Second Supplements to the 2nd Edition of

RODD'S CHEMISTRY OF CARBON COMPOUNDS

ELSEVIER SCIENCE B.V.
Sara Burgerhartstraat 25
P.O. Box 211, 1000 AE Amsterdam, The Netherlands

ISBN: 0-444-82416-2

© 1996 ELSEVIER SCIENCE B.V. All rights reserved.

No part of this publication may be reproduced, stored in a retrieval system or transmitted in any form or by any means, electronic, mechanical, photocopying, recording or otherwise, without the prior written permission of the publisher, Elsevier Science B.V. , Copyright & Permissions Department, P.O. Box 521, 1000 AM Amsterdam, The Netherlands.

Special regulations for readers in the U.S.A. - This publication has been registered with the Copyright Clearance Center Inc. (CCC), 222 Rosewood Drive Danvers, MA 01923. Information can be obtained from the CCC about conditions under which photocopies of parts of this publication may be made in the U.S.A. All other copyright questions, including photocopying outside of the U.S.A., should be referred to the copyright owner, Elsevier Science B.V., unless otherwise specified.

No responsibility is assumed by the publisher for any injury and/or damage to persons or property as a matter of products liability, negligence or otherwise, or from any use or operation of any methods, products, instructions or ideas contained in the material herein.

This book is printed on acid-free paper.

Printed in The Netherlands.

Second Supplements to the 2nd Edition of

RODD'S CHEMISTRY OF CARBON COMPOUNDS

VOLUME I

ALIPHATIC COMPOUNDS
★

VOLUME II

ALICYCLIC COMPOUNDS
★

VOLUME III

AROMATIC COMPOUNDS
★

VOLUME IV

HETEROCYCLIC COMPOUNDS
★

VOLUME V

MISCELLANEOUS
GENERAL INDEX
★

Second Supplements to the 2nd Edition of

RODD'S CHEMISTRY OF CARBON COMPOUNDS

A modern comprehensive treatise

Edited by
MALCOLM SAINSBURY
*School of Chemistry, The University of Bath,
Claverton Down, Bath BA2 7AY, England*

Second Supplement to

VOLUME III AROMATIC COMPOUNDS

Part A: General Introduction; Mononuclear Hydrocarbons and
Their Halogeno Derivatives; and Derivatives with
Nuclear Substituents Attached through Nonmetallic Elements from
Group VI of the Periodic Table

1996
ELSEVIER
Amsterdam – Lausanne – New York – Oxford – Shannon – Tokyo

Contributors to this Volume

G.C. BARRETT

School of Biological and Molecular Sciences, Oxford Brookes University, Gipsy Lane, Headington, Oxford OX3 0BP, U.K.

R.W. BROWN

ICI Agricultural Products, Western Research Center, 1200 S 47th Street, Box 4023, Richmond, CA 94804-0023, U.S.A.

N. FURUKAWA

Department of Chemistry, Tsukuba University, Tsukuba, Ibaraki, Japan

H. HEANEY

Department of Chemistry, University of Loughborough, Leicester LE11 3TU, U.K.

S. SATO

Department of Chemistry, Tsukuba University, Tsukuba, Ibaraki, Japan

D.J. SIMMONDS

Department of Chemistry, Sheffield City University, Pond Street, Sheffield S1 1WB, U.K.

P.G. STEEL

Department of Chemistry, University of Durham, Science Laboratories, South Road, Durham DH1 3LE, U.K.

J.H.P. TYMAN

Department of Chemistry, Brunel University, Uxbridge, Middlesex UB8 3PH, U.K.

Preface to Volume III A

Volume III of Rodd's "Chemistry of the Carbon Compounds" deals with Aromatic Compounds and, in this the first sub-volume, the topic is introduced by reference to the chemistry and synthesis of aromatic hydrocarbons, their halogen, hydroxy, sulphur, selenium and tellurium derivatives.

It is the intention of the Supplements to select all that is of importance in the development of Organic Chemistry since the publication of the Second Edition of Rodd. The order of the chapters and the topics should thus run from one presentation to the other. The current work covers all that which has been added to our knowledge of the topics noted above in the period 1983–1994. A daunting task, considering the fact that all of the authors are busy chemists working full time either in industry or in academia.

It is particularly pleasing that two of the contributors, Professor Harry Heaney and Dr. John Tyman, were responsible for the same chapters in the First Supplement to Rodd, Volume III A. Their experience has been very valuable in bringing continuity into the series. I am also happy to acknowledge the skills shown by the other contributors to this volume: Professor Furukawa, and Drs. Barrett, Brown, Simmonds and Steel, for the way they have presented most readable accounts of such a huge mass of new chemical discoveries. Finally, I would like to thank all the authors for their patience during the time it has taken to compile this volume.

Malcolm Sainsbury　　　　　　　　　　　　　　　　　　　　　　　　　　　　　Bath
　　　　　　　　　　　　　　　　　　　　　　　　　　　　　　　　　　　October 1995

Contents
Volume III A

Aromatic compounds: General introduction: Mononuclear hydrocarbons and their halogeno derivatives; and derivatives with nuclear substituents attached through non-metallic elements from Group VI of the Periodic Table

List of contributors vi
Preface vii
List of common abbreviations and symbols used xvii

Chapter 1. Benzene and its Homologues
by D.J. SIMMONDS

1. Introduction 1
   (a) General Introduction 1
   (b) Aromaticity 2
   (c) Homoaromaticity 5
2. Annulenes and related systems 8
   (a) Introduction 8
   (b) Homologous series of annulenes 8
   (c) C_2–C_5 systems 10
   (d) Benzene rings and aromatization 15
   (e) 8π and 10π systems 22
   (f) 12π and 14π systems 26
   (g) Higher annulenes 31
   (h) Carbon-only systems 35
   (i) Heterocyclic systems 36
3. Benzenoid compounds 43
   (a) Benzene isomers 43
   (b) Benzynes 45
   (c) Non-condensed systems 49
      (i) Cyclophanes, 49 – (ii) Calixarenes and related systems, 55 – (iii) Phenylenes, 57 – (iv) Iptycenes, 58 –
   (d) Fused systems 61
      (i) Oligoarylenes, 61 – (ii) Series of fused benzenoids, 62 – (iii) Cyclacenes, circulenes and related compounds, 63 – (iv) Fullerenes, 66 –
4. Substitution reactions 73
   (a) Electrophilic substitution 73
      (i) General, 73 – (ii) Wheland intermediates, 74 – (iii) Electron transfer, 75 –
   (b) Nucleophilic substitution 79
      (i) General, 79 – (ii) Mechanism of substitution, 80 – (iii) Vicarious nucleophilic substitution (VNS), 84 –

(c) Organometallic methods 85
   (i) General, 85 – (ii) Reactions with electrophiles, 86 – (iii) Reactions with nucleophiles, 88 – (iv) Radical reactions, 92 –
5. Formation and loss of the benzene nucleus 101
   (a) Formation 101
      (i) General, 101 – (ii) Enediyne and related cyclizations, 102 – (iii) Metal mediated alkyne cyclizations, 105 –
   (b) Loss of the benzene ring 109
      (i) General, 109 – (ii) Photoaddition of the benzene ring with alkenes, 110 – (iii) Other photochemical processes, 112 –

Chapter 2. Mononuclear Hydrocarbons: Benzene and its Homologues
by H. HEANEY

1. Nuclear magnetic resonance spectroscopy 115
2. The formation of benzene derivatives with built-in functionality 116
   (a) The formation of benzene derivatives by cyclisation reactions 116
      (i) [3+3] Cyclisation reactions, 116 – (ii) [4+2] Cyclisation reactions, 117 – (iii) [2+2+2] Cyclisation reactions, 119 –
3. Benzo- small ring compounds 123
   (a) Benzocyclopropenes 123
   (b) Benzocyclobutenes 125
4. The metallation of arenes and arylation reactions using arylmetallic compounds 127
   (a) Aryllead tricarboxylates 128
   (b) Arylbismuth reagents 131
   (c) Arylpalladium reagents 134
5. Electrophilic addition-with-elimination reactions 137
   (a) Protonation reactions 137
   (b) Group (IV) electrophiles 138
      (i) Alkylation reactions, 139 – (ii) Mannich reactions, 146 –
   (c) Acylation reactions 147
      (i) General principles, 147 – (ii) Catalysts, 149 – (iii) Reactions involving mixed anhydrides, 151 – (iv) Other catalyst systems, 153 – (v) Reactions using stoichiometric catalysts, 155 – (vi) Intramolecular acylation reactions, 157 – (vii) Metal catalysed ketone syntheses, 158 –
   (d) Formylation reactions 160
      (i) Formylation reactions using formyl cation equivalents, 161 – (ii) Vilsmeier formylation reactions, 162 –
   (e) Group (V) electrophiles 164
      (i) Nitration reactions, 164 – (ii) Nitrosation reactions, 170 –
   (f) Group (VI) electrophiles 171
      (i) Sulfonation reactions, 171 – (ii) Hydroxylation reactions, 174 –
   (g) Group (VI) electrophiles 175

Chapter 3. Halogenobenzenes
by P.G. STEEL

Introduction 178
1. Nuclear halogen derivatives 179
   (a) Methods of preparation 179
   (i) Replacement of an aromatic hydrogen by halogen, 179 – (ii) Replacement of the carboxylic acid function by a halogen, 188 – (iii) Replacement of an amino group by a halogen, 188 – (iv) Replacement of nitro, chlorosulphonyl and substituted amino groups by a halogen, 190 – (v) Replacement of one halogen by another, 191 –
   (b) Reactions and properties 192
   (i) Reactions with metals, 192 – (ii) Dehalogenation, 198 – (iii) Electrophilic reactions, 205 – (iv) Nucleophilic substitution, 206 – (v) Photochemical reactions, 212 – (vi) π-Haloarene metal complexes, 214 –
2. Alkylbenzenes with halogen in the side chain 217

Chapter 4. Hydroxy Aromatics
by J.H.P. TYMAN

1. Monohydric Phenols 225
   (a) Synthesis 226
   (i) The hydroxylation of arenes, 227 – (ii) The conversion of phenolic esters and phenolic ethers to phenols, 228 – (iii) Formation of phenols by replacement of boron, alkyl, oxoalkyl, carboxyl, formyl, halogeno, hydroperoxide, lithio, nitro, phosphate, sulpho, silano, triazeno, and thallic groups, 229 – (iv) Formation of phenols from cycloaliphatic precursors, 232 – (v) Formation of phenols from an acyclic precursor, 234 – (vi) Formation of phenols from reactions of two acyclic precursors, 235 – (vii) Formation of phenols and naphthols from transition metal intermediates, 236 – (viii) Phenols formed by alkylations and arylations, 239 – (ix) formation of phenols by rearrangements, 240 – (x) Formation of polycyclic phenols, 241 –
   (b) Reactions of phenols 241
   (i) Esterification, 241 – (ii) Diaryl carbonates, 243 – (iii) Thio and thionecarbonates, xanthates and carbamates, 244 – (iv) Phosphites, phosphates and aryloxylamines, 246 – (v) Sulphates and sulphonyl esters, 247 – (vi) Replacement of the hydroxyl group, 247 – (vii) Replacement by hydrogen, 249 – (viii) Protection of, and by, the phenolic group, 250 –
   (c) Reactions of the aromatic ring 250
   (i) Deuteration, 250 – (ii) Diels-Alder reaction, 250 – (iii) Oxyalkyl and oxycarboxyalkyl compounds by substitution, 250 – (iv) Alkyl side chains containing S or Si atoms, 252 –
   (d) Oxidations 253
   (i) Oxidative coupling, 253 – (ii) Cyclohexadienone formation, 255 – (iii) Quinone formation, 259 – (iv) Carbonylic and other transformation products, 261 –

(e) Reactions of alkylphenols 262
   (i) Esterification and hydrogenation, 262 – (ii) Aldol and Mannich reactions, 263 – (iii) Cyclisation of o-alkenylphenols, 265 – (iv) Dimerisation and arylation, 267 – (v) Calixarenes, 269 – (vi) Crypto- and hindered phenols, 269 –
(f) Phenolic aldehydes, ketones, acids and esters 273
   Synthesis
   (i) 2-Hydroxyaldehydes, 273 – (ii) 4-hydroxyaldehydes, 274 – (iii) 2-hydroxyaryl alkyl ketones, 274 – (iv) 2- and 4-hydroxybenzophenones, 276 – (v) Polycyclic hydroxyketones, 276 – (vi) Phenolic acids, 277 – (vii) Phenolic esters, 278 – (viii) Alkyl 4-hydroxybenzoates, 280 – (ix) Dialkyl hydroxyphenyldicarboxylates, 280 – (x) Phenolic nitriles, amides, ketoacids and phthalides, 281 –
   Reactions
   (i) Phenolic aldehydes, 282 – (ii) Phenolic ketones, 284 – (iii) Phenolic acids, their ethers and phenolic esters, 286 –
(g) Phenolic ethers 287
   Synthesis of phenolic ethers
   (i) Alkyl ethers from phenolic intermediates by alkylation under catalytic or acidic conditions, 288 – (ii) Alkyl ethers from phenoxide intermediates, 288 – (iii) Naphthyl ethers from naphthoxide intermediates, 290 – (iv) Substituted alkyl ethers from phenoxide intermediates, 290 – (v) Phenolic ethers from transetherification reactions, 293 – (vi) Phenolic ethers formed by the displacement of nitro, chloro and other groups, 293 – (vii) Phenolic ethers from alicyclic and acyclic precursors, 296 –
   Reactions of phenolic ethers
   (i) Dealkylation and dealkoxylation, 298 – (ii) Substitution of the phenolic ether group, 300 – (iii) Substitution in the aryl ring of phenolic ethers, 300 – (iv) Reactions in the alkyl side chain of phenolic ethers, 304 – (v) Displacement of nuclear chlorine, bromine or fluorine from a phenolic ether, 305 – (vi) Cyclisation reactions of phenolic ethers, 306 – (vii) Reactions of keto derivatives of phenolic ethers, 307 – (viii) Reactions of carboxy, amido, formyl and nitro derivatives of phenolic ethers, 309 – (ix) Phenolic ether coupling, 310 – (x) Miscellaneous reactions of phenolic ethers, 310 –
(h) Halogeno, nitro, amino, azo, sulpho and thio derivatives of phenols 311
   Synthesis of halogen, nitro, amino, azo and thio derivatives
   (i) Fluorophenols, 311 – (ii) Chlorophenols, 312 – (iii) Bromophenols, 313 – (iv) Iodophenols, 315 – (v) Nitrophenols, 315 – (vi) Aminophenols, 316 – (vii) Hydroxyazo and hydroxydiaryl compounds, 319 – (viii) Thio derivatives and thiophenols, 319 –
   Reactions of halogeno, nitro, amino, and thio derivatives of phenols
   (i) Fluorophenols, 321 – (ii) Chlorophenols, 321 – (iii) Bromophenols, 323 – (iv) Iodophenols, 324 – (v) Nitrophenols, 325 – (vi) Aminophenols, 326 – (vii) Phenolsulphonic acids, 329 –

2. Dihydric phenols 330
   (a) Synthesis
      (i) 1,2-Dihydroxybenzenes (catechols) and their derivatives, 330 –
      (ii) Mono O-alkyl and acyl derivatives of catechol, 331 – (iii) 1,3-Dihydroxybenzenes (resorcinols) and their derivatives, 332 –
      (iv) 1,4-Dihydroxybenzenes (hydroquinones) and their derivatives, 337 –
   (b) Reactions of dihydric phenols
      (i) 1,2-Dihydroxybenzenes (catechols) and their derivatives, 340 –
      (ii) Mono O-alkyl derivatives of 1,2-dihydroxybenzenes, 341 –
      (iii) Di-O-alkyl derivatives of 1,2-dihydroxybenzenes, 343 –
      (iv) Methylenedioxy derivatives of 1,2-dihydroxybenzenes, 346 –
      (v) Dioxin derivatives, 347 – (vi) 1,3-Dihydroxybenzenes (resorcinols) and their derivatives, 347 – (vii) Mono O-alkoxy derivatives of 1,3-dihydroxybenzenes, 352 – (viii) Mixed unsymmetrical O-substituted derivatives of 1,3-dihydroxybenzenes, 354 – (ix) 1,3-Dialkoxy derivatives of 1,3-dihydroxybenzenes, 354 – (x) 1,4-Dihydroxybenzenes (hydroquinones) and their derivatives, 357 –
      (xi) Mono O-alkyl derivatives of 1,4-dihydroxybenzenes, 360 –
      (xii) Mono O-acyl derivatives of 1,4-dihydroxybenzenes, 362 –
      (xiii) 1,4-Dialkoxy derivatives of 1,4-dihydroxybenzenes, 363 –
      (xiv) Diacyl derivatives of 1,4-dihydroxybenzenes, 364 – (xv) Naphthalenic and polycyclic dihydric systems, 365 –
3. Polyhydric phenols 366
   (i) 1,3,5-Trihydroxybenzene and its alkoxy derivatives, 368 – (ii) 1,2,3-Trihydroxybenzenes and alkoxy derivatives, 371 – (iii) 1,2,4-Trihydroxybenzene and alkoxy derivatives, 373 – (iv) Polyalkoxybenzenes, tetraacylbenzenes and polyalkoxy naphthalenes, 375 –

Chapter 5. Hydrocarbons carrying substituents attached through divalent sulphur, selenium, or tellurium: thiophenols, sulphides, etc.
by G.C. BARRETT

1. Introduction 377
2. Organization of this chapter 377
3. Reference sources and textbooks 378
4. Properties of thiols and sulphides 378
5. Preparation of arenethiols 380
   (a) Preparation by reduction of arenesulphonyl halides 381
   (b) Preparation by reduction of arenesulphonic acids 381
   (c) Reduction of thiolsulphonates 381
   (d) Preparation by reduction of arenesulphonic acids 382
   (e) Preparation by reduction of arenesulphonates 382
   (f) Preparation by reduction of disulphides 382
   (g) Preparation from phenols 382
   (h) Preparation from arenes 384
   (i) Preparation from aryl halides using H₂S, NaSH or MSR; or using other simple inorganic reagents 386
   (j) Preparation from arenesulphenyl halides 387

(k) Ring opening of sulphur heterocycles [see also Section 9(a)(1)] 387
(l) Preparation by cleavage of sulphides 388
6. Thiolesters 389
7. Reactions of thiols 391
   (a) Replacement of SH by H 391
   (b) Oxidation of thiols 391
   (c) Substitution reactions 392
   (d) Synthetic uses: heterocyclic ring formation 393
   (e) Acylation 393
8. Protection of the SH group in synthesis 394
   (a) Preparation of sulphides 396
9. Saturated alkyl, aryl, and diaryl sulphides 396
10. Reactions of sulphides and selenides and their analogues 404
    (a) Oxidation 404
11. Synthetic uses of sulphides 405
12. Thioacetals and related compounds 409
13. Unsaturated sulphides 411
    (a) Preparation 411
    (b) Synthetic uses of unsaturated sulphides 411

Chapter 6. Benzenes carrying sulfonyl, sulfinyl, or sulfenyl functional groups, but excluding those compounds bearing functionalized side chains
by R.W. BROWN

1. Benzenesulfonic acids and their derivatives 413
   (a) Benzenesulfonic acids 413
       (i) Preparation, 413 – (ii) Reactions, 415 –
   (b) Benzenesulfonyl halides 416
       (i) Preparation, 416 – (ii) Reactions, 416 –
   (c) Benzenesulfonic anhydrides 418
   (d) Benzenesulfonyl peroxides 419
   (e) Benzenesulfonates 420
       (i) Preparation, 420 – (ii) Reactions, 421 –
   (f) Benzenesulfonamides 424
       (i) Preparation, 424 – (ii) Reactions, 424 –
   (g) Benzenesulfonyl hydrazides and hydrazones 427
   (h) Benzenesulfonyl azides 431
   (i) Benzenesulfonyl isocyanates 432
   (j) Benzenesulfonyl oxaziridines 433
   (k) Biological activity of benzenesulfonic acids and their derivatives 434
2. Benzenesulfinic acids and their derivatives 437
   (a) Benzenesulfinic acids 437
       (i) Preparation, 437 – (ii) Reactions, 438 –
   (b) Benzenesulfinates 444
       (i) Preparation, 444 – (ii) Reactions, 445 –
   (c) Benzenesulfinyl chlorides 448
   (d) Benzenesulfinamides 449
3. Benzenesulfenic acids and their derivatives 453
   (a) Benzenesulfenic acids 453

(b) Benzenesulfenates 454
(c) Benzenesulfenyl halides 459
(d) Benzenesulfenamides 462
   (i) Preparation, 463 – (ii) Reactions, 465 –
(e) Aryl benzenethiolsulfonates 467

Chapter 7. Mononuclear hydrocarbons carrying nuclear substituents containing selenium, or tellurium
by N. FURUKAWA and S. SATO

1. Introduction 469
2. Se(II) and Te(II) derivatives 471
   (a) Arylselenols, diaryl diselenides and their derivatives 471
   (b) Arylselenides 474
      (i) Preparation, 474 – (ii) Structures and reactions, 477 –
   (c) Aryltellurols, ditellurides, tellurides and related derivatives 492
      (i) Preparation, 492 – (ii) Reactions of dicoordinate aryltellurium compounds, 495 –
3. Se(III) and Te(III) derivatives 496
   (a) Selenoxides and telluroxides 496
   (b) Optically active selenoxides 497
   (c) Selenoimines and telluroimines 499
   (d) Selenonium, telluronium salts and their ylides 499
4. Se(IV) and Te(IV) compounds: selenones and tellurones 502
5. Hypervalent compounds of Se(IV), Te(IV) 503
   (a) General introduction and historical background of hypervalent compounds of chalcogens 503
   (b) Detection of unstable chalcogenuranes as intermediates 507
      (i) Detection of tellurium and selenium ate-complexes [10-M-3(C3)] (M: Se, Te) using Li–Te and Li–Se exchange reactions, 507 – (ii) Organic synthesis via tellurium ate-complexes, 509 – (iii) Ligand exchange and coupling reactions of chalcogenide oxides with organometallic reagents, 510 – (iv) Formation of tetraarylchalcogenuranes [10-M-4(C4)] (M: S, Se, Te) and ligand coupling reactions, 512 –
   (c) Miscellaneous hypervalent compounds of sulfur, selenium and tellurium 517

Guide to the index 521
Index 523

List of Common Abbreviations and Symbols Used

A	acid
Å	Ångström units
Ac	acetyl
a	axial
as, asymm.	asymmetrical
at.	atmosphere
B	base
Bu	butyl
b.p.	boiling point
c, C	concentration
CD	circular dichroism
conc.	concentrated
crit.	critical
D	Debye unit, 1×10^{-18} e.s.u.
D	dissociation energy
D	dextro-rotatory; dextro configuration
d	density
dec., decomp	with decomposition
deriv.	derivative
E	energy; extinction; electromeric effect
E	entgegen (opposite) configuration
$E1$, $E2$	uni- and bi-molecular elimination mechanisms
E1cB	unimolecular elimination in conjugate base
ESR	electron spin resonance
Et	ethyl
e	nuclear charge; equatorial
f.p.	freezing point
G	free energy
GLC	gas liquid chromatography
g	spectroscopic splitting factor, 2.0023
H	applied magnetic field; heat content
h	Planck's constant
Hz	hertz
I	spin quantum number; intensity; inductive effect
IR	infrared
J	coupling constant in NMR spectra
J	Joule
K	dissociation constant
k	Boltzmann constant; velocity constant
kcal	kilocalories
L	laevorotatory, laevo configuration
M	molecular weight; molar; mesomeric effect
Me	methyl

m	mass; mole; molecule; *meta-*
m.p.	melting point
Ms	mesyl (methanesulphonyl)
[M]	molecular rotation
N	Avogadro number; normal
NMR	nuclear magnetic resonance
NOE	Nuclear Overhauser Effect
n	normal; refractive index; principal quantum number
o	*ortho-*
ORD	optical rotatory dispersion
P	polarisation; probability; orbital state
Pr	propyl
Ph	phenyl
p	*para-*; orbital
PMR	proton magnetic resonance
R	clockwise configuration
S	counterclockwise configuration; entropy; net spin of incompleted electronic shells; orbital state
$S_N 1, S_N 2$	uni- and bi-molecular nucleophilic substitution mechanism
$S_N i$	internal nucleophilic substitution mechanism
s	symmetrical; orbital
sec	secondary
soln.	solution
symm.	symmetrical
T	absolute temperature
Tosyl	*p*-toluenesulphonyl
Trityl	triphenylmethyl
t	time
temp.	temperature (in degrees centigrade)
tert	tertiary
UV	ultraviolet
v	velocity
Z	zusammen (together) configuration
α	optical rotation (in water unless otherwise stated)
$[\alpha]$	specific optical rotation
ϵ	dielectric constant; extinction coefficient
μ	dipole moment; magnetic moment
μ_B	Bohr magneton
μg	microgram
μm	micrometer
λ	wavelength
ν	frequency; wave number
χ, χ_d, χ_μ	magnetic; diamagnetic and paramagnetic susceptibilities
$\sim$	about
(+)	dextrorotatory
(−)	laevorotatory
−	negative charge
+	positive charge

Second Supplements to the 2nd Edition of Rodd's Chemistry of Carbon Compounds, Vol.III A, edited by M. Sainsbury
© 1996 Elsevier Science B.V. All rights reserved.

Chapter 1 Benzene and its Homologues

D. J. SIMMONDS

1. INTRODUCTION

(a) **General Introduction**

This second supplement (cf. "Rodd's Chemistry of Carbon Compounds", 2nd Edition, Supplement to III A, M. F. Ansell (Ed)) reflects massive activity in the design and synthesis of aromatic structures, backed up by enhanced hardware and software for theoretical investigations, and almost routine availability of instrumentation (especially NMR techniques) for spectroscopic scrutiny. Although the basic outline of this supplement matches the first, the areas of emphasis are very different. The bulk of this chapter provides a detailed survey of structural aspects of aromaticity, homoaromaticity and anti-aromaticity encompassing benzenoid and non-benzenoid systems (benzene being just one of a series of homologous annulenes).

The reactivity of benzene is dominated by substitution reactions (cf. Chapter 2). Electrophilic behaviour remains the predominant mechanism for substitution but the use of organometallic reagents and catalysts has animated developments in nucleophilic and homolytic processes also. The involvement of single electron transfer (SET) has been identified in both electrophilic and nucleophilic substitutions. In this second supplement substitutions, and other reactions, involving organometallic species provide a particular focus for attention.

The stability of the benzene ring dictates the course of many reactions but, under photochemical conditions particularly, useful and interesting transformations are possible. Many photochemical reactions are distributed throughout this chapter and Section 5 b summarizes those that lead to a loss of aromaticity. On the other hand some important reactions have come to light that lead to the formation of benzene rings. In particular the cyclotrimerization of alkynes (Vollhardt cyclisation), the ring-closure of enediynes (Bergman cyclisation), and the elaboration of transition metal carbene complexes have become powerful synthetic procedures (Section 5 a).

Finally, the properties of the benzene ring and some related structures are relevant in materials chemistry. Secondary bonding forces in appropriately substituted oligophenyls, aryl benzoates etc, lead to stable mesophases from which liquid crystal devices can be fabricated [*eg.* G. W. Gray (Ed.), "Thermotropic Liquid Crystals", Wiley (UK), 1987], and the LC phenomenon can be transposed to polymeric materials with a wide range of possible applications in optical and engineering technologies [*eg.* A. A. Collyer (Ed.], "Liquid Crystal Polymers; from Structures to Applications", Elsevier, 1992). On the other hand the electronic and topological properties of aromatic species assist the formation of solid state materials with "metallic" band structures suitable for fabrication into conductors, semi-conductors and even superconductors (review: M. R. Bryce, Chem. Soc. Rev., 1991, **20**, 355). One of the most intriguing of the molecular superconductors derives from fullerene C_{60} ("Buckminsterfullerene"), the "new allotrope" of carbon. Fullerenes, and other carbon-only molecules, are introduced in Sections 2 and 3 as members of a burgeoning class of aromatic compounds of which benzene is just one example, albeit a supremely important example.

(b)   **Aromaticity**

The concept of aromaticity is the subject of various books (*eg.* P. J. Garratt, "Aromaticity", John Wiley and Sons, 1986; D. Lloyd, "The Chemistry of Conjugated Cyclic Compounds", Wiley, New York, 1989) and reviews (*eg.* M. V. Gorelik, Russ. Chem. Rev., 1990, **59**, 116; P. J. Garratt in "Comprehensive Organic Chemistry", Vol. 1, J. F. Stoddart (Ed.), Pergamon, 1979). Despite the long history of the concept (review: A. T. Balaban, Pure Appl. Chem., 1980, **52**, 1409) the precise origin of aromatic stabilisation remains controversial. Benzene, as the archetypal arene, has been exposed to detailed theoretical calculations (*eg.* J. E. Boggs, F. Pang, P. Pulay, J. Comput. Chem., 1982, **3**, 344; C. W. Bock, M. Trachtman and P. George, *ibid*, 1985, **6**, 592; V. Melissas, K. Faegri and J. Almlof, J. Amer. Chem. Soc., 1985, **107**, 4640) but complete and general agreement has not yet resulted.

Until recently molecular orbital theory held sway; the structure of benzene was assumed to result from the delocalisation of six π electrons in three bonding MOs leading to a structure of optimum stability. Huckel's rule then predicted that all annulenes having (4n + 2) π electrons would display aromatic stability (at least up to n = 5), while 4n π annulenes would, if planar, be anti-aromatic. Thus valence-bond treatments were considered redundant and Kekulé (and other) resonance structures could safely be superceded by a single, planar, hexagonal structure involving a delocalised π-system. A discussion of various criteria for aromaticity/anti-aromaticity, based mainly on the MO approach was published in 1985 (M. N. Glukhovtsev, B. Y. Simkin and V. I. Minkin, Russ. Chem. Rev., Engl. Transl., 1985, **54**, 54).

However, using a new valence bond approach, Hiberty, Shaik and co-workers asserted that the π-bonding energies of benzene were actually weaker in the regular, hexagonal geometry than in a distorted, Kekulé-type arrangement. *Ab initio* calculations revealed that the hexagonal structure results from two opposing influences, a distortive π system and a symmetrizing σ framework; *ie* the regular structure is driven by that σ framework with π overlap and delocalisation being the outcome and not the driving force (P. C. Hiberty *et al*, J. Org. Chem., 1985, **50**, 4657; S. S. Shaik and P. C. Hiberty, J. Amer. Chem. Soc., 1985, **107**, 3089). Despite challenges (*eg.* N. C. Baird, J. Org. Chem., 1986, **51**, 3907), the new ideas were vigorously defended by their originators (P. C. Hiberty *et al*, *ibid*, 1986, **51**, 3908; S. S. Shaik *et al,* J. Amer. Chem. Soc., 1987, **109**, 363) and by others (K. Jug and A. M. Koster, *ibid*, 1990, **112**, 6772). Other workers dislike the σ/π conflict and argue that π delocalisation stabilizes a strongly symmetric benzene in accord with classical resonance theory (E. D. Glending *et al*, *ibid*, 1993, **115**, 10952). A review summarizes the σ framework ideas (P. C. Hiberty *et al*, Pure Appl. Chem., 1993, **65**, 35).

One drawback of classical valence-bond methods as applied to delocalised systems is that a truly comprehensive description requires a large number of resonance structures. In the case of benzene a total of 175 structures, 2 Kekulé, 3 Dewar and 170 ionic, are needed (J. Gerratt, Chem. Brit., 1987, **23**, 327). Using a more sophisticated *ab initio* method, spin-coupled valence bond theory, Gerratt and co-workers have remodelled benzene (D. L. Cooper, J. Gerratt and M. Raimondi, Nature (London), 1986, **323**, 699). The preliminary result associates the six π electrons with six non-orthogonal orbitals that are highly localised, and stability arises from the symmetric coupling of the electron spins around the ring. Further refinement was then used to calculate electron exchange interactions around the ring leading to a complete picture including "ring currents" and consistent with the MO based (4n + 2) rule and the phenomenon of aromatic stabilisation. A series of responses to these new ideas appeared in the journal Nature (London) throughout 1987, and the originators have since extended their calculations to other benzenoid and heteroaromatic molecules, claiming results that correlate well with empirical data (review: D. L. Cooper, J. Gerratt and M. Raimondi, Chem. Rev., 1991, **91**, 929).

Irrespective of the origin of ring currents in aromatic molecules, the associated magnetic anisotropy remains the most convenient property whereby aromaticity or anti-aromaticity can be assessed, using NMR spectroscopy (see Section 2a). Routine methods now include ^{13}C-NMR, and a lanthanum shift reagent method has been advocated (R. J. Abraham et al, J. Chem. Soc., Chem. Commun., 1982, 998). By using high field PMR an alternative method, based on magnetic polarizability anisotropies, is possible for probing aromaticity (P. C. M. Van Zijl et al, J. Amer. Chem. Soc., 1986, **108**, 1415).

Aromatic stabilisation can also be exploited, eg. by investigating the reversibility of protonation of a test structure either by experiment (B. Capon and S. Q. Lew, Tetrahedron, 1992, **48**, 7823) or by calculation (J. M. Bofill et al, J. Org. Chem., 1988, **53**, 5148). Alternatively stabilisation can be verified by calculating an energy profile for distorting the planarity of a test molecule; distortion will be endothermic for aromatic, and exothermic for anti-aromatic, systems (B. L. Podlogar et al, ibid, 1988, **53**, 2127). Finally the concept of "relative hardness", based on the HOMO-LUMO gap of a test molecule relative to that in acyclic reference structures, has been proposed as a means of assessing aromaticity (Z. Zhou and R. G. Parr, J. Amer. Chem. Soc., 1989, **111**, 7371). Other tests that have been applied are mentioned when they arise below.

(c)   **Homoaromaticity**

Homoaromaticity provides a stabilising mechanism in some structures despite the presence of one (homoaromatic) or more (bishomoaromatic etc) sp^3 atom in the molecular periphery. The most familiar example, obtained by protonation of cyclooctatetraene, is the homotropylium ion (1) (see R. F. Childs *et al*, J. Org. Chem., 1983, **48**, 1431; R. C. Haddon, J. Amer. Chem. Soc., 1988, **110**, 1108):-

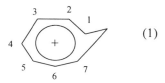

Planarity encompassing C_1 - C_7 allows sufficient interaction of orbitals on C_1 and C_7 for delocalisation of the 6π electrons and positive charge (review: R. F. Childs, Acc. Chem. Res., 1984, **17**, 347). Homoaromaticity has also been demonstrated in the 2-hydroxyhomotropylium ion (R. F. Childs *et al*, J. Amer. Chem. Soc., 1982, **104**, 2452) and thiatropylium ion (K. Yamamoto, S. Yamazaki and I. Murata, Angew. Chem., Int. Ed. Engl., 1985, **24**, 576) and, in the former case, X-ray showed homoaromaticity in the solid state also. The homoconjugation gap needs to be small, *eg.* C_1 - C_7 is *ca.* 1.6A° in (1) (L. T. Scott and M. M. Hashemi, Tetrahedron, 1986, **42**, 1823). In 2-ethoxyhomotropylium salts the distance is 2.28A° and the system is not aromatic (despite PMR evidence of a ring current) (R. F. Childs *et al*, J. Amer. Chem. Soc., 1986, **108**, 3613).

Confirmation of homoaromaticity can be difficult and NMR evidence is not always sufficient. X-ray crystallography is useful for corroboration if stable salts are available, and Haddon advocates the use of sophisticated theoretical treatment (R. C. Haddon and K. Raghavachari, *ibid*, 1983, **105**, 118). Using POAV-3D HMO calculations he confirmed homoaromaticity in (1) but the stabilisation is rather small (R. C. Haddon, *ibid*, 1988, **110**, 1108; Acc. Chem. Res., 1988, **21**, 243). The bishomotropylium ion (2) seems homoaromatic by NMR (C. J. Abelt and H. D. Roth, J. Amer. Chem. Soc., 1986, **108**, 2013) but calculations have still not provided definite corroboration (D. Cremer *et al*, *ibid*, 1993, **115**, 7457).

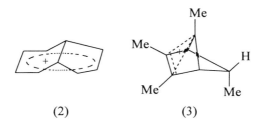

(2)   (3)

2π Homoaromaticity is quite common. The ion (3) contains a bishomoaromatic feature (R. F. Childs and D. L. Mulholland, *ibid*, 1983, **105**, 96), while the dication (4) is a sandwich of two such moieties (G. K. S. Prakash *et al*, *ibid*, 1987, **109**, 911), and (5) neatly combines bishomoaromatic and allyl cations (G. A. Olah, M. Arvanaghi and G. K. S. Prakash, Angew. Chem., Int. Ed. Engl., 1983, **22**, 712). The 2π adamantanediyl dication (6) provides a dramatic example of three-dimensional homoaromaticity (P. von R. Schleyer *et al*, *ibid*, 1987, **26**, 761).

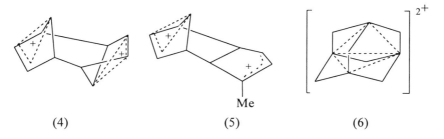

(4)   (5)   (6)

Homoaromaticity may be restricted to cations although radical cations can probably be included also, *eg.* the bishomoaromatic species (7) obtainable (S. Dai, J. T. Wang and F. Williams, *ibid*, 1990, **112**, 2835) from semibullvalene (*cf.* H. D. Roth and C. J. Abelt, *ibid*, 1986, **108**, 2013):-

(7)

Anionic and neutral systems remain controversial. Separate calculations regarding anion (8) have favoured both the localised allylic form (8a) (P. von R. Schleyer et al, ibid, 1981, **103**, 1375) and the homoaromatic form (8b) (J. M. Brown, R. J. Elliott and W. G. Richards, J. Chem. Soc., Perkin Trans. 2, 1982, 485):-

Empirical evidence in favour of the homoaromatic ion (8b) has been accrued using both deuterium isotope ^{13}C-NMR (M. Christe, H. Leininger and D. Bruckner, J. Amer. Chem. Soc., 1983, **105**, 118) and pKa measurements (W. N. Washburn, J. Org. Chem., 1983, **48**, 4287). The latter approach was also used to identify the extended anion (9) as the most convincing candidate so far (A. Tuncay et al, J. Chem. Soc., Chem. Commun., 1988, 1590):-

Amongst neutral molecules, only triquinacene (10) has received strong support, as a 6π trishomoaromatic system, on the basis of thermochemical data (L. A. Paquette et al, J. Amer. Chem. Soc., 1986, **108**, 8267). However theoretical analysis of (10) was not in agreement (M. J. S. Dewar and A. J. Holder, ibid, 1989, **111**, 5384); when the relative strain of (10) and its hydrogenated derivatives is taken into account then any homoaromatic effect becomes vanishingly small (J. W. Storer and K. N. Houk, *ibid*, 1992, **114**, 1165).

2.   ANNULENES AND RELATED SYSTEMS

(a)   **Introduction**

The power of modern synthetic chemistry has given access to a wealth of molecules pertinent to the concept of aromaticity. Huckel's analysis, having implications for the magnetic properties expected for aromatic systems, is amenable to the techniques of NMR spectroscopy as established elsewhere (*eg.* P. J. Garratt in "Comprehensive Organic Chemistry", *op. cit.*). The routine availability of instrumentation makes PMR an attractive empirical probe for aromaticity although UV/visible spectroscopy and chemical reactivity (especially electrophilic substitution) provide corroboration in many instances.

Aromatic molecules, with cyclic structures containing 4n + 2 delocalised π-electrons, are diatropic; protons outside the perimeter are shifted downfield, and those inside are shifted upfield, relative to comparable but atropic structures due to the operation of a ring current. Paratropic effects, with a reversal of the outer-inner proton shifts noted for diatropic systems, are expected when 4n π-electron systems are constrained to planarity since delocalisation in those circumstances leads to anti-aromatic electronic structures. In the absence of conformational constraint such systems can avoid an anti-aromatic state as non-planar polyalkenes where single-double bond alternation occurs; PMR spectra in such cases reveal atropicity of protons on the π-periphery. Thus NMR provides a simple method for assessment of the many ingenious synthetic products designed as either diatropic (aromatic) or paratropic (anti-aromatic) species.

(b)   **Homologous series of annulenes**

Annulenes are cyclic alternant hydrocarbons (*ie.* those whose structures can be represented by alternating single and double bonds in a carbocyclic framework) corresponding to the general formula $[(C_2H_2)_n, (n > 1)]$ (1). They form an homologous series (including geometric isomers for large n) and can be named accordingly so that benzene (n = 3), as the six carbon homologue, would be [6]annulene (see P. J. Garratt, "Aromaticity", *op. cit.*, Ch. 4).

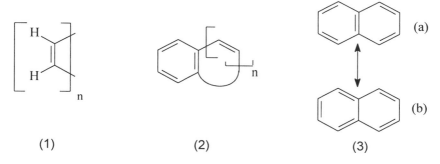

(1)   (2)   (3)

Related homologous series are also significant. For example naphthalene is one of a series of benzannelated annulenes (2), ie. benzo[6]annulene (canonical form 3a). It is also a bicyclic alternant hydrocarbon (canonical form 3b) with an isomer, azulene (4), that implies other homologous series

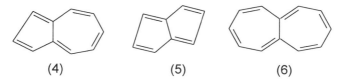

(4)   (5)   (6)

of condensed rings. For example azulene and pentalene (5) both contain five membered rings while azulene and heptalene (6) both contain seven membered rings; in either pairing membership of a homologous family is implied. Furthermore, bicyclic systems like (3)-(6) have as isomers monocyclic dehydroannulenes that contain a triple bond or a cumulene in place of a double bond without prejudice to π-delocalisation, leading to further series of structures.

If the π-system of an annulene is bridged by additional atoms then planarity and aromatic (or anti-aromatic) delocalisation may still be possible, indeed bridging has been used as a constraint against a tendency to distort from planarity. Bridged structures of that sort are discussed here in the context of their parent annulenes.

Finally, the ability of appropriate cyclic systems to attain (4n + 2) delocalised electrons by ion formation, leads us to include some functionalised molecules such as diketones or quinones (7),as parents for ionic species of interest.

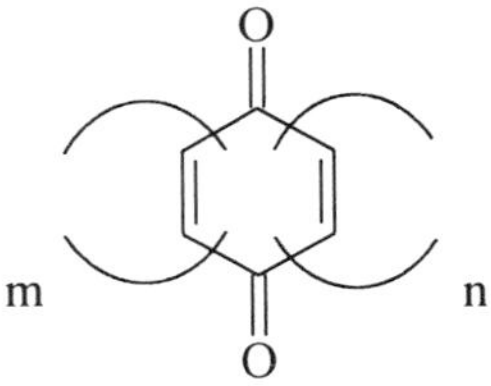

(7)

Earlier work on annulenes and their derivatives has been collated and reviewed (A. T. Balaban, M. Banciu and V. Ciorba, "Annulenes, Benzo- , Hetero- , Homo-Derivatives and their Valence Isomers", Vol. 1-3, CRC Press, Florida, 1987), and a good deal was summarized (according to π-electron numbers in the first Rodd supplement (P. J. Garratt in M. F. Ansell, *op. cit.*).

(c) **C_2 - C_5 systems**

The smallest aromatic rings are 2π cyclopropenium ions of which many stable examples are known (W. E. Billups and A. W. Moorehead, "The Chemistry of the Cyclopropyl Group", Part 2, Z. Rappoport (Ed), Wiley N. Y., 1987), indeed the tricyclopropylcyclopropenium ion is stable in water (R. A. Moss *et al*, J. Amer. Chem. Soc., 1986, **108**, 134). *Ab initio* calculations indicate that the cyclopropenyl carbene (8) will be nucleophilic due to contribution by the zwitterIonic cyclopropenium form (Y. Apeloig *et al*, Tetrahedron Lett., 1980, **21**, 411). An aromatic ylide structure (9) was

(8)

deduced for the singlet ground state of cyclopropenylidene, obtained by cycloreversion and stable in an argon matrix at 10K (H. P. Reisenauer *et al*, Angew. Chem., Int. Ed. Engl., 1984, **23**, 641).

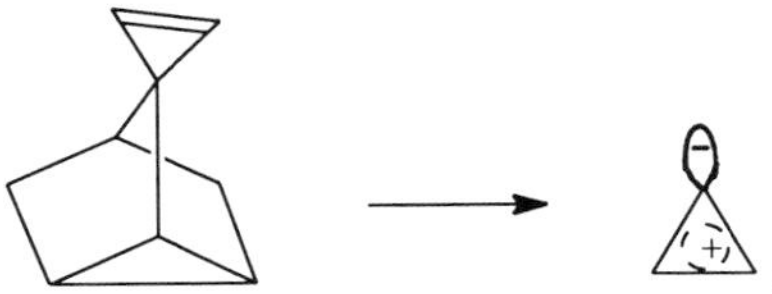

(9)

At -30°C the PMR spectrum of 1, 4-di-t-butylcyclopropene contains signals at 4.02 and 7.09δ suggesting charge separation into the aromatic form (W. E. Billups and L-J. Lin, Tetrahedron Lett., 1983, **24**, 1683):-

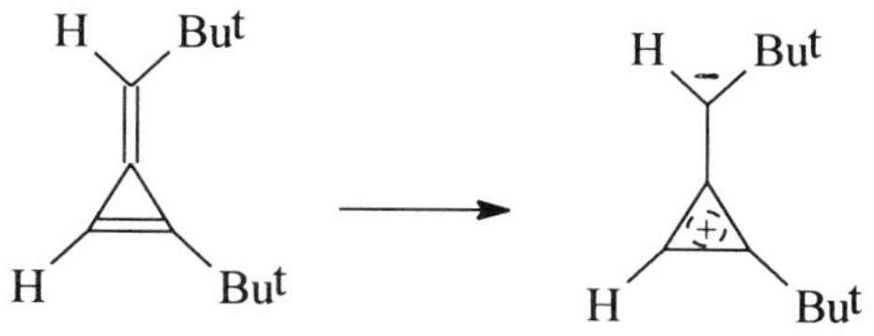

Methylenecyclopropene itself is available by elimination of halogen acids or sulphenic acid or *via* pyrolysis of a diazoketone (W. E. Billups, L-J. Lin and E. W. Casserly, J. Amer. Chem. Soc., 1984, **106**, 3698; S. W. Slaley and T. D. Norden, *ibid*, 1984, **106**, 3699; G. Maier *et al*, Tetrahedron Lett., 1984, **25**, 5645). It is an unstable molecule whose PMR contains two triplets at 3.6 and 8.18δ suggestive of the diatropic, charge-separated form. The benzannelated derivative is, however, a stable solid (B. Halton, C. J. Randall and P. J. Stang, J. Amer. Chem. Soc., 1984, **106**, 6108).

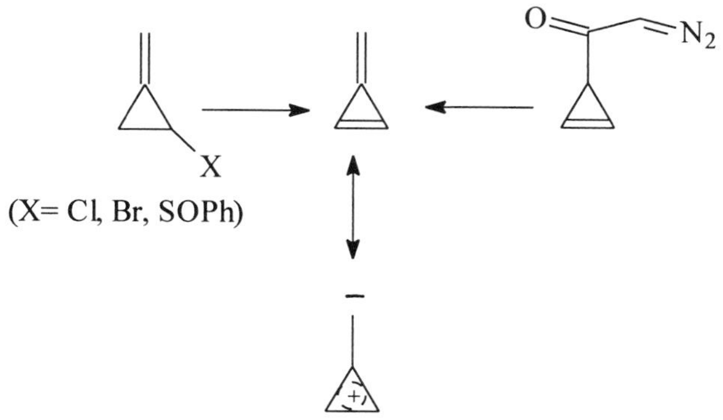

(X= Cl, Br, SOPh)

Cyclopropene itself can be iodinated to give cyclo-C_3I_4, an explosive compound that seems to exist as triiodocyclopropenium iodide (10) (R. Weiss *et al*, Angew Chem., Int. Ed. Eng., 1986, **25**, 103).

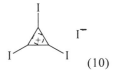
(10)

The well-known triphenylcyclopropenium ion emerges spontaneously by dissociation of the dicyano-p-nitrobenzyl compound (11) upon dissolution in polar solvents (E. M. Arnett *et al*, J. Amer. Chem. Soc., 1983, **105**, 6172):-

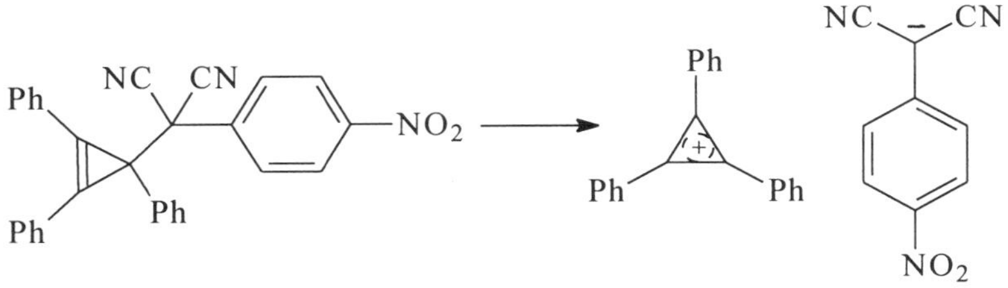

(11)

Cycloproparene chemistry has been reviewed (W. E. Billups, W. A. Rodin and M. M. Haley, Tetrahedron, 1988, **44**, 1305). Cyclobutadiene (review: G. Maier, Angew Chem., Int. Ed. Engl., 1988, **27**, 309) is the smallest of the annulenes having 4π electrons. As such the square planar, delocalised form would be anti-aromatic and an interconverting, rectangular diene structure has been proposed (R = H) (M. J. S. Dewar, K. M. Merz and J. J. P. Stewart, J. Amer. Chem. Soc., 1984, **106**, 4039):-

Two isomers (R = D) have been identified in apparent corroboration (D. W. Whitman and B. K. Carpenter, *ibid*, 1980, **102**, 4272). X-ray studies of sufficiently long-lived cyclobutadiene derivatives confirm the presence of single-double bond alternation (H. Irngartinger *et al*, Chem. Ber., 1988, **121**, 673).

Benzocyclobutadiene has low stability (see R. Boese and D. Blaser, Angew. Chem., Int. Ed. Engl., 1988, **27**, 304). Valence Bond calculations using quantitative resonance theory favour the benzenoid (12) rather than quinoid form (13) (J. M. Schulman and R. L. Disch, J. Amer. Chem. Soc., 1993, **115**, 11153). The parent hydrocarbon is only isolable at 20K but more stable derivatives were found to be non-aromatic or anti-aromatic (see P. J. Garratt, "Aromaticity", op. cit., Ch. 9). Double benzannelation

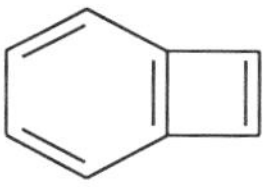

(12) 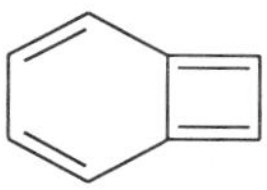   (13) 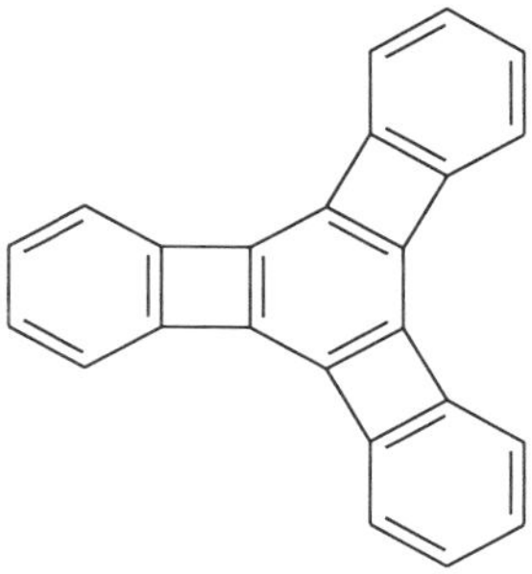   (14)

gives interesting derivatives, the phenylenes, that are discussed later (Section3c).The branched phenylene, tris(benzocyclobutadienato)-benzene (14), is typical in that the central ring corresponds to a bond-localised cyclohexatriene having bond lengths of 1.33 and 1.50 A° (R. Diercks and K. P. C. Vollhardt, J. Amer. Chem. Soc., 1986, **108**, 3150).

Both the dication of cyclobutadiene (a 2π-system) (G. K. S. Prakash, T. N. Randah and G. A. Olah, Angew. Chem., Int. Ed. Engl., 1983, **22**, 390) and its dianion (6π) obey Huckel's rule but MO calculations suggest that neither is actually subject to aromatic stabilisation (B. A. Hess, C. S. Ewig and L. J. Schaad, J. Org. Chem., 1985, **50**, 5869). Transition metal complex

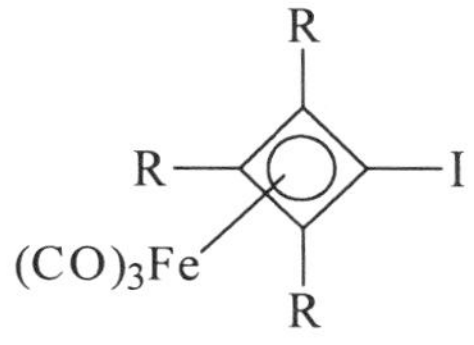

(15) R=H

(16) R=I

derivatives are however fairly stable, eg (15) (J. E. C. Wiegelmann and U. H. F. Bunz, Organometallics, 1993, **12**, 3792) and (16) (U. H. F. Bunz and V. Enkelmann, Angew. Chem., Int. Ed. Engl., 1993, **32**, 1653), and are formal dications of the cyclobutadiene system. The tetraphenyl salt (17) and other apparent dianions (see P. J. Garratt, *op. cit.*, Ch. 5) involve charge-density in substituents rather than on the cyclobutadiene core [*eg.* the phenyl groups in (17)].

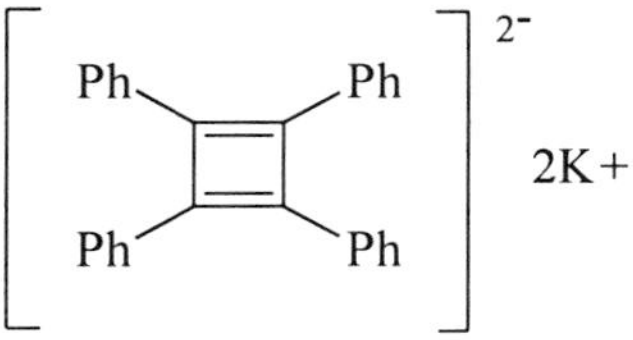

$$\left[\begin{array}{c} Ph \diagdown \diagup Ph \\ \square \\ Ph \diagup \diagdown Ph \end{array} \right]^{2-} \quad 2K+$$

(17)

By Huckel's rule cations of the cyclopentadiene system are anti-aromatic (4 π), but the annelated 9-fluorenyl cation (18) is not subject to appreciable destabilisation (T. L. Amyes, J. P. Richard and M. Novak, J. Amer. Chem. Soc., 1992, **114**, 8032). It can be generated photochemically and used in Friedel-Crafts alkylations of benzene and alkylbenzenes (F. Cozens *et al*, Angew. Chem., Int. Ed. Engl., 1992, **31**, 743).

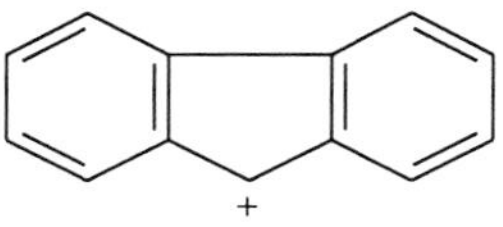

(18) 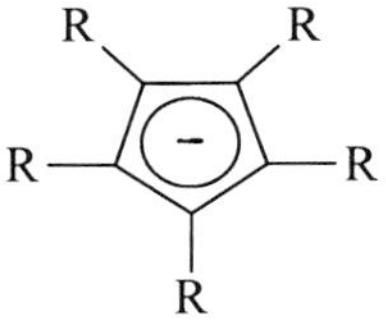   (19)

The 6π cyclopentadienyl ion (19, R = H) is diatropic and aromatic with resonance energy *ca.* 100-113 KJ mol^{-1} (F. G. Bordwell, G. E. Drucker and H. E. Fried, J. Org. Chem., 1981, **46**, 632). The pentakis (trifluoromethyl) derivative (19, R = CF$_3$) dissociates spontaneously from the parent cyclopentadiene (pK_a ≤ -2) and requires concentrated sulphuric acid for reprotonation (E. D. Laganis and D. M. Lemal, J. Amer. Chem. Soc., 1980, **102**, 6633).

Tetraphenylcyclopentadiene is converted to the anion by dark cathodic reduction with H atom homolysis in a procedure that can be used for subsequent alkylation reactions (M. A. Foxe and R. C. Owen, *ibid*, 1980, **102**, 6559). Alternatively photoanodic oxidation of the tetraphenylcyclopentadienyl anion leads to a radical which dimerises:-

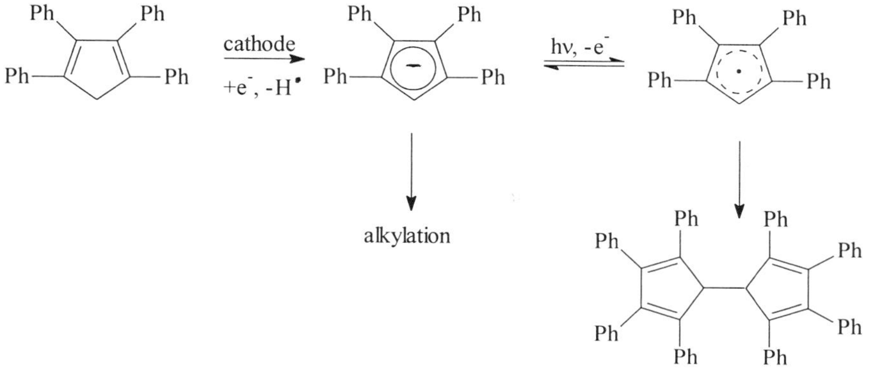

The familiar metallocene sandwich compounds (20) (reviews:D. A. Lemenovskii, V. P. Fedin, Russ. Chem. Rev., 1986, **55**, 127; D. J. Sikora, D. W. Macomber and M. D. Rausch, Adv. Organomet. Chem., 1986, **25**, 317) have been extended into series of multidecker sandwich compounds containing 6π-cyclopentadienide systems (review: W. Siebert, Angew. Chem., Int. Ed. Engl., 1985, **24**, 943).

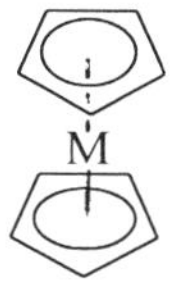

(20) (M = Fe, Co, Ni etc)

(d) **Benzene rings and aromatization**

The stability of the 6π benzene ring system is attested to by the ease of many reactions leading to its formation. The Diels-Alder dimer (22) reverted to benzene with free energy of activation of *ca.* 15.7 kcal mol^{-1} (A. Bertsch, W. Grimme and G. Reinhardt, Angew. Chem., Int. Ed. Engl., 1986, **25**, 377) after photolytic generation from the trimer (21). Under thermal conditions the trimer is stable at 50°C while the dimer reverts to benzene at -20°C. This was explained by the production of only one benzene ring on decomposition of (21) while (22) reverts to two benzene rings (W. Grimme and G. Reinhardt, *ibid*, 1983, **22**, 617):-

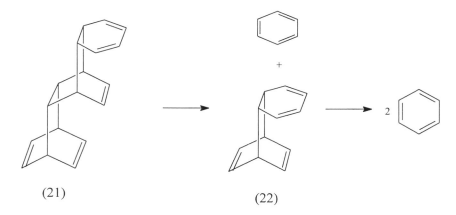

(21)   (22)

Adducts of benzene with anthracene and naphthalene were more stable but showed similar trends (A. Bertsch, W. Grimme and G. Reinhardt, *op. cit.*, but see N. C. Yang and X. Yang, J. Amer. Chem. Soc., 1987, **109**, 3804). For comparable observations in naphthalene chemistry see L. A. Paquette, H. Jendralla and G. Lelucca, *ibid*, 1984, **106**, 1518.

Cycloreversion reactions from adducts (23) releasing benzene (X = Y = H) or naphthalene [X = H, Y = $(CH)_4$] or anthracene [X = Y = $(CH)_4$] provide a

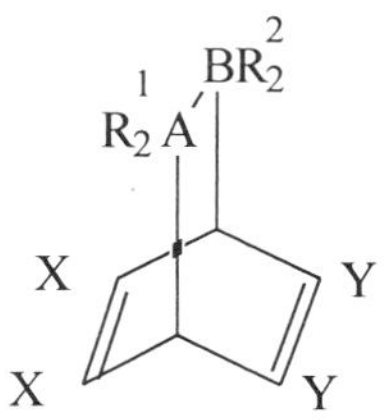

(23)

useful means of generating sensitive molecules (R^1_2 A = B R^2_2), eg. $Me_2Si = SiMe_2$ (Y. Nakadaira, T. Otsuka and H. Saturai, Tetrahedron Lett., 1981, **22**, 2417, 2421), $H_2Si = CH_2$ (G. Maier, G. Mihm and H. P. Reisenauer, Angew. Chem., Int. Ed. Engl., 1981, **20**, 597), and enediols (A = B = C; R^1, R^2 = H or OH) (M-C. Lasne and J-L. Ripoll, Tetrahedron Lett., 1982, **23**, 1587). The same strategy was used in the unmasking of conductive polyacetylene in the "Durham route" preparation (D. C. Bott, J. H. Edwards and W. J. Feast, Polymer, 1984, **25**, 395) and similar approaches are evident in the generation, from benzenoid adducts, of eg. carbon monoxide (D. M. Birney, K. B. Wiberg and J. A. Berson, J. Amer. Chem. Soc., 1988, **110**, 6631), dimethylgermylene (J. Kocher and W. P. Neumann, ibid, 1984, **106**, 3861) and phenylsulphenylnitrene (R. S. Atkinson, M. Lee and J. R. Malpass, J. Chem. Soc., Chem. Commun., 1984, 919).

Aromatic stabilisation of the product was deemed responsible for the unusual base-catalysed migration of a methyl group from a 1,3-cyclohexadiene ring (B. Miller and A. K. Bhattacharya, J. Amer. Chem. Soc., 1980, **102**, 2450):-

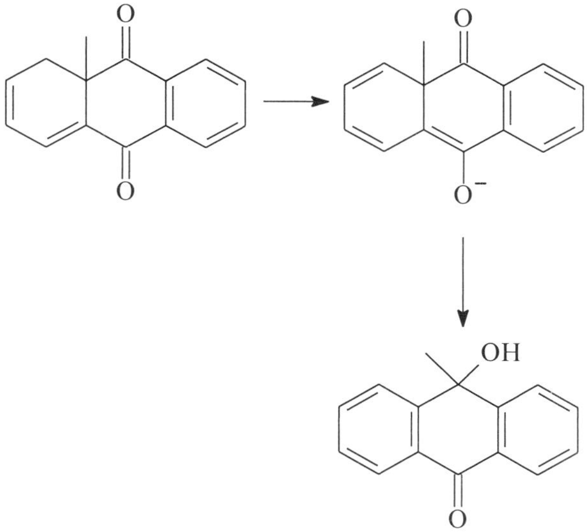

and aromatisation was the driving force in the production of an unusual oxygen cation (trapped by phenolic coupling) (K. Shudo et al, ibid, 1981, **103**, 943):-

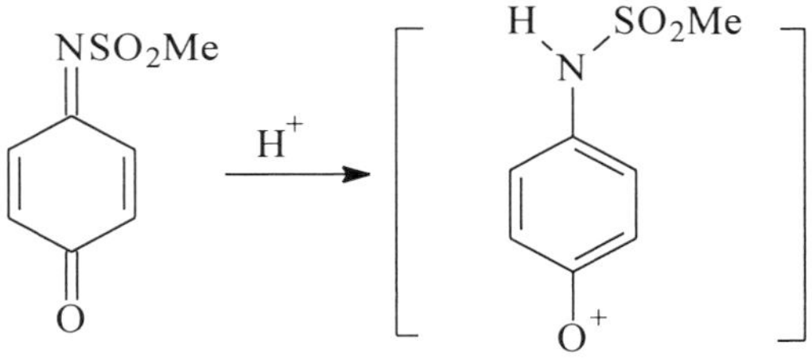

Huckel-type delocalisation of the benzene π-system presupposes a planar ring, but in many benzenoids (*eg.* the cyclophanes, Section 3c) the ring is distorted and properties become modified. The interrelationships between steric strain and aromatic stabilisation have been illustrated by several theoretical and experimental investigations, and the transition state for dehydrogenation of 1,4-cyclohexadiene into benzene has been shown to be boat-like (R. J. Rico, M. Page and C. Doubleday Jr., *ibid*, 1992, **114**, 1131). Some highly crowded benzene derivatives are planar *eg.* hexahalogenobenzenes (J. Almlof and K. Faegri, *ibid*, 1983, **105**, 2965), hexaisopropylbenzene (J. S. Siegel *et al*, *ibid*, 1986, **108**, 1569), and hexakis(dimethylsilyl)benzene (I. I. Schuster, W. Weissensteiner and K. Mislow, *ibid*, 1986, **108**, 6661). Additional crowding, however, leads to distortion as in the puckered hexa-t-butyl, hexakis(trimethylgermyl) and hexakis(trimethylsilyl) derivatives (K. Mislow *et al*, *ibid*, 1986, **108**, 6664). Systems as strained as these tend to be unstable with respect to valence isomerisation (see Section 3a) or ring-opening *eg.* (H. Sakurai *et al*, *ibid*, 1990, **112**, 1799):-

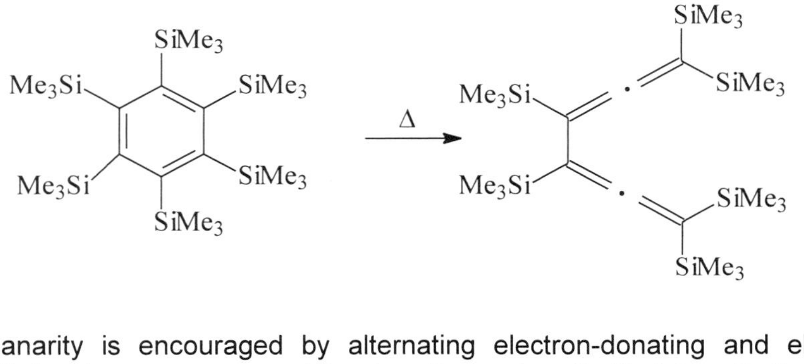

Planarity is encouraged by alternating electron-donating and electron-withdrawing groups around the ring, the push-pull effect (see P. C. Hiberty and G. Ohanessian, *ibid*, 1984, **106**, 6963). However the steric strain imposed in (24) is too great and the molecule exists as a boat or twist boat

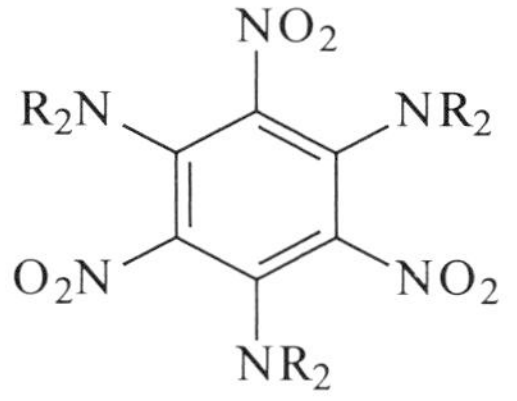

R_2N — benzene ring with NO$_2$, NR$_2$, NO$_2$, NR$_2$, O$_2$N substituents

(24) R = Et
(25) R = H

in solution (J. S. Siegel *et al*, *ibid*, 1989, **111**, 5940). In the less hindered (25) planarity was demonstrated by X-ray crystallography and peripheral H-bonding (predicted from calculations) was a further stabilising influence (K. K. Baldridge and J. S. Siegel, *ibid*, 1993, **115**, 10782).

A variety of ionic and radical species is derived from the benzene ring. Those that arise, *eg.* as "σ-complexes", during substitution reactions

(26)   (27)

receive fuller coverage elsewhere. The spiro or *ipso* species (26) (A. Effio *et al*, *ibid*, 1980, **102**, 6063) and (27) (S. Fornarini and V. Muraglia, *ibid*, 1989, **111**, 873), were both observed during studies of arylethyl systems. Phenyl radicals can arise *via* anaerobic hydrogen atom abstraction from benzene using hydroxyl radicals (P. Muldur and R. Louw, Tetrahedron Lett., 1982, **23**, 2605). Theoretical studies reveal good relative stability for Ph•, which should be available by the cyclisation of dienyne radicals, *eg.* in soot nucleation (M. J. S. Dewar *et al*, J. Amer. Chem. Soc., 1987, **109**, 4456):-

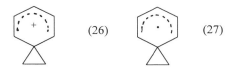

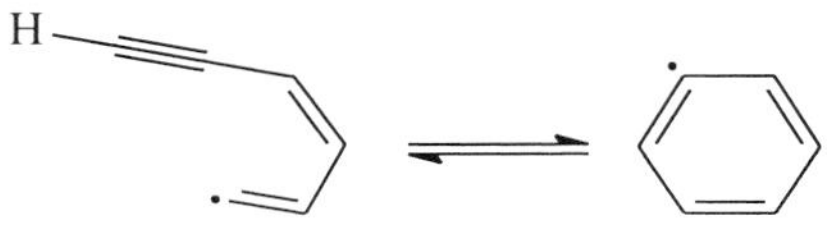

Calculations predict bond shortening to the carbon bearing the unpaired electron in the phenyl radical (J. Pacansky, B. Liu and D. DeFrees, J. Org. Chem., 1986, **51**, 3720):-

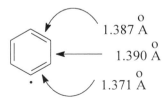

Phenyl cations result from β-decay of 1,4-ditritiobenzene (G. Angelini *et al*, *ibid*, 1980, **45**, 3291), while S_N1 solvolysis of the hindered 2,6-bis(trimethylsilyl)- , or 2,6-di-t-butyl-phenyl triflates produces the corresponding aryl cations (Y. Himeshima, H. Koboyashi and T. Sonada, J. Amer. Chem. Soc., 1985, **107**, 5286). Aryl cations are available indirectly *via* ring closure reactions (M. Hanack *et al*, J. Chem. Soc., Chem. Commun., 1985, 1487; Angew. Chem., Int. Ed. Engl., 1979, **18**, 870).

Aryl anions resulting directly from deprotonation are rare. The negative charge in the anion of 2,3,4,5,6-pentamethylbenzene arises by deprotonation of a methyl group, and involves one resonance stabilised methylene anion site with no fluxion around the other substituents (F. M. Menger and M. G. Banaszczyk, Tetrahedron Lett., 1992, **33**, 301).

Aryl radical anions and cations are important in the fabrication of molecular electronics materials (review: M. R. Bryce, Chem. Soc. Rev., 1991, **20**, 355). The parent radical cation, $C_6H_6^{+\cdot}$, is advocated as a reagent gas for chemical ionisation mass spectrometry (C. Allgood, Y. C. Ma and B. Munson, Anal. Chem., 1991, **63**, 721). Radical cations are readily available from sulphur-substituted arenes, *eg*. (28) (R. Lapouyade and J. P. Morand, J. Chem. Soc., Chem. Commun., 1987, 223) and (29) (N. Boden *et al*, Tetrahedron Lett., 1991, **32**, 6195):-

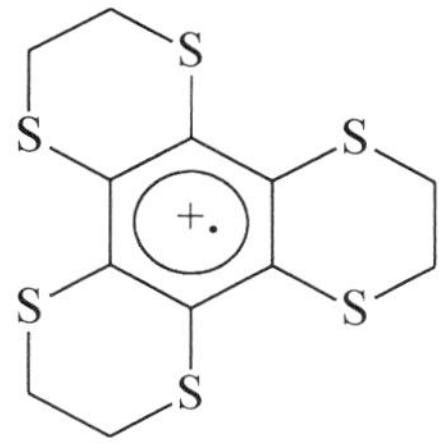

(28) 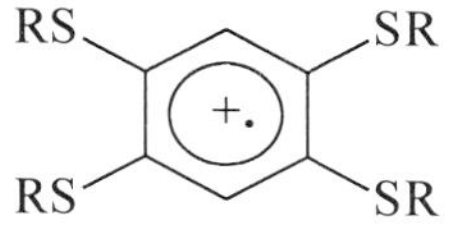 (29)

Arene biradicals arise from Bergman cyclisation of appropriate enediynes as covered in Section 5a. The benzene dication is a 4π system. *ab initio* MO calculations show that it should exist in both a singlet state (chair-like C_{2h} symmetry) and a triplet state (planar D_{6h} symmetry) and not as a delocalised (and anti-aromatic) carbenium species (K. Krogh-Jespersen, J. Amer. Chem. Soc., 1991, **113**, 417).

Dianion salts (30) and (31) have been obtained by reduction of the parent arenes with lithium in THF (A. Sekiguchi *et al*, *ibid*, 1991, **113**, 1464 and 7081):-

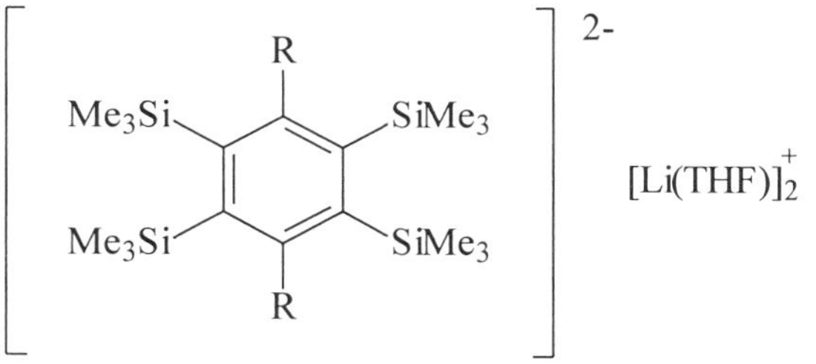

$[Li(THF)]_2^+$ (30) R = SiMe$_3$

(31) R = H

The hindered hexasilyl ion (30) exists in a boat form but X-ray crystallography reveals that the anion in (31) is a planar, 8π-electron species and thus anti-aromatic.

Hexaiodobenzene forms a triflate salt in which the arene occurs as a dication (32). ^{13}C-NMR reveals an upfield shift of 42.6ppm with respect to hexaiodobenzene itself, and the dication is claimed to be an example of σ-aromaticity. σ-Type overlap of p-orbitals containing lone pairs on each iodine gives, in the dication, a 10π-electron system delocalised around the outer "σ-periphery". The inner, carbon periphery is then subject to a diatropic, upfield shift revealing aromaticity in the iodine "ring". The effect is <u>not</u> observed in fluoro-pentaiodobenzene dications (D. J. Sagl and J. C.

Martin, *ibid*, 1988, **110**, 5827; *cf.* V. I. Minkin *et al*, J. Org. Chem. USSR, (Engl. Transl.), 1988, **24**, 1, and D. Cramer, Tetrahedron, 1988, **44**, 7427).

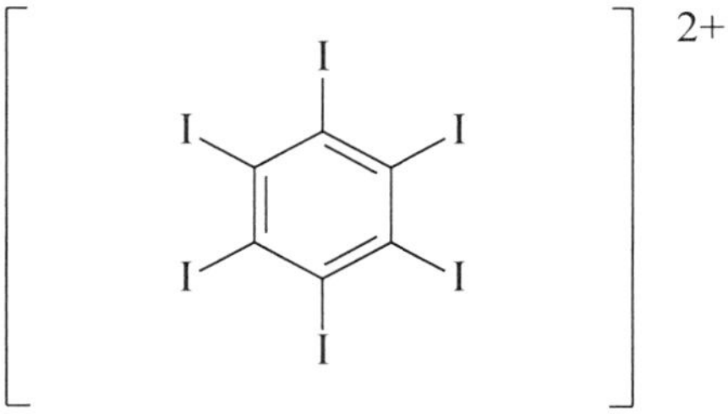

(32)

(e) **8π and 10π systems**

Cyclooctatetraenes (8π) and their 6π-dications were discussed in the first supplement. Benzannelation reduces ring currents, *eg.* paramagnetic currents in antiaromatic systems, as exemplified by the condensed species (33) (as well as its dehydro-derivative having two triple bonds in the central ring) (H. N. C. Wong and F. Sondheimer, Tetrahedron, 1981, **37**, Supp. 1, 99), by (34) (H. Durr *et al*, Angew. Chem., Int. Ed. Engl., 1983, **22**, 332), and by (35) (H. N. C. Wong *et al*, J. Amer. Chem. Soc., 1990, **112**, 7790) which are reasonably stable, planar compounds despite their strain and their 4n π structures. Dehydro[8]annulenes have been reviewed (N. Z. Huang and F. Sondheimer, Acc. Chem. Res., 1982, **15**, 96).

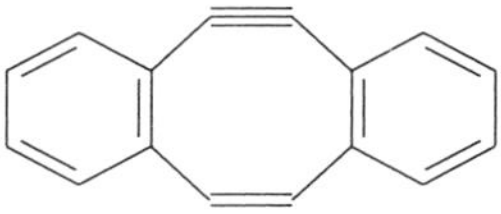

(33) 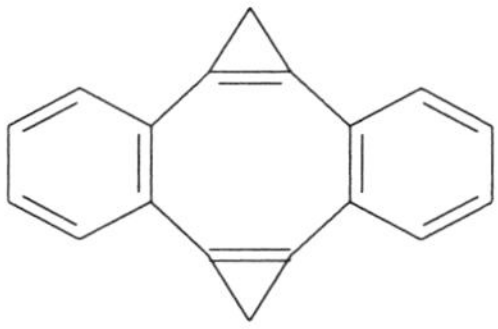   (34) 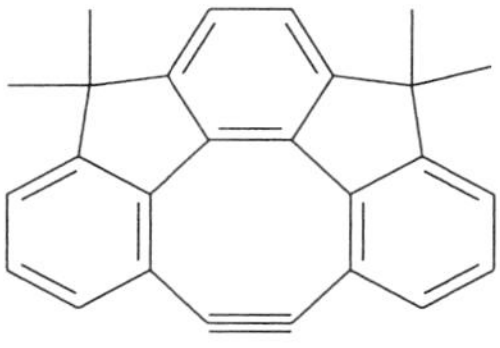   (35)

Planar monocyclic [10]annulenes are unstable by angle-strain unless there are two <u>trans</u> double bonds but even then planarity is prevented by steric hindrance of the inner protons from the <u>trans</u> double bonds. Bridged (homonapthalene) structures can approach planarity however and the parent hydrocarbon 1,6-methano[10]annulene (36, X = Y = H) is

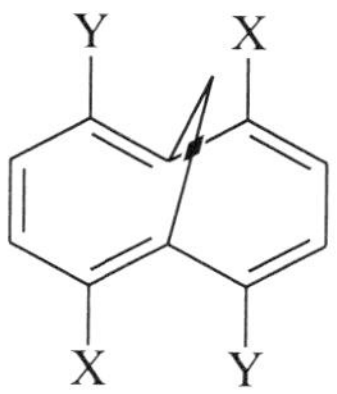

(36)

aromatic (R. Niedlin et al, Angew. Chem., Int. Ed. Engl., 1985, **24**, 587). Chiral derivatives have been reported, eg. (36, X = D, Y = H) (A. Meyer and K. Schogl, Monatsh. Chem., 1992, **123**, 465), and efficient syntheses devised (eg. D. G. Barrett, G. B. Liang and S. H. Gellman, J. Amer. Chem. Soc., 1992, **114**, 6915). However large substituents disrupt the planarity, eg. a derivative (36, X = $SiMe_3$, Y = H) shows bond alternation and conformational isomerism (R. Niedlin et al, op. cit). Fragility of the aromatic system is reflected by a tendency toward valence tautomerism with a bridged tricyclic naphthene structure (D. Cremer and B. Dick, Angew. Chem., Int. Ed. Engl., 1982, **21**, 865; cf. R. Neidlin and U. Kux, ibid, 1993, **32**, 1324):-

The doubly benzannelated derivative indeed, has no extended aromaticity but exists as a bis-styrene system (R. K. Hill, C. B. Giberson and J. V. Silverton, J. Amer. Chem. Soc., 1988, **110**, 497):-

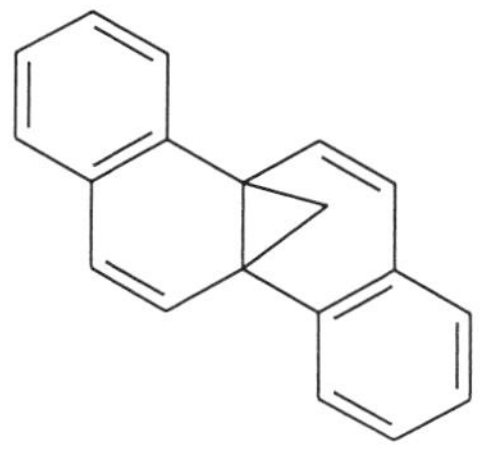

Functionalisation of the bridging methylene can lead to spontaneous isomerisation to the benznorcaradiene system (E. Vogel et al, Angew. Chem., Int. Ed. Engl., 1982, **21**, 869):-

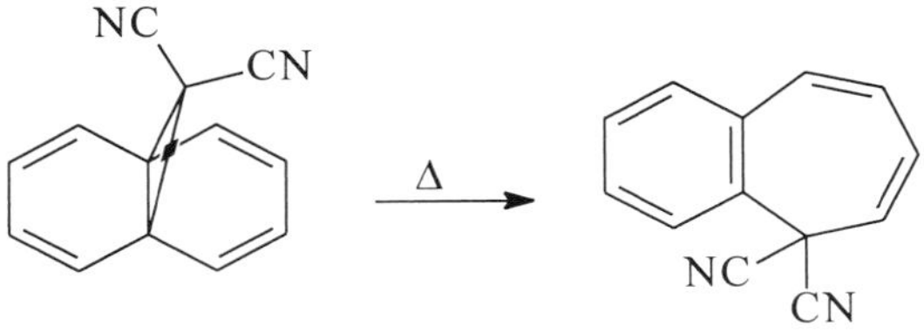

1,5-Methano[10]annulene is also prone to isomerisation; a di-π-methane mechanism was proposed for the thermal process leading to its tricyclic homoazulene isomer (and then to further products) (L. T. Scott and I. Erden, J. Amer. Chem. Soc., 1982, **104**, 1147):-

The methanobridged annulenes are subject to reduction either electrochemically, leading to 10π-dianions (W. Huber et al, Angew. Chem., Int. Ed. Engl., 1982, **21**, 301), or by metal reduction leading to paratropic but reasonably stable 12π-dianions (D. Schmalz and H. Gunther, *ibid*, 1988, **27**, 1692).

Methano-bridged [10]-annulenes are then rather borderline as aromatic structures, aromatic stabilisation being in a precarious balance with other factors (*eg.* steric) that tell against planarity. Direct bridging, involving the sp^2 carbons only, leads to naphthalene isomers. Azulene (37) is a stable

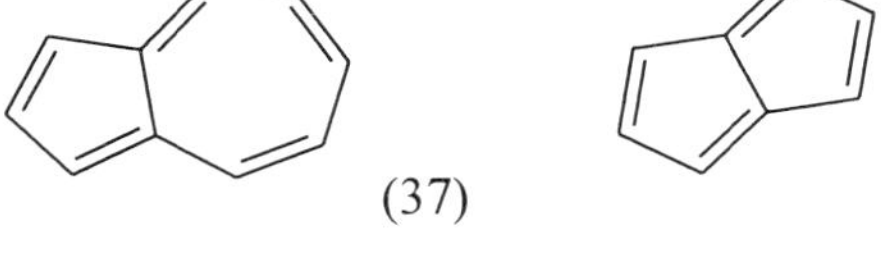

(37)   (38)

aromatic compound for which efficient syntheses are available (*eg.* L. T. Scott, M. A. Minton and M. A. Kirms, J. Amer. Chem. Soc., 1980, **102**, 6311). The isoelectronic 10π-dianion of pentalene (38) has also been prepared and characterised (see P. von R. Scheyer *et al*, J. Chem. Soc., Chem. Commun., 1985, 1263). A valence isomer of azulene, methoxy-"Dewar"-azulene has been prepared and shown to rearrange to methoxyazulene on heating (Y. Sugihara, T. Sugimura and I. Murata, J. Amer. Chem. Soc., 1982, **104**, 4295):-

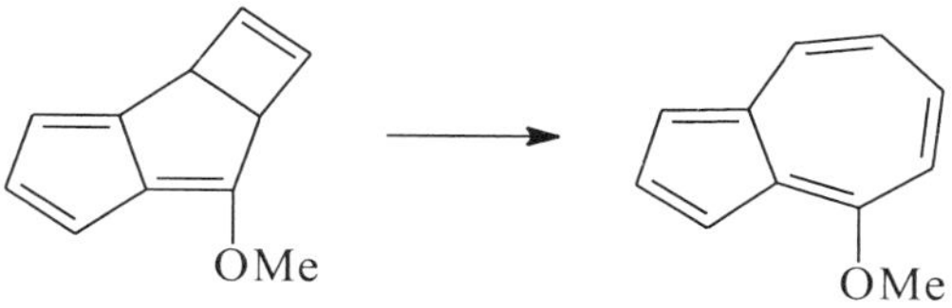

The stability of the azulene system is attested to by the room temperature, Cope-type rearrangement of a suitable dihydroazulene (K. Hafner, J. Hartung and C. Syren, Tetrahedron, 1992, **48**. 4879):-

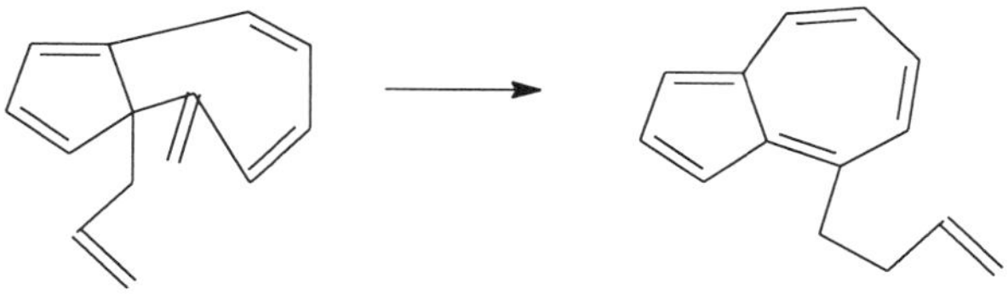

Azulene has a dipole moment suggesting contribution from a charge-separated tropylium cyclopentadienide form, and calculations suggest that azuloquinones should be stable, *eg.* 1,3-azuloquinone (39) (L. T. Scott *et al*, J. Amer. Chem. Soc., 1980, **102**, 5169):-

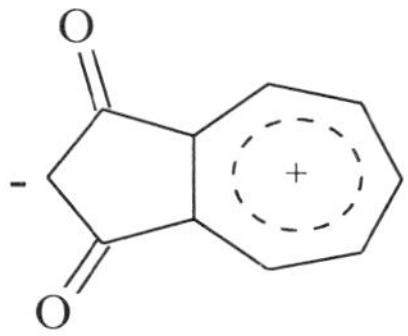

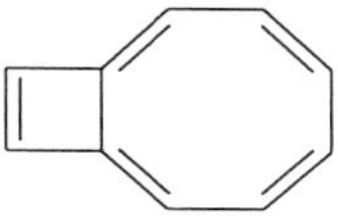

The azulene isomer, bicyclo[6.2.0]decapentene (40), shows a small positive resonance energy (by heats of hydrogenation) (W. D. Roth *et al*, Chem. Ber., 1986, **119**, 837) and does not behave as two fused 4n π systems. Calculations suggest a planar 10π, weakly aromatic periphery (D. Cremer, T. Schmidt and C. W. Bock, J. Org. Chem., 1985, **50**, 2684).

The tricyclic 1,4,7-methino[10]annulene (41, R = X = H) is strongly

(41)   (42)

aromatic (R. C. Haddon and L. T. Scott, Pure Appl. Chem., 1986, **58**, 137). Aromaticity remains in the more strained methyl derivative (41, R = Me, X = H) (C. W. Rees *et al*, J. Chem. Soc., Chem. Commun., 1982, 497, 499) but in the phenolic compound (41, R = Me, X = OH), the less strained keto-tautomer (42) is more stable (R. McCague, C. J. Moody and C. W. Rees, *ibid*, 1982, 622).

(f)   **12π and 14π systems**

Anti-aromatic destabilisation is not sufficient to prevent the preparation of certain 12π-annulenes *eg.* the paratropic (43) (*eg.* δ_{Me} = 4.75) (K. Hafner and V. Kuhn, Angew. Chem., Int. Edn. Engl., 1986, **25**, 632), the monocyclic (44) (J. Anthony, C. B. Knobler and F. Diederich, *ibid*, 1993, **32**, 1148), and the optically active twisted hexamethylheptalene (45) (K. Hafner and G. L. Knaup, Tetrahedron Lett., 1986, **27**, 1665).

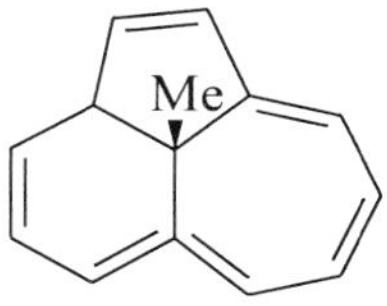

(43) 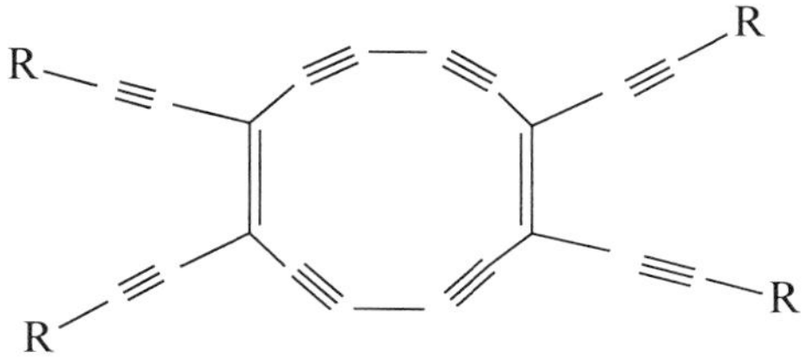 (44) 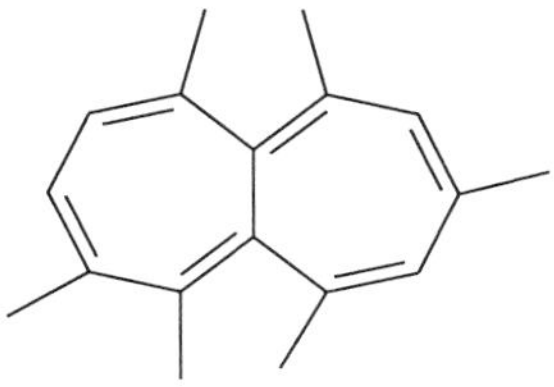 (45)

Surprisingly calculations reveal a small positive resonance energy for planar cis-[12]annulene (R. C. Haddon, Pure Appl. Chem., 1986, **58**, 129). Methano[12]annulenes have been prepared and benzannelation reduces the paratropicity, eg. (46) (L. T. Scott et al, J. Amer. Chem. Soc., 1983, **105**, 1372):-

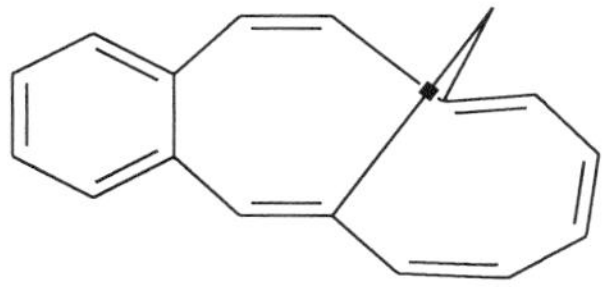

(46)

Aromatic stabilisation of [14]-annulene by electron delocalisation is predicted by electron-pair correlation theory (K. Jug and E. Fasold, ibid, 1987, **109**, 2263). However steric interference of the internal protons reduces stability even in the pyrene-shaped form which, accordingly, is not planar although bond alternation is not apparent by X-ray (H. Rottele and G. Schroder, Chem. Ber., 1982, **115**, 248). But the π-periphery of [14]-annulene is evident in dihydropyrene derivatives (47) (review: R. H. Mitchell, Adv. Theor. Interesting Mol., 1989, **1**, 135):-

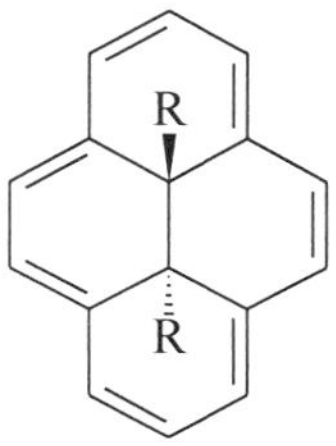

[structure 47 with two R groups on dihydropyrene]   (47)

Trans-15,16-dimethyldihydropyrene (47, R = Me) is a diatropic 14π aromatic compound, and some instructive, annelated derivatives have been prepared. Benzannelation reduces the diatropicity of the annulene leading to bond localisation (R. H. Mitchell *et al*, J. Amer. Chem. Soc., 1982, **104**, 2544 and succeeding papers), while the complex formed between the annulene and $Fe_2(CO)_9$, which is antiaromatic, is stabilised by benzannelation (*ie.* paratropicity AND diatropicity are both reduced) (R. H. Mitchell and P. Zhou, Angew. Chem. Int. Ed. Engl., 1991, **30**, 1013). However annelation involving the "olefinic" 9,10 bond of phenanthrene produces an annulene with a strong diamagnetic ring current (*eg.* δ_{Me} = -3.31) (Y-H. Lai, J. Amer. Chem. Soc., 1985, **107**, 6678):-

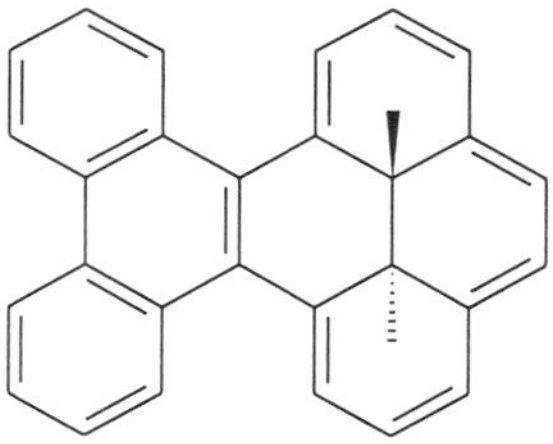

The demonstration of aromaticity in azupyrene (48) (A. G. Anderson Jr., G. M. Masada and G. L. Kao, J. Org. Chem., 1980, **45**, 1312) probably depends on its 14π periphery (in a 16π molecule). Reductive alkylation generates aromatic, bridged 14π annulenes (K. Mullen *et al*, Chem. Ber., 1992, **125**, 505; *cf.* E. Vogel *et al*, Chem. Lett., 1987, 33):-

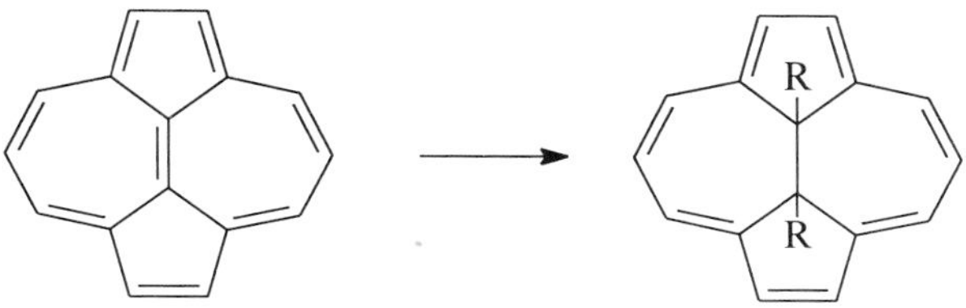

(48)

Peracid oxidation of isopyrene yields a quinone in which resonance stabilisation involves charge separation (E. Vogel et al, Angew. Chem., Int. Edn. Engl., 1991, **30**, 681):-

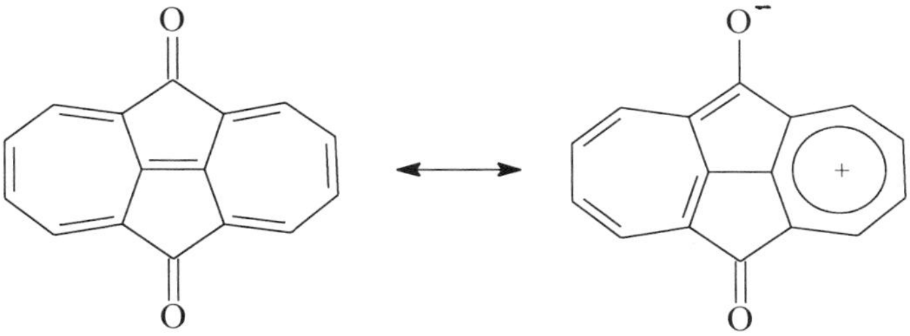

Bridged [14]-annulenes have been prepared that correspond to homoanthracenes (49) and homophenanthrenes (50); two bridges are required to replace all inner protons that would cause steric strain. The *syn*-isomer is shown in each case but the corresponding *anti*-isomers are also known for the hydrocarbons (49, X = Y = CH_2) and (50). In either

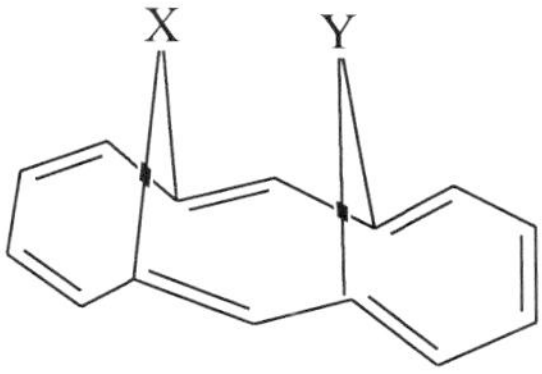

(49) 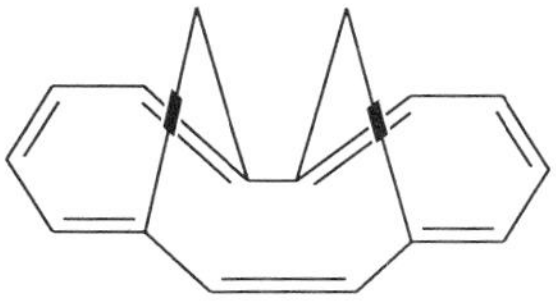   (50)

case only the *syn* forms are diatropic due to poor π overlap in the geometry of the *anti*-isomers (see A. P. Laws and R. Taylor, J. Chem. Soc., Perkin Trans. 2, 1987, 1691; E. Vogel *et al*, Angew. Chem., Int. Ed. Engl., 1986, **25**, 720, 723 and references cited). *Syn*-isomers of a corresponding diether (49, X = Y = O) (E. Vogel *et al*, *ibid*, 1966, **5**, 734), diamine (49, X = Y = NH) (E. Vogel *et al*, J. Amer. Chem. Soc., 1983, **105**, 6982), and amino-ether (49, X = NH, Y = O) (J. A. Marco and J. F. Sanz, Tetrahedron Lett., 1990, **31**, 999) have also been prepared and shown to be diatropic.

Other 14π systems have been designed and prepared. Some (like azulene) contain only sp^2 carbons at their core, *eg.* (K. Hafner, G. F. Thiele and C. Mink, Angew. Chem. Int. Edn. Engl., 1988, **27**, 1191):-

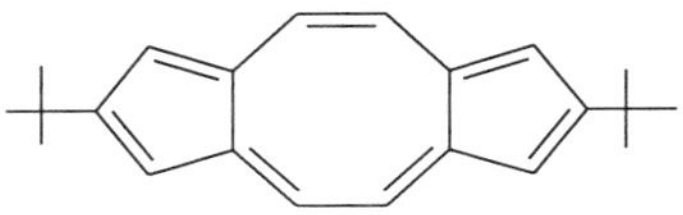

In others an alternant 14π periphery surrounds sp^3 bridging features (*cf.* the dihydropyrenes above) leading to strongly diatropic structures *eg.* (51) (G. Neumann and K. Mullen, J. Amer. Chem. Soc., 1986, **108**, 4105), and (52) (K. Mullen, Pure Appl. Chem., 1986, **58**, 177):-

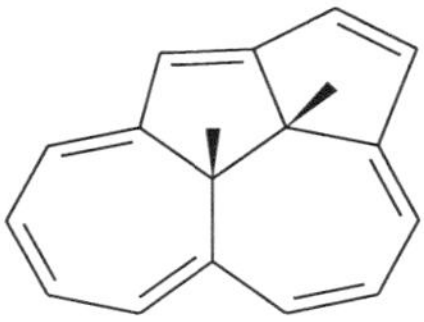

(51) 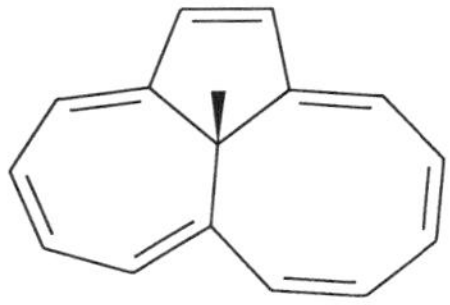 (52)

Cyclopenta[a]azulene derivatives provide interesting structures. In the fulvene (53), cross conjugation disrupts delocalisation; the molecule has no alternant canonical form and behaves as separate azulene and fulvene entities (M. Yasunami *et al*, Bull. Chem. Soc. Jpn., 1993, **66**, 2273).

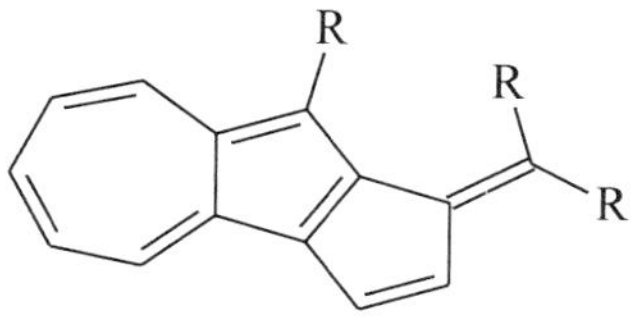

Either isomer of the parent hydrocarbon (*ie.* no exocyclic, fulvene unit) leads, on deprotonation, to the anion (54) which localises as shown and behaves as isolated azulene and cyclopentadienide moeities (Y. Kitamuri *et al*, *ibid*, 1992, **65**, 1527).

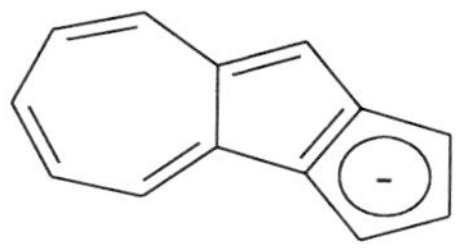

(g) **Higher annulenes**

1,6:9,14-Bismethano[16]annulene (55) is formally anti-aromatic but leads to strongly diatropic ions, a 14π-dication with antimony pentafluoride and an 18π-dianion with activated lithium (E. Vogel *et al*, Tetrahedron Lett., 1985, **26**, 3087, 3091). Similar planning led to the [24]-annulene (56) which is paratropic (J. Ojima *et al*, *ibid*, 1986, **27**, 975).

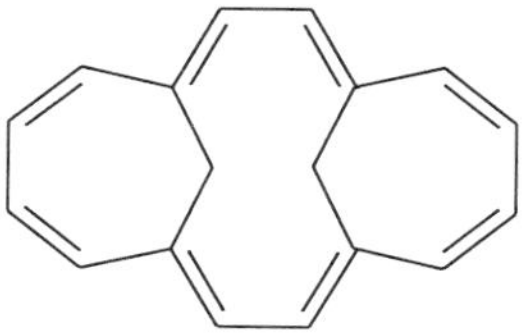

It has been demonstrated that unbridged [26]-annulene (57) is not aromatic (see T. Wessel *et al*, Angew. Chem., Int. Edn. Engl., 1993, **32**, 1148). This confirms a long-standing prediction based on Huckel's approach (M. J. S. Dewar and G. J. Gleicher, J. Amer. Chem. Soc., 1965, **87**, 685) that resonance stabilisation in (4n + 2) rings would be sufficient to sustain planarity only as far as [22]-annulene, which is indeed diatropic (F. Sondheimer *et al*, J. Chem. Soc., Chem. Commun., 1971, 338).

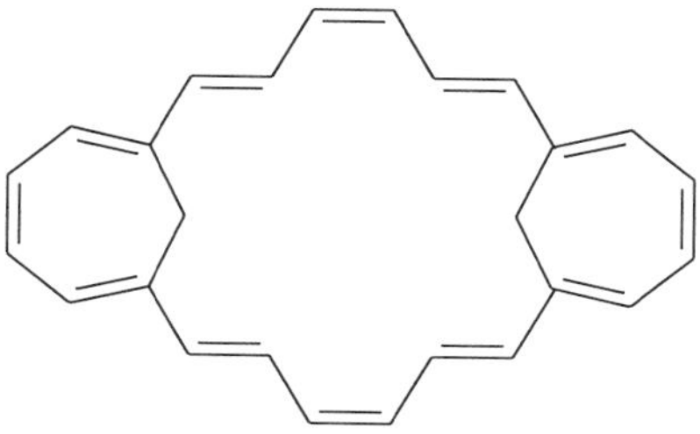

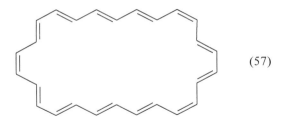

(57)

However, the use of methano-bridging and triple bonds or odd-numbered cumulenes in molecular design has led to a wealth of synthetic annulene derivatives that are (approximately) planar as judged from NMR and other spectra. Using Wittig-reaction methodology, Ojima and co-workers have prepared every member of the series shown in (58) corresponding to a sequence of tetradehydromethano[18]annulene up to tetradehydromethano[38]annulene, initially excepting the [36]annulene (J. Ojima *et al*, J. Chem. Soc., Perkin Trans. 1, 1988, 385 and references cited):-

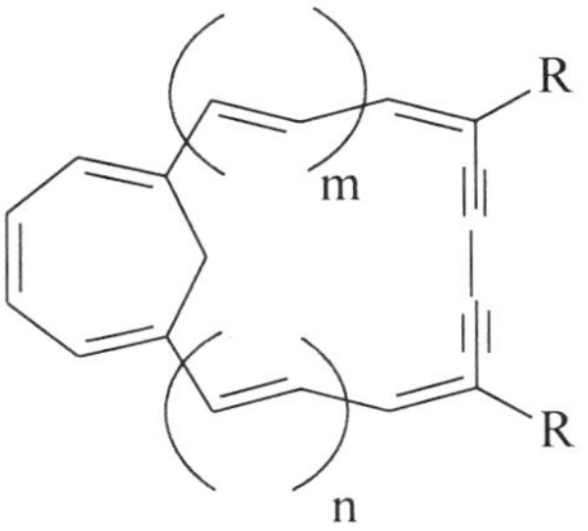

(58)

NMR of the annulenes (58) revealed diatropic behaviour for the $(4n + 2)\pi$ [18]- , [22]- , [26]- , [30]- , and [34]-annulenes. On the other hand paratropicity was demonstrated for the $4n\pi$ [20]- , [24]- , and [28]-annulenes. The [32]- and [38]-annulenes were atropic. Subsequent work provided the missing, corresponding [36]-annulene which was also atropic (J. Ojima *et al*, *ibid*, 1993, 983). Clearly, attention to molecular design can deliver annulene derivatives for whom diamagnetic or paramagnetic ring currents are sustainable in ring sizes much greater than those of the parent annulenes.

Using aldol chemistry and copper coupling of terminal alkynes, Ojima and co-workers have also prepared diketone analogues of the above annulenes (58). Methano-bridged tetradehydroannulenediones (59) were

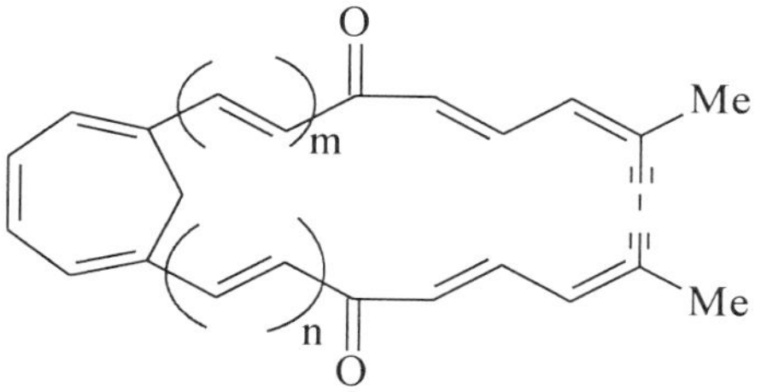

(59)

produced corresponding to [20]- , [24]- , [26]- , [28]- , [30]- , and [32]-annulenediones (J. Ojima *et al, ibid*, 1993, 89), and subsequently the [22]-annulenedione (*idem, ibid*, 1994, 1579). The neutral diones were atropic apart from the [24]-annulenedione (59, m = n = 1) which was weakly diatropic (presumably due to polarisation of the two carbonyls); lack of planarity was considered responsible for the atropicity. However, dissolution in D_2SO_4 led to NMR spectra revealing diatropic dications from both the [20]- and [24]-annulenediones (18π and 22π respectively) and a paratropic 20π dication from the [22]-annulenedione. The higher homologues decomposed before NMR data could be obtained.

The cumulenes (60), methano-bridged dichlorodidehydro-[16]- , [20]- , and [24]-annulenediones also afforded diatropic dications in D_2SO_4 (14π, 18π and 22π respectively). Surprisingly, and unlike systems (58) and (59) the diatropic effect increased with ring size (J. Ojima *et al, ibid*, 1994, 1453):-

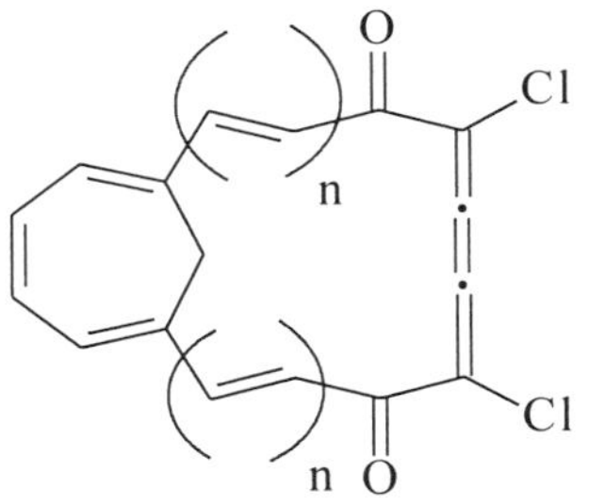

(60) n=1,2,3

Modification of the aldol/copper coupling protocols led to a series of monocyclic compounds (61), tetramethyloctadehydrodihydro-[26]- , [28]- , [30]- , [32]- and [34]-annulenediones (J. Ojima *et al*, *ibid*, 1994, 1957):-

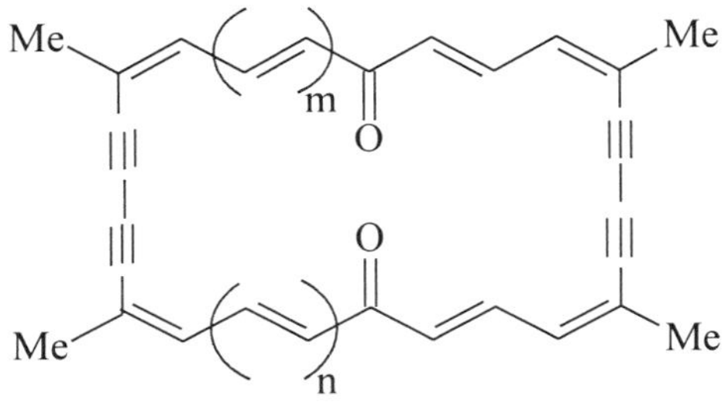

(61)

All five annulenediones were reasonably stable, but effectively atropic species in neutral conditions. In D_2SO_4 however, the [26]-annulenedione (61, m = n = 1) generated a paratropic 24π dication and the [28]-annulenedione (61, m = 1, n = 2) a diatropic 26π dication, the largest monocarbocyclic tropic systems so far (but see Section 2i).

Greater tropicity is expected in ionic rather than comparable neutral species; indeed the bicyclic annulene (62) behaves as two separate annulene rings when neutral but as a single (peripheral) ring when ionised (W. Huber *et al*, Helv. Chim. Acta., 1986, **69**, 949):-

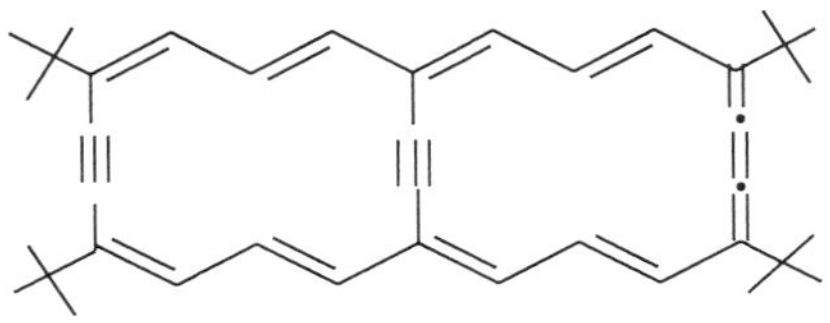

(62)

(h)   **Carbon-only systems**

Rings of sp-hybridised carbon atoms are called cyclo[n]carbons (review: F. Diederich and Y. Rubin, Angew. Chem., Int. Edn. Engl., 1992, **31**, 1101). Calculations on the [18]-annulene, cyclo[18]carbon (63), identify either the alkyne (63a, having D_{9h} symmetry) or the cumulene (63b, having D_{18h} symmetry) as the more stable form depending on how the calculations are performed (see F. Diederich et al, Science, 1989, **245**, 1088, and V. Parasuk, J. Almlof and M. W. Feyereisen, J. Amer. Chem. Soc., 1991, **113**, 1049):-

(63)

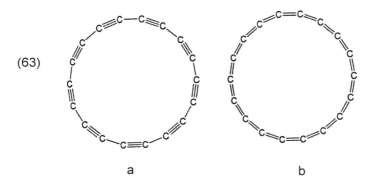

a b

Both calculations predict room-temperature stability for (63) and, at planarity, the angle strain would be no worse than other, known cycloalkynes (review: R. Gleiter, Angew. Chem., Int. Edn. Engl., 1992, **31**, 27). The resultant aromaticity would involve two delocalised 18π electron systems perpendicular to each other.

Synthetic approaches to (63) were based on Sondheimer's method for the planar, diatropic hexadehydro[18]annulene (64a) (W. H. Okamura and F. Sondheimer, J. Amer. Chem. Soc., 1967, **89**, 5991), and non-planar, atropic octadehydro[24]annulene (64b) (R. M. McQuilkin, P. J. Garratt and F. Sondheimer, ibid, 1970, **92**, 6682). Minor modification of the approach has provided the aromatic compound (64c) (J. Anthony, C. B. Hobbler and F. Diederich, Angew. Chem., Int. Edn. Engl., 1993, **32**, 406), but the use of groups X that provide stable, extrudable species on the periphery promises eventual access to (63).

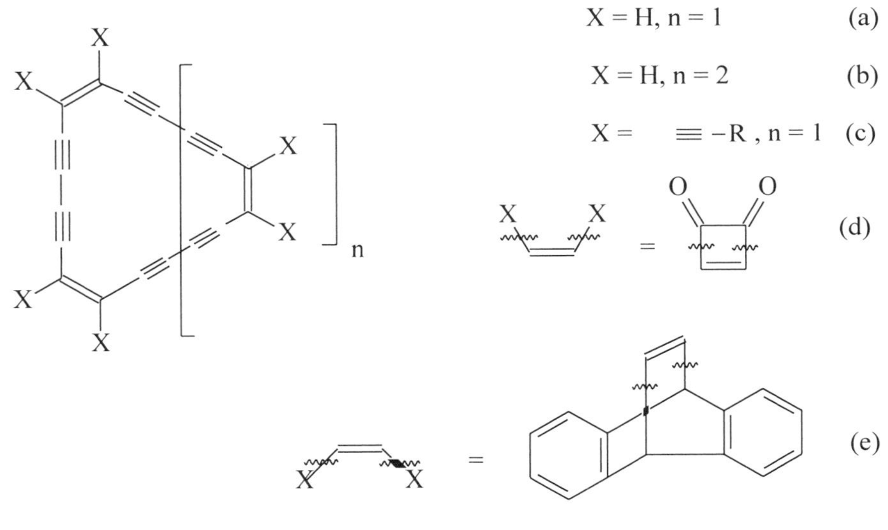

	X = H, n = 1	(a)
	X = H, n = 2	(b)
	X = ≡–R , n = 1	(c)

(d)

(e)

(64)

Extrusion of six CO molecules from (64d, n = 1) would generate (63). Acetals of (64d) (n = 1, 2 and 3) have been prepared and shown to be planar for n = 1 or 2, and diatropic (n = 1) or paratropic (n = 2) as expected. The pentamer (n = 3) was atropic (F. Diederich et al, J. Amer. Chem. Soc., 1989, **111**, 6870; ibid, 1990, **112**, 1607 and 1618). The corresponding cyclobutene-1,2-diones (64d) arose by deprotection in concentrated sulphuric acid but were too unstable for detailed study; their laser desorption FTMS spectra do, however, display prominent ions for C_{18}, C_{24} and C_{30} respectively (Y. Rubin et al, ibid, 1991, **113**, 495).

Cyclo[18]carbon (63) would also result from extrusion of three anthracene molecules from adduct (64e, n = 1), a planar diatropic molecule that is very stable but generates C_{18} as a prominent species in resonant two photon ionisation MS (see F. Diederich and Y. Rubin, op. cit.).

(i) **Heterocyclic systems**

Among the small-ring heterocycles, boracyclopropene (65) is calculated to have ca 70% of the resonance energy of the cyclopropenium cation (P. von R. Schleyer, J. Amer. Chem. Soc., 1981, **103**, 1375) whereas no significant stabilisation is expected for the silacyclopropenium cation (66) (or the silacyclopentadienyl anion) (M. S. Gordon, P. Boudjouk and F. Anwari, *ibid*, 1983, **105**, 649).

$$\underset{(65)}{\overset{B}{\triangle}} \quad \underset{(66)}{\overset{Si}{\triangle}}$$

Resonance has been demonstrated in the diphosphirenium cation (67) isolated as its tetrafluoroborate salt (G. Bertrand *et al*, *ibid*, 1991, **113**, 8160):-

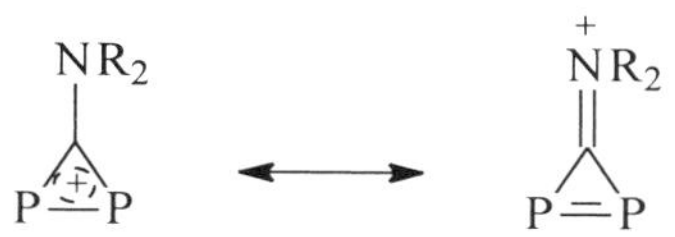

$$\underset{P\overset{(+)}{=}P}{\overset{NR_2}{\triangle}} \longleftrightarrow \underset{P=P}{\overset{\overset{+}{N}R_2}{\triangle}} \quad (67)$$

A general synthesis of metal "oles" (68, E = As, Sb, Bi), using zirconium transfer, is available that can be extended *eg.* for E = Ga, In, Sn, Se (P. J. Fagan and V. A. Nugent, J. Amer. Chem. Soc., 1988, **110**, 2310). In other work the bismolyl anion has been generated and bibismole (69) isolated (A. J. Ashe and F. J. Drone, Organometallics, 1984, **3**, 495). Dewar-furan and Dewar-thiophene have both been identified in matrix photolysis experiments (W. A. Rendall *et al*, J. Amer. Chem. Soc., 1986, **108**, 1691; *cf.* I. G. Pitt, R. A. Russell and R. N. Warrener, *ibid*,1985, **107**, 7176).

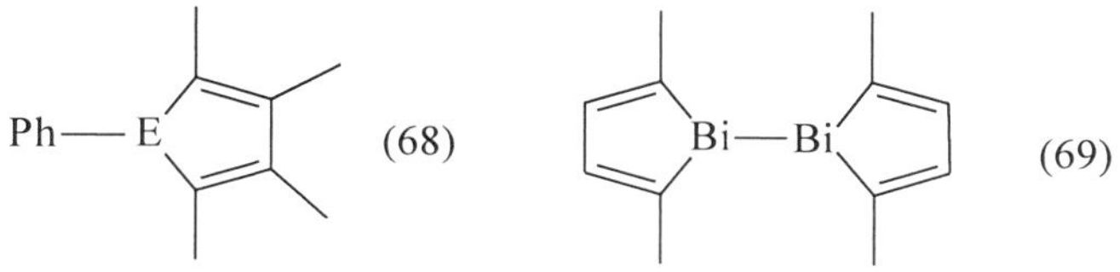

A variety of heterobenzenes (70) has been studied. Silabenzene (X = SiH) is expected to be aromatic (H. B. Schlegel, B. Coleman and M. Jones, *ibid*, 1978, **100**, 649); it and the silatoluene (X = SiMe) have been identified spectroscopically (G. T. Burns *et al*, *ibid*, 1980, **102**, 429), and a reasonably stable derivative (71) isolated (G. Markl and W. Schlosser, Angew. Chem., Int. Ed. Engl., 1988, **27**, 963). Both phosphabenzenes (or phosphinines) (70, X = P) (G. Markl and C. Dorges, *ibid*, 1991, **30**, 106; S. Holand, L. Ricard and F. Mathey, J. Org. Chem., 1991, **56**, 4031), and arsabenzenes (70, X = As) (A. J. Ashe *et al*, *ibid*, 1981, **46**, 881), have been prepared and studied. Both are aromatic but arsenic allows a lesser degree of aromaticity than does phosphorus (or nitrogen) (review: C. W. Bird, Tetrahedron, 1990, **46**, 5697). Indeed a variety of six-membered phosphorus compounds has been made including a 1,3,5-triphosphabenzene (E. Fluck *et al*, Angew. Chem., Int. Ed. Engl., 1986, **25**, 1002), a phosphaphenol (G. Markl and A. Kallmunzer, Tetrahedron Lett., 1989, **30**, 5245), and the all-heteroatom ring (72) which is planar and has equal B-P bond lengths (partial double bonds) (H. V. R. Dias and P. P. Power, Angew. Chem., Int. Ed. Engl., 1987, **26**, 1270).

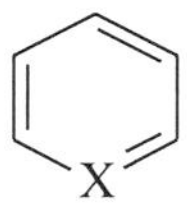

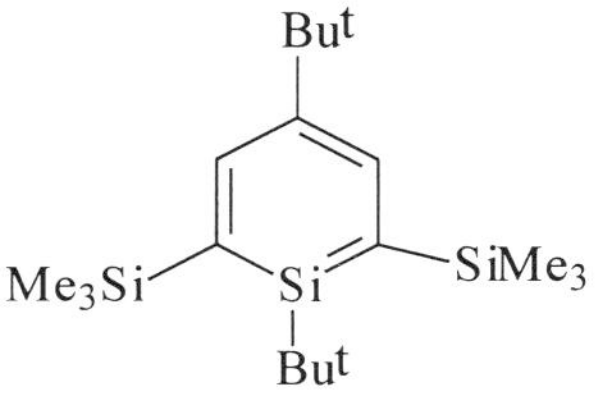

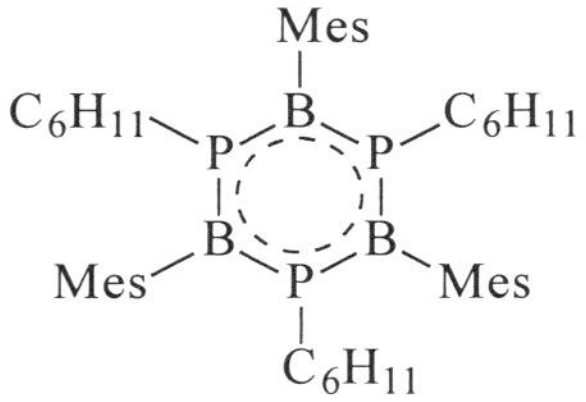

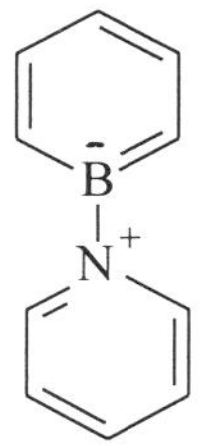

Ab initio calculations predict that, while hexaazabenzene should be planar and stable (P. Saxe and H. F. Schaefer III, J. Amer. Chem. Soc., 1983, **105**, 1760), hexaphosphabenzene would be unstable with respect to its valence isomers (D. S. Warren and B. M. Gilmore, *ibid*, 1992, **114**, 5378). Using π-complex isolation methods Scherer and co-workers have studied both hexaphosphabenzene and the P_5 analogue (O. J. Scherer *et al*, Angew. Chem., Int. Ed. Engl., 1985, **24**, 351 and 1986, **25**, 363). Nitrogen is involved in two fascinating compounds, the aromatic borabenzene-pyridine complex (73) (G. Maier *et al*, Chem. Ber., 1985, **118**, 1644), and the extraordinary bicyclic compounds (74, X = As or Sb) whose planar, delocalised structures were rationalised by contribution from tripolar canonical forms eg. (A. J. Arduengo III et al, J. Amer. Chem. Soc., 1985, **107**, 1089 and 5543):-

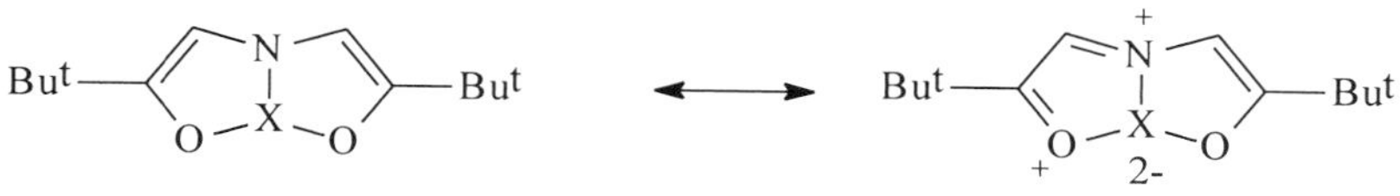

(74)

The seven-membered heterocycle (75, X = CH) is a planar, 10π system (S-N bonds having partial double bond character) that gives classical aromatic reactions (C. W. Rees et al, J. Chem. Soc., Chem. Commun., 1985, 396 and 398). Its aza-analogue (75, X = N) is also a stable, 10π aromatic compound (P. J. Dunn, J. L. Morris and C. W. Rees, J. Chem. Soc., Perkin Trans. 1, 1988, 1745).

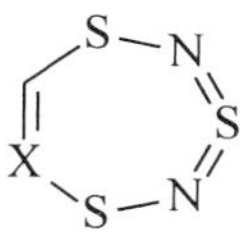

(75) 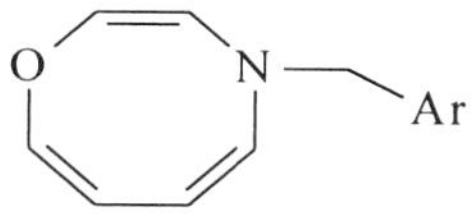   (76)

Oxazine (76) is a stable, planar molecule with little bond-alternation (X-ray) suggesting a 10π delocalised structure (H. Prinzback et al, Angew. Chem., Int. Ed. Engl., 1984, **23**, 309). An analogous (but 8π) diazine (77) generates a stable, 10π dianion (C. Schneiders et al, ibid, 1985, **24**, 576), as do the cyanoheterocycles (78) with base (H. Prinzbach et al, Tetrahedron Lett., 1987, **28**, 2517).

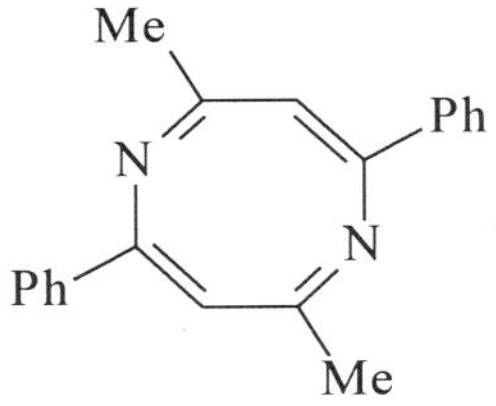

(77) [structure with Me, Ph, N groups]

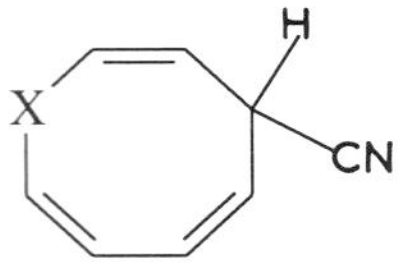

(78) (X = NR , O)

Ojima and co-workers have produced azaannulenes (79, m = 1 - 3, n = 1 - 3) with 14- to 22-membered rings showing the expected diatropic $(4n + 2)\pi$ - , and paratropic $(4n\pi)$-behaviour (J. Ojima et al, J. Chem. Soc., Perkin Trans. 1, 1986, 933).

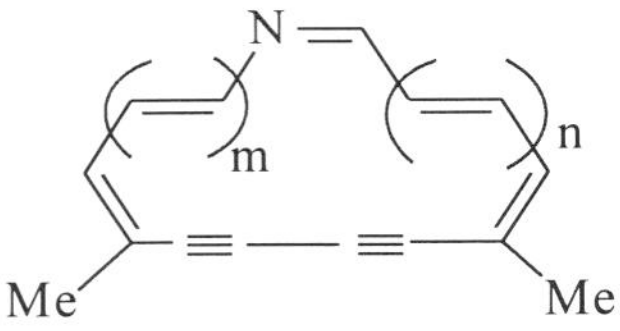

(79)

Methanobridged heteroannulenes have also been prepared. The 10π thiaannulene (80) arose via a fragmentation reaction (H. Kato and S. Toda, J. Chem. Soc., Chem. Commun., 1982, 510) while a more general approach to methanobridged thiaannules has been devised (J. Ojima et al, J. Chem. Soc., Perkin Trans. 1, 1990, 333). 1,6-methanoaza[10]annulene (81), despite apparent delocalisation (by NMR and X-ray), shows little

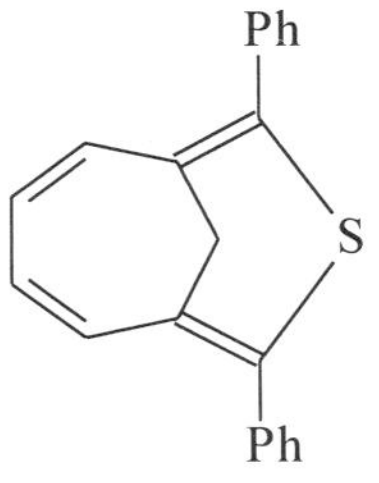

(80)

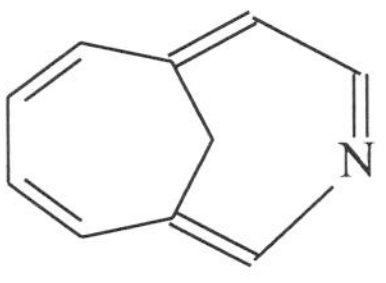

(81)

aromatic stability (E. Vogel et al, Angew. Chem., Int. Ed. Engl., 1985, **24**, 592), and the diaza[22]annulene (82) was atropic in NMR investigations (J. Ojima et al, J. Chem. Soc., Perkin Trans. 1, 1993, 975).

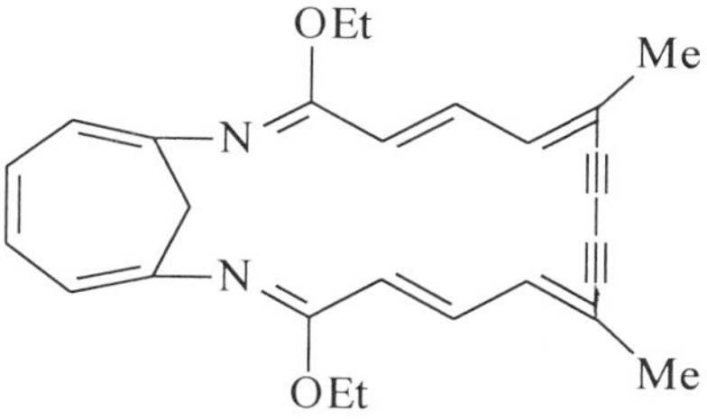

The incorporation of furan or pyrrole units has been used to produce macrocyclic annulenes; eg. the bisfuran (83) is a 14π, aromatic analogue of tropolone (H. Ogawa et al, Tetrahedron Lett., 1985, **26**, 5567). The 20π carbon periphery in (84) derives from diprotonated porphyrin (X = NH) which generates the familiar 18π aromatic system associated with the natural porphyrin pigments. The aromaticity of other analogues eg. dications from (84, X = O, S, Se) and of expanded, vinylogous macrocycles, has been reviewed (E. Vogel, Pure Appl. Chem., 1993, **65**, 143). Franck and co-workers have exploited the planar constraints imposed in tetrapyrrolic macrocycles to prepare and study large aromatic molecules. The 26π annulene (85, n = 1) is strongly diatropic with ca 25 ppm difference between inner and outer protons (M. Gosmann and B. Franck, Angew. Chem., Int. Ed. Engl., 1986, **25**, 1100). Other 26π expanded porphyrins have been reported also (eg. B. Franck et al, ibid, 1993, **32**, 1148; J. L. Sessler, T. Morishima and V. Lynch, ibid, 1991, **30**, 977). However the 34π analogue (85, n = 2), with an even stronger diatropic effect (Δδ ca 30 ppm), is the largest individual aromatic system encountered so far (G. Knubel and B. Franck, ibid, 1988, **27**, 1179).

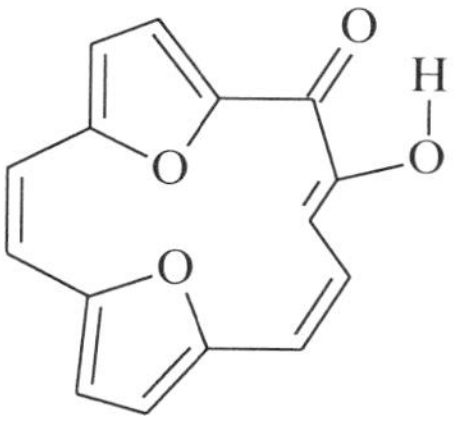

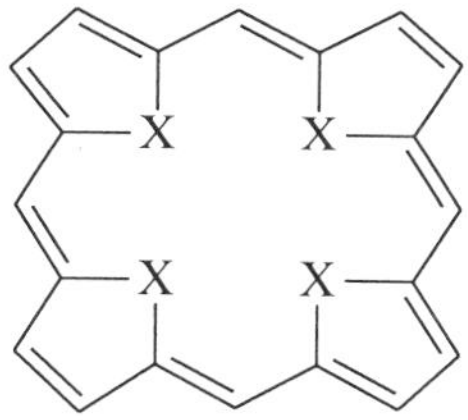

(85)

3. BENZENOID COMPOUNDS

Aspects of the benzene ring, related to its place among the annulene series, were discussed earlier (Section 2 d). Here we consider isomerism of benzene and of dehydrobenzene, and survey the larger molecules obtained when several benzene rings are connected together.

(a) Benzene isomers

Two isomeric species, 1,2,3-cyclohexatriene (1) and cyclohexen-3-yne (2) have been prepared using F^- elimination of TMS and triflate from appropriate trimethylsilyl, triflate precursors. The unstable products can be trapped by dienophiles (W. C. Shakespeare and R. P. Johnson, J. Amer. Chem. Soc., 1990, **112**, 8578), or isolated as zirconocene complexes (J. Yin, K. A. Abboud and W. M. Jones, ibid, 1993, **115**, 8859).

(1)    (2)

Benzene itself automerizes at sufficiently high temperatures. [1,2-^{13}C]Benzene scrambles to generate 1,2:1,3:1,4 homomer mixtures (ratio 72:24:4), and the process can be rationalised via thermal benzene-benzvalene rearrangements (L. T. Scott, N. H. Roelofs and T-H. Tsang, ibid, 1987, **109**, 5456; but cf. K. M. Merz and L. T. Scott, J. Chem. Soc., Chem. Commun., 1993, 412):-

(3)

Benzvalene (3) is one of four recognised valence isomers of benzene, the others being Dewar benzene (4), prismane (5) and bicyclopropenyl (6). Although a prefulvene biradical has also been identified on the theoretical $(CH)_6$ energy surface (S. Oikawa et al, J. Amer. Chem. Soc., 1984, **106**,

6751), the involvement of fulvene (7) as a direct photochemical valence isomer is unconfirmed.

(4)   (5)   (6)   (7)

The valence isomers have been reviewed (F. Bickelhaupt and W. H. de Wolf, Recl. Trav. Chim. Pays-Bas, 1988, **107**, 459) and an X-ray structure of bicyclopropenyl was obtained recently (W. E. Billups *et al*, J. Amer. Chem. Soc., 1993, **115**, 743). Interconversions between benzene and the valence isomers (3-6) can be realised photochemically using suitable derivatives *eg*. (a. F. Bokisch *et al*, Chem. Ber., 1991, **124**, 1831; b. S. Miki *et al*, Tetrahedron Lett., 1992, **33**, 953; c. S. Miki *et al*, *ibid*, 1992, **33**, 1619):-

Benzvalene (review: M. Christl, Angew. Chem., Int. Ed. Engl., 1981, **20**, 529) has been used as a precursor for ring-opening metathesis polymerisation producing conjugated polyenes related to polyacetylene (T. M. Swager, D. A. Dougherty and R. H. Grubbs, J. Amer. Chem. Soc., 1988, **110**, 2973):-

(b) Benzynes

The three possible isomers of C_6H_4 are usually called benzynes [review: T. L. Gilchrist in "The Chemistry of Functional Groups, Supplement C", S. Patai and Z. Rappoport (Eds), Wiley (N. Y.), 1983], although only o-benzyne (8) contains a formal triple bond. Heat of formation calculations match experimental values for o-benzyne only, while m- and p-benzyne are anomalous (A. Nicolaides and W. T. Borden, J. Amer. Chem. Soc., 1993, **115**, 11951; S. U. Wierschke, J. J. Nash and R. R. Squires, ibid, 1993, **115**, 11960). p-Benzyne is now established as a diradical species (9) (T. P. Lockhart, C. B. Mallon and R. G. Bergman, ibid, 1980, **102**, 5976) that can be generated by pyrolitic, Bergman cyclisation of cis-enediynes (Section 5 a) (T. P. Lockhart and R. G. Bergman, ibid, 1981, **103**, 4082, 4091):-

    (8)   (9)

Other calculations confirm a diradical structure for both m- and p-benzyne (M. J. S. Dewar, G. P. Ford and C. H. Reynolds, ibid, 1983, **105**, 3162).

Relatively stable o-benzyne complexes have been reported with titanocenes (E. G. Berkovich et al, Chem. Ber., 1980, **113**, 70) and especially zirconocenes, eg. (S. L. Buchwald, R. T. Lumm and J. C. Dewan, J. Amer. Chem. Soc., 1986, **108**, 7441):-

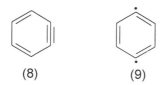

o-Benzyne analogues have been obtained from phenanthrene (H-F. Grutzmacher and U. Straetmans, Tetrahedron, 1980, **36**, 807), and from various heteroarenes (review: M. G. Reinecke, *ibid*, 1982, **38**, 427). Highly strained arynes have also been prepared as judged from the isolation of furan adducts, *eg.* from bisarynes (H. Hart *et al*, J. Amer. Chem. Soc., 1980, **102**, 6649), and from cyclopropynes (B. Halton and C. J. Randall, *ibid*, 1983, **105**, 3162).

The generation of benzynes in the gas phase has been reviewed (R. F. C. Brown and F. W. Eastwood, SYNLETT, 1993, 9). Usually *o*-benzyne is prepared in solution and used *in situ*, most commonly by treating a suitable halogenobenzene with strong base (H. Hart *et al*, J. Org. Chem., 1983, **48**, 4357), or from *o*-aminobenzoic acids by elimination [see structure (10) below]. Other methods include fluoride-assisted generation from trimethylsilylaryl triflates (K. Shankaran and V. Snieckus, Tetrahedron Lett., 1984, **25**, 2827), decomposition of N-aminobenzotriazoles using lead tetraacetate (H. Hart and D. Ok, *ibid*, 1984, **25**, 5415), and photolysis of benzocyclobutenedione (H. V. Linnert and J. M. Riveros, J. Chem. Soc., Chem. Commun., 1993, 48):-

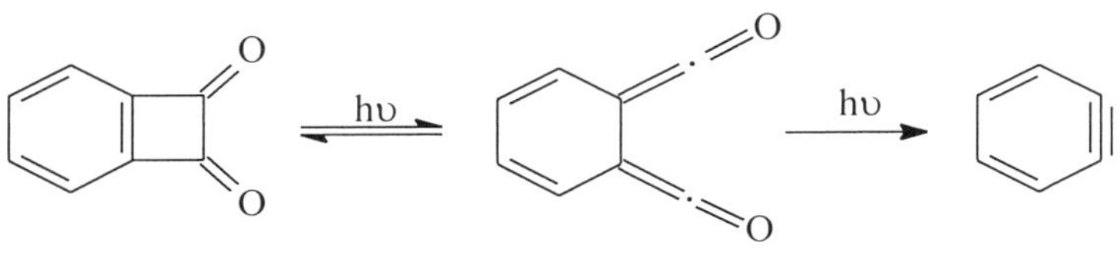

o-Benzyne is useful for the synthesis of a wide range of benzene ring containing molecules. It is susceptible to nucleophiles, to cycloaddition and to dipolar addition. It is a very reactive substrate for [4 + 2] cycloaddition especially with furans (H. Hart and Y. Takehira, J. Org. Chem., 1982, **47**, 4370), but also with less obvious dienes *eg.* (L. Castedo *et al*, Tetrahedron Lett., 1982, **23**, 457):-

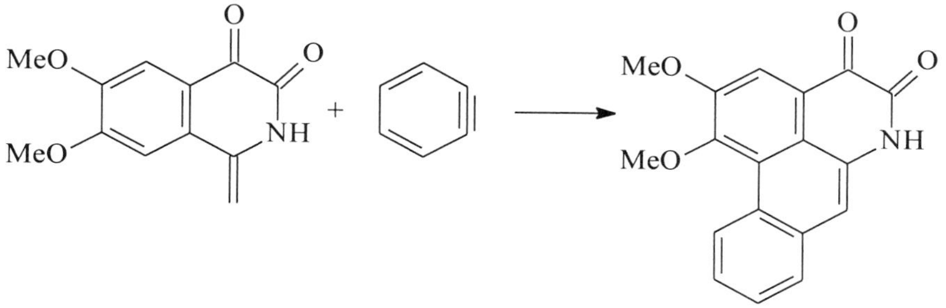

Intramolecular [4 + 2] reactions can be used to construct complex molecules, *eg.* the decomposition and cycloaddition of the aminobenzoic acid (10) (W. M. Best and D. Wege, *ibid*, 1981, **22**, 4877):-

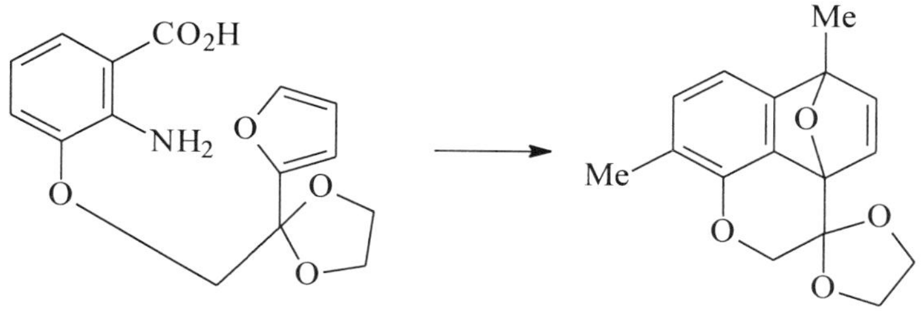

(10)

o-Benzyne also undergoes [2 + 2] cycloaddition with alkenes *eg.* to form benzocyclobutenes (R. V. Stevens and G. S. Bisacchi, J. Org. Chem., 1982, **47**, 2393; *cf.* M-C. Carre, B. Gregoire and B. Caubere, *ibid*, 1984, **49**, 2050), or reactive intermediates for further elaboration (*eg.* C. W. G. Fishwick, R. C. Gupta and R. C. Storr, J. Chem. Soc., Perkin Trans. 1, 1984, 2827):-

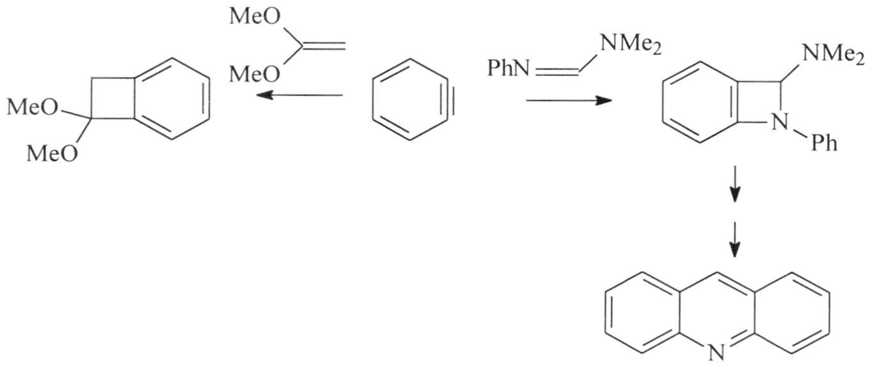

o-Benzyne is trapped by diazofluorene in a 1,3-dipolar addition reaction (K. Hirakawa, Y. Minami and S. Hayashi, *ibid*, 1982, 577):-

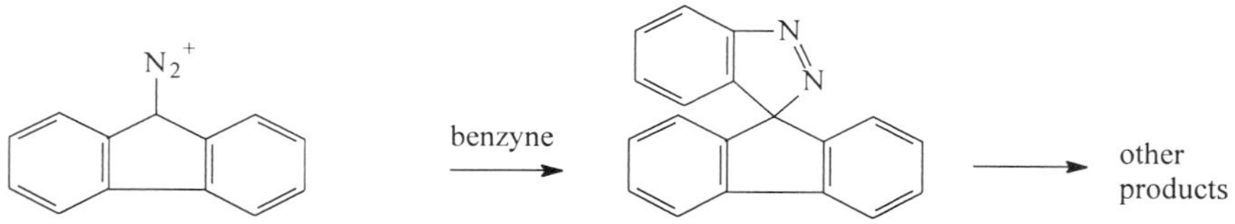

Certain oxygen and sulphur nucleophiles add to *o*-benzyne leading to aryl anions and thence to addition products *eg.* (J. Nakayama, S. Takene and M. Hoshino, Tetrahedron Lett., 1984, **25**, 2679; *cf.* R. B. Bates and K. D. Janda, J. Org. Chem., 1982, **47**, 4374; and K. Shankaran and V. Snieckus, *op. cit.*):-

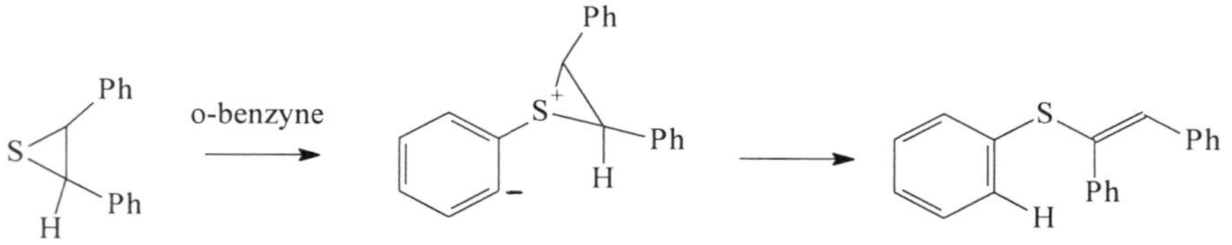

The generation of zirconium-benzyne complexes can be followed by alkyne insertion and metal exchange *eg.* in the synthesis of a stilbaindole (S. L. Buchwald, R. A. Fisher and B. M. Foxman, Angew. Chem., Int. Ed. Engl., 1990, **29**, 771):-

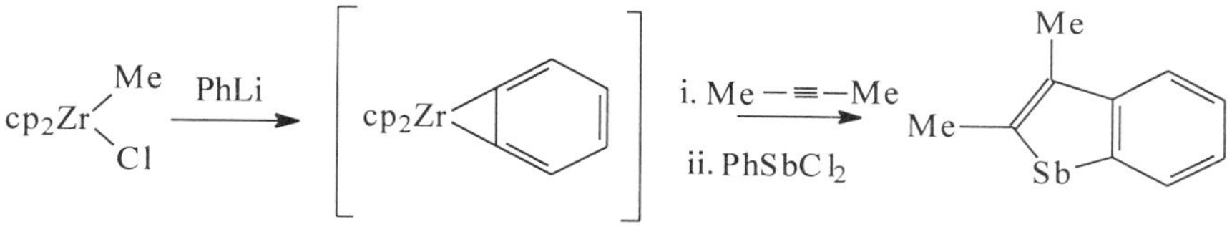

(c) Non-condensed systems

(i) Cyclophanes

Benzocyclo(n)alkenes are "ortho-analogues" of [n]-metacyclophanes (11) and [n]-paracyclophanes (12). The smallest reported examples (unstable compounds generated briefly from their corresponding Dewar isomers) are [4]-metacyclophane (G. B. M. Kostermans *et al*, J. Amer. Chem. Soc., 1987, **109**, 7887) and [4]-paracyclophane (T. Tsuji and S. Nishida, J. Chem. Soc., Chem. Commun., 1987, 1189). Corresponding dibenzo compounds are the [n, m]-metacyclophane (13) and [n, m]-paracyclophane (14) double decker molecules.

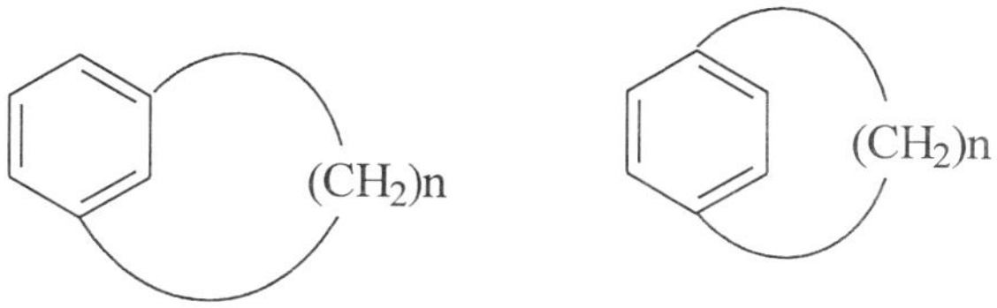

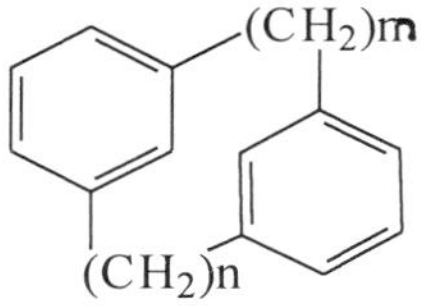

(13) 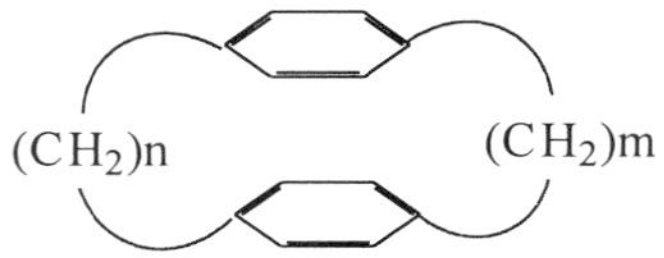   (14)

In the latter systems smaller bridges can be accommodated (*eg.* n = m = 2) and multiple bridging can be used, for example [2.2.2.2.2.2] (1,2,3,4,5,6)cyclophane (15) has been prepared and studied (Y. Sekine and V. Boekelheide, J. Amer. Chem. Soc., 1981, **103**, 1777).

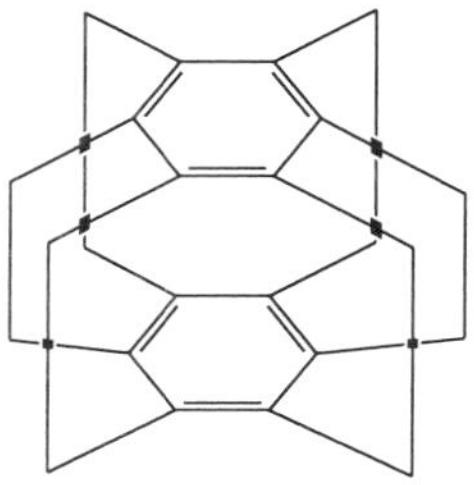

(15) 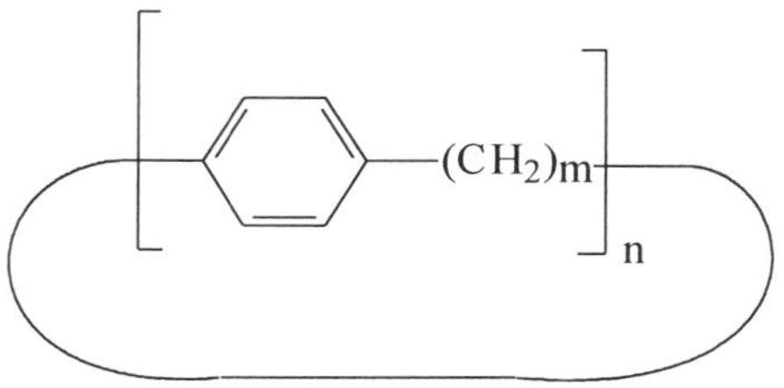   (16)

Alternatively, cyclic oligomers (16, or the meta-isomers, calixarenes) can be made, *eg.* [1.1.1.1.1.1]-paracyclophane (16, n = 6 m = 1) (G. W. Gribble and C. F. Nutaitis, Tetrahedron Lett., 1985, **26**, 6023), and furthermore unsaturated linkages can be used [*eg.* (16) but with (CH = CH)$_m$ rather than (CH$_2$)$_m$] (see K. Mullen *et al*, J. Amer. Chem. Soc., 1984, **106**, 7514).

Cyclophanes are important compounds and a detailed monograph has been published (F. Diederich, "Cyclophanes", R. S. C. Monographs in Supramolecular Chemistry, R. S. C. (Cambridge), 1991). A sizeable family of cyclophane relatives has been prepared that includes pyrrolecyclophanes (F. Muller and J. Mattay, Angew. Chem., Int. Ed. Engl., 1992, **31**, 209), pyridinecyclophanes (H. C. Kang and V. Boekelheide, *ibid*, 1981, **20**, 571), metallocenecyclophanes (S. Kamiyama *et al*, Bull. Chem. Soc. Jpn., 1981, **54**, 2079), azulenocyclophanes (Y. Fukazawa, M.

Sobukawa and S. Ito, Tetrahedron Lett., 1982, **23**, 2129), cyclophanequinones (Y. Miyahara, T. Inazu and T. Yoshino, *ibid*, 1982, **23**, 2189), troponecyclophanes (Y. Fujise *et al*, *ibid*, 1982, **23**, 1601), furanocyclophanes (H. Hopf and B. Witulski, Pure Appl. Chem., 1993, **65**, 47), naphthaleno- and anthracenocyclophanes (Y. Tobe *et al*, J. Amer. Chem. Soc., 1990, **29**, 191).

Variations to the linkages between benzene rings has resulted in *eg.* porphyrin-bridged cyclophanes (F. Diederich *et al*, Angew. Chem., Int. Ed. Engl., 1990, **29**, 191), benzannelated bridges (A. de Meijere *et al*, *ibid*, 1990, **29**, 1418), crown-ether bridges (R. Leppkes and F. Vogtle, Chem. Ber., 1983, **116**, 215), diene bridges (useful for further elaboration) (B. Thulin and O. Wennerstrom, Acta Chem. Scand., Ser. B, 1983, **37**, 297, 589), sulphur bridges (L. R. Hanton and H. Sikanyika, J. Chem. Soc., Perkin Trans. 1, 1994, 1883), selenium bridges (H. Higuchi and S. Misumi, Tetrahedron Lett., 1982, **23**, 5571), a series of S and N containing bridges (F. Vogtle *et al*, J. Chem. Soc., Chem. Commun., 1989, 1757 and 1994, 1361), an adamantyl bridge (R. Lemmerz, M. Nieger and F. Vogtle, *ibid*, 1993, 1168), alkyne bridges (W. J. Youngs *et al*, *ibid*, 1994, 1257), dimethylgermyl bridges (A. Sekiguchi *et al*, J. Organomet. Chem., 1990, **390**, C27) and dimethylsilyl bridges (W. Ando, T. Tsumuraya and Y. Kabe, Angew. Chem., Int. Ed. Engl., 1990, **29**, 778).

Other cyclophanes have been designed and prepared with materials applications in mind. These include a cyclophane propeller compound (W. Kissener and F. Vogtle, *ibid*, 1985, **24**, 222), a donor-acceptor charge-transfer[2,2]-metacyclophane (H. A. Staab, A. Dohling and C. Krieger, Tetrahedron Lett., 1991, **32**, 2215), and a metacyclophane polymer (S. Mizogami and S. Yoshimura, J. Chem. Soc., Chem. Commun., 1985, 1736).

Interest in cyclophanes is widespread. Among the most intriguing are those with the greatest strain, *eg.* (11), (12) and (14) with n and m being small. Synthetic methods for small cyclophanes have been reviewed (V. V. Kane, W. H. de Wolf and F. Bickelhaupt, Tetrahedron, 1994, **50**, 4575), and methods involving either extrusion of links from longer spacer chains, or valence isomerisation of Dewar-benzene precursors are particularly useful for difficult targets. Strain is one of the most interesting features of [n]-cyclophanes (11) and (12). The n = 4 examples, cited above, could not be isolated but [5]-metacyclophane (11, n = 5) can be studied; it has a highly distorted benzene ring but is aromatic even so (P. C. M. van Zijl *et al*, J. Amer. Chem. Soc., 1986, **108**, 1415). [5]-Paracyclophane (12, n = 5) is barely isolatable and is also strongly distorted though aromatic (L. W. Jenneskens *et al*, *ibid*, 1985, **107**, 3716); substituted derivatives show improved stability (G. B. M. Kostermans, W. H. de Wolf and F. Bickelhaupt, Tetrahedron Lett., 1986, **27**, 1095). [6]-Paracyclophane (11, n = 6), while still strained, can be isolated and stored; efficient syntheses have been devised (*eg.* J. Liebe, C. Wolff and W. Tochtermann, *ibid*, 1982, **23**, 171; Y. Tobe *et al*, Chem. Lett., 1983, 1645).

Strained [n]-cyclophanes contain benzene rings distorted into boat conformations (17) with distortion angles, θ, estimated at 19.4° for [6]-paracyclophane (Y. Tobe *et al*, J. Org. Chem., 1987, **52**, 2639), 20.5° for [6]-paracycloph-3-ene (*ie.* *cis*-alkene bridge) (Y. Tobe *et al*, J. Amer. Chem. Soc., 1987, **109**, 1136), and 23° for [5]-paracyclophane which is calculated to be 78Kcal mol^{-1} less stable than benzene (J. E. Rice *et al*, *ibid*, 1987, **109**, 2902). Although aromaticity is retained these distorted

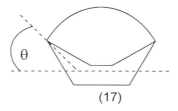

(17)

benzene rings exhibit anomalous reactivity, *eg.* they are prone to nucleophilic attack (P. A. Kraakman *et al*, *ibid*, 1990, **112**, 6638; Y. Tobe *et al*, J. Org. Chem., 1993, **58**, 5883), to Diels-Alder reactions (L. A. M. Turkenburg *et al*, Angew. Chem., Int. Ed. Engl., 1982, **21**, 298) and other additions (Y. Tobe *et al*, Tetrahedron, 1986, **42**, 1851). There is a tendency to undergo skeletal rearrangement (L. W. Jenneskens *et al*, J. Amer. Chem. Soc., 1990, **112**, 8941), often involving [n]-para- to less strained [n]-metacyclophanes (but see G. B. Kostermans *et al, ibid*, 1987, **109**, 2471). [6]-Paracyclophane rearranges thermally to the spirotriene (18) (Y. Tobe *et al*, 1986, *op. cit.*), a reaction whose reverse can be used to prepare less strained paracyclophanes *eg.* (11, n = 7, 8) (L. W. Jenneskens, W. H. de Wolf and F. Bickelhaupt, Tetrahedron, 1986, **42**, 1571):-

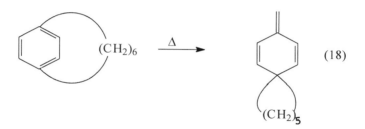

Face to face [n, m]-cyclophanes can also exhibit ring distortion effects but ring-ring interactions are, perhaps, more interesting. Interactions can be detected spectroscopically (*eg.* S. Mataka *et al*, J. Org. Chem., 1987, **52**, 2653; L. Ernst, S. Eltamany and H. Hopf, J. Amer. Chem. Soc., 1982, **104**, 299), and can even lead to cycloaddition reactions between the rings (*eg.* H. Prinzbach *et al*, *ibid*, 1987, **109**, 4626; H. Higuchi *et al*, Tetrahedron Lett., 1982, **23**, 671). There are numerous examples showing the influence of one ring on the other causing either enhanced reactivity (*eg.* J. Nishimura, Y.Okada and A. Oku, J. Org. Chem., 1986, **51**, 1838) or reduced reactivity (*eg.* R. Filler, G. L. Cantrell and E. W. Chloe, *ibid*, 1987, **52**, 511). Different π-electron densities in the two rings lead to transannular reactions, *eg.* acid catalysed alkyl migration of the [2, 2]-paracyclophane (19) (J. Kleinschroth, S. El-Tamany and H. Hopf, Tetrahedron Lett., 1982, **23**, 3345) and Lewis-acid catalysed ring closure of the tert-butyl [2, 2]-metacyclophane ether (20) (Y. Yamamoto *et al*, J. Chem. Res. (S), 1993, 272):-

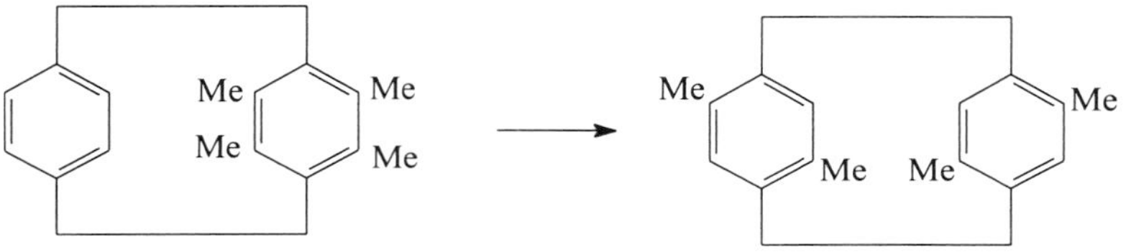

Cyclophanes have interesting isomer possibilities. [2, m]-Metacyclophanes (m ≠ 2) can be chiral due to helicity (F. Vogtle *et al*, J. Chem. Soc., Chem. Commun., 1986, 1248), and chirality has been demonstrated in [n][n]-paracyclophanes, *eg.* [12][12], structure (21) (T-L. Chan *et al*, *ibid*, 1994, 1971), and in 1,5-naphthaleneophanes *eg.* (22) (M. H. Chang and D. A. Dougherty, J. Amer. Chem. Soc., 1983, **105**, 4102):-

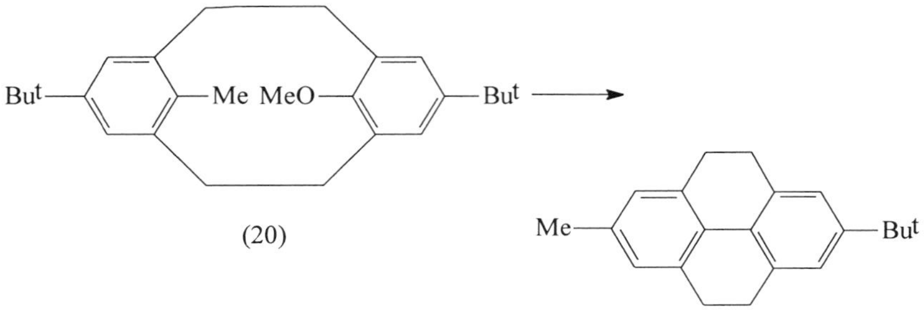

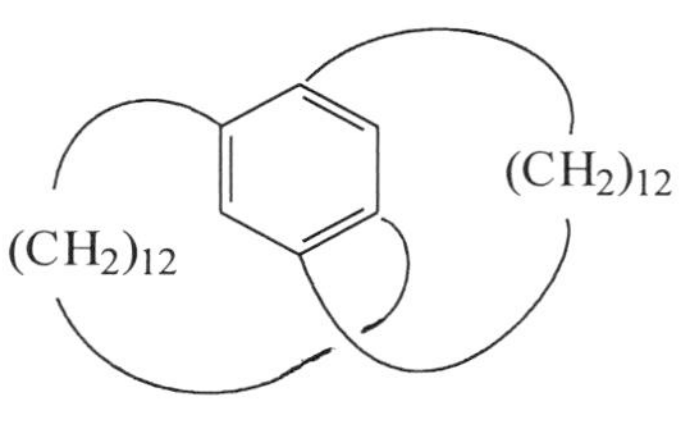

[2.2]-metacyclophanes can occur in *syn* (23) or *anti* (24) forms. The *anti* conformer is more stable (*eg.* in toluene derivatives; S. Ito *et al*, Tetrahedron Lett., 1986, **27**, 2907; R. H. Mitchell, Pure Appl. Chem., 1986, **58**, 2907) and, of the parent compounds, *syn*-[2.2]-metacyclophane (23)

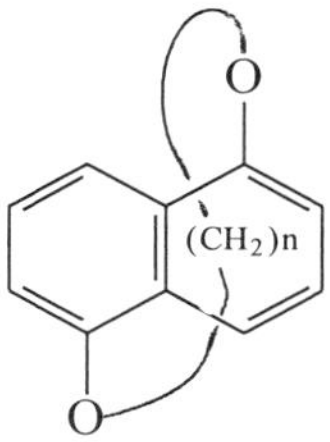

was the last to be obtained (R. H. Mitchell, T. K. Vinod and G. W. Bushnell, J. Amer. Chem. Soc., 1990, **112**, 3487).

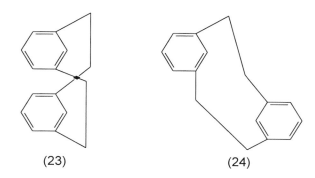

(23)   (24)

The long term importance of cyclophanes may rest with their applications which encompass electrical conductors (*eg.* V. Boekelheide, Pure Appl. Chem., 1986, **58**, 1) and semi-conductors (*eg.* K. Mullen *et al*, Angew. Chem., Int. Ed. Engl., 1986, **25**, 443, 444), charge-storage (O. Reiser *et al*, J. Amer. Chem. Soc., 1993, **115**, 3511), organic ferromagnetism (A. Izuoka *et al*, *ibid*, 1987, **109**, 2631), and especially catalysis (*eg.* A. McCurdy *et al*, *ibid*, 1992, **114**, 10314) and other phenomena associated with host-guest chemistry and molecular recognition (see monograph by F. Diederich, 1991, *op. cit.*).

(ii)   Calixarenes and related systems

A particular class of cyclic oligo-[1]-metacyclophanes, often the tetramers (25, n = 4), corresponds to the family of calixarenes (review: S. Shinkai, Tetrahedron, 1993, **49**, 8933). Their structures organise into cone-shaped

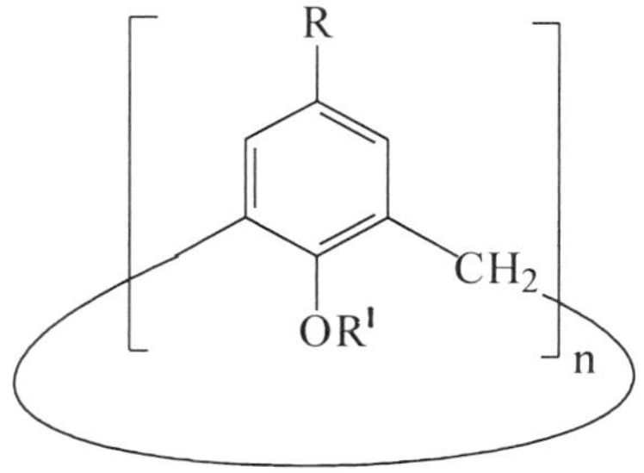

[Structure (25): a phenol unit with R at para position, OR^1 at the hydroxyl, and CH_2 bridge, repeated n times, drawn above a cone/cavity]

(25)

'cavitands' (*ie.* cavity within the cone) (J. R. Moran, S. Karback and D. J. Cram, J. Amer. Chem. Soc., 1982, **104**, 5826; C. D. Gutsche and I. Alam, Tetrahedron, 1988, **44**, 4689) that remain conformationally rigid at normal temperatures (C. D. Gutsche *et al*, Tetrahedron, 1983, **39**, 409). The parent, calix[4]arene (25, n = 4, R = R^1 = H) is commercially available along with some higher oligomers (n = 5 - 8) and some substituted forms with varying solubilities (25, n = 4, R = various alkyl, R^1 = H). Water-soluble versions have been prepared also, *eg.* (25, n = 4, R = $CH_2N^+Me_3$, R^1 = propyl) [S. Arimori, T. Nagasaki and S. Shinkai, J. Chem. Soc., Perkin Trans. 1, 1993, 887]. The cavitands bind metal and other ions, and have important uses in host-guest chemistry (monograph: C. D. Gutsche, "Calixarenes", R. S. C. Monographs in Supramolecular Chemistry, R. S. C. (Cambridge), 1992). Calixarenes can also be adapted for use in analytical chemistry, *eg.* by incorporating chromogenic substituents (R) (H. M. Chawla and K. Srinivas, J. Chem. Soc., Chem. Commun., 1994, 2593).

Other useful host molecules are based on cyclophane-like architecture. They include spherands [similar to (25) but with no bridging CH_2 groups (n = 6 or more)] (D. J. Cram *et al*, J. Amer. Chem. Soc., 1988, **110**, 2554; 1989, **111**, 8662; 1990, **112**, 5837), hemispherands which combine a spherand moiety with a crown ether *eg.* (26) (G. M. Lein and D. J. Cram, *ibid*, 1985, **107**, 448), and cryptophanes where two cyclophane cyclic oligomers are connected face to face *via* complexing ether bridges (review: A. Collet, Tetrahedron, 1987, 43, 5725).

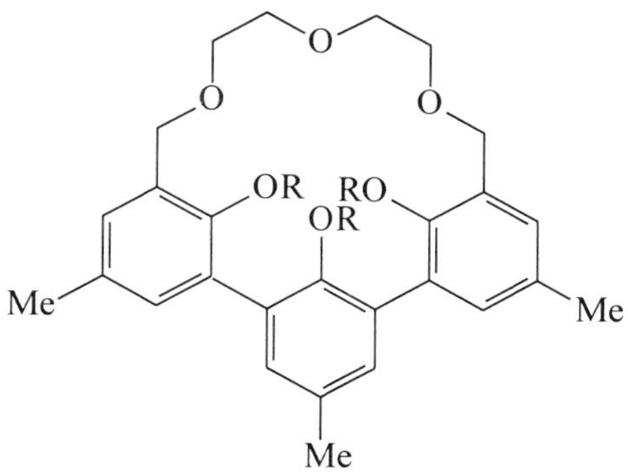

(26)

(iii)　　Phenylenes

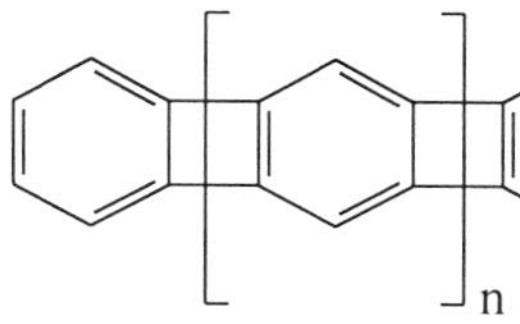

(27)

In phenylenes, benzene rings alternate with cyclobutadiene rings (27) (review: K. P. C. Vollhardt, Pure Appl. Chem., 1993, **65**, 153). They can be prepared by cobalt catalysed alkyne-cyclisation (K. P. C. Vollhardt *et al*, J. Amer. Chem. Soc., 1985, **107**, 5670), which can, by using bis(trialkyltin)alkynes, be iterative (M. Hirthammer and K. P. C. Vollhardt, *ibid*, 1986, **108**, 2481):-

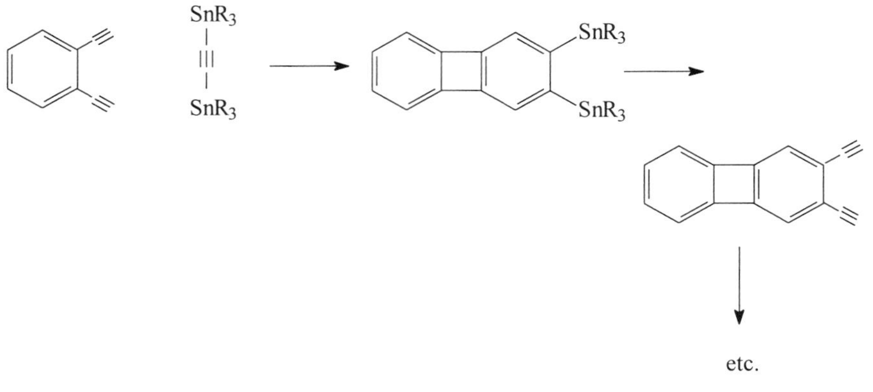

However the upper limit appears to be the pentamer (27, n = 3) and the hexamer failed to form (K. P. C. Vollhardt et al, Angew. Chem., Int. Ed. Engl., 1987, **26**, 1246). Variations to the method provided access to angular phenylenes (K. P. C. Vollhardt et al, J. Amer. Chem. Soc., 1986, **108**, 3150; ibid, 1992, **114**, 9713) and to naphthylene analogues (H. E. Helson, K. P. C. Vollhardt and Z-Y. Yang, Angew. Chem., Int. Ed. Engl., 1985, **24**, 114).

The phenylenes are of particular interest due to their alternating $6\pi/4\pi$ aromatic/antiaromatic rings. It seems that terminal benzene rings tend to maximise their aromaticity at the expense of internal benzene rings, which approximate to bond-alternating cyclohexatrienes. Certainly internal benzene rings are highly reactive; in [3]-phenylene (27, n = 1) the central ring undergoes facile base-catalysed alkylation (G. H. Hovakeemian and K. P. C. Vollhardt, ibid, 1983, **22**, 994), while the central ring of [5]-phenylene (27, n = 3) is very susceptible to cycloaddition reactions (K. P. C. Vollhardt et al, ibid, 1987, **26**, 1246).

(iv)   Iptycenes

o-Benzyne (Section 3 b) reacts with anthracene to give triptycene (28) by cycloaddition. The preparation may be achieved using o-dibromobenzene and BuLi to generate the benzyne. Hart and co-workers extended the method by generating dibenzynes from tetrabromobenzenes (29) and preparing pentiptycenes, *eg.* (30), after double cycloaddition with anthracene (H. Hart, S. Shamovilian and Y. Takehira, J. Org. Chem., 1981, **46**, 4427); 1,2,4,5-tetrabromobenzene (29, $R_1 = R_3 = H$, $R_2 = Br$) gave the "*para*-pentiptycene" shown (30), while (29, $R_1 = Br$, $R_2 = R_3 = H$) led to the corresponding "*ortho*-pentiptycene".

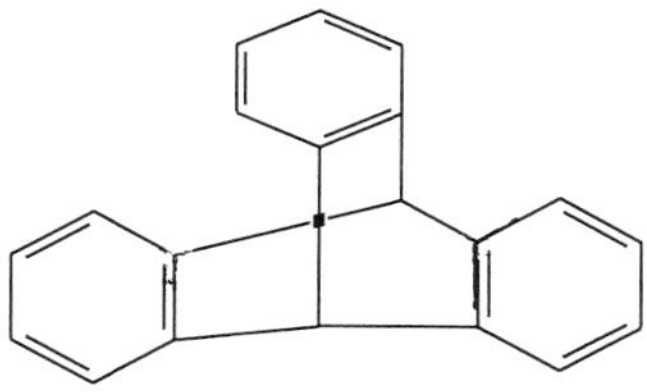

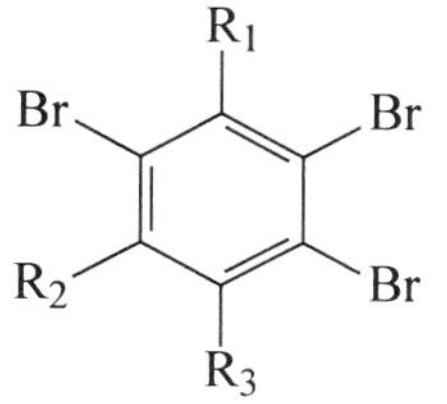

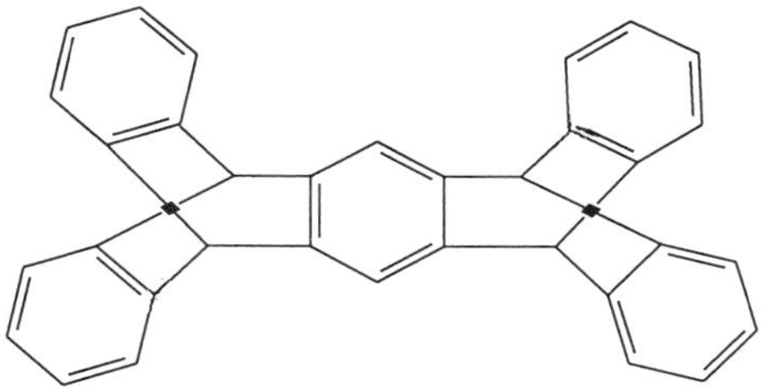

Iptycenes, then, are arene-containing, propeller-like compounds named by prefixes to show the number of benzene rings. By modification of the preparative method so that a central benzene ring is constructed during synthesis (K. Shahlai and H. Hart, J. Amer. Chem. Soc., 1988, **110**, 7136), larger iptycenes have been produced; the largest to date being the nonadecaiptycene (31) (S. B. Singh and H. Hart, J. Org. Chem., 1990, **55**, 3412):-

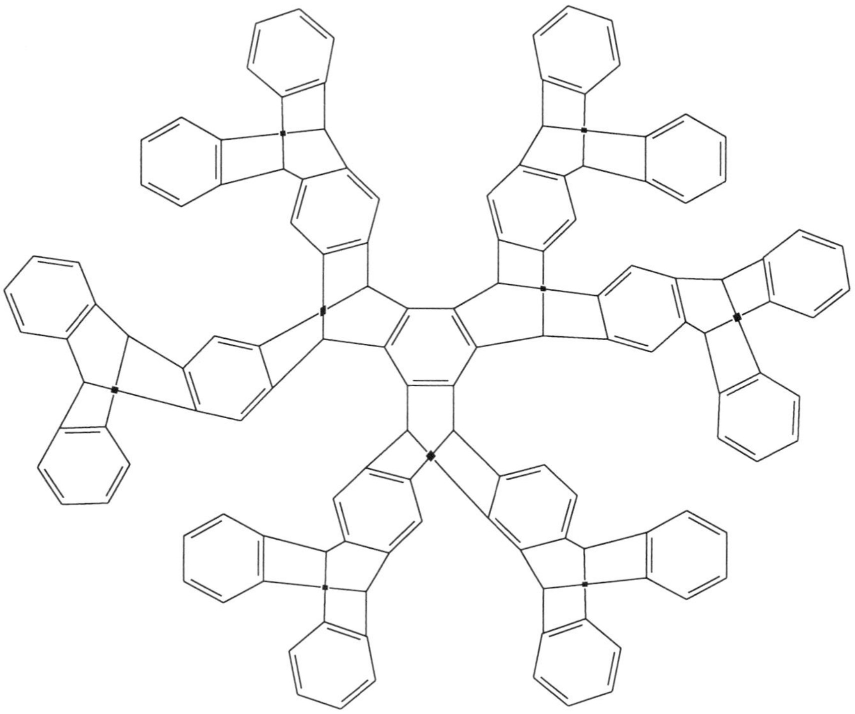

(31)

There are actually two types of iptycene structure, those having a propellane core like triptycene itself (28), and those like (30) and (31) that have a central arene ring and may be centrosymmetric, *eg.* (30). Other variants that have been investigated include those containing fused acene rings, or those displaying helical chirality (H. Hart *et al*, Tetrahedron, 1986, **42**, 1641), and those (not yet achieved) in which the iptycene framework is itself cyclic (H. R. Karfunkel and T. Dressler, J. Amer. Chem. Soc., 1992, **114**, 2285)

Iptycenes exhibit great thermal stability and can be studied in some depth (review: H. Hart, Pure Appl. Chem., 1993, **65**, 27). They are of interest theoretically and spectroscopically (partly due to their symmetry), and larger specimens contain well defined cavities which may be useful in host-guest chemistry.

(d) Fused systems

(i) Oligoarylenes

Oligoarylenes have been prepared as analogues of benzene-ring oligomers and polymers such as polyphenyls (32) and poly(paraphenylenevinylene)s (33) which display interesting electrical and

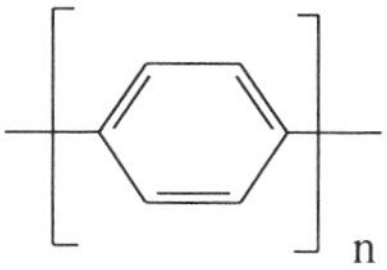

(32)

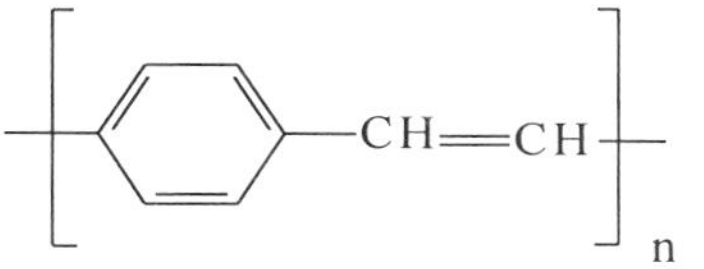

(33)

magnetic properties. Fused benzenoids are more reactive than benzene itself (*eg.* to electroreduction) and Mullen and co-workers have made series of oligonaphthylenes (34 a) (K. Mullen *et al*, Angew. Chem., Int. Ed. Engl., 1989, **28**, 904) and oligoanthrylenes (34 b) (K. Mullen *et al, ibid*, 1991, **30**, 1003; *ibid*, 1992, **31**, 448) together with arylenevinylene analogues (*cf.* 33) (H-P. Weitzel and K. Mullen, Makromol. Chem., 1991, **191**, 2837). Suitable oligonaphthylenes (34 a) could also be converted into ladder-structures (35) (K. Mullen *et al*, Angew. Chem., Int. Ed. Engl., 1990, **29**, 525).

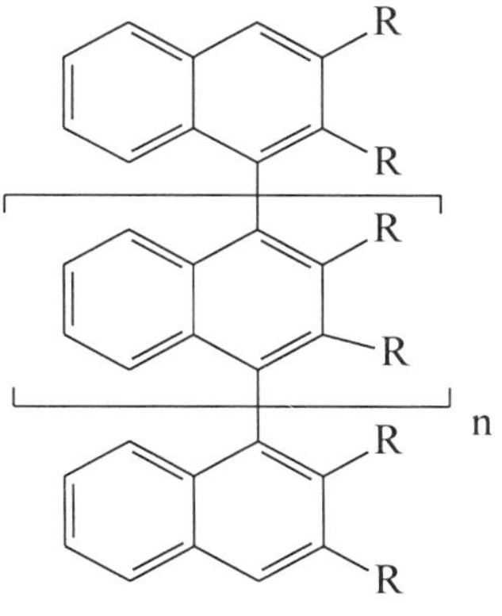

(34) a. R = H
b. R,R = -(CH_4)-

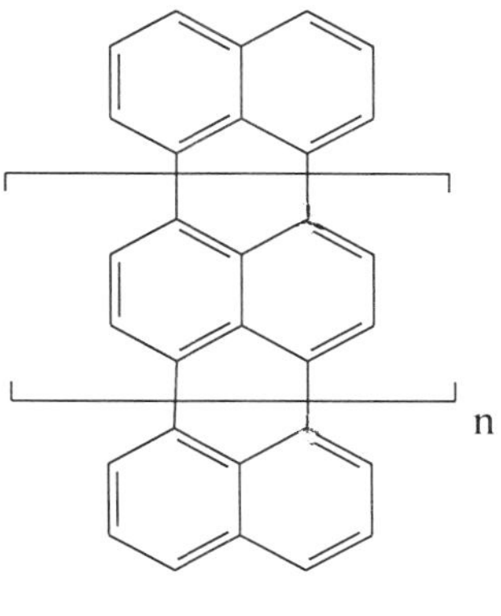

(35)

Some electrical properties of the anthrylenes (34 b) have been reviewed. There is little interaction between the anthracene units, partly because of torsional orthogonality from layer to layer (although vinyl linkages do not greatly increase the interactions). However, the compounds are good electron-storage media and may be candidates for the development of organic ferromagnets (review: K. Mullen, Pure Appl. Chem., 1993, **65**, 89).

(ii)   Series of fused benzenoids

Graph theory can be applied to polycyclic benzenoid hydrocarbons (monograph: N. Trinajstic, "Chemical Graph Theory", CRC Press, Florida, 1992). Benzenoids can be represented by dualist graphs whose vertices are the centres of the hexagons and whose edges are lines connecting those vertices, as seen in figures (36) - (39). Three classes of benzenoid result; the catafusenes with acyclic graphs, *eg.* anthracene (36) and phenanthrene (37), the highly condensed perifusenes with graphs

containing three-membered rings like perylene (38) and graphite, and the coronoids with graphs containing larger rings such as [7]-circulene (39) as well as cyclacenes and corannulene (following Sections).

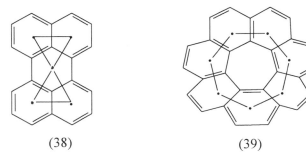

Graph theory is useful in the recognition of patterns among the huge array of known and possible polycyclic benzenoids. For example, in the catafusene series (review: A. T. Balaban, Pure Appl. Chem., 1993, **65**, 1) the most stable isomers having formula C_nH_m will be those containing the most Kekule rings, the K_{max} structures (eg. 37 rather than 36), as predicted by classical valence-bond theory. Computer generation, on that basis, of all the Kmax structures amongst all the possible hydrocarbons from one hexagon up to sixty hexagons, has been performed (D. J. Klein and X. Liu in A. T. Balaban, op. cit.). As a result, periodicity became apparent among the structures. Other workers have also produced periodic classifications of benzenoids using different analytical approaches (eg. J. D. Dias, Chem. Brit., 1994, **30**, 384). The exercise is not just a theoretical one. The results help to interpret the outcomes of eg. automerization reactions of catafusenes and may have implications for the detoxification of certain carcinogenic PAHs (A. T. Balaban, op. cit.).

(iii)   Cyclacenes, circulenes and related compounds

Cyclacenes are belt-like structures comprising [a, d]-fused benzenoid rings (ie. anthracene-like annelation) coiled round to form bracelets of benzenoid rings (40) (review: J. F. Stoddart et al, Pure Appl. Chem., 1993, **65**, 119). The architecture contains drawbacks; the belt arrangement is subject to considerable strain (R. W. Alder and R. B. Sessions, J. Chem. Soc., Perkin Trans. 2, 1985, 1849), and the electronic

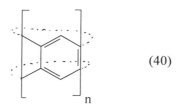

(40)

structure is flawed in two ways as follows. Firstly, classical theory predicts a rapid loss of stability in oligo[a, d]catafusenes as more rings are incorporated. While this instability could be offset by extra, stabilising delocalisation around the macrocycle of rings in the bracelet, so called "superaromaticity" (eg. J. Cioslowski, P. B. O'Connor and E. D. Fleischmann, J. Amer. Chem. Soc., 1991, **113**, 1086), the extra delocalisation might in fact be de-stabilising (anti-aromatic type).

[12]-Cyclacene (40, n = 12) can be envisaged as two annulene rings, an inner ring and an outer ring, connected by the benzenoid fusion bonds (41) whereupon it is apparent that both "rings", the inner and outer peripheries as shown, are 4n 24π systems. Indeed it proved impossible to

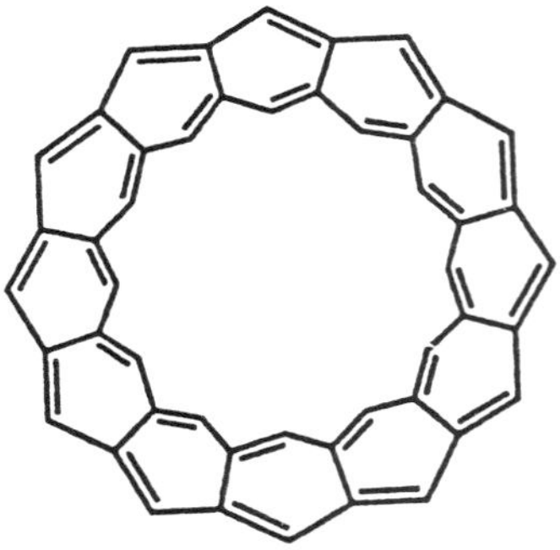

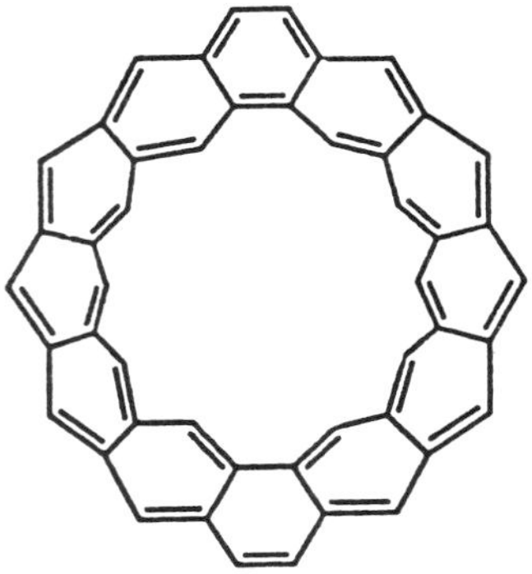

(41)   (42)

generate [12]-cyclacene by dehydrogenation of its precursor [12]-collarene, a belt of six benzene rings alternating with six saturated rings (see P. R. Ashton et al, ibid, 1992, **114**, 6330, and J. F. Stoddart, 1993, op. cit.). This second problem could be overcome in alternative structures. Stoddart's group has identified the constitutional isomer (42), having two [a, c] fused rings, as a new synthetic target containing 4n + 2 peripheries (26π outer and 22π inner) (see J. F. Stoddart op. cit.).

Further light is thrown on the electronic structures of circulenes, and their susceptibility to "superaromatic" delocalisation, by the 48π [12]-circulene molecule kekulene (43). The resonance canonical form shown is a K_{max} structure (previous section) whose predominance is supported by PMR analysis (H. A. Staab et al, Chem. Ber., 1983, **116**, 3487, 3504). In an alternative, quinonoid form superaromatic delocalisation would be possible via outer and inner peripheries of 30 and 18π electrons respectively (J. Cioslowski, P. B. O'Connor and E. D. Fleischmann, op. cit.). A similar situation was noted in the [10]-circulene analogue of kekulene (43, but with "naphthyl" rather than "anthryl" vertical edges) where a K_{max} form predominates over a 26π/14π superaromatic bis-annulene form (D. J. H. Funhoff and H. A. Staab, Angew. Chem., Int. Ed. Engl., 1986, **25**, 742). The superaromaticity concept is, accordingly, somewhat controversial (J. Aihara, J. Amer. Chem. Soc., 1992, **114**, 865).

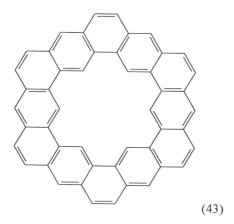

(43)

In cyclacenes (if not kekulenes) there is considerable distortion to the benzenoid rings. However several "helicenes" have been prepared wherein the coil of rings is not closed off into a macrocycle so that strain is greatly reduced even in the tightly coiled [7]-helicenes (*eg.* A. Sudhakar and T. J. Katz, Tetrahedron Lett., 1986, **27**, 2231). Alternatively the ends of the helix can be connected by alkyl or alkenyl bridges without excessive strain (*eg.* H. A. Staab, F. Diederich and V. Caplar, Liebigs Ann. Chem., 1983, 2262). The helical twist of such molecules raises interesting possibilities for optically active helicene systems *eg.* (A. Sudhaker, T. J. Katz and B-W. Yang, J. Amer. Chem. Soc., 1986, **108**, 2790).

In small circulenes the distortion imposed by annulation results in pronounced bowl shapes, *eg.* [7]-circulene (39) (K. Yamamoto *et al, ibid,* 1988, **110**, 3578). In the highly strained circulene, corannulene (44), bowl

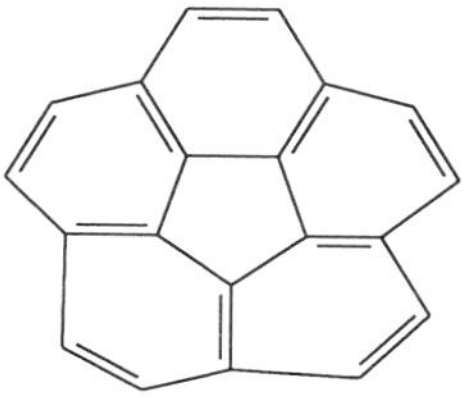

(44)

to bowl inversion has been demonstrated (L. T. Scott, M. M. Hashemi and M. S. Bratcher, ibid, 1992, **114**, 1921; A. Sygula and P. W. Rabideau, J. Chem. Soc., Chem. Commun., 1994, 1497). Corannulene corresponds to a segment of Buckminsterfullerene, and certain derivatives are synthons for fullerene synthesis (next section).

(iv)   Fullerenes

A football-like structure with formula C_{60} was predicted as early as 1970 but its actual existence, in laser-ablated graphite, was first demonstrated by Kroto, Smalley and colleagues in 1985 (H. W. Kroto et al, Nature (London), 1985, **318**, 162). An efficient procedure for obtaining macroscopic quantities was developed (W. Kratschmer et al, ibid, 1990, **347**, 354) and a convenient bench-top generator has been described (A. S. Koch, K. C. Khemani and F. Wudl, J. Org. Chem., 1991, **56**, 4543). The material obtained, fullerite (which is commercially available), is actually a mixture of C_{60} and C_{70} (ca. 85:15%) (H. Ajie et al, J. Phys. Chem., 1990, **94**, 8630), from which the pure fullerenes can be obtained by column chromatography (A. D. Darwish et al, J. Chem. Soc., Chem. Commun., 1994, 15).

Despite the recency of these discoveries there is already a formidable literature associated with fullerenes. The list includes some valuable reviews (H. W. Kroto, A. W. Allaf and S. P. Balm, Chem. Rev., 1991, **91**, 1213; R. F. Curl and R. E. Smalley, Sci. Am., 1991, **265**, 54; J. F. Stoddart, Angew. Chem., Int. Ed. Engl., 1991, **30**, 70; F. Diederich and R. L. Whetten, ibid, 1991, **30**, 678; H. W. Kroto, ibid, 1992, **31**, 111; H. Schwarz, ibid, 1992, **31**, 293; R. Taylor and D. R. M. Walton, Nature, 1993, **363**, 685), and an entire edition of "Accounts of Chemical Research" was devoted to fullerene papers (Acc. Chem. Res., 1992, **25**, (no. 3), 97-175). Papers from a Royal Society symposium in 1992 (proceedings: Phil. Trans. R. Soc. Lond. A, 1993, **343**, no. 1667) were later published as a book ["The Fullerenes" (New Horizons for the Chemistry, Physics and Astrophysics of Carbon), H. W. Kroto and D. R. M. Walton (Eds), Cambridge University Press, 1993].

C_{60} and C_{70} are the most accessible members of a family, the "fullerenes" (H. W. Kroto, Nature (London), 1987, **329**, 529). A fullerene is a 'geometrically closed trivalent polyhedral network in which n carbon atoms are arranged in 12 pentagonal and ($\frac{1}{2}n - 10$) hexagonal rings' (P. W.

Fowler in "The Fullerenes", 1993, *op. cit.*). So far, apart from C_{60} and C_{70}, only C_{76} and C_{78} (three isomers) have been obtained pure (R. Taylor *et al*, Pure Appl. Chem., 1993, **65**, 135). Many others have been predicted (*eg.* F. Diederich and R. L. Whetten, Acc. Chem. Res., 1992, **25**, 119; H. W. Kroto, A. W. Allaf and S. P. Balm, *op. cit.*, 1991) including some, like C_{76}, that may be chiral (D. L. D. Caspar in "The Fullerenes", 1993, *op. cit.*). Comprehensive experimental characterisation has, so far, been restricted to C_{60}.

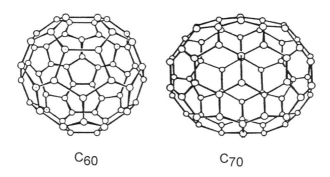

C_{60} C_{70}

Fullerene-C_{60}, or Buckminsterfullerene or [60]-fullerene, is a fairly stable, rather insoluble purple-brown solid that gradually decomposes in light and is thermally unstable in air above 200°C. X-ray crystallography (*eg.* J. M. Hawkins *et al*, Science, 1991, **252**, 312) reveals a virtually spherical molecule (diameter *ca.* 10 A°). ^{13}C-NMR gives a single resonance (142.68 ppm in benzene) showing that all 60 carbon are in identical environments, at the apex of two hexagons and one pentagon. The structure comprises 20 hexagons and 12 pentagons. UV/visible spectra contain intense bands at λ_{max} 213, 257 and 329 nm and IR spectroscopy reveals four characteristic absorptions at 1429, 1183, 577 and 528 cm^{-1} (spectroscopic data from H. W. Kroto, A. W. Allaf and S. P. Balm, 1991, *op. cit.*).

Early expectations for C_{60} assumed a high degree of aromaticity. A total of 12,500 resonance structures is possible (D. J. Klein *et al*, J. Amer. Chem. Soc., 1986, **108**, 1301) in superaromatic C_{60}. There is, however, just one structure corresponding to minimum strain in benzannelated systems (Mills-Nixon effect). In that structure no double bonds appear in any pentagonal rings and the hexagonal rings exhibit bond alternation (as seen in corannulene also). C_{60} does indeed behave as a giant, closed-cage alkene with very restricted electron delocalisation (R. Taylor and D. R. M. Walton, 1993, *op. cit.*). In C_{60} every hexagon has another hexagon fused across each double bond (a 6,6 bond) with a pentagon fused across each single bond (6,5-bond) (partial structure 45), and every pentagon has a bond fixated hexagon across each bond as in corannulene (44) and partial structure (46) (showing one of the subsequent pentagons).

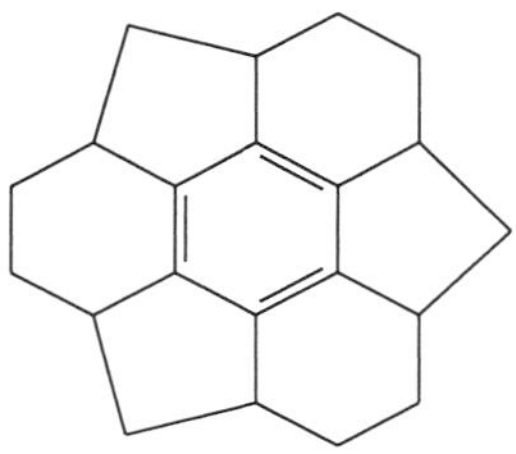

(45) 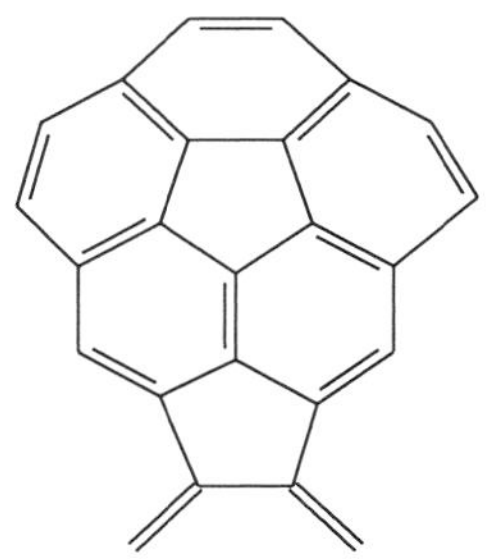 (46)

The ubiquitous, electron-withdrawing sp^2 carbons generate a strong electron affinity (R. C. Haddon, L. E. Brus and K. Raghavachari, Chem. Phys. Lett., 1986, **125**, 459). The crystal structure (face-centred cubic) provides three interstitial sites per C_{60} molecule, *eg.* for three alkali metal cations (R. C. Haddon *et al*, Nature, 1991, **350**, 321). The molecular structure sub-divides into six units related to pyracyclene (47), but benzannelated as in partial structure (46). Pyracyclene can accommodate

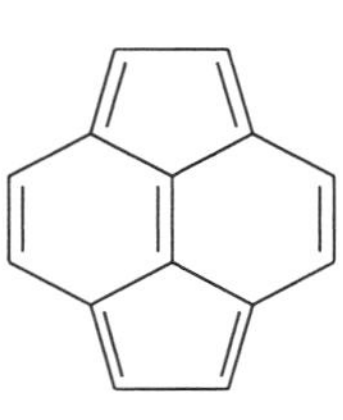

(47) 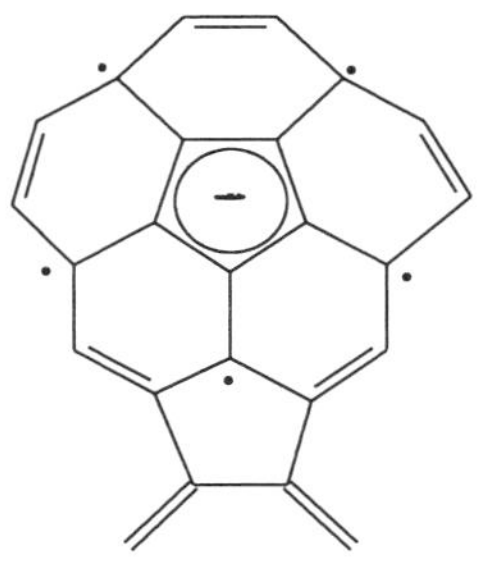 (48)

two extra electrons (*eg.* electrochemical doping) and generate two 6π cyclopentadienyl rings, but in C_{60}, one-electron doping leading to anion (48) is preferred so that the molecule can accommodate up to six electrons. When C_{60} is doped by alkali metals A (*eg.* K, Rb), two phases predominate corresponding to A_3C_{60} and A_6C_{60} (R. C. Haddon, Pure Appl. Chem., 1993, **65**, 11). A face-centred cubic structure is retained by A_3C_{60} but a more open body-centred cubic lattice is required to accommodate six interstitial cations in A_6C_{60}. The band structure of A_3C_{60} contains a half-filled conduction band and the lattice is conductive. A_6C_{60}, however, contains a filled band leading to an insulating state. Furthermore the former, conductive lattices undergo low temperature transition to a superconductive state (Tc = 18K and 28K for A = K and Rb respectively), and C_{60} may lead to useful organic superconductors after further development (R. C. Haddon in "The Fullerenes", 1993, *op. cit.*, p 53).

In other C_{60} complexes bonding may arise *via* charge transfer, *eg.* with thiafulvalene donors (A. Izuoka *et al*, J. Chem. Soc., Chem. Commun., 1992, 1472), or with ferrocene (J. D. Crane *et al*, *ibid*, 1992, 1764), or with metalloporphyrins (F. Wudl *et al*, J. Amer. Chem. Soc., 1991, **113**, 6698).

Alternatively covalent bonding between metals and fullerene is observed, involving π electrons from the C_{60} bonds that are members of two hexagons (*ie.* those connecting the apices of two pentagons, analogous to the central 6,6-double bond in pyracylene structure (47). η^2- Complexes of this sort have been reported, *eg.* for osmium (J. M. Hawkins, Acc. Chem. Res., 1992, **25**, 150), for platinum (P. J. Fagan, J. C. Calabrese and B. S. Malone, *ibid*, 1992, **25**, 134), and for iridium (R. S. Koefod, M. F. Hudgens and J. R. Shapley, J. Amer. Chem. Soc., 1991, **113**, 8957). In these examples the metals bridge across a fullerene C-C bond, whereas gas-phase studies with *eg.* Fe, Co, Ni suggest the formation of metallocene-type bonding implicating the pentagonal sites (*eg.* Y. Huang *et al*, *ibid*, 1991, **113**, 6298, 8186, 9418).

All the foregoing complexes are exohedral (metal or donor outside the fullerene cage) and many are somewhat unstable with respect to reversion to C_{60}. Free space inside the fullerene cage has a diameter of *ca.* 3.5 A°, and mass-spectrometric evidence has been presented that suggests the formation of endohedral complexes, with encapsulated La, K and Cs, that are stable until the fullerene cage itself fragments (R. E. Smalley, Acc. Chem. Res., 1992, **25**, 98).

The organic chemistry of C_{60} is dominated by addition reactions since in carbon-only molecules, in the absence of leaving groups, substitution would involve rupture of the cage. Adducts are often unstable, reverting to C_{60} quite readily (*eg.* during MS characterisation of reaction products!). Substitution protocols are therefore possible *via* displacement from an initial adduct by an appropriate nucleophile (although neither S_N1 nor S_N2 mechanism is feasible). Cycloadditions have been reported including [4 + 2], Diels-Alder reactions (*eg.* M. L. H. Green *et al*, J. Chem. Soc., Chem. Commun., 1994, 1641), or [3 + 2] reactions *eg.* with trimethylenemethane (T-Y. Luk *et al*, *ibid*, 1994, 647), or [2 + 2] additions *eg.* with benzyne (S. H. Hoke *et al*, J. Org. Chem., 1992, **57**, 5069). Hydroboration and alkaline work up leads to water soluble fullerols (N. S. Schneider *et al*, J. Chem. Soc., Chem. Commun., 1994, 463). As before, the central 6,6-bonds are implicated as the reaction sites and up to six adducts can be formed per C_{60} molecule (review: R. Taylor and D. R. M. Walton, Nature, 1993, **363**, 685).

The electron-deficient C_{60} is susceptible to attack by nucleophiles, including arenes undergoing electrophilic substitution by fullerene under Friedel-Crafts conditions *eg.* benzene/$FeCl_3$ (A. G. Avent *et al*, J. Chem. Soc., Chem. Commun.,1994, 1463). Once again, initial attack by nucleophile Nu⁻ occurs at a 6,6-junction bond leading to intermediates related to (49) with aromatic pentagonal rings. In some reactions (*eg.* the

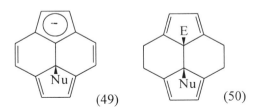

foregoing arylation) final products result after union with electrophiles, and addition protocols have been developed using Grignard reagents or organolithiums to generate intermediate anions that are quenched with alkyl or aryl electrophiles (E^+) giving *eg.* (50) or isomeric products (see below). However bridged products, *eg.* epoxides and homofullerenes (by addition of methylenes), are also observed. In these cases 6,6- or 5,6-addition may occur (51 a or b). When X is carbon both product types have been obtained, *eg.* a 5,6-product (51b) where X = CH_2 (F. Wudl, Acc. Chem. Res., 1992, **25**, 106), or a 6,6-product (51a) (a homofullerene or "fulleroid") where X = CAr_2 (J. Osterodt, M. Nieger and F. Vogtle, J. Chem. Soc., Chem. Commun., 1994, 1607). In C_{60} epoxides the oxidoannulene structure (51b, X = O) is more stable by calculation (K. Raghavachari, Chem. Phys. Lett., 1992, **195**, 221), but the 6,6-adduct (51a, X = O) is actually predominant despite its extra strain (R. Taylor and D. R. M. Walton, 1993, op. cit.).

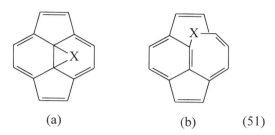

(a)   (b)   (51)

Additions that introduce two groups across a 6,6 double bond have been reviewed (R. Taylor in "The Fullerenes", 1993, op. cit., p 87). The regiochemical outcome depends largely on steric factors. Partial structures (52) indicate that addition can lead to either 1,2-adducts (52a) or 1,4-adducts (52b). 1,2-Addition avoids Mills-Nixon instability since no

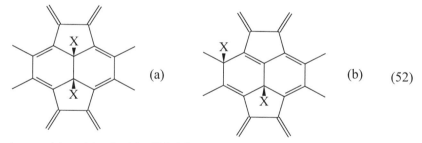

(a)   (b)   (52)

pentagonal bond is double (51a) but, since the two X groups are eclipsed there is steric strain if X is large. That strain is avoided by 1,4-addition (52b) but at the cost of locating a double bond in a pentagon. 1,2-Addition is observed with small X while bulkier addends lead to 1,4-addition.

Reactions with X_2, hydrogen or halogens, can lead to the uptake of up to $6X_2$ in sites like (52) and then further uptake can occur, whose course depends on X. Birch reduction leads to $C_{60}H_{36}$, the structure probably corresponds to a supertetrahedron containing four isolated benzene rings, one at each corner, with saturation in all 12 five-membered and the other six-membered rings, the structure arising by 1,2-addition (52a). Only 24 groups can be added to C_{60} without any two becoming adjacent and, apart from hydrogen addition, only fluorine adducts exceed this limit presumably due to the small size of F atoms. Thus adducts corresponding to $C_{60}F_{36}$ (H. Selig et al, J. Amer. Chem. Soc., 1991, **113**, 5475) and even $C_{60}F_{60}$ have been reported (J. H. Holloway et al, J. Chem. Soc., Chem. Commun., 1991, 966). With larger halogens 1,4-addition occurs leading to adducts $C_{60}Cl_{12}$ (F. N. Tebbe et al, J. Amer. Chem. Soc., 1991, **113**, 9900) and $C_{60}Cl_{24}$ (G. A. Olah et al, ibid, 1991, **113**, 9385), or to adducts $C_{60}Br_6$ and $C_{60}Br_8$ (P. R. Birkett et al, Nature (London), 1992, **357**, 479) and $C_{60}Br_{24}$ (F. N. Tebbe et al, Science, 1992, **256**, 822). Methoxydehalogenation has been used to convert some of these halides into methoxyfullerene derivatives (G. A. Olah et al, 1991, op. cit.).

Knowledge about the chemical properties of C_{60} and other fullerenes is increasing rapidly; reviews by Taylor and by Taylor and Walton (already cited) give further information. C_{60} also gives stable products in radical reactions (P. J. Krusic et al, Science, 1991, **254**, 1183), and dissolution in strong acids generates cation radicals that will react with electrophiles eg. SO_3 (G. P. Miller et al, J. Chem. Soc., Chem. Commun., 1994, 1549).

The total synthesis of C_{60} may not be far away. The relevance of corannulene (44) as a building block is clear by comparison with partial structure (46). Rapid bowl-bowl inversions in corannulene could be a problem but Rabideau and co-workers have prepared a derivative (53)

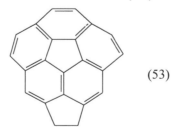 (53)

(cf. 46) that is conformationally rigid and may open the way towards synthetic C_{60} (P. W. Rabideau et al, Tetrahedron Lett., 1993, **34**, 6351; J. Amer. Chem. Soc., 1994, **116**, 7891). Other routes under exploration are discussed by Diederich (F. Diederich and Y. Rubin, Angew. Chem., Int. Ed. Engl., 1992, **31**, 1101).

4. SUBSTITUTION REACTIONS

The organisation of this section is somewhat different to the equivalent part of the first supplement, reflecting the primary literature since 1979. Radical species are now widely implicated throughout the gamut of aromatic substitutions and no section has been devoted entirely to homolytic processes (for a review on aromatic photosubstitutions see C. Parkanyi, Pure Appl. Chem., 1983, **55**, 331). There has, however, been prolific activity with metal-mediated substitutions requiring a full section here that includes some important radical processes.

(a) Electrophilic substitution

(i) General

A comprehensive monograph is available that covers all aspects of electrophilic aromatic substitution up to 1988 ("Electrophilic Aromatic Substitution", R. Taylor, John Wiley and Sons, 1990). Nitration remains the most studied transformation and, for the specialist, two books are available ("Nitration: Methods and Mechanisms", G. A. Olah, R. Malhotra and S. C. Narang, VCH (New York), 1989; "Aromatic Nitration", K. Schofield, Cambridge University Press, 1980). Reviews have appeared discussing the use of Hammett/Brown selectivity relationships (V. A. Koptyug, N. F. Salakhutdinov and A. N. Detsina, J. Org. Chem. USSR, 1984, **20**, 1039; G. A. Olah, J. A. Olah and T. Ohyama, J. Amer. Chem. Soc., 1984, **106**, 5284; F. P. DeHaan et al, J. Org. Chem., 1986, **51**, 1591), and a comprehensive listing of σ constants has been published (C. Hansch, A. Leo and R. W. Taft, Chem. Rev., 1991, **91**, 165).

With the arrival of chemical ionisation mass spectrometry, gas-phase analogues of electrophilic substitutions in solution can be investigated (M. D. Bezoari, P. Kovacic and A. R. Gagneux, Org. Mass Spectrom., 1982, **17**, 493). Studies have concentrated on alkylation (F. Cacace and G. Ciranni, J. Amer. Chem. Soc., 1986, **108**, 887) and nitration (A. P. Laws and R. Taylor, J. Chem. Soc., Perkin Trans. 2, 1987, 591; M. Attina and F. Cacace, Gazz. Chim. Ital., 1988, **118**, 241), and gas-phase processes seem rather similar to solution processes insofar as solvent effects in the latter can be disregarded.

In the previous Rodd supplement, J. H. Ridd featured ipso attack in particular. Although that topic is not discussed further here, its importance has been emphasised in reviews (*eg.* J. G. Traynham, J. Chem. Ed., 1983, **60**, 937), and many examples have been discussed (see R. Taylor, 1990, *op. cit.*). Ridd also described growing evidence for electron transfer as a mechanistic component of some electrophilic substitutions. In this, second supplement we concentrate on mechanistic intermediates, not just Wheland type arenium ions but also the electron transfer complexes that Ridd alluded to.

(ii)   Wheland intermediates

Following the classical work of Olah and his colleagues, many arenium complexes have been isolated then shown to give the expected substitution products *eg.* on heating (review: V. A. Koptyug, Top. Curr. Chem., 1984, **122**, 1). The benzenonium ion itself was prepared using super-acid protonation, allowing ^{13}C-NMR characterisation of charge-distributions (G. A. Olah *et al*, J. Amer. Chem. Soc., 1978, **100**, 6299). Wheland arenium intermediates are now accepted as ubiquitous features of electrophilic aromatic substitutions. But isolated arenium salts have additional uses in studies of mechanistic details of these reactions.

In a long series of investigations, Effenberger and co-workers have prepared arenium derivatives (1) from 1,3,5-tris(dialkylamino)benzenes (review: F. Effenberger, Acc. Chem. Res., 1989, **22**, 27). When NR_2 corresponded to 1-pyrrolidinyl, stable derivatives (1) were isolable (especially with X = BF_4^- or ClO_4^-, as obtained by anion exchange). They

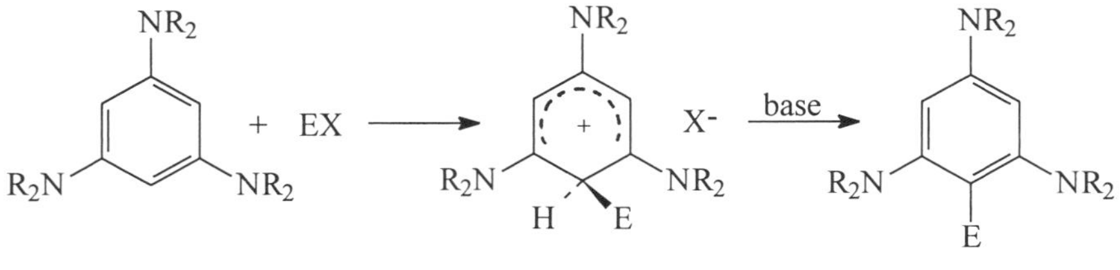

could be characterised fully by X-ray crystallography then deprotonated to substitution products (2). Using this approach alkylations, acylations and sulphonations were studied in depth, although halogen electrophiles gave rather unstable complexes (1).

By comparing empirical results [eg. X-ray structures of complexes (1)] with MO calculations of comparable Wheland intermediates, insights can be gained especially concerning stereoelectronic effects in the rearomatisation of those intermediates (F. Effenberger et al, J. Amer. Chem. Soc., 1987, **109**, 882). For example in Wheland intermediates (3)

(a)　　　　　　　　　(b)　　　　　　　　　(c)　　(3)

a planar conformation (3 b) is preferred in some cases (eg. $R = R^1 = H$ or CH_3). With non-identical substituents ($R \neq R^1$) bent complexes are preferred however (empirical and calculated data). In the electrophilic substitution of substrate $Ar - R^1$ by electrophile R^+ the Wheland intermediate must lose substituent R^1 for a successful outcome, and the geometry of complex (3) helps determine the energetics of that step since the favoured leaving group in bent complexes (3 a or 3 c) is quasi-axial so that the desired product is obtained from structure (3 c) rather than (3 a). Since the use of X-ray structures of isolated complexes gives details of the preferred geometry (3) of analogues (1), predictions can be made about the feasibility and energetics of substitutions. For example when converting Ar - H to Ar - X, the product is obtained from an intermediate of type (3a, $R = H$, $R^1 = X$) which may require a rate-determining 'ring-flip' from 3c to 3a, and a good estimate of the energetics of that process can be made. Similarly, in potential ipso substitutions via (3, $R \neq R^1$, neither = H) the approach described can be used to explain the isolation of addition products when substitution is not feasible due to stereoelectronic factors in the intermediate. These and other examples are discussed in the review (F. Effenberger, 1989, op. cit.).

(iii)　　Electron transfer

While the involvement of Wheland-type σ complexes in electrophilic substitutions is well established the interactions between substrate and reagent (and other possible species), that leads to those intermediates, are far from clear. In some processes at least, the involvement of arene radical cations, formed by one electron transfer as a preliminary to σ-complex formation, is now accepted.

In many reactions, mechanisms are proposed as shown:-

$$Ar-H + E^+ \longrightarrow [complex] \longrightarrow \left[Ar \begin{array}{c} H \\ E \end{array} \right]^+ \longrightarrow Ar-E$$

Sometimes "[complex]" is defined to be a π-complex (eg. O. M. E. El-Dusouqui, K. A. M. Mahmud and Y. Sulfat, Tetrahedron Lett., 1987, **28**, 2417), yet often its nature is ill-defined and corresponds to some form of encounter complex or even a sequence of transient species (review: A. S. Morkovnik, Russ. Chem. Rev., 1988, **57**, 144). The bonding in π-complexes involves partial charge transfer, but many electron-donor-acceptor complexes of arenes (EDA) formed by full, one electron transfer are stable, isolable species so that EDA structures are viable reaction intermediates (J. K. Kochi, Acta Chem. Scand., 1990, **44**, 409). Thus the overall mechanism could include an EDA intermediate, involving a substrate radical cation, leading on to the σ-complex Wheland intermediate that aromatises to product.

In a key experiment the naphthalene radical cation (hexafluorophosphate complex) was prepared and isolated by electrochemical oxidation , then reacted with NO_2 in dichloromethane. A high yield of nitro naphthalenes was obtained demonstrating the feasibility of radical cation intermediates in electrophilic substitution (L. Eberson and F. Radner, Acta Chem. Scand. Ser. B, 1980, **34**, 739). Furthermore, electron-rich arene donors visibly form coloured EDA complexes by admixture with familiar acceptors like tetracyanoethene but also with acceptors like nitronium salts, NO^+ (eg. S. Brownstein et al, J. Chem. Soc., Chem. Commun. 1984, 1566), and tetranitromethane, (TNM) (S. Sankararaman, W. A. Haney and J. K. Kochi, J. Amer. Chem. Soc., 1987, **109**, 5235, 7824) that are sources of electrophiles. Using picosecond time-resolved spectroscopy, Kochi and co-workers have investigated nitrations involving [$ArH^{+\cdot}$ $NO_2^{\cdot}$] radical pairs generated by irradiation at the appropriate charge-transfer band frequency. It seems that on irradiation an initial complex undergoes charge transfer to an ion pair which rapidly generates some encounter complex involving an ion radical pair and trinitromethyl anion (J. K. Kochi et al, ibid, 1986, **108**, 1126):-

$$\text{ArH} + \text{C(NO}_2)_4 \rightleftharpoons \left[\text{ArH, C(NO}_2)_4\right]$$

$$\left[\text{ArH, C(NO}_2)_4\right] \xrightarrow{h\upsilon} \left[\text{ArH}^{+\cdot}, \text{C(NO}_2)_4^{-\cdot}\right]$$

$$\left[\text{ArH}^{+\cdot}, \text{C(NO}_2)_4^{-\cdot}\right] \xrightarrow{\text{fast}} \left[\text{ArH}^{+\cdot}, \text{NO}_2^{\cdot}\ \text{C(NO}_2)_3^{-}\right]$$

Now the participation of radical cations had been mooted in 1945 (by G. W. Kenner) as one of two possible entries to the Wheland intermediate for nitration (*ie.* path a as well as conventional ionic addition *via* path b):-

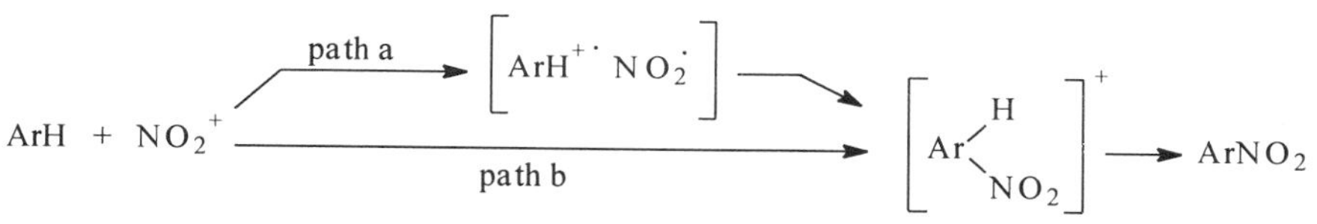

$$\text{ArH} + \text{NO}_2^{+} \begin{array}{c} \xrightarrow{\text{path a}} \left[\text{ArH}^{+\cdot}\ \text{NO}_2^{\cdot}\right] \searrow \\ \xrightarrow{\hspace{2cm}\text{path b}\hspace{2cm}} \end{array} \left[\text{Ar}\begin{array}{c}\text{H}\\ \text{NO}_2\end{array}\right]^{+} \longrightarrow \text{ArNO}_2$$

Although Kenner's proposal for proton loss by the path a complex to generate Ar· is unlikely (*eg.* L. Eberson and F. Radner, Acta Chem. Scand. Ser. B, 1986, **40**, 71), many studies confirm the involvement of an electron transfer step in some substitution reactions. Using CIDNP as a probe for single-electron events, Ridd and co-workers have studied nitrous acid catalysed nitrations (J. H. Ridd *et al*, J. Chem. Soc., Perkin Trans. 2, 1984, 1659, 1667). As a result of these and other experiments a more detailed mechanism was proposed for some nitrations (corresponding to path a) which has become established (review: J. H. Ridd, Chem. Soc. Rev., 1991, **20**, 149) :-

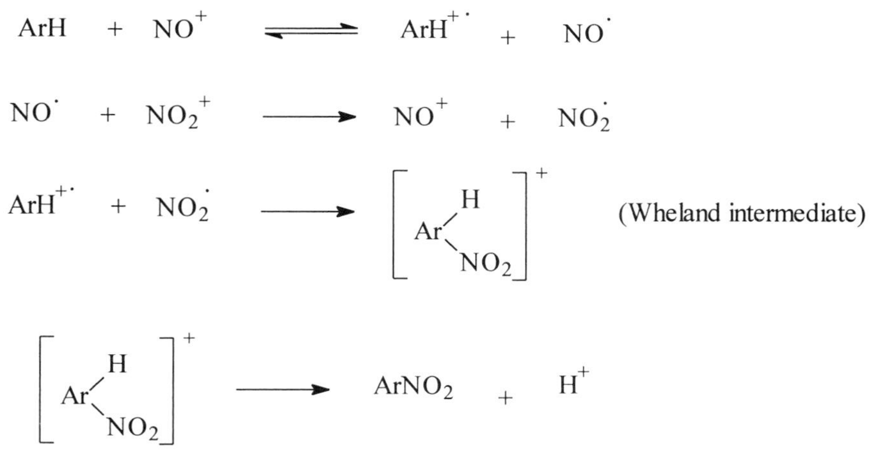

$$ArH + NO^+ \rightleftharpoons ArH^{+\cdot} + NO^{\cdot}$$

$$NO^{\cdot} + NO_2^+ \longrightarrow NO^+ + NO_2^{\cdot}$$

$$ArH^{+\cdot} + NO_2^{\cdot} \longrightarrow \left[Ar \begin{matrix} H \\ NO_2 \end{matrix} \right]^+ \quad \text{(Wheland intermediate)}$$

$$\left[Ar \begin{matrix} H \\ NO_2 \end{matrix} \right]^+ \longrightarrow ArNO_2 + H^+$$

The mechanism is consistent with Kochi's results (above) using charge-transfer band irradiation, assuming that encounter complexes formed thereby are precursors for Wheland intermediate formation in the third step of this mechanism (ie. $ArH^{+\cdot}$ with $NO_2\cdot$). It is also important that the reverse of the third step is unfavourable otherwise CIDNP observations could result from homolysis of a Wheland intermediate (formed by ionic addition) in a side reaction (see L. Eberson and F. Radner, Acta Chem. Scand. Ser. B, 1985, **39**, 343, 357; J. H. Ridd *et al*, J. Chem. Soc., Perkin Trans. 2, 1985, 1227).

Thus some electrophilic substitutions, especially those involving oxidising reagents or additives, involve charge-transfer as a significant step towards a Wheland intermediate that forms by reaction of an arene cation radical with a radical (*eg.* $NO_2\cdot$) (for a general review see Z. V. Todres, Tetrahedron, 1985, **41**, 2771). An outstanding problem remains with the detailed structure of the intermediate ion radical pair prior to Wheland intermediate formation. For example, in purely practical terms such pairs are still very difficult to detect unless there is reasonable separation between the components, so that *eg.* CIDNP observations may fail (J. H. Ridd, 1991, *op. cit.*). More fundamentally however, the distance apart and relative orientation of the components in an ion radical pair are bound to affect the product distribution when substituted benzenes, napthalenes etc are the substrates. Eberson and Radner have applied Marcus theory to help rationalize electronic interactions in the transition state for electron transfer, and so explain the orientation within the radical ion pair once formed (review: L. Eberson and F. Radner, Acc. Chem. Res., 1987, **20**, 53). When applied to nitrations or nitrosations in oxidative conditions the application of Marcus theory supports an electron transfer mechanism in line with experimental outcomes of regiochemistry and product distribution (J. P. B. Sandall, J. Chem. Soc., Perkin Trans. 2, 1992, 1689).

While electron transfer may be restricted to comparatively few electrophilic substitutions it is common in nitrations (L. Main, R. B. Moodie and K. Schofield, J. Chem. Soc., Chem. Commun., 1982, 48), including ipso nitrodeaminations of arylamines (J. H. Ridd *et al*, J. Chem. Soc., Perkin Trans. 2, 1985, 1217), nitrations of phenols using nitric acid catalysed by nitrous acid (M. Ali and J. H. Ridd, *ibid*, 1986, 327), nitrations using nitronium salts in dinitrogen pentoxide (R. B. Moodie, A. J. Sanderson and R. Willmer, *ibid*, 1991, 645), and ozone-mediated nitrations with nitrogen dioxide (H. Suzuki, T. Murashima and T. Mori, J. Chem. Soc., Chem. Commun., 1994, 1443). Single electron transfer is also implicated in certain acetoxylations and benzoyloxylations (see Z. V. Todres, 1985, *op. cit.*) and in the thallation of toluene (W. Lau and J. K. Kochi, J. Amer. Chem. Soc., 1984, **106,** 7100).

(b) Nucleophilic substitution

(i) General

In the previous supplement two sections dealt with nucleophilic substitution (N. B. Chapman) and homolytic substitution (G. H. Williams) respectively. Subsequent discoveries have complicated the picture to the extent that a different approach is taken here. The intervention of organometallic reagents and catalysts has been significant enough to require a separate section containing some reactions that could appear here (*eg.* nucleophilic attack on arene-chromium complexes), and others that could be classed as arene radical processes (*eg.* using aryl halides with tin hydrides). On the other hand the involvement of arene radical anions and radicals in nucleophilic substitutions is much more widespread than was once apparent. Accordingly no separate section on aryl radical processes appears here but, instead, a survey of important organometallic protocols is presented later (Section 4c), while the menu of nucleophilic substitution mechanisms given emphasises the significance of anion radical intermediates.

(ii) Mechanism of substitution

Three types of mechanism are of general importance in nucleophilic substitution reactions of benzene derivatives, the S_NAr mechanism, the S_{RN} mechanism and the benzyne mechanism. S_N1 reactions of diazonium compounds, and $S_NANRORC$ reactions of heterocyclic arenes are not general processes. Vicarious substitutions (Section iii below) provide a special case in the latter stages of S_NAr-type processes.

Activated arene substrates, in particular those featuring nitro substituents ortho and/or para to a leaving group, are susceptible to nucleophilic substitution by the **S_NAr mechanism** in which formation of a σ-complex (usually rate determining) is followed by departure of a leaving nucleophile to give the product. The presence of at least one nitro group is important for activation and a detailed monograph covers all aspects of reactions of this type (F. Terrier, "Nucleophilic Aromatic Displacement: the Influence of the Nitro Group", VCH (New York), 1991).

The σ-complex intermediates (4) are generally known as Meisenheimer complexes. They are often sufficiently stable for isolation and certainly the physico-chemical parameters associated with their formation, structure and properties are amenable to detailed study (reviews: F. Terrier, Chem. Rev., 1982, **82**, 77; G.A. Artamkina, M. P. Egorov and I. P. Beletskaya, *ibid*, 1982, **82**, 427). Their participation in S_NAr reactions may be summarised as follows, where at least one Electron-Withdrawing Group must be ortho or para to X :-

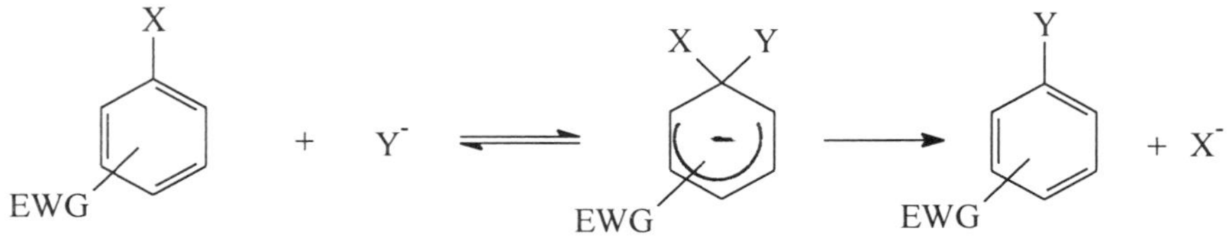

(4)

When, as is common with at least one nitro EWG, formation of the σ-complex (4) is rate limiting then the charge density around the substrate ring is important and the influence of X on the rate is generally as follows (C. E. Peishoff and W. L. Jorgensen, J. Org. Chem., 1985, **50**, 1056) :-

NR_3^+, SR_3^+ > NO_2, F > OSO_2Ar > Cl, Br, I > SOR, SO_2R, N_3 > OR, OAr > SR, SAr > NR_2

In protic solvents the influence, on the overall rate, of attacking nucleophiles Y is generally as follows (*idem, ibid*) :-

SO_3^{2-} > ArS^- > R_3S^- > RO^- > R_2NH,RNH_2 > $ArO^-,ArNH_2$ > NH_3 > I^- > Br^- > Cl^- > H_2O,ROH

Often the use of polar aprotic solvents leads to enhanced rates (*eg.* E. R. de Vargas and R. H. de Rossi, *ibid*, 1984, **49**, 3978), and beneficial effects have been reported for the use of phase-transfer catalysts (*eg.* F. P. Schmidtzen, Chem. Ber., 1984, **117**, 725, 1287) and micellar catalysts (C. A. Bunton *et al*, J. Chem. Soc., Perkin Trans. 2, 1986, 1799; 1987, 547). However particularly active substrates may undergo facile uncatalysed substitution under ambient conditions in conventional solvents (*eg.* W. M. Koppes *et al*, J. Chem. Soc., Perkin Trans. 1, 1981, 1815; F. Terrier *et al*, J. Chem. Res. (s)., 1979, 272).

Useful synthetic transformations can be achieved. In a synthesis of substituted dibenzo[1,4]dioxines (5) using catechol dianion as the nucleophile in HMPA solvent, effects of leaving groups and substituents are apparent (Ho H. Lee and W. A. Denny, J. Chem. Soc., Perkin Trans. 1, 1990, 1071):-

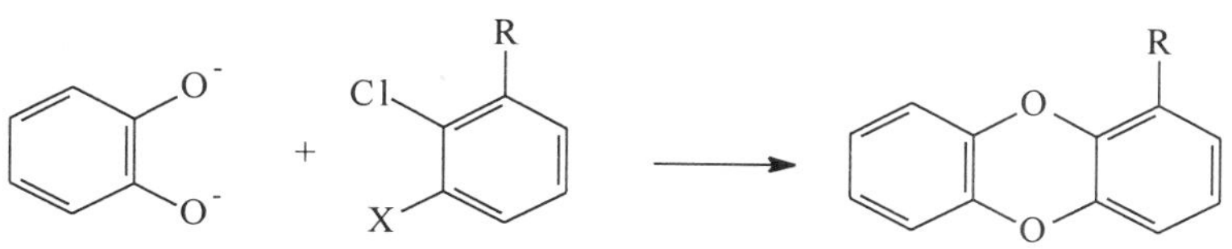

(5)

Under comparable conditions, substituents R of H, CO_2Me, Cl, NO_2 gave (5) in yields of 1, 15, 27 and 96% respectively. On the other hand (5, R = CO_2Pr^i) was obtained in 59% with X = NO_2, but only 35% with X = Cl (allowing regiochemical control *via* different leaving groups).

In many transformations attention must be paid to the reversibility of σ-complex formation, *eg.* 2,4,6-trinitroanisole with methoxide leads initially to the kinetic complex (6) while at equilibrium the predominant, thermodynamic complex is (7):-

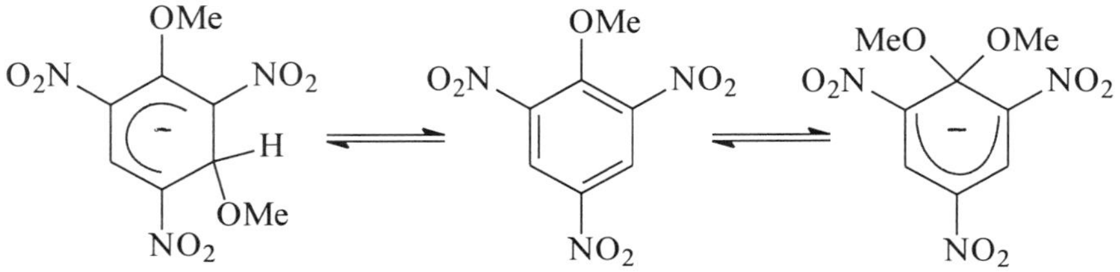

(6)   (7)

Moreover stereoelectronic factors in nucleophiles RO^-, affect the competitive rates. Clean transformations are best achieved by allowing the systems to reach thermodynamic equilibrium (E. Buncel *et al*, J. Amer. Chem. Soc., 1992, **114**, 5610).

The discovery, by J. F. Bunnett, of the **S$_{RN}$1 mechanism** was described in the first supplement (N. B. Chapman, *op. cit.*). Aryl halides (unactivated by EWG), after initiation giving anion radicals, undergo substitution by nucleophiles Y^- *via* chain propagation (review: J-M. Savéant, Tetrahedron, 1994, **50**, 10117):-

$$[ArX] \longrightarrow [ArX]^{-\bullet}$$

$$[ArX]^{-\bullet} \longrightarrow Ar^{\bullet} + X^-$$

$$Ar^{\bullet} + Y^- \longrightarrow [ArY]^{-\bullet}$$

$$[ArY]^{-\bullet} + ArX \longrightarrow ArY + [ArX]^{-\bullet}$$

Initiation may be photochemical or *via* solvated electrons (see J-M. Savéant, *op. cit.*, 1994, p. 10140), or electrochemical (J. M. Savéant, *op. cit.* and Acc. Chem. Res., 1980, **13**, 323), or *via* radical initiators like AIBN (see J. F. Bunnett, *ibid*, 1992, **25**, 2) or even thermal (*eg.* J. E. Swartz and J. F. Bunnett, J. Org. Chem., 1979, **44**, 340).

Since Bunnett's discovery, charge-transfer encounter complexes have also been detected en route to Meisenheimer complexes in S_NAr reactions of active substrates (*eg.* J. Hayami *et al*, Chem. Lett., 1987, 739), and electron transfer was implicated as a mechanistic step in other S_NAr transformations (N. J. Bunce *et al*, J. Org. Chem., 1987, **52**, 4214). MNDO calculations have been used to propose that one electron charge transfer is a rather general step towards σ-complex formation which leads to a Meisenheimer complex only when appropriate EWGs are present to stabilise it, otherwise S_{RN} occurs instead (S. K. Dotterer and R. L. Harris, *ibid*, 1988, **53**, 777; *cf.* C. A. Bunton *et al*, J. Amer. Chem. Soc., 1988, **110**, 3495, 3503, 3512). Experimental support for that proposal was provided by EPR spectroscopic analysis of many S_NAr reactions (L. Grossi, Tetrahedron, 1992, **33**, 5645) although counter evidence was presented in hydrolysis reactions (using hydroxide) of polynitrohalogenobenzenes (M. R. Crampton *et al*, J. Chem. Soc., Perkin Trans. 2, 1989, 675) and of 2,4-dinitrophenyl ethers (E. B. de Vargas *et al*, J. Org. Chem., 1993, **58**, 7364), where no evidence for single electron transfer could be found.

The utility of S_{RN} reactions for transformations of unactivated aryl halides is clear (review: J. F. Bunnett, Acc. Chem. Res., 1992, **25**, 2). There is however, controversy about the mechanism, centred on the fate of radical anions [Ar X]⁻· In the $S_{RN}1$ mechanism (above) unimolecular dissociation leads to aryl radicals Ar· , but some radical anions are rather stable toward such dissociation (D. B. Denney, D. Z. Denney and A. J. Perez, Tetrahedron, 1993, **49**, 4463) and are more likely to undergo attack by nucleophiles (Y⁻) in an alternative, bimolecular propagation, the **$S_{RN}2$ mechanism** (D. B. Denney and D. Z. Denney, *ibid*, 1991, **47**, 6577):-

$$[ArX]^{-\bullet} + Y^- \longrightarrow [ArY]^{-\bullet} + X^-$$

$$[ArY]^{-\bullet} + ArX \longrightarrow ArY + [ArX]^{-\bullet}$$

The $S_{RN}2$ proposal has been criticised on empirical kinetic grounds (R. A. Rossi and S. M. Palacios, Tetrahedron, 1993, **49**, 4485) and through a consideration of transition states (J. F. Bunnett, *ibid*, 1993, 4477). A weighing-up of the pros and cons is provided by Savéant (J-M. Savéant, *op. cit.*, 1994).

In the **benzyne mechanism** (see N. B. Chapman in the first supplement) aryl halides (Ar X) eliminate HX with strong base then undergo nucleophilic addition to give overall substitution as described in Section 3b (above). Benzyne is an intermediate in some reactions of aryl halides with HO^-, rather than simple ipso hydroxydehalogenation, unless copper salts are present (M. Zoratti and J. F. Bunnett, J. Org. Chem., 1980, **45**, 1769, 1776). Yet with dialkylamide bases protiodehalogenation can occur by an S_{RN} mechanism especially when benzyne formation is prevented by ring substituents ortho to the halogen (J. Winiarski and J. F. Bunnett, J. Amer. Chem. Soc., 1985, **107**, 5271). Positive evidence for benzyne participation is often provided by the isolation of cine- as well as ipso-substituted products.

(iii)   Vicarious Nucleophilic Substitution (VNS)

In VNS kinetic σ-complexes (6) are intercepted by using attacking carbanions containing a leaving nucleophile bonded to the C^- site. Protonation of the resulting anion (8) gives the VNS product in which formal carbodeprotiation has occurred (review: M. Makosza and J. Winiarski, Acc. Chem. Res., 1987, **20**, 282), *eg*.:-

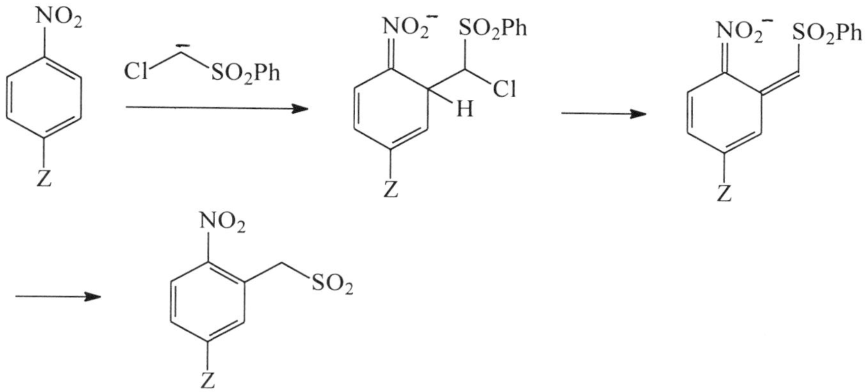

The VNS procedure does seem to require at least one nitro group in the substrate (but, for a possible exception, see R. Caputo, M. DeNisco and G. Palumbo, Tetrahedron, 1993, **49**, 11383) but reactions are clean, some regiocontrol is possible and the method is useful for introducing new functionality onto benzene rings. Further developments using other nucleophiles have led to VNS syntheses of nitroarylethylene derivatives (M. Makosza and A. Tyrala, Synthesis, 1987, 1142), dichloromethyl- and thus formyl-arenes (M. Makosza and Z. Owczarczyk, Tetrahedron Lett., 1987, **28**, 3021), arylacetonitriles (M. Makosza, W. Danikiewicz and K. Wojciechowski, Liebigs Ann. Chem., 1988, 203), alkylaminoarenes (A. R. Katritzky and K. S. Laurenzo, J. Org. Chem., 1988, **53**, 3978), trihalogenomethylarenes and thus carboxylates (M. Makosza and Z. Owczarczyk, *ibid*, 1989, **54**, 5094), nitroanilines (M. Makosza and M. Bialecki, *ibid*, 1992, **57**, 4784), benzosultans (K. Wojciechowski and M. Makosza, Synthesis, 1992, 571) and isocyanomethylarenes (M. Makosza, A. Kimowski and S. Ostrowski, *ibid*, 1993, 1215).

(c)   Organometallic methods

(i)   General

Complexes involving aryl carbon to metal bonding are common and many have uses in transformations of arenes. A brief and selective survey is given here, concentrating on processes that have become widely used. Unified texts (*eg.* A. J. Pearson, "Metallo-organic Chemistry", John Wiley and Sons, 1985), and more focused texts (*eg.* S. G. Davies, "Organotransition Metal Chemistry: Applications to Organic Synthesis",

Pergamon, 1982) are also somewhat selective and the interested reader really needs to trawl the specialist organometallic literature for detailed information. Helpful elementary texts are available in the series "Oxford Chemistry Primers" (M. Bochmann, "Organometallics 1: Complexes with transition metal-carbon σ-bonds", and "Organometallics 2: Complexes with transition metal-carbon π-bonds", Oxford University Press, 1994). Other monographs and reviews are cited below were relevant.

The methods and reactions covered here are all capable of bringing about substitution reactions on benzene rings. The sub-divisions correspond to formal electrophilic substitutions (section ii), nucleophilic substitutions (section iii) and radical substitutions (section iv) respectively. In many cases the transformations would be difficult or impossible to achieve without the use of organometallic chemistry.

(ii)   Reactions with electrophiles

Lithiation of benzene derivatives is a well known strategy for generating nucleophilic arenes either by halogen displacement (H. R. Rogers and J. Houk, J. Amer. Chem. Soc., 1982, **104**, 522), or proton displacement in which case up to three sites in all can be lithiated and quenched by electrophile *eg.* (S. Cabbidu *et al*, Tetrahedron, 1991, **47**, 9279):-

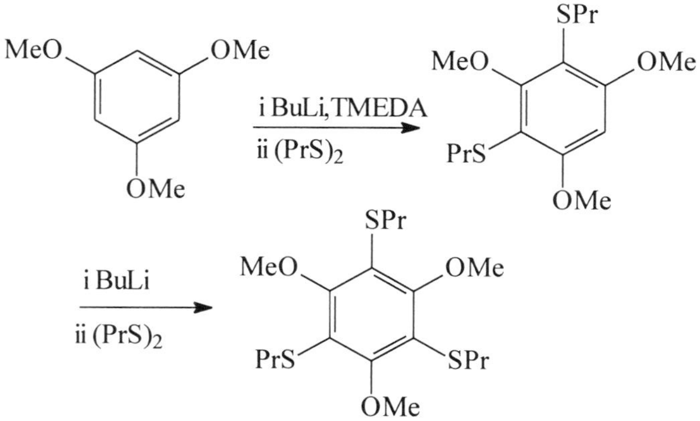

Lithiation offers regiospecific substitution by electrophiles with none of the isomer mixtures that sometimes dog conventional substitutions. With substituted benzenes therefore, it is important to achieve regiospecific lithiation. The strategy of directed ortho-metallation (DOM) has become well established (review: V. Snieckus, Chem. Rev., 1990, **90**, 879). Using this approach pre-association of the lithium base by an appropriate ring substituent directs the Li specifically to the ortho position, with N-alkyl benzamides being particularly effective eg.:-

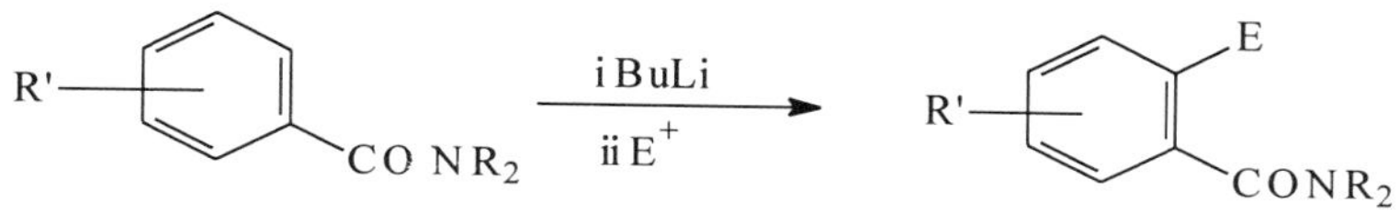

$$R'-\text{C}_6H_3-\text{CONR}_2 \xrightarrow[\text{ii E}^+]{\text{i BuLi}} R'-\text{C}_6H_3(\text{E})-\text{CONR}_2$$

π-Deficient arenes may undergo nucleophilic addition by alkyllithiums but if a directing group is present specific ortho-lithiation occurs (V. Snieckus et al, Adv. Met. Chem., 1991, **52**, 189). The scope of DOM has been extended by the discovery of new procedures for lithiation of N-benzylamides (Y. Simig and M. Schlosser, Tetrahedron Lett., 1988, **29**, 4277), phenylamines (A. R. Katritzky, W-Q. Fan and K. Akutagawa, Tetrahedron, 1986, **42**, 4027), phenyl ethers (J. Morey et al, J. Org. Chem., 1990, **55**, 3902), fluorobenzenes (D. C. Furlano et al, ibid, 1988, **53**, 3145; A. J. Bridges et al, Tetrahedron Lett., 1992, **33** 7495, 7499), t-butylsulphoxides (V. Snieckus et al, ibid, 1992, **33**, 2625), thiols (D. M. Giolando and K. Kirschbaum, Synthesis, 1992, 451), thiolates (S. Masson et al, ibid, 1993, 485) and aldimines (L. A. Flippin et al, J. Org. Chem., 1993, **58,** 2462). Reaction conditions can be quite mild allowing chiral synthesis without racemization eg. (S. Matsui et al, J. Chem. Soc., Perkin Trans, 1, 1993, 701):-

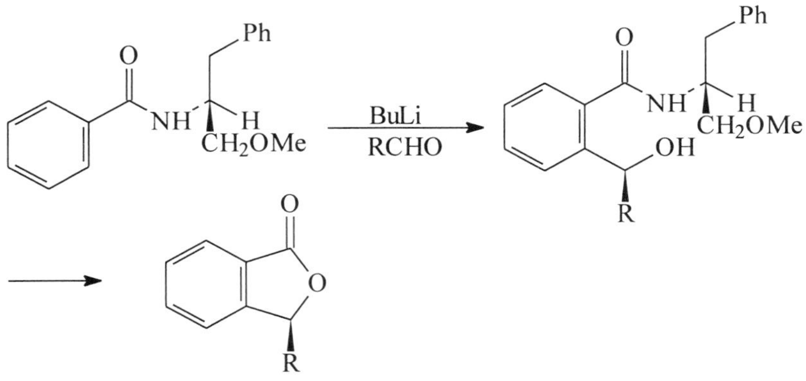

(iii) Reactions with nucleophiles

The use of copper-assisted procedures for nucleophilic displacement of halogen from unactivated aryl halides is well known and of broad scope. Methods for displacement by N- , O- , S- , Hal- , P- and C-nucleophiles have been reviewed (J. Lindley, Tetrahedron, 1984, 1433). Cyanodehalogenation can be a useful transformation for synthetic applications and the original, inconvenient Rosenmund-von Braun conditions are no longer necessary. High yields can be obtained using CuCN in pyridine with aryl bromides. Alternatively, alkali metal cyanides can be used with Pd or Ni catalysts giving high yields with aryl bromides, chlorides and also fluorides (review: G. P. Ellis and T. M. Romney-Alexander, Chem. Rev., 1987, **87**, 779).

A particularly dramatic development has centred on the η^6-complexes of arenes with transition metals, expecially Cr. Arene chromium tricarbonyl complexes (9) are available from the arene and $Cr(CO)_6$ either thermally or photochemically. The complexes are stable and easily handled but have very useful reactivity [M. F. Semmelhack in "Comprehensive Organic Synthesis", (Ed. B. M. Trost and I. Fleming), Vol. 4, Pergamon, 1991, Ch. 2.4]. Eventual products are recoverable by direct or oxidative decomplexation usually by iodine (*eg.* in ether) but other oxidants are also effective (see A-M. Lluch *et al*, Tetrahedron Lett., 1991, **32**, 5629).

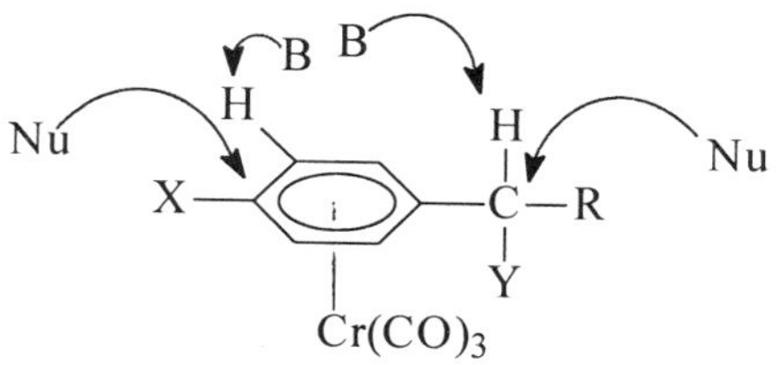

$$\text{(Figure 9)} \qquad (9)$$

The complexes have uses beyond the scope of this chapter since benzylic protons have enhanced acidity, benzylic leaving groups (Y) are prone to nucleophilic attack, and both benzylic processes are subject to chiral induction due to facial hindrance by the chromium tricarbonyl (Figure 9). However the same reactivity applies to the arene rings; protons are rather acidic, the rings are subject to nucleophilic addition and leaving groups (X) to nucleophilic displacement (reviews: M. F. Semmelhack, Pure Appl. Chem., 1981, **53**, 2379; M. F. Semmelhack *et al*, Tetrahedron, 1981, **37**, 3957). The dual susceptibility to bases and to nucleophiles was used to investigate a possible benzannelation procedure (M. Ghavshou and D. A. Widdowson, J. Chem. Soc., Perkin Trans. 1, 1983, 3065):-

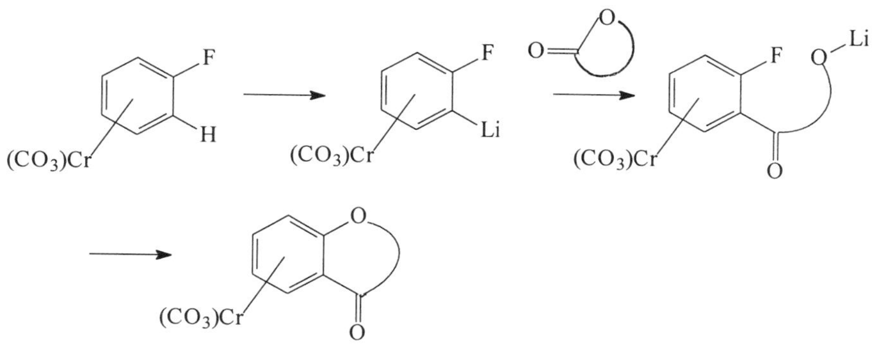

[Other examples of base treatment and reaction with electrophiles include benzylation (T. V. Lee, A. J. Leigh and C. B. Chapleo, Tetrahedron Lett., 1989, **30**, 5519) and metallation (*eg.* Ti, Au) (P. H. van Rooyen *et al*, Organometallics, 1992, **11**, 1104)].

In reactions with nucleophiles ring substituents tend to be meta directing irrespective of their identity (*eg.* OR, R, Hal). If the meta position is occupied then ortho attack occurs, para attack being rare. When a good leaving group is present the initial meta adduct tends to rearrange to an ipso complex from which the leaving group departs, *eg.*:-

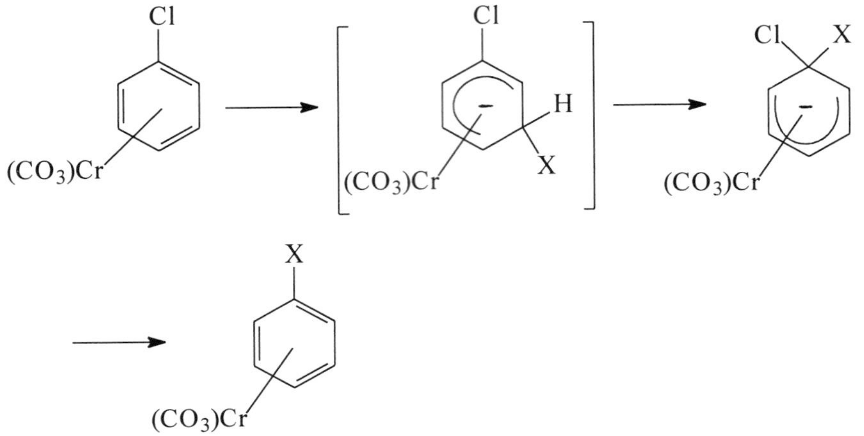

In the absence of a suitable leaving group, nucleophilic addition leads to an anionic complex which can be protonated and demetallated to the cyclohexadiene addition product, or demetallated with oxidation to give the arene corresponding to nucleophilic substitution of hydride:-

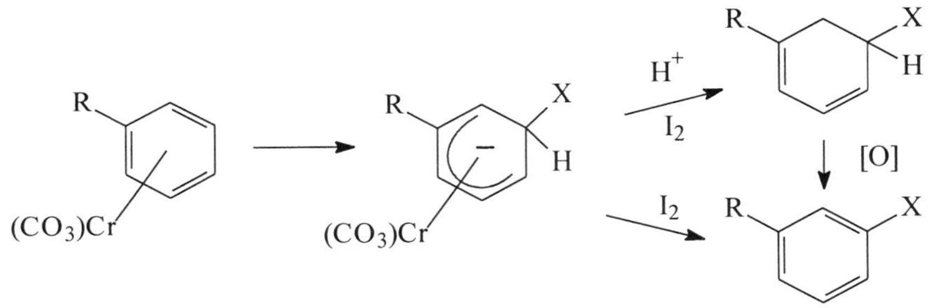

Cyclohexadienes are also available by an iterative sequence of addition-elimination followed by a second addition (*eg.* F. Rose-Munch, L. Mignon and J. P. Souchez, Tetrahedron Lett., 1991, **32**, 6323):-

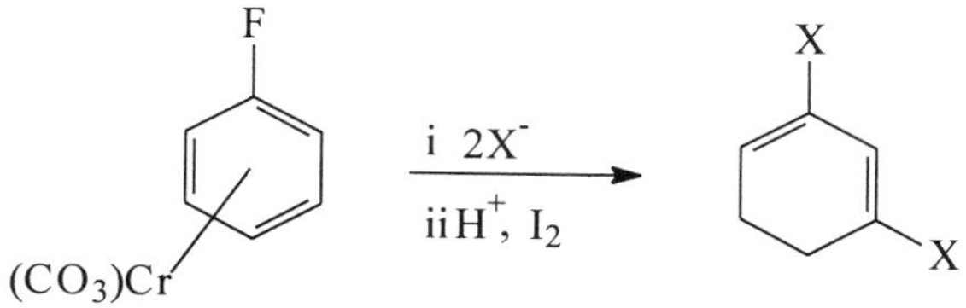

$$\text{(CO}_3\text{)Cr-C}_6\text{H}_4\text{F} \xrightarrow[\text{ii H}^+, \text{I}_2]{\text{i } 2X^-} \text{X-C}_6\text{H}_6\text{-X}$$

When the addition product is quenched not with H$^+$ but with RX/CO then the alkyl electrophile is delivered as an acyl group eg. (E. P. Kundig et al, J. Amer. Chem. Soc., 1991, **113**, 9676):-

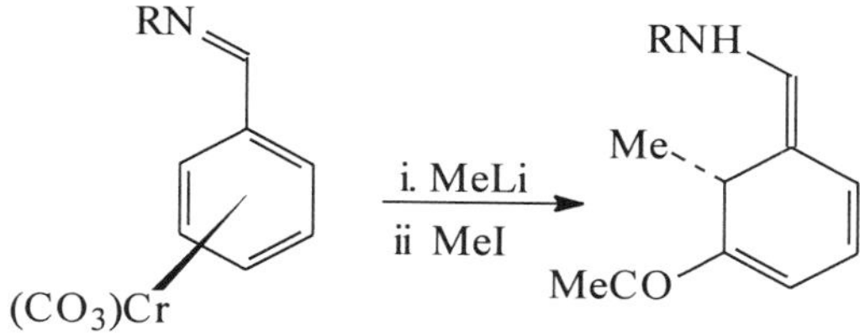

$$\text{(CO}_3\text{)Cr-C}_6\text{H}_4\text{-N=CHR} \xrightarrow[\text{ii MeI}]{\text{i. MeLi}} \text{product with RNH, Me, MeCO substituents}$$

This strategy has been extended to procedures involving nucleophiles such as carboxylic or phosphoric amides (C. Baldoli et al, Tetrahedron Lett., 1992, **33**, 4049), amino esters and nitriles (F. Rose-Munch et al, J. Organomet. Chem., 1992, **415**, 223).

The addition step can be carried out with a wide range of nucleophiles including alkoxides and phenoxides (F. Hossner and M. Voyle, ibid, 1988, **347**, 365) and oximes (A. Alemagna et al, Synthesis, 1987, 192) as O-nucleophiles. Most commonly C-nucleophiles are used however, with nitrile anions (see H. Kunzer and M. Thiel, Tetrahedron Lett., 1988, **29**, 1135 for a cyanomethyldemethoxylation), reactive enolates (M. Chaari et al, J. Organomet. Chem., 1991, **401**, C10) and sulphonates (M. F. Semmelhack, et al, Tetrahedron Lett., 1993, **34**, 5051) being useful for further elaboration. The utility of arene-chromium complexes is apparent from their increasing application in the synthesis of complex natural products such as indole alkaloids (eg. M. F. Semmelhack et al, 1993, op. cit.: M. F. Semmelhack and H. Rhee, ibid, 1993, **34**, 1395, 1399).

While Cr(CO)$_3$ complexes are the most commonly used, arene η^6-complexes with iron (review: D. Astruc, Tetrahedron, 1983, **39**, 4027) or manganese are also useful. In each case the complexes are cationic (rather than neutral), more reactive with a wider range of nucleophiles than the chromium counterparts, and more likely to react with two nucleophiles (since the initial adducts are not anionic). Ring substituents tend to be orthodirecting. The η^6-complexes are conveniently prepared by Lewis acid catalysed ligand exchange and eventual decomplexation to arene products occurs with oxidation (*eg.* cerium IV or DDQ) or merely by irradiation in acetonitrile. Cyclohexadienes are obtainable by aerobic decomplexation *eg.* (B. C. Roell Jr. *et al*, Organometallics, 1993, **12**, 224):-

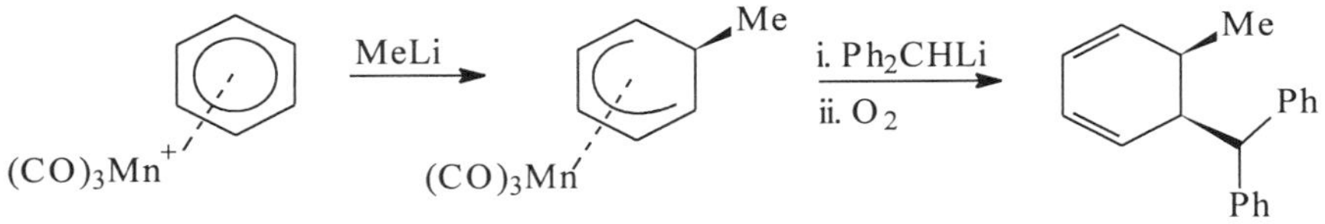

A useful monograph on organoiron chemistry includes a section, with experimental details, on arene η^6-complexes (A. J. Pearson, "Iron Compounds in Organic Synthesis", Academic Press, 1994). Sandwich complexes are used in which the second ligand is either η^5-cyclopentadienyl (giving monocationic complexes) or η^6-aryl (giving dication complexes) although the former are better known. A wide range of nucleophiles add to $^+$FeCp complexes, mainly C but also H, N and O nucleophiles, and DDQ in acetonitrile is a convenient medium for arene recovery after these reactions *eg.* (D. A. Brown, W. K. Glass and K. M. Kreddan, J. Organomet. Chem., 1991, **413**, 233):-

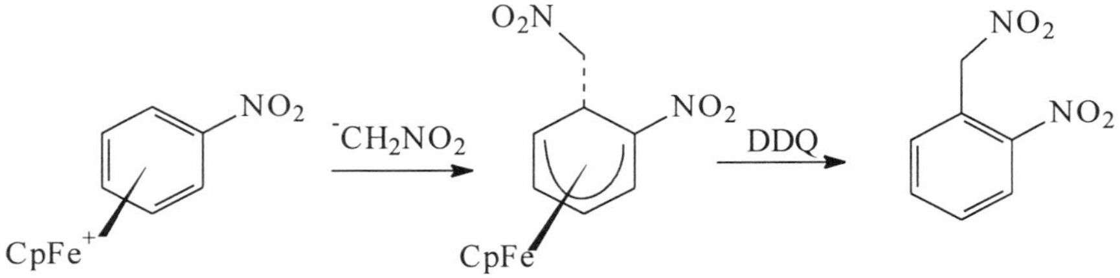

(iv) Radical reactions

Aryl halides react with many metal complexes and compounds with, at least formally, homolysis of the arene-halogen bond. The arene radicals or σ-metal complexes have useful reactivity in synthetic chemistry. In some cases σ-complexes can be elaborated by ligand exchange and/or insertion of CO in quite complicated protocols, *eg.* (M. Tanaka, Bull. Chem. Soc. Jpn., 1981, **54**, 637):-

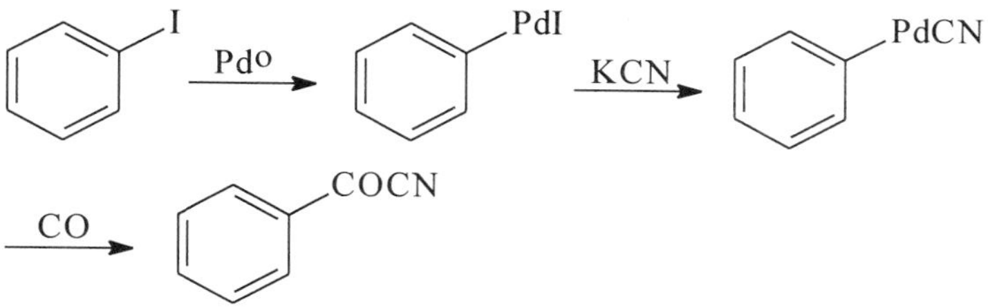

Ph–I $\xrightarrow{Pd^0}$ Ph–PdI $\xrightarrow{KCN}$ Ph–PdCN

$\xrightarrow{CO}$ Ph–COCN

This carbonylation approach has been used to prepare amides from halides, *eg.* in anthramycin synthesis (M. Ishikuru *et al*, J. Chem. Soc., Chem. Commun., 1982, 741), and in a general preparation of arenecarboxylic anhydrides from halides (Y. Fujiwara *et al, ibid*, 1982, 132). However, two classes of transformation have tended to dominate the application of this type of chemistry, the coupling of arenes to sp^2 carbons (arenes or alkenes), and the addition of aryl radicals to alkenes. The remainder of this survey concentrates on those transformations.

In the **Heck Reaction** an aryl halide is coupled to an alkene in the presence of base and a Pd catalyst, in a procedure of broad scope (review: R. F. Heck, Org. React., 1982, **27**, 345; monograph: R. F. Heck, "Palladium Reagents in Organic Synthesis", Academic Press, 1985):-

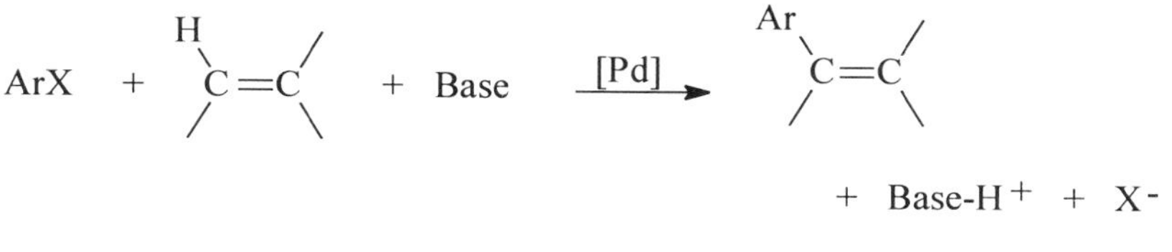

ArX + H\C=C/ + Base $\xrightarrow{[Pd]}$ Ar\C=C/

+ Base-H$^+$ + X$^-$

The aryl halide is normally bromide or iodide, but chloride can be used (J. J. Bozell and C. E. Vogt, J. Amer. Chem. Soc., 1988, **110**, 2655) and triflate reacts well (W. Cabri *et al*, J. Org. Chem., 1991, **56**, 5796). The catalyst normally involves palladium (II) chloride or acetate (PdY$_2$ below) as a bis(triarylphosphine) complex (L_2 below) and a two-stage mechanism, reduction of Pd(II) to Pd(o) then catalytic coupling, has been proposed (R. F. Heck, 1982, *op. cit.*):-

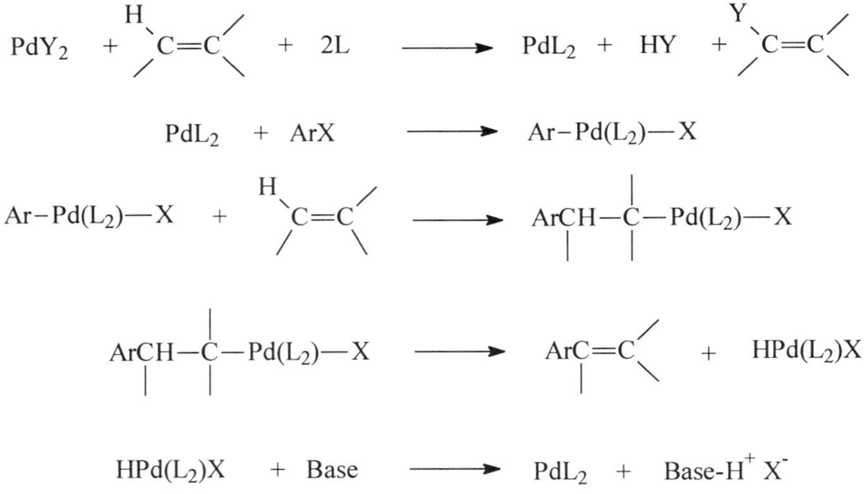

$$PdY_2 + \;\;^H\!\!\!>\!\!C=C\!\!<\;\; + \;2L \longrightarrow PdL_2 + HY + \;\;^Y\!\!\!>\!\!C=C\!\!<$$

$$PdL_2 + ArX \longrightarrow Ar-Pd(L_2)-X$$

$$Ar-Pd(L_2)-X + \;\;^H\!\!\!>\!\!C=C\!\!<\;\; \longrightarrow ArCH-C-Pd(L_2)-X$$

$$ArCH-C-Pd(L_2)-X \longrightarrow ArC=C\!\!<\;\; + HPd(L_2)X$$

$$HPd(L_2)X + Base \longrightarrow PdL_2 + Base\text{-}H^+\;X^-$$

The most common bases are tertiary amines or alkali metal acetates, carbonates or bicarbonates. Secondary amines can be used although they occasionally become incorporated into the product (*eg.* with diene rather than alkene substrates). Regiospecificity and rate effects of alkenes are dominated by steric factors. Ethene reacts fastest but monosubstituted or terminal alkenes are good substrates also. Electron-withdrawing substituents direct coupling to the opposite end of the alkene bond while electron donating groups give mixtures mainly under steric control. The arene reactant can be quite highly hindered but still give good yields, albeit at a slow rate. Similar couplings are also possible between aryl halides and alkynes (*eg.* R. Singh and G. Just, J. Org. Chem., 1989, **54**, 4453), and trimethylsilylketene acetals (C. Carfagna *et al*, *ibid*, 1991, **56**, 261).

Normally the alkene double bond does not migrate unless migration leads to conjugation, *eg.* (Y. Zhang, B. O'Connor and E. Negishi, *ibid*, 1988, **53**, 5588). When the alkene is an allylic alcohol (or even homoallylic) then Pd catalysed migration can occur giving carbonyl products, but some control is now possible allowing the isolation of either arylated allylic alcohols or ketones/aldehydes (T. Jeffrey, Tetrahedron Lett., 1991, **32**, 2121).

The involvement of chiral ligands (*eg.* BINAP) can lead to high enantiomeric excess in the product, *eg.* (F. Ozawa *et al*, *ibid*, 1992, **33**, 1485; J. Organomet. Chem., 1992, **33**, 1485; *cf.* B. M. Trost, D. L. Van Vranken and C. Bingel, J. Amer. Chem. Soc., 1992, **114**, 9327):-

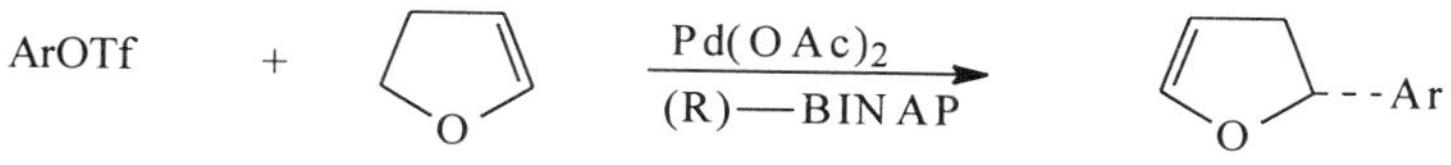

$$\text{ArOTf} + \underset{O}{\bigcirc\!\!=} \quad \xrightarrow[\text{(R)—BINAP}]{\text{Pd(OAc)}_2} \quad \underset{O}{\bigcirc\!\!-}\text{-}\text{-Ar}$$

Aryl iodides are the most reactive halides, and selective control of aryl dihalides is possible *eg.* (A-S. Carlstrom and T. Frejd, J. Org. Chem., 1991, **56**, 1289):-

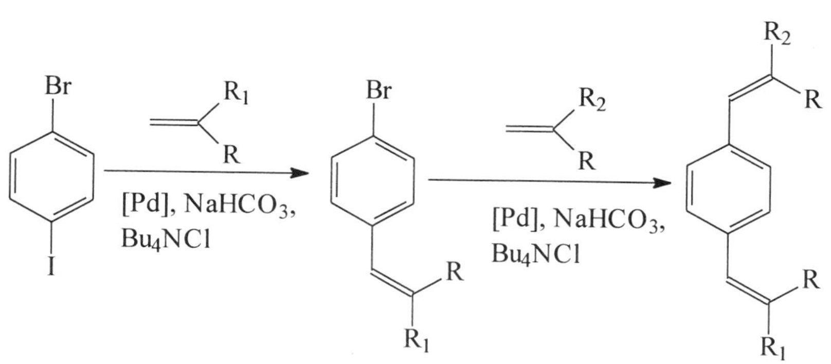

Other advances in the control of Heck reactions have been made (see R. Grigg *et al*, Tetrahedron Lett., 1991, **32**, 687; Tetrahedron, 1991, **47**, 9703). Intramolecular couplings are feasible using such mild conditions that the Heck method is finding increasing application in the synthesis of complex natural products, *eg.* taxol (S. J. Danishefsky *et al*, Tetrahedron Lett., 1993, **34**, 7253) and polycyclic alkaloids (F. Yokokawa, Y. Hamada and T. Shiori, *ibid*, 1993, **34**, 6559). The active Heck intermediate, Ar-Pd(L_2)-X, can be prepared from a large number of arene starting materials greatly extending the utility of palladium-mediated arylations (see R. A. Abramovitch, D. H. R. Barton and J-P. Finet, Tetrahedron, 1988, **44**, 3039, esp. p. 3056).

Palladium catalysts are also used in aryl couplings *eg.* with organotin compounds (review: T. N. Mitchell, Synthesis, 1992, 803) or with vinyl- or arylboronic acids and esters. In the **Stille coupling** procedure aryl halides or triflates are coupled with vinyl- or aryl-trialkyltins, normally using Pd(Ph$_3$P)$_4$ (J. K. Stille et al, J. Org. Chem., 1987, **52**, 422; J. Amer. Chem. Soc., 1987, **109**, 5478). Selectivity between halide and triflate substituents is possible eg. (J. K. Stille et al, ibid, 1988, **53**, 1170):-

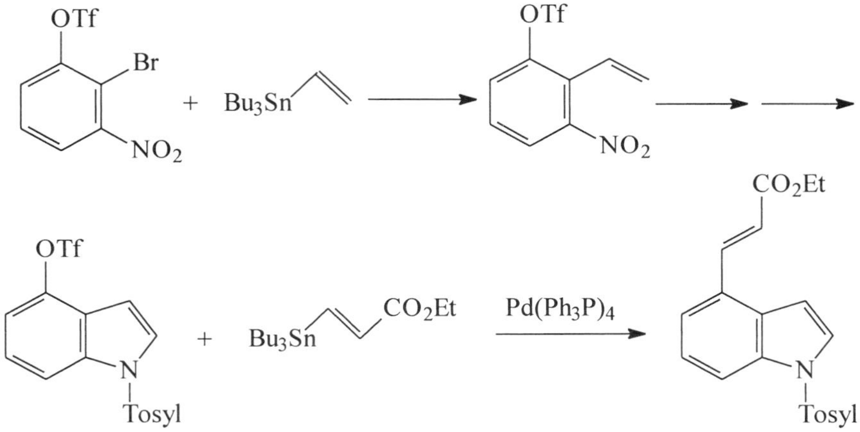

The reaction has been applied to benzannelation strategies (eg. R. J. P. Corriu, B. Geng and J. J. E. Moreau, ibid, 1993, **58**, 1443) and to synthesis related to natural products (eg. T. Sakamoto et al, Heterocycles, 1993, **36**, 2597; H-C. Zhang, M. Brakta and G. D. Daves Jr., Tetrahedron Lett., 1993, **34**, 1571). Minor modifications allow coupling to allylic tin alkyls (Y. Yamamoto, S. Hatsuya and J. Yamada, J. Chem. Soc., Chem. Commun., 1988, 86), and aryl triflates give aryl ketones (ArCOR) with CO and R$_3$SnH (A. M. E. Echavarren and J. K. Stille, J. Amer. Chem. Soc., 1988, **110**, 1557). The reverse coupling, of aryltrialkyl tins with vinylic halides is also practicable (eg. Y. Yamamoto, T. Seko and H. Nemoto, J. Org. Chem., 1989, **54**, 4734).

In **Suzuki coupling** an aryl halide and an aryl- or vinyl-borate ester or boronic acid are coupled normally using Pd(Ph$_3$P)$_4$ (N. Miyaura, T. Yanagi and A. Suzuki, Synth. Commun., 1981, **11**, 513). Aryl triflates are also effective coupling partners (T. Ohne, N. Miyaura and A. Suzuki, J. Org. Chem., 1993, **58**, 2201). With some vinylboron substrates a competitive Heck reaction seems kinetically preferred but regiocontrol is possible by modifying the reaction conditions (A. R. Hunt, S. K. Stewart and A. Whiting, Tetrahedron Lett., 1993, **34**, 3599):-

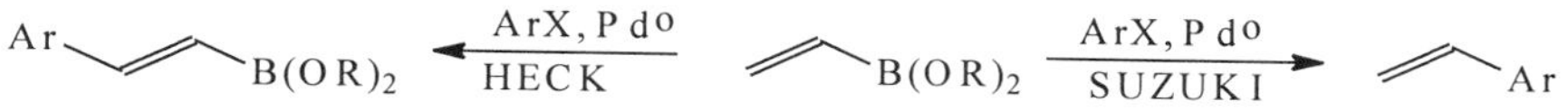

$$Ar\diagdown\!\!=\!\!\diagup^{B(OR)_2} \xleftarrow[\text{HECK}]{ArX, Pd^o} \diagup\!\!=\!\!\diagdown_{B(OR)_2} \xrightarrow[\text{SUZUKI}]{ArX, Pd^o} \diagup\!\!=\!\!\diagdown_{Ar}$$

However in other cases clean Suzuki coupling has been achieved with no evident Heck contamination *eg.* (F. Jin, Y. Xu and B. Jiang, J. Fluorine Chem., 1993, **65**, 111):-

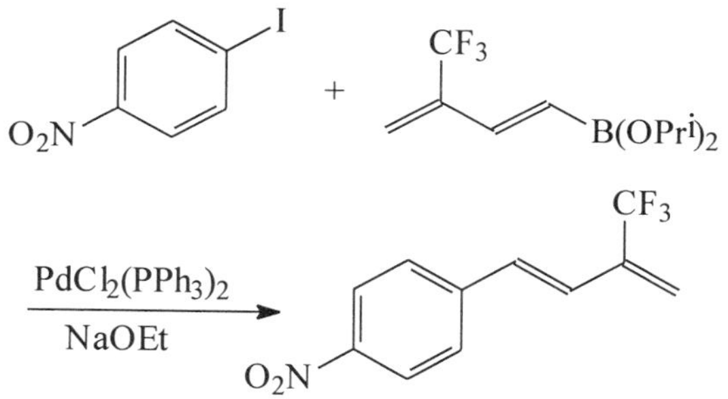

Suzuki coupling has been used during natural product synthesis (*eg.* P. Rocca *et al*, Tetrahedron Lett., 1993, **34**, 7917), but seems particularly well suited to materials synthesis (*eg.* J-P. Sauvage *et al*, *ibid*, 1993, **34**, 2933; C-S. Chan, C. C. Mak and K. S. Chan, *ibid*, 1993, **34**, 5125), and to poly(arylene) polymerization (T. I. Wallow and B. M. Novak, J. Amer. Chem. Soc., 1991, **113**, 7411). It has become a method of choice for oligoarene liquid crystals *eg.* (J. W. Goodby *et al*, J. Mater. Chem., 1993, **3**, 821):-

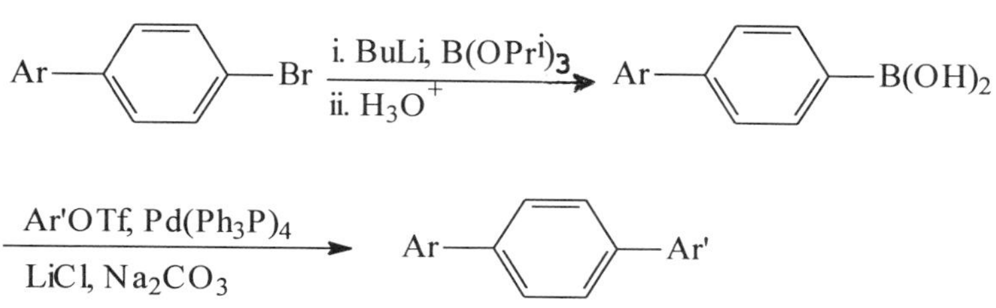

The Suzuki method provides efficient access to **unsymmetrical biaryls** (E. M. Campi *et al*, J. Chem. Soc., Chem. Commun., 1994, 2395). The Stille approach can also be used (T. R. Bailey, Tetrahedron Lett., 1986, **27**, 4407) and arylpalladium complexes also give biaryls with arylmagnesium and arylmercury derivatives (D. A. Widdowson and Y. Zhang, Tetrahedron, 1986, **42**, 2111; N. A. Bumagin, P. G. More and I. P. Beletskaya, J. Organomet. Chem., 1989, **364**, 231). Other successful methods for biaryl synthesis involve organobismuth or organolead chemistry (J-P. Finet, Chem. Rev., 1989, **89**, 1487; D. M. X. Donnelly *et al*, J. Chem. Soc., Perkin Trans. 2, 1994, 2921). A detailed review describes many of these methods (R. A. Abramovitch, D. M. R. Barton and J-P. Finet, Tetrahedron, 1988, **44**, 3039). Useful procedures for aryl coupling were reviewed in 1980 (M. Sainsbury, Tetrahedron, 1980, **36**, 3327).

Tin mediated **radical addition** of arenes to double bonds is a useful procedure especially for intramolecular ring closure where, typically a suitable aryl iodide moiety adds to a nearby double bond when exposed to Bu$_3$SnH/AIBN. A series of investigations of oxygen heterocycle synthesis, by Whiting and co-workers, is instructive. While Heck coupling (see above) gave the expected benzpyrene (10) (P. C. Amos and D. A. Whiting, J. Chem. Soc., Chem. Commun., 1987, 510), radical addition of a similar substrate led to a benzpyran product (11) *via* 6-exo-trig cyclization favouring *cis*-ring closure as shown (S. A. Ahmad-Junan and D. A. Whiting, *ibid*, 1988, 1160). On the other hand 6-endo-trig cyclization produced a *trans*-ring junction in the peltogynol derivative (12) after stereospecific reduction (*idem*, J. Chem. Soc., Perkin Trans. 1, 1990, 418):-

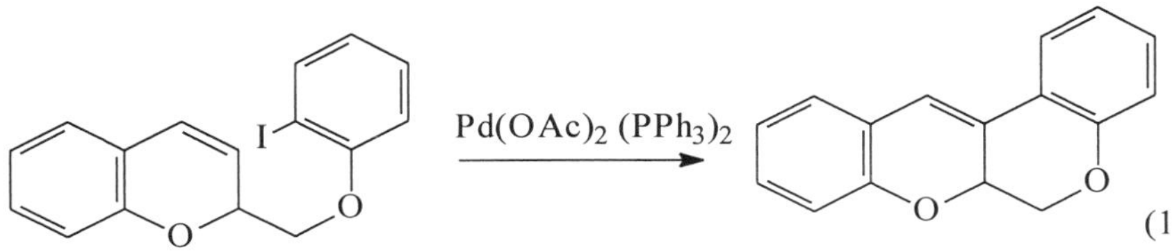

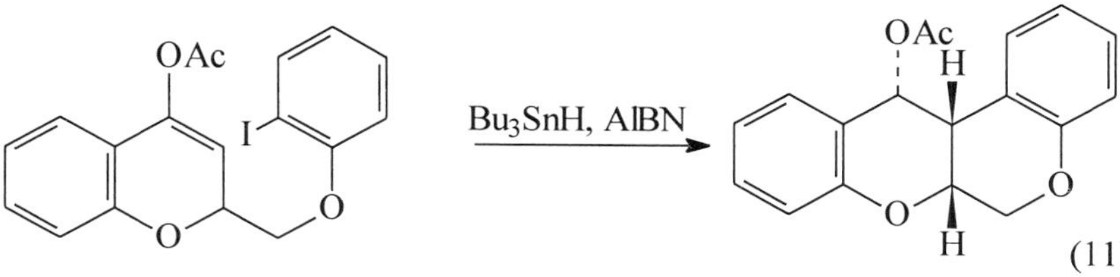

Baldwin's rules are usually obeyed although a 5-endo-trig cyclization onto an azo group was reported (C. P. A. Kunka and J. Warkentin, Can. J. Chem., 1990, **68**, 575):-

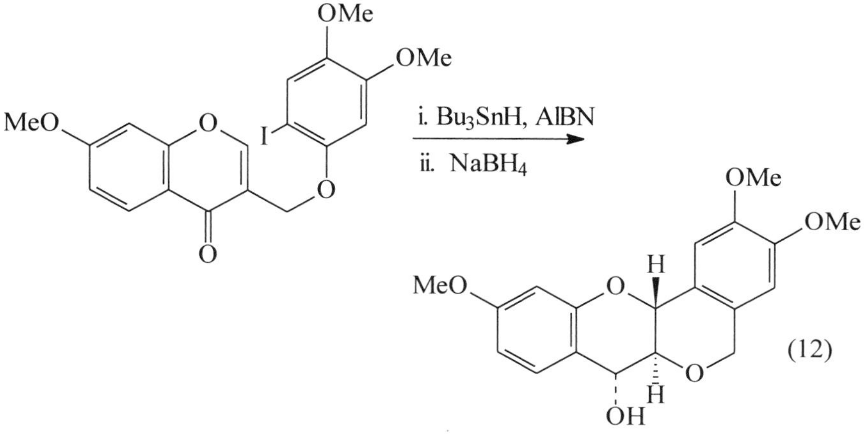

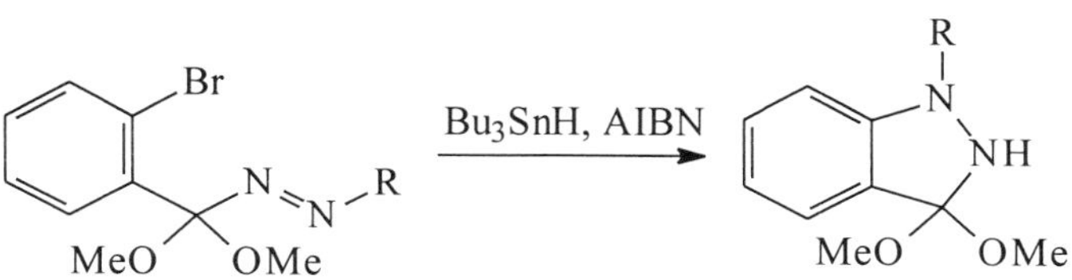

Sometimes mixtures are obtained when both 6-endo and 5-exo pathways are available (*eg.* M. J. Tomaszewski and J. Warkentin, Tetrahedron Lett., 1992, **33**, 2123; *cf.* S. Takano *et al, ibid*, 1990, **31**, 2315):-

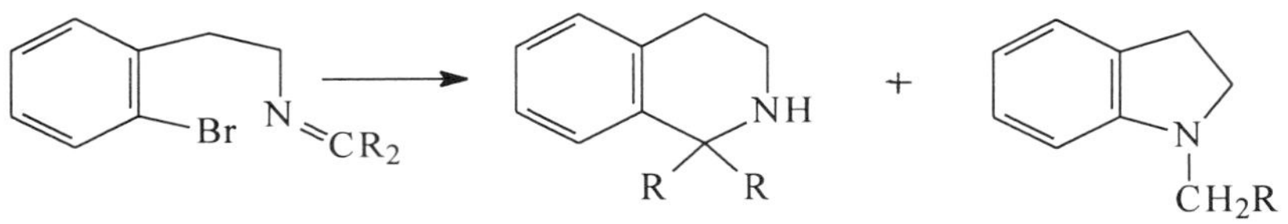

Arene to alkene cyclizations are the most common and enamines are useful substrates in alkaloid synthesis where complex functionality and stereochemistry elsewhere in the molecule survives unscathed (*eg.* A. G. Schultz, M. A. Holoboski and M. S. Smyth, J. Amer. Chem. Soc., 1993, **115**, 7904). Ingenious cascade cyclizations have been devised, most dramatically in a triple cyclization leading to the lysergic acid system (D. E. Cladingboel and P. J. Parsons, J. Chem. Soc., Chem. Commun., 1990, 1547):-

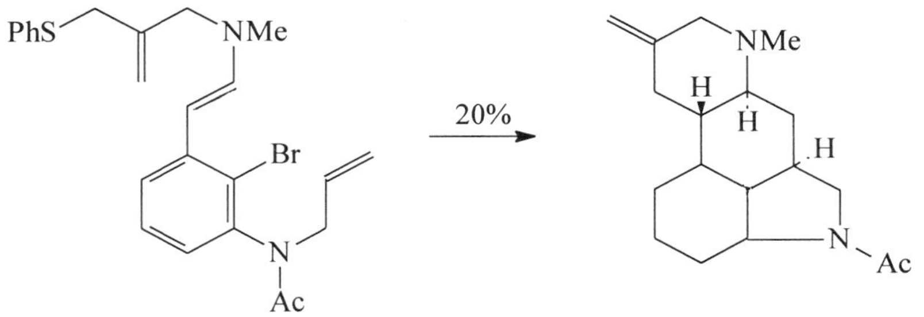

Tandem cyclizations, however, deliver higher yields as exemplified by double cyclizations leading (as above but in 74% yield) to lysergic acid derivatives (Y. Ozlu, D. E. Cladingboel and P. J. Parsons, SYNLETT, 1993, 357) and to dihydroisocodeine (D. P. Curran and H. Liu, J. Amer. Chem. Soc., 1992, **114**, 5863).

5. FORMATION AND LOSS OF THE BENZENE NUCLEUS

(a) Formation

(i) General

The main theme of Section 5a involves benzene ring-forming cyclization of alkynes. Other routes to the benzene nucleus are discussed in Chapter 2, while a detailed review discusses general routes to simple benzene derivatives (P. Bamfield and P. F. Gordon, Chem. Soc. Rev., 1984, **13**, 441).

Excluded from the rest of the section are Diels-Alder type cycloadditions. However numerous examples of such methodology can be found for the preparation of simple benzene derivatives (eg. A. D. Buss, G. C. Hirst and P. J. Parsons, J. Chem. Soc., Chem. Commun., 1987, 1836), of monoarylcatechols (eg. T. V. Lee, A. J. Leigh and C. B. Chapleo, SYNLETT, 1989, 30), of benzannelated compounds (eg. S. Gronowitz, G. Nikitidis and A. Hallberg, Acta Chem. Scand., 1991, **45**, 632; K. Kanematsu and I. Kinoyama, J. Chem. Soc., Chem. Commun., 1992, 735), and of polycyclic aromatics (eg. H. Fujihara et al, J. Org. Chem., 1993, **58**, 5291). Benzyne can be used as the 2π component in naphthoquinone synthesis (eg. M. A. Brimble and S. J. Phythia, Tetrahedron Lett., 1993, **34**, 5813). Electrocyclic reactions of suitable cyclobut-2-enones (R. Danheiser and S. K. Gee, J. Org. Chem., 1984, **49**, 1672; A. Gurski and L. S. Liebskind, J. Amer. Chem. Soc., 1993, **115**, 6101), or of corresponding 2-exomethylenecyclobutenols (J. E. Ezcurra and H. W. Moore, Tetrahedron Lett., 1993, **34**, 6177) also lead via eventual [4 + 2] cycloaddition to benzenes or phenols eg. (R. L. Danheiser, S. K. Gee and J. J. Perez, J. Amer. Chem. Soc., 1986, **108**, 806):-

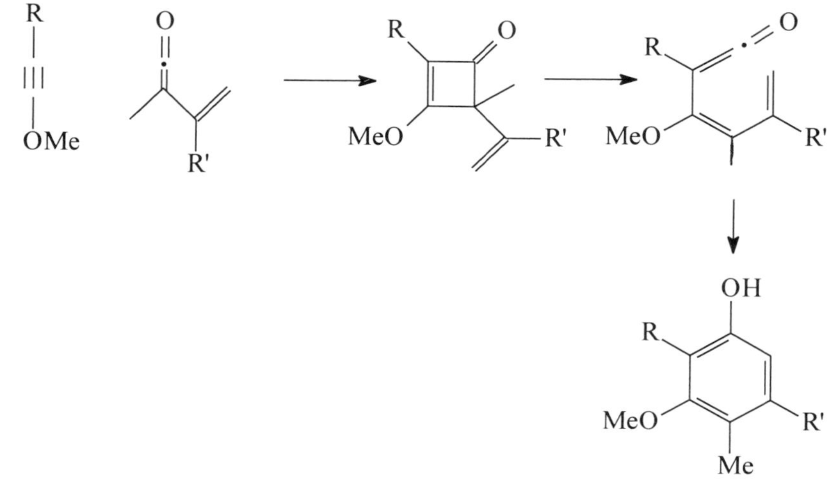

(ii)   Enediyne and related cyclizations

The ring-forming rearrangement of 3-ene-1,5-diynes (13) is called Bergman cyclization (R. R. Jones and R. G. Bergman, J. Amer. Chem. Soc., 1972, **94**, 660) and leads to benzene 1,4-diradicals (review: F. E. Ziegler, Chem. Rev., 1988, **88**, 1423). A related process (Myers cyclization) generates tolyl diradicals from the corresponding en-yn-allenes (14) (A. G. Myers et al, J. Amer. Chem. Soc., 1989, **111**, 8057, 9130):-

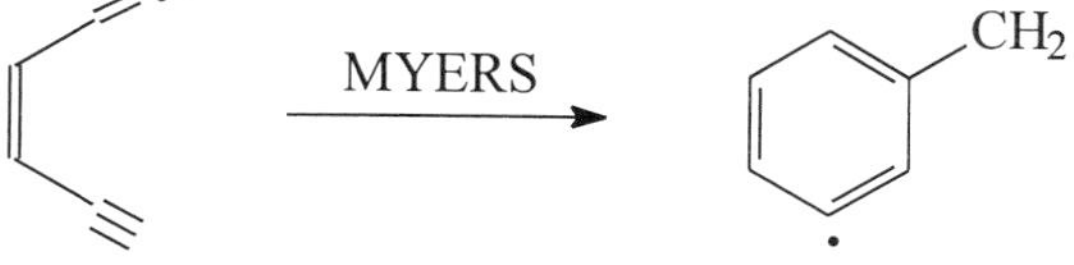

Despite the similarity the Myers process has lower energy of activation and is more exothermic (N. Koga and K. Morokuma, *ibid*, 1991, **113**, 1907). In either case the cyclizations are of broad scope and it is uncertain whether the key factor in reaction design is the relief of strain (*eg.* P. Magnus *et al*, *ibid*, 1990, **112**, 4986; J. P. Snyder, *ibid*, 1990, **112**, 5367) or the distance apart of the reaction termini C_1 and C_6 (*eg.* K. C. Nicolaou *et al*, Angew. Chem., Int. Ed. Engl., 1989, **28**, 1272). In some cases at least the thermal cyclization is remarkably facile *eg.* (A. G. Meyers and N. S. Finney, J. Amer. Chem. Soc., 1992, **114**, 10986):-

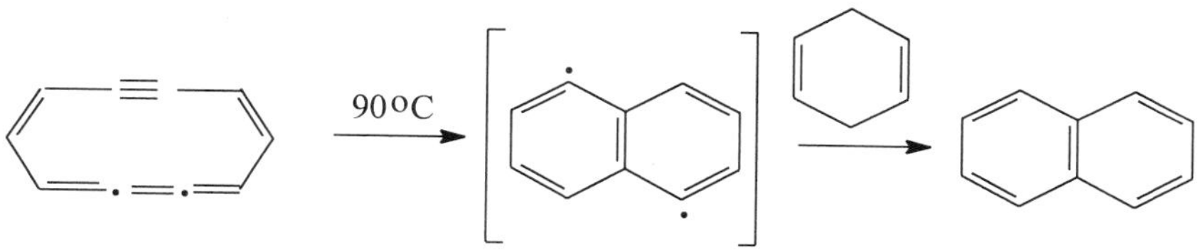

The initial diradical products are highly reactive and will quench *via* further reaction and/or an added source of hydrogen atoms (as above) or *via* solvent homolysis *eg.* (R. G. Bergman *et al*, *ibid*, 1992, **114**, 3120):-

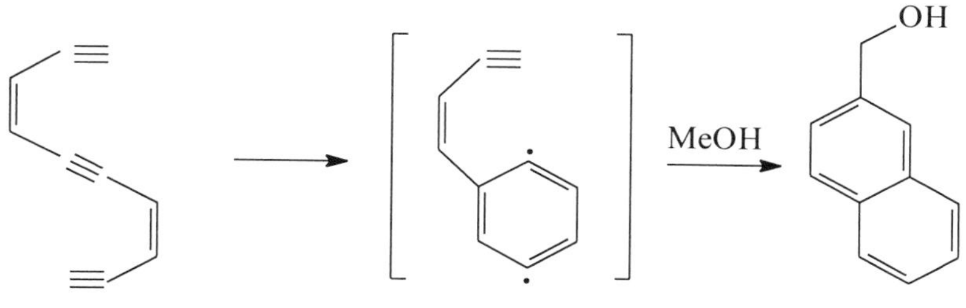

Suitable pendant alkenes can be used for tandem cyclizations to introduce functionality into the final products *eg.* (J. W. Grissom, T. L. Calkins and H. A. McMillen, J. Org. Chem., 1993, **58**, 6556):-

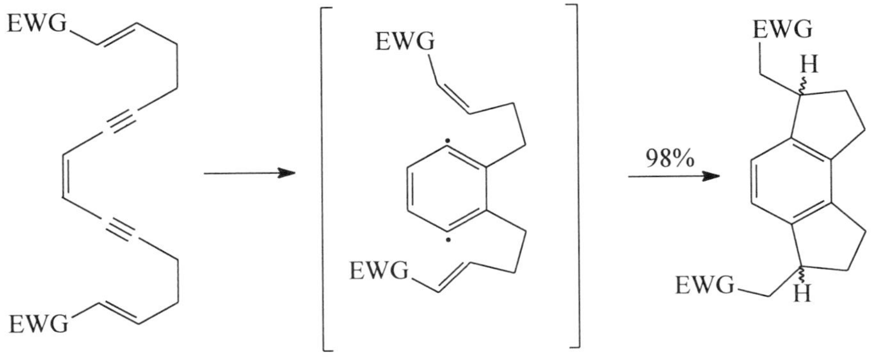

The cyclization can also be used for benzannelation but reaction is inhibited if the double bond is part of an existing benzene ring although it is accelerated if the ene bond comes from a quinone ring eg. (M. F. Semmelhack et al, Tetrahedron Lett., 1992, **33**, 3277; cf. K. C. Nicolaou et al, J. Amer. Chem. Soc., 1992, **114**, 9279):-

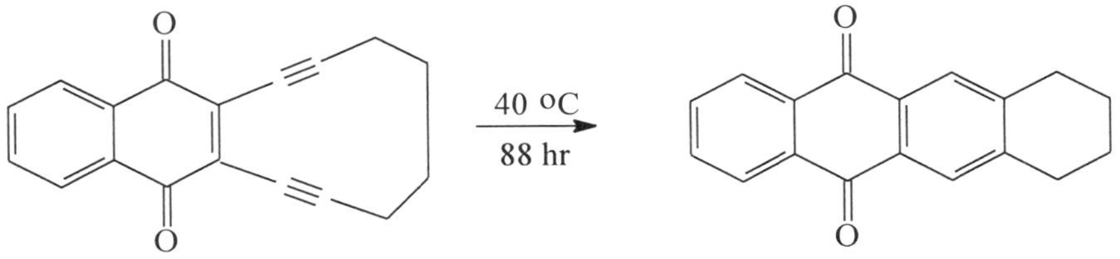

In addition to their synthetic potential, the cyclization processes are crucial to the biological activity of natural enediyne antibiotics such as calicheamicin γ (15), neocarzinostatin and dynemicin A (for reviews on these important compounds see K. C. Nicolaou and W. M. Dai, Angew. Chem., Int. Ed. Engl., 1991, **30**, 1387; I. H. Goldberg, Acc. Chem. Res., 1991, **24**, 191; M. D. Lee, G. A. Ellstad and D. B. Borden, ibid, 1991, **24**, 191; K. C. Nicolaou, Chem. Brit., 1994, **30**, 33). Their anti-tumour activity

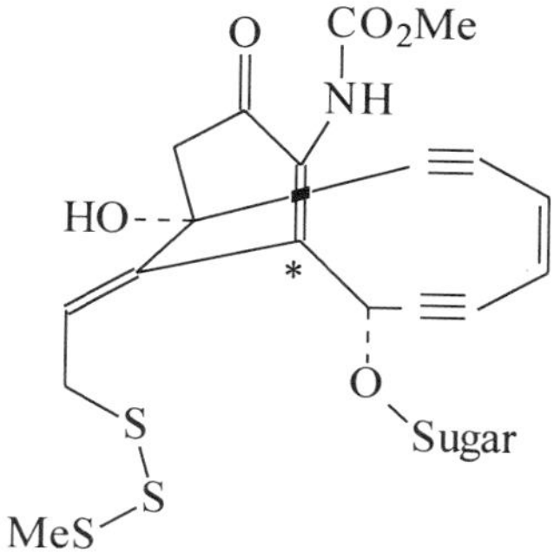

(15)

probably involves activation by nucleophilic attack (*eg.* at the trisulphide moeity of 15) followed by conjugate addition of S⁻ to the enone double bond to unlock the conformation of the enediyne functionality, which then undergoes cycloaromatisation (Bergman in the case of 15). The resulting, highly reactive diradical can then quench by attacking double stranded DNA (K. C. Nicolaou *et al*, J. Amer. Chem. Soc., 1988, **110**, 7247) leading to cleavage and cell death.

(iii)   Metal mediated alkyne cyclizations

Benzenoid compounds can be made *via* the cyclotrimerization of alkynes, [2 + 2 + 2] reactions mediated by one of several possible transition metal catalysts. The Vollhardt cyclization, using $CpCo(CO)_2$, was introduced earlier as an entry to various phenylenes (Section 3ciii) but its scope is broad (for one step cyclization to a ring B aromatic steroid see K. P.C. Vollhardt *et al*, J. Org. Chem., 1986, **51**, 5496). By using two alkynes with one alkene it is possible to isolate cyclohexadienes *eg.* (E. P. Johnson and K. P. C. Vollhardt):-

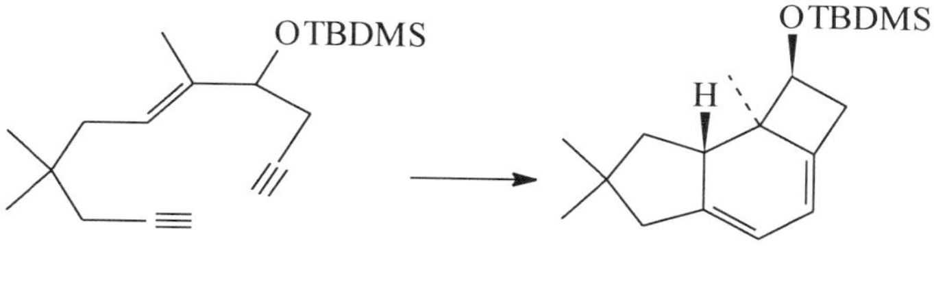

Variation to the alkene component and to the particular cobalt catalyst used affects the feasibility of isolating a cyclohexadiene before it aromatises *eg.* (Z. Zhou, M. Costa and G. P. Chiusoli, J. Chem. Soc., Perkin Trans. 1, 1992, 1399):-

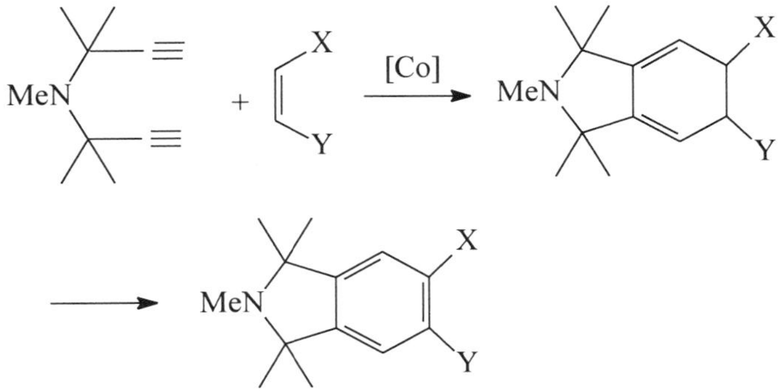

Preformed cobalt-vinylketene complexes react directly with alkynes to give phenols (S. H. Cho, K. R. Wirtz and L. S. Liebeskind, Organometallics, 1990, **9**, 3067).

Alkyne cyclotrimerization can also be performed using alkynes complexed with niobium (A. C. Williams *et al*, *ibid*, 1989, **8**, 1566) or with ruthenium carbonyl (E. Lindner *et al*, *ibid*, 1989, **8**, 2355), and RhCl(PPh$_3$)$_3$ is useful for catalytic cyclotrimerizations (R. Grigg, R. Scott and P. Stevenson, J. Chem. Soc., Perkin Trans. 1, 1988, 1357). Nickel (0) has been used for mild cyclizations with concomitant functional group transformation *eg.* (P. Bhatarah and E. H. Smith, *ibid*, 1992, 2163):-

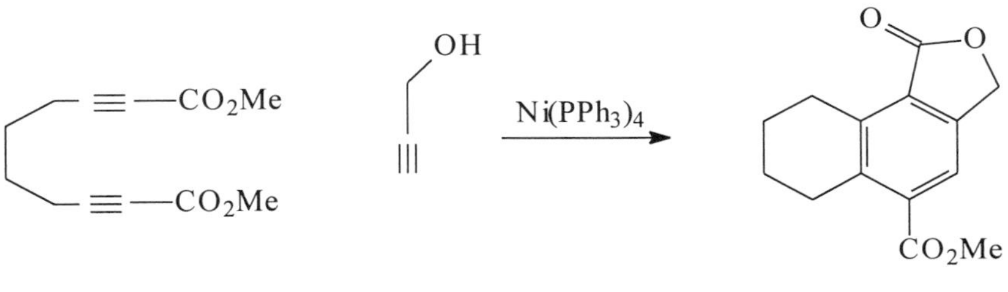

Using a silicon trapping agent, two molecules of CO were cyclised successfully with a diyne to give a catechol silyl ether with ruthenium carbonyl catalysis (N. Chatani et al, J. Amer. Chem. Soc., 1993, **115**, 11614):-

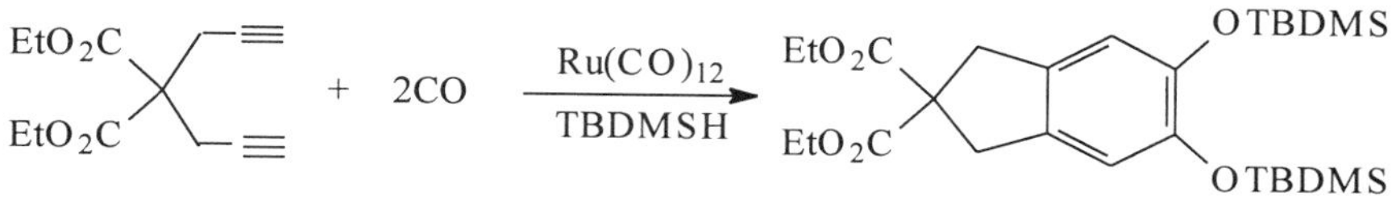

Palladium catalysts are also effective. The trialkyne (16a) cyclised in acidic solution (HOAc/MeCN) while the corresponding vinyl bromide (16b) required basic conditions (Et_3N/MeCN) (E. Negishi et al, Tetrahedron Lett., 1992, **33**, 3253):-

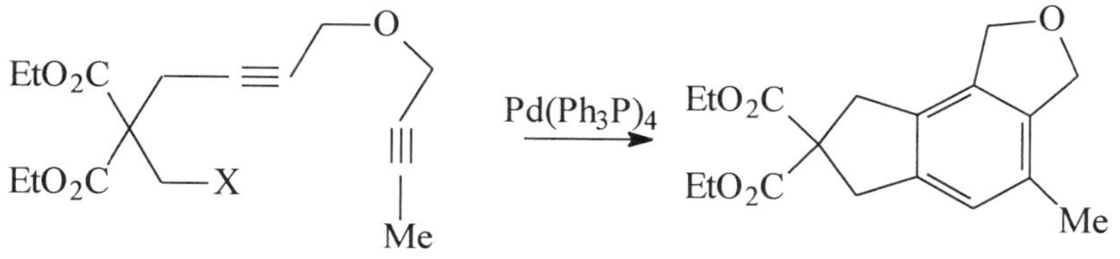

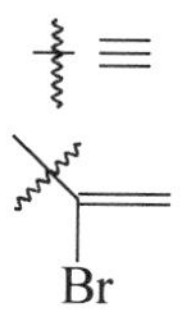

Benzene rings are also formed in the reactions of Fischer carbene complexes (usually Cr) with alkynes in the presence of CO which, via incorporation into the ring, provides phenolic products eg. (W. D. Wulff and C. D. Jung, J. Amer. Chem. Soc., 1984, **106**, 7565):-

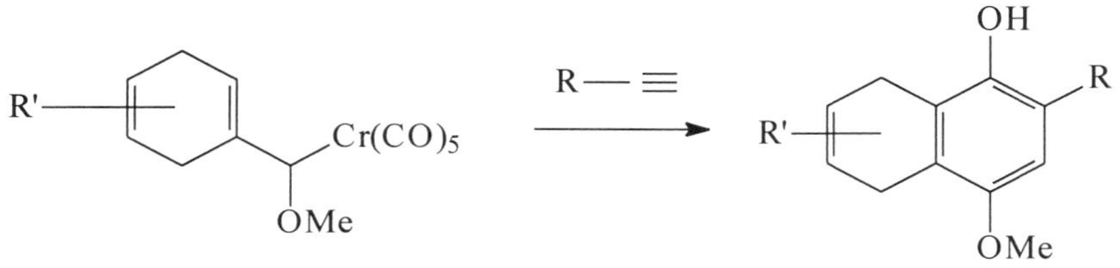

The course of the reaction has been studied in depth (W. D. Wulff *et al*, *ibid*, 1993, **115**, 10671). The initial product, an η^6-Cr(CO)$_3$ complex is not normally isolated but it can be intercepted by a nucleophile (see Section 4ciii) in favourable cases *eg*. (S. Chamberlain and W. D. Wulff, *ibid*, 1992, **114**, 10667):-

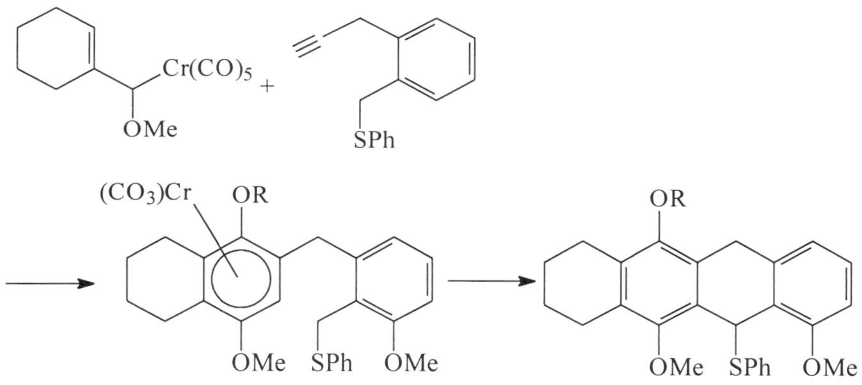

Tungsten carbenes can also be used, most notably in a tandem cyclization of two open chain precursors to a ring A - ring C aromatic steroid in 62% yield (W. D. Wulff *et al*, *ibid*, 1991, **113**, 9873). Modification of the reactions is possible. If an isonitrile is present then it can provide amine rather than phenol functionality (C. A. Merlic *et al*, *ibid*, 1992, **114**, 8722), and if an amine nucleophile is also present then a second amino group is also introduced (R. Aumann, Chem. Ber., 1993, **126**, 1867):-

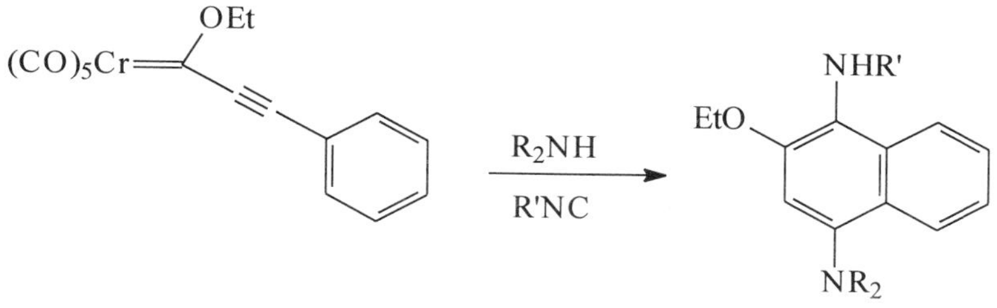

If the conjugation of the carbene is extended by an alkene group then no alkyne is required and an ortho-dioxygenated product is formed (C. A. Merlic and D. Xu, J. Amer. Chem. Soc., 1991, **113**, 7418):-

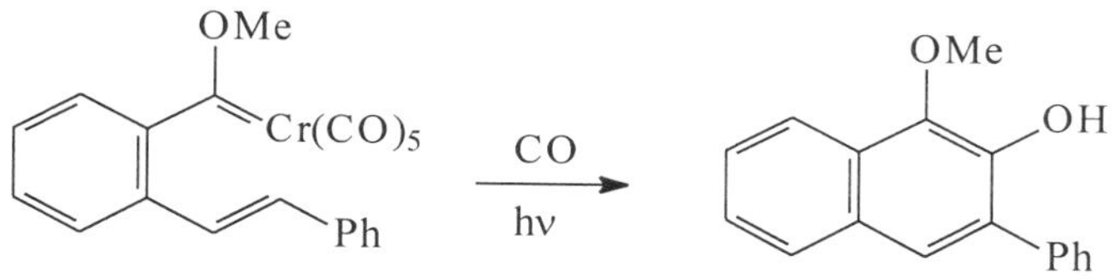

(b) Loss of the benzene ring

(i) General

This section (5b) concentrates on photochemical processes, especially the important meta-photocyclization reactions with alkenes. However, reduction of the benzene ring is important also. It is commonly achieved by Birch reduction involving a metal (usually, but not necessarily, sodium) dissolving in liquid ammonia (review: P. W. Rabideau, Tetrahedron, 1989, **45**, 1579). An electrochemical variant can be used instead of the dissolving metal approach (J. Chaussard et al, Tetrahedron Lett., 1987, **28**, 1173). Birch-type reactions have a long history and have been of use in numerous natural product syntheses (review: J. M. Hook and L. N. Mander, Nat. Prod. Rep., 1986, **3**, 35).

Oxidative dearomatisation of benzenes can be brought about using osmium tetroxide (J. M. Wallis and J. K. Kochi, J. Org. Chem., 1988, **53**,

1679; J. Amer. Chem. Soc., 1988, **110**, 8207). However a microbial method is available using the dioxygenase enzymes of *Pseudomonas putida* mutants (review: H. A. J. Carless, Tetrahedron: Asymmetry, 1992, **3**, 795) leading to *cis*-1,2-dihydrocatechols. A wide variety of ring substituents is tolerated (D. R. Boyd *et al*, J. Amer. Chem. Soc., 1991, **113**, 666) and, in the case of styrene, substituents (R) can affect the product distribution (T. Hudlicky *et al*, SYNLETT, 1992, 391; J. Org. Chem., 1989, **54**, 4239):-

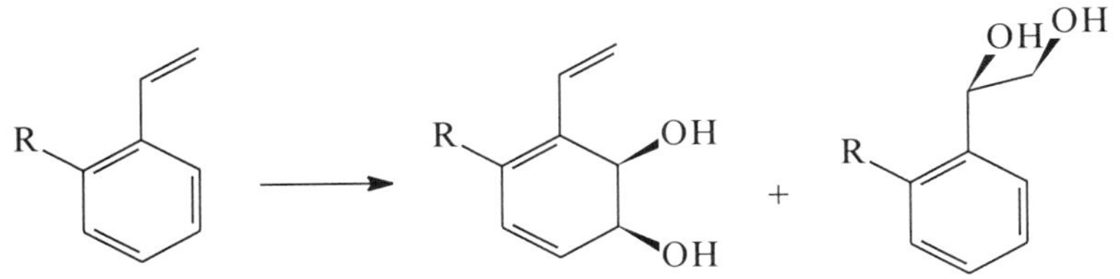

(ii) Photoaddition of the benzene ring with alkenes

Benzene can add to alkenes in a [2 + 2] process to give 1,2-adducts (a) or, via [4 + 2] addition, to give either 1,3-adducts (b) or 1,4-adducts (c):-

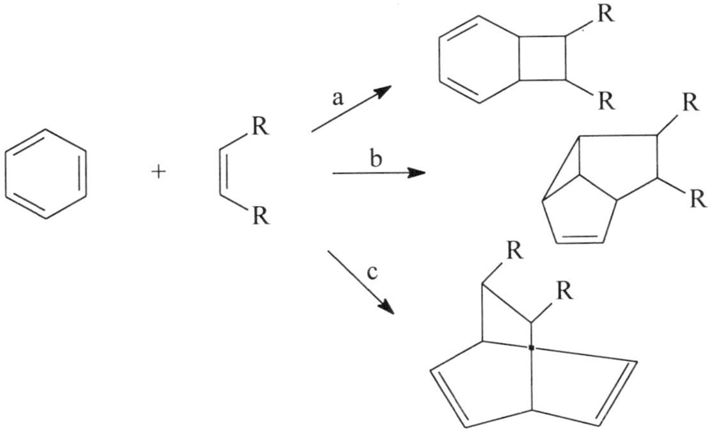

Of the three, 1,4-addition is rare although favoured in certain intramolecular additions, *eg.* (A. Gilbert and G. Taylor, J. Chem. Soc., Chem. Commun., 1978, 129):-

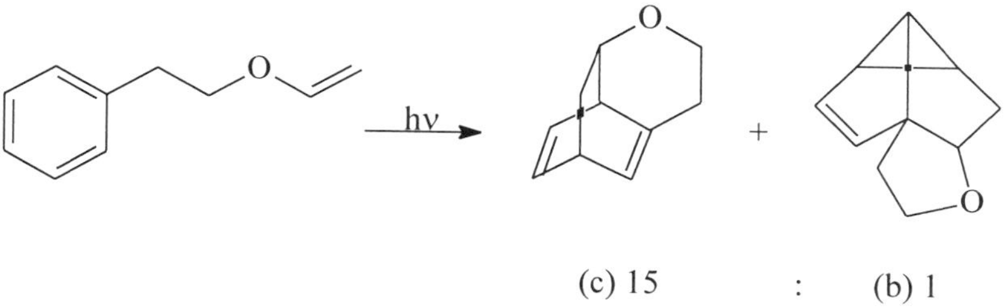

(c) 15 : (b) 1

1,2-Addition reactions are quite common especially in intermolecular reactions with alkenes containing electron withdrawing groups (or electron donating groups if the benzene ring is π-deficient), *eg* (A.Gilbert and P.Yianni, Tetrahedron, 1981, **37**, 3275):-

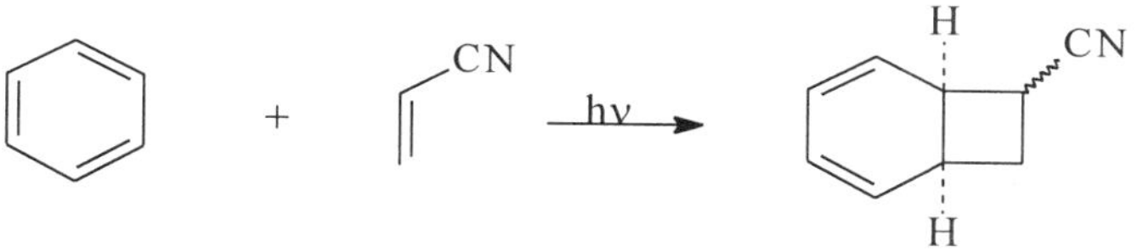

Intramolecular 1,2-cycloadditions, while less common, have been achieved using ω-alkenyl aryl ethers (*eg* A.Gilbert *et al*, J.Chem.Soc.Perkin Trans. 1,1992, 1145, but see *idem, ibid,* 1992, 2265) and even using a styrene derivative (H. Aoyama, Y. Arata and Y. Omote, J.Chem.Soc., Chem.Commun.,1990,736).

The most important mode is the 1,3- or meta-photocycloaddition. It seems that addition by the alkene to a prefulvene or prebenzvalene biradical (section iii) is not normally involved, but rather that the mechanism involves initial formation of an exciplex, with some zwitterionic character, between the alkene and photexcited arene (P. de Vaal, G.Lodder and J.Cornelisse, Tetrahedron, 1986, **42**, 4585):-

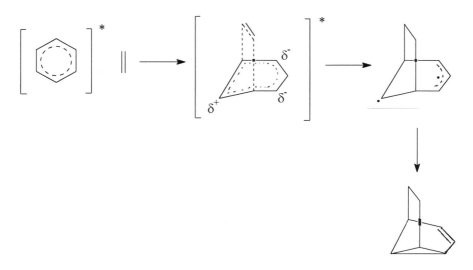

With substituted benzenes the regiochemistry of addition depends on whether the the substituent is electron donating or withdrawing. The stereochemical outcome (endo or exo) also depends on electronic factors with exo addition being restricted mainly to reactions involving electron-rich arenes or benzene itself; electron withdrawing substituents lead to endo adducts (review : J.J.McCullough, Chem. Rev., 1987, **87**, 811).

The importance of the meta-photocycloaddition lies not just with its relatively common occurrence. The reaction involves reasonably mild conditions, is tolerant to several functional groups and often leads to the creation of several asymmetric carbon atoms. It has accordingly been exploited in natural product synthesis, especially by Wender and coworkers (*eg.* P.A. Wender, T.W. von Geldern and B.H.Levine, J.Amer.Chem.Soc., 1988, **110**, 4858).

(iii) Other photochemical processes.

Detailed reviews of the photochemistry of benzene derivatives appeared in the late 1970s (D. Bryce-Smith and A.Gilbert, Tetrahedron, 1976, **32**, 1309; 1977, **33**, 2459). A subsequent review sought to provide systematic classification of the reactions (G.Kaupp, Angew. Chem.,Int.Ed.Engl., 1980, 19, 243). A photochemical variant of the Birch reduction results from photoexcitation of a substrate in the presence of an electron acceptor (*eg.* 1,3-dicyanobenzene, DCB) followed by hydride reduction (*eg.* $NaBH_4$) (M.

Yasuda, C. Pac and H. Sakurai, J.Org.Chem., 1981, **46**, 788). The product cyclohexadiene emerges after protonation of the intermediate radical and recovery of the DCB catalyst (G.A.Epling and E.Florio, Tetrahedron Lett., 1986, **27**, 1469):-

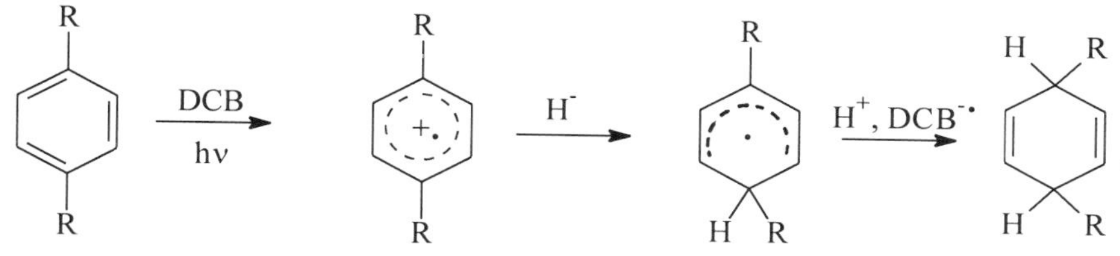

In most cases the photoreduction, unlike conventional Birch reductions, leaves substituents R on sp^3 carbon atoms in the product (M.Kropp and G.B. Schuster, *ibid*, 1987, **28**, 5295)

Photoisomerisation of benzene (liquid or in solution) often leads to the diradical (17) and thus to benzvalene (D. Bryce-Smith and A. Gilbert, *op.cit.*). Fulvene can then arise by H-migration and ring contraction :-

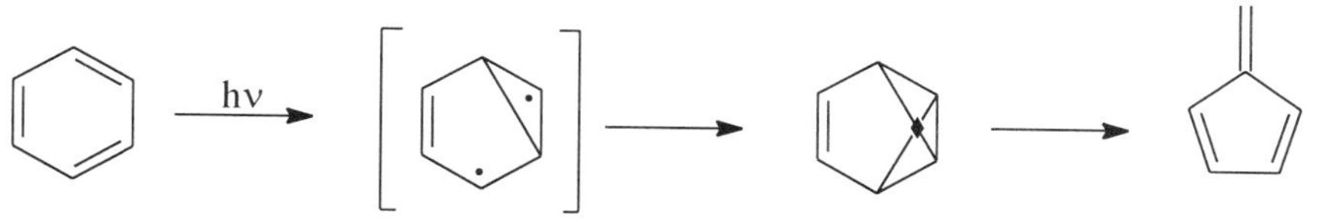

(17)

At shorter wavelengths isomerisation to Dewar benzene is favoured (see G.Kaupp, 1980, *op.cit.*). The valence isomers were surveyed in section 3a above. Hexakis(trimethyl)silyl)benzene readily forms the Dewar isomer (H. Sakurai *et al*, J.Amer.Chem.Soc, 1990, **112**, 1799) and Dewar forms have been prepared that are isomeric with thiophenes (W.A. Rendall, M. Torres and O. P. Strausz, *ibid*, 1985, **107**, 723), furans (I.G.Pitt, R.A.Russell and R.N.Warrener, *ibid*, 1985, **107**, 7176; *idem*, J.Chem.Soc., Chem.Commun., 1984, 1464, 1466), and pyridines (G.Maas *et al*, *ibid*, 1990, 1456).

Photorearrangement of azatriptycene (18, X = N) at 77K leads *via* trapping of nitrene (19, X = N̈:) to a product (20, X = N) (T. Sugawara *et al*, J. Amer.Chem. Soc., 1983, 105, 858), that is analogous to the carbene insertion product (20, X = CH) from triptycene itself (see G. Kaupp, 1980, *op.cit.*).

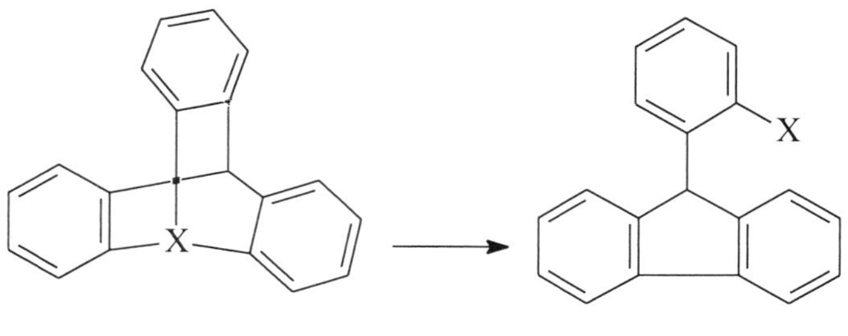

(18)

(19)

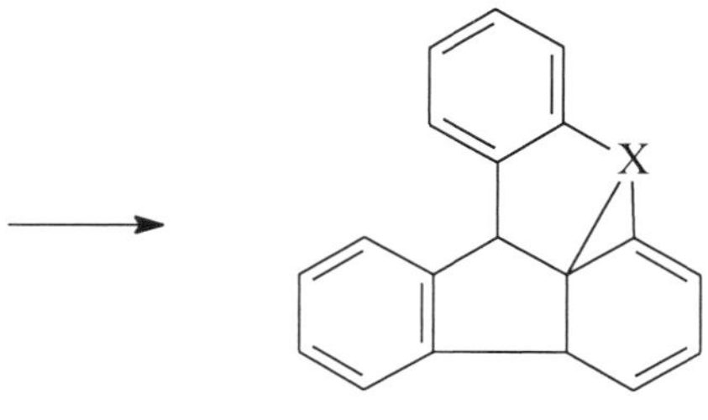

(20)

A similar product (20, X = N) is formed by laser flash photoysis of the azide (19, X = N_3) (S. Murata *et al*, Tetrahedron Lett., 1984, **24**, 1933), but direct treatment of dialkylbenzenes with methoxycarbonylnitrene leads to ring fission. (T. Kumagai *et al*, *ibid*, 1983, **24**, 2275):-

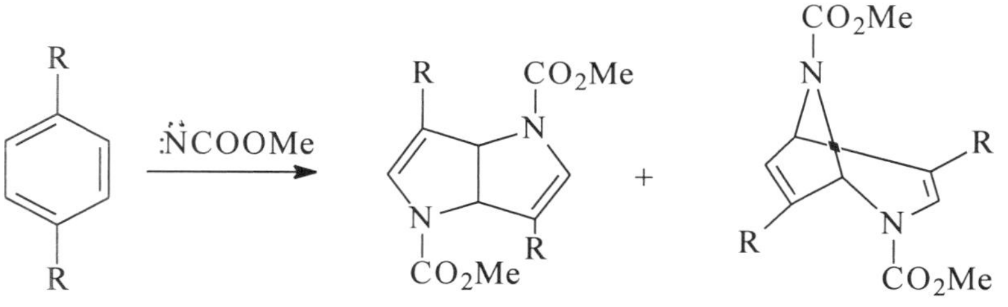

Chapter 2

MONONUCLEAR HYDROCARBONS: BENZENE AND ITS HOMOLOGUES

H. HEANEY

1. *Nuclear Magnetic Resonance Spectroscopy*

The main advances in the use of nuclear magnetic resonance spectroscopy relate to the use of pulse sequences that allow easier interpretation as a result of two-dimensional techniques. Both homonuclear and heteronuclear correlation spectroscopy (COSY) procedures are used in structure determination of carbocyclic aromatic compounds. A number of books have been published that provide good introductions (see Derome, A.D. *'Modern N.M.R. Techniques for Chemistry Research'*, Pergamon, Oxford, **1987**; Sanders, J.K.M.; Hunter, B.K. *'Modern NMR Spectroscopy. A Guide for Chemists'*, Oxford University Press, Oxford, **1987**; and Bax, A. *'Two Dimensional Nuclear Magnetic Resonance in Liquids'*, Delft University Press, Dordrecht, **1982**). The methods available for the generation and study of arenium ions in solution have been comprehensively reviewed (Koptyug, V.A. *Topics Current Chem.*, **1984**, *122*, 1). ^{1}H, ^{13}C, and ^{19}F Nuclear magnetic resonance spectroscopy are among the methods discussed; ^{1}H and ^{13}C nmr spectra have been recorded for a range of benzocyclopropenyl cations (Halton, B. *et al. J. Chem. Soc., Perkin Trans. 2*, **1976**, 258). ^{1}H, ^{2}H, and ^{13}C Nuclear magnetic resonance spectroscopy have all been used to study the mechanisms of reactions involving the transformation of benzenoid derivatives; some of the results are discussed in later sections. The influence of the alignment of molecules on their nmr spectra is well known. Orientation effects have been studied using electric field nmr experiments on deuteriated benzene derivatives (van Zijl, P.C.M. *et al. Accounts Chem. Res.*, **1984**, *17*, 172). The concept of a linear relationship between ^{13}C chemical shift and electron density has been studied with respect to cyclohexadienyl anions and cations (Hallden-Abberton, M.; Fraenkel, G. *Tetrahedron*, **1982**, *38*, 71). Although the values vary with the medium and the counterion the overall conclusion is that there is a linear correlation of charge with shift.

2. The Formation of Benzene Derivatives with Built-in Functionality

The construction of benzene rings from acyclic precursors with functionality already in place is a valuable strategy and useful alternative to functional group interconversion. Annelation strategies are inherently convergent and hence allow the efficient assembly of highly substituted benzene derivatives. The dehydrogenation of cyclohexane derivatives is well known but there have been some useful improvements to the technology involved in some venerable reactions. The Semmler-Wolf aromatisation of cyclohexenone oximes, which can be accessed by the Michael-Aldol strategy, is one such example. The conversion of the oxime derived from 3,5-dimethylcyclohexenone into 3,5-dimethylacetanilide as shown (equation [1]) is achieved under much milder reaction conditions (Tamura, Y. *et al. Synthesis*, **1980**, 483) than those formerly used.

$$\underset{\text{Me}}{\overset{\text{N}^{\text{OH}}}{\bigcirc}}\text{Me} \quad + \quad \text{CH}_2\text{=C=O} \quad \xrightarrow[\text{MeCN, 70-80 °C}]{\text{TsOH}} \quad \underset{\text{Me}}{\overset{\text{HN}^{\text{Ac}}}{\bigcirc}}\text{Me} \qquad [1]$$

61%

2. (a) The Formation of Benzene Derivatives by Cyclisation Reactions

Cyclisation reactions may be conveniently classified under three main sub-headings relating to the number of carbon atoms in the sub-units.

2. (a) (i) *[3 + 3] Cyclisation reactions*

The formation of highly functionalised benzene derivatives has been reviewed in detail (Bamfield, P.; Gordon, P.F. *Chem. Soc. Revs.*, **1984**, *13*, 441). A large number of phenolic compounds have been prepared based on procedures that require two three-atom components. This method requires that one component has two donor sites, for example the methylene groups α-to a carbonyl group. The other component requires two acceptor sites, frequently a β-dicarbonyl compound (Chan, T.H.; Brownbridge, P. *Tetrahedron*, **1981**, *37, suppl. 1*, 387). Reactions of malonyl dichloride with acyclic enol ethers afford phloroglucinol derivatives (equation [2]), together in some cases with 4-hydroxy-α-pyrones (Effenberger, F.; Schönwälder, K.-H.; Stezowski, J.J. *Angew. Chem., Int. Ed. Engl.*, **1982**, *21*, 871; Effenberger, F.; Schönwälder, K.-H. *Chem. Ber.*, **1984**, *117*, 3270). The latter compounds can be converted into benzene derivatives easily. When enol ethers that are derived from cyclic ketones are used *meta*-cyclophane derivatives are accessed (equation [**3**]).

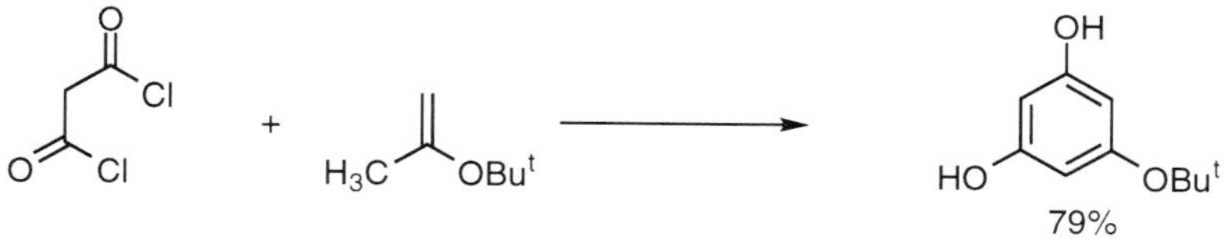

[2]

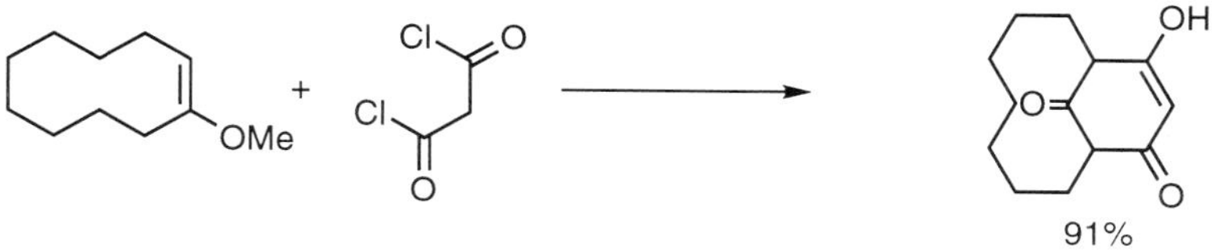

[3]

The condensation of enamines derived from acyclic ketones with 4-trimethylsilyl-3-dialkylamino crotonate esters afford benzenoid derivatives under acid catalysis as shown in equation [**4**] (Chan, T.H.; Kang, G.J. *Tetrahedron Lett.*, **1983**, *24*, 3051).

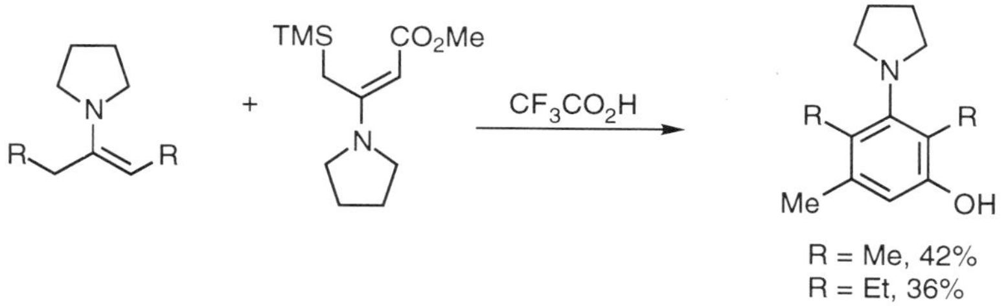

[4]

R = Me, 42%
R = Et, 36%

2. (a) (ii) *[4 + 2] Cyclisation reactions*

The Robinson annelation method which leads to the formation of cyclohexenone derivatives can be recognised as a model for four atom plus two atom syntheses. The cyclohexenones can be converted into phenols or, as exemplified in the introduction to this section, into aniline derivatives. The condensation of 4-trimethylsilyl-3-dialkylamino crotonate esters, shown above (*loc. cit.*), with enamines derived from cyclic ketones take part in four atom plus two atom syntheses as indicated in equation [**5**]. The cyclisation of enynones, which may be prepared from acetylenes and an acyl halides, have been reported to undergo cyclisation to phenols at high temperatures in the presence of collidine *p*-toluenesulfonate (Jacobi, P.A.; Kravitz, J.I. *Tetrahedron Lett.*, **1988**, *29*, 6873).

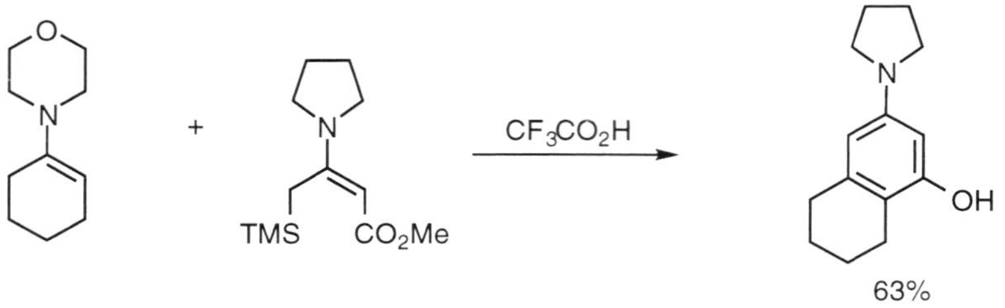

[Reaction scheme 5: morpholine enamine of cyclohexanone + pyrrolidine enamine with TMS and CO₂Me groups, CF₃CO₂H, gives tetrahydronaphthalenol with pyrrolidine and OH substituents, 63%] [5]

The Diels-Alder approach to benzene ring synthesis is perhaps the best known of the four atom plus two atom syntheses and has been exploited on many occasions (Tsuge, O. *et al. Chem. Lett.*, **1984**, 3221; 3234). The majority of examples require an oxidation stage, but by using an acetylenic dienophile and a diene that contains a group that easily eliminates, the correct oxidation level is already present within the two components. The cheletropic ejection of a small thermodynamically stable molecule also serves the same purpose and the two strategies can be combined (Boger, D.L.; Mullican, M.D. *J. Org. Chem.*, **1984**, *49*, 4033). In addition, the correct choice of substituents on the diene and on the dienophile makes it possible to construct substituted benzene derivatives in a single step that would be difficult to access by alternative strategies (Danishefsky, S. *Accounts Chem. Res.*, **1981**, *14*, 400). For example the reactions of 3,4-di-t-butylthiophene-1,1-dioxide with a variety of alkynes provides a route (Nakayama, J. *et al. J. Am. Chem. Soc.*, **1988**, *110*, 6598) to a number of *o*-di-t-butylbenzene derivatives. Part of a synthesis of the antibiotic lasiodiplodin (equation [**6**]) also illustrates this method (Danishefsky, S.; Etheredge, S.J. *J. Org. Chem.*, **1979**, *44*, 4716).

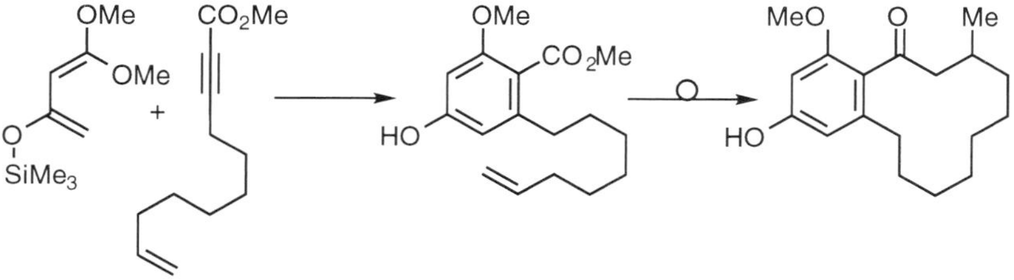

[Reaction scheme 6: Danishefsky diene + alkyne dienophile gives substituted benzene intermediate, then cyclized to lasiodiplodin-type macrocycle] [6]

The alternative strategy in which the dienophile contains a group that is eliminated in the final aromatisation step is exemplified (Krauss, G.A. *et al. Tetrahedron Lett.*, **1988**, *29*, 1879) by the synthesis of 2-nitrotoluene in 95% yield by the reaction between 2-benzoyloxynitroethene and *E*-penta-1,3-diene. In the final step it was found that potassium t-butoxide was the best base.

2. (a) (iii) *[2 + 2 + 2] Cyclisation reactions*

The trimerisation of acetylenes to afford benzene derivatives have been studied using a number of different catalyst systems since the Reppe approach was first published. The topic has been covered as part of a more general review (Schore, N.E. *Chem. Rev.*, **1988**, *88*, 1081). The cyclotrimerisation of acetylene on palladium single crystals proceeds under both ultra high vacuum and atmospheric pressure conditions. Under UHV conditions the different crystal faces have different catalytic activities (Logan, M.A. *et al. J. Phys. Chem.*, **1986**, *90*, 2703). The co-ordination of three acetylenes to a divalent titanium centre has been shown to afford benzene derivatives (Meijer-Veldman, M.E.E.; Meijer, H.J.de L. *J. Organometal. Chem.*, **1984**, *260*, 199). In particular the co-oligomerisation of α,ω-diacetylenes with a monoacetylene that does not itself trimerise has been exploited in a number of interesting ways. In the well known cyclopentadienyl cobalt dicarbonyl method (Vollhardt, K.P.C. *Angew. Chem., Int. Ed. Engl.*, **1984**, *23*, 539) the intermediacy of cobalta-cyclopentadienes has been suggested (Wakatsuki, Y. *et al. J. Am. Chem. Soc.*, **1983**, *105*, 1907). The further elaboration of the arylsilane, formed when an acetylenic component is substituted by a trialkylsilyl group, by the interaction with electrophiles, including cycloacylation by Friedel-Crafts methodology (Gesing, E.R.F.; Sinclair, J.A.; Vollhardt, K.P.C. *J. Chem. Soc., Chem. Commun.*, **1980**, 286) will be discussed in a later section. The methodology has also been used in a concise synthesis of a benzo[3,4]cyclobuta[1,2*b*]biphenylene from 1,2,4,5-tetraiodobenzene (Berris, B.C.; Hovakeemian, G.H.; Vollhardt, K.P.C. *J. Chem. Soc., Chem. Commun.*, **1983**, 502). Palladium catalysed coupling with trimethylsilylacetylene followed by the cyclopentadienylcobalt dicarbonyl catalysed bis(cyclotrimerisation) using bis(trimethylsilyl(acetylene gave the expected product in a 71% yield. The parent hydrocarbon, that is obtained by protiodesilylation, shows little aromatic character in the outer rings. (η^5-Cyclopentadienyl) rhodium complexes have been used to study the cyclotrimerisation of for example hex-3-yne with dimethyl acetylenedicarboxylate (Abdulla, K.; Booth, B.L.; Stacey, C. *J. Organometal. Chem.*, **1985**, *293*, 103). The Wilkinson catalyst [RhCl(PPh)$_3$] has also been used and shown to afford benzene derivatives very rapidly using low concentrations of the catalyst. In the example shown in equation **[7]** the reaction was carried out using 1 mol % of the catalyst at 79 °C for 10 min (Grigg, R.; Scott, R.; Stevenson, P. *Tetrahedron Lett.*, **1982**, *23*, 2691).

It has also been shown that cyclopentadienylcobalt stabilised cyclobutadiene derivatives interact with triply bonded species, including acetylenes, in the latter case to afford substituted benzenes (Gleiter, R.; Kratz, D. *Angew. Chem., Int. Ed. Engl.*, **1990**, *29*, 276). The cyclisation reactions of allenediynes have also been studied by the cobalt methodology and diastereomeric mixtures of cobalt cyclopentadienyl complexes obtained. In the example shown in equation [8] the decomplexation on silica gel led to the partial aromatisation of the initial product (Aubert, C.; Lierena, D.; Malacria, M. *Tetrahedron Lett.*, **1994**, *35*, 2341).

[8]

Bis(cycloocta-1,5-diene)nickel [0] in combination with a trialkylphosphine catalyses the cycloaddition reactions of 1,6- and 1,7-diynes with carbon dioxide to afford mixtures of the α-pyrone and the dimer, a substituted benzene derivative, (Tsuda, T.*et al. J. Org. Chem.*, **1988**, *53*, 3140). The yield of the α-pyrone, which can be used, in principle in the synthesis of other benzene derivatives, is dependent on the structure of the phosphine that is used.

An extension of Heck reactions (Heck, R.F. *Accounts Chem. Res.*, **1979**, *12*, 146; *Org. React.*, **1982**, *27*, 345) involving palladium catalysis in the coupling of haloalkenes with alkynes has been shown to lead to regioselective benzene ring formation. Palladium catalysis has been used in the cyclisation of skipped triacetylenes that result in the formation of benzenes that also carry two additional alicyclic rings (Negishi, E.-i. *et al. Tetrahedron Lett.*, **1992**, *33*, 3253). The carbometallation of skipped halodienes with a suitably substituted alkyne was shown to proceed slowly (Zhang, Y.; Negishi, E.-i. *J. Am. Chem. Soc.*, **1989**, *111*, 3454). On the other hand the formation of benzene derivatives using haloenynes together with mono-acetylenes proceeds rapidly and with good regioselectivity (Equation [9]) (Negishi, E.-i.; Ay, M.; Sugihara, T. *Tetrahedron*, **1993**, *49*, 5471).

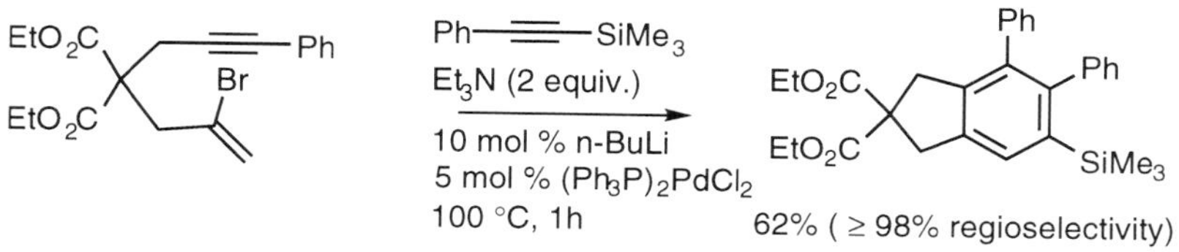

[Scheme for equation 9: EtO₂C/EtO₂C-substituted bromoalkene-alkyne with Ph + Ph-≡-SiMe₃, Et₃N (2 equiv.), 10 mol % n-BuLi, 5 mol % (Ph₃P)₂PdCl₂, 100 °C, 1h → indane-benzene product with Ph, Ph, SiMe₃ substituents, 62% (≥ 98% regioselectivity)] [9]

A similar approach (Torii, S.; Okumoto, H.; Nishimura, A. *Tetrahedron Lett.*, **1991**, *32*, 4167) provided evidence for the sequence of reactions involved. That the first step involves the intramolecular insertion of the palladium into the acetylenic bond is indicated by the isolation of the monocyclic product when the reaction shown in equation [**10**] was carried out at room temperature. The benzene derivative was only isolated when the reaction was carried out under reflux.

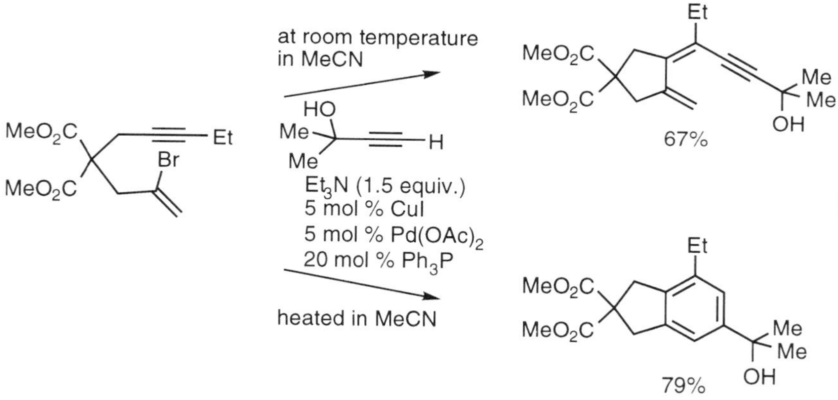

[Scheme for equation 10: MeO₂C/MeO₂C bromoalkene-alkyne-Et + HO-CMe₂-C≡C-H, Et₃N (1.5 equiv.), 5 mol % CuI, 5 mol % Pd(OAc)₂, 20 mol % Ph₃P. At room temperature in MeCN → monocyclic product 67%. Heated in MeCN → tricyclic benzene product 79%.] [10]

The synthesis of highly functionalised tricyclic benzene derivatives by suitably positioned bromoalkenes linked to two other alkyne residues has also been achieved by tandem cyclisations using Heck reaction conditions. The principle is based on earlier work (Meyer, F.E. *et al. J. Chem. Soc., Chem. Commun.*, **1992**, 390; Meyer, F.E.; Parsons, P.J.; de Meijere, A. *J. Org. Chem.*, **1991**, *56*, 6487). In the examples shown in equation [**11**] oxidative addition of the carbon bromine bond onto palladium (0) is presumed to be followed either by way of two successive n-exo-*dig* ring closure reactions or an n-exo-*dig* reaction followed by electrocyclisation (Meyer, F.E.; de Meijere, A. *Synlett*, **1991**, 777). Reductive elimination from the final σ-alkylpalladium species or hydride elimination would then yield the benzene derivative.

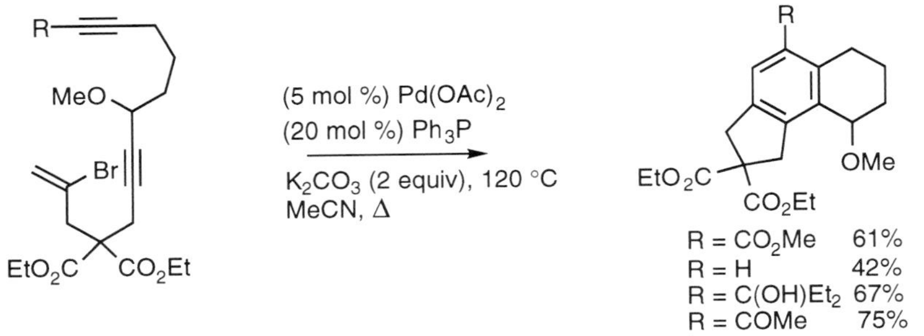

[Scheme for equation 11]

R = CO₂Me   61%
R = H   42%
R = C(OH)Et₂   67%
R = COMe   75%

[11]

Double intramolecular two-acetylene annelations of Fischer carbene complexes that incorporate the carbene ligand have been developed into a benzene ring synthesis as exemplified in equation [**12**] (Bao, J. *et al. J. Am. Chem. Soc.*, **1991**, *113*, 9873).

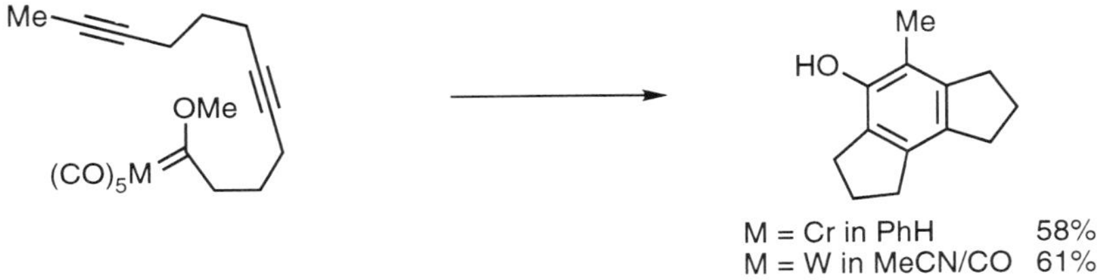

[Scheme for equation 12]

M = Cr in PhH   58%
M = W in MeCN/CO   61%

[12]

A new general route to phenols (Danheiser, R.L. *et al. Tetrahedron Lett.*, **1988**, *29*, 4917; *J. Am. Chem. Soc.*, **1990**, *112*, 3093) has been established. The key step involves the photochemical Wolff rearrangement of an α,β-unsaturated diazoketone to a vinyl ketene followed by reaction with an acetylene which affords a cyclobutenone. This then undergoes a four electron electrocyclic cleavage to a dienyl ketene which can take part in an electrocyclic ring closure to a cyclohexadienone and hence to the phenol. This protocol is exemplified (equation [**13**]) in a synthesis of the host defence stimulant maesanin (Danheiser, R.L.; Cha, D.D. *Tetrahedron Lett.*, **1990**, *31*, 1527) in which the benzenoid ring was assembled with all of the required functional groups in place.

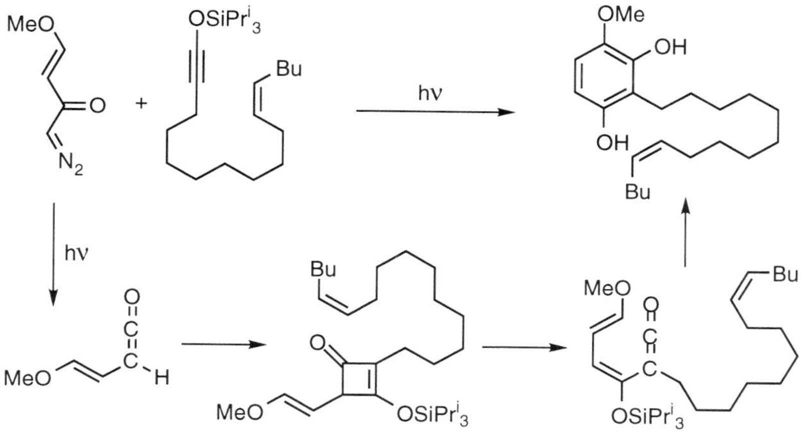

[Scheme showing equation [13]: photochemical reaction of MeO-substituted diazo ketone with alkyne/alkene bearing OSiPr$^i_3$ and Bu groups, proceeding via ketene intermediate through cyclobutenone to give methoxy-dihydroxyphenyl macrocycle with Bu-vinyl side chain]

The palladium catalysed cross coupling of 4-chloro-2-cyclobutenones with vinylstannanes or the related organozirconium compounds also provides a route to dienylketenes and hence to phenols (equation [14]) (Krysan, D.K.; Gurski, A.; Liebeskind, L.S. *J. Am. Chem. Soc.*, **1992**, *114*, 1412).

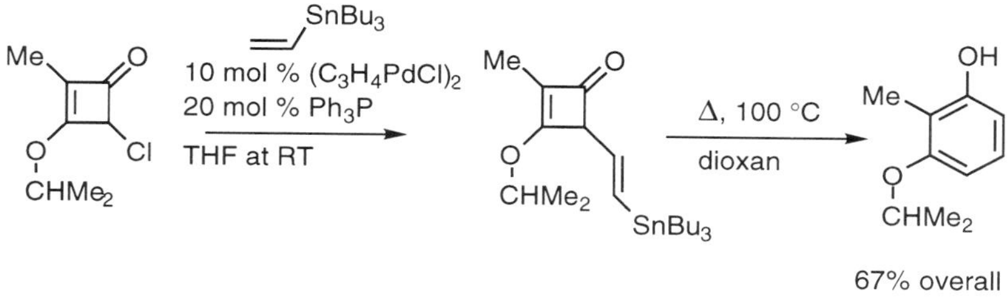

[Scheme for equation [14]: 4-chloro-cyclobutenone with Me and OCHMe$_2$ substituents + CH$_2$=CHSnBu$_3$, 10 mol% (C$_3$H$_4$PdCl)$_2$, 20 mol% Ph$_3$P, THF at RT → vinyl-substituted cyclobutenone → Δ, 100 °C, dioxan → methyl isopropoxy phenol, 67% overall]

3. Benzo- Small Ring Compounds

3.(a) Benzocyclopropenes

The checked preparation of benzocyclopropene in good yield by the dehydrochlorination of 7,7-dichlorobicyclo[4.1.0]hept-3-ene (Billups, W.E.; Blakeney, A.J.; Chow, W.Y. *Org. Synth.*, **1976**, *55*, 12) has made the study of its reactions much more accessible. The versatility of the base induced dehydrochlorination of dichlorocarbene adducts of cyclohexadiene and its analogues is indicated by the synthesis of the rather unstable 1,6-methano[10]annulene derivative (Vogel, E.; Sombroek, J. *Tetrahedron Lett.*, **1974**, 1627) shown in equation [15].

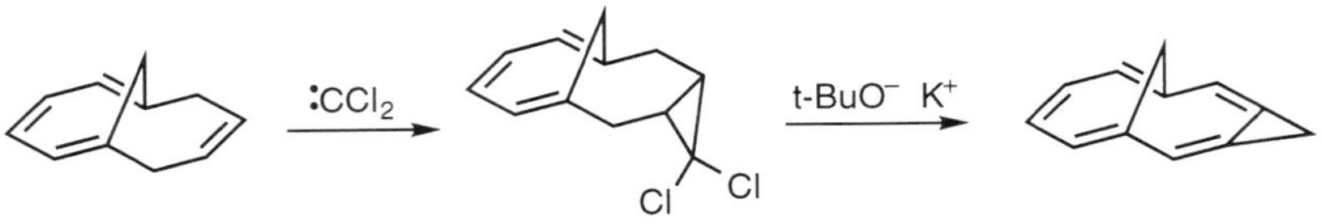

[Scheme 15]

An early review (Halton, B. *Chem. Rev.*, **1973**, *73*, 113) indicated that the chemistry is dominated by processes that involve the cleavage of the three-membered ring. More recent reviews (Billups, W.E.; Rodin, W.A.; Haley, M.M. *Tetrahedron*, **1988**, *44*, 1305; Halton, B. *Chem. Rev.*, **1989**, *89*, 1161) have been published. Benzocyclopropene is estimated to be less stable than cyclopropene by 18.3 k cal. mol^{-1}. Thus the addition of dibromo- or dichloro- carbene results in the formation of the corresponding 7,7-dihalo-benzocyclobutene in very high yields (Kagabu, S.; Saito, K. *Tetrahedron Lett.*, **1988**, *29*, 675) and α,β-unsaturated carbonyl compounds afford 1,3-dihydroisobenzofuran derivatives (Neidlein, R.; Krämer, B. *Chem. Ber.*, **1991**, *124*, 353) in yields that are increased by the addition of Yb(fod)$_3$. The irradiation of a solution of benzocyclopropene and thiocyanogen in benzene leads to the formation of 1,6-dithiocyanatocyclo-heptatriene in 61% yield (Okazaki, R. *et al. Angew. Chem., Int. Ed. Engl.*, **1981**, *20*, 799) and thermal reactions with sulfonyl isocyanates afford cycloaddition products that also result from the fragmentation of the three-membered ring (Kagabu, S. *et al. Bull. Chem. Soc. Jpn.*, **1991**, *64*, 106). Furan reacts with benzocyclopropene in a 2π + 2σ-type cycloaddition reaction in which the three-membered ring breaks as the 2σ component (Saito, K.; Ishihara, H.; Kagabu, S. *Bull. Chem. Soc. Jpn.*, **1987**, *60*, 4141). Inverse electron demand Diels-Alder reactions in which benzocyclopropene functions as the dienophile are exemplified (equation [**16**]) by the reactions with 1,2,4-triazines (Martin, J.C.; Muchowski, J.M. *J. Org. Chem.*, **1984**, *49*, 1040).

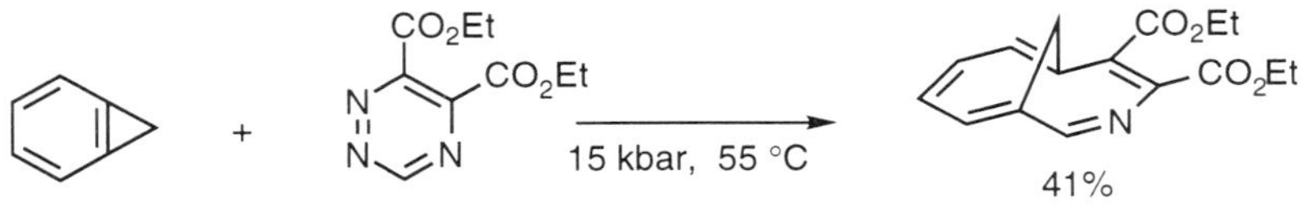

[Scheme 16]

An improved synthesis of cyclopropa[4,5]benzocyclobutene (2.7 to 32%) based on the cyclopentadienyl cobalt dicarbonyl methodology (Saward, C.J.; Vollhardt, K.P.C. Tetrahedron Lett., **1975**, 4539) used a tributyl-stannylpropargyl trimethylsilyl ether as shown in equation [**17**] (McNichols, A.T.; Stang, P.J. *Synlett*, **1992**, 971).

3. (b) Benzocyclobutenes

The use of benzocyclobutenes, as the ring opened *o*-quinodimethanes, in synthesis has been reviewed (Oppolzer, W. *Synthesis*, **1978**, 793). An improved synthesis (Hoey, M.D.; Dittmer, D.C. *J. Org. Chem.*, **1991**, *56*, 1947) of the sultine 1,4-dihydro-2,3-benzoxathiin from *o*-xylylene dihalides and sodium hydroxymethane sulfinate (83% yield) makes the parent *o*-quinodimethane precursor readily available. *o*-Quiodimethane has also been generated in high yield, for use in Diels-Alder reactions, from *o*-xylylene dibromide using chromium (II) chloride (Stephen, D.; Gorgues, A.; Le Coq, A. *Tetrahedron Lett.*, **1984**, *25*, 5649), using activated nickel in dimethoxyethane at room temperature (Inaba, S. *et al. J. Org. Chem.*, **1988**, *53*, 339), and also from analogues (Rubottom, G.M.; Wey, J.E. *Synthetic Commun.*, **1984**, *14*, 507) using a zinc-silver amalgam. The latter method was found to be more efficient than when using zinc. Proton induced 1,4-elimination from δ-hydroxyalkyl trialkylstannanes (Sano, H.; Ohtsuka, H.; Migita, T. *J. Am. Chem. Soc.*, **1988**, *111*, 2014) is another new route to *o*-quinodimethanes.

It is of interest to note that naturally occurring benzocyclobutene derivatives have been isolated and syntheses have been recorded using established routes (Rawal, V.H.; Cava, M.P. *Tetrahedron Lett.*, **1983**, *24*, 5581; Honda, T.; Toya, T. *Heterocycles*, **1992**, *33*, 291). The use of diisobutylaluminium hydride in the demethylation of 4,5-dimethoxybenzocyclobutene gave the best yield of the related dihydric phenol. Oxidation with DDQ then gave the stable quinone, benzocyclobutene-4,5-dione (Sato, M.; Katsumata, N.; Ebine, S. *Synthesis*, **1984**, 685). A number of routes to 1-substituted benzocyclobutenes have been investigated. The optimum conditions, particularly temperatures, have been evaluated for the flash vacuum pyrolysis of a number of benzocyclobutene precursors (Schiess, P.; Rutschmann, S.; Toan, V.V. *Tetrahedron Lett.*, **1982**, *23*, 3665); in particular the flash vacuum pyrolysis of α-cyano- and α-chloro- 2-methylbenzyl chlorides have been shown to be efficient precursors to 1-cyano- and 1-chloro- benzocyclobutene (Schiess, P.; Rutschmann, S.; Toan, V.V. *Tetrahedron Lett.*, **1982**, *23*, 3669). An interesting entry to 1-bromobenzocyclobutene (equation [**18**]) in yields varying from 18 to 45% allows the ready conversion into a number of other 1-substituted derivatives (DeCamp, M.R.; Viscogliosi, L.A. *J. Org. Chem.*, **1981**, *46*, 3918). A high temperature is required for the reaction but unfortunately this also results in the decomposition of the desired product, presumably, at least in part, by way of the related bromo-*o*-quinodimethane. When the reaction was conducted in the presence of dimethyl acetylenedicarboxylate, dimethyl naphthalene-2,3-dicarboxylate was isolated, which corroborates the suggestion of the possible involvement of the quinodimethane. Reactions of arynes with enolates derived from the monoacetals of 1,2-diketones have been used to prepare interesting functionalised benzocyclobutenes, as exemplified in equation [19] (Gregoire, B.; Carre, M.-C.; Caubere, P. *J. Org. Chem.*, **1986**, *51*, 1419).

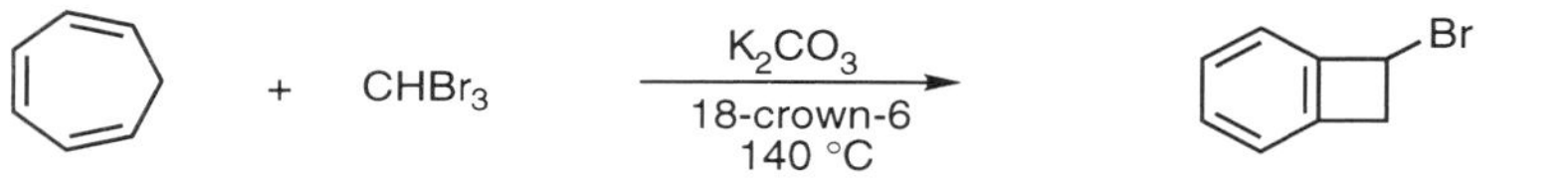

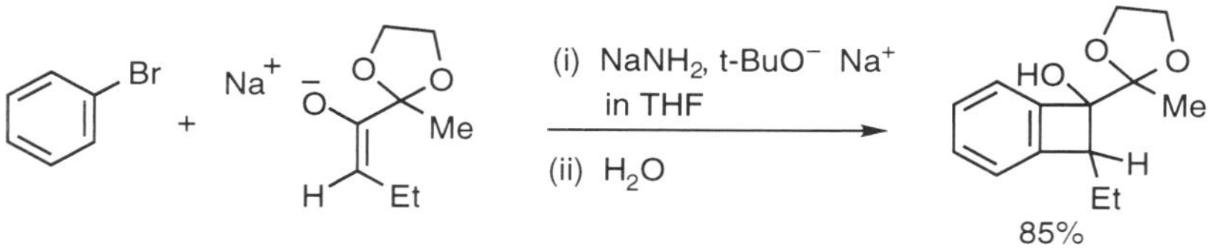

The generation of 1,3-cyclooctadiene-5-yne from thermolysis of the selenodiazole (equation [20]), which can be trapped by dienes used in conventional benzyne chemistry, rearranges to benzocyclobutene in the absence of a co-reactant (Meier, H.; Hanold, N.; Kolshorn, H. *Angew. Chem., Int. Ed. Engl.*, **1982**, *21*, 66).

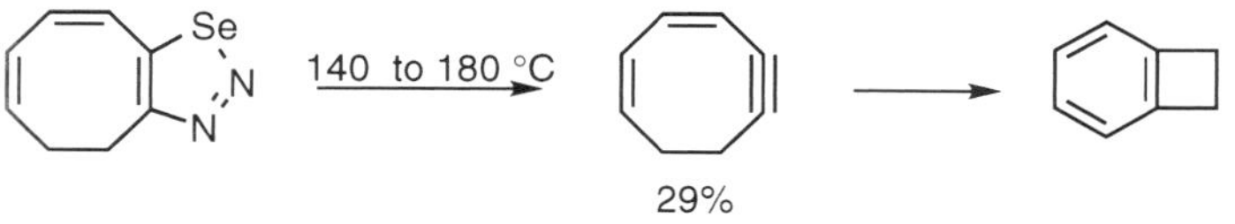

The kinetics of the photochemical decomposition of the ketone shown in equation [21] to *o*-quinodimethane (E_a = 25.6 k cal.mol^{-1}) (Roth, W.R.; Scholz, B.P. *Chem. Ber.*, **1981**, *114*, 3741) and its cyclisation to benzocyclobutene (E_a = 26.9 k cal.mol^{-1}) were measured using a shock tube technique. Temperature dependant measurements on the equilibrium between *o*-quinodimethane and benzocyclobutene gave a value of 60.8 k cal.mol^{-1} for the enthalpy of formation of *o*-quinodimethane. In more recent measurements (Roth, W.R.; Ebbrecht, T.; Beitat, A. *Chem. Ber.*, **1988**, *121*, 1357) activation parameters have been redetermined using dioxygen trapping and gave a value of 26.3 k cal.mol^{-1} for the cyclisation of *o*-quinodimethane to benzocyclobutene. The addition of the carbene benzocyclobutenylidene to benzene (O'Leary, M.A.; Richardson, G.W.; Wege, D. *Tetrahedron*, **1981**, *37*, 813) gives a spirocycloheptatriene derivative which undergoes ring opening to the related *o*-quinodimethane derivative. At 71.8 °C the ring opening is 2.7×10^6 times faster than that of benzocyclobutene itself.

[21]

Benzocyclobutene undergoes thermal ring opening and isomerisation to styrene at temperatures higher than those required to produce *o*-quinodimethane. The processes involved have been studied by nmr spectroscopy using ^{13}C (Chapman, O.L.; Tsou, U.-P. E. *J. Amer. Chem. Soc.*, **1984**, *106*, 7974; Chapman, O.L.; Tsou, U.-P. E.; Johnson, J.W. *J. Amer. Chem. Soc.*, **1987**, *109*, 553) and in the case of ^2H labelling, using both ^1H and ^2H nmr techniques (Trahanovsky, W.S.; Scribner, M.E. *J. Amer. Chem. Soc.*, **1984**, *106*, 7976). The major pathways were shown to involve a series of arylcarbene-cycloheptatetraene interconversions. Minor pathways that were detected in the ^{13}C nmr study involve the interconversion of o-quinodimethane and *o*-methylphenylcarbene and onwards to styrene labelled in the *o*-, *m*-, and *p*- positions. The distribution of the ^{13}C label showed that two initial pathways distributed the label in approximately equal amounts. The styrene was labelled β- (48%), *o*- (30%), α- (14%), *m*- (4%), and *p*- (4%). In a related study (Chapman, O.L.; McMahon, R.J.; West, P.R. *J. Amer. Chem. Soc.*, **1984**, *106*, 7973) *o*-, *m*-, and *p*-tolyldiazomethanes were photolysed in an argon matrix at 10-15 °K. The carbenes that were generated were shown to be triplets by esr spectroscopy. In a complex series of equilibria the carbenes were seen to be interconverted to benzocyclobutene and *o*-quinodimethane. Thus the interconversions proceed by a mechanistic pathway that operates over a 1200 degree range. Photolysis of the parent benzocyclobutene at 254 nm in pentane (Turro, N.J. *et al. Tetrahedron Lett.*, **1988**, *29*, 2543) affords two dihydropentalene derivatives.

4. *The Metallation of Arenes and Arylation Reactions using Arylmetallic Compounds*

In the first supplement to volume IIIA it was pointed out that a distinction may be drawn between those arylation reactions that involve organometallic reagents such as aryllithium species, which function as anions, and those which function as nucleophiles. Some of the latter take part in *ipso*-electrophilic addition-with-elimination reactions. Arylation reactions involving arenediazonium species are known to proceed by a variety of mechanistic pathways, for example the Gomberg-Bachman-Hey reactions with arenes involve radical processes while the thermal decomposition of arenediazonium fluoroborates may involve aryl-cations (Zollinger, H. *Accounts Chem. Res.*, **1973**, *6*, 335). A comparison of the arylation reactions of arylmetallic reagents with arenediazonium or diaryliodonium ions (Beringer, F.M. *et al. J. Am. Chem. Soc.*, **1953**, *75*, 2708) is appropriate. The kinetics and mechanism of aromatic thalliation has been studied (Lau, W.; Kochi, J.K. *J. Am. Chem. Soc.*, **1984**, *106*, 7100) and competing electrophilic and electron-transfer pathways identified. Some of the reagents, for example arylthallium species, may react with nucleophiles, formally as aryl-cation equivalents. In this chapter we will consider reactions where a hydrogen atom in the co-reactant is formally replaced by an aryl residue. The major developments that have been published since the last supplement have been concerned with arylation reactions using derivatives of lead (Pinhey, J.T. *Austral. J. Chem.*, **1991**, *44*, 1353), bismuth, (Abramovitch, R.A.; Barton, D.H.R.; Finet, J.-P. *Tetrahedron*, **1988**, *44*, 3039), and palladium (Heck, R.F. *Org. React.*, **1982**, *27*, 345; Trost, B.M.; Verhoeven, T.R. in *Comprehensive Organometallic Chemistry*, Eds. Wilkinson, G.; Stone, F.G.A.; Abel, E.W.; Vol. 8, p 799, Pergamon Press, Oxford, **1982**; Heck, R.F. in *Comprehensive Organic Chemistry*, Eds. Trost, B.M.; Fleming I.; Vol. 4, Vol. Ed. Semmelhack, M.F., p 833, Pergamon, Oxford, **1991**). Mechanistic tests have been used to probe the nature of the arylation processes.

4. (a)   *Aryllead tricarboxylates*

A checked procedure has been published for the direct plumbylation of anisole and the conversion of *p*-methoxyphenyllead triacetate into 1-(*p*-methoxyphenyl)-2-oxocyclohexanecarboxylate (Kozyrod, R.P.; Pinhey, J.T. *Org. Synth.*, **1984**, *62*, 24). Direct plumbylation with monosubstituted benzene derivatives normally occurs at the *para*-position. In reactions of the halobenzenes using lead(IV) acetate in trifluoroacetic acid the *p*-halophenyl-lead tris(trifluoroacetate) is formed, but with benzene derivatives that are less nucleophilic than the halobenzenes an indirect method, for example involving the replacement of mercury, thallium, or silicon, must be used. The replacement of tin by lead (Kozyrod, R.P.; Morgan, J.; Pinhey, J.T. *Austral. J. Chem.*, **1985**, *38*, 1147) is highly favoured. The known boron-mercury exchange reactions involving arylboronic acids (Larock, R.C. "*Organometallic Compounds in Organic Synthesis*", Springer-Verlag, Berlin, **1985**) led to the development of analogous reactions using lead(IV) acetate. The reactions were studied using nmr spectroscopy and it was shown that a catalytic amount of mercury(II) acetate leads to a supression of diaryllead diacetate formation (Morgan, J.; Pinhey, J.T. *J. Chem. Soc., Perkin Trans. 1*, **1990**, 715). High yields of aryllead triacetate can thus be obtained in two steps from the aryl halide *via* the organolithium reagent. The direct plumbylation of *iso*butylbenzene with lead(IV) acetate followed by reaction with the methyl derivative of Meldrum's acid proceeds in a quantitative yield (equation [22]) (Pinhey, J.T.; Rowe, B.A. *Tetrahedron Lett.* **1980**, *21*, 965). Hydrolysis and decarboxylation of the malonic acid derivative affords the antiimflammatory agent ibuprofen.

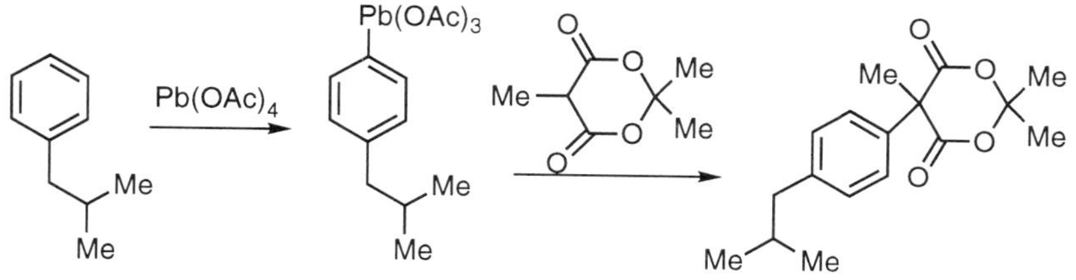

[22]

The majority of the early arylation reactions using aryllead tricarboxylates involved soft carbon nucleophiles, as exemplified by the reaction shown in equation [22]. In these reactions the aryllead tri-carboxylate functions as an aryl cation equivalent. Biaryls were obtained in reactions involving a number of aromatic compounds (Bell, H.C. et al. *Austral. J. Chem.*, **1979**, *32*, 1531). When the co-reactant is reactive towards electrophiles the yields are high. Reactions of *p*-fluorophenyllead triacetate with *p*-xylene, durene, and mesitylene in trifluoroacetic acid afford the expected biaryls in 64, 67, and 88% yields respectively. Although the reactions clearly suggest that an electrophilic substitution reactions is involved a *free* aryl cation is probably not involved in reactions of the more nucleophilic arenes. A suggested mechanism involves the formation of a complex between the π-donor aromatic component and the aryllead bis(trifluoroacetate) cation which then collapses to a Wheland intermediate after the loss of lead bis(trifluoroacetate) and reorganisation of an aryl σ-cation π-complex. Aryllead triacetates, for example the *p*-methoxyphenyl derivative, react with phenols to give predominantly the *C*-arylation products. The reaction involving mesitol is an exception to the above generalisation; the diaryl ether was obtained in that case. We may note here that some reactions involving arylbismuth reagents are more selective in their reactions with phenols. In contrast to the above reactions it has been suggested that *free* aryl cations may be involved in reactions where biaryls are formed in reactions that are catalysed by boron trifluoride diethyl etherate. The arylation of 3-allyloxycarbonylchroman-4-ones (Donnelly, D.M.X.; Finet, J.-P.; Rattigan, B.A. *J. Chem. Soc., Perkin Trans. 1*, **1993**, 1729) allows high yield syntheses of isoflavanones and isoflavones after selective catalytic de-allyloxycarbonylation. The potential pharmacological importance of α-arylglycine derivatives prompted the use of oxazolones as the nucleophilic component. In particular, 4-ethoxycarbonyl-2-phenyloxazol-5-one is fully enolised in deuteriochloroform and was shown to be efficiently arylated by using either pre-formed lead(IV) derivatives or those formed *in-situ* from the arylboronic acid (Koen, M.J.; Morgan, J.; Pinhey, J.T. *J. Chem. Soc., Perkin Trans. 1*, **1993**, 2383). The reactions involved in the synthesis of *N*-benzoyl-2,4-dimethoxyphenylglycine are shown in equation [**23**].

[23]

There are a number of interesting reactions of aryllead tricarboxylates with non-carbon nucleophiles. Thus high yields of aryl azides can be obtained in reactions involving sodium azide in DMSO at room temperature (Huber, M.-L.; Pinhey, J.T. *J. Chem. Soc., Perkin Trans. 1*, **1990**, 721). The *N*-arylation of amide-type nitrogen has been achieved in high yields in a number of cases. The sodium salts of the amides react, for example with p-tolyllead triacetate, in the presence of a catalytic quantity of copper(II) acetate (López-Alvarado, P.; Avendaño, C.; Menéndez, J.C. *Tetrahedron Lett.*, **1992**, *33*, 6875). A versatile route to arylamines, including diarylamines, under mild neutral conditions involves reactions of an aryllead triacetate with, for example, an arylamine catalysed by copper diacetate (Barton, D.H.R. *et al. J. Chem. Soc., Perkin Trans. 1*, **1991**, 2095). In the proposed mechanism it was suggested that the aryl group is transferred to copper by oxidative addition, forming a copper(III) species, followed by ligand coupling and reductive elimination of the catalytic copper(I) species. Arylamines that are readily oxidised, for example aniline, give low yields. The arylation of indole-nitrogen and phenol-oxygen cannot be achieved by this method. The use of arylbismuth reagents (see below) again provides an interesting comparison.

The ready arylation of iodide, azide ions, and the salts of nitroalkanes together with the failure of aryllead triacetates to react with other carbon nucleophiles in the absence of copper catalysis suggested the possible involvement of free radicals or an $S_{RN}1$ process in these reactions. However, reactions designed to check those mechanisms allowed such processes to be excluded, see equation [**24**] (Morgan, J.; Pinhey, J.T. *J. Chem. Soc., Perkin Trans. 1*, **1993**, 1673). The formation of dihydrobenzofuran derivatives would be expected from the intervention of aryl radicals.

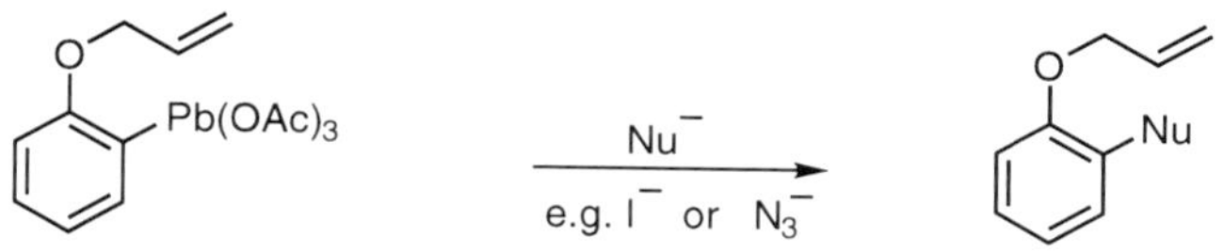

[Scheme 24: o-allyloxyphenyl-Pb(OAc)₃ + Nu⁻ (e.g. I⁻ or N₃⁻) → o-allyloxyphenyl-Nu] [24]

Reactions of aryllead triacetates with trimethylsilylenol ethers in the presence of boron trifluoride diethyl etherate lead to the formation of relatively unstable derivatives (Morgan, J.; et al. J. Chem. Soc., Perkin Trans. 1, **1993**, 1677) as exemplified in equation [25]. The ligand coupling products, the desoxybenzoin derivatives, were formed when the initial products were heated in chloroform. The involvement of a radical intermediate was also excluded in these latter reactions by the use of a lead derivative with an o-allyloxy- group which did not afford a dihydrobenzofuran derivative.

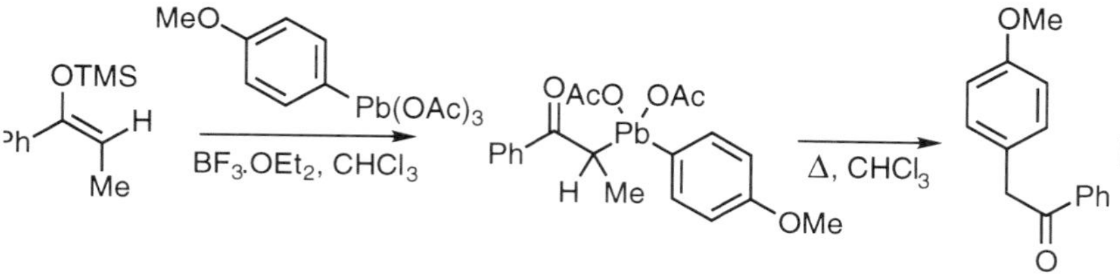

[Scheme 25: Ph(OTMS)C=CHMe + 4-MeO-C₆H₄-Pb(OAc)₃, BF₃·OEt₂, CHCl₃ → PhC(O)CH(Me)Pb(OAc)₂(C₆H₄-OMe) →(Δ, CHCl₃) 4-MeO-C₆H₄-CH(Me)C(O)Ph] [25]

The intervention of an aryl cations has also been suggested in reactions where aryllead triacetates react with boron trifluoride to afford aryl fluorides (De Meio, G.; Morgan, J.; Pinhey, J.T. *Tetrahedron*, **1993**, *49*, 8129). Reactions of triarylboroxines, electron rich aryltrimethylsilanes, and some arenes that are known to be converted directly into aryllead triacetates, when stirred with lead tetraacetate in boron trifluoride diethyl etherate, are converted directly into the corresponding aryl fluoride.

4. (b)   *Arylbismuth reagents*

Triarylbismuthines have found relatively little use in organic synthesis but they are readily available by the interaction of bismuth trichloride with aryl Grignard reagents and they can then be converted to the more useful bismuth (V) reagents by oxidation to triarylbismuth dichlorides using sulfuryl chloride and hence to a range of other compounds (Barton, D.H.R.; et al. *Tetrahedron*, **1981**, *37*, Supp. 1, 73; Barton, D.H.R.; et al. *Helv. Chim. Acta*, **1984**, *67*, 586). Heteroatom- and C-phenylation reactions will be considered in this section. In the case of reactions using ambident nucleophiles as co-reactants it is possible to effect some control over the position to be arylated.

In some cases effective reactions are achieved using copper catalysts. For example aromatic and aliphatic amines are N-phenylated using triphenylbismuth bis(carboxylates) in the presence of copper (0) powder. Aniline is converted into diphenylamine in high yield using the diacetate but the reaction is more rapid when using the bis(trifluoroacetate) (Barton, D.H.R.; Finet, J.-P.; Khamsi, J. *Tetrahedron Lett.*, **1986**, *27*, 3615). Amines also react with triphenylbismuth in the presence of copper (II) acetate (equation [26]); in this latter case in rather better yield than when using triphenylbismuth bis(trifluoroacetate) (Barton, D.H.R.; Finet, J.-P.; Khamsi, J. *Tetrahedron Lett.*, **1987**, *28*, 887).

[26]

The phenylation of indole with tetraphenylbismuth trifluoroacetate gave mainly 3-phenylindole together with a small amount of 1,3-diphenylindole (Barton, D.H.R.; et al. *J. Chem. Soc., Perkin Trans. 1*, **1985**, 2667); the expected result of a reaction involving a positive phenylating agent. A reaction of the indolyl anion gave as the major product 3,3-diphenyl-3H-indole. N-Phenylation of indole is known to be rather difficult to achieve but if the 3-position is blocked triphenylbismuth bis(trifluoroacetate) in the presence of low concentrations of either copper (II) acetate or copper (0) gives the N-substitution product (Barton, D.H.R.; Finet, J.-P.; Khamsi, J. *Tetrahedron Lett.*, **1988**, *29*, 1115). 3-Methylindole gave 3-methyl-1-phenylindole in a 21% yield but carbazole gave N-phenylcarbazole in an 84% yield.

Although simple alcohols are oxidised by a number of bismuth (V) reagents glycols are monophenylated (David, S.; Thieffry, A. *J. Org. Chem.*, **1983**, *48*, 441) using triphenylbismuth diacetate. The phenylation of one alcoholic function is only possible if there is another hydroxyl group that is suitably located in the same molecule. There is a strong preference for the phenylation of axial hydroxyl groups in six-membered rings. For example, benzyl 4,6-O-benzylidene-α-D-mannopyranoside gives a 42% yield of the axial product and 11% of the equatorial product and *trans*-4-t-butyl-*cis*-cyclohexane-1,2-diol gave the products shown in equation [27].

[27]

The *C*-phenylation of enols and enolate ions has been studied (Barton, D.H.R.; *et al. J. Chem. Soc., Perkin Trans. 1*, **1985**, 2667) using a number of bismuth reagents. Under basic conditions even hindered ketones are perphenylated using, for example triphenylbismuth carbonate. 1,3-Dicarbonyl compounds are *C*-phenylated using pentaphenylbismuth, triphenylbismuth carbonate under neutral conditions, and triphenylbismuth dichloride under basic conditions. The phenylation of cyclohexanone (equation [**28**]) has been reported (Barton, D.H.R.; *et al. J. Am. Chem. Soc.*, **1985**, *107*, 3607) using triphenylbismuth carbonate in the presence of an excess of potassium hydride in tetrahydrofuran.

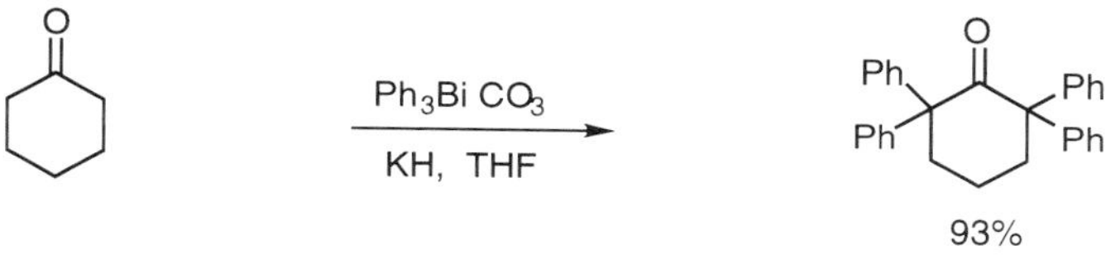

Pentavalent bismuth reagents react with phenols under basic conditions to afford *C*-arylated products by way of an intermediate that is formed by ligand exchange. When 2-naphthol was allowed to react with triphenylbismuth under basic conditions, using tetramethyl-2-t-butylguanidine or sodium hydride, 1-phenyl-2-naphthol (equation [**29**]) was obtained in good yield (Barton, D.H.R.; *et al. J. Chem. Soc., Perkin Trans. 1*, **1985**, 2657). In some cases the ligand exchange intermediate was detected by ^{1}H nmr spectroscopy. Under neutral or acidic conditions *O*-phenylation (equation [**30**]) was observed. It was suggested that *O*-phenylation occurs if there is sufficient positive charge on the aryl carbon for the oxygen to displace the bismuth in a concerted process. Benzene was observed as the other product in an nmr experiment (Barton, D.H.R.; Charpiot, B.; Motherwell, W.B. *Tetrahedron Lett.*, **1982**, 23, 3365) during the phenylation of 2-naphthol using tetraphenylbismuth trifluoroacetate.

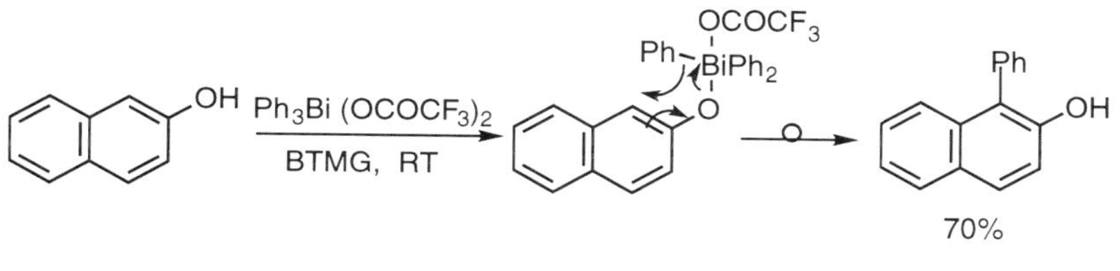

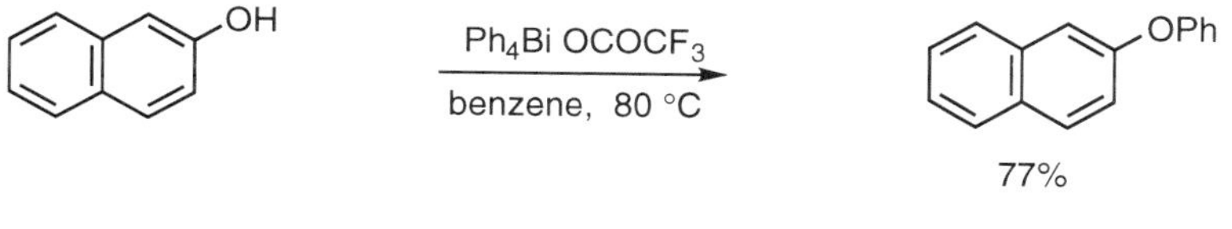

4. (c)   *Arylpalladium reagents*

There are three methods that are most frequently used for generating arylpalladium reagents. These are transmetallation reactions, for example, using arylmercury (II) reagents, the direct palladation of an arene which involve the use of a palladium (II) catalyst, and reactions involving aryl halides and pseudohalides requiring a palladium (0) catalyst.

The direct palladation of benzenoid derivatives (Moritani, I.; Fujiwara, Y. *Synthesis*, **1973**, 524) can be achieved using palladium (II) acetate in acetic acid. Electron donating substituents direct attack to the *ortho-* and *para-* positions as expected of an electrophilic addition-with-elimination reaction. However, although substituents on the benzene ring influence the direction of substitution, the effect is smaller than with more typical electrophiles. Competition studies, although showing that the more electron rich compounds are more reactive, also show smaller differences than may have been expected (Fujiwara, Y.; *et al. J. Org. Chem.*, **1976**, *41*, 1681). The arylpalladium species then take part in Heck reactions as exemplified in equation [**31**]. Benzene reacts with styrene to afford *trans*-stilbene in a 95% yield and 1,1-diphenylethene gave a 72% yield of triphenylethene. That the reactions are strongly influenced by steric effects is indicated by the fact that *trans*-stibene only gave a 28% yield of triphenylethene under analogous reaction conditions.

$$\text{PhOMe} \xrightarrow[\text{AcOH}]{\text{Pd(OAc)}_2} \text{ArPd(OAc)} \xrightarrow{\text{Ph-CH=CH}_2} \text{Ar-CH=CH-Ph} \quad [31]$$

o- = 30%; *m-* = 5%; *p-* = 48%

The regiochemistry observed on direct palladation can be strongly influenced by substituents that are present on the benzene ring. In a series of *N,N*-diethylbenzylamine derivatives it was shown that palladation normally occurs exclusively at the less hindered position (equation [**32**]) whereas when more strongly co-ordinating sulfur atoms were present reaction occurred only at the more hindered position. The sulfur containing complexes were found to be particularly stable and unreactive (Holton, R.A.; Davis, R.G. *J. Am. Chem. Soc.*, **1977**, 99, 4175).

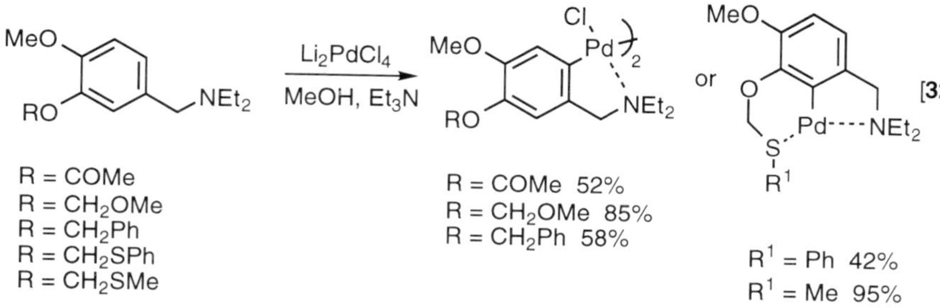
MeO
RO
NEt2
Li2PdCl4
MeOH, Et3N
Cl
Pd
2
or
S
R1
Pd
R = COMe
R = CH2OMe
R = CH2Ph
R = CH2SPh
R = CH2SMe
R = COMe 52%
R = CH2OMe 85%
R = CH2Ph 58%
R1 = Ph 42%
R1 = Me 95%

[**32**]

When direct palladation proceeds inefficiently the yields obtained in Heck reactions are inevitably lower as compared to the results obtained when transmetallation is involved.

The treatment of dicyclopentadiene with *p*-methoxyphenylmercury (II) chloride and a stoichiometric amount of lithium tetrachloropalladate (equation [**33**) results in arylation at the more reactive norbornene double bond (Kasahara, A.; *et al. Bull. Chem. Soc. Jpn.*, **1974**, *47*, 1967). Coordination of the residual double bond to the palladium stabilises the intermediate which can be reduced as shown. Where the intermediate is not stabilised the overall result of the reaction is insertion by *cis*-addition (Heck, R.F. *J. Am. Chem. Soc.*, **1969**, *91*, 6707) followed by *cis*-elimination of a hydridopalladium species. Thus the reaction of methyl acrylate and diphenylmercury with lithium trichloropalladium (II) afforded methyl cinnamate in an 88% yield (Heck, R.F. *J. Am. Chem. Soc.*, **1968**, *90*, 5518). The reactions can become catalytic in the palladium species when an oxidant such as copper (II) chloride is included in the regent system in order to convert the palladium (0) species back to the palladium (II) salt.

HgCl
OMe
+
H
H
Li2PdCl4
Pd
Cl
NaBH4
42%
65%

[**33**]

The ease with which a number of arenes are thalliated provides a good route to arylpalladium intermediates by transmetallation. This is exemplified by the high yield preparation of ethyl cinnamate shown in equation [**34**] (Uemura, S. *et al. Bull. Chem. Soc. Jpn.*, **1980**, *53*, 553).

$$\text{C}_6\text{H}_6 \xrightarrow[\text{CCl}_3\text{CO}_2\text{H}]{\text{Tl}_2\text{O}_3} \text{PhTl(OCOCCl}_3)_2 \xrightarrow[\text{AcO}^-\ \text{Na}^+,\ \text{AcOH},\ 117\,°\text{C},\ 1\text{h}]{\text{CH}_2=\text{CHCO}_2\text{Et},\ \text{PdCl}_2} \text{PhCH=CHCO}_2\text{Et} \quad [34]$$

66%   91%

Despite the above considerations the use of aryl halides in the Heck arylation procedure provides the most practical approach to the generation of the required arylpalladium intermediates. Although a palladium (0) catalyst is required it is most frequently generated *in situ* from a palladium (II) salt, normally the reactive acetate, and two equivalents of a phosphine as shown in equation [35]. The nucleophilic nature of the attack on the benzene ring is indicated by the rate acceleration that was observed on substitution by an electron withdrawing group. The converse was observed when electron releasing groups were present (Dieck, H.A.; Heck, R.F. *J. Am. Chem. Soc.*, **1974**, *96*, 1133).

$$\text{4-MeO}_2\text{C-C}_6\text{H}_4\text{-Br} + \text{CH}_2=\text{CHCO}_2\text{Me} \xrightarrow[\text{Ph}_3\text{P, 2 mol\%},\ n\text{-Bu}_3\text{N, 110 °C}]{\text{Pd(OAc)}_2,\ 1\ \text{mol\%}} \text{4-MeO}_2\text{C-C}_6\text{H}_4\text{-CH=CH-CO}_2\text{Me} \quad [35]$$

81%

A number of *B*-alkylboranes are known to couple with aryl halides in the presence of dichloro{1,1'-[bis(diphenylphosphine)]ferrocene} palladium-(II). The B-alkyl -9-BBN derivatives have been the most thoroughly investigated. The reaction of 2-bromoallylbenzene was shown to afford indane by this method in an 86% yield (Miyaura, M. *et al. J. Am. Chem. Soc.*, **1989**, *111*, 314). The conversion of arylboronic acids into biaryls by the interaction with aryl bromides in the presence of tetrakis(triphenylphosphine)palladium proceeds efficiently (Sharp, M.J.; Cheng, W.; Snieckus, V. *Tetrahedron Lett.*, **1987**, *28*, 5093; 3097) and consistently high yields of ketones are obtained in the coupling reactions of aryl triflates with organostannanes in the presence of carbon monoxide using the dichloro{1,1'-[bis(diphenylphosphine)] palladium(II) catalyst (Echavarren, A.M.; Stille, J.K. *J. Am. Chem. Soc.*, **1988**, *110*, 1557). Aryl iodides are more reactive than the corresponding bromides as indicated in the internal competition reaction shown in equation [36] (Plevyak, J.E.; Dickerson, J.E.; Heck, R.F. *J. Org. Chem.*, **1979**, *44*, 4078). The reactions are remarkably chemoselective and almost all common functional groups do not interfere and sulfur does not poison the catalyst. The reaction shown in equation [37] was carried out under solid-liquid phase transfer conditions (Citation, P.G.; Ortar, G. *Synthesis*, **1986**, 70).

$$\text{Br-C}_6\text{H}_4\text{-I} + \text{CH}_2=\text{CH-CO}_2\text{Me} \xrightarrow[\text{Et}_3\text{N, 100 °C, 5 h}]{\text{Pd(OAc)}_2, \text{ 1 mol \%}} \text{Br-C}_6\text{H}_4\text{-CH=CH-CO}_2\text{Me}$$

68%   [36]

$$\text{THP-O-CH}_2\text{-CH=CH-CO}_2\text{Me} + \text{Ph-I} \xrightarrow[\substack{\text{NaHCO}_3\text{, HMPA} \\ 60 \text{ °C}}]{\text{Pd(OAc)}_2\text{, n-Bu}_4\text{N}^+ \text{ Cl}^-} \text{THP-O-CH}_2\text{-C(CH}_2\text{Ph)=CH-CO}_2\text{Me} \xrightarrow[\text{MeOH, 45 °C, 2 h}]{\text{Dowex 50W x 8, H}^+} \text{Ph-substituted butenolide}$$

71 %   [37]

5. *Electrophilic Addition-with-Elimination Reactions*

The tendency to undergo electrophilic addition-with-elimination reactions is a widely used criterion for aromaticity. The problem of the competition between the elimination of (normally) a proton and the addition of a nucleophile has been addressed by considering the ability of the methylene group, that is formed on protonation of an aromatic hydrocarbon, to either dissociate a proton, or function as a source of hyperconjugative electron release (Aihara, J.-i. *Bull Chem. Soc. Jpn.*, **1981**, *54*, 2268). Hyperconjugation leads to an aromatic like resonance energy while generating a large positive charge density at the methylene group. This effect promotes the elimination of a proton to regenerate the aromatic system. The earlier suggestion (Stock, L.M. "*Aromatic Substitution Reactions*", Prentice-Hall, Englewood Cliffs, N.J., **1968**, Chapter 2) that overall substitution might occur *via* an aromatic transition state is thus justified and generalised in terms of hyperconjugative resonance energy.

5. (a)   *Protonation Reactions*

The protonation of simple aromatic compounds in superacids (Olah, G.A.; Prakash, G.K.S.; Sommer, J. "*Superacids*", Wiley, **1985**, New York) has been reviewed (Farcasiu, D. *Accounts Chem. Res.*, **1982**, *15*, 46). The protonation of benzene can therefore be used in acidity measurements for various protic acid-Lewis acid mixtures. For example the degree of protonation of benzene in $HF\text{-}SbF_5$, $HBr\text{-}AlBr_3$, and $HF\text{-}TaF_5$ has been investigated by ^{13}C nmr spectroscopy (Farcasiu, D. *et al. J. Org. Chem.*, **1982**, *47*, 453). Both of the former two acids are stronger than $HF\text{-}TaF_5$. Benzene and alkylbenzenes dissolved in trifluoromethanesulfonic acid (triflic acid) have been shown to be completely protonated when present at low concentrations (Bakoss, H.J. et al. *Tetrahedron*, **1982**, *38*, 623). The position of protonation of alkylbenzenes (mainly *para*) and the position of the dealkylation-realkylation equilibrium has also been discussed. Protonation equilibria of mesitylene in concentrated solutions of triflic acid have been studied by ultraviolet spectroscopy and ^{13}C nmr chemical shifts (Marziano, N.C.; Tortato, C.; Bertani, R. *J. Chem. Soc., Perkin Trans. 2*, **1992**, 955) and the uv data used to evaluate the basicity constant. A thermochemical cycle has been used to evaluate the pKa's of a number of aromatic hydrocarbons (Nagaoka, T. et al. *J. Org. Chem.*, **1990**, *55*, 3707). A value of – 24 was suggested in the case of the benzenium ion in acetonitrile. The ability to compute nmr chemical shifts (Gauss, J. *Chem. Phys. Letters*, **1992**, *191*, 614) has allowed the evaluation of MP2/6-31G *ab initio* geometry calculations (Sieber, S.; Schleyer, P.v.R.; Gauss, J. *J. Am. Chem. Soc.*, **1993**, *115*, 6987) of phenonium and benzenium ions. The symmetrically bridged structure was confirmed to be the transition state for hydrogen migration as indicated in equation [**38**].

[38]

5. (b)   *Group (IV) Electrophiles*

Friedel-Crafts and related reactions still remain as some of the most important synthetic carbon-carbon bond forming procedures involving benzene and its homologues. The more important recent developments have been concerned with the development of (i) new mild reagents, (ii) new catalysts, particularly those that can be recovered and reused and are therefore environmentally friendly, and (iii) control of stereochemistry where that is potentially important. A balance between the electrophilicity of the reagent system and the nucleophilicity of the benzenoid derivative is required in order to achieve successful reactions. Not all benzene derivatives will undergo reactions with the relatively weak electrophiles involved, for example in Vilsmeier formylation and Mannich reactions, where the electrophile replaces hydrogen. It is worth mentioning at this point that the use of *s*-trioxan, tin (IV) chloride, and chlorotrimethylsilane has been recommended as a replacement for chloromethyl methyl ether in chloromethylation reactions (Itsuno, S.; Uchikoshi, K.; Ito, K. *J. Am. Chem. Soc.*, **1990**, *112*, 8187).

5. (b) (i) *Alkylation reactions*

Although the well known industrially important preparations of styrene and cumene (*iso*propylbenzene) use the alkenes ethene and propene, the majority of Friedel-Crafts alkylation reactions (Roberts, R.M.; Khalaf, A.A *"Friedel-Crafts Alkylation Chemistry"*, Dekker, New York, **1984**; Olah, G.A.; Krishnamurti, R.; Prakash, G.K.S. in *"Comprehensive Organic Synthesis"*, Eds. Trost, B.M.; Fleming I.; Vol. 3, Vol. Ed. Pattenden, G., p 293, Pergamon, Oxford, **1991**) use an alkyl halide, usually the chloride or bromide, as the alkylating agent. The catalyst system is normally a Lewis acid, most frequently aluminium chloride or aluminium bromide, together with a co-catalyst. It is frequently assumed that anhydrous conditions are required to carry out Friedel-Crafts reactions. However, it is very difficult to obtain a Lewis acid in a completely anhydrous state and traces of moisture are found to accelerate the reactions that are catalysed by some Lewis acids such as aluminium chloride. In fact water functions as a co-catalyst. The effectiveness of a catalyst depends on its purity and it is not unusual to find that that what is apparently the same catalyst can lead to different yields of products when the catalyst is purchased from different suppliers. The reaction mixtures are normally added to water which presents a disposal problem in addition to the fact that the catalyst cannot be reused. New reagent systems continue to be developed, for example dicyclohexylcarbodiimide (Kim, J.N.; Chung, K.H.; Ryu, E.K. *Tetrahedron Lett.*, **1994**, *35*, 903) and N,N´-dicyclohexylurea (Chung, K.H.; Kim, J.N.; Ryu, E.K. *Tetrahedron Lett.*, **1994**, *35*, 2913) have been used as cyclohexyl cation equivalents. The catalytic activity of zinc chloride has been reported to be enhanced for Friedel-Crafts benzylation reactions by the addition of polar solvents (Hayashi, E. *et al. Bull. Chem. Soc. Jpn.*, **1993**, *66*, 3520). A considerable amount of work has centred on the development of new catalysts. Lanthanide trihalides have been shown to act a catalysts in the alkylation of aromatic compounds (Mine, N.; Fujiwara, Y.; Taniguchi, H. *Chem. Lett.*, **1986**, 357) and they can be isolated and reused without an apparent loss of catalytic activity. The reaction of benzyl chloride with benzene in the presence of dysprosium (III) chloride gave diphenylmethane in a 74% yield which was unchanged when using recovered catalyst. The late lanthanides were shown to have particularly high activity, for example thulium (III) chloride gave a high yield of diphenylmethane after a very short reaction time. Polystyrene cross-linked with 5-7% of divinylbenzene has been shown to form a stable complex with titanium tetrachloride (Ran, R.; Jiang, S.; Shen, J. *Yingyong Huaxue*, **1985**, *2*, 29; *Chem. Abs.*, **1985**, *103*, 27,939) which can be used in Friedel-Crafts alkylation reactions and reused four or five times. Similarly polystyrene-divinylbenzene co-polymer beads can be coated with gallium (III) chloride (Ran, R.C.; Jiang, S.J.; Shen, J. *J. Macromol. Sci., Chem.*, **1987**, *A24*, 669) which then function as an efficient catalyst that is easily separated from reaction mixtures for reuse. Sulfonated polystyrene-divinylbenzene co-polymers are also efficient solid acids. Although the conversion of benzene into cumene has been carried out for many years using supported catalysts such as phosphoric acid the protonation of ethene occurs far less easily than the protonation of propene.The availability of perfluorinated resinsulfonic acids [Nafion-H®] (Olah, G.A.; Iyer, P.S.; Prakash, G.K.S. *Synthesis*, **1986**, 513) has allowed that problem to be addressed. It has been shown (Olah, G.A.; Kasp, J.; Bukala, J. *J. Org. Chem.*, **1977**, *42*, 4187) that benzene undergoes alkylation with both ethene and propene in the gas phase at atmospheric pressure with short contact times and temperatures as low as 110 °C. Natural and synthetic zeolites and clays have been used as catalysts. Various zeolites exist with characteristic pore and channel sizes. The ZSM-5 zeolite, for example, has interconnecting channels that allow certain benzene derivatives to fit quite closely and to diffuse in and out of the pores. Molecules with dimensions larger than those of 1,3,5-trimethylbenzene are excluded. The high *para*-selectivity in the alkylation of toluene by methanol that is catalysed, for example by the zeolite ZSM-5, results from such an effect (Chen, N.Y.; Kaeding, W.W.; Dwyer, F.G. *J. Am. Chem. Soc.*, **1979**, *101*, 6783), reaction occurring primarily within the internal pore structure. The efficient benzylation of benzene and a number of substituted derivatives has been reported (Barlow, S.J. *et al. J. Chem. Research (S)*, **1991**, 74) using catalyst systems of zinc-, copper (II)-, and magnesium- chlorides supported on montmorillonite K10. The catalyst is removed at the end of the reaction by simple filtration and is clearly environmentally friendly. The use of a clay-impregnated zinc chloride catalyst (Clayzic) showed, rather surprisingly, that toluene is more reactive than mesitylene towards benzylation using benzyl chloride. However, in a competition reaction (Cornélis, A. *et al. Tetrahedron Lett.*, **1991**, *32*, 2901) a switch in reactivity and a substantial increase in the rate was reported. Mesitylene was then fifteen times as reactive as toluene. Similarly (Cornélis, A. *et al. Tetrahedron Lett.*, **1991**, *32*, 2903) a switch of reactivity was also observed when a mixture of benzyl alcohol and benzyl chloride were allowed to compete for toluene in the presence of clayzic. At 80 °C all of the alcohol was consumed before reaction involving the chloride was initiated. The benzylation of benzene and substituted benzene derivatives is carried out efficiently (Yamato, T. *et al. J. Org. Chem.*, **1991**, *56*, 2089) using benzyl alcohols and Nafion-H® and the catalyst may be filtered off at the end of the reaction. The reactions proceed in good yields to give clean products and the water that is produced in the reactions does not deactivate the catalyst. Intramolecular cyclisation reactions were also reported together with trimerisation and tetramerisation reactions of methoxybenzyl alcohols, as exemplified in equation [**39**]. Nuclear magnetic resonance evidence has been reported (Wu, T.-T.; Speas, J.R. *J. Org. Chem.*, **1987**, *52*, 2330) in favour of a partial cone structure.

The nature of the intermediates involved in Friedel-Crafts alkylation reactions depends on the structures of the alkylating agents involved. Methyl halides do not give rise to carbenium ions in solution and so the intermediates in those cases are either tight donor-acceptor complexes or cations derived by alkylation of the solvent. It has been shown (Olah, G.A. *et al. J. Am. Chem. Soc.*, **1972**, *94*, 156) that the solvents sulfur dioxide and sulfuryl chloride fluoride are *O*-methylated by methyl fluoride-antimony pentafluoride and that the solutions are very powerful methylating systems. We may note here that the methylation of toluene by dimethylhalonium ions and the trimethyloxonium ion in the presence of magic acid give isomer distributions that are typical of reactions involving very reactive electrophiles (Olah, G.A. *et al. J. Am. Chem. Soc.*, **1974**, *96*, 884). At the other extreme, *t*-butyl halides and other tertiary halides interact with a range of Lewis acids to form carbenium ions. The rearrangement of primary alkyl halides to secondary carbenium ions and secondary alkyl halides to the isomeric tertiary carbenium ions leads to the well known formation of rearranged products. It is interesting to note here that the *t*-butyl group is valuable as a positional protective group (Tashiro, M. *Synthesis*, **1979**, 921). Not only can the group be introduced easily but also it can be removed with facility by protio-dealkylation or by way of a transalkylation reaction (Lewis, N.; Morgan, I. *Synthetic Commun.*, **1988**, *18*, 1783).

An interesting variant on the conventional alkylation reaction has been reported (Nutaitis, C.F.; Gribble, G.W. *Synthesis*, **1985**, 756) where arenes such as *m*-xylene react with sodium borohydride in trifluoroacetic acid to afford 1,1,1-trifluoro-2,2-diarylethanes. It was proposed that an initial condensation gave an intermediate carbinol (equation [**40**]) which also acts as an alkylating agent in a reaction with a second molecules of the arene.

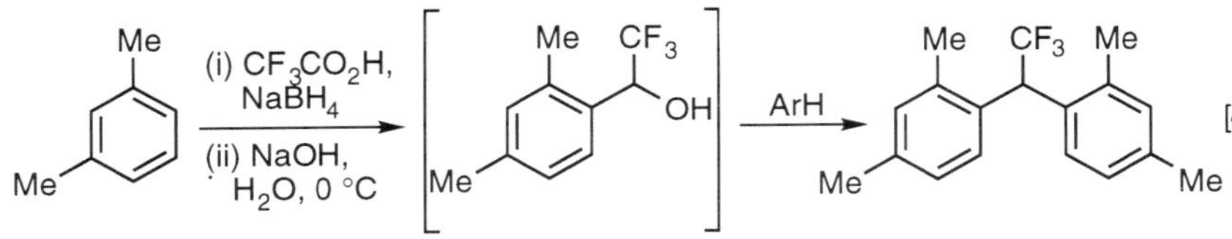

[Scheme 40: m-xylene + (i) CF$_3$CO$_2$H, NaBH$_4$; (ii) NaOH, H$_2$O, 0 °C → intermediate alcohol with CF$_3$ and OH, then ArH → diaryl CF$_3$ product]

Bridgehead halides interact with aluminium chloride or boron trifluoride to form strongly polarised complexes that are in equilibrium with highly energetic carbenium ions. The strained bridgehead cations cannot completely flatten out and so the empty orbital must have a high sp^3 component. The alkylation of benzene (48% yield) and toluene (60% yield) were reported together with m-xylene (equation [41]) which was shown (Olah, G.A. et al. J. Am. Chem. Soc., **1993**, *115*, 10728) to afford preferentially the thermodynamic (*meta*-substitution) product.

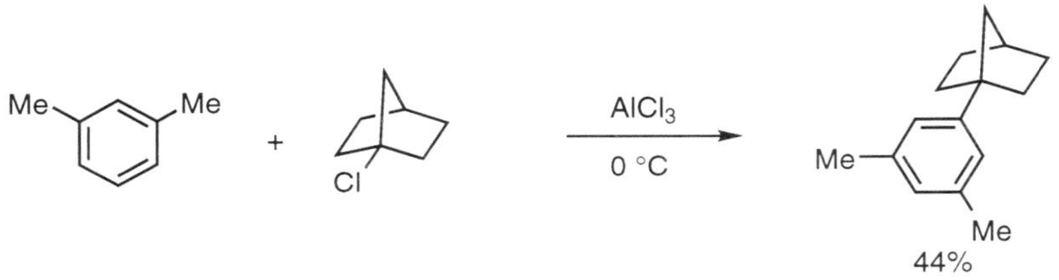

[Scheme 41: m-xylene + norbornyl chloride, AlCl$_3$, 0 °C → norbornyl-substituted m-xylene, 44%]

Among the various cycloalkylation reactions that have been reported the double Friedel-Crafts alkylation reactions involving allylic sulfones is interesting and can be carried out in a single operation in low yield (Trost, B.M.; Ghadiri, M.R. J. Am. Chem. Soc., **1984**, *106*, 7260). In the first step the sulfone group is displaced and the second step proceeds after protonation of the residual double bond as shown in equation [42]. By carrying out the sequence in two steps, first by attenuating the Lewis acidity of aluminium chloride by means of diethyl ether, and secondly by using a specific protic acid, good yields may be obtained.

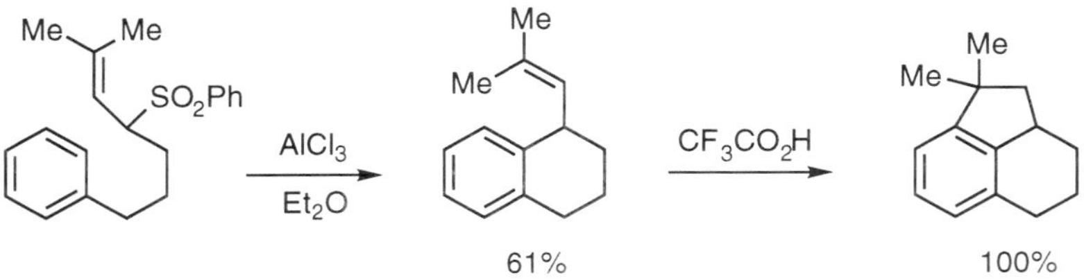

[Scheme 42: allylic sulfone substrate, AlCl$_3$/Et$_2$O → tetralin derivative (61%), CF$_3$CO$_2$H → fused tricyclic product (100%)]

The stereochemical outcome of Friedel-Crafts reactions have been reported including those that proceed *via* cations as well as those that involve S_N2 like transition states.

The first observation of a highly stereoselective Friedel-Crafts reaction involved the alkylation of benzene by γ-valerolactone where it was concluded (Brauman, J.I.; Pandell, A.J. *J. Am. Chem. Soc.*, **1967**, *89*, 5421) that the high degree of inversion of configuration resulted from a tight ion-pair involving the oxygen-stabilised cyclic cation. Racemisation also occurs in that the alkylation competes with return from the cation to starting material. The overall effect is that the reaction proceeds with a 40% net inversion of configuration. In the case of reactions that involve the ring opening of cyclic ethers the smaller the ring size the easier is the alkylation of arenes. In reactions with, for example propene oxide (Nakajima, T. *et al. Bull. Chem. Soc. Jpn.*, **1967**, *40*, 2980), 1,2-epoxybutane (Nakajima, T.; Nakamoto, Y.; Suga, S. *Bull. Chem. Soc. Jpn.*, **1975**, *48*, 960), *cis*-2,3-epoxybutane (Inoue, M. *et al. Bull. Chem. Soc. Jpn.*, **1980**, *53*, 458), and 2-methyltetrahydrofuran (Brauman, J.I.; Solladié-Cavallo, A. *J. Chem. Soc., Chem. Commun.*, **1968**, 1124), it was also shown that the alkylation reactions occur with predominant inversion of configuration. In the case of the reactions of *cis*-2,3-epoxybutane it was shown that although the reactions proceed with inversion of configuration when using aluminium chloride the stereoselectivity was significantly reduced when using aluminium bromide. The alkylation of benzene by (+)-2-methyloxetane (Segi, M. *et al. Bull. Chem. Soc. Jpn.*, **1982**, *55*, 167) affords a 3-phenyl-1-butanol as a result of ring opening. The stereochemistry of the ring opening process was studied and the reaction was found to proceed with inversion of configuration via an S_N2 like transition state using aluminium chloride, titanium tetrachloride, and tin (IV) chloride. The optical yields varied from 20% using aluminium chloride to 55 and 62% using titanium tetrachloride and tin (IV) chloride respectively. Neither titanium tetrachloride or tin (IV) chloride were found to act as catalysts in attempted reactions of 3-chloro-1-butanol with benzene and so the higher optical yields may be interpreted as involving a lower degree of racemisation when using those Lewis acids. This is in contrast to the almost complete racemisation observed (Burwell, R.L. Jr. *et al. J. Am. Chem. Soc.*, **1952**, *74*, 4570) in Friedel-Crafts reactions involving acyclic ethers. On the other hand it has been shown (Piccolo, O. *et al. J. Org. Chem.*, **1985**, *50*, 3945) that the alkylation of benzene with optically pure (S)-methyl 2-(chlorosulfonyloxy)propionate or the related mesylate using aluminium chloride as catalyst gave (S)-methyl 2-phenylpropionate in excellent optical yield (>97%), involving inversion of configuration, and good chemical yields (50-80%). Similar results were obtained (Piccolo, O. *et al. J. Org. Chem.*, **1991**, *56*, 183) when using (R)-alkyl 3-(sulfonyloxy)butanoates and also when using the N-benzoyl derivative of 3-(S)-methanesulfonyloxyproline (Kronenthal, D.R. *et al. Tetrahedron Lett.*, **1990**, *31*, 1241).

A rather smaller number of studies of the stereochemical results of arene alkylation reactions involving cationic intermediates have been reported. High diastereoselectivity was observed (Snider, B.B.; Jackson, A.C. *J. Org. Chem.*, **1983**, *48*, 1471) in the Prins reaction of paraformaldehyde with a 1,4-diaryl-1-butene catalysed by an aluminium halide catalyst that was prepared by mixing methylaluminium dichloride with an equimolar amount of dimethylaluminium chloride. Some diastereoselectivity (4.4 : 1) was also observed in the cycloalkylation reaction shown in equation [**43**]. The major diastereomer was shown (Angle, S.R.; Louie, M.S. *J. Org. Chem.*, **1991**, *56*, 2853) to have both substituents in pseudo-equatorial positions and this was presumably the controlling factor involved in the transition state leading to the major product. Intramolecular cyclisation of benzylic cations favours six-membered ring formation as compared with reactions that lead to the formation of five- or six-membered rings.

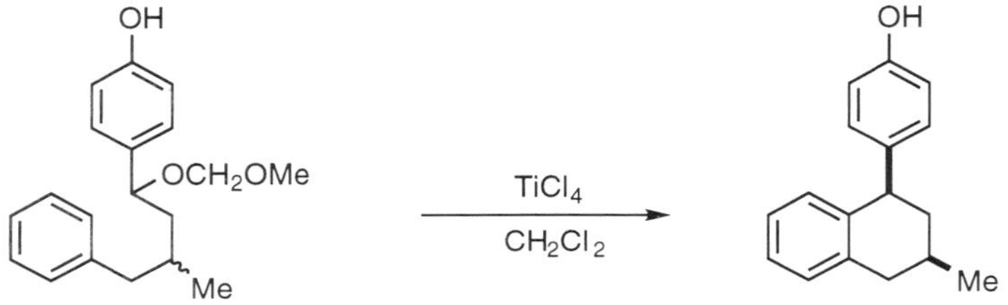

[43]

Considerably better control was observed (Effenberger, F.; Weber, T. *Chem. Ber.*, **1988**, *121*, 421) in reactions of derivatives of threonine methyl ester as shown in equation [**44**]. The alkylation of, for example, benzene and mesitylene proceed in rather low chemical yields but with almost complete retention of configuration at C-3. The reactions evidently proceed *via* a free cation and the stereochemical control results from the shielding of one side of the cation that is related to restriction of rotation of the phthaloyl residue.

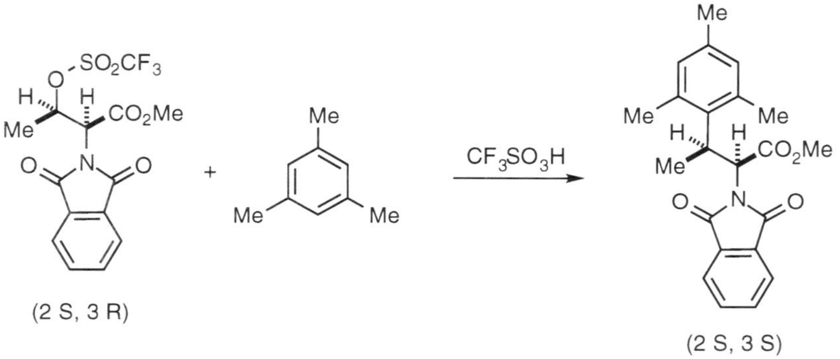

[44]

(2 S, 3 R)

(2 S, 3 S)

The shielding of one face of a planar cation has been used in the highly diastereoselective syntheses of 2-hydroxymandelic esters (Bigi, F. *et al. Tetrahedron Asymmetry*, **1990**, *1*, 861) and 4-hydroxymandelic esters (equation [45]) (Bigi, F. *et al. Tetrahedron Asymmetry*, **1993**, *4*, 2411) in which excellent diastereoselectivity was observed when using 8-phenylmenthyl glyoxylate.

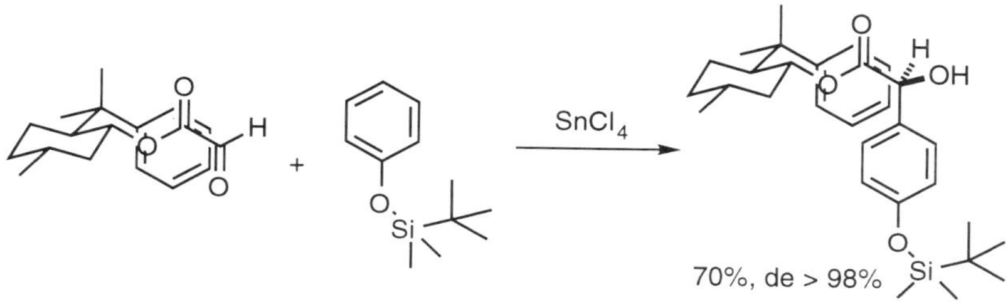

[45]

70%, de > 98%

The shielding of one face of a cation has also been invoked (Matsuda, S.; Nakajima, T.; Suga, S. *Bull. Chem. Soc. Jpn.*, **1983**, *56*, 1089) to explain the stereochemical control that was observed in the alkylation reactions of (S)-(−)-2-chloro-1-phenylpropane and (S)-(+)-1-chloro-2-phenylpropane. Both result in the formation of (R)-(−)-1,2-diphenylpropane (equation [46]): the former compound gave the product with retention of configuration whereas the latter compound reacted with inversion of configuration. The optical yields were very high when using aluminium chloride as catalyst but dropped rapidly when the temperature was raised, particularly when using iron (III) chloride as the catalyst.

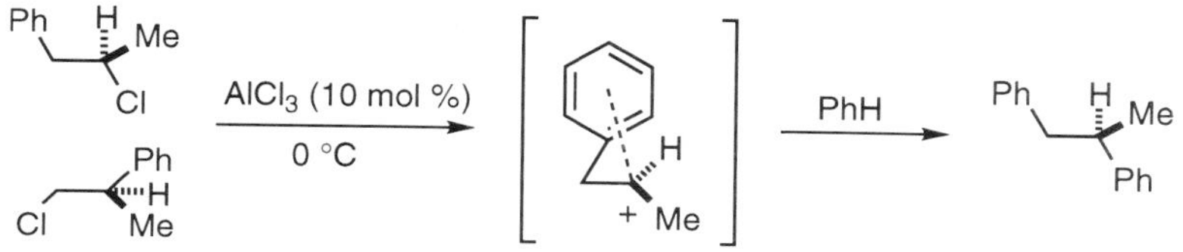

[46]

The synthesis of optically pure, non racemic, non proteinogenic α-amino acids has been investigated in considerable detail. Among these, arylglycines are of pharmacological interest because of the possibility of incorporation into a range of new drugs. Among the many new α-amino acid syntheses that have been developed, glycinyl cation equivalents have been investigated as Friedel-Crafts alkylation reagents. The well known use of anions derived from bislactim ethers has been extended to the generation of cations from chloro-derivatives. An example (Schöllkopf, U. *et al. Angew. Chem.,*

Int. Ed. Engl., **1987**, *26*, 683) is shown in equation [**47**]. After hydrolysis the α-amino acid was isolated with an ee >95% . The well known propensity with which arylglycines racemise and the very high enantiomeric excess obtained illustrates that the method has very great potential.

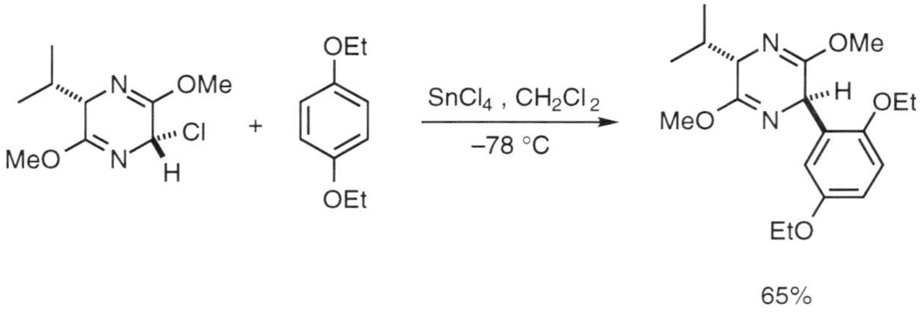

[47]

65%

5. (b) (ii) *Mannich Reactions*

Aminomethylation reactions of aromatic compounds (Heaney, H. in "*Comprehensive Organic Synthesis*", Eds. Trost, B.M.; Fleming I.; Vol. 2, Vol. Ed. Heathcock, C.H., p 953, Pergamon, Oxford, **1991**) only occurs with highly nucleophilic substrates. The reactions are important for a number of reasons. The amine function in the form of a salt confers water solubility and in addition the fact that the quaternary ammonium salts undergo nucleophilic substitution reactions easily means that the use of Mannich reactions is popular with groups who are interested in studying the biological activity of products. Reactions with phenols have been widely investigated but are not within the pervue of this chapter. Because the Mannich reaction involves relatively weak electrophiles there are few benzenoid compounds that are sufficiently nucleophilic to react. Anisole is not nucleophilic enough but *m*-dimethoxybenzene is. Even with the latter compound the classical aqueous reaction conditions cannot be used. Reactions of 1,3-dimethoxybenzene and 1,3,5-trimethoxybenzene were the first examples (Böhme, H.; Eichler, D. *Arch. Pharm. (Weinheim, Ger.)* , **1967**, *300*, 679) where preformed methyleneiminium salts in aprotic solvents were used successfully. In the example shown in equation [**48**] the product was characterised as its hydrobromide salt.

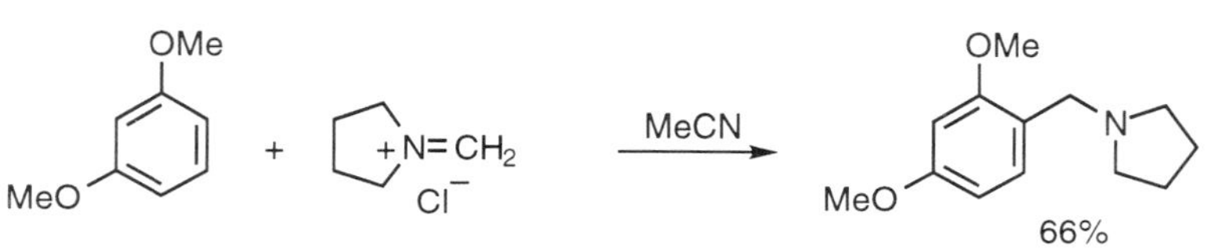

[48]

66%

The problem can also be circumvented by using the increased electron density at the *ipso*-carbon of certain arylmetallic compounds. Aryltrialkylstannanes have been shown to be particularly useful . *N*,*N*-Dimethylmethyleneiminium chloride reacts with phenyltrimethylstannane to give *N*,*N*-dimethylbenzylamine (Cooper, M.S. et al. *Tetrahedron*, **1989**, *45*, 1155) in a 65% yield and with *p*-tolyltributylstannane (equation [**49**]) to afford *N*,*N*-dimethyl-4-methylbenzylamine in a 67% yield.

$$\underset{\text{Me}}{\text{C}_6\text{H}_4\text{-SnBu}_3} + \underset{\text{Me}}{\overset{\text{Me}}{\text{N}}}{=}\text{CH}_2 \ \text{Cl}^- \xrightarrow[\Delta, 24\text{ h}]{\text{MeCN}} \underset{\text{Me}}{\text{C}_6\text{H}_4\text{-CH}_2\text{NMe}_2} \quad 67\% \qquad [49]$$

5. (c)   *Acylation Reactions*

5. (c) (i)   *General principles*

The majority of the same general considerations that were outlined in the section on Friedel-Crafts alkylation reactions also apply to the acylation reactions (Heaney, H. in "*Comprehensive Organic Synthesis*", Eds. Trost, B.M.; Fleming I.; Vol. 2, Vol. Ed. Heathcock, C.H., 733; 753, Pergamon, Oxford, **1991**).

The importance of the introduction of an acyl residue into an aromatic system relates to a number of important synthetic transformations that can be carried out, as well as to the inherent importance of aromatic ketones themselves. The conversion of ketones into amines using the Beckmann rearrangement, into aryl esters and hence phenols *via* the Baeyer-Villiger rearrangement, and into the related hydrocarbon by a number of reductive dihydrodeoxo-disubstitution reactions, is well established. The latter transformations overcome a number of drawbacks that are observed when Friedel-Crafts alkylation reactions are carried out, even though extra steps are involved. The preparation of alkylarenes in a one pot procedure in which arenes were reacted with acyl chlorides, aluminium chloride, and triethylsilane in dichloromethane (Jaxa-Chamiec, A.; Shah, V.P.; Kruse, L.I. *J. Chem. Soc., Perkin Trans. 1*, **1989**, 1705). The alkylated arenes are formed by the deoxygenation of the intermediate acylated products and the yields obtained were high.

The alkyl substituted aromatic compounds are more nucleophilic than their precursors and so overalkylation can be a problem. Proto-dealkylation is frequently a low energy process and reversibility can lead to a lack of control over the regiochemical position of alkyl groups in an aromatic system.

In addition, where alkylation proceeds *via* a carbocation, rearrangement of the alkyl residue is not unusual. On the other hand, the aromatic ketones are less nucleophilic than their precursors and so it is usually more difficult to introduce more than one acyl group into an aromatic ring and over-acylation is not normally a problem. The acylation of mesitylene is an exception; diacetylmesitylene is formed easily and undergoes acetyl exchange reactions (Gore, P.H.; Hoskins, J.A. *J. Chem. Soc. C*, **1970**, 517; Andreou, A.D. *et al. J. Chem. Soc., Perkin Trans, 2*, **1981**, 830), presumably because the first formed ketone complex is prevented from attaining co-planarity with the ring by the adjacent methyl groups.

The other important feature that confers an advantage to the Friedel-Crafts acylation reaction is that rearrangement of the carbon chain of the acylating agent is not observed. However, where the product of the reaction is soluble in the reaction medium, for example when using boron trifluoride catalysts; acylation of the side chain can occur which leads to the formation of 1,3-diketones (Walker, H.G.; Sanderson, J.J.; Hauser, C.R. *J. Am. Chem. Soc.*, **1953**, *75*, 4109). Toluene reacts with an excess of acetic anhydride in the presence of boron trifluoride to give 4-methylbenzoylacetone in a 43% yield. It is also of interest to note that boron tris(trifluoroacetate) can function as a trifluoroacetylating agent (Briody, J.M.; Marshall, G.L. *Synthesis*, **1982**, 939). The reactions are summarised in equation [**50**]. Only in the case of a reaction using mesitylene did the reaction stop at the first stage to afford 2,4,6-trimethyltrifluoroacetophenone. We may assume that this is related to the steric effects of the methyl groups in the 2- and 6- positions. With benzene 1,1,1-trifluoro-2,2,2-triphenylethane was formed in low yield, but with the more nucleophilic anisole, 1,1,1-trifluoro-2,2,2-tris(*p*-methoxyphenyl)ethane was obtained in 90% yield. With 1,3-dimethoxybenzene, the 1,1-diaryl-2,2,2-trifluoroethanol was isolated in an 82% yield.

$$ArH + B(OCOCF_3) \longrightarrow \begin{cases} \text{if ArH is } Me_3C_6H_3 \\ \text{if ArH is } MeOC_6H_5 \\ \text{if ArH is } 1,3\text{-}(MeO)_2C_6H_4 \end{cases}$$

[50]

Ar = 4-MeOC$_6$H$_4$–

Ar = 2,4-(MeO)$_2$C$_6$H$_3$–

The intermediates involved in acylation reactions are more reactive and therefore less accessible than those involved in the related alkylation reactions. The result is that when lactones are used as reagents in Friedel-Crafts reactions the alkylation step precedes acylation. The reactions of ethyl cyclopropanecarboxylate, exemplified in equation [51], using aluminium chloride and hydrogen chloride as the catalyst system, provide an interesting variation and also follow the same sequence. The reaction involving benzene was shown to proceed *via* ethyl 3-chloro-2-methylpropionate and results in the formation of 2-methylindanone in 93% yield (Pinnick, H.W. *et al. J. Org. Chem.*, **1981**, *46*, 3758) and with toluene 2,6-dimethylindanone was formed in a 67% yield.

$$\text{Me-C}_6\text{H}_5 + \triangleright\text{-CO}_2\text{Et} \xrightarrow{\text{AlCl}_3,\ \text{HCl}} \text{2,6-dimethylindanone}\quad 67\%\qquad [51]$$

5. (c) (ii) *Catalysts*

The majority of the investigations have reported Friedel-Crafts acylation reactions using acyl chlorides or carboxylic anhydrides together with a slight excess over a stoichiometric amount of a Lewis acid, normally aluminium chloride or bromide. The fact that the ketonic product and a carboxylic acid coordinate to a mole of the Lewis acid also means that whereas one equivalent of the Lewis acid is involved in this way when using an acyl halide, two equivalents of the Lewis acid are tied up when using a carboxylic anhydride. As we noted in connection with Friedel-Crafts alkylation reactions, ongoing research is concerned with attempts to find truly catalytic systems, especially those that are easily separated from the reaction mixtures and are reusable. Such systems are categorised as involving "clean chemistry".

A wide variety of protic acids have been used to catalyse acylation reactions. It is assumed that the reactions involve the generation of acylium ions. An electrochemically generated acid catalyst has been shown to be successful in the acetylation of a number of electron rich carbocyclic aromatic compounds (Gatti, N. *Tetrahedron Lett.*, **1990**, *31*, 3933). For example, the reaction of acetic anhydride with veratrole (in the anode compartment) gave an 84% yield of 4-acetylveratrole when lithium perchlorate was electrolysed in a divided cell in a mixture of dichloromethane and an excess of acetic anhydride. No product was detected without electrolysis.

Polymeric reagents such as Nafion-H® have been used; for example 2-fluorobenzoyl chloride and toluene give the benzophenone derivatives with an *ortho-* : *para-* ratio of 4:81 (Olah, J.A. *et al. Synthesis*, **1978**, 672).

Intramolecular reactions of diphenylalkane-2-carboxylic acids (equation [52]) have also been carried out using Nafion-H® but the reactions proceed much more rapidly when the carboxylic acid is first converted into the corresponding acyl chloride (Yamato, T. *et al. J. Org. Chem.*, **1991**, *56*, 3955). It is assumed that the reactions proceed via mixed carboxylic-perfluoroalkane-resin sulfonic anhydrides. The ease of isolation of the reaction products and the ready regeneration of the catalyst provide obvious potential to this method.

$$\text{diphenylmethane-CO}_2\text{H} \xrightarrow[\text{p-xylene, }\Delta,\text{ 12 h}]{\text{Nafion-H}^\circledR\text{ 30 wt \%}} \text{dibenzosuberone} \quad 90\% \quad [52]$$

Mixtures of methanesulfonic acid and phosphoric anhydride have been used (Eaton, P.E.; Carlson, G.R. *J. Org. Chem.*, **1973**, *38*, 4071) but methanesulfonic acid itself appears to be superior to polyphosphoric acid in intramolecular acylation reactions though not for intermolecular reactions (Premasagar, V.; Palaniswamy, V.A.; Eisenbraun, E.J. *J. Org. Chem.*, **1981**, *46*, 2974). γ-Phenylbutyric acid gave α-tetralone in a 90% yield after being heated for two hours at 90 °C in methanesulfonic acid. We may note that acylureas have been used as acylating agents in conjunction with polyphosphoric acid as shown in equation [53] (Majumdar, M.P.; Kudav, N.A. *Chem. Ind. (London)*, **1976**, 1069).

$$\text{PhOMe} + \text{MeCON(H)CONH}_2 \xrightarrow{\text{PPA, 95 °C}} \text{4-MeO-C}_6\text{H}_4\text{-COMe} \quad 55\% \quad [53]$$

Trimethylsilyl polyphosphate also proves to be an efficient and convenient reagent for intramolecular Friedel-Crafts acylation reactions involving carboxylic acids (Berman, E.M.; Showalter, H.D.H. *J. Org. Chem.*, **1989**, *54*, 5642). A Y-faujasite-type zeolite exchanged with cerium (III) shows high regioselectivity in the acylation of toluene (Chiche, B. *et al. J. Org. Chem.*, **1986**, *51*, 2128) giving at least 94% *para*- substitution. This effect presumably reflects the fact that the acylation reaction occurs within the cavity of the zeolite where the dimensions involved in *para*-disubstitution can be accommodated much more easily than the dimensions involved in forming the other two isomers. Short chain carboxylic acids gave low yields but the yields with long chain acids were excellent (equation [54]).

$$\text{Me-C}_6\text{H}_5 + \text{n-C}_{11}\text{H}_{23}\text{COOH} \xrightarrow{\text{zeolite, 150 °C}} \text{4-Me-C}_6\text{H}_4\text{-CO-C}_{11}\text{H}_{23} \quad 96\%$$

[54]

Silica supported Keggin-type heteropoly acids such as 12-molybdophosphoric acid have been used in acylation reactions, as well as in alkylation reactions, for example in the benzoylation of *p*-xylene using benzoyl chloride. It was suggested (Izumi, Y. *et al. Bull. Chem. Soc. Jpn.*, **1989**, *62*, 2159.) that the high catalytic activity was due to the ability of the heteropoly anion to stabilise acylium ions.

Iron oxides that are formed by calcining iron (III) nitrate or the iron (III) hydroxides that are obtained by base induced precipitation from a solution of a salt show very high catalytic activity in the benzylation, *t*-butylation and the acetylation of toluene at room temperature (Arata, K.; Hino, M. *Chem. Lett.*, **1980**, 1479). A heterogeneous catalyst that may be filtered off at the end of reactions can be prepared by first calcining iron (III) sulfate at 800 °C followed by activation using benzyl chloride (Arata, K.; Hino, M. *Bull. Chem. Soc. Jpn.*, **1980**, *53*, 446). No catalytic activity was observed in the absence of the activation step. The relative rates for the acetylation of toluene and benzene (k_T/k_B) using acetyl chloride and the catalyst was *ca.* 100 with quite high positional selectivity in favour of attack at the *para-* position in toluene. A high value for k_T/k_B is normally taken to reflect the relatively weak electrophilicity of the acetylium ion. The value for k_T/k_B of 65.2 using the iron sulfate (800 °C) system and acetyl bromide suggests that the reactions do not just involve iron chlorides bound to the catalyst surface. It is interesting to note that trifluoromethoxybenzene is acylated exclusively in the *para-* position by acetyl chloride in nitromethane using iron (III) chloride as the catalyst (Olah, G.A. *et al. J. Am. Chem. Soc.*, **1987**, *109*, 3708).

5. (c) (iii) *Reactions involving mixed anhydrides*

In the first supplement we highlighted reactions that proceed *via* mixed anhydrides of trifluoromethanesulfonic acid (Effenberger, F.; Epple, G. *Angew. Chem., Int. Ed. Engl.*, **1972**, *11*, 299; 300). Similar results have been obtained using nonafluorobutanesulfonic acid (Effenberger, F.; Sohn, E.; Epple, G. *Chem. Ber.*, **1983**, *116*, 1195) and trifluoroacetyl trifluoromethanesulfonate is capable of uncatalysed trifluoroacetylation of a variety of aromatic rings (Forbus, T.R. Jr.; Martin, J.C. *J. Org. Chem.*, **1979**, *44*, 313). Kinetic studies (Roberts, R.M.; Sadri, A.R. *Tetrahedron*, **1983**, *39*, 137) involving reactions of carboxylic acids in triflic acid suggest that protonated mixed anhydrides are the probable precursors to the electrophiles involved in the acylation reactions. There have been a number of interesting developments since the early results were published. For example, 2-trifluoromethanesulfonyloxypyridine together with a carboxylic acid and trifluoroacetic acid affords very high yields of ketones (Keumi, T. *et al. Chem. Lett.*, **1977**, 1099) as illustrated in equation [**55**]. In addition, aryl ketones have been prepared (Effenberger, F.; König, G.; Klenk, H. *Chem. Ber.*, **1981**, *114*, 926) using alkyl and aryl carboxylic dichlorophosphoric anhydrides in reactions with electron rich compounds such as anisole (equation [**56**]).

[55]

[56]

As a development of work involving catalyst systems generated from mixtures of Lewis acids and other salts, similar systems have been investigated where mixed anhydrides are implicated. Reactions involving 4-trifluoromethylbenzoic anhydride and either another free carboxylic acid or the related trimethylsilyl ester, it is adduced that a mixed anhydride is first formed (Mukaiyama, T.; Suzuki, K. *Chem. Lett.*, **1992**, 1751; Suzuki, K.; Kitagawa, H.; Mukaiyama, T. *Bull. Chem. Soc. Jpn.*, **1993**, 66, 3729). In these reactions the acylating species, an acylium ion, is generated by interaction with an active catalyst that is formed from silicon tetrachloride and silver perchlorate. A reaction between equimolar amounts of anisole and the mixed anhydride formed from 4-trifluoromethylbenzoic anhydride and trimethylsilyl acetate gave 4-methoxyacetophenone in 68% yield using a 1:3 mixture of silicon tetrachloride and silver perchlorate as the catalyst system. Evidence concerning the involvement of mixed anhydrides was obtained during an investigation of the active titanium (IV) catalyst that is generated by the interaction of silver triflate with half an equivalent of titanium tetrachloride (Miyashita, M.; Shiina, I.; Mukaiyama, T. *Bull. Chem. Soc. Jpn.*, **1994**, 67, 210). The mixed anhydride was detected by ^{1}H nmr spectroscopy when a mixture of trimethylsilyl acetate and 4-trifluoromethylbenzoic anhydride was treated with a catalytic amount of tin (II) triflate.

The possibility of carrying out reactions using derivatives of optically pure α-amino acids has been explored by a number of groups. The amino group must be protected in order to preclude racemisation. The involvement of a mixed anhydride is implicated in the example shown in equation [**57**] (Mukaiyama, T. *et al. Chem. Lett.*, **1991**, 1059).

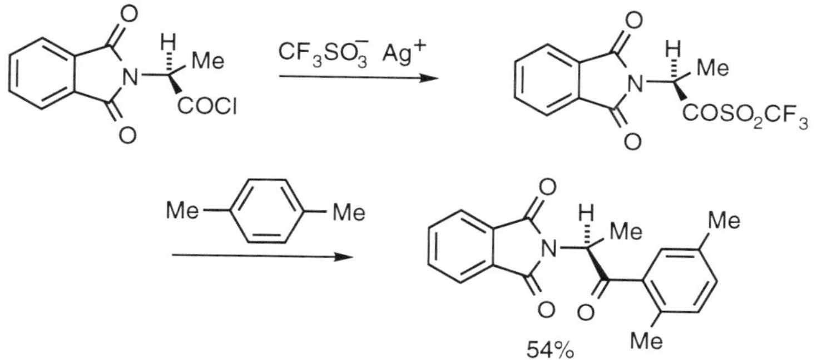

[57]

5. (c) (iv) *Other catalyst systems*

Dichlorogallium (III) perchlorate can be generated by the reaction of silver perchlorate and gallium (III) chloride (Harada, T. *et al. Synthesis*, **1991**, 1216; Mukaiyama, T. *et al. Chem. Lett.*, **1991**, 1059) and is an effective catalyst in the reaction of anisole with, for example, hexanoic anhydride, as shown in equation [**58**]. The catalyst is effective at a concentration of 10 mol% of gallium (III) chloride and the yield in the reaction with anisole only increases from 85% to 91% when the amount of silver perchlorate is increased from 10 mol% to 20 mol%. Silver hexafluoroantimonate has also been used (Mukaiyama, T. *et al. Chem. Lett.*, **1991**, 1059) in conjunction with gallium (III) chloride in high yield reactions involving a range of carboxylic anhydrides and anisole, veratrole, and 2-methylanisole.

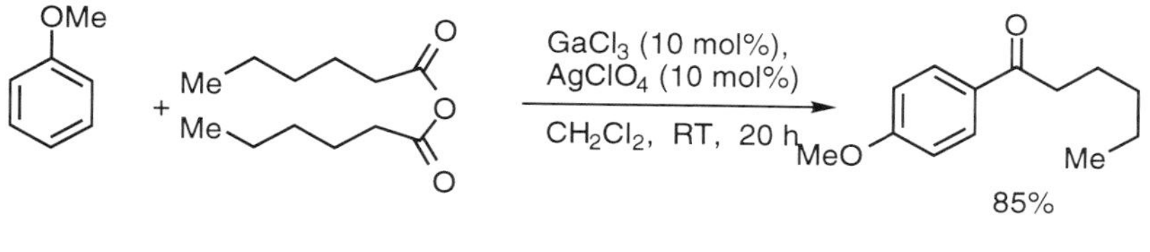

[58]

Reactions using 1-10 mol% of iron (III) chloride as the catalyst has also been used with N-phthaloyl-α-amino acid chlorides and reactive aromatic compounds such as p-dimethoxybenzene (Effenberger, F.; Steegmüller, D. *Chem. Ber.*, **1988**, *121*, 117). These reactions also proceed in good yields without racemisation as indicated in equation [**59**].

[Reaction scheme [59]: phthalimide-N-CH(Me)-COCl + 1,4-dimethoxybenzene, FeCl₃, 10 mol%, DCE – HCl, 50 °C, 61%]

Indium (III) chloride has been used (Pivsa-Art, S. *et al. J. Chem. Soc., Perkin Trans 1*, **1994**, 1703) at the level of 5 mol% in the benzoylation of 2-methoxynaphthalene. The regiochemical result of such acylation reactions are strongly dependent on the catalyst used. A catalyst system consisting of antimony pentachloride (4 mol%) and lithium perchlorate (100 mol%) has been shown to be an effective catalyst in reactions involving carboxylic anhydrides (Mukaiyama, T. *et al. Chem. Lett.*, **1992**, 435). For example a reaction of acetic anhydride with mesitylene gave 2,4,6-trimethylacetophenone in a 76% yield. The diarylboryl hexachloroantimonates are another series of catalysts that hold out promise in special cases (Mukaiyama, T. *et al. Chem. Lett.*, **1986**, 165). Acyl transfer is effected using diphenylboryl hexachloroantimonate in dichloromethane under almost neutral reaction conditions. In the reaction of propenylbutyrate with veratrole, 3,4-dimethoxybutyrophenone was isolated in a 65% yield. The decarbonylation of the pivaloylium ion is well known and as a result the introduction of a *t*-butyl group often complicates reactions designed to introduce the pivaloyl group. The reaction of anisole with pivaloyl chloride is therefore a good test for new Friedel-Crafts acylation reagent systems and is shown in equation [**60**]

[Reaction scheme [60]: anisole + pivaloyl chloride, (MeO-C₆H₄)₂B SbCl₆, CH₂Cl₂, → 4-methoxy-*t*-butyl phenyl ketone, 70%]

The use of lanthanide triflates promises to establish a series of very valuable Lewis acid catalysts in acylation reactions. The catalysts are easily recovered from the reaction mixtures and may be reused without a reduction in catalytic activity. The reaction between anisole and acetic anhydride catalysed by 20 mol% of ytterbium triflate gave an almost quantitative yield of 4-methoxyacetophenone (Kawada, A.; Mitamura, S.; Kobayashi, S. *J. Chem. Soc., Chem. Commun.*, **1993**, 1157) and scandium triflate gives higher yields in the acetylation of mesitylene than does ytterbium triflate (Kawada, A.; Mitamura, S.; Kobayashi, S. *Synlett*, **1994**, 545).

5. (c) (v) *Reactions using stoichiometric catalysts*

The use of an acyl chloride together with aluminium chloride still constitutes the most frequently used Friedel-Crafts protocol and a very wide range of ketones have been prepared in good to excellent yields. Acylation with some half-ester half-acid chlorides and some diacid chlorides proceeds satisfactorily. The reaction of ethyl oxalyl chloride with phenetole (equation [**61**]) (Johnson, W.M.P.; Holan, G. *Aust. J. Chem.*, **1981**, *34*, 2363) gives a good yield of the expected product. In a variation on the Vilsmeier formylation reaction, methyl *N,N*-dialkyloxamates, for example methyl morpholinylglyoxylate, have been shown to give glyoxylic esters by interaction with pyrophosphoryl chloride and electron rich aromatic compounds such as *m*-dimethoxybenzene (Downie, I.M. *et al. Tetrahedron*, **1993**, *49*, 4015).

We mentioned earlier the use of the *t*-butyl group in positional protection (Tashiro, M. *Synthesis*, **1979**, 921) and use has been made of that strategy, for example in the benzoylation of 4-*t*-butylanisole which leads to the formation of 2-hydroxybenzophenone. It is not unusual to find that dealkylation of an ether function occurs when an *o*-acyl group is introduced. This effect is exemplified in equation [**62**] (Rizzi, G.P. *Synth. Commun.*, **1983**, *13*, 1173). Advantage of this effect was also taken in a synthesis of the naturally occur-ring phenol aurentiacin (Schiemenz, G.P.; Schmidt, U. *Liebigs Ann. Chem.*, **1982**, 1509).

The acylation of phenyl methyl sulfide using 100 mol% of aluminium chloride gives the maximum yield and *para* selectivity (Pines, S.H. *J. Org. Chem.*, **1976**, *41*, 884). As has been indicated in a number of the other examples in this section, the acylation of ethers is widely used. However, it has been shown that acylation reactions of ethers are arrested when a large excess of catalyst is present (Buckley III, T.F.; Rapoport, H. *J. Am. Chem. Soc.*, **1980**, *102*, 3056). The details of that effect were studied using aluminium bromide because that catalyst allowed a more precise control of the stoichiometry than when using aluminium chloride. Reactions using acetyl chloride in *o*-dichlorobenzene were almost quantitative using 100 mol% of aluminium bromide, but the inhibiting effect of an excess of the catalyst was so great that the solvent then became the reactant. In a competition reaction an equimolar mixture of veratrole and benzene was allowed to react at room temperature with 200 mol% of acetyl chloride and 400 mol% of aluminium bromide in *o*-dichlorobenzene. After 1 h more than 95% of the benzene had been consumed and acetophenone was the only product detected by gas chromatography. Methylenedioxybenzene and benzene gave similar results. On the other hand, the acetal formed between catechol and 2,4-dimethylpentan-3-one is acylated particularly easily and does not suffer the inhibition effect. The inhibition effect is evidently due to coordination of the Lewis acid with the methyl ether and methylenedioxy functions.

The presence of halogens in the alkyl residue of the acyl halide may cause no significant problem and may prove to be useful as indicated in equations [63] (Kulinkovich, O.G.; Tishchenko, I.G.; Masalov, N.V. *J. Org. Chem. USSR (Engl. Transl.)*, **1982**, *18*, 859.) {see equations [51] and [64]} (Mahato, S.B.; Podder, G.; Maitra, S.K. *Indian J. Chem., Sect. B.*, **1986**, *25*, 1263).

In connection with the introduction of more than one function in a "one-pot" Friedel-Crafts acylation reaction we may note that enol lactones have been used in the preparation of 1,4-diketones (Chiron, R.; Graff, Y. *C. R. Hebd. Seances Acad. Sci.*, **1973**, *276*, 1207). In the example shown in equation [65] an oxazolone (azlactone) was used and gave an acylamino ketone (Balaban, A.T. *et al. Tetrahedron*, **1963**, *19*, 2199; El-Hashash, M.A. *Synthesis*, **1981**, 798). Friedel-Crafts bis(acylation) reactions occur using phthaloyl chloride and hydroquinones or their dimethyl ethers and have been used in anthracyclinone synthesis (Sartori, G. *et al. Tetrahedron Lett.*, **1987**, *28*, 1533).

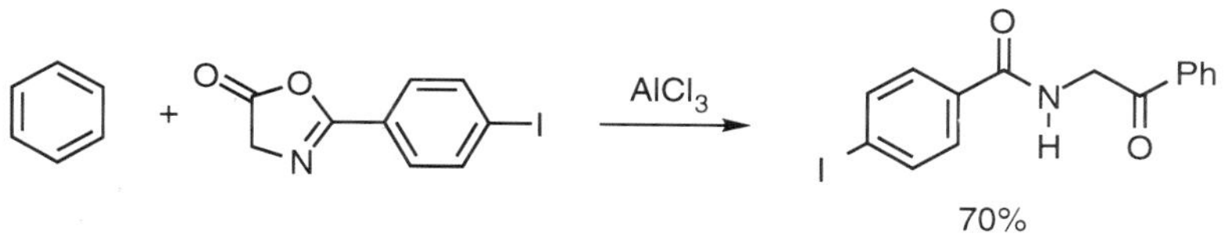

[Reaction scheme [65]: benzene + oxazolone-aryl iodide, AlCl₃, giving aryl ketone amide product, 70%]

5. (c) (vi) *Intramolecular acylation reactions*

A wide variety of cyclo-acylation reactions have been reported and summarised (Heaney, H. in *"Comprehensive Organic Synthesis"*, Eds. Trost, B.M.; Fleming I.; Vol. 2, Vol. Ed. Heathcock, C.H., p. 753, Pergamon, Oxford, **1991**). We will therefore concentrate on aspects not covered in that review. The possibility of building a highly functionalised benzene derivative involving as its final step an intramolecular acylation reaction was mentioned in the section on benzene ring syntheses. The cobalt diyne-yne methodology has been exploited in this connection as indicated in equation [**66**] (Gesing, E.R.F.; Sinclair, J.A.; Vollhardt, K.P.C. *J. Chem. Soc., Chem. Commun.*, **1980**, 286) where the final acylation step involved the *ipso*-acylation involving the increased electron density at the carbon atom that is substituted by silicon. In this connection we may note that attempted Mannich reactions are not successful and lead to the conclusion that the electrophiles involved in acylation reactions are significantly stronger than methyleneiminium salts.

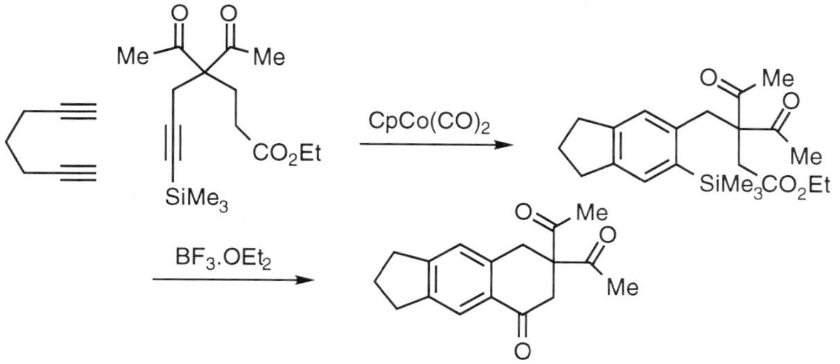

[Reaction scheme [66]: diyne-yne with SiMe₃ and CO₂Et groups, CpCo(CO)₂, then BF₃·OEt₂, giving indane intermediate then tricyclic diketone product]

The retention of chirality in intramolecular acylation reactions involving α-amino acids such as phenylalanine depends on having the amino group suitably protected. The formation of α-amino ketones *via* azlactones works well for the formation of five-, six-, or seven-membered rings. However, their known rapid racemisation due to the high acidity of the hydrogen at C-5 invalidates their use in the synthesis of enantiomerically pure α-amino ketones. On the other hand, carbamates derived from α-amino acids are known to show greater optical stability as compared with azlactones and this suggested a study of alkoxycarbonyl protected α-amino acids. Attempts to use *N*-benzyloxycarbonyl (Cbz) protection were unsuccessful, presumably due to the ease of formation of a benzyl cation under the reaction conditions. Cyclisation reactions of *N*-methoxycarbonyl-protected phenylalanine and homophenylalanine derivatives have been reported (McClure, D.E. *et al. J. Org. Chem.*, **1981**, *46*, 2431) and enantiomeric excesses better than 96% can be achieved. An example is shown in equation [**67**].

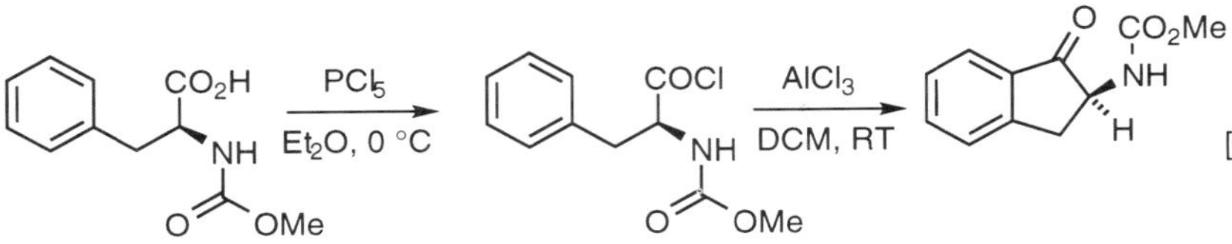

[67]

Similar results were obtained (Effenberger, F. *et al. Chem. Ber.*, **1988**, *121*, 125) using *N*-phthaloyl protection for phenylalanine and homophenylalanine derivatives. Cyclisation of the acid chlorides to the tetralone using either aluminium chloride or iron (III) chloride proceeds without racemisation whereas when using iron (III) chloride cyclisation to the indanone did result in some racemisation. The yields were good using aluminium chloride but lower when using iron (III) chloride. The conversion of (S)-*O*-methyl-*N*-phthaloyltyrosine to the mixed anhydride *via* the acid chloride using silver triflate (equation [**68**]) gave the indanone derivative without racemisation.

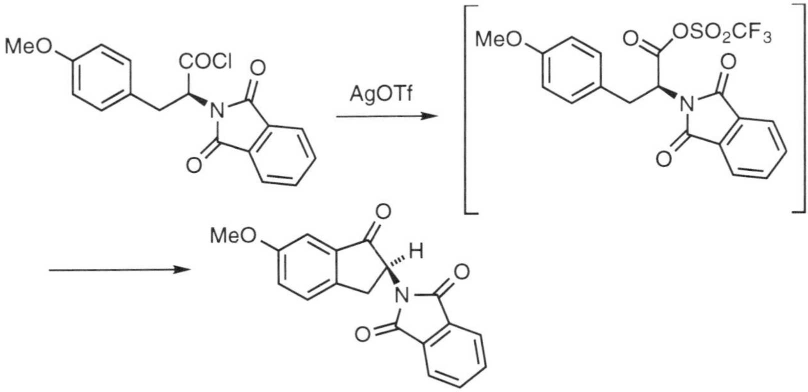

[68]

5. (c) (vii) *Metal catalysed ketone syntheses*

The formation of ketones by the reactions of acyl halides with Grignard reagents or organolithium compounds is normally complicated by the formation of unwanted carbinols as a result of secondary reactions. On the other hand, other kinds of organometallic reagents have been successfully employed. Organocopper regents can be used, but low temperatures are generally required in order to suppress the undesired reactions (Posner, G.H. *Org. React.*, **1975**, *22*, 253). It has been shown that the conversion of acid chlorides into ketones can be carried out efficiently in the presence of the catalyst tris(acetylacetonate)iron (III) at or just below room temperature using Grignard reagents (Fiandanese, V. *et al. Tetrahedron Lett.*, **1984**, *25*, 4805). When using a 1:1 ratio of phenyl Grignard to acid chloride isobutyrophenone was formed in a 92% yield at 0 °C; and with pivaloyl chloride and benzoyl chloride the yields of the expected ketones were 80 and 75% respectively.

Benzylchlorobis(triphenylphosphine)palladium (II) has been shown to be an efficient catalyst in the conversion of acid chlorides into ketones by coupling with organotin compounds (Milstein, D.; Stille, J.K. *J. Am. Chem. Soc.*, **1978**, *100*, 3636; *J. Org. Chem.*, **1979**, *44*, 1613). The yields obtained are high and the reactions are tolerant to a wide variety of other functional groups, including esters and even aldehydes. The reactions are accelerated by oxygen but deactivated by triphenylphosphine. The synthetic method is quite general with respect to the acid chloride and the organotin reagent and catalytic turnovers of 1000 have been achieved (Stille, J.K. *Angew. Chem., Int. Ed. Engl.*, **1986**, *25*, 508). Only one of the four groups on tin is transferred easily. In the case of an aryltrialkyltin the alkyl partners on tin function as anchoring groups and are not transferred. The result is that phenyltributylstannane functions as a phenyl anion equivalent. 4-Nitrobenzoyl chloride gave 4-nitrobenzophenone in a 97% yield (equation [**69**]) (Labadie, J.W.; Tueting, D.; Stille, J.K. *J. Org, Chem.*, **1983**, *48*, 4634).

The mechanism involved in the reaction of phenyltributylstannane with benzoyl chloride has been studied by ^{31}P nmr spectroscopy after the addition of benzylchlorobis(triphenylphosphine)palladium (II) (Labadie, J.W.; Stille, J.K. *J. Am. Chem. Soc.*, **1983**, *105*, 6129). The disappearance of the spectrum due to the catalyst was mirrored by the appearance of the spectrum due to the presence of benzoylchlorobis(triphenylphosphine)palladium (II). Benzophenone was formed in that reaction. The results were interpreted in terms of a transmetallation reaction in which electrophilic cleavage of the carbon-tin bond takes place with the benzoylpalladium (II) complex functioning as the electrophile.

Reactions of acid chlorides with benzyl bromide and a number of derivatives using a combined palladium and zinc catalyst system affords hydroxybenzoin derivatives in good yields (Sato, T. *et al., Chem. Lett.*, **1981**, 1135). Thus the reaction of benzyl bromide with benzoyl chloride in the presence of bis(triphenylphosphine)palladium dichloride and zinc powder gave desoxybenzoin in an 83% yield, presumably *via* the intervention of benzyl-zinc bromide. Alkyl halides have also been used in conjunction with di-sodium tetracarbonylferrate which form complexes that contain triphenylphosphine and react with aryl iodides in the presence of tetrakis(triphenylphosphine)palladium and zinc chloride as a co-catalyst. (R)-2-methyl-1-phenyl-1-octanone was prepared in good optical yield by that method starting from (S)-2-bromooctane (Koga, T.; Makinouchi, S.; Okukado, N. *Chem. Lett.*, **1988**, 1141).

The palladation of a number of *N,N*-dialkylbenzylamines has been shown to result in the formation of dimeric *ortho*-palladated species. Benzoyl and acetyl chloride have been shown to afford *o*-dialkylaminomethylaryl ketones in excellent yields as shown in equation [70] (Holton., R.A.; Natalie, K.J. Jr. *Tetrahedron Lett.*, **1981**, 22, 267). Other examples have extended the scope of the reaction using acid chlorides such as cinnamoyl-, phenylacetyl-, and diphenylacetyl- chloride (Clark, P.W. *et al., J. Organomet. Chem.*, **1983**, *253*, 399). No insertion was observed using pivaloyl- and triphenylacetyl- chloride.

5. (d) *Formylation Reactions*

In this section we will consider Gattermann and related formylation reactions (Miethchen, R.; Kroger, C.-F. *Z. Chem.*, **1975**, *15*, 135; Olah, G.A.; Ohannesian, L.; Arvanaghi, M. *Chem. Rev.*, 1987, 87, 671) and the Vilsmeier reaction (Simchen, G. in "*Houben-Weyl*", 4th ed., p. 36, Thieme, Stuttgart, 1983: Meth-Cohn, O.; Stanforth, S.P.in "*Comprehensive Organic Synthesis*", Eds. Trost, B.M.; Fleming I.; Vol. 2, Vol. Ed. Heathcock, C.H., p. 777, Pergamon, Oxford, **1991**). We note that formylation using the Reimer-Tiemann methodology (Wynberg, H.; Meijer, E.W. *Org. React.*, **1982**, *28*, 1; Wynberg, H. in "*Comprehensive Organic Synthesis*", Eds.

Trost, B.M.; Fleming I.; Vol. 2, Vol. Ed. Heathcock, C.H., p.769, Pergamon, Oxford, **1991**) only applies to phenols and is therefore outside the scope of this chapter.

5. (d) (i)   *Formylation reactions using formyl cation equivalents*

A number of variations on the Gattermann-Koch and Gattermann formylation reactions are available. The formylation of benzene and toluene using the Gattermann-Koch conditions shows the highest selectivity (k_T/k_B 155-860) and the highest positional selectivity towards attack at the *para*-position in toluene (Olah, G.A. *et al. J. Am. Chem. Soc.*, **1976**, *98*, 296). The lowest selectivity was observed in the formylation reaction using carbon monoxide catalysed by HF-SbF$_5$ in sulfurylchloride fluoride and the highest using carbon monoxide catalysed by HF-BF$_3$ in an excess of the aromatic hydrocarbons. *Para*-substitution is favoured by a transition state that is late; that is when it is like the Wheland intermediate. This is because the *para*-methyl group is better able to stabilise the transition state than an *ortho*-methyl group and, of course, much more than a *meta*-methyl group. Formylation using carbon monoxide and triflic acid also gives reasonable yields of aldehydes; for example *p*-xylene gave 2,5-dimethylbenzaldehyde in a 54% yield (Olah, G.A.; Laali, K.; Farooq, O. *J. Org. Chem.*, **1985**, *50*, 1483). Formyl fluoride forms a complex with boron trifluoride which can be used as a form-ylating agent. Boron trifluoride is introduced into a solution of the aromatic compound and formyl fluoride. Benzene gave benzaldehyde and mesitylene gave 2,4,6-trimethylbenzaldehyde in 58% and 70% yields respectively. An improved and convenient method for the preparation of acyl fluorides includ-ing formyl fluoride involves the reaction between the carboxylic acid and cyanuric fluoride (equation [71]) (Olah, G.A.; Nojima, M.; Kerekes, I. *Synthesis*, **1973**, 487).

$$3\ \text{HCO}_2\text{H} + \text{C}_3\text{N}_3\text{F}_3 \longrightarrow \text{HCOF} + \text{C}_3\text{N}_3(\text{OH})_3 \qquad [71]$$

It may be noted that whereas the system CO-HF-SbF$_5$ shows no primary kinetic isotope effect, the formyl fluoride - boron trifluoride complex shows a k_H/k_D of 2.68. Formylation using dichloromethyl methyl ether and aluminium chloride or titanium tetrachloride (de Haan, F.P. *J. Org. Chem.*, **1984**, *49*, 3963) is less selective than the classical Gattermann-Koch method as also is the zinc cyanide aluminium chloride method (Olah, G.A. *et al. J. Am. Chem. Soc.*, **1976**, *98*, 296). There are a number of recent examples of the use of zinc cyanide-hydrogen chloride formylation reactions (for example Chen, K.-M.; Joullié, M.M. *Org. Prep. Proc. Int.*, **1986**, *18*, 109).

Reactions that result in the formation of amides are related to formylation processes. Benzamide is formed in a 63% yield (equation [72]) (Firing, A.E. *J. Org. Chem.*, **1976**, *41*, 148) and *N,N*-diethylchloroformamide gave *N,N*-diethylbenzamides with a number of benzene derivatives in yields ranging from 69 - 95% in the presence of tin (IV) chloride (Naumov, Y.A. *et al. Zh. Org. Khim.*, **1975**, *11*, 370; *Chem. Abs.*, **1975**, *82*, 125018m).

$$\text{C}_6\text{H}_6 + \text{H}_2\text{N-CHO-F} \xrightarrow[100\ °\text{C}]{\text{HF}} \text{C}_6\text{H}_5\text{CONH}_2 \quad 63\%$$ [72]

5. (d) (ii) *Vilsmeier formylation reactions*

It is normally assumed that the reactive intermediate involved in Vilsmeier formylation reactions is a *N,N*-disubstituted chloromethyleneiminium salt although convincing arguments have been presented (Simchen, G. in "*Houben-Weyl*", 4th ed., p. 36, Thieme, Stuttgart, 1983) in favour of a series of equilibria as indicated in equation [73].

$$\text{H-C(=O)NR}_2 + \text{Cl-X} \rightleftharpoons \text{H-C(O-X)(}^+\text{NR}_2\text{)}\ \text{Cl}^- \rightleftharpoons \text{H-C(Cl)(}^+\text{NR}_2\text{)}\quad \text{O=P(O}^-\text{)Cl}_2$$

$$X = \text{COCl, COCOCl or SO}_2 \downarrow$$

$$\text{H-C(Cl)(}^+\text{NR}_2\text{)}\ \text{Cl}^-$$ [73]

Excellent yields of carboxaldehydes can be isolated with very electron rich aromatic compounds using the whole range of electrophiles indicated in equation [73], even using pre-formed *N,N*-disubstituted chloromethyleneiminium chlorides. The major evidence in favour of the normally accepted view is that attempts to obtain convincing ^{1}H or ^{31}P nmr evidence when using phosphoryl chloride as the electrophile have failed. On the other hand reactions using pyrophosphoryl chloride lend support to the proposal that a series of equilibria are involved (Downie, I.M. *et al. Tetrahedron*, **1993**, *49*, 4015). A series of reactions using the weakly nucleophilic anisole were carried out using *N,N*-dimethylchloromethyleneiminium chloride, and *N,N*-dimethylformamide with phosphoryl chloride or pyrophosphoryl chloride. In the latter reactions we may assume that the leaving group is the phosphorodichloridate anion and hence the Vilsmeier intermediate is not phosphorus free (equation [74]). The yields obtained using the three reagent systems with the other reaction conditions constant showed a wide variation (<5 : 38 : 75 %). It is reasonable to conclude that *N,N*-dimethylchloromethyleneiminium chloride is at lower energy than the phosphorus containing species and that the proposed equilibria do describe the events involved in Vilsmeier reactions. A similarly reactive intermediate (equation [**75**]) is formed by the interaction of *N,N*-dimethylformamide and triflic anhydride (Martínez, A.G. *et al. J. Chem. Soc., Chem. Commun.*, **1990**, 1571).

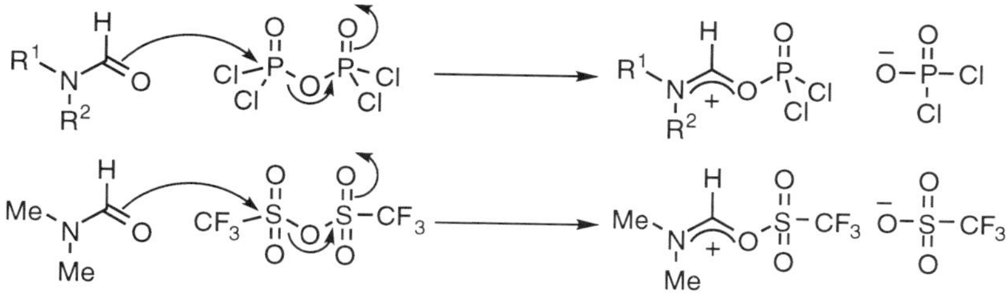

[74]

[75]

Vilsmeier reactions of a number of phloroglucinol ethers where one of the functions is a group that can be deprotonated easily have been shown to give benzofuran derivatives. For example 3,5-dimethoxyphenoxyacetophenone was shown to give 2-benzoyl-4,6-dimethoxybenzofuran (Hirota, T. *et al. Heterocycles*, **1986**, *24*, 771) and 3,5-dimethoxyphenoxyacetonitrile gave 2-cyano-4,6-dimethoxybenzofuran in a 61% yield together with 3,5-dimethoxy-4-formylphenoxyacetonitrile in 8.5% yield (Hirota, T. *et al. J. Heterocycl. Chem.*, **1986**, *23*, 1347). In the example shown in equation [**76**] the acetal function was used to produce a formyl group at the 2-position in the final benzofuran (Hirota, T. *et al. J. Heterocycl. Chem.*, **1986**, *23*, 1715).

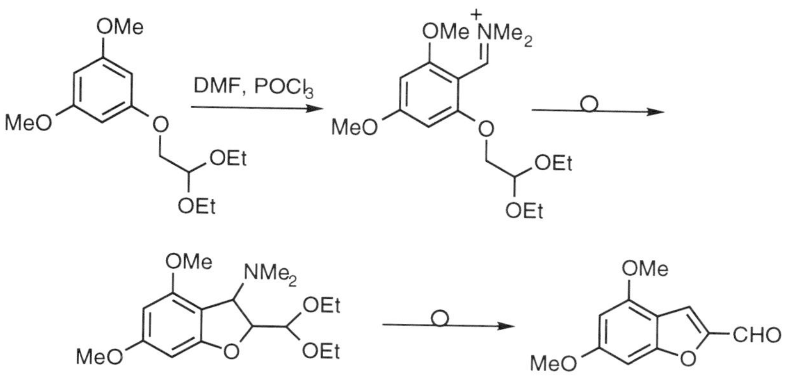

[76]

5. (e) *Group (V) Electrophiles*

5. (e) (i) *Nitration reactions*

Important books (Olah, G.A.; Malhotra, R.; Narang, S.C. *"Nitration - Methods and Mechanisms"*, VCH, New York, **1989**; Taylor, R. *"Electrophilic Aromatic Substitution"*, Wiley, New York, **1990**) and reviews, (Eberson, L.; Radner, F. *Accounts Chem. Res.*, **1987**, *20*, 53.; Ridd, J.H. *Chem. Soc. Revs.*, **1991**, *20*, 149 ; Kochi, J.K. *Accounts Chem. Res.*, **1992**, *25*, 39) have been published: the latter are principally concerned with electron transfer reactions that are involved in some nitration processes. The major advances that have been made in studies of nitration reactions relate to two areas. These are attempts to devise new and improved "nitronium carriers" and secondly studies aimed at a better understanding of the mechanistic aspects of the processes involved. In this latter context the use of pre-formed nitronium salts in the presence of very strong acids has been re-investigated (Olah, G.A. *Angew. Chem., Int. Ed. Engl.*, **1993**, *32*, 767). The protonation of the linear nitronium ion in fluorosulfuric acid generates a non-linear dication which is highly reactive and accounts for the ability of that system to nitrate electron depleted systems such as *m*-dinitrobenzene. Even the trityl cation is nitrated by the nitronium ion in triflic acid (Olah, G.A. *et al. J. Org. Chem.*, **1993**, *58*, 5017).

5. (e) (i/i) *Reagents and methods for the nitration of benzene and its derivatives*

Although mixtures of nitric acid and strong protic acids such as sulfuric and fluorosulfuric acid produce highly efficient nitrating agents, they do suffer from the fact that the more nucleophilic benzene derivatives may be sulfonated or oxidised under the reaction conditions. Non oxidising protic acids that have been used in conjunction with nitric acid include triflic acid and polyphosphoric acid, that is phosphoric acid to which phosphoric anhydride has been added. It is of interest to note that in the nitration of toluene the *ortho-* : *para-* ratio decreases as the amount of phosphoric anhydride in the phosphoric acid is increased, presumably because the electrophile involved is more sterically demanding. Nitric acid together with triflic anhydride affords a very efficient nitrating agent that affords high yields of products at low temperatures after short reaction times (Olah, G.A.; Reddy, V.P.; Prakash, G.K.S. *Synthesis*, **1992**, 1087). The predominantly covalent nature of trifluoromethanesulfonyl nitrate is indicated by the ^{15}N nmr chemical shift data.

The requirement that spent acid has to be neutralised and removed at the end of reactions is a major environmental problem and some attempts have been made to avoid the problem and to devise recyclable catalysts. One approach that is being used on an industrial scale and which we will return to later involves the use of nitrogen dioxide in the presence of ozone. Nitric acid adsorbed on silica gel has been shown to be an efficient reagent for the nitration of activated systems such as anisole (Tapia, R.; Torres, G.; Valderrama, J.A. *Synth. Commun.*, **1986**, *16*, 681). As mentioned in the first supplement, the availability of perfluorinated resinsulfonic acids [Nafion-H®] and the ease of isolation for reuse prompted a study of the nitration of a number of arenes using nitric acid, together with the resin acid under conditions where the water produced in the reactions was removed azeotropically (Olah, G.A.; Malhotra, R.; Narang, S.C. *J. Org. Chem.*, **1978**, *43*, 4628). For example benzene and mesitylene gave the expected nitro-derivatives in 77 and 79% yields respectively.

Nitration reactions have been carried out by using a number of metal nitrates in the presence of a Lewis acid. Silver nitrate has been used together with boron trifluoride to effect the mononitration of a number of arenes as exemplified in equation [77] (Olah, G.A. et al. *J. Org. Chem.*, **1981**, *46*, 3533). The larger ratio of *ortho* : *para* nitration of anisole suggests that the reactions proceed *via* a polarised complex rather than by way of the free nitronium ion. Boron trifluoride monohydrate is rather easier to handle than the gaseous boron trifluoride and has been used in conjunction with potassium nitrate to give nitroarenes in good yields (Olah, G.A. et al. *Synthesis*, **1994**, 468).

[77]

Trifluoroacetic anhydride has been used together with ammonium nitrate (Crivello, J.V. *J. Org. Chem.*, **1981**, 46, 3056). Trifluoroacetate is a better leaving group than acetate and so the nitration of relatively electron deficient benzenoid derivatives proceed to afford good yields. Benzoic acid gave *m*-nitrobenzoic acid in 98% yield using ammonium nitrate and trifluoroacetic anhydride. It is reasonable to assume that trifluoroacetyl nitrate is involved in these reactions. More recently trifluoroacetic anhydride has been used with tetrabutylammonium nitrate in a number of different solvents (Masci, B. *Tetrahedron*, **1989**, *45*, 2719). In the latter study, homogeneous reactions involved compounds with nucleophilicities that varied from benzene to mesitylene. Relative rates and partial rate factors for the nitration of a number of benzene derivatives have been determined using trifluoroacetic anhydride with tetrabutylammonium nitrate. Substrate and positional selectivities were observed to change when large ligands such as 18-crown-6 were added; it was assumed that they produce a greater sensitivity to electronic effects and larger crowding in the transition state leading to nitration (Masci, B. *J. Org. Chem.*, **1985**, *50*, 4081). Tetra-butylammonium nitrate has also been used together with triflic anhydride to produce anhydrous nitronium triflate (Adams, C.M.; Sharts, C.M.; Shackelford, S.A. *Tetrahedron Lett.*, **1993**, *34*, 6669). A wide range of acylnitrates have been generated by the interaction of silver nitrate with acyl chlorides and used together with other nitrating agents in relative rate and selectivity studies involving, for example toluene, (Olah, G.A. *et al. Proc. Natl. Acad. Sci. USA*, **1978**, *75*, 1045). A high, and reasonably constant *ortho-* to *para-* selectivity, was observed with all of the reagents. Metal nitrates such as copper (II) nitrate have also been used with acetic anhydride, for example in the nitration of [2.2]paracyclophane derivatives. It was assumed that acetyl nitrate was involved in the reactions shown in equation [**78**] where transannular π-interactions lead predominantly to skeletal rearrangement and the formation of cyclohexadienones (Horita, H.; Sakata, Y.; Misumi, S. *Tetrahedron Lett.*, **1976**, 1509).

A series of investigations using metal nitrates, acetic anhydride and an acidic clay provide interesting results. The nitration of toluene in a dilute solution in carbon tetrachloride with copper (II) nitrate impregnated in K10 montmorillonite and acetic anhydride gave a very high yield of a mixture of products with an unusually high *para-* content (Cornélis, A. *et al. Tetrahedron Lett.*, **1988**, *29*, 5657). A more practical method uses the generation of acetyl nitrate continuously in apparatus containing the substrate, acetic anhydride, and the montmorillonite clay. Nitric acid is extracted into the apparatus using a Dean-Stark trap in reverse (Cornélis, A.; Gerstmans, A.; Laszlo, P. *Chem. Lett.*, **1988**, 1839). The four-membered ring in benzocyclobutene is labile and so makes the preparation of its nitro-derivative difficult. However, the latter method has provided a significantly improved procedure (equation [**79**]) (Thomas, P.J.; Pews, R.G. *Synth. Commun.*, **1993**, *23*, 505).

A number of "nitronium carriers" were discussed in the first supplement and of those acetone cyanhydrin nitrate is probably the most useful. It is significantly more stable than, for example alkyl nitrates, and yet the cleavage of the O-N bond is relatively easy when the reagent is coordinated to a Lewis acid. Nitryl halides such as nitrosyl chloride function as nitrating agents in conjunction with a Lewis acid. High yields of nitroarenes have been obtained in reactions where the nitrosyl chloride was generated by the interaction of chlorotrimethylsilane with sodium nitrate (Olah, G.A. et al. *Synthesis*, **1994**, 468). The first formed intermediate is trimethylsilyl nitrate which then forms nitrosyl chloride and hexamethyldisiloxane by reaction with a second equivalent of chlorotrimethylsilane. The activation of dinitrogen pentoxide in sulfur dioxide appears to be particularly useful for the nitration of electron depleted aromatic systems. The nitration of, for example, dimethyl isophthalate (equation [80]) proceeds in high yield (Bakke, J.M.; Hegborn, I. *Acta Chem. Scand.*, **1994**, *48*, 181).

$$\text{MeO}_2\text{C-C}_6\text{H}_4\text{-CO}_2\text{Me} \xrightarrow{N_2O_5,\ SO_2} \text{MeO}_2\text{C-C}_6\text{H}_3(\text{NO}_2)\text{-CO}_2\text{Me}\quad 80\%$$

[80]

5. (e) (i/ii) *Mechanisms of nitration reactions*

It has often been assumed that all there is to be known about "the mechanism" of electrophilic aromatic substitution reactions has been reported long ago. It is remarkable how frequently we have had to changed our interpretations. Perhaps this is particularly the case with respect to the nitration of arenes.

There appear to be differences in the details involved in the mechanism of nitration reactions that depend on a number of factors. The nucleophilicity of the substrate and the precise nitrating reagent system both have an influence. In the case of arenes such as 1,2,4-trimethylbenzene the interaction with the nitronium ion proceeds by way of an encounter complex before the Wheland intermediate is reached. It is perhaps just a semantic argument to try to distinguish between an encounter complex and a charge transfer interaction. In any event there is a requirement that two intermediates are involved under some circumstances. Whereas the nitronium ion is linear the Wheland intermediate has a carbon-nitrogen bond that requires the -NO$_2$ group to be non linear. So the transition state leading to the Wheland intermediate must involve some expenditure of energy involved in bending the linear nitronium ion. Presumably this is why activation by the interaction with a very strong acid is required when nitronium salts react with substrates such as *m*-dinitrobenzene. The activation of "nitronium carriers" such as nitrosyl chloride require an excess of the Lewis acid in order to generate a "bent ion" (equation [**81**]) (Olah, G.A. *et al. J. Org. Chem.*, **1993**, *58*, 5017).

$$Cl-NO_2 + 3\ AlCl_3 \longrightarrow \underset{\underset{O-AlCl_3}{\diagdown}}{\overset{Cl_3Al\diagdown}{O=N^+}} \quad AlCl_4^- \qquad [\mathbf{81}]$$

Calculation of the reaction profiles suggest that the unsolvated nitronium ion reacts with benzene to give the Wheland intermediate without an energy barrier (Szabó, K.J.; Hörnfeldt, A.-B.; Gronowitz, S. *J. Am. Chem. Soc.*, **1992**, *114*, 6827). On the other hand, a solvated nitronium ion, for example protonated methyl nitrate, reacts *via* an energy barrier that is substituent dependent and also depends on the solvating species. In addition to these considerations, recent research that involves new nitrating reagent systems has shown that cation radicals are involved in some cases.

The nitronium ion and "nitronium carriers" such as acetyl nitrate, dinitrogen pentoxide, and nitrosyl chloride are all electron deficient and are therefore capable of acting as electron acceptors, as also are other nitro compounds. Comparative product studies and time-resolved spectroscopy following charge-transfer excitation of aromatic complexes formed with "nitronium carriers" has demonstrated that [Ar$^{+\cdot}$, $\cdot$NO$_2$] is a reasonable intermediate *en route* to the normally considered first intermediate - the Wheland intermediate or σ-complex (Kochi, J.K. *Accounts Chem. Res.*, **1992**, *25*, 39). A consideration of electron transfer mechanisms also comes to similar conclusions (Eberson, L.; Radner, F. *Accounts Chem. Res.*, **1987**, *20*, 53). Studies using *N*-nitropyridinium and substituted cations have demonstrated that transient charge transfer interactions can be involved in aromatic nitration reactions. The collapse to Wheland intermediates and hence to the products of nitration proceed at rates that are related to the HOMO-LUMO gap of the [ArH, X-PyNO$_2^+$] complexes (Kim, E.K.; Lee, K.Y.; Kochi, J.K. *J. Am. Chem. Soc.*, **1992**, *114*, 1756).

Although the nitrosonium ion is not sufficiently electrophilic to nitrosate benzene and various methylbenzenes it has been shown that nitration can be effected using nitrosonium salts in a number of polar solvents following the introduction of dioxygen (Kim, E.K.; Kochi, J.K. *J. Org. Chem.*, **1989**, *54*, 1692). Brightly coloured solutions are obtained when arenes and nitrosonium salts are mixed; in the case of benzene the charge transfer complex has λ_{max} = 346nm. The colour is discharged rapidly when dioxygen is passed through the solutions and nitroarenes can be isolated in good yields. *p*-Xylene gave 1,4-dimethyl-2-nitrobenzene in 75% yield. The suggested scheme leading to the Wheland intermediate is indicated in equation [**82**]. Nitrous acid has also been shown to give 4-nitro-*N*,*N*-dimethylaniline with a number of 4-substituted *N,N*-dimethylanilines (Colonna, M.; Greci, L.; Poloni, M. *J. Chem. Soc. Perkin Trans. 2*, **1984**, 165). It was proposed that an electron transfer process is involved in the initial *ipso*-attack.

$$\text{p-xylene} \xrightarrow{NO^+} [\text{ArH}\cdots NO^+] \xrightarrow{O_2} [\text{ArH}^{\cdot+} \quad \cdot NO_2] \longrightarrow \text{Wheland intermediate} \quad [82]$$

Aromatic hydrocarbons such as mesitylene are readily nitrated by nitrogen dioxide in dichloromethane at room temperature or below without irradiation. The transient red colours are due to the metastable charge transfer complex [ArH, NO$^+$] NO$_3^-$. Deliberate irradiation at the wavelength of the charge transfer band, even at –78 °C, resulted in rapid nitration (Bosch, E.; Kochi, J.K. *J. Org. Chem.*, **1994**, *59*, 3314). The photochemical reaction was shown to proceed by way of the radical cation.

The involvement of radical intermediates has also been demonstrated in nitration reactions of naphthols, phenols, and arylamines, that use nitrocyclohexadienones (Lemaire *et al. Synthesis*, **1989**, 761). ^{15}N Nuclear magnetic resonance studies involving the nitration of phenol showed a CIDNP effect with an enhancement of >200 (Coombes, R.G.; Ridd, J.H. *J. Chem. Soc., Chem. Commun.*, **1992**, 174). A radical process was invoked involving the phenoxyl radical and the ·NO$_2$ radical that had escaped from the radical pair in which it was formed.

An interesting series of papers involving the activation of nitrogen dioxide by ozone has shown that a wide variety of substrates can be nitrated in an environmentally friendly way by that method. It was suggested that an electron transfer mechanism operates by way of nitrogen trioxide (Suzuki, H.; Murashima, T.; Mori, T. *J. Chem. Soc., Chem. Commun.*, **1994**, 1443). Substrates such as acetophenone (Suzuki, H. *et al. Chem. Lett.*, **1993**, 1421), benzoic acid and its alkali and alkaline earth salts (Suzuki, H.; Tomaru, J.-i; Murashima, T. *J. Chem. Soc., Perkin Trans. 1*, **1994**, 2413), and alkylbenzenes (Suzuki, H. et al. *J. Chem. Soc., Perkin Trans. 1*, **1993**, 1591) have all been successfully used. In the latter case the addition of a protic acid increases the rate of the reaction and can lead to polynitration.

The results of studies involving *ipso*-nitration reactions were discussed in the first supplement and are covered in detail in one of the recent books (Olah, G.A.; Malhotra, R.; Narang, S.C. "*Nitration - Methods and Mechanisms*", VCH, New York, **1989**). We will only mention one recent example involving nitro-de-*t*-butylation reactions. The possibility of functionalising the upper rim of calixarenes has been addressed by carrying out nitration reactions on a number of phenolic ethers derived from 4-*t*-butylcalix[4]arene using 100% nitric acid in acetic acid. The conformational inversions that are frequently observed can be frozen in favour of the cone conformation by a suitably sized alkyl ether group and all four *t*-butyl groups can be replaced as shown in equation [**83**] (Verboom, W. et al. *J. Org. Chem.*, **1992**, *57*, 1313).

$$\text{calix[4]arene tetra-propyl ether} \xrightarrow{HNO_3} \text{tetra-nitro calix[4]arene tetra-propyl ether} \quad [83]$$

5. (e) (ii) *Nitrosation Reactions*

The chemistry of nitrosation reactions has been summarised in a recent book (Williams, D.H.L.. *"Nitrosation"*, Cambridge University Press, **1988**).

Although the direct introduction of a nitroso-group by electrophilic aromatic substitution has been thought to be restricted to very electron rich substrates such as *N,N*-dimethylaniline and phenol, recent results have shown that aromatic ethers can be *C*-nitrosated without concomitant *O*-de-alkylation (Radner, F.; Wall, A.; Loncar, M. *Acta Chem. Scand.*, **1990**, *44*, 152). Anisole was converted into 4-nitrosoanisole using sodium nitrite in a mixture of dichloromethane and trifluoroacetic acid (equation [**84**]).

$$\text{MeO-C}_6\text{H}_5 \xrightarrow[\text{CH}_2\text{Cl}_2,\ \text{CF}_3\text{CO}_2\text{H}\ (3:1\ \text{v/v})]{\text{Na NO}_2} \text{MeO-C}_6\text{H}_4\text{-NO} \quad (57\%) \quad [84]$$

Anisole has also been shown to give 4-nitrosoanisole in 87% yield by reaction with nitrosonium fluoroborate after 0.5 h. in a reaction carried out at room temperature and mesitylene (equation [**85**]) a 78% yield of 2,4,6-trimethylnitrosobenzene after 5 h (Bosch, E.; Kochi, J.K. *J. Org. Chem.*, **1994**, *59*, 5573). The difference in the behaviour of the nitrosonium ion as compared with reactions of the nitronium ion was ascribed to the rate limiting deprotonation of the Wheland intermediate. No primary kinetic isotope effect is observed in nitration reactions using the nitronium ion.

```
Me        Me                              Me        Me
  \  /\  /         NO⁺ BF₄⁻                \  /\  /
   ||  ||       ─────────────►              ||  ||          [85]
  /  \/  \         CH₃CN, 5 h, RT          /  \/  \NO
    Me                                       Me
                                            78%
```

Tertiary arylamines undergo *N*-deakylation-*N*-nitrosation with a range of nitrosating agents (Verado, G.; Giumanini, A.G.; Strazzolini, P. *Tetrahedron*, **1990**, *46*, 4303). The dealkylation process involves the intermediacy of iminium ions. These results should be contrasted with the well known *C*-nitrosation reactions that occur when reactions are carried out under acidic conditions. For example, the interaction of boron trifluoride with dinitrogen trioxide in nitroethane affords a quantitative yield of 4-nitroso-*N*,*N*-dimethylaniline (Bachman, G.B.; Hokama, T. *J. Am. Chem. Soc.*, **1957**, *79*, 4370).

5. (f) *Group (VI) Electrophiles*

5. (f) (i) *Sulfonation reactions*

Sulfonation reactions have been reviewed on a number of occasions (see for example: Gilbert E.E. "Sulfonation and Related Reactions", Wiley, New York, 1965; Cerfontain, H. *Recl. Trav. Chim. Pays-Bas*, **1985**, *104*, 153). The majority of the work that has been published since the first supplement has been concerned with sulfonation reactions using sulfur trioxide or the addition compounds formed between sulfur trioxide and Lewis bases. The Lewis bases include triphenylphosphine (Galpin, I. *et al. J. Chem. Soc., Chem. Commun.*, **1981**, 789), sulfolane (Hayashi, T.; Iida, H.; Ogata, I. *J. Appl. Chem. Biotechnol.*, **1976**, 513), and nitromethane (Cerfontain, H.; Koeberg-Telder, A. *Recl. Trav. Chim. Pays-Bas*, **1970**, *89*, 569).

In addition to the normal acidic sulfonating agents, sulfuric acid and fuming sulfuric acid ($H_2SO_4.SO_3$), methanesulfonic acid together with sulfur trioxide (methanepyrosulfonic acid) (Koeberg-Telder, A; Cerfontain, H. *Recl. Trav. Chim. Pays-Bas*, **1971**, *90*, 193) and arenepyrosulfonic acids have been used (van Albada, M.P.; Cerfontain, H.; Koeberg-Telder, A. *Recl. Trav. Chim. Pays-Bas*, **1972**, *91*, 33). Acetylsulfonic acid and acetylselenic acid have also been investigated (Ris, C.; Cerfontain, H. *J. Chem. Soc. Perkin Trans. 2*, **1973**, 2129). Whereas acetylsulfonic acid sulfonates benzene the selenium analogue, formed by the interaction of selenic acid with acetic anhydride, does not react with benzene. With more nucleophilic arenes acetylselenic acid gives isomer distributions that are similar to those obtained in the related sulfonation reactions.

Reactions of sulfate esters such as trimethylsilyl chlorosulfonate with arenes have been investigated and shown to afford trimethylsilyl arenesulfonates (equation [**86**]) or, depending on the work up procedure used, sulfonic acids (Hofmann, K.; Simchen, G. *Liebigs Ann. Chem.*, **1982**, 282).

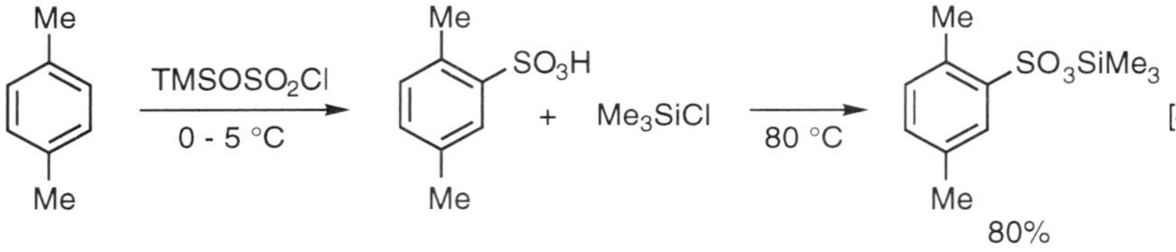

The sulfonation of arenes by sulfur trioxide has been shown to be considerably more complex than is normally described. The observed kinetics implicate the involvement of more than one mole of sulfur trioxide in complexing solvents such as nitromethane and dioxane (Cerfontain, H. *Recl. Trav. Chim. Pays-Bas*, **1985**, *104*, 153). The equations shown below describe the processes involved.

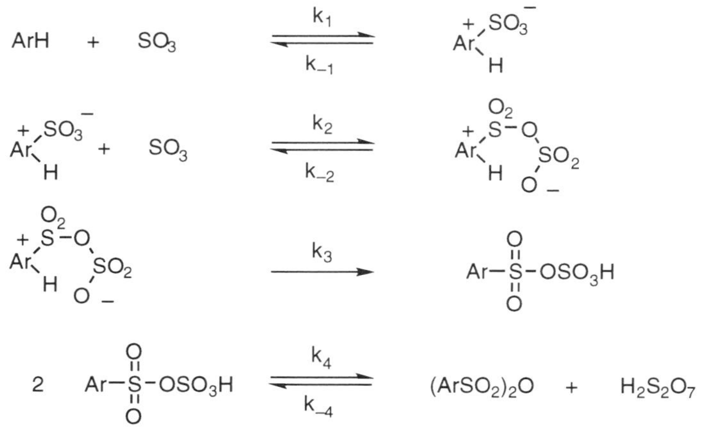

$$ArH + SO_3 \underset{k_{-1}}{\overset{k_1}{\rightleftharpoons}} Ar\overset{+}{\underset{H}{\diagdown}}SO_3^- \quad \text{[i]}$$

$$Ar\overset{+}{\underset{H}{\diagdown}}SO_3^- + SO_3 \underset{k_{-2}}{\overset{k_2}{\rightleftharpoons}} Ar\overset{+}{\underset{H}{\diagdown}}\overset{O_2}{\underset{O^-}{S}}\diagdown SO_2 \quad \text{[ii]}$$

$$Ar\overset{+}{\underset{H}{\diagdown}}\overset{O_2}{\underset{O^-}{S}}\diagdown SO_2 \overset{k_3}{\longrightarrow} Ar-\overset{O}{\underset{O}{\overset{\|}{S}}}-OSO_3H \quad \text{[iii]}$$

$$2 \; Ar-\overset{O}{\underset{O}{\overset{\|}{S}}}-OSO_3H \underset{k_{-4}}{\overset{k_4}{\rightleftharpoons}} (ArSO_2)_2O + H_2S_2O_7 \quad \text{[iv]}$$

The sulfonation of 1,6-methano[10]annulene with sulfur dioxide occurs, as expected, exclusively at the α-position (equation [**87**]) (Cerfontain, H. *et al. J. Org. Chem.*, **1984**, *49*, 3097).

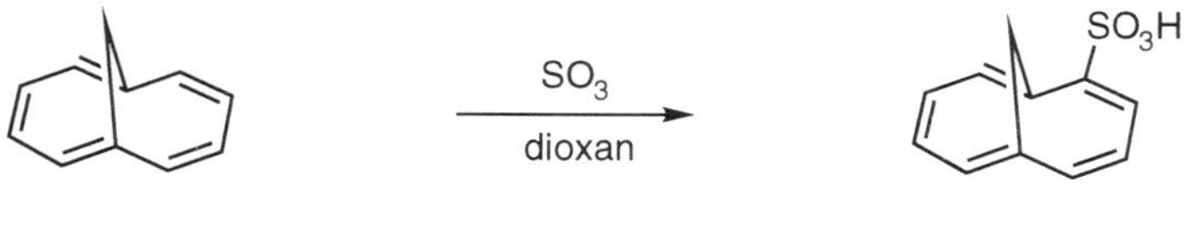

It has been shown that benzene and a wide range of its derivatives react with fluorosulfonic acid in the presence of antimony pentafluoride to give diarylsulfones as the major products together, in some cases, with small amounts of the arylsulfonyl fluoride. *p*-Xylene gave the diarylsulfone shown (2,5-2,5) (equation [**88**]) as the major product together with (2,5-2,4; and 2,4-2,4) {ratio 9.3 : 0.5 : 0.2} (Tanaka, M.; Souma, Y. *J. Org. Chem.*, **1992**, *57*, 3738).

[88]

The transfer sulfonation of anisole has been studied using polymethylbenzenesulfonic acids, catalysed by triflic acid. The results show that, in addition to the electronic activation by a methyl group *para*- to the sulfonic acid residue, the degree of overcrowding around the sulfonic acid group is important (Koeberg-Telder, A.: Cerfontain, H. *Recl. Trav. Chim. Pays-Bas*, **1990**, *109*, 41). The sulfonation of sterically unhindered aryl ethers using sulfur trioxide in dichloromethane results in an increase in the amount of *ortho*- sulfonic acid as compared with reactions carried out in nitromethane or dioxane (Ansink, H.R.W.; Cerfontain, H. *Recl. Trav. Chim. Pays-Bas*, **1992**, *111*, 183). It was suggested that coordination of two molecules of sulfur trioxide delivered the sulfur trioxide to the *ortho*- position. The sulfonation of the isomeric dimethoxybenzene derivatives has been studied using sulfur trioxide and the results were similar to those obtained when using sulfuric acid (Cerfontain, H.; Coenjaarts, N.J.; Koeberg-Telder, A. *Recl. Trav. Chim. Pays-Bas*, **1989**, *108*, 7). An investigation involving a number of catechol derivatives, including benzo-15-crown-5, showed that the rates were significantly larger when using sulfur trioxide than when using sulfuric acid (Ansink, H.R.W.; Cerfontain, H. *Recl. Trav. Chim. Pays-Bas*, **1989**, *108*, 395). The results were discussed in terms of the Curtin-Hammett principle. Both methylenedioxy- and ethylenedioxy- benzene gave predominantly 4-substitution as the initial monosubstitution product. Reactions of sulfur dioxide with benzene derivatives that contain electron withdrawing groups lead, as expected, to the formation of 1,3-disubstitution products. In some cases the 1,3,5-disulfonic acids and the 3,3'-diarylsulfones are also produced (Cerfontain, H.; Zon, Y.; Bakker, B.H. *Recl. Trav. Chim. Pays-Bas*, **1994**, *113*, 403). In the case of benzene derivatives that contain an oxygen containing electron withdrawing substituent, for example trifluoromethoxybenzene (equation [**89**]) attack occurs at the 4-position (Ansink, H.R.W.; Cerfontain, H. *Recl. Trav. Chim. Pays-Bas*, **1992**, *111*, 215). This result complements those obtained in nitration, halogenation, acylation, and alkylation reactions (Olah, G.A. *et al. J. Am. Chem. Soc.*, **1987**, *109*, 3708).

5. (f) (ii) *Hydroxylation reactions*

The hydroxylation of arenes can be carried out by three different procedures. The oxidation of organometallic reagents is one obvious way and is exemplified by reactions of aryllithium reagents with bis(trimethylsilyl)peroxide (Taddei, M.; Ricci, A. *Synthesis*, **1986**, 633). Much effort has also been expended into an investigation of the free radical oxidation of arenes, involving Fenton type reactions. Some recent synthetic work involving hydroxyl radicals has been reported. The reaction of benzene with hydrogen peroxide and iron (III) in the presence of electron transfer agents have proved useful: phenol was obtained in 80% yield in the presence of *N,N,N',N'*-tetramethyl-*p*-phenylenediamine (Tamagaki, S.; Sasaki, M.; Tagaki, W. *Bull. Chem. Soc. Jpn.*, **1989**, *62*, 153). Hydroxyl radicals are involved in a number of other reactions involving hydrogen peroxide and metal catalysts (see: Brook, M.A. *et al. J. Chem. Soc. Perkin Trans 2*, **1982**, 687; Eberhardt, M.K.; Ramirez, G.; Ayala, E. *J. Org. Chem.*, **1989**, *54*, 5922).

A fundamental problem that is associated with the hydroxylation of arenes is that once hydroxylated the ring becomes more susceptible to further oxidation. The use of hydrogen peroxide in the presence of superacid media was highlighted in the first supplement (Olah, G.A.; Ohnishi, R. *J. Org. Chem.*, **1978**, *43*, 865). The electrophile is assumed to be the protonated form of hydrogen peroxide. In these reactions the phenolic hydroxy group in the product is protonated and so the ring is protected from further electrophilic oxidation by the electron withdrawing effect of the cationic functional group. The method has been extended by investigations involving a range of substituted benzene derivatives. Benzaldehyde and aromatic ketones are hydroxylated and Baeyer-Villiger oxidation does not occur because the carbonyl groups are protonated under the reaction conditions. Acetophenone gave 3-acetylphenol in a 31% yield using hydrogen peroxide in the presence of hydrogen fluoride and antimony pentafluoride (Gesson, J.-P. *et al. Tetrahedron Lett.*, **1983**, *24*, 3095). Reactions involving amides such as acetanilide (Berrier, C. *et al. Bull. Chim. Soc. Fr.*, **1987**, 158) and esters, for example phenyl acetate, (Jacquesy, J.-C. et al. *Bull. Chim. Soc. Fr.*, **1994**, 658), have been studied. In the case of phenyl acetate a mixture of the isomeric monohydroxylated products were obtained in a reaction carried out at –40 °C; the *meta*- (44%) and *ortho*- (37%) isomers predominated.

Although the use of superacids prevents over-oxidation the systems are not easy to handle and recovery of the acid is difficult. The use of 30% aqueous hydrogen peroxide with hydrogen fluoride and boron trifluoride at –78 to –60 °C has been advocated as an alternative and benzene was converted into phenol in 37% yield by that method (Olah, G.A.; Fung, A.P.; Keumi, T. *J. Org. Chem.*, **1981**, *46*, 4305). Hydrogen peroxide has also been used in the presence of pyridinium polyhydrogen fluoride (Olah, G.A.; ; Keumi, T.; Fung, A.P. *Synthesis*, **1979**, 536).

Peroxytrifluoroacetic acid in the presence of boron trifluoride has been used in the oxidation of arenes (Hart, H. *Accounts Chem. Res.*, **1971**, *4*, 337) and other peroxyacids have also been investigated more recently. Peroxymonophosphoric acid has been shown to have a reactivity similar to that of peroxytrifluoroacetic acid and mesitylene is converted into mesitol (equation [**89**]) in good yield by that method (Ogata, Y. *et al. Tetrahedron*, **1981**, *37*, 1485).

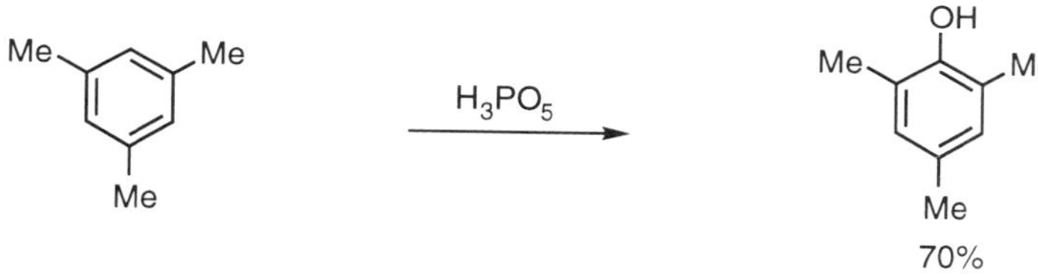

$$\text{Mesitylene} \xrightarrow{H_3PO_5} \text{Mesitol} \quad [89]$$

70%

Triflic acid has been used together with bis(trimethylsilyl)peroxide (Olah, G.A.; Ernst, T.D. *J. Org. Chem.*, **1989**, *54*, 1204) in the electrophilic hydroxylation of arenes, and mesitylene was converted into mesitol with sodium perborate - triflic acid in 73% yield (Prakash, G.K.S. *et al. Synlett*, **1991**, 39). The reactions of the perborate - triflic acid system with *ortho*- and *para*- xylenes give rise to products resulting from intramolecular methyl migrations (equation [**90**]) as also happen in reactions using *t*-butyl hydroperoxide in the presence of aluminium chloride (Apatu, J.; Chapman, D.C.; Heaney, H. *J. Chem. Soc., Chem. Commun.*, **1981**, 1079).

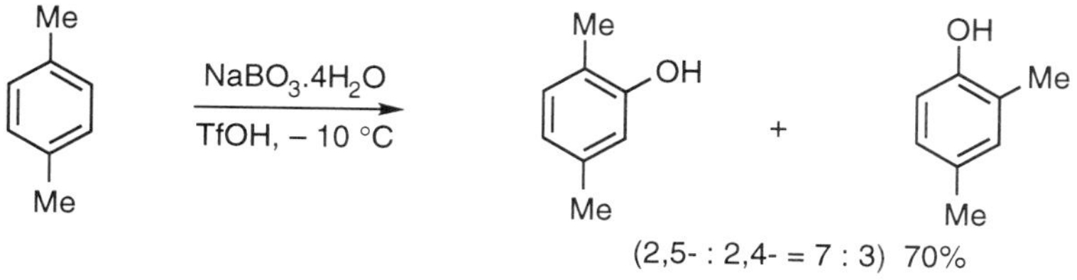

$$p\text{-xylene} \xrightarrow[\text{TfOH}, -10\,°C]{NaBO_3 \cdot 4H_2O} \text{product A} + \text{product B} \quad [90]$$

(2,5- : 2,4- = 7 : 3) 70%

5. (g)   *Group (VI) Electrophiles*

A number of mild and selective halogenating agents have been reported since the last supplement was published. In addition further mechanistic investigations have also been undertaken. The involvement of charge transfer complexes of bromine and chlorine and their relevance to aromatic bromination and chlorination reactions have been discussed (Fukuzumi, S.; Kochi, J.K. *J. Org. Chem.*, **1981**, *46*, 4116). Spectra of arene charge transfer complexes with chlorine, bromine, and iodine have also been compared with arene - halogen atom complexes (Radner, K.D.; Lusztyk, J.; Ingold, K.U. *J. Phys. Chem.*, **1989**, *93*, 564). Iodine monochloride and other interhalogen compounds are known to form charge transfer complexes and iodine monochloride is known to iodinate a range of functionalised benzene derivatives. Electron transfer occurs from the arene iodine monochloride complex to give an arene radical cation, chloride ion, and atomic iodine (Hubig, S.M.; Jung, W.; Kochi, J.K. *J. Org. Chem.*, **1994**, *59*, 6233). The cation radical has been detected spectroscopically in the case of 1,4-dimethoxybenzene at –78 °C in dichloromethane. The cation radical is quenched by chloride ion to give the chloroarene and by atomic iodine to give the iodoarene. The ratio of the two products can be varied by changing the solvent polarity or by added salt. Benzeneselinyl chloride together with aluminium chloride has been shown to be a regioselective chlorinating agent, or with aluminium bromide a brominating agent. In the example shown in equation [91] the intervention of chlorine monobromide was invoked (Kamigata, N. *et al. Bull. Chem. Soc. Jpn.*, **1988**, *61*, 2226).

$$Me_2N-C_6H_5 \xrightarrow[AlBr_3, -10\,°C, 3h]{PhSeOCl} Me_2N-C_6H_4-Br \quad 99\% \quad [91]$$

Charge transfer interactions and hydrogen bonding between phenol and substituted phenols and hexachlorocyclohexadienones were invoked to explain the regioselective chorination reactions involving those reagents (Lemaire, M.; Guy, A. Guetté, J.-P. *Bull. Soc. Chim. Fr.*, **1985**, 477; Guy, A.; Lemaire, M.; Guetté, J.-P. *Tetrahedron*, **1982**, *38*, 2339).

Bromodimethylsulfonium bromide and the related chloro-compound have been used to effect the regioselective bromination and chlorination of nucleophilic aromatic compounds. Phenols and aryl ethers react efficiently in dichloromethane at temperatures between – 25 and 0 °C to afford essentially pure *para*-substituted derivatives (Olah, G.A.; Ohannesian, L.; Arvanaghi, M. *Synthesis*, **1986**, 868). The regioselectivity (equation [92]) was ascribed to a combination of the bulky nature of the reagent and that the transition state leading to the Wheland intermediate was "late".

$$MeO-C_6H_5 \xrightarrow[8h]{Me_2\overset{+}{S}Br\ Br^-} MeO-C_6H_4-Br \quad 94\% \quad [92]$$

It is noteworthy that no halogenation was observed with *p*-cresol and its derivatives. The chlorination and bromination of trifluoromethoxybenzene in the presence of iron (III) salts gives predominantly *para*-substitution and chlorination in the presence of iodine is even more *para*-selective (Olah, G.A. et al. *J. Am. Chem. Soc.*, **1987**, *109*, 3708). These results should be compared with acetylation which gives exclusively the *para*- product. The oxidation of potassium bromide or iodide using *meta*-chloroperoxybenzoic acid in the presence of 18-crown-6 leads to electrophilic halogenating species that have been used with a number of phenols and aryl ethers (Srebnik, M.; Mechoulam, R.; Yona, I. *J. Chem. Soc., Perkin Trans. 1*, **1987**, 1423). Alumina supported copper (II) halides have been used to afford nuclear halogenated products free from side chain derivatives with a number of polymethylbenzenes (Kodomari, M.; Satoh, H.; Yoshimoto, S. *Bull. Chem. Soc. Jpn.*, **1988**, *61*, 4149; *J. Org. Chem.*, **1988**, *53*, 2093). 1,3-Dibromo-5,5-dimethylhydantoin has been used as a brominating agent in dichloromethane catalysed by strong acids (Eguchi, H. et al. *Bull. Chem. Soc. Jpn.*, **1994**, *67*, 1918) and catalytic amounts of 70% perchloric acid have been used to initiate the halogenation of electron rich aromatic compounds using *N*-chloro- and *N*-bromo- succinimide (Goldberg, Y.; Alper, H. *J. Org. Chem.*, **1993**, *58*, 3072). Mesitylene gave the bromo- derivative (equation [93]) and the chloro- derivative in 81% yield. Primary and secondary amines have been used in conjunction with *N*-bromosuccinimide in the bromination of phenol in dichloromethane (Fujisaki, S. et al *Bull. Chem. Soc. Jpn.*, **1993**, *66*, 1576).

[93]

A number of benzyltrimethylammonium salts have been used as sources of electrophilic halogenating agents (Kajigaeshi, S. et al. *Bull. Chem. Soc. Jpn.*, **1988**, *61*, 597; **1990**, *63*, 941). For example the selective bromination of aromatic ethers was reported using benzyltrimethylammonium tribromide in either dichloromethane - methanol or acetic acid - zinc chloride (Kajigaeshi, S. et al. *J. Chem. Soc., Perkin Trans. 1*, **1990**, 897).

Chapter 3

HALOGENOBENZENES

P. G. STEEL

Introduction

As in the previous update to this chapter the sections and subsections have the same titles and numbering as those used in Chapter 3, Volume IIIA of the Second Edition. Not all the topics are repeated, whilst a new section on the chemistry of transition metal π-complexes of halobenzenes appears as section 1(b)(ix).

Although as a sequel to the first supplement the present report covers the literature from ~1980 onwards, earlier references are included where it is felt that they contribute to the discussion. Since the previous supplement was published a number of major review works have appeared and these are referenced throughout the chapter. Similarly a number of new books containing material of relevance to this chapter have been published. Reflecting the rise in importance of organofluorine compounds, aspects of the chemistry of fluoroarenes are covered in the following texts; 'Biomedical Aspects of Fluorine Chemistry', Eds. R. Filler and Y. Kobayashi, Nodansha/Elsevier, New York, 1982; 'New Fluorinating Agents in Organic Synthesis' Ed. L. German and S. Zemskov, Springer Verlag, Berlin, 1989; and 'Synthetic Fluorine Chemistry' Eds. G. A. Olah, R. D. Chambers and G. K. S. Prakash, J. Wiley, Chichester, 1992. In addition a reissue of the revised second edition of '"Chemistry of Organic Fluorine Compounds "- a practical laboratory manual" M. Hudlicky, Ellis Horwood/PTR Prentice Hall, Chichester, 1992 has appeared. Other monographs containing relevant material have been published, including 'The Organic Chemistry of Polycoordinated Iodine', A. Varvoglis, VCH Publishers, Cambridge, 1992. This last subject has also been comprehensively reviewed by G. F. Koser in 'The Chemistry of Functional Groups Supplement D' ed S. Patai, Wiley, New York, 1993, pp774-806.

1. Nuclear Halogen Derivatives

(a) Methods of Preparation

(i) Replacement of an aromatic hydrogen by halogen
Major advances in this area have been in the development of direct fluorination procedures. Work has appeared on the use of dilute solutions of fluorine in nitrogen or the inert gases and the use of this reagent has been reviewed (S. T. Purrington, B. S. Kagen and T. B. Patrick, *Chem. Rev.*, 1986, **86**, 997; and S. Rozen, *Accs. Chem. Res.*, 1988, **21**, 307). Selectivity is the key issue and in this respect the use of low reaction temperatures with arylorganometallic substrates in hydrocarbon ether solvents has provided some success (J. DeYoung, H. Kawa and R. J. Lagow, *J. Chem. Soc., Chem. Commun.*, 1992, 811; and D. Naumann and H. Lange, *J. Fluorine Chem.*, 1983, **23**, 37), as has the use of Lewis Acids (S. T. Purrington and D. L. Woodard, *J. Org. Chem.*, 1991, **56**, 142). However, despite this progress far greater emphasis has been placed on alternative more convenient reagents. Many of these are significantly less reactive and require the prior generation of an intermediate aryl metal species.

Reagents that directly react with the aryl unit via a traditional electrophilic aromatic substitution pathway include xenon difluoride which continues to be investigated. The early applications of this reagent have been comprehensively reviewed (R. Filler, *Isr. J. Chem.*, 1978, **17**, 71; M. Zupan 'Xenon Halide Halogenations' in *The Chemistry of Functional Groups, Supplement D. Chemistry of Halides, Pseudohalides and Azides*, S. Patai and Z. Rappoport Eds, Wiley, Chichester, 1983; M. Zupan, *Vestn. Slov. Kem. Drus. (Suppl.)*, 1984, **31**, 151). A more recent report, describing reactions of xenon difluoride with aryl silanes, has suggested that a radical based mechanistic pathway is more appropriate than one based on simple electrophilic substitution (A. P. Lothian and C. R. Ramsden, *Synlett*, 1993, 753). The use of trifluoromethylhypofluorite is less common owing to the fact that it is no longer commercially available and also that its use frequently yields polyfluorinated products (T. B. Patrick, G. L. Cantrell and C. Chang, *J. Am. Chem. Soc.*, 1979, **101**, 7434).

Acetylhypofluorite has been introduced by Rozen and collaborators as a simpler and more convenient method for the introduction of fluorine than the classical Balz-Schiemann reaction, particularly radiolabelled fluorine for which the latter reaction is very inefficient (O. Lerman *et al.*, *J. Org. Chem.*, 1984, **49**, 806). Interestingly, this reagent provides an unusually high *ortho* to *para* ratio of fluorinated products in low temperature reactions with activated aromatics when using freon™ solvents. This observation has been rationalised in terms of an addition/elimination mechanism. Support for this hypothesis was obtained from the reaction with piperonal. This combination produced the adduct (**1**) which is sterically constrained from eliminating the elements of acetic acid required to regenerate the aromatic nucleus. The nature of the solvent appears to be critical, since, in a study on the application of this reagent Visser and co-workers, carried out reactions in acetic acid (G. W. M. Visser *et al., J. Org. Chem.,* 1986, **51**, 1886) and consequently achieved an efficient fluorination of phenol. This is in contrast to the result obtained using freon™ solvents in which oxidation products predominated.

$$\text{OHC-}\underset{}{\bigcirc}\text{-O} \xrightarrow[55\%]{CH_3COOF} \text{OHC-}\underset{F}{\bigcirc}\text{-O (OAc)}$$
(1)

Furthermore, the products from the use of acetic acid as the solvent do not show the enhanced *ortho* to *para* ratios observed in the less polar freon™ solvents. These results have led the authors to suggest that an alternative radical based mechanism is operative under these conditions.

The other well established reagents that continue to find applications in this area are fluorosulphates, notably the cesium salt (D. P. Ip *et al., J. Am. Chem. Soc.,* 1981, **103**, 1964; and T. B. Patrick and D. L. Darling, *J. Org. Chem.,* 1986, **51**, 3242). Again, recent work with these reagents is concerned with the extension to other arenes; chemistry which is beyond the scope of this chapter; although mechanistic data is surveyed. Studies indicate that, as with acetylhypofluorite, there is a mechanistic divergence. The reaction either proceeds via electrophilic attack (this is itself unusual as the attacking species is an anion), or by the generation of a radical cation (E. H. Appelman, L. J. Basile and R. Hayatsu, *Tetrahedron,* 1984, **40**, 189). Being a much milder reagent, cesium fluorosulphate smoothly fluorinates phenol. This outcome is in contrast to the analogous reaction using acetylhypofluorite which yields products of oxidative polymerisation.

One problem common to almost all the O-F reagents is that they are toxic, hygroscopic, unstable (occasionally explosive) and consequently require specialised equipment to handle safely. More recently milder more selective reagents which contain an N-F bonded species have come to the fore. This has resulted in the introduction of N-fluoropyridone (S. T. Purrington and W. A. Jones, *J. Fluorine Chem.,* 1984, **26**, 43); N-fluoro-N-alkylarene-sulphonamides (W. E. Barnette, *J. Am. Chem. Soc.,* 1984, **106**, 452) and related species which are easy to handle. Unfortunately, this stability results in their use being limited to the fluorination of activated arenes. Recent developments have lead to reagents with enhanced reactivity such as N-fluoroperfluoroalkylsulphonamides (**2**) (S. Singh *et al., J. Am. Chem. Soc.,* 1987, **109**, 7194) and N-fluoro-N'-alkyl-1,4-diazoniabicyclo-[2. 2.2]octane salts (**3**) (R. E. Banks *et al., J. Chem. Soc., Chem. Commun.,*

1992, 595) which fluorinate unsubstituted arenes in moderate to excellent yield.

$R_fSO_2N(F)SO_2R_f'$

(2)

$R_f, R_f' = CF_3, C_4F_9, C_6F_{13}, (CF_2)_n$ [n=2-4]

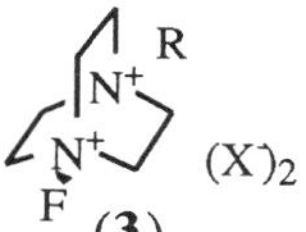

(3)

R= CH_3, CH_2Cl, CH_2CF_3
X= OTf, BF_4

However, in general and certainly for non-activated systems, the N-fluoropyridyl triflates (4) developed by Umemoto and coworkers appear to be the most popular reagents for this task (T. Umemoto et al., *J. Am. Chem. Soc.*, 1990, **112**, 8563).

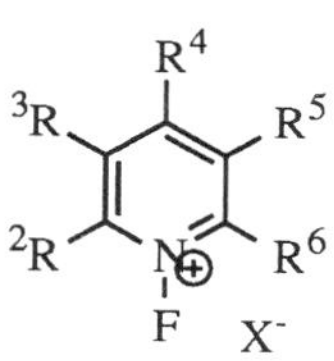

(4)

X= OTf, $OSO_2C_4F_9$, BF_4, SbF_6, ClO_4
R^1-R^5 =H, Me, ^tBu, Cl, CO_2Me, CH_2OMe, CH_2OCOMe, SO_3^-

The principal feature of these reagents is that by varying the nature of the counter ion or pyridine substituents it is possible to tune their reactivity. The power of the reagent increases with the number of electron withdrawing pyridine substituents. This is a σ-bond effect as shown by the fact that the 2,6-dichloro reagent is more reactive than its 3,5- analogue. Other control elements that can be used include the counter ion and/or the solvent. The use of a triflate anion in conjunction with a non-polar solvent such as dichloromethane provides the most powerful reagent. This combination fluorinates nitrophenol in 73% yield and also produces respectable amounts of polyfluorination of 4-tolylurethane. The regioselectivity of these reagents is modest showing a slight preference for *ortho* - substituted products, although the incorporation of the the counterion into the molecule as in the zwitterion (5) affords complete *ortho*- selectivity in its reaction with phenol.

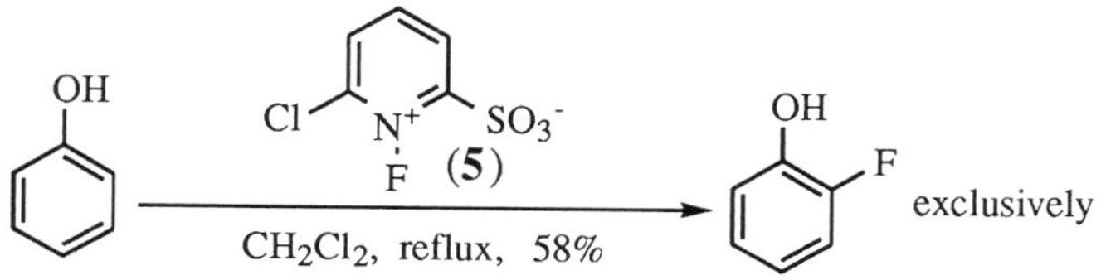

OH + (5) $\xrightarrow{CH_2Cl_2, reflux, 58\%}$ 2-fluorophenol   exclusively

Although these reagents seem to be a simple source of electrophilic fluorine (K. O. Christie, *J. Fluorine Chem.*, 1983, **22**, 519; O. Lerman *et al.*, *J. Org. Chem.*, 1984, **49**, 806 ref. 12b therein), it appears that these reactions proceed via a one electron transfer process through an intermediate π-complex, followed by fluorine atom transfer to generate a fluorinated carbocation.

An alternative strategy, for the specific incorporation of fluorine into less reactive, electron deficient, arenes is to increase their nucleophilicity by the generation of an aryl metal species. Metalation is generally achieved either directly or via metal halogen exchange. A large number of different metal units have been employed in combination with an equally diverse number of positive halogen equivalents (N-fluoropyridone-PhMgCl, S. T. Purrington and W. A. Jones, *J. Fluorine Chem.*, 1984, **26**, 43; RSO_2NFR' -ArLi W. E. Barnette, *J. Am. Chem. Soc.*, 1984, **106**, 452; $CsSO_4F$ - $ArSnR_3$ M. R. Bryce *et al., J. Chem. Soc., Chem. Commun.*, 1986, 1623; CH_3CO_2F - ArM [M= Sn, Si, Ge, Pb, Hg, Tl] M. J. Adam, *J. Fluorine Chem.*, 1984, **25**, 329; F_2 - $ArSiMe_3$ P. DiRaddo, M. Diksic, D. Jolly, *J. Chem. Soc., Chem. Commun.*, 1984, 159 and M. Speranza *et al., J. Fluorine Chem.*, 1985, **30**, 97). The high regiospecifity of the last two procedures has been of considerable use in the preparation of specific radiolabelled compounds where the relatively short halflife of ^{18}F ($t_{1/2}$=110min) demands both rapid and regioselective incorporation.

One final preparation of monofluorinated aromatics by the displacement of hydrogen is the treatment of N-aryl-N-hydroxyamides with DAST as reported by Kikugawa (Y. Kikugawa *et al., J. Chem. Soc., Chem. Commun.*, 1992, 921) which proceeds with almost complete regioselectivity.

R=Me (71%), Ph (83%)

Unlike all of the procedures mentioned so far this is believed to be a nucleophilic fluorination process. Consistent with this is the fact that the hydroxylamino function is essential for selectivity. To account for these observations a mechanism has been postulated in which initial reaction of

DAST with the N-hydroxy group is followed by a heterolytic cleavage of the N-O bond to generate a conjugated amidyl cation. This is then trapped by the fluoride anion at the site of highest charge density, namely the *para-* position.

Although there are many existing methods for the preparation of chlorine and bromine substituted benzenoid species, new work continues to appear. This is mainly associated with two features, the development of reagents for the direct introduction of halogen into non-activated aromatic rings and the control of regioselectivity.

In work directed towards the first of these objectives Harrison and coworkers (J. J. Harrison, J. P. Pellegrini and C. M. Selwitz, *J. Org. Chem.,* 1981, **46**, 2169) have used potassium bromate in the presence of 68% sulphuric acid to achieve the bromination of deactivated aromatic rings in moderate to excellent yields (50-97%). One problem with this reagent combination is the interference of other functional groups elsewhere in the molecule. Similar problems of a lack of control also befall the use of bromine-mercuric oxide (V. N. Shishkin, *Zh. Org. Khim.,* 1991, **27**, 1486) and dichlorine monoxide (F. D. Marsh et al., *J. Am. Chem. Soc.,* 1982, **104**, 4680). This latter reagent in the presence of either sulphuric or trifluoromethanesulphonic acid rapidly chlorinates all aromatic nuclei with polysubstitution being common.

$$\text{4-O}_2\text{N-C}_6\text{H}_4\text{-CH}_3 \xrightarrow[\text{rt, 3h, 97\%}]{\text{Cl}_2\text{O, CF}_3\text{CO}_2\text{H}} \text{4-O}_2\text{N-C}_6\text{Cl}_4\text{-CH}_3$$

The degree of substitution is controlled by the amount of Cl_2O added and, experimentally, it is easiest to achieve either mono- or perchlorination. Given that the oxide readily dissolves in water to form hypochlorous acid (HOCl) it is suggested that the active species in this reaction is the acylhypochlorite. Interestingly, if the reaction is carried out in the absence of a protic acid (carbon tetrachloride solution at room temperature) the principal products are those of benzylic substitution. Again polysubstitution readily occurs, although in this case the mechanism is believed to be via ClO• hydrogen atom abstraction. Rozen has introduced the use of bromine trifluoride as an efficient brominating reagent for electron deficient arenes. In the presence of a slight excess of bromine this produces a low concentration of bromine monofluoride, an exceptionally powerful electrophilic bromine source. Electron rich substrates give much poorer yields as they intercept the original bromine trifluoride, a source of electrophilic fluorine (S. Rozen and O. Lerman, *J. Org. Chem.,* 1993, **58**, 239).

In the search for selectivity it has been reported that, in the presence of a catalytic amount of diphenyldisulphide and a Lewis acid, sulphuryl chloride chlorinates phenols and similarly substituted arenes in a highly *para-* selective fashion (W. D. Watson, *J. Org. Chem.*, 1985, **50**, 2145). This contrasts with the uncatalysed reaction in which sulphuryl chloride has been reported to react with phenols to give the monochlorinated products as a mixture of *ortho* - and *para-* isomers (D. Masilamani and M. H. Rogic, *J. Org. Chem.*, 1981, **46**, 4486). The reactive species in this later modification appears to be be the complex $Ph_2SCl_2^+\cdots AlCl_3$. This increased selectivity is suggested to be due to an attractive interaction between the phenolic OH and the reagent (H-bonding or coordination). Evidence for this hypothesis comes from a comparison of a series of 2-substituted phenols in which the *para-* to *ortho-* selectivity decreases as the steric bulk of the alkyl group is increased.

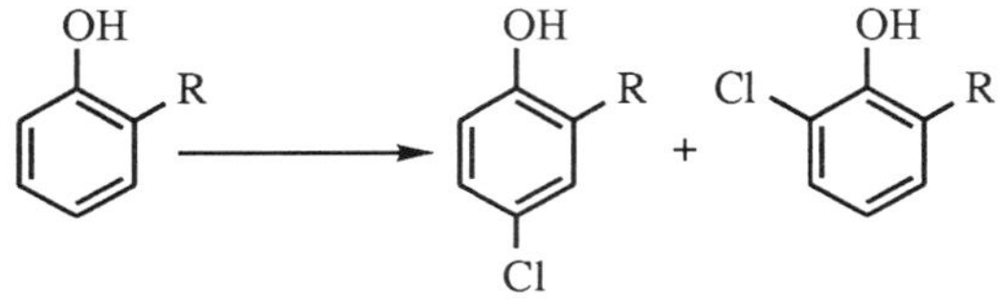

R = CH$_3$ (94%) k_{rel} = 1 $p:o$ = 19.0 : 1
R = ^{t}Bu (80%) k_{rel} = 2 $p:o$ = 4.6 : 1

Unfortunately, with less reactive species the more vigorous conditions required result in decomposition of the complex generating free chlorine and consequently non-selective polysubstitution results. Hydrogen bonding has also been invoked to acount for the selectivity of chlorinations using hexachlorocyclohexadienone (A. Guy, M. Lemaire and J. -P. Guetté, *Tetrahedron*, 1982, **38**, 2339).

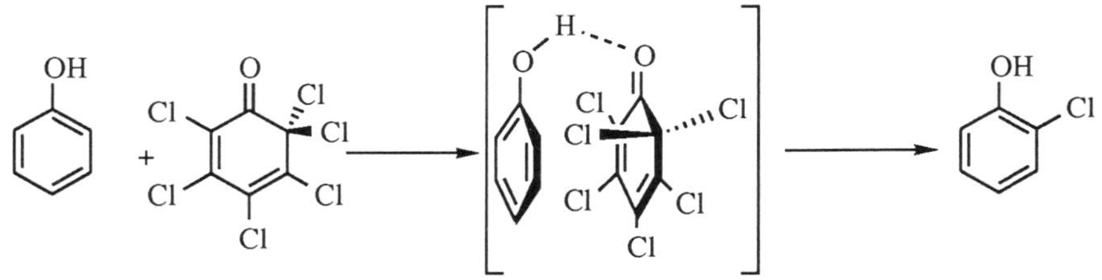

Interestingly, the 2,4 dienone shows enhanced *ortho-* selectivity whilst a *para-* selective reaction occurs with the 2,5 dienone isomer. These reaction are believed to proceed via hydrogen-bond tethered charge-transfer complexes. Consistent with this theory is the fact that using the analogous anisoles, in which hydrogen bonding in the complexes is precluded, exclusive *para*- substituted products result.

Other attempts to control regioselectivity have been developed. These include the development of new reagents, such as morpholine-N-chloride which is reported to monochlorinate electron rich substrates in the *para*-position to the activating group (J. R. Lindsay-Smith, L. C. McKeer and J. M. Taylor, *Org. Synth.*, 1989, **67**, 222) and the application of new technologies. For example, zeolite catalysis has been demonstrated to give products with high *para*- / *ortho*- ratios (T. Nakamura, K. Shinoda, K. Yasuda, *Chem. Lett.*, 1992, 1881 and references therein) whilst carrying out the halogenation reaction on a hplc column has been shown to produce moderate *para*-selectivity (D. A. Jaeger *et al.*, *Tetrahedron Lett.*, 1987, 4805). Alternatively, the use of micelle bound alkylhypochlorite to solubilise and orientate phenolic substrates towards chlorination has been suggested (S. M. Onyiruika and C. J. Suckling, *J. Chem. Soc., Chem. Commun.*, 1982, 833). Although hypohalites are normally relatively poor electrophiles the addition of Lewis acids can generate a highly reactive electrophilic species (V. L. Heasley, *J. Org. Chem.*, 1983, **48**, 3195).

In this respect, the use of quaternary alkyl perbromides has continued to receive much attention with interest focusing on the nature of the active species. Suggestions vary from electrophilic attack using a Br_3^- anion (J. Berthelot *et al., Can. J. Chem.*, 1989, **67**, 2061) to a simple bromonium ion generated by the presence of an additional Lewis acid e.g. $Br^+[BnMe_3]^+$ $[ZnCl_2Br_2]^{2-}$ (S. Kajigaeshi *et al., J. Chem. Soc., Perkin Trans. I*, 1990, 897). An interesting contribution to this area has appeared in which the reaction is carried out in the presence of hydrogen peroxide (J. Dakka and Y. Sasson, *J. Chem. Soc., Chem. Commun.*, 1987, 1421). In contrast to classical direct halogenation, in which half the halogen used is wasted, the hydrogen bromide byproduct is, in this modification, recycled via oxidation by hydrogen peroxide. This simple procedure works perfectly satisfactorily for activated aromatics such as phenol. However, with unactivated nuclei a Lewis acid catalyst is necessary which, in addition to the differing solvent requirements, can also cause problems of hydrogen peroxide decomposition. This can be circumvented by the use of quaternary ammonium salts which act as both Lewis acid catalysts and also phase-transfer reagents for the extraction of hydrogen bromide and hydrogen peroxide into the organic layer.

Unlike most of the procedures outlined above, iodination of arenes is frequently reversible and this can lead to complex mixtures of products and competing reactions. The subject has been reviewed, ('Advances in the Synthesis of Iodoaromatic Compounds' E. B. Merkushev, *Synthesis*, 1988, 923) and it has been suggested from a comparative survey of a number of methods that all proceed via a common reactive intermediate namely the iodonium cation, I^+ (C. Galli, *J. Org. Chem.*, 1991, **56**, 3238). A kinetic study on the iodination of phenols in water (R. H. deRossi and A. V. Veglia,

Tetrahedron Lett., 1986, 5983) reveals that the product distribution seems to vary widely depending on the conditions. Low pH and buffer concentrations favour formation of *ortho* - products, this has been attributed to the different rates of protonation of the cyclohexadienone intermediates. Consequently, in order to obtain satisfactory yields it is necessary to remove the HI produced from the reaction mixture. In attempts to achieve this several strategies have been developed. These include the use of more reactive iodinating reagents such as iodine monofluoride (S. Rozen and D. Zamir, *J. Org. Chem.*, 1990, **55**, 3552), which yields the products typically expected of electrophilic aromatic substitution. Unlike many of the earlier techniques, this is suitable for a wide range of substituted benzenes; the only problem being the occurrence of polysubstitution of activated substrates particulary at higher temperatures and prolonged reaction times required for complete conversion of starting material. Another new reagent system that has been developed is an iodine / bis(trifluoro-acetoxy)iodobenzene combination.

i. $PhI(OCOCF_3)_2 + I_2 \rightleftharpoons PhI + 2CF_3CO_2I$

ii. $CF_3CO_2I + ArH \longrightarrow ArI + CF_3CO_2H$

This is the first reported use of a polyvalent iodine compound in aromatic iodination reactions (E. B. Merkushev, N. D. Simakhina and G. M. Koveshnikova, *Synthesis*, 1980, 486). The reaction is again believed to be a standard electrophilic aromatic substitution in which the active species is trifluoracetylhypoiodite generated *in situ*.

An alternative to the use of these more powerful reagents is to carry out the reaction in the presence of an oxidant. This rapidly reacts with the hydrogen iodide produced, removing it from the equilibrium mixture and in the process regenerating iodine. The use of sodium vanadate under an atmosphere of oxygen as a co-oxidant (A. Shimizu, K. Yamataka and T. Isaya, *Bull. Chem. Soc. Jpn.*, 1985, **58**, 1611) is one such example; whilst the action of an alkali metal iodide in combination with ceric ammonium nitrate (CAN) also proves satisfactory. Although deactivated aromatics do not react, more reactive substrates such as polyalkyl or polyalkoxy arenes only yield the products of aromatic ring substitution (T. Sugiyama, *Bull. Chem. Soc. Jpn.*, 1981, **54**, 2847). This implies that ceric ammonium nitrate, apart from recycling the hydrogen iodide, has a role in producing the reactive iodonium cation in association with a metallo-iodide species. Similar reagent combination systems have subsequently been reported in which CAN has been replaced by either chloramine-T (T. Kometani, D. S. Watt and T. Ji, *Tetrahedron Lett.*, 1985, 2043) or *tert* -butylhypochlorite (T. Kometani, D. S. Watt, T. Ji and T. Fitz, *J. Org. Chem*, 1985, **50**, 5384). These work equally well with yields being in the range 20-97%. Either mono or polysubstitution can be achieved by varying the stoichiometry of the reagents.

One other strategy that has been examined in order to overcome this lack of permanent reactivity has been to incorporate the halogenating reagent onto a solid support. A particularly simple technique involves the adsorption of iodine onto activated chromatography alumina (R. Boothe et al., Tetrahedron Lett., 1986, 2207). This system is believed to act in several ways: the aluminium ions functioning as Lewis acids, polarising and activating the iodine, whilst the oxygen anions are converted to a bound hypoiodite reactive species as well as reacting with the HI generated thus preventing the reverse reaction from occurring.

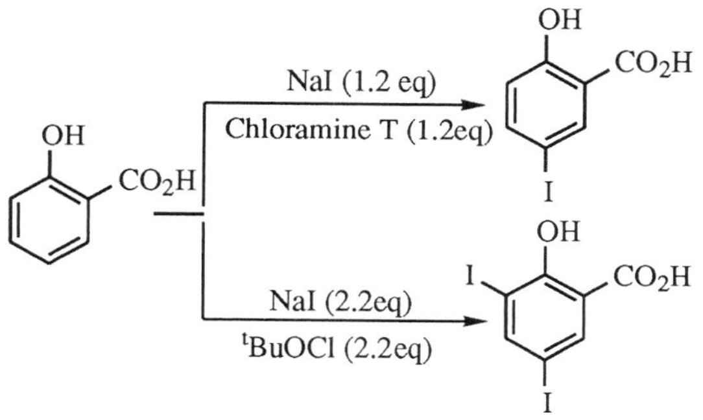

More specific polymers have also been developed. In one such case, iodine dichloride has been bound to the nitrogen of an alternating copolymer of styrene and 4-vinylpyridine. The resulting reagent efficiently iodinates a variety of activated arenes, e.g. 3-aminobenzoic acid is converted into 3-amino-2,4,6-triiodobenzoic acid in 77% yield.

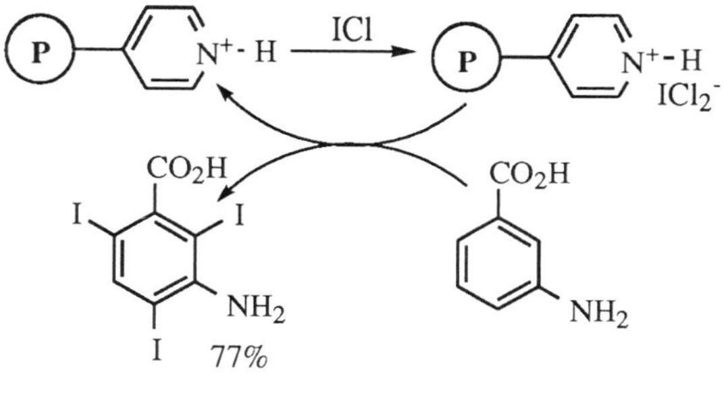

The principal advantage of these reagents is their practical simplicity. Stirring the polymeric complex in a solution of the substrate followed by filtration and washing of the residue affords good yields of the iodinated product and permits recycling of the basic polymer for reactivation. Unfortunately, these reagents only react efficiently with activated aromatics (B. Skeet, P. Zupet and M. Zupan, *J. Chem. Soc., Perkin Trans 1*, 1989, 2279).

(ii) Replacement of the carboxylic acid function by a halogen

Much of the work in this field has been concerned with the development of milder and more efficient versions of the Hunsdiecker reaction. In this respect the treatment of arylradicals generated from the corresponding acylpyridylthione ester with a halogen atom source, e.g. bromotrichloro-methane, is noteworthy (D. H. R. Barton, B. Lacker and S. Z. Zard, *Tetrahedron*, 1987, **43**, 4321). The procedure is particularly useful as it can be carried out in a single step without isolation of the intermediate thione ester. The reaction simply involves the addition of the acid chloride to a mixture of the sodium salt of N-hydroxypyridine, a catalytic amount of AIBN and either carbon tetrachloride, bromotrichloromethane, or iodoform depending upon which haloarene is required.

Alternatives to the use of rhodium complexes in the decarbonylation of aroyl chlorides have been reported. Treatment of acid chlorides with either catalytic palladium on carbon, palladium dichloride, or tetrakistriphenyl-phosphinepalladium(0) produces aryl chlorides in good to moderate yields (J. W. Verbick Jr., B. A. Dellacolette and L. Williams, *Tetrahedron Lett.* 1982, 371).

One further reaction which, although not a decarbonylation process does produce a carbonyl compound as the by-product, is the directed electrophilic fluorination of *para*-substituted benzylic alcohols using cesium fluorosulphate. The substituent must be activating; neither unsubstituted nor trifluoromethyl analogues react (S. Stavber, I. Kosir and M. Zupan, *J. Chem. Soc., Chem. Commun.*, 1992, 274).

Me(O)CHN—⟨ ⟩—CH(OH)— $\xrightarrow[\text{MeCN, rt}]{\text{CsSO}_4\text{F}}$ Me(O)CHN—⟨ ⟩—F + CH$_3$CHO

(iii) Replacement of an amino group by a halogen

Despite the requirement for the prior regiospecific synthesis of the starting aryl amine the Sandmeyer reaction continues to be used in synthesis [see for example D. L. J. Clive, A. G. Angoh and S. M. Bennet, *J. Org. Chem.*, 1987, **51**, 1339; and K. L. Rinehart *et al.*, *J. Am. Chem. Soc.*, 1987, **109**, 3378]. In the search for greater control recent developments have resulted in the preparation of stable diazonium ion equivalents which may be isolated and purified prior to reaction with the halogen source. Treatment of an arylamine with nitrous acid and pyrrolidine produces the stable triazene (**6**).

PhNH$_2$ + HNO$_2$ + pyrrolidine(NH) → Ph-N=N-N(pyrrolidine) →[KI*, H$^+$] Ph-I*

(6)

After purification, this is subsequently converted to the desired aryl halide by treatment with an appropriate alkali metal (usually potassium) halide in the presence of an acid. The isolation of the triazene intermediate renders this a particularly attractive strategy for the preparation of radiolabelled iodides (N. I. Foster *et al.*, *Synthesis*, **1980**, 572).

Triazene intermediates are also postulated (H. Ku and J. R. Barrio, *J. Org. Chem.*, 1981, **46**, 5239) to form in a procedure in which arylamines are nitrosated in the presence of a secondary amine base and trimethylsilylchloride. Subsequent nucleophilic displacement then provides the desired aryl halide.

Ph-NH$_2$ →[HNO$_2$, Et$_2$NH / Me$_3$SiCl] Et$_2$(Et)N$^+$(Me$_3$Si)-N=N-Ph →[LiBr or NaI] Ph-X

X = Br, I

Other workers in this field have developed new methods to introduce the halogen atom in a single step obviating the need to isolate the intermediate diazonium salt (S. Oae, K. Shinhama and Y. H. Kim, *Bull. Chem. Soc. Jpn.*, 1980, **53**, 1065). The introduction of chlorine or bromine is achieved by treatment of the arylamine with *tert*-butylthionitrite in the presence of either copper(I) chloride or bromide. The same workers have subsequently reported that changing the diazonium generating species to *tert*-butylthionitrate additionally permits the synthesis of the analogous iodo compounds (S. Oae, K. Shinhama and Y. H. Kim, *Bull. Chem. Soc. Jpn.*, 1980, **53**, 2023). In this latter variation the halogen source can be either carbon tetrachloride, bromoform, or molecular iodine depending on which haloarene is required. Under the reaction conditions the intermediate diazonium species, $ArN_2^+OH^-$, decomposes to generate an aryl radical which then abstracts a halogen atom from one of the halogen sources listed above. Similarly, the use of alkylnitrites to achieve the diazotisation in the presence of HF provides for an efficient 'one pot' conversion of arylamines to the corresponding fluoroarenes (Y. Yoda, H. Hokonohara and T. Yamaguchi, *Chem. Abstr.*, 1990, **112**, 197812).

Other enhancements to this transformation include the addition of various additives to increase the boiling point of the reaction mixture and hence accelerate the dediazonation step e.g. water and alkali metal fluorides

(*Chem. Abstr.*, 1985, **102**, 95,343) or various organic bases (T. Fukuhara *et al.*, *Synth. Commun.*, 1991, **51**, 299). With this latter group of additives, the ratio of base to hydrogen fluoride has been found to be critical for high yields (T. Fukuhara *et al.*, *Bull. Chem. Soc. Jpn.*, 1990, **63**, 2058). Yoneda has also reported a modified procedure involving photochemical dediazoniation which has the added advantage of avoiding the need to isolate the intermediate diazonium tetrafluoroborate salt (N. Yoneda *et al.*, *Synth. Commun.*, 1989, **19**, 865). In contrast to the thermal reaction, the photochemical dediazoniation of arene diazonium salts having amino, halogen, methoxy, or nitro groups in the 2-position proceeds in good yield with the absence of significant by-products. Finally, the application of high pressure is reported to improve both the yield and scope of the process (M. S. Howarth and D. M. Tomkinson, EP. 258,985, *Chem. Abstr.*, 1989, **109**, 229723e).

The Balz-Schiemann reaction and variants thereof continue to be used for the incorporation of fluorine into aromatic substrates. However, the inefficient use of the fluorine component coupled with a high temperature requirement have made alternative procedures more attractive. One significant advance has been the introduction of NOBF$_4$ which removes the need to isolate the intermediate diazonium salt. However, the decomposition step still requires the use of relatively high boiling solvents such as 2-dichlorobenzene (D. J. Milner, *Synth. Commun.*, 1992, **22**, 73). A recent technological innovation that permits the reaction to be achieved under much milder conditions is the use of ultrasound (A. Muller *et al.*, *Zeitschrift Chem.*, 1986, **26**, 169).

(iv) Replacement of nitro, chlorosulphonyl and substituted amino groups by a halogen

Heating nitro- or fluorosulphonylarenes, containing an electron withdrawing group in the *meta* position,with mixture of potassium fluoride and tetraphenylphosphonium bromide at 150-200°C affords the products of fluorodenitration or fluorodesulphonation respectively (N. Yazawa *et al.*, *Chemistry Lett.*, 1989, 2213). Clark and coworkers have also reported a similar procedure, see below. which represented the first publication dealing with the introduction of fluorine into an aromatic ring in this manner. However, these and the other related fluorodenitration procedures reported since, are frequently accompanied by other products, notably phenols and diaryl ethers. These are believed to be due to reactions of the displaced nitrite ion. Consequently Scorrano and co-workers have introduced the use of nitrite ion scavengers such as phthaloyl difluoride or tetrafluorophthaloyl difluoride and these provide for an efficient procedure of general applicability (M. Maggini *et al., J. Org. Chem.*, 1991, **56**, 6406). An alternative solution to this problem is the use of anhydrous tetramethylammonium fluoride as reported by Boechat and Clark where the large cation inhibits the initial steps in the side reactions [nitrite attack and decomposition to the oxide] (N. Boechat and J. H. Clark, *J. Chem. Soc., Chem. Commun.*, 1993, 921).

(v) Replacement of one halogen by another

Owing to difficulties in the direct preparation of fluoro and iodo compounds their production from the more readily available chloro- and bromoarenes continues to be a common option. Without exception, the preparation of the fluorinated compounds are all nucleophilic aromatic substitution processes and consequently require an appropriate electron withdrawing group attached to the aromatic ring in either the *ortho-* or the *para-* position. The current emphasis is concerned with development of more convenient, milder and more selective reagents. Treatment of aryl chlorides with the salt, tetraphenylphosphonium dihydrogenfluoride, leads to a smooth displacement reaction (S. J. Brown and J. H. Clark, *J. Fluorine Chem.*, 1985, **30**, 251). This reagent has also been shown to be capable of displacing a nitro group (see above). A procedure utilising an analogous ammonium fluoride and HMPA as the solvent has also been shown to be a viable process (P. Bosch *et al., Tetrahedron Lett.*, 1987, 4733). A recent modification is to prepare the active species, tetraphenylphosphonium fluoride, *in situ* via the displacement of bromide from tetraphenylphosphonium bromide with potassium fluoride (J. H. Clark and D. J. MacQuarrie, *Tetrahedron Lett.*, 1987, 111). The addition of 18-crown-6 to the reaction mixture enhances the reactivity and extends the scope to include certain benzaldehyde (Y. Yoshida and Y. Kimura, *Chemistry Lett.*, 1988, 1355) and benzonitrile derivatives (N. Yazawa *et al., Chemistry Lett.*, 1989, 2213; S. Kumai and R. Seki, *Asahi Garasu Kenkyu Hokoku*, 1989, **39**, 317), although the temperature requirement (~200°C) is still too high for sensitive substrates. In a search to achieve substitution under milder conditions Yoshioka and coworkers have prepared tetrabutylphosphonium hydrogendifluoride [Bu$_4$PF•HF] and dihydrogentrifluoride [Bu$_4$PF•(HF)$_2$] (Y. Uchibori *et al., Synlett*, 1992, 345). These reagents react with a variety of halogenated aromatic, substrates in relatively low boiling non polar solvents, such as toluene and xylene with reaction temperatures ranging from 80-140°C to produce the fluorinated products in good yields (75->95%). The hydrogendifluoride is also an efficient reagent for fluoro-denitration reacting competitively with 2-chloro-6-nitrobenzonitrile in xylene at 80°C to produce 2-chloro-6-fluorobenzonitrile in 94% yield.

In contrast the dihydrogentrifluoride reagent will not displace the nitro group and is, in general, less reactive requiring more vigorous conditions.

Interestingly, the monofluoride [Bu$_4$PF], although highly active in the fluorination of aliphatic substrates shows no reactivity towards arenes. No explanation has been proposed to account for this observation.

The corresponding iodides are considerably less reactive than the fluorides. Consequently, it is difficult to obtain aryl iodides by such direct displacement methods. Furthermore the products are often labile under the reaction conditions and, accordingly, alternative strategies for their preparation have been developed. For example, electrolysis of a mixture of an arylbromide, potassium iodide and nickel(II)bromide affords the corresponding aryl iodide in good yield (G. Meyer, Y. Rollin and J. Perichon, *Tetrahedron Lett.*, 1986, 3497). The reaction involves an initial electrochemical reduction giving Ni(0) which then oxidatively inserts into the Ar-Br bond. Subsequent halogen exchange then affords the desired product. In a related process the aryl bromide is stirred in a suspension of potassium iodide, iodine and metallic nickel in anhydrous DMF (S. H. Yang, C. S. Li and C. H. Cheng, *J. Org. Chem.*, 1987, **52**, 691). It is postulated that the reaction involves a nucleophilic displacement of a π-adsorbed aromatic halides bound to the nickel surface. Interestingly, unlike most nucleophilic aromatic displacement reactions there is a decrease in rate with increasing substitution by electron withdrawing groups. This is consistent with the mechanistic picture in which π-absorption onto the metal surface will be slower and weaker with electron poor arenes. The use of supported reagents has also been demonstrated by Clark and Jones (J. H. Clark and C. W. Jones, *J. Chem. Soc., Chem. Commun.*, 1987, 1409) who treated a variety of aryl chlorides and bromides with cuprous iodide adsorbed onto either alumina or activated carbon to provide the desired iodides in good yields.

Preparation of chloro and bromoarenes is also possible by halide displacement. For example, treatment of bromoaromatics with aqueous hypochlorite in the presence of a phase transfer catalyst provides good yields of the corresponding chloroaromatic (J. T. Arnold *et al.*, *J. Org. Chem.*, 1992, **57**, 391). Surprisingly, subjecting iodobenzene to the same reaction conditions produces only modest yields of chlorobenzene. However, under carefully monitored conditions, pH=8-9 and prolonged reaction times, oxidation to the iodylbenzene, followed by chloride displacement permits the efficient replacement of iodine by chlorine (T. O. Bayraktaroglu *et al.*, *J. Org. Chem.*, 1993, **58**, 1264).

(b) Reactions and Properties

(i) Reactions with Metals

The most frequent reaction in this category is the metal halogen exchange with either lithium or magnesium. However, owing to the basicity of the resulting organometallic reagents, other metals, notably zinc, have been explored. Organozincs are normally prepared from aryl lithiums via transmetalation, although direct reaction of aryl iodides with metallic zinc occurs for aryl iodides bearing electron withdrawing groups (K. Takagi, N. Hayama and S. Inokawa, *Bull. Chem. Soc. Jpn.*, **1980**, *53*, 3691).

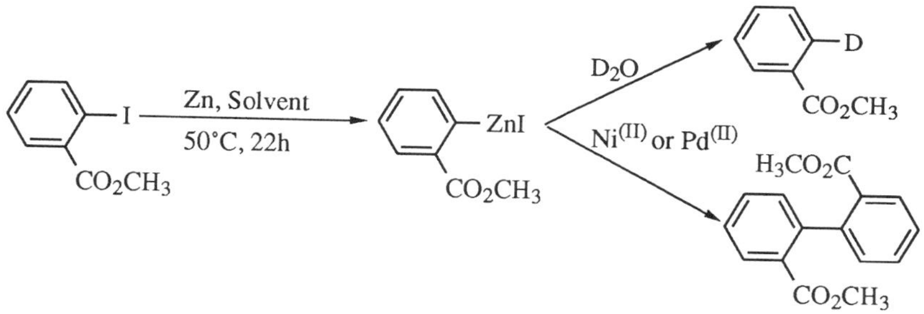

Other transition metals, such as nickel and palladium, have also been used in coupling reactions of haloarenes. These offer the advantage of circumventing the vigorous conditions, notably the high temperature requirement, of the traditional copper based Ullmann reaction for the synthesis of biaryls via the cross coupling of aryl halides (for a recent review on this aspect of target synthesis see G. Bringmann, R. Walter and R. Weinrich, *Angew. Chem. Int. Ed. Engl.*, **1990**, *29*, 977). Much of the early work has been reviewed (P. W. Jolly "Nickel Catalysed Coupling of Organic Halides and Related Reactions" in Comprehensive Organometallic Chemistry, Eds G. Wilkinson, F. G. A. Stone and E. W. Abel, Pergamon Press, Oxford 1982, vol. 8 p713). More recent work focuses on increasing the efficiency and range of the reaction and also elucidation of the mechanism.

The copper catalysed reaction is limited to aryl iodides and in some cases aryl bromides, with chlorides and fluorides being progressively more inert. The same order of reactivity applies to the nickel catalysed reaction with only aryl fluorides being completely unreactive (K. Tagaki, N. Hayama and S. Inokawa, *Chem. Lett.*, 1979, 917).

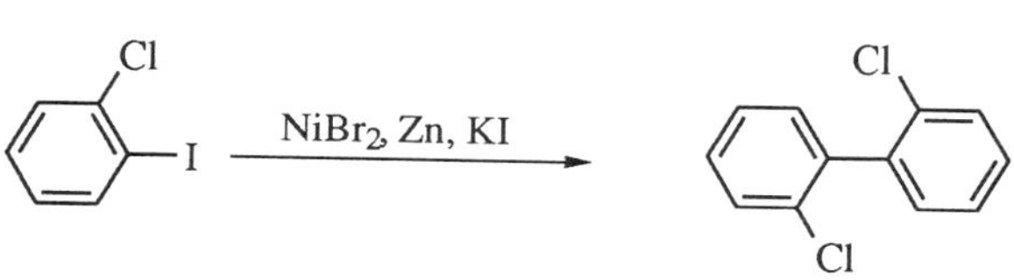

The degree of reactivity can vary dramatically with the nature of the complex. For example; in an early example, Semmelhack and co-workers, using a previously synthesised Ni(0) species, in the form of air sensitive bis(1,5-cyclo-octadienyl)nickel(0), could only achieve the cross coupling of chlorobenzene in 14% yield. In contrast Ni(0) generated *in situ* by a zinc initiated reduction allows the isolation of biphenyl in 98% yield (I. Colon and D. R. Kelsey, *J. Org. Chem.,* 1986, **51**, 2627).

$$\text{Ph-Cl} \xrightarrow[\text{DMAC, N}_2, 80°C]{\text{NiCl}_2, \text{PPh}_3} \text{Ph-Ph}$$
$$>98\%$$

One further example is the reaction of the nickel reagents generated from sodium hydride (60mmol), nickel (II) acetate (40 mmol) and 2,2'-bipyridine (20mmol) as devised by Caubere *et al.* These, unusually, react far more efficiently with aryl chlorides than with the equivalent bromides or iodides. The various products obtained implicate a radical pathway but decisive mechanistic evidence has not been forthcoming (R. Vanderesse, J. J. Brunet and P. Caubere, *J. Orgmet. Chem.,* 1984, **264**, 263; and P. Caubere *et al.,* 1987, **54**, 4840).

Mechanistically these process all commence with the oxidative addition of the aryl halide to the metal species and of the three possibilities for this initial step electron transfer is the favoured pathway (T. T. Tsou and J. K. Kochi, *J. Am. Chem. Soc.,* 1979, **101**, 6319). The subsequent steps of the cross coupling depend very much on the precise reaction conditions. Kochi and Tsou suggest that biaryl formation is a radical chain process in which Ni(II) and Ni(III) species are reactive intermediates. (T. T. Tsou and J. K. Kochi, *J. Am. Chem. Soc.,* 1979, **101**, 7547) However, this has been disputed by Semmelhack and colleagues (M. F. Semmelhack *et al., J. Am. Chem. Soc.,* 1981, **103**, 6460). A more recent conclusion by Rieke and co-workers, based on studies using highly reactive nickel powder generated by the reduction of nickel halides with lithium and naphthalene, invokes the disproportionation of the aryl nickel(II) halide as outlined in equations (1)-(3).

$$[Ni] + ArX + L_n \longrightarrow ArNi^{II}XL_n \quad (1)$$

$$2\ ArNi^{II}XL_n \longrightarrow Ar_2Ni^{II}L_n + Ni^{II}X_2 + L_n \quad (2)$$

$$Ar_2Ni^{II}L_n \longrightarrow ArAr + [Ni] + L_n \quad (3)$$

Evidence for this proposal is found in the isolation of the intermediate nickel(II) species, (**7**) and (**8**) as their phosphine complexes (H. Matsumoto, S.-i. Inaba and R. D. Rieke, *J. Org. Chem.,* **1983**, *48*, 840). In this case, the presence of the electron withdrawing perfluorosubstituted arene prevents the final reductive elimination steps allowing the isolation of the these products.

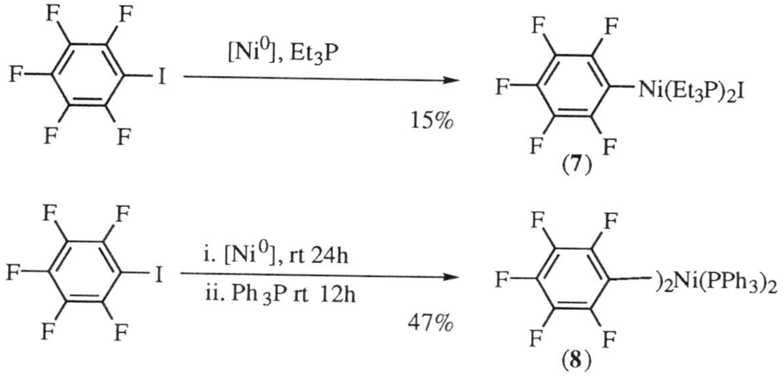

[Scheme showing pentafluoroiodobenzene + [Ni⁰], Et₃P → C₆F₅-Ni(Et₃P)₂I (7), 15%]

[Scheme showing pentafluoroiodobenzene: i. [Ni⁰], rt 24h; ii. Ph₃P rt 12h → (C₆F₅-)₂Ni(PPh₃)₂ (8), 47%]

Other polyhalogenated species have also been studied. For example 1,2-dibromobenzene on treatment with activated nickel powder affords moderate yields of the cyclic oligomers (**9**) and (**10**). Similar treatment of other 1,3-dibromobenzenes provide varying amounts of the linear polyphenyls (**11**) (C. S. Chao, C. H. Cheng and C. T. Chang, *J. Org. Chem.*, 1983, **48**, 4904; Z.-h. Zhou and T. Yamamoto, *J. Orgmet. Chem.*, 1991, **414**, 119).

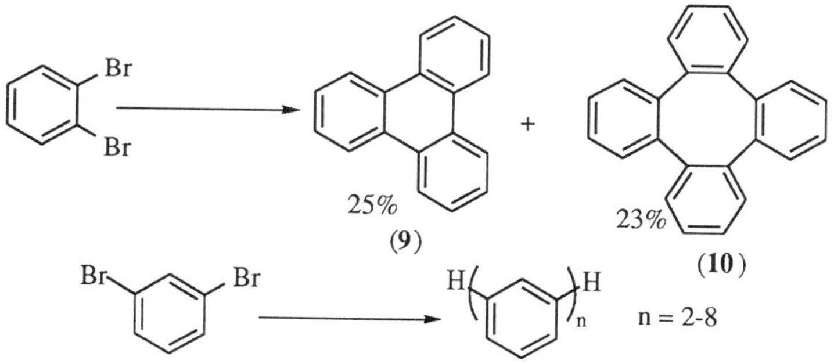

[Scheme: 1,2-dibromobenzene → triphenylene (9) 25% + tetraphenylene (10) 23%]

[Scheme: 1,3-dibromobenzene → H-(C₆H₄)ₙ-H (11), n = 2-8]

Similar products can be obtained from the equivalent aryl chlorides, although in this case the presence of potassium iodide is essential and this process is believed to proceed via a facile nickel promoted halogen exchange. Other transition metals have also been used to insert into the aryl halogen bond. Photolysis of an aryl halide with dicobaltoctacarbonyl, under an atmosphere of carbon monoxide and in the presence of tetrabutylammonium bromide as a phase-transfer catalysis leads to carbonylation of the aromatic nucleus (J. J. Brunet, C. Sidot and P. Caubére, *J. Org. Chem.*, 1983, **48**,

1166). The reaction is believed to proceed via photostimulation of a charge-transfer complex between the halide and metal carbonyl species. This excited intermediate then decays to give an aryl free radical closely associated with a cobalt carbonyl radical anion. An intracage $S_{RN}1$ reaction between these two species then generates an aryl cobalt intermediate, $ArCo(CO)_4$, which undergoes carbon monoxide insertion to afford a cobalt acyl derivative (equations 4-8).

$$[ArX, Co(CO)_4^-, M^+] \longrightarrow [ArX, Co(CO)_4^-, M^+]^* \quad (4)$$
c-t complex

$$[ArX, Co(CO)_4^-, M^+]^* \longrightarrow [Ar\bullet \bullet Co(CO)_4, X^-M^+] \quad (5)$$

$$[Ar\bullet \bullet Co(CO)_4, X^-M^+] \xrightarrow{\text{Cage constrained } S_{RN}1} Ar\text{-}Co(CO)_4 \quad (6)$$

$$Ar\text{-}Co(CO)_4 \longrightarrow Ar(CO)Co(CO)_4 \quad (7)$$

$$ArCOCo(CO)_4 \xrightarrow{\text{NuH}} ArCONu \quad (8)$$

Subsequent trapping with a nucleophile then yields the observed carbonylated products. The fact that the whole process takes place within an attractive charge transfer complex (cage) is suggested as an explanation for the extremely high ratios observed of CO insertion to reduction of the intermediate radical. In the case in which the starting aryl halide has a pendant amino or alcohol group this reaction provides easy access to the corresponding lactam or lactone.

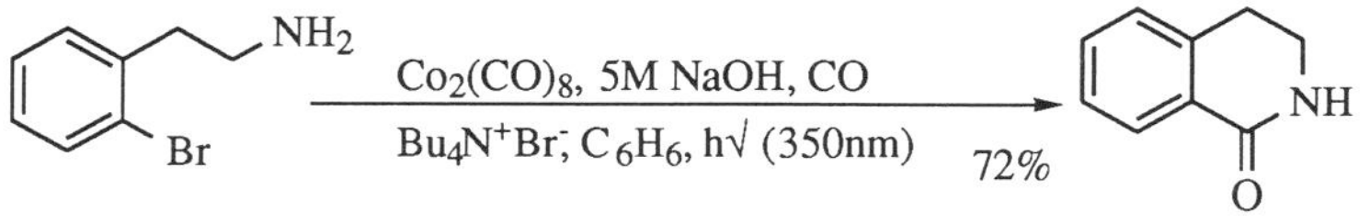

Treatment of an aryl halide with various cobalt(I) complexes [NaCo(salen) or NaCo(dmgH)$_2$] leads to a very rapid insertion, probably by a single electron-transfer mechanism, to produce the aryl cobalt(III) complexes [cobaloximes and cobalamines respectively]. These are highly reactive, readily undergoing atom transfer radical reactions as in the following example drawn from the work of Pattenden (V. F. Patel and G. Pattenden, *Tetrahedron Lett.*, 1987, 1451).

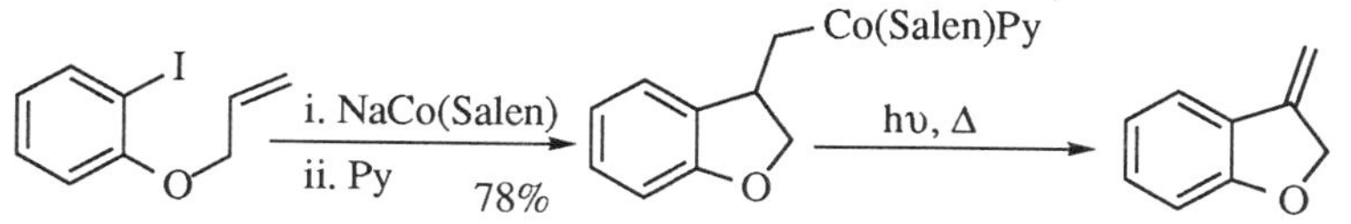

The increasing use of the lanthanide metals has extended into this area. The reactions of the series Yb, Sm, Eu, Ce have been examined with a range of aryl halides. The reactions and their products all bear analogy to the equivalent Grignard processes (A. B. Sigalov *et al.*, *Izv. Akad. Nauk. SSSR.*, *Ser. Khim.*, 1983, 2615; L. F. Rybakova *et al.*, *ibid.*, 1984, 1413; and O. P. Syutkina *et al.*, *J. Orgmet. Chem.*, 1985, **280**, C67).

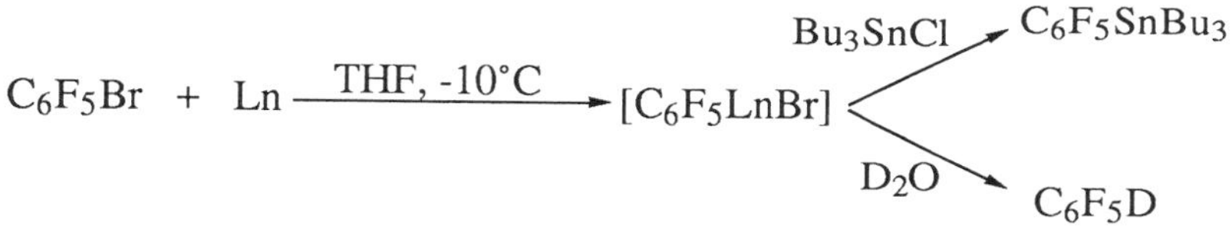

The reactions of halogenated benzenes with aryl Grignards has been revisited (K. Harada, H. Hart and C. F. J. Du, *J. Org. Chem.*, 1985, 52, 5524). The ease of metal halogen exchange increases with the extent of halogenation as the -I effect more efficiently stabilises the resultant anion. The anions then undergo well documented elimination to the corresponding arynes which on subsequent reaction with further equivalents of the Grignard reagent give rise to the tetraphenylarenes in good yield (65%).

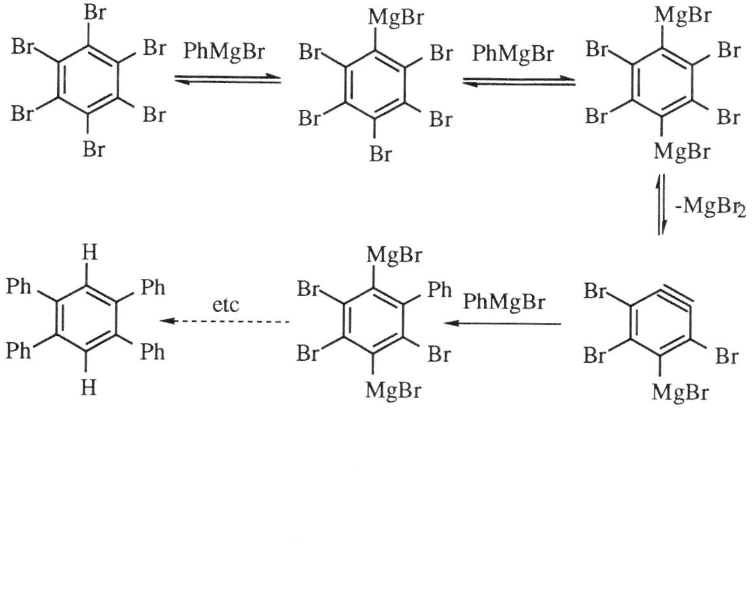

Notably, complete regioselectivity is observed in the reaction between perbromobenzene and phenylmagnesiumbromide with only 1,2,4,5-tetraphenylbenzene being formed. Again chloro and fluoro species are inert, 1,4-dibromo-2,3-5,6-tetrachlorobenzene undergoing simple reductive dehalogenation.

$$\underset{\text{Br}}{\underset{\text{Cl}}{\text{Cl}}}\underset{\text{Br}}{\overset{\text{Br}}{\text{C}_6\text{Cl}_4}}\underset{\text{Cl}}{\text{Cl}} \xrightarrow{\text{PhMgBr}} \underset{\text{H}}{\underset{\text{Cl}}{\text{Cl}}}\underset{\text{H}}{\overset{\text{H}}{\text{C}_6\text{Cl}_4}}\underset{\text{Cl}}{\text{Cl}}$$

High regioselectivity is also observed in the deprotonation of monofluorinated benzenes using lithium dialkylamide bases. Under carefully controlled conditions, a single fluorine atom can be one of the most selective *ortho* directing substituents available (A. J. Bridges *et al.*, *Tetrahedron Lett.*, 1992, 7495). Similar directing effects have also been noted for the corresponding chloroarenes (M. Iwao, *J. Org. Chem.*, 1990, **55**, 3622). Similar o-haloarylorganometallic species can be obtained through the selective metallation of 1,2-dihalogenoarenes at very low temperatures. The requirement for such low temperatures can be obviated through the use of highly activated copper (G. W. Ebert *et al.*, *Tetrahedron Lett.*, 1993, 2279).

(ii) Dehalogenation

Much of the earlier work in this area has been reviewed (A. R. Pinder, *Synthesis*, 1980, 425). Many recent developments involve the use of transition metal complexes. These reactions can broadly be divided into two classes; metal hydride reductions and catalysed hydrogenolyses. In the former class a variety of different transition metals have been studied, the most common being iron, cobalt, nickel and titanium. In a survey of the various possible combinations of di- and tri-halide salts with lithium aluminium hydride (1 : 1 ratio) Ashby and Lin found that the nickel (II) species produced the most reactive reagent (E. C. Ashby and J. J. Lin, *J. Org. Chem.*, 1978, **43**, 1263). However, Dams *et al.* have reported an equally efficient process using a low valent titanium species similarly generated by reduction of titanium tri- and / or tetrachloride with lithium aluminium hydride (R. Dams, M. Malinowski and H. J. Giese, *Recl. Trav. Chim. Pays-Bas*, 1982, **101**, 112) The mechanisms of these transition metal assisted hydride reductions has been reviewed (R. C. Wade, *J. Mol. Cat.*, 1983, **18**, 273). More recently, Ganem and co-workers using cobaltous chloride as the co-reductant deduced that the key step was a halogen atom transfer from the organic halide to an aryl radical bound to the aluminium surface (J. O. Osby, S. W. Heinzman and B. Ganem, *J. Am. Chem. Soc.*, 1986, **108**, 67).

The use of lithium aluminium hydride is frequently unsuitable for more functionalised aryl halides and alternative reducing systems have been examined. Sodium borohydride in combination with bistriphenylphosphinenickel(II) chloride proves to be a satisfactory alternative provided that the reaction is run in DMF. This reducing mixture does, however, require a somewhat higher reaction temperature (~70°C) than the aluminium based alternative (S. T. Lin and J. A. Roth, *J. Org. Chem.*, 1979, **44**, 309). Under these more vigorous conditions, problems can arise due to reaction of the borohydride with the DMF solvent and this is particulary the case for titanium based reagents. These difficulties can be avoided through the use of dimethylacetamide as the solvent, although this requires longer reaction times. Interestingly this change of solvents appears to result in a change of mechanism from nucleophilic hydride displacement to a single electron transfer process (Y. Liu and J. Schwartz, *J. Org. Chem.*, 1994, **59**, 5005). Alternatively, other hydride sources may be employed, for example the use of alkoxyborohydrides enable the efficient reduction of pentachlorobenzene in THF (S.-M. H. Tabaei and C. U. Pittman, Jr., *Tetrahedron Lett.*, 1993, 3263). Other co-reductants have also been developed, for example metallic zinc has been used in conjunction with triphenylphosphine and sodium iodide (I. Colon, *J. Am. Chem. Soc.*, 1982, **47**, 2622). The last reagent is believed to have a role in the reduction of the nickel salts although the generation of a more reactive intermediate aryl iodide cannot be ruled out. As with all the reagents discussed so far, the order of reactivity is ArI > ArBr > ArCl with aryl fluorides being unreactive. This reagent will, however, reduce polychlorinated aromatics.

2,4,5-trichlorobenzene → benzene

NiCl$_2$, Ph$_3$P, NaI, Zn
H$_2$O, DMF, 50°C, 4h

> 90%

A complex reducing agent generated from sodium hydride, sodium alkoxide and a transition metal salt has been demonstrated to reduce all the aryl halides, albeit with the reduction of aryl fluoride species being significantly slower (G. Guillaumet, L. Mordenti and P. Caubére, *J. Organomet. Chem.*, 1975, **92**, 43).

X—C$_6$H$_4$—OMe → H—C$_6$H$_4$—OMe

NaH - ^{t}AmONa - Ni(OAc)$_2$ (40:20:10)
65°C, THF

X = F 12h 97%
X = Br 1h 92%

By varying the nature of the transition metal, it is possible to tune the reactivity of these systems. Using cuprous salts reduction of only iodo- and bromobenzenes is possible, with cobalt all but fluorobenzenes are labile whilst with certain nickel based reagents all the halogenobenzenes are reduced (see above). With these differences in reactivity, it is possible to observe some selectivity in the reduction of polyhalogenated benzenes,. For example, 2-chlorobromobenzene is cleanly converted to chlorobenzene. However, with 1,2-dichlorobenzene the reduction of the second chlorine bond is competitive and a complex mixture of products and starting material results.

$$\text{1,2-Cl}_2\text{C}_6\text{H}_4 \longrightarrow \text{PhCl (42\%)} + \text{PhH (45\%)} + \text{1,2-Cl}_2\text{C}_6\text{H}_4 (10\%)$$

In the majority of these cases the active species is believed to be an alkoxy bridged Ni species containing metal hydride units. However, as indicated above, an electron transfer mechanism cannot be discounted.

Finally under the classification of metal hydride reductions, it has been observed that the use of ultrasound presents some advantages particularly in the case of electron rich aryl halides. Reduction of 4-bromoanisole with lithium aluminiumhydride in THF afforded only a 17% yield of anisole. Repetition of the reaction in DME with the reaction vessel immersed in a standard laboratory ultrasound cleaning bath afforded the desired product in a much improved 70% yield. The choice of the solvent appears to be important although the reason why remains to be elucidated (B. H. Han and P. Boudjouk, *Tetrahedron Lett.*, 1982, 1643).

The alternative to hydride based reagents is transition metal catalysed hydrogenolysis and refinements have been forthcoming in this area. Treatment of aryl halides (not fluorides) with formic acid and a catalytic quantity of palladium on charcoal permits a smooth reductive dehalogenation; aromatic halides being reduced in preference to their aliphatic counterparts (P. N. Pandey and M. L. Purkayastha, *Synthesis*, 1982, 876).

$$\text{Cl-C}_6\text{H}_4\text{-CH}_2\text{CH}_2\text{Cl} \xrightarrow[\Delta, 5h]{\text{Pd-C, HCO}_2\text{H, DMF}} \text{C}_6\text{H}_5\text{-CH}_2\text{CH}_2\text{Cl} \quad (85\%)$$

In a move away from heterogeneous catalysis Helquist and co-workers reported that treatment of an aryl bromide with sodium formate in DMF at 100°C in the presence of 5mol% of tetrakistriphenylphosphinepalladium(0) smoothly afforded the hydrogenolysed product. The reaction proceeds via oxidative addition of the aryl bromide to the palladium complex. Metathetical substitution of bromide by formate followed by decarboxylation of the latter leads to an aryl metal hydride. Subsequent reductive elimination affords the desired product (P. Helquist, *Tetrahedron Lett.*, 1978, 1913).

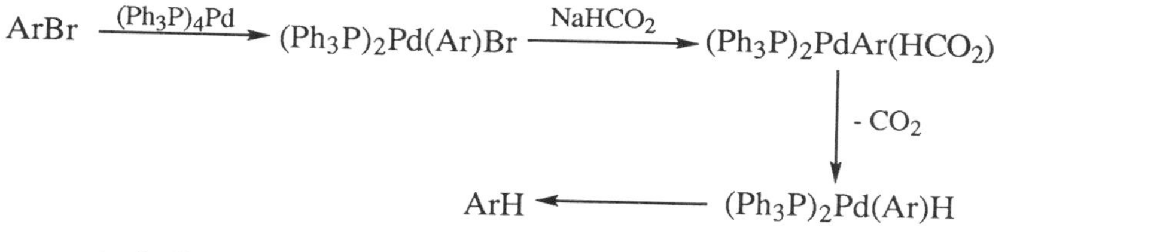

$$ArBr \xrightarrow{(Ph_3P)_4Pd} (Ph_3P)_2Pd(Ar)Br \xrightarrow{NaHCO_2} (Ph_3P)_2PdAr(HCO_2)$$

$$\downarrow -CO_2$$

$$ArH \longleftarrow (Ph_3P)_2Pd(Ar)H$$

A similar result can be achieved using methoxide as the hydrogen source. This is converted *in situ* to formaldehyde and hence to formate (A. Zask and P. Helquist, *J. Org. Chem.*, 1978, **43**, 1619). Polyhalogenated species, e.g. pentahalophenyl ethers, can also be reduced using this methodology (T. Okamoto and S. Oka, *Bull. Chem. Soc. Jpn.*, 1981, **54**, 1265).

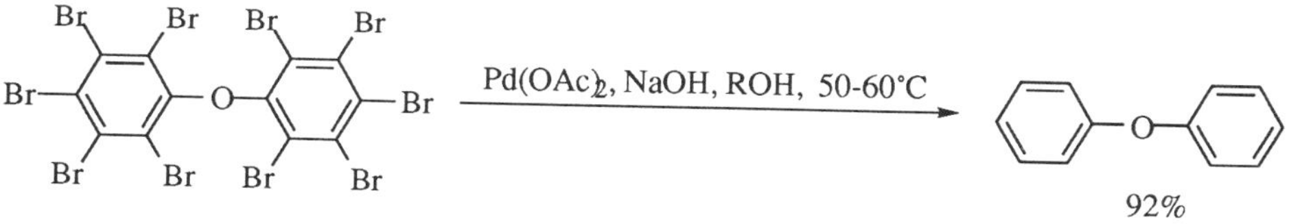

(Br₈-diphenyl ether) $\xrightarrow{Pd(OAc)_2, NaOH, ROH, 50-60°C}$ Ph-O-Ph   92%

Other hydrogen transfer reagents have also been used including sodium hypophosphite in conjunction with palladium on carbon (S. K. Boyer *et al.*, *J. Org. Chem.*, 1985, **50**, 3409; C. A. Marques, M. Selva and P. Tundo, *J. Chem. Soc. Perkin Trans. I*, 1993, 529); and poly(methyl)hydrosiloxane (I. Pri-Bar and O. Buckman, *J. Org. Chem.*, 1986, **51**, 734). The latter is a soluble alternative to sodium formate, compatible with a large range of functional groups, which when used with palladium(0) catalysts, e.g. (Ph₃P)₄Pd, provides for a very efficient and selective reduction.

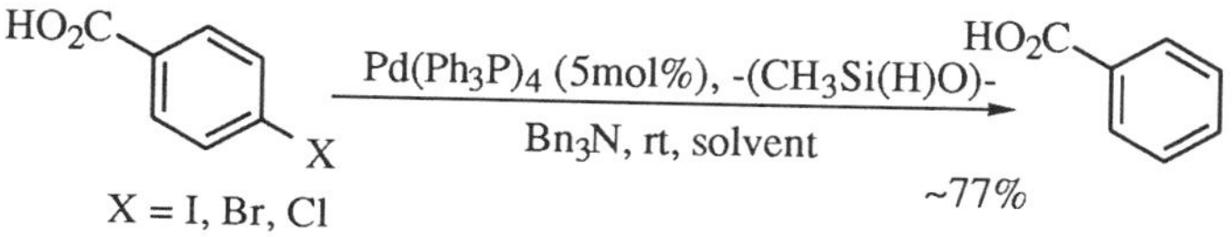

$$HO_2C\text{-}C_6H_4\text{-}X \xrightarrow[Bn_3N, rt, solvent]{Pd(Ph_3P)_4 (5mol\%), -(CH_3Si(H)O)-} HO_2C\text{-}C_6H_5$$

X = I, Br, Cl   ~77%

An NADP(H) mimic has also been employed (S. Yasui *et al.*, *J. Org. Chem.*, 1985, **50**, 3283). The combination of the dihydronicotinamide

(BNAH) with either Wilkinsons catalyst [RhCl(PPh$_3$)$_3$] or palladium (II) acetate creates an efficient reducing system. The latter catalyst is somewhat less selective leading in certain cases to biaryl products.

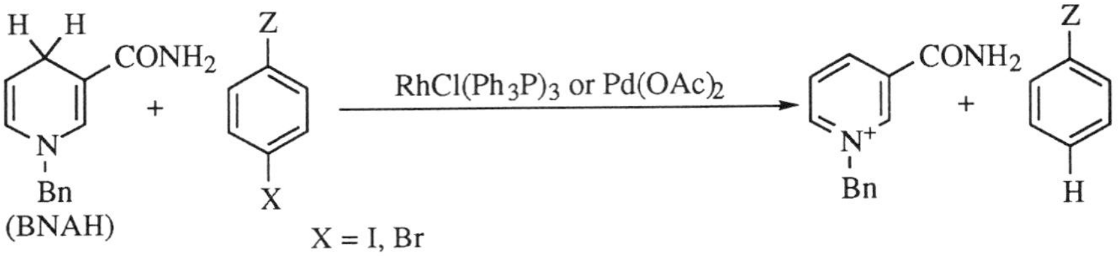

$$\text{BNAH} + \text{ArX} \xrightarrow{\text{RhCl(Ph}_3\text{P)}_3 \text{ or Pd(OAc)}_2} \text{Bn-pyridinium-CONH}_2 + \text{ArH}$$

X = I, Br
Z = NO$_2$, COMe, CO$_2$Me, NH$_2$, CH$_3$, Br

Electron deficient arenes react more readily. Thus, when Z = NO$_2$ 100% conversion occurs within 3h, whereas when Z = NH$_2$ only 72% conversion is achieved after 23h. This is consistent with a simple oxidative addition of the aryl halide to the metal catalyst. It is then postulated that the remaining steps are (i) electron transfer from the dihydronicotinamide, (ii) free radical formation via homolytic cleavage of the aryl metal complex and (iii) hydrogen atom abstraction from the (BNAH) radical cation.

Free radical intermediates have also been identified in the peroxide initiated reduction of aryl bromides and iodides using sodium borohydride. With allyl-2-iodophenyl ether as a substrate Beckwith obtained the dihydrobenzofuran (12) in addition to the simple reduction product. This is the expected product of a 5-exo-trig cyclisation of an aryl radical onto the terminal alkene of the side chain. (A. N. Abeywickream and A. L. J. Beckwith, *Tetrahedron Lett.*, 1986, 109).

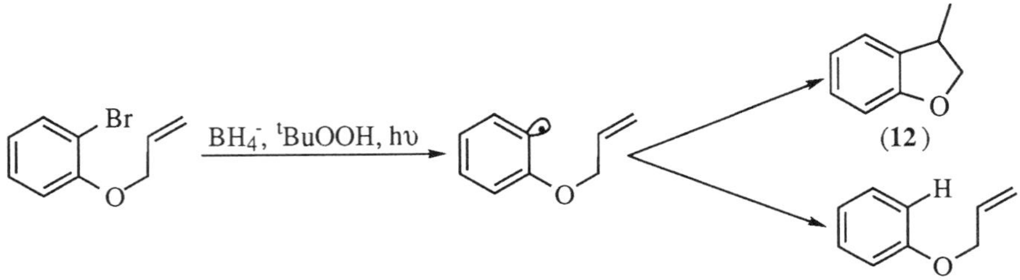

Similar products can be obtained from treatment of aryl halides with tributyltin hydride. The relative amounts of reduction and cyclisation observed depending on the precise reaction conditions (Y. Ueno, K. Chino and M. Okarawa, *Tetrahedron Lett.*, 1982, 2575). Analogous reactions using lithium aluminium hydride have also been investigated (A. L. J. Beckwith and S. H. Goh, *J. Chem. Soc., Chem. Commun.*, 1983, 905, 907). This reagent cleaves aryl fluoride bonds although its use is limited to substrates lacking other reducible functionality. Detailed mechanistic studies indicate that with this reagent there are two competing pathways depending on whether or not oxygen is present. If the reaction is run in the absence of oxygen an aryl metal intermediate is postulated as supported by the high incorporation of deuterium on quenching with D_2O. This is believed to involve reaction on the surface of the aluminium as there is no evidence for any direct S_N2 displacement of the halogen atom. On addition of peroxide a free radical pathway operates, equations (9) - (11). The initial step is now believed to be of the S_H2 type providing an aryl radical. Small substituent effects observed imply that this is a halogen atom transfer not an electron transfer process.

$$^tBuO^\bullet + AlH_4^- \longrightarrow {}^tBuOH + AlH_3^{\bullet -} \quad (9)$$

$$Ar\text{-}X + AlH_3^{\bullet -} \longrightarrow Ar^\bullet + AlH_3X^- \quad (10)$$

$$Ar^\bullet + AlH_4^- \longrightarrow ArH + AlH_3^{\bullet -} \quad (11)$$

Lastly, aryl halogen bonds have been cleaved electrolytically. The early work has been comprehensively reviewed (A. J. Bard and H. Lund (Eds.) "Encyclopaedia of Electrochemistry of the Elements" vol XIV, Marcel Decker, NY, 1980, p179). More recent work has provided much mechanistic detail and reports continue to appear in the literature with an emphasis on the competition between reductive dehalogenation and cross coupling to generate biaryls. For example, Toni and co-workers studied the electrolysis of a series of substituted benzenes in the presence of a palladium(0) complex (S. Toni, H. Tanaka and K. Moriski, *Tetrahedron Lett.,* 1985, 1655). It was found that the presence of bulky *ortho* substituents completely inhibit the cross coupling reaction. In the absence of the palladium catalyst only reduction was observed, albeit at a much reduced rate.

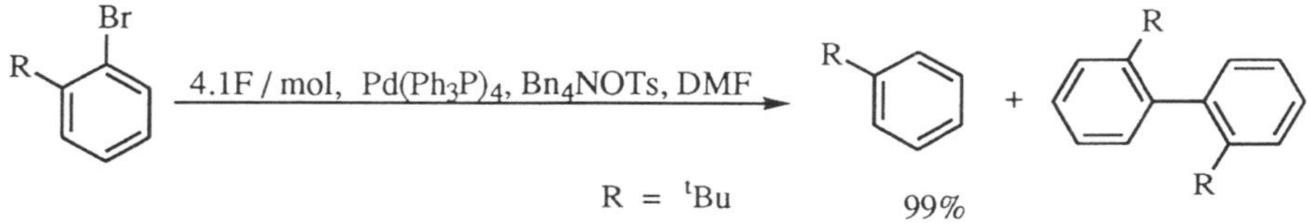

R		
^{t}Bu	99%	
OMe	96%	
H		99%
H (no Pd)	57%	

Reaction conditions: 4.1F / mol, Pd(Ph$_3$P)$_4$, Bn$_4$NOTs, DMF

(iii) Electrophilic Reactions

Electrophilic aromatic substitution continues to be an area of interest. A comprehensive survey of the reaction in general has been published (R. Taylor, 'Electrophilic Aromatic Substitution' J. Wiley, Chichester, 1990). In relation to this chapter, current interest is focused on the reactions of polyhalogenobenzenes, notably fluoroaromatics. Replacement of hydrogen by fluorine activates the reaction, pentafluorobenzene undergoing rapid substitution of the hydrogen by almost any electrophile (V. D. Steingarts, 'Synthetic Aspects of Electrophilic Ipso Reactions of Polyfluoroarenes' in "Synthetic Fluorine Chemistry" Eds. G. A. Olah, R. D. Chambers and G. K. S. Prakash, J. Wiley, Chichester, 1992, Chapter 12, p 259). With perfluorinated aromatics loss of cationic fluorine is unfavourable. The Meisenheimer complexes, formed by the addition of electrophiles to perfluoroarenes, can be trapped to provide an efficient route to fluorinated cyclohexadienones. In general, the principal difference between perfluoroarenes and their hydrocarbon analogues is the directing effect of the fluorine substituents. This disfavours the reaction at a carbon *meta* to a fluorine substituent and consequently pentafluorotoluene preferentially reacts *meta* to the methyl group. Strong electron donors overcome this effect in which case the normal regiochemistry is observed. For example, treatment of fluoroalkyl or perfluorophenyl ethers with a nitric / hydrofluoric acid mixture provides a mixture of the 2- and 4-alkoxy- or aryloxynitrocyclohexa-2,5-dienes.

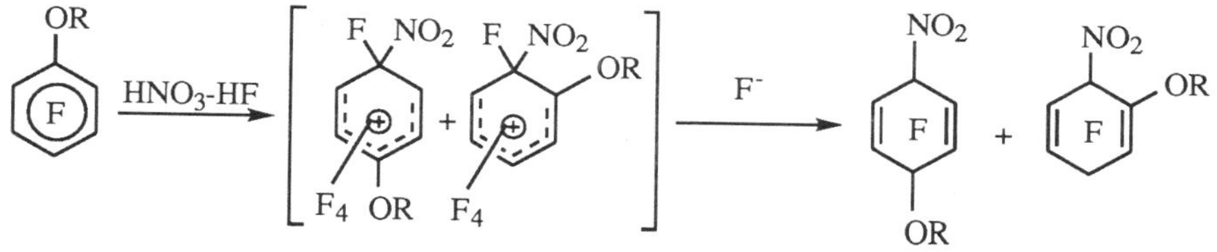

Other electrophiles have been studied, in particularly those involving antimony pentafluoride catalysis. Treatment of tetrafluoro-4-xylene with MeF-SbF$_5$ at -90°C initially produces the gem-dimethyl substituted cation (**13**).

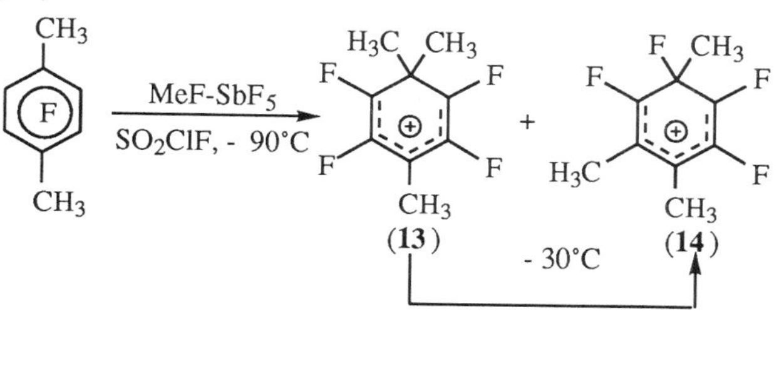

This is the kinetic product and on warming to -30°C undergoes a methyl shift to provide the stable arenium ion (**14**) (P. N. Dobronravov and V. D. Steingarts, *Zh. Org. Khim.*, 1983, **19**, 995).

Similar stable ions are obtained from the equimolar combination of perfluorobenzenium ions and one equivalent of pentafluorobenzene. Carrying out the same reaction with an excess of pentafluorobenzene provides an efficient entry to polynuclear perhalogenated species typified by perfluoro-1,3,5-triphenylbenzene. The formation of this product reflects both a steric requirement and also the propensity for electrophiles to attack *meta* to the non-fluorinated positions in polyfluorinated aromatics (V. D. Steingarts, *loc. cit.* and references therein).

(iv) Nucleophilic Substitution

Nucleophilic aromatic substitution reactions have been comprehensively reviewed (*Comprehensive Organic Chemistry*, Eds I. Fleming and B. M. Trost, Pergamon Press, Oxford, 1991, Vol. 4 chapters 2. 1 - 2. 4 inclusive). Many of the reactions covered elsewhere in this chapter are included in the mechanistic discussion contained in this mammoth reference source. In addition the mechanisms involed in the various steps of the $S_{RN}1$ reaction pathway have been the subject of a monograph (R. A. Rossi and R. H. de Rossi, *A. C. S. Monograph # 178*, 1983). However, recently it has been suggested that many of these reactions involve the direct participation of the arylhalide radical anion in an $S_{RN}2$ type process, equations (12) and (13), without the need for fragmentation to an aryl radical as originally proposed (D. B. Denney and D. Z. Denney, *Tetrahedron*, 1991, **47**, 6577). This suggestion has provoked considerable response (see for example: D. B. Denney, D. Z. Denney and A. J. Perez, *Tetrahedron*, 1993, **49**, 4463; J. F. Bunnet, *ibid.*, 1993, **49**, 4477; R. A. Rossi and S. M. Palacios, *ibid.*, 1993, **49**, 4485).

$$ArX^{-\bullet} + Nu^- \longrightarrow ArNu^{-\bullet} + X^- \quad (12)$$

$$ArNu^{-\bullet} + ArX \longrightarrow ArNu + ArX^{-\bullet} \quad (13)$$

Displacement of fluorine from activated fluoroaromatics via either S$_N$Ar or a nucleophilic hydride displacement process occur in moderate yields under relatively mild conditions. Carrying out the reaction at elevated temperatures can lead to the products of multiple substitution. For example, the tertiary amine (**15**) is the major product when 4-cyanoaniline and 4-fluoronitrobenzene are coupled (J. H. Gorvin, *J. Chem. Soc. Chem. Commun.*, 1985, 238).

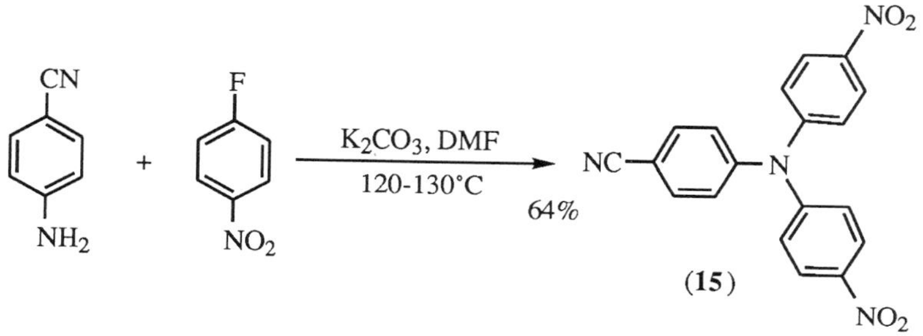

In general these displacement reactions are limited to fluoro or chlorobenzenes with *ortho* and / or *para* electron-attracting groups and relatively unhindered nucleophiles. It has been reported that certain haloarenes undergo nucleophilic substitution with tertiary alkylamines with concomitant dealkylation (K. Matsumoto, S. Hashimoto, S. Otani, *J. Chem. Soc., Chem. Commun.*, 1991, 306). An EPR study of this process provided evidence for radical intermediates and this is consistent with an initial electron-transfer step in this and other S$_N$Ar reactions (L. Grossi, *Tetrahedron Lett.*, 1992, 5645). Alternatively, in reactions involving nitrogen nucleophiles where there is a large negative volume of activation (K. R. Brower, *J. Am. Chem. Soc.*, 1959, **81**, 3504), the application of high pressures provide aryl amines in good yield from activated aryl halides derived from any halogen.

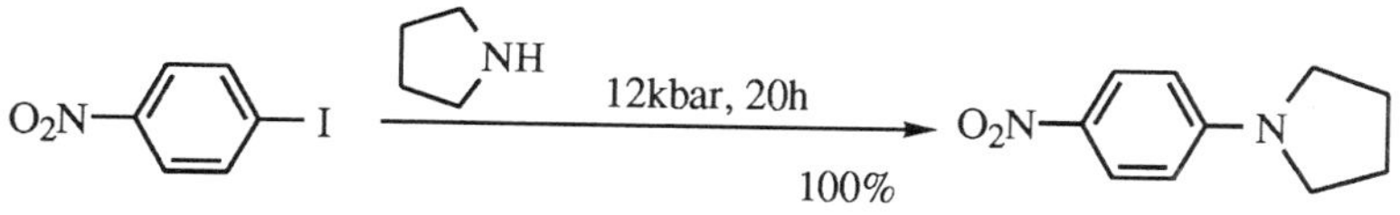

The equivalent reactions at normal pressure do not proceed. However, this acceleration is restricted by the steric requirements of the nucleophile to primary amines and certain cyclic species (T. Ibata, Y. Isogami and J. Toyoda, *Chemistry Lett.*, 1987, 1187).

A similar enhancement of reactivity is observed when the reaction is carried out under phase-transfer conditions (G. Barak, Y. Sasson, *J. Chem. Soc. Chem. Commun.,* 1987, 1187). In many of these cases the catalyst is an onium salt, e.g. tetrabutylammonium bromide which provides good yields in alkoxide displacements (S. M. Andrews, C. Konstantinou and W. A. Feld, *Synth. Commun.,* 1987, **17**, 1041). Some selectivity can be observed, for example the reaction of 1,2,3,5-tetrachlorobenzene and octane thiolate provides the sulphide (**16**) as the major product. This selectivity is attributed to the absence, in this position, of a *para-* chlorine substituent which is known to retard nucleophilic displacement reactions,*vide infra*.

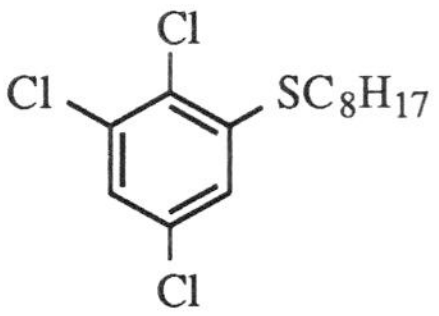

(**16**)

Tricyclohexyldodecylphosphonium bromide is another effective catalyst permitting the selective substitution of the bromine atom from 3-chlorobromobenzene with thiol nucleophiles (D. J. Brunelle, *J. Org. Chem.,* 1984, **49**, 1309).

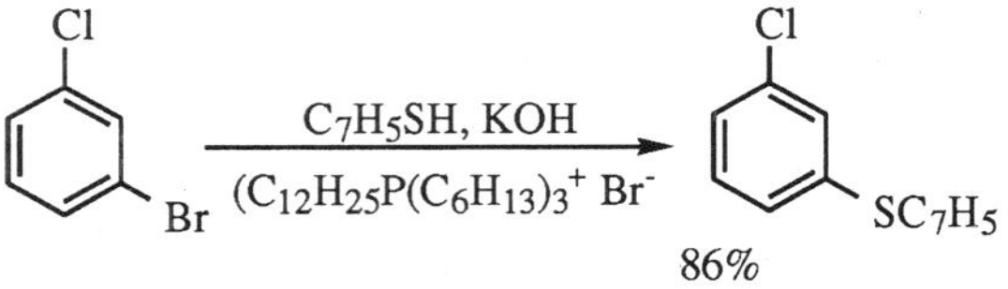

However, in the same report Brunelle notes that a general limitation with the use of phosphonium or ammonium salts is their lability under severe reaction conditions; e.g. tetrabutylammonium bromide has been shown to be rapidly degraded by both alkyl thiols and sodium phenoxide. In an approach to circumvent these problems new more stable catalysts such as the pyridinium salt (**17**) have been developed (D. J. Brunelle and D. A. Singleton, *Tetrahedron Lett.,* 1984, 3384).

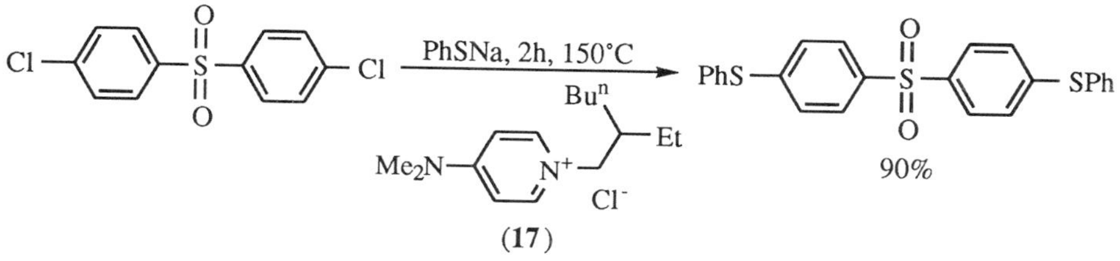

Scheme showing reaction of 4,4'-dichlorodiphenyl sulfone with PhSNa, 2h, 150°C, with Bu^n and Me$_2$N-pyridinium chloride catalyst, giving the bis(phenylthio) product in 90% yield.

(17)

In highly polar solvents, such as HMPA, unactivated halobenzenes are also prone to direct substitution (J. E. Shaw, D. C. Kunerth and S. B. Swanson, *J. Org. Chem.*, 1976, **41**, 732) and in this medium the relative reactivity was observed to be I>F>Br>Cl (P. Cogolli *et al.*, *J. Org. Chem.*, 1979, **44**, 2642). With polychlorinated benzenes the initial reaction with methoxide is moderately regioselective attempts to achieve polysubstitution produce complex mixtures of mono- and disubstituted chlorophenols. The phenols are formed via a nucleophilic demethylation pathway (L. Testaferri *et al.*, *Tetrahedron*, 1983, **39**, 193).

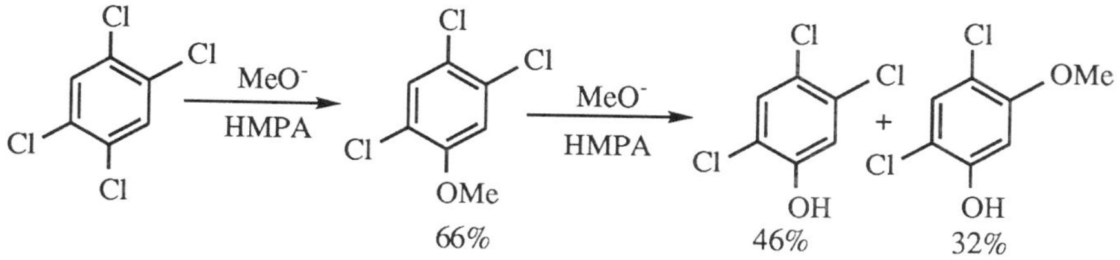

Scheme: 1,2,4,5-tetrachlorobenzene + MeO⁻/HMPA → 2,4,5-trichloroanisole (66%) → MeO⁻/HMPA → mixture of 2,4,5-trichloro-3-methoxyphenol-type products (46% + 32%).

Sulphur based nucleophiles behave in a similar fashion, but with demethylation competing more efficiently (L. Testaferri, M. Tingoli and M. Tieco, *Tetrahedron Lett.*, 1980, 3099). In contrast to chlorinated substrates fluorinated aromatics show considerable regioselectivity. In a study of the kinetics of substitution of fluorinated aromatics, Chambers and co-workers have shown that the effect of introducing fluorine is to activate the ring towards substitution at the *ortho* and *meta* positions by a factor of 30-100 and to slightly deactivate the *para* position; 0. 2-0. 5 x the rate for the equivalent H compound (R. D. Chambers, D. Close and D. L. H. Williams, *J. Chem. Soc. Perkin Trans. II*, 1980, 778; and R. D. Chambers, *Dyes and Pigments*, 1982, **3**, 183). These ratios are dependent upon the conditions. The rate constants for the similar reaction of difluorobenzene with sodium methoxide were found to vary in the ratio $p : o : m$ 1 : 8 : 48 . This anomalous activating influence of the ortho fluorine substituent is a common observation and is attributed to the reaction proceeding through an early transition state in which the electrophilic nature of the carbon atom under attack is enhanced by the strong electron inductive effect of the *ortho* fluorine atom. This is illustrated in the following examples in which the incoming nucleophile attacks at a position with the optimim number of *ortho-* and *meta-* fluorine atoms.

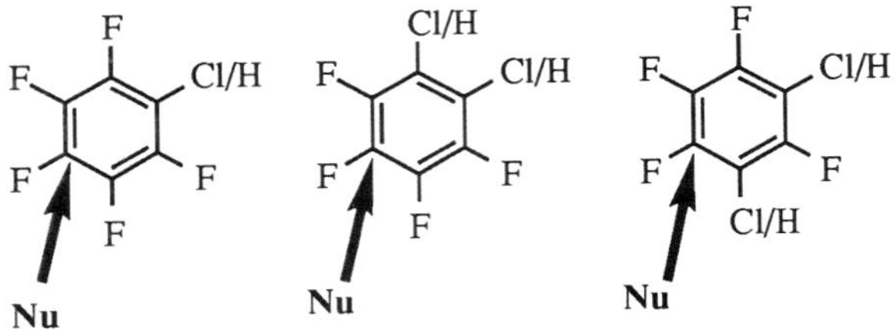

This effect is accentuated with increasing reactivity of the aromatic system, emphasising the close relationship between the starting material and the transition state involved. A compilation of kinetic studies which contains many references to relatively inaccessible Russian literature has been published (P. P. Rodionov and G. G. Furin, *J. Fluorine Chem.*, 1990, **47**, 361). The ease of substitution increases with increasing halo substitution: 1,2,3-trichlorobenzene undergoes complete substitution on reaction with excess isopropylthiol at 100°C in HMPA whilst the same reaction using hexachlorobenzene must be carried out at or below 20°C to ensure good yields (L. Testaferri, M. Tingoli and M. Tiecco, *J. Org. Chem.*, 1980, **45**, 4376). Similarly, a change from chlorine to fluorine substitution increases the reactivity of the arene towards nucleophiles, e.g. hexafluorobenzene undergoes the analogous reaction below 0°C. Owing to this high reactivity perfluorobenzenes have been used to prepare pentafluorophenyl esters as masking groups in peptide chemistry (L. Kisfaludy and I. Schön, *Synthesis*, 1983, 325). The analogous ethers are too difficult to cleave (A. H. Haines and K. C. Symes, *J. Chem. Soc., Perkin Trans. 1.*, 1973, 53). However, substitution of the *para-* fluorine by a trifluoromethyl group removes this hurdle and consequently, it is possible for perfluorotoluene to be used as a protecting groups for alcohols (M. Jarman and R. McCaque, *J. Chem. Soc. Chem. Commun.*, 1984, 125). Regeneration of the original alcohol is achieved by treatment with sodium methoxide. Interestingly, the by-product of this step is 2,6-difluoro-3,5-dimethoxy-4-trifluoromethylphenol (**18**), the formation of which is consistent with the directing effects discussed above.

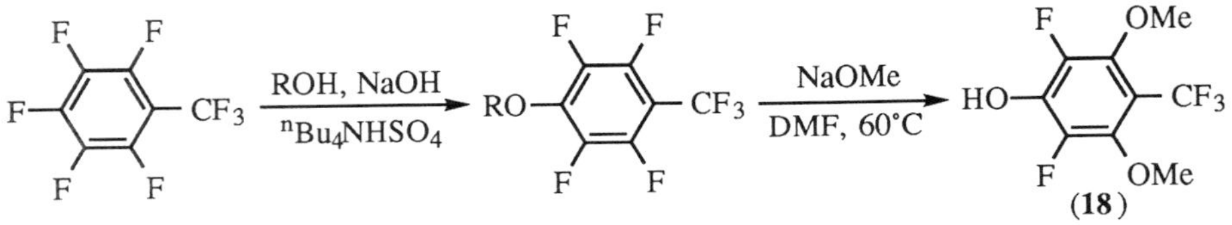

Aryl iodides and bromides only rarely undergo direct nucleophilic substitution. However in the presence of transition metal salts the effective products of nucleophilic substitution may be observed. For example, substitution of aryl iodides with a variety of pseudohalides and chalcogens occurs in good yield in the presence of cuprous salts dissolved in HMPA (H. Suzuki, *Bull. Chem. Soc. Jpn.*, 1980, **53**, 1765). The use of copper assisted nucleophilic aromatic substitution of halogen atoms has been reviewed (J. Lindley, *Tetrahedron*, 1984, **40**, 1433). Again these reactions can be affected by the nature of other substituents on the benzene ring. For example, in the trifluoromethylation of a series of chlorobenzenes using trifluoromethylcopper it was found that only those compounds in which a chelated transition state was possible reacted efficiently (J. H. Clark *et al., J. Fluorine. Chem.*, 1990, 50, 411). Thus 2-nitrobenzene affords the trifluoromethylated product in 59% yield via transition state (**19**) whilst the *meta*- isomer was inert and the *para*- form (reflecting the non chelated reaction) gave only 5% of the analogous product. Further results which support this hypothesis come from the failure of both the trifluoromethyl and cyano- analogues to undergo substitution. The trifluoromethyl species cannot complex the cuprate reagent and, although, chelation is possible with the nitrile the geometry of the complex (**20**) is not correct for reaction.

(19)   (20)

In the $S_{RN}1$ reaction the order of reactivity is reversed i. e. I > Br > Cl » F and this has been confirmed by a series of competition experiments using disubstituted species. Diiodobenzenes afford only products of disubstitution, no monosubstituted intermediates are observed and this has been rationalised as being indicative of preferential dissociation of IArNu•⁻ rather than electron transfer to another molecule of starting material. In contrast, fluorine containing substrates only give the products of monosubstitution. Consequently bromo and iodo species tend to be the most synthetically useful, although when the reaction is initiated electrochemically (J. M. Savéant, *Acc. Chem. Res.*, 1980, **13**, 323) the use of chlorinated substrates has been advocated to avoid problems of over reduction (C. Amatore *et al., J. Am. Chem. Soc.*, 1979, **101**, 6062). The reactivity is also significantly affected by the presence of other substituents notably in the *ortho* position (J. F. Bunnett, E. Mitchell and C. Galli, *Tetrahedron*, 1985, **41**, 4119). The photostimulated reaction of the enolate of ᵗbutylmethylketone with 2-

trifluoromethyliodobenzene provides the annulated heterocycle (**21**) as the major product along with a small amount of the expected product of $S_{RN}1$ substitution. Resubjecting the latter to the reaction conditions provides more of the benzopyran (**21**). The formation of this compound is postulated to proceed via the formation of the difluoro-orthoquinone methide (**22**) followed by the addition of a second equivalent of the enolate and subsequent cyclisation. Support for this hypothesis arises since similar products are obtained from 4-trifluoromethyliodobenzene (J. F. Bunnet and C. Galli, *J. Chem. Soc., Perkin Trans. I*, 1985, 2515).

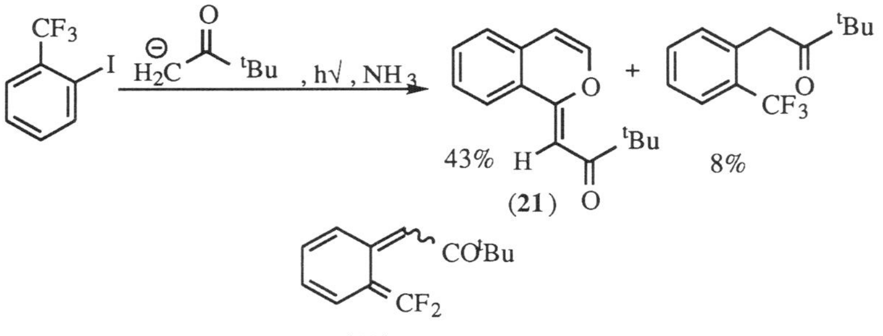

Finally this methodology has found applications in polymer synthesis; for example, the preparation of poly(1,4-phenylene sulphide) from 4-dichlorobenzene and sodium sulphide (V. Z. Annenkova *et al.*, *Chem. Abstr.*, 1986, **105**, 6848).

(v) Photochemical Reactions

Owing to the growing concern regarding the environmental effect of halogenated aromatics the photochemistry of these compounds is an area of some interest. The use of tertiary amine additives as electron donors has been shown to accelerate the dehalogenation process via an electron-transfer mechanism (R. A. Beecroft, R. S. Davidson and D. Goodwin, *Tetrahedron Lett.*, 1983, 5873; and N. J. Bunce, *J. Org. Chem.*, 1982, **47**, 1948). Fluorinated aromatics were found to be resistant to dehalogenation giving instead products of amine addition (A. Gilbert and S. Krestonosich, *J. Chem. Soc., Perkin Trans. I*, 1980, 1703). More recently, the presence of anthracene, or a related compound, as a photocatalyst has permitted the dechlorination of biphenyls using near uv-visible light (Y. Tanaka *et al.*, *J. Chem. Soc. Chem. Commun.*, 1987, 1703).

Irradiation of chlorophenoxide anions in the presence of amines [or any other suitable nucleophile] leads to the products of substitution which form via the α-ketocarbene (L. M. Tolbert, 'The Photochemistry of Resonance

Stabilised Anions' in "Comprehensive Carbanion Chemistry" Eds. E. Buncel and T. Durst, Elsevier, Oxford, 1978, p262 and reference 67 therein).

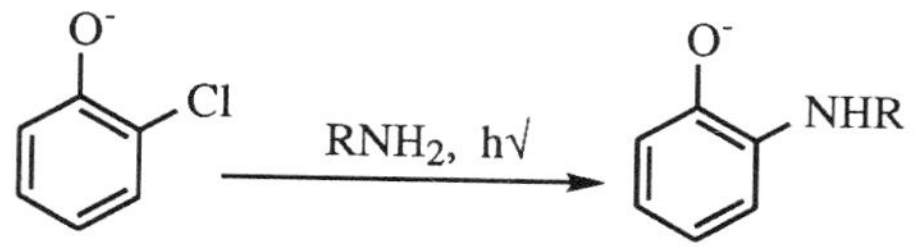

Evidence for carbene intermediates has been deduced from the irradiation of 2-chlorophenol in dilute aqueous alkalis. This affords the tricyclic dicarboxylic acid (**25**) in 75% yield (C. Guyon, P. Boule and J. Lemaire, *Nouv. J. Chim.*, 1984, **8**, 685; and *Tetrahedron Lett.*, 1982, 1581). The formation of this product is rationalised via the Wolff rearrangement of the carbene (**23**) to provide the cyclopentadienyl ketene (**24**) which, after hydrolysis, undergoes spontaneous dimerisation.

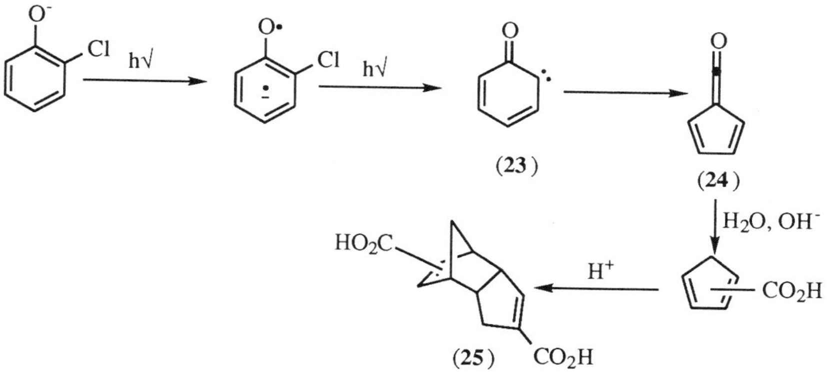

An account of the early photocycloaddition chemistry of perfluorobenzene with a range of olefins has been published (M. Zupan and B. Sket, *Isr. J. Chem.*, 1978, *17*, 92). More recently, Lemal and co-workers have examined the photocycloaddition of perfluorobenzene with a variety of halogenated olefins in a study of the perfluorobenzeneoxide-perfluorooxepin equilibrium (N. E. Takenaka, R. Hamlin and D. M. Lemal, *J. Am. Chem. Soc.*, 1990, **112**, 6715 and references therein).

Photocycloaddition of 1,2-dichlorodifluoroethene with hexafluorobenzene provides a 45% yield of isomeric dichloroperfluorotricyclooctenes.

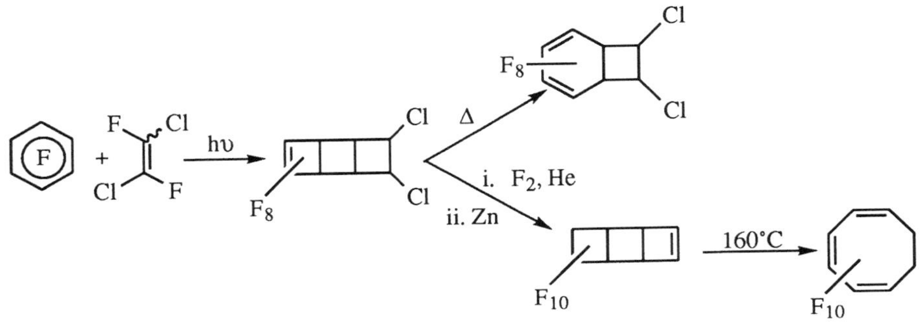

These stable but strained tricycles can be easily converted into a variety of analogues, thermolysis of which provides routes to a variety of fluorinated benzocyclobutanes and cyclooctatrienes and tetraenes. The tetrahydro analogue (**26**) undergoes a rapid pericyclic rearrangement to the benzocyclobutane (**27**) which, in the presence of acid, spontaneously cleaves to generate the alkylperfluorobenzene (**28**) (M. M. Rahmann et al., *J. Am. Chem. Soc.*, 1990, **112**, 5986).

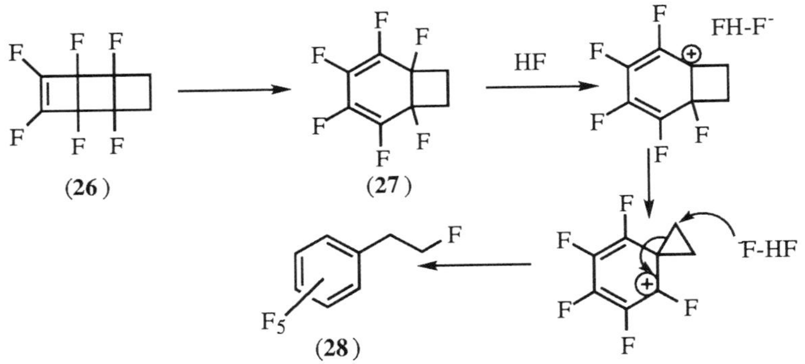

(vi) π-Haloarene Metal Complexes

These compounds are easily and routinely prepared by treatment of haloarenes with a suitable $ML_3L'_3$ type complex [L' = CO, Ph$_3$P, MeCN, etc] (see for example, C. A. L. Mahaffy and P. L. Pauson, *Inorg. Synth.*, 1979, **19**, 154). The early chemistry of these complexes has been reviewed (M. F. Semmelhack, *Pure Appl. Chem.*, 1981, **53**, 2379; and M. F. Semmelhack et al., *Tetrahedron*, 1981, **37**, 3957). In general, the complexation of an aryl halide with a metal carbonyl or equivalent results in enhanced susceptibility towards nucleophilic aromatic substitution and directed deprotonation.

The order of reactivity towards aromatic nucleophilic substitution of several arene metal complexes relative to some electron deficient aryl halides is as follows (W. E. Watts in *Comprehensive Organometallic Chemistry*, Eds G. Wilkinson, F. Abel and G. A. Stone, Pergamon Press, Oxford, 1982, vol 8, p1013):-

$[ArenMn(CO)_3]^+$ > 2,4-$(NO_2)C_6H_3Cl$ > $[areneFeCp]^+$ » 4-$(NO_2)C_6H_4Cl$ > areneCr$(CO)_3$

The most widely studied of these species are the more stable arene chromium tricarbonyl complexes. All the monohalo derivatives are known although the bromo and iodo complexes fail to undergo S_NAr processes.

In a study on the mechanism of this reaction, Knipe and co-workers have found that, when using methoxide as the nucleophile, the fluorobenzene complexes are some 2000x more reactive than their chlorobenzene analogues (A. C. Knipe, J. McGuinness and W. E. Watts, *J. Chem. Soc. Chem. Commun.*, 1979, 842). In this case, the reaction proceeds via a fast and reversible nucleophilic attack at a C-H position on the complex (see below), followed by rate limiting formation of the Meisenheimer intermediate (**29**) (i.e. $k_{-1} > k_1 > k_2$). Halide loss then affords the products of *ipso* substitution. With amine nucleophiles, a detailed mechanistic study suggests that halide loss is now rate limiting (i.e. $k_{-2} > k_3$) (J. F. Bunnet and H. Hermann, *J. Org. Chem.*, 1971, **36**, 4081).

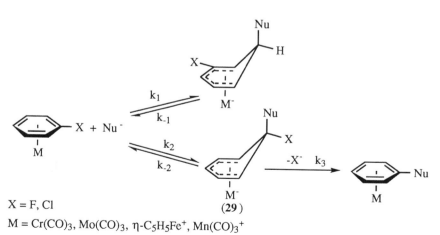

X = F, Cl (**29**)
M = Cr$(CO)_3$, Mo$(CO)_3$, η-$C_5H_5Fe^+$, Mn$(CO)_3^+$

This mechanistic dependence upon the nucleophile is found to be a general phenomenon. Stabilised carbanions, such as the malonate ion, undergo complete *ipso* substitution via reversible attack on the starting complex. With more reactive nucleophiles the mechanism is one of fast substitution at a C-H position followed by rearrangement and subsequent loss of the halide. These equilibration and rearrangement steps are both solvent and temperature dependent (M. F. Semmelhack in Comprehensive Organic Synthesis, Eds. B. M. Trost and I. Fleming, Pergamon Press, Oxford, 1992, vol. 4, p. 517).

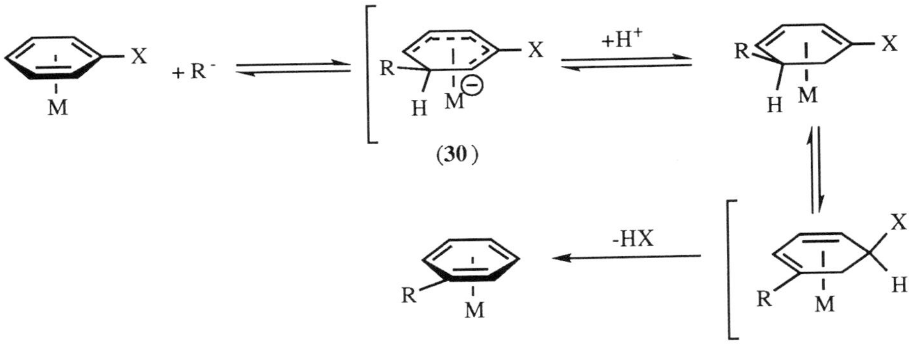

For example, in the reaction of lithiobutyronitrile with the chlorobenzene chromium tricarbonyl complex complete consumption of starting material to give the anion (**30**, M = Cr(CO)$_3$, X = Cl, R = (CH$_3$)$_2$CH(CN)CH-) occurs rapidly at -78°C, but the appearance of the the substituted product does not occur until the reaction is warmed to room temperature. Quenching of the anion at -78°C by addition of trifluoroacetic acid, followed by iodine initiated decomplexation affords a complex mixture of dienic and aromatic products.

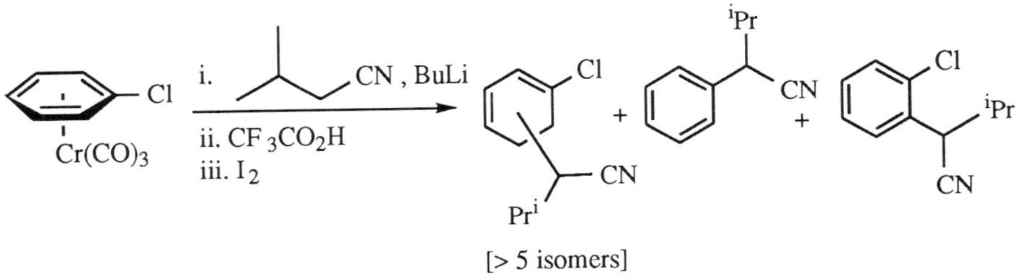

[> 5 isomers]

Highly reactive carbanions (pK$_a$ >20) add irreversibly and some selectivity can be observed in the regiochemistry of these reactions. As with the uncomplexed haloarenes, fluorinated substrates show the greatest regioselectivity, reacting primarily at the *meta* position. *Ortho* and *para* substitution can become significant with the chloro complexes although other factors, such as steric effects, can overide these weakly directing electronic effects (F. Rose-Munch *et al.*, *J. Orgmet. Chem.*, 1988, **353**, 53; 1989, **363**, 297; and F. Rose-Munch, E. Rose and A. Semra, *J. Chem. Soc., Chem. Commun.*, 1987, 942).

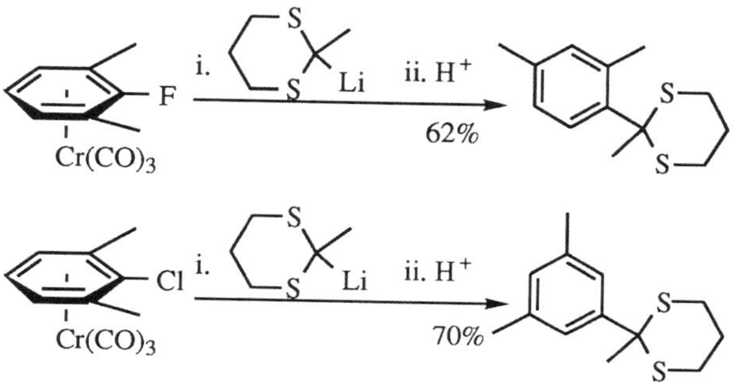

With the exception of the iodobenzene complex, which undergoes transmetalation, treatment of the other halobenzene complexes with strong base leads to deprotonation of the aromatic nucleus. This is entirely analogous to the uncomplexed species (R. J. Carol and W. S. Trahanovsky, *J. Org. Chem.*, 1980, **45**, 2560). As with nucleophilic substitution, a strong directing effect can be observed, particularly with the fluorobenzene complex (M. Ghavshou and D. Widdowson, *J. Chem. Soc., Perkin Trans. I*, 1983, 3065). Addition of the resultant anion to butyrolactone provides an efficient synthesis of the benzannulated heterocycle (**31**) via a tandem alkylation-aromatic substitution process. Thus the halobenzene complexes can behave as a source of complexed benzyne, although the latter species has never been observed directly.

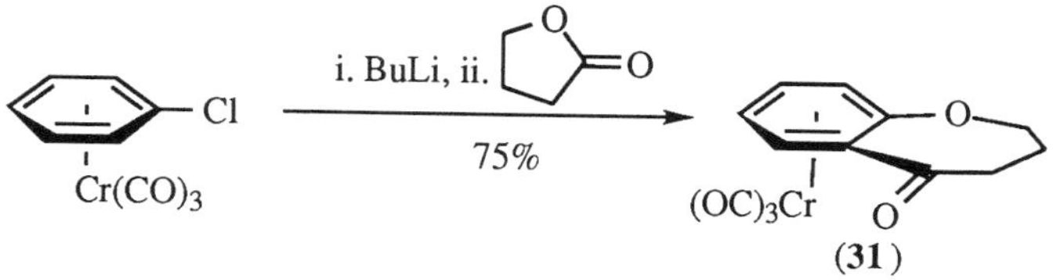

2. Alkylbenzenes with halogen in the side chain

Benzyl iodides can be readily generated (~68-75%) via a pyrolytic S_N2 displacement from 4,6-diphenyl-2-mercaptopyridinium iodide (**33**) generated *in situ* from the corresponding amine by reaction with 4,6-diphenylpyran-2-thione and alkylation of the resultant pyridinethione (**32**) with methyl iodide (A. Lorenzo, P. Molina and M. J. Vilaplana, *Synthesis*, 1980, 853).

$$ArCH_2NH_2 + \underset{S}{\underset{\|}{O}}\!\!\!\diagdown\!\!\!\diagup\!\!\!\overset{Ph}{\diagdown}\!\!\!\diagup\!\!-Ph \longrightarrow ArCH_2-\underset{S}{\underset{\|}{N}}\!\!\!\diagdown\!\!\!\diagup\!\!\!\overset{Ph}{\diagdown}\!\!\!\diagup\!\!-Ph \xrightarrow{MeI} ArCH_2-\underset{MeS}{\overset{Ph}{\underset{+}{N}}}\!\!\!\diagdown\!\!\!\diagup\!\!-Ph \;\; I^-$$

(32)   (33)

$$\downarrow \Delta$$

$$ArCH_2I$$

In a related process, Katritzky and colleagues have used 2,4,6-triphenylpyridinium salts to prepare benzylic iodides, bromides, chlorides and fluorides (A. R. Katritzky, A. Chermprapai and R. C. Patel, *J. Chem. Soc., Perkin Trans. I*, 1980, 2901; A. R. Katritzky *et al.*, *ibid.*, 1979, 433, 436).

Synthesis of ß-haloalkylbenzenes can be achieved by deamination of aromatic amines using either a thionitrite or thionitrate salt in the presence of the requisite cupric halide and an electrophilic olefin. It is postulated that the reaction proceeds via nucleophilic addition of an aryl radical, generated *in situ*, to the olefin followed by capture of the alkyl radical with the copper salt (S. Oae, K. Shinhama and Y. H. Kim, *Bull. Chem. Soc. Jpn.*, 1980, **53**, 1065). Treatment of simple arenes with 1,1,1-trifluoropropanone and aluminium chloride yields the corresponding 1,1,1-trifluoro-2-chloro-2-arylpropane accompanied by the corresponding 2,2-diaryl-1,1,1-trifluoropropane. The ratio of the two depends on the reaction conditions, e. g. the use of low temperature in dichloromethane solvents favours the formation of the chlorinated compound (D. Bonnet-Delpon and M. Charpentier-Morize, *Bull. Soc. Chim. Fr.*, 1986, 933).

Halomethylation continues to be an area of intense activity. Problems with chloromethylation, notably the hazards associated with many of the reagents [both chloromethyl(methyl)ether and particularly bis(chloromethyl)ether are serious carcinogens]; as well as the standard hurdles of selectivity have been addressed.

A variety of Lewis acids have been explored in attempts to find solutions to these issues. One recent example is the use of trivalent lanthanide salts as weak but hard Lewis acids. These favour chloromethylation over Friedel Crafts alkylation although the major products are diarylmethanes (D. V. Davydov, S. A. Vinogradov and I. P. Beletskaya, *Bull. Acad. Sci. U. S. S. R.*, 1990, **39**, 627).

McKillop and co-workers introduced the use of methoxyacetylchloride as an alternative to chloromethyl(methyl)ether. In the presence of a

Lewis acid catalyst, this reagent permits the chloromethylation of electron rich aromatics, although with more reactive substrates over-reaction leading to diarylmethane formation is still a problem (A. McKillop, F. A. Madjdabadi and D. A. Long, *Tetrahedron Lett.*, 1983, 1933).

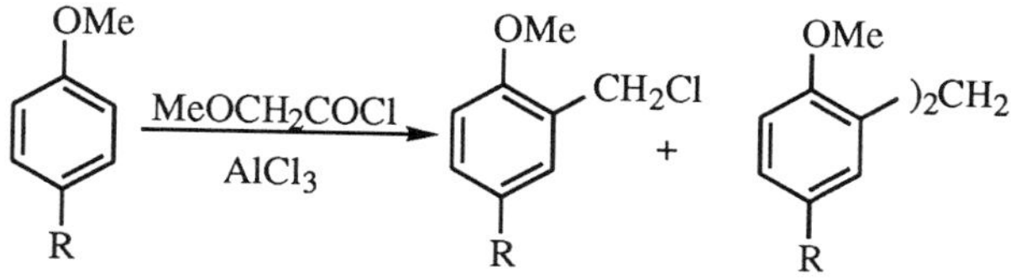

![Reaction scheme: 4-R-anisole + MeOCH$_2$COCl / AlCl$_3$ → 4-R-2-(chloromethyl)anisole + bis(4-R-2-methoxyphenyl)methane. R = NO$_2$: 89%, 0. R = Br: trace, 70%.]

It is proposed that on complexation with aluminium chloride the acid chloride decomposes to give the methoxymethyl cation as the reactive species and carbon monoxide. In support of this, the latter gas has been detected on mixing the two components. Subsequent reaction with the aromatic substrate gives the aryl methoxymethyl ether which then undergoes a further displacement by either chloride ion or a second aromatic molecule. An alternative method for avoiding the hazardous byproducts has been to use longer chain alkylhaloalkylethers, ROCH$_2$X, where R is at least heptyl (A. Warshawasky, A. Deshe and R. Gutman, *Br. Polym. J.*, 1984, **16**, 234). These reagents have been found to produce almost no bis(chloromethyl)ether during their use, although the higher boiling alcohol generated in the reaction can create difficulties in the purification stages. A further refinement in this area, which obviates this problem, has been the development of a polymeric halomethylating agent, 2-bromomethoxyethylsulphonamidopolystyrene (**34**) (A. Warshawsky and N. Shoef, *J. Polym. Sci., Polym. Chem. Ed.*, 1985, **23**, 1843).

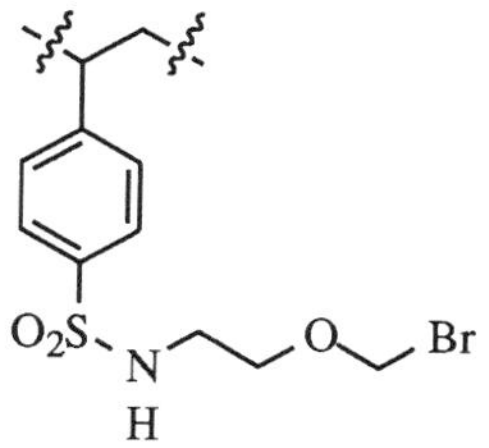

![Structure of polymeric reagent 34: polystyrene-bound aryl with para-SO$_2$NH-CH$_2$CH$_2$-O-CH$_2$Br substituent.]

(**34**)

Although these procedures do not use and/or produce hazardous materials the problems of diarylmethane formation remains. This has been circumvented by the use of milder reaction conditions provided by the use of phase-transfer conditions and aqueous media. Selective monochloromethylation has been achieved by an Italian group using aqueous formaldehyde and sodium chloride in sulphuric acid in the presence of an ambiphilic surfactant such as hexadecyltrimethylammonium bromide which provides a high boundary surface area between the two phases. However, with substituted benzenes the regioselectivity is not good, showing only a small preference for *para* over *ortho* substitution (M. Selva, F. Trotta and P. Tundo, *Synthesis*, 1991, 1003). Similarly, Mitchell and Lyer have reported that the treatment of a variety of aromatic hydrocarbons with a 1,3,5 trioxane and 48% aqueous hydrobromic acid combination in glacial acetic acid using tetradecyltrimethylammonium bromide as the phase-transfer catalyst afforded very high yields of the monobromomethylated product. In contrast to the chloromethylation reported above, almost complete regioselectivity was obtained with no diarylmethane products being observed. For example, toluene afforded 2-bromomethyltoluene in 96% isolated yield (R. H. Mitchell and V. S. Lyer, *Synlett*, 1989, 55). The deactivating effect of the halomethyl group coupled with the moderating effect of the aqueous media appear to prevent polysubstitution. Carrying out the reaction under anhydrous conditions, with longer reaction times and higher temperatures, allows the di- and tri- bromomethylation of mesitylene and other alkyl benzenes. Again no evidence for any diarylmethane formation is found. Less reactive substrates such as benzene only undergo monobromomethylation whilst deactivated aromatics such as chlorobenzene are inert (A. W. van der Made and R. H. van der Made, *J. Org. Chem.*, 1993, **58**, 1262). The higher reactivity of the products coupled with the greater efficiency (yield and selectivity) observed in bromomethylation should make this the favoured halomethylation procedure (See reference 4 in A. W. van der Made and R. H. van der Made, *loc. cit.*).

The selective introduction of the bromomethyl groups can also be achieved by functionalisation of aromatic methyl groups. Treatment of 2,3-dimethylanisole with various different stoichiometries of N-bromosuccinimide under photothermal conditions allows the selective introduction of up to four bromine atoms (V. G. S. Box and G. P. Yiannikous, *Heterocycles*, 1990, **31**, 1261).

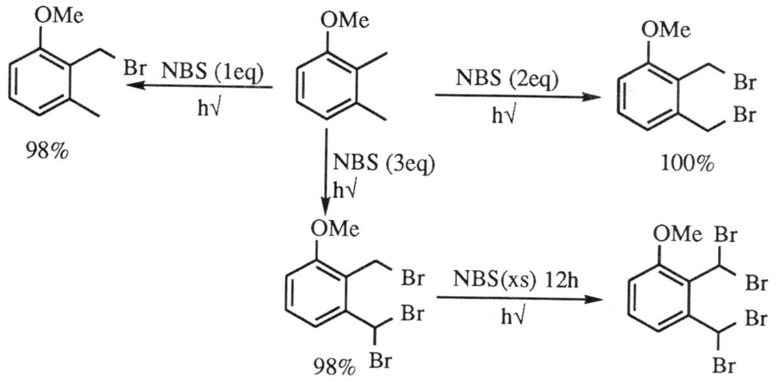

[Scheme: 2-methylanisole reactions with NBS/hν]

- With NBS (1 eq), hν → 2-(bromomethyl)-3-methylanisole-type product (98%)
- With NBS (2 eq), hν → bis(bromomethyl)anisole (100%)
- With NBS (3 eq), hν → ArCH₂Br and ArCHBr₂ substituted anisole (98%); then NBS (xs), 12 h, hν → ArCHBr-/ArCHBr₂ further brominated product

Fluoroalkylarenes have been recognised as valuable industrial chemicals and this has led to significant advances in synthetic methodology in this area (see for example Organofluorine Compounds and Their Industrial Applications, Ed. R. E. Banks, Ellis Harwood ltd., Chichester, 1979; and Biomedical Aspects of Fluorine Chemistry, Eds. R. Filler and Y. Kobayashi, Nodansha/Elsevier, New York, 1982). Traditionally, the perfluoroalkyl group has been introduced via the thermolysis of the appropriate perfluoroalkyliodide and the optimum conditions for this process have been surveyed (A. B. Cowell and C. Tamborski, *J. Fluorine Chem.*, 1981, **17**, 345).

More recently, milder alternative procedures have been developed. The combination of xenon difluoride and perfluoroalkanoic acids gives moderate yields of fluoroalkylated products with electron poor aromatics (Y. Tanabe, N. Matsuo and N. Ohno, *J. Org. Chem.*, 1988, **53**, 4585). A somewhat cheaper and simpler procedure involves the use of perfluoroalkylperoxides, $(R_fCOO)_2$, as the radical source (M. Yoshida, *J. Fluorine Chem.*, 1990, **49**, 1). Although the lower peroxide homologues are stable at room temperature, they all react readily at elevated temperatures to deliver the perfluoralkylated products in moderate to excellent yield. Better yields are obtained with the higher homologues; for example, benzene undergoes trifluoromethylation in 50% yield, whereas perfluoroheptylation occurs in 97% yield. The mechanism of the reaction is suggested to involve initial electron transfer from the aromatic substrate to the peroxide, followed by radical coupling (H. Sawada *et al.*, *J. Fluorine Chem.*, 1990, **46**, 423; and M. Yoshida *et al.*, *J. Chem. Soc. Chem. Commun.*, 1985, 234).

$(R_fCOO)_2$ + ⌬ ⟶ $(R_fCOO)_2^{-\bullet}$ + [⌬$^{+\bullet}$] $\xrightarrow{-CO_2}$ $R_f^\bullet$ + $R_fCO_2^-$ + [⌬$^{+\bullet}$]

⌬–R_f + R_fCO_2H ⟵ [⌬$^{\oplus}$(H)(R_f)] + $R_fCO_2^-$

The other principal method for the introduction of the perfluoroalkyl group is the use of perfluoro-organometallic reagents (D. J. Burton and Z. Y. Yang, *Tetrahedron*, 1992, **48**, 185) and in particular the use of trifluoromethyl copper species has received much attention. Trifluoromethylation in general has been comprehensively reviewed (M. A. McClinton and D. A. McClinton, *Tetrahedron*, 1992, **48**, 6555). With less reactive arenes, the carbenoid like nature of these reagents has been revealed through the isolation of products of higher perfluoroalkylation (Y. Kobayashi and I. Kumadaki, *J. Chem. Soc. Perkin Trans. I*, 1980, 661).

The trifluoromethyl group can also be introduced indirectly via the displacement of other functional groups. Treatment of orthothioesters with either N-bromosuccinimide or 1,3-dibromo-5,5-dimethylhydantoin in the presence of HF / pyridine complex leads to the corresponding trifluoromethyl arenes in moderate to good yields [34-67%] (D. P. Matthews, J. P. Witten and J. R. McCarthy, *Tetrahedron Lett.*, 1986, 4861).

Reaction of arenes with carbon tetrachloride and hydrogen fluoride provides a route to the trifluoromethylated products, presumably through the intermediacy of the corresponding trichloromethylated aromatic, in good overall yields (≥ 90%) (A. Marhold and E. Klauke, *J. Fluorine Chem.*, 1981, **18**, 281). In a similar fashion treatment of polychlorinated benzenes with a mixture of fluorotrichloromethane and aluminium trichloride provides the trifluoromethylarene. On treatment with additional aluminium chloride in carbon disulphide this product is subsequently converted to the trichloromethyl analogue (J. Castaner *et al.*, *J. Org. Chem.*, 1991, **56**, 103). Addition of more aluminium chloride and fluorotrichloromethane can reverse this second conversion.

Cl_n–⌬ $\xrightarrow{CCl_3F, AlCl_3}$ Cl_n–⌬–CF_3 $\xrightleftharpoons[AlCl_3, CCl_3F]{(xs) AlCl_3, CS_2}$ Cl_n–⌬–CCl_3

The rationalisation of this sequence is based on the knowledge that, in the presence of aluminium trichloride, fluorotrichlormethane decomposes to a mixture of carbon tetrachloride, difluorodichloromethane and chlorotrifluoromethane. Although the last named compound can directly give the observed product, via a highly reactive CF_3^+ ion adsorbed onto the aluminium chloride surface, it is more likely that the mixed haloalkylbenzene is formed. This subsequently undergoes halogen exchange via benzylic cation intermediates in an equilibrium controlled by the concentration of the aluminium chloride. With trichlorobenzenes it is possible to obtain the product of bis(trihalomethylation).

The other effective electrophilic perfluoroalkylation method reported concerns the use of Umemoto's perfluoroalkylphenyliodonium triflate salts. Treatment of aromatic substrates with this R_f^+ carrier reagent smoothly affords the desired perfluoroalkylarenes (T. Umemoto, Y. Kurro and H. Shuyama, *Chem. Lett.*, 1981, 1633; and T. Umemoto and Y. Gotoh, *Bull. Chem. Soc. Jpn.*, 1987, **60**, 3307).

As with the trifluoromethyl group, trichloromethylation may be achieved indirectly via alkylarenes. Recent developments in this field include free radical chlorinations with dichlorine monoxides, reactions which show enhanced selectivities (D. J. Sam *et al.*, in 'Current Trends in Organic Synthesis', Ed. H. Nozaki, Pergamon Press, New York, 1983, pp 413-422). Similar results are obtained from the treatment of benzaldehyde dithioacetals with sulphuryl chloride (W. F. Goure, US 4,575,565, *Chem. Abstr.*, 1986, **105**, 6305m) and the related conversion of benzylsulphides using chlorine gas (A. Marhold and E. Klauke, *Synthesis*, 1982, 951). One example of the latter process is the conversion of bis(3-methylbenzyl)sulphide into two equivalents of 3-methylbenzotrichloride. Difluorodiarylmethanes can similarly be obtained from the corresponding diaryldithiolanes on treatment with sulphurylchloride and pyridinium polyhydrogen fluoride. Competitive and non selective ring chlorination of dialkyl and alkylarylketones limits this process to diarylketones (G. K. S. Prakash *et al.*, *Synlett*, 1993, 691).

Strongly electron deficient arenes, e.g. 2-nitrotrichlorobenzene, are not accessible by these procedures. However, treatment of benzyl or benzal chlorides with aqueous base and a perchloroalkane under phase transfer conditions provides an efficient synthesis of these and other trichloromethylarenes (J. P. Chupp *et al.*, *Synthesis*, 1986, 224). The process is experimentally simpler if carbon tetrachloride is the halogen source. However, the use of hexachloroethane is more efficient and produces less waste since the byproduct, tetrachloroethene, can easily be recycled.

Direct nucleophilic trichloromethylation is a much less studied reaction and many of the procedures developed are limited, either in the range of possible substrates or in their selectivity. Complexation of an arene with a transition metal enhances the reactivity of the system allowing reaction at lower temperatures and hence, provides for the clean production of mono(trichloromethyl)arenes after *in situ* oxidative decomplexation with iodine (R. G. Sutherland, C. Zhang and A. Piorko, *Tetrahedron Lett.*, 1990, 6831).

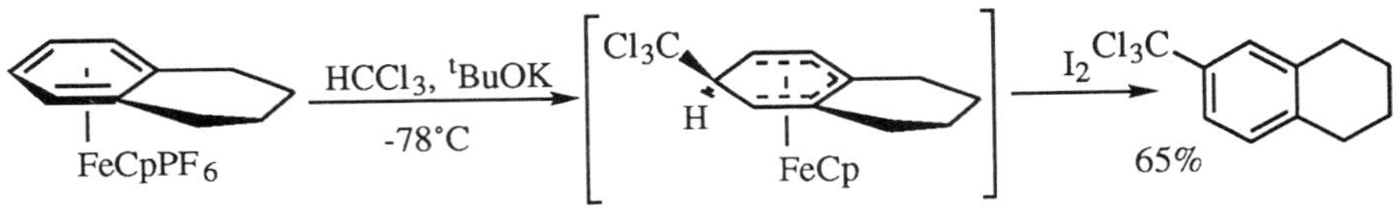

Finally, treatment of perfluorobenzocycloalkanes with antimony pentafluoride at elevated temperatures produces di(perfluoroalkyl)benzenes in good yield. This fragmentation process is unique to the fluorinated series. (V. M. Karpov *et al.*, *Bull. Soc. Chim. Fr.*, 1986, 980).

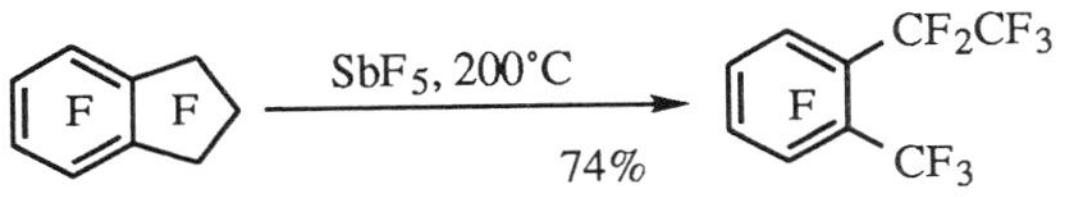

Second Supplements to the 2nd Edition of Rodd's Chemistry of Carbon Compounds, Vol.III A, edited by M. Sainsbury

Chapter 4

HYDROXY AROMATICS

J.H.P. Tyman

1. Monohydric Phenols

Introduction

Several theoretical studies concerned with basic properties of the phenol structure have been carried out. An *ab initio* conformational analysis has been made of the effect of hydroxyl rotation in phenol and 4-nitrophenol (P. Politzer and N. Sukumar, Theor. Chem., 1988, *48, 439)*. MO calculations have been reported for the ή electron-transfer to or from substituents for 20 monosubstituted benzenes, including phenol and anisole (S. Marriott, A. Silvestro and R.D. Topsom, J. Chem. Soc., Perkin Trans 2, 1988, 457). Electron-acceptor properties of substituents in aromatic molecules having a variety of groups, including the hydroxyl, have been calculated by the CNDO/2 method for the ring carbon atoms in PhR (A.N. Sarat, Zh. Obsch. Khim., 1988,*58*,2780). The separation of inductive from mesomeric effects in aromatic systems has been examined and the calculated acidities of phenols then analysed in this context (J. Niwa, Bull. Chem. Soc., Japan, 1989, *62*, 226). Substituent effects on the stabilities of phenoxyl radicals and the acidities of phenoxyl radical cations have been examined in a study of the oxidation potentials of 35 phenoxide and 3 naphthoxide ions (F.G. Bordwell and J. Cheng, J. Am. Chem. Soc., 1991, *113*, 1736). The thermodynamics of the dissociation of various protonic acids in aqueous organic water-rich media have been studied. The K_A values, enthalpies of neutralisation, and the ΔG, ΔH and ΔS values for phenol, acetic, benzoic and valeric acids, and anilinium and pyridinium cations in aqueous t-BuOH, DMSO and Me_2CO have been calculated (H. Gillet, J. Chim. Phys.

Phys.-Chim. Biol., 1987, 87, 475). The study of keto tautomerism in phenol continues. Measurements of the gas-phase heats of formation of the linearly conjugated dienone (1) show this to be more stable by 4 Kcal mol^{-1} than its isomer (2)(C.S Shiner, P.E. Vorndam and S.R. Kass, J. Am. Chem. Soc., 1986, 108, 5699). MO calculations in a CNDO/S3 study have concerned quinone methide, phenol and phenoxide ion structures (S.G. Semenov and S.M. Shevchenko, Croat. Chem. Acta, 1988, 61, 113).

Aspects of the UV photoelectron and absorption spectra of substituted phenols (J.P. LaFemina, Int. J. Quantum Chem., 1989, 36, 563), infrared spectra (T.N. Pliev, Izv. Vyssh. Uchebn. Zaved., Khim. Khim. Technol., 1987, 30, 29), and the C-13 NMR spectra (C.J. O'Connnor et al., Austral. J. Chem., 1987, 40, 677) of certain phenols have been described.
The cumulative indexes of the J. Chromatography for volumes 251-350 (pp.351-2), volumes 351-400 (pp.303-4), volumes 401-450, (pp.582-4), and volumes 451-500, (pp.336) list many separatory and analytical studies of phenols concerned with synthetic, environmental, biological and medicinal chemistry.

(a) Synthesis

No fundamentally new methods have been introduced for the synthesis of phenolic rings with the exception of certain strategies using transition metal intermediates. Five other basic methodologies have been exhaustively investigated during the previous decade. These comprise the hydroxylation of arenes, the replacement of substituents by the hydroxyl group, the formation of phenolic rings by way of alicyclic precursors, the production of phenols from a single acyclic precursor and from two acyclic components with the same methodology. In this account preparations of naphthols and some hydroxypolycyclic compounds have been included. The methods described essentially relate predominantly to phenol, and homologous compounds including alkylene relatives. The section includes mention of the synthesis of alkylphenols from phenol which are conveniently considered here although they are equally examples of reactions of the phenolic aromatic ring.

(i) The Hydroxylation of Arenes

The mechanism of the catalytic hydroxylation of aromatic hydrocarbons has been reviewed (E.A. Kharakhanov, S.Y. Narin and A.G. Dedov, App. Organomet. Chem., 1991, 5, 445). The hydroxylation of benzene or toluene by peroxydiphosphate (K. Tomizawa and Y. Ogata, J. Org. Chem., 1981, 46, 2107) and by peroxydisulphate (M.K. Eberhardt, J. Am. Chem. Soc., 1981, 103, 3876) are similar. Mesitylene (1,3,5-trimethylbenzene) in acetone solution with a small proportion of peroxymonophosphoric acid affords a 74% yield of mesitol (1,3,5-trimethylphenol). An aqueous mixture of benzene, hydrogen peroxide, ferric nitrate and a catalytic proportion of hexadecyltrimethylammonium bromide at 35-50°C gives an 82% yield of phenol, although in the absence of the phase-transfer catalyst the yield drops to 8-10% (SU 1268562 Moscow Lomonosov Univ.). Phenol is available if silica gel containing adsorbed iron(III) catechol complex in benzene is treated with 35% hydrogen peroxide (S. Tamagaki, K. Hotta and W. Tagaki, Chem Letters, 1982, 651). A 56% yield of phenol results from a mixture of benzene and a vanadium(V) catalyst in acetonitrile (H. Mimoun, et al., J. Amer. Chem. Soc., 1983, 103, 3101).

Trifluoromethanesulphonic acid solutions of various arenes are converted to phenols under a variety of conditions. The electroreduction of dioxygen (R. Onishi and A. Armata, J. Chem. Soc. Chem. Commun., 1986, 1630), and the generatinn of hydroxyl radicals in benzene solution (A.Kuni, S, Hata, S. Ito and K. Sasaki, J. Amer, Chem. Soc., 1986, 108, 6012) gives phenol, while diazonium fluoroborate solutions give phenyltriflates by thermal, or photochemical decomposition (N. Yoneda, T. Fukuhara, T. Mizokami and A. Suzuki, Chem. Letters, 1991,459). Toluene in trifluoromethanesulphonic acid undergoes electrophilic o-hydroxylation by the addition of sodium borate giving 2-methylphenol in 66% yield, (G.K.S. Prakash, N. Krass, Qi Wang and G.A. Olah, Synlett., 1991, 1, 39). o-Hydroxylation of the t-butyldimethylsilyl ether of benzyl methyl ketone is effected in 66% yield by photooxidation in the presence of tetraphenylporphine followed by reduction with sodium borocyano hydride or lithium aluminium hydride (I. Saito, et al., Tetrahedron Letters, 1982, 23, 1717) a procedure which can be extended to the synthesis of other o-alkylphenols.

By contrast with the preceding methods, 4-hydroxylation has been achieved by the use of the fungus *Beauverya sulphurescens* (B. Vigne, A Archelas, J.D. Fourneron and R Furstoss, Tetrahedron, 1986, *42*, 2451). 2-Phenoxypropionic acid has been converted to the 4-hydroxy compound, 2-(4-hydroxyphenoxy)propionic acid in 97% yield with the use of *Streptomyces hygroscopicus* in a solution containing corn steep liquor and inorganic phosphates over 3-7 days at 28°C (BASF WO-9011362).

(ii) The Conversion of Phenolic Esters and Phenolic Ethers to Phenols

1-Naphthyl benzoate with 1-butylamine affords a 97% yield of 1-naphthol whereas alkyl and allylic analogues are unaffected (K.H. Bell, Tetrahedron Letters, 1986, *27*, 2263). 5,6,7,4'-Tetrabenzoyloxyflavone undergoes selective hydrolysis in pyridine with phenol and silver carbonate at ambient temperature to give 5,6-dibenzoyloxy-7,4'-dihydroxyflavone (P.Ramesh, C.R. Yuvarajan and V. Narayanan, Curr. Sci., 1989, *58*, 29).

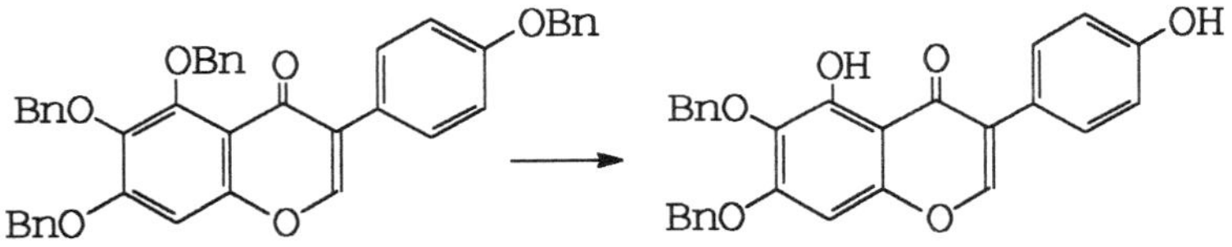

Sodium hydrogen telluride (from tellurium powder and sodium borohydride *in situ*) reacts with ethyl phenyl carbonate in ethanol containing deoxygenated acetic acid as a buffer, to give a 98% yield of phenol (N. Shobana *et. al.*, Indian J. Chem., 1988, *27B*, 965). The demethylation of anisole affords phenol in quantitative yield when reacted with aluminium iodide and tetra-n-butyl iodide (S. Andersson, Synthesis, 1985, 437). When mixed ethers are present preferential cleavage can occur and the reagent is milder than aluminium chloride and boron tribromide. A freshly-prepared solution of the reagent in acetonitrile converts allyloxybenzene to phenol in 87% yield (M.V. Bhatt and J.R. Babu, Tetrahedron Letters, 1984, *25*, 3437).

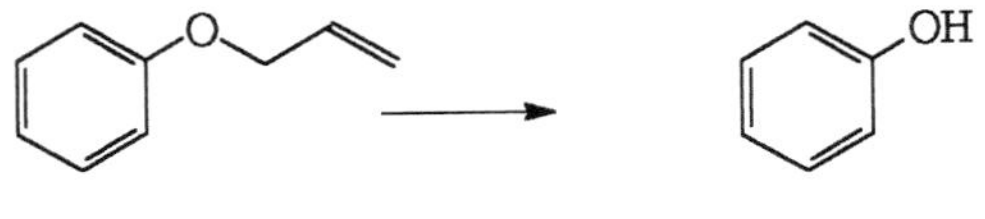

By contrast 2-naphthyl benzyl ether with sodium formate and formic acid in methanol or ethanol containing a little 10% Pd-C is converted by heating to 2-naphthol in 79% yield, (H.G. Krishnamurty, S. Ghosh and S. Sathyanarayana, Indian J. Chem., 1986, *25B*, 1253). Selective effects are possible as in the removal of the protective azidomethyl group in the presence of methoxyl and ester groups with an equimolar proportion of stannous chloride (B. Loobinoux, S. Tabbache, P. Gerardin and J. Miazimbakana, Tetrahedron Letters, 1988, *44*, 6055).

MeCO$_2$CH$_2$-C$_6$H$_3$(OCH$_2$N$_3$)(OMe) → MeCO$_2$CH$_2$-C$_6$H$_3$(OH)(OMe)

Microbiological demethylation in 90% yield is effected by addition of anisole and reduced nicotinamide adenine dinucleotide phosphate to a cell-free extract of methane-grown *methylosinus tricosporium* OB 3b, in a shaker-incubator (S.G. Jezequel, B. Kaye and L.J. Higgins, Biotech. Letters, 1984, *6*, 567).

(iii)Formation of Phenols by Replacement of Boron, Alkyl, Oxoalkyl, Carboxyl, Formyl, Halogeno, Hydroperoxide, Lithio, Nitro, Phosphate, Sulpho, Silano, Triazeno, and Thallic Groups
Phenylboronic acid and also triazenes are primarily used for obtaining ^{17}O-labelled phenol. Thus a solution of phenylboronic acid in ether-ethanol (10:1) when treated with ^{17}KO$_2$H gives an 86% yield of phenol with 13.8% enrichment (S. Bank and K.L. Longley, J. Labelled Compounds Radiopharm., 1990, *28*, 41). Both ^{17}O and ^{18}O-labelled phenols are obtained in 88% yield from triazenes in acetonitrile solution by their dropwise addition to a boiling aqueous suspension of bio-rad AG 50W-X12 acid resin in ^{18}O-enriched water (N. Satyamurthy et al., Tetrahedron Letters, 1990, *31*, 4409).
Benzoic acid in a mixture of steam and air (1:38:0.6)is converted to phenol over Cu-ZrO-K$_2$O on alumina at 280-310°C at 90-110psig. (Lummus Co., USP 4277630). The process is reminiscent of the pyrolysis of copper(II) benzoate by the Dow method (W. W. Kaeding, J. Org. Chem., 1962, *27*, 3551; A. A. Durrani and J. H. P. Tyman, J. Chem. Soc., Perkin Trans, 1979, 2069).

C$_6$H$_5$CO$_2$H → C$_6$H$_5$OH

3-Formylbenzocyclobutene when added to monoperoxyphosphoric acid in acetonitrile affords a 79% yield of 3-hydroxybenzocyclobutene (Dow Chemical Co., USP 5120884).

1,3-Dinitrobenzenes and nitronaphthalenes are converted to the corresponding m-nitrophenols by reaction with alkoxides and with benzyl alcohol containing tetramethylurea, followed by a debenzylation (F. Effenberger, M. Koch and W. Streicher, Chem. Ber., 1991, *124*, 163).

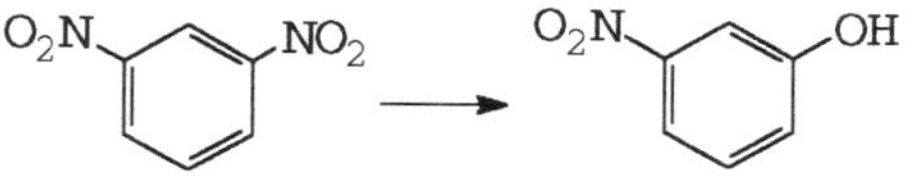

Naphthalene-1,5-disulphonamide reacts by a Bucherer-type procedure with 20% aqueous sodium hydroxide in an autoclave at 220°C during 10-15 hours to afford a 89% yield of 5-hydroxy-1-naphthalenesulphonamide (Bayer AG, DE 3633906).

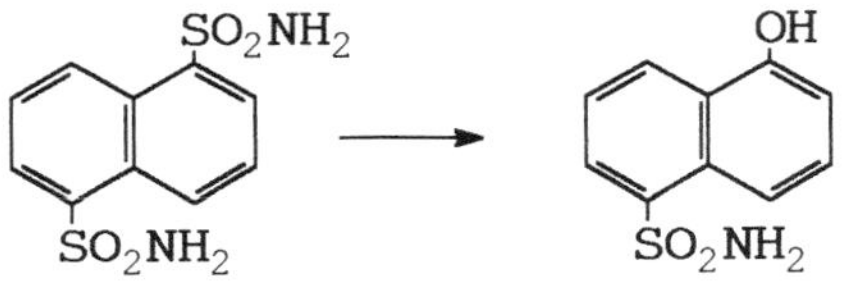

Iodobenzene stirred in a sealed reactor with potassium hydroxide and a little cuprous oxide in deionised water for 4 hours at 180°C gives a 61% yield of phenol (Montedipe Spa, EP 204271).

Phenols are derived, albeit in a circuitous manner, from the

oxidation of aryltriethoxysilanes (by way, initially, of Grignard reagents and tetrahalogenosilanes) in methanol with 3-chloroperbenzoic acid (A. Hosomi et al., Chem. Letters, 1981, 243).
1-Acetyl-7-hydroxyindoline is obtained in 62% yield by the oxidation of (1-acetyl-2,3-dihydroindol-7-yl)thallium bis trifluoroacetate in dimethylformamide/water (1:1) at 125°C, containing copper sulphate. These conditions are mild to those normally used (Nippon Steel Chem., KK, JO 2091059).

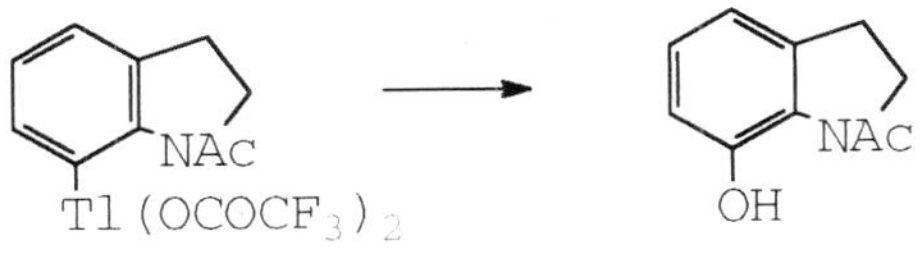

Aromatic aldehydes (having alkyl, or two or more alkoxy groups) in dichloromethane containing a small proportion of (2,4-dinitrophenyl)selenide upon treatment with excess 30% hydrogen peroxide, and hydrolysis with methanolic potassium hydroxide give the corresponding phenols in high yield (L. Syper, Synthesis, 1989, 167).

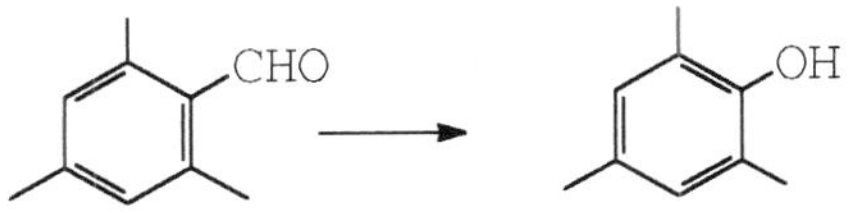

2-(1-Hydroxy-1-methylethyl)-3-methoxy-5,6,7,8-tetrahydronaphthalene upon treatment with 90% hydrogen peroxide, followed by stirring with a little 4-toluenesulphonic acid afforded, by way of the hydroperoxide, a 85% yield of 2-hydroxy-3-methoxy-5,6,7,8-tetrahydronaphthalene (D.L. Boger and R.S. Coleman, J. Org. Chem., 1986, *51*, 5436), in a procedure reminiscent of the cumenehydroperoxide rearrangement.
In this connection the catalyic reduction with hydrogen under slight pressure over Pd-alumina of other arylhydroperoxides can also be used to afford phenols (Mitsui Toatsu Chem. Inc., J59 110639).

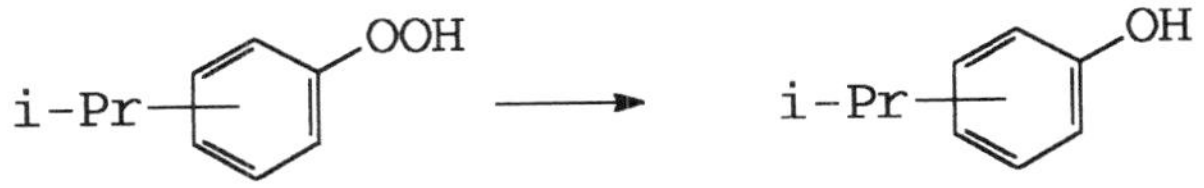

Tri-(2-hydroxyphenyl)phosphine is obtained in 87% yield from the reaction of triphenylphosphate with lithium diisopropylamide (B. Dhawan and D. Redmore, Synth. Commun., 1987, *17*, 465).

(iv) Formation of Phenols from Cycloaliphatic Precursors

In this approach a number of saturated monocyclic and bicyclic compounds are employed as intermediates for the synthesis of phenols and naphthols. Cyclohexanone and 2 moles of propanal at 150°C in the presence of bis(cyclopentadienyl zirconium dichloride) afford 2,6-di-n-propylphenol in 69% yield (T. Nakano et al., J. Org. Chem., 1988, *53*, 5181).

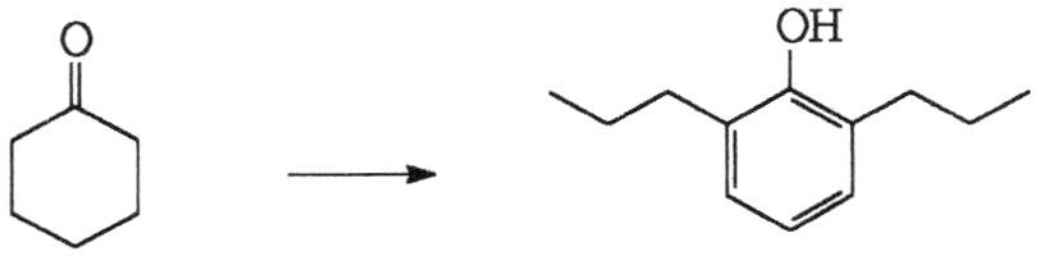

Cyclohex-2-enone, in tetrahydrofuran containing lithium diethylamide at -78°C, is converted to the 6-phenylselenide by the action of phenylselenylchloride. The isolated intermediate upon treatment with either 3-chloroperbenzoic acid or 30% hydrogen peroxide in tetrahydrofuran at -15°C, and warming of the mixture to ambient temperature affords phenol in 55% yield. The procedure was ineffective for 1-tetralone (M.I. Al-Hassan, Synth. Commun., 1989, *19*, 453).

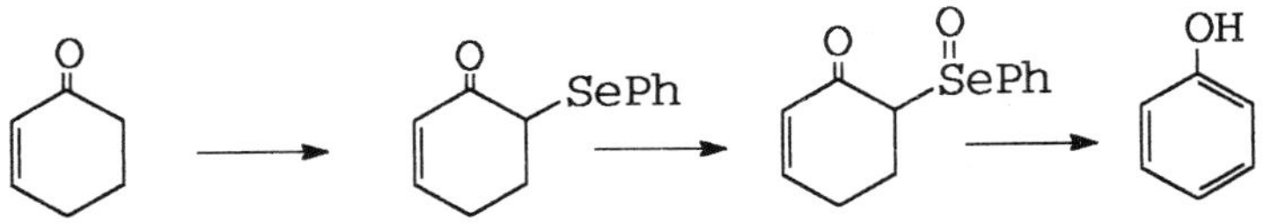

Thermolysis of 2-cyclopropylcyclohex-2-enone yields 2-n-propylphenol (G. Pattenden and D. Whybrow, J. Chem. Soc., Perkin I, 1981, 3147).

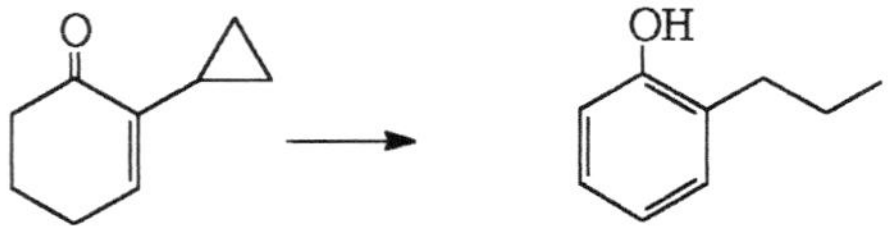

When isophorone is heated at 400-600°C over chromia-alumina 3,5-xylenol is obtained in 87% yield (K.V. Ramanamurthy et al., Adv. Catal., Proc. 7th Natl. Symp. Catal., 621; Chem. Abs., *104*, 186059).

The trimethylsilyl ether of 1-tetralone in dichloromethane containing 2,6-dimethylpyridine can be oxidised by trityl perchlorate to give 1-naphthol in 93% yield (M.T. Reetz and W. Stephan, Ann. Chem., 1980, 533).

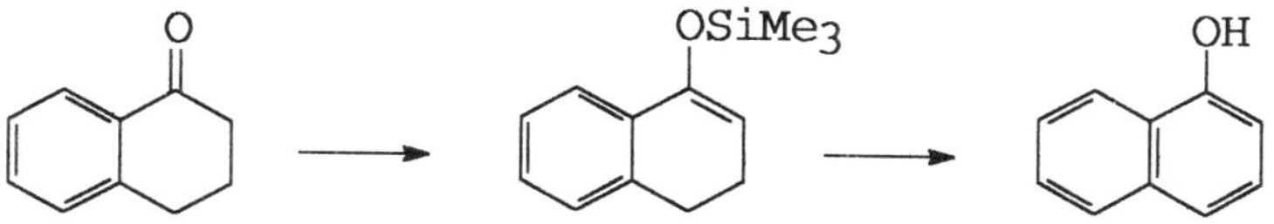

The cyclohexenocyclohexen-1,4-dione prepared, by Diels-Alder addition from 2-acetamido-3-iodo-1,4-benzoquinone and 1-methoxy-2-methyl-3-trimethylsiloxy-1,3-butadiene, upon treatment with 2M hydrochloric acid, followed by aerial oxidation gives the corresponding naphthaquinone, 2-acetamido-3-iodo-6-hydroxy-7-methylnaphtho-1,4-quinone in 78% yield (A.P. Kozikowski, K. Sugiyama and J.P. Springer, Tetrahedron Letters, 1980, *21*, 3257).

Certain bridged benzocyclohexanones afford substituted 1-naphthols merely by thermal or irradiation treatment. Thus, 3,4-benzo-6-methylenebicyclo[3.1.0]hex-3-ene-2-one in 1,2-ethanediol at 150°C for 1 hour gives a quantitative yield of 3-hydroxymethyl-1-naphthol (M. Rule et al., Tetrahedron, 1982, *38*, 787).

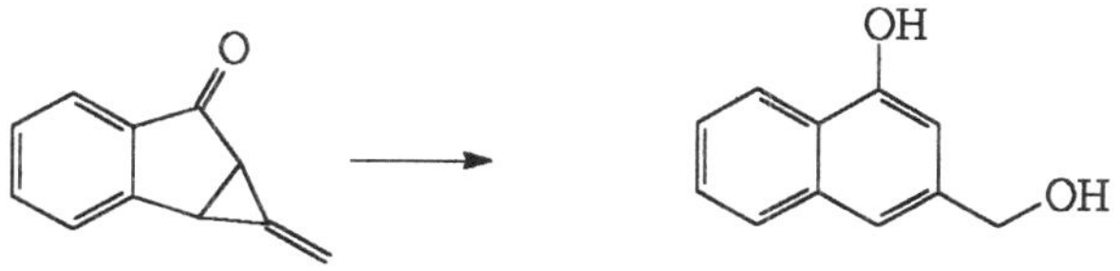

1,2-Xylylene dibromide as its monophenylsulphonyl derivative, reacts with diethyl methylmalonate in tetrahydrofuran containing excess sodium hydride to effect both alkylation and cyclisation. The product obtained in 80% yield upon desulphonylation, hydrolysis, decarboxylation and aromatisation affords 3-methyl-2-naphthol (E. Ghera and Y. Ben-David, Tetrahedron Letters, 1983, *24*, 3533).

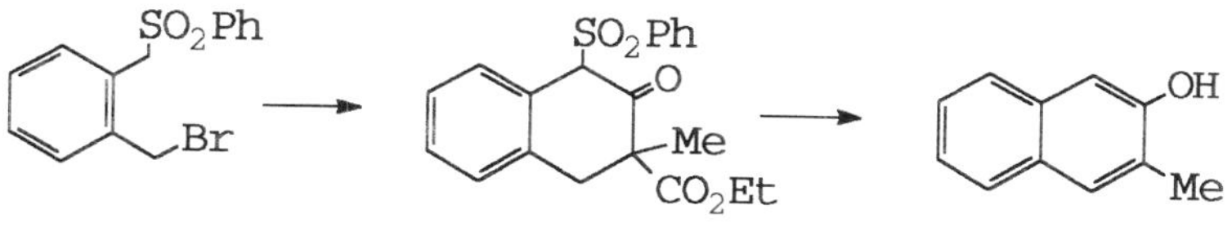

(v) Formation of Phenols from an Acyclic Precursor

A review has appeared on this topic (P. Bamfield and P.F. Gordon, Chem. Soc. Rev., 1984, *13*, 441). Very little work has been reported on the transformation of C_6 acyclic intermediates and homologues to phenols, although in the described examples below great ingenuity has been demonstrated. The synthesis of phenol has been described in 93% yield by the flash-vacuum pyrolysis of 1-(E,E-hexa-2,4-dienoyl)-3,5-dimethylpyrazole at 650°C or at 80°C under 10^{-3} mm. (J. Besida, et al., Austral. J. Chem., 1982, *35*, 1373).

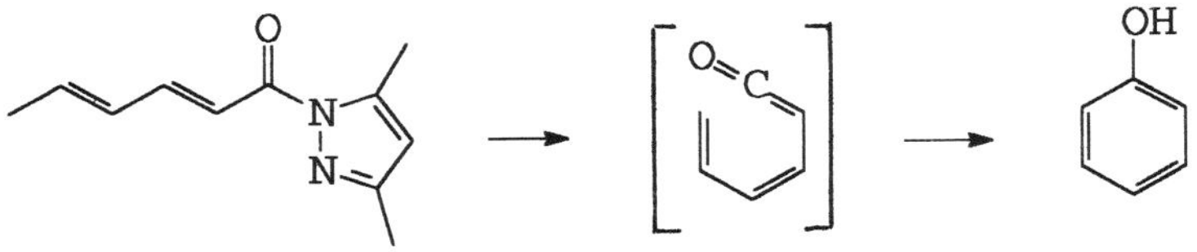

6-Ethyloct-5-ene-2-yn-4-one on heating with 2,4,6-trimethylpyridinium 4-toluenesulphonate in toluene solution affords an 82% yield of 5-ethyl-3,4-dimethylphenol (R.A. Jacobs and J.I. Kravitz, Tetrahedron Letters, 1988, *29*, 6873).

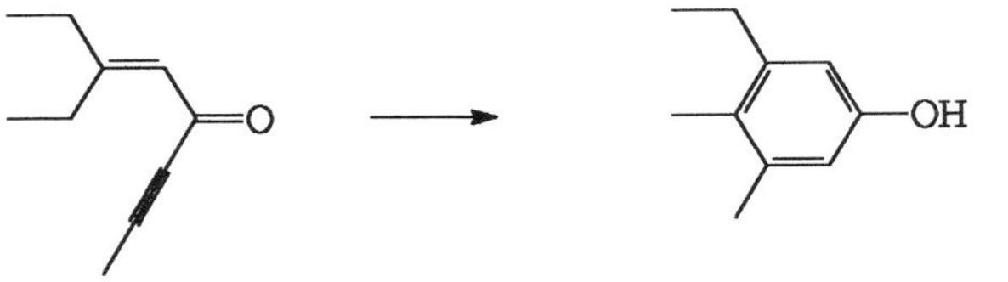

3-Methyl-6-(6-methyl-5-hepten-2-yl)-2,4-hexadienoic acid can be cyclised to 5-methyl-2-(6-methyl-5-hepten-2-yl)phenyl acetate in 70% yield by heating in acetic anhydride containing sodium acetate (D. Murali and G.S.K. Rao, Synthesis, 1987, 254).

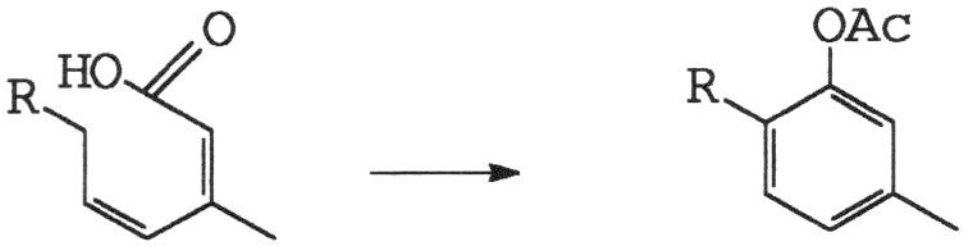

A related acid, 3-methoxycarbonyl-5-methylhexa-3,5-dienoic acid with N-methylmorpholine and diphenylphosphonic acid, after treatment with triethylamine and completion of reaction at ambient temperature, affords methyl 3-hydroxy-5-methylbenzoate in 50% yield. The procedure is similar to that of the Horner-Emmons methodology (K. Clinch et al., Tetrahedron, 1989, *45*, 239).

Methyl 3-oxo-6-methylhept-6-enoate in acetic acid containing excess manganese(III) acetate and some lithium chloride is oxidatively cyclised to produce methyl 5-methylsalicylate in 71% yield (B.B. Snider and J.J. Patricia, J. Org. Chem., 1989, 54, 38).

In the last two examples hydrolysis and decarboxylation can be employed to obtain the respective parent phenols.
4-Benzyl-1,2,5-trimethyl-1H-imidazo[4,5-c]-pyridinium iodide upon heating with aqueous ethanolic potassium hydroxide undergoes hydrolysis at the 4,5 N=C bond, followed by alkylation/recyclisation to give 4-hydroxy-5-phenyl-1,2-dimethylbenzimidazole in 95% yield (Y.M. Yutilov and A.G. Ignatenko, Khim. Geterotskil. Soedin, 1987, 995).

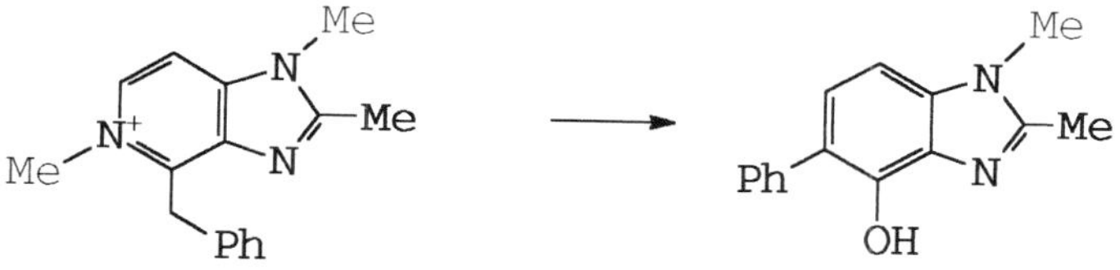

(vi) Formation of Phenols from Reactions of two Acyclic Precursors

A variety of procedures has been adopted in this somewhat neglected approach to the synthesis of phenols.
3-Phenylphenol results in 83% yield by reaction of the aminoketonic salt from 2-chloroacetone and pyridine with the Mannich base from acetophenone, formaldehyde and dimethylamine in hot ethanol containing triethylamine (K. Eichinger and P. Nussbaumer, Synthesis, 1991, 663).

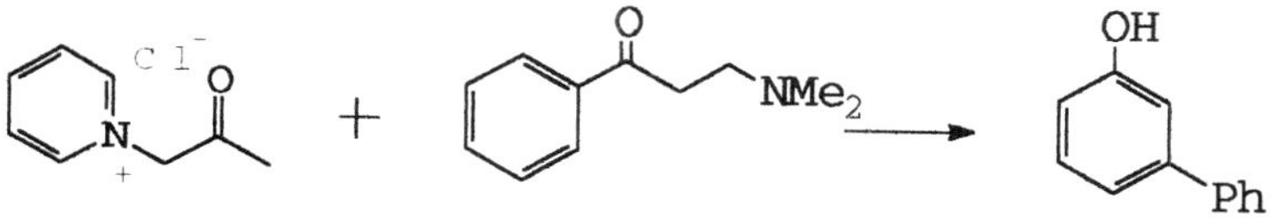

2,6-Diarylphenols are obtained from 2,6-diarylcyclohexanones itself by dehydrogenation. The latter are obtained from dibenzylketones and 1,3-dibromopropane in the presence of sodium hydroxide and chlorobezene containing tetrabutylammonium bromide (D.E. Dana and A.S. May, Synthesis, 1982, 164).

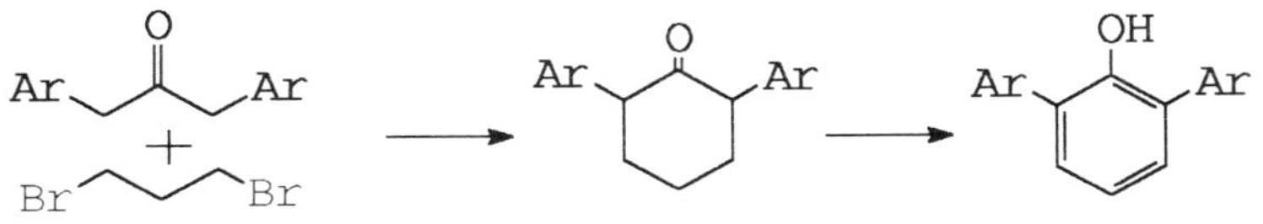

Michael addition reactions are employed to obtain 5,6,7,8-tetrahydronaphthol. Thus the lithium enolate of cyclohexanone in tetrahydrofuran reacted with 3-phenylthiobut-3-en-2-one giving an adduct which was transformed to the final product in 74% yield by prolonged heating with 4-toluenesulphonoic acid in boiling benzene solution (K. Takaki et al., J. Chem. Soc. Chem. Comun., 1980, 1183).

A variation on the preceding scheme consists in the reaction of the carbanion from 2-(phenylsulphinyl)cyclohexanone with methyl vinyl ketone (D.L. Boger and M.D. Mullican, J. Org. Chem. 1980, 45, 5002).

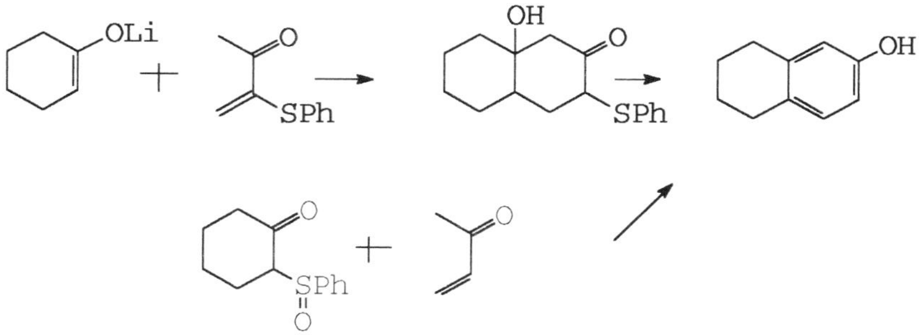

Acetylenic precursors have also found a place in the synthesis of phenols. For example, 2,3-dimethylbut-3-enoyl chloride reacts with propyne to give 2,3,5-trimethylphenol, whereas but-2-yne affords 2,3,5,6-tetramethylphenol (M. Karpf, Tetrahedron Letters, 1982, 47, 4923).

(vii)Formation of Phenols and Naphthols from Transition Metal Intermediates

5-Hydroxy-6-methylindane is synthesised in 57% yield by a regiospecific double ring closure involving the reaction of 1,6-heptadiyne with a chromium carbene complex in deoxygenated tetrahydrofuran (W. Wulff et al., J. Am. Chem. Soc., 1985, 107, 1060).

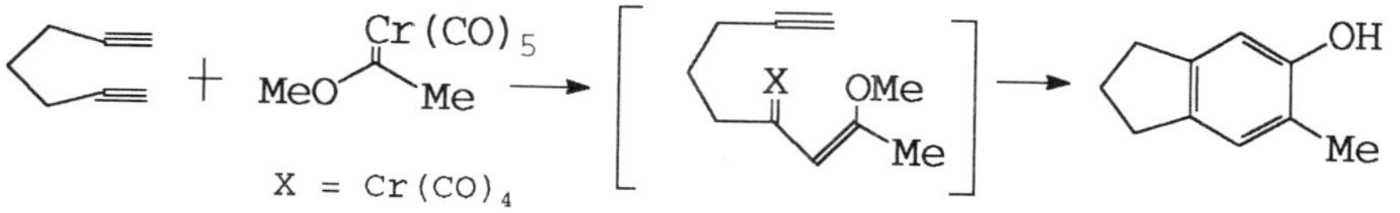

Substituted 1,7-heptadiynes are used in similar reactions (D. Plouin and R. Glenat, Bull. Soc. Chim. Fr., 1975, 336), and this methodology can be extended to the synthesis of steroidal compounds.

Phenols can be produced by the general nickel(0)-promoted reactions of cyclobutenones and alkynes (M.A. Huffman and L.S. Liebeskind, J. Am. Chem. Soc., 1990, 113, 2771). Thus, 2,3-diethyl-5-phenylphenol is obtained in 75% yield from 3-phenylcyclobutenone and 2 moles of 3-hexyne in diethyl ether containing 0.1 mole Ni(COD)$_2$, [bis(1,5-cyclooctadiene)nickel].

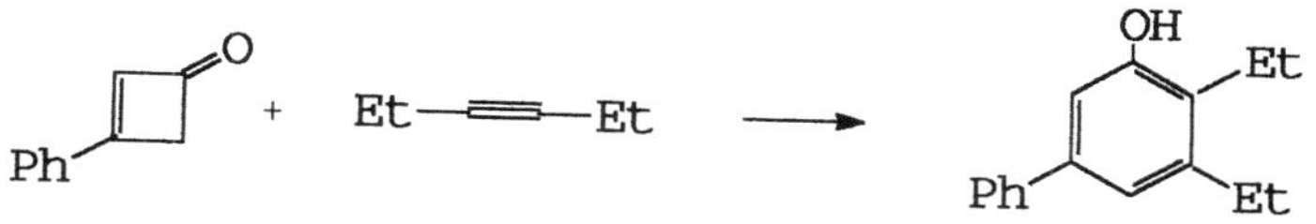

Several different organometallic reactants have been utilised for the formation of polysubstituted phenols. For example, vinylcyclopropenes in benzene solution containing a catalyic amount of tetracarbonyldichlororhodium undergo carbonylation and rearrangement at 80°C with carbon monoxide under 1 atmosphere pressure. In this way 5,6-diethyl-3-phenyl-2-propylphenol is obtained in 82% yield (S.H. Cho and L.S. Liebeskind, J. Org. Chem., 1987, 52, 2631).

6-Butyl-2-phenyl-4-propylphenol is synthesised in quantitative yield from 3-benzoyl-2-propylcycloprop-1-ene and 1-hexyne in dichloromethane upon treatment with the rhodium complex, [Rh(CO)$_2$Cl$_2$]$_2$. An intermediate oxepin is formed which gives the final product upon acidification with 1 drop of hydrochloric acid (A. Padwa and S.L. Xu, J. Am. Chem. Soc., 1992, 114, 5881).

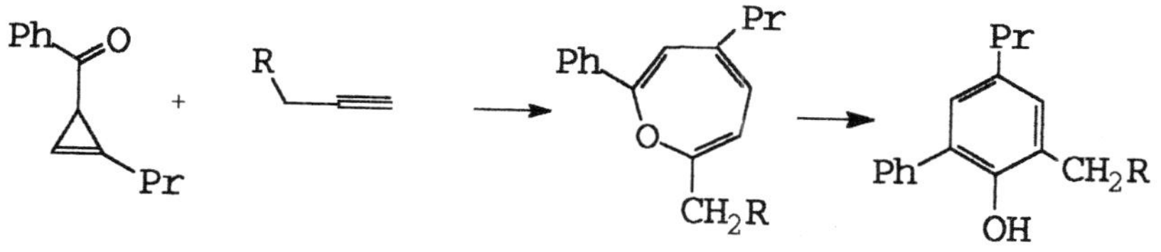

Cobalt complexes of vinylketene react with alkynes affording moderate yields of phenols (M.A. Huffman and L.S. Liebeskind, J. Am. Chem. Soc., 1990, *112*, 8617).
Palladium-catalysed insertion reactions have proved useful for both bicyclic and polycyclic compounds. By this methodology 3-(2-furyl)allyl acetate with a little $PdCl_2(PPh_3)_2$ in benzene at 170°C containing triethylamine and acetic anhydride (1:1) when reacted in an autoclave pressured (70Kg/cm^2) with carbon monoxide gave an 85% yield of 4-acetoxybenzofuran (M. Iwasaki et al., Tetrahedron Letters, 1989, *30*, 95). Similarly, 7-acetoxybenzofuran can be derived from 3-(3-furyl)allyl acetate.

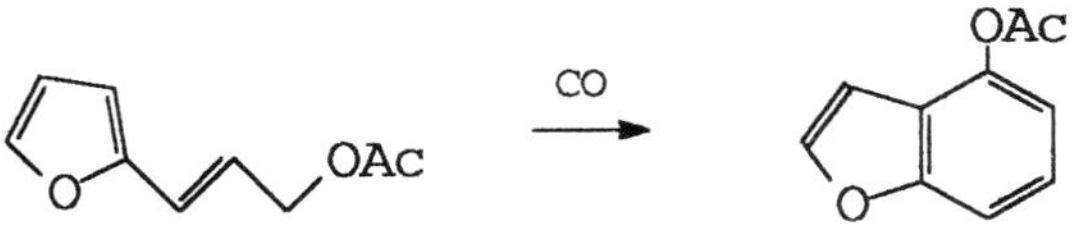

Dienoid systems have also been investigated and, for example, 5-phenylpenta-2,4-dienyl acetate reacts with palladium(II) chloride bis(triphenylphosphine), triethylamine and acetic anhydride under carbon monoxide pressure at 140°C to afford 2-phenylphenyl acetate in 69% yield (Y. Ishii et al., J. Chem. Soc. Chem. Commun., 1991, 695).

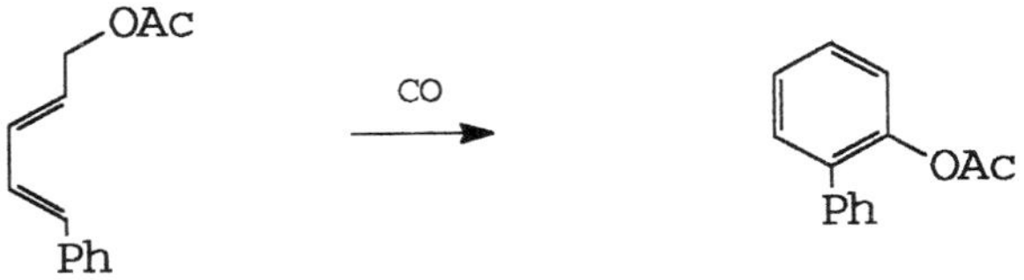

1,2-Aryldialdehydes and alkynes undergo niobium-catalysed regioselective cycloadditions to afford 1-naphthols (J.B. Hartung and S.F. Pedersen, J. Am. Chem. Soc., 1989, *111*, 5468).

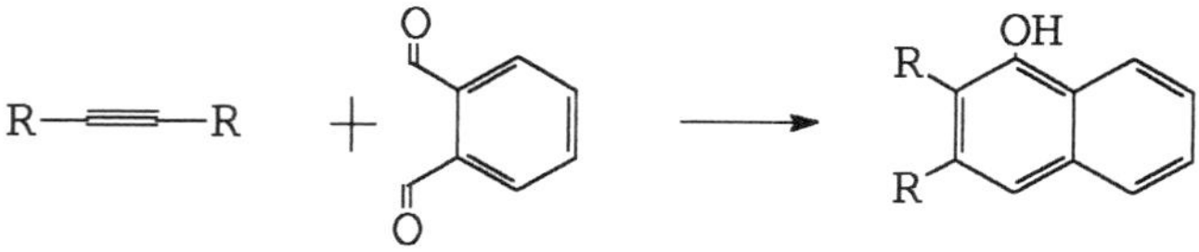

Niobium complexes of alkynes, such as 1-dodecyne, also furnish 1-naphthols in low yields along with other products. Thus phthalaldehyde in tetrahydrofuran with 1-dodecyne and niobium pentachloride/dimethoxyethane in benzene give 2-decylnaphthalene (14%), 2-decyl-1-naphthol (14%) 1-dodecene (24%) and 2-decyl-1,4-dihydroxynaphthalene (2%)(Y. Kataoka et al., Tetrahedron Letters, 1990, *31*, 369).

(viii) Phenols formed by Alkylations and Arylations

Organometallic compounds, alkyl halides, alkenes, alcohols, acetals and ethers have been employed for the synthesis of alkyl and aralkylphenols and a notable feature of many of these processes is their ortho regiospecificity.

2-Methylphenol is formed in 50% yield by the addition of phenol in toluene solution to ethyl iodomethyl zinc at 0°C (E.K. Lehnert, J.S. Sawyer and T.C. Macdonald, Tetrahedron Letters, 1989, *30*, 5215). 2-Isopropylphenol is produced in 92% yield with an 85% conversion from the passage of propan-2-one and phenol (5:1) through a zinc oxide/ferric oxide catalyst at 355°C followed by hydrogen (Mitsui Petrochem. Ind., KK, J5 9025342). 2-(1,1,3,3-Tetramethylbutyl)phenol is formed in 64% yield from the reaction of 2,4,4-trimethylpent-1-ene and phenol (8:1) at 100-110°C in the presence of aluminium turnings (J.A.M. Laan and J.P. Ward, Chem. and Ind., (London), 1987, 34). (E)-2-(3,7-Dimethyl-2,6-octadienyl)phenol is synthesised in 86% yield by treatment of potassium phenate with zinc chloride and (E)-3,7-dimethylocta-2,6-dienylchloride (F. Bigi et al., Synthesis, 1981, 310).

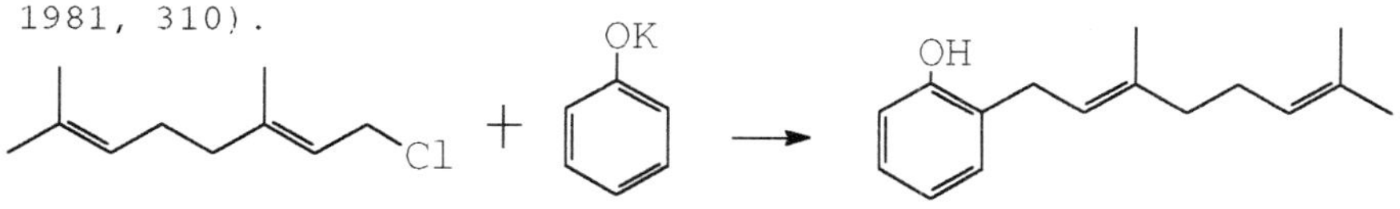

1-(3-Methylbut-2-enyl)-2-naphthol results in 60% yield from the sequential addition of sodium iodide, isoprene and trimethylchlorosilane to 2-naphthol in acetonitrile (D.N. Sharma et al., Tetrahedron Letters, 1984, *25*, 5581). By contrast, the reaction of isoprene, more particularly with resorcinol, in the presence of magnesium iodide/diethylether affords chromans (P.K. Chowdhury, J. Chem. Res., (S), 1990, 390).

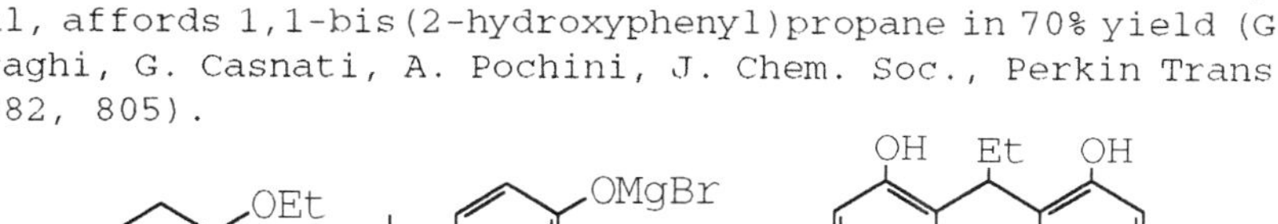

Phenoxymagnesium bromide when reacted with propanal diethyl acetal, affords 1,1-bis(2-hydroxyphenyl)propane in 70% yield (G. Casiraghi, G. Casnati, A. Pochini, J. Chem. Soc., Perkin Trans. I, 1982, 805).

2,6-Dimethylphenol in carbon tetrachloride, containing silica gel and sodium carbonate, upon treatment with tert-butylbromide affords 2,6-dimethyl-4-tert-butylphenol in 85% yield (Y. Kamitori et al., J. Org. Chem., 1984, *49*, 4161).
2,2-bis(4-Hydroxyphenyl)propane ('bis phenol A'), is synthesised in 82% yield by the sequential addition of telluriun tetrachloride and 2-bromopropene to phenol at 0°C (Rhone-Poulenc Chim., FR 2646418).

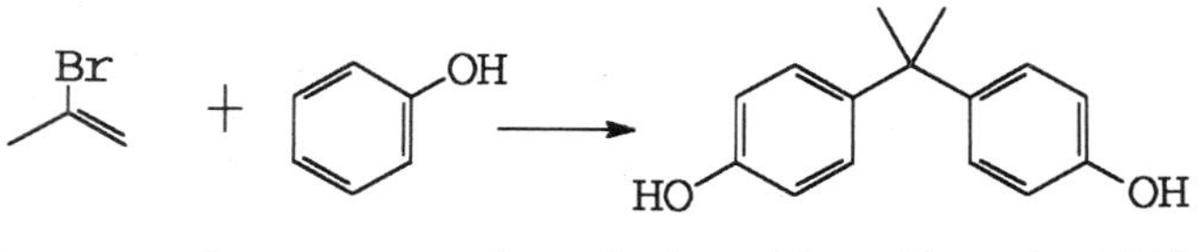

Grignard reagents are employed for the direct arylation of phenols (H. Jendralla and L.-J. Chen., Synthesis, 1990, 827). 2-Benzylphenol is produced in 88.5% yield from phenol and γ-alumina (or magnesium oxide) treated with benzyl alcohol and reacted at 200°C (Mitsubishi Chem. Ind., KK, J60 032740), or in 84% yield from phenol and dibenzyl ether in boiling toluene containing γ-alumina, while 4-chloro-2-benzylphenol is obtained in 92% yield from 4-chlorophenol, benzyl alcohol and sodium-Y-zeolite by heating at 200°C (Bayer AG, DE 370917).

(ix) **Formation of phenols by Rearrangements**
The methodology of the well-known Claisen rearangement is extended by the use of Montmorillonite clays (W.G. Dauben, J.M. Cogan and V. Behar, Tetrahedron Letters, 1990, *31*, 3241).

Rearrangment of certain naphthyl ethers affords naphthols. Thus 4-chloro-1-naphthyl tetrahydropyranyl ether when treated with boron trifluoride etherate in dichloromethane at 0°C gives a 65% yield of 4-chloro-2-(tetrahydropyranyl)-1-naphthol. (T. Kometani, H. Kondo and Y. Fujimori, Synthesis, 1988, 1005).

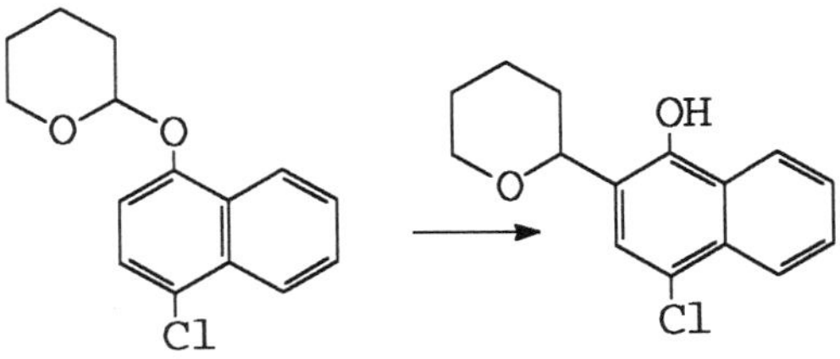

(x) Formation of Polycyclic Phenols

A number of reactions for the formation of tricyclic compounds are described.
2-Methylbiphenyl-2'-N,N-diethylcarboxamide when treated with excess lithium diethylamide in tetrahydrofuran at 0°C gives a 92% yield of 9-phenanthrol by a process reminiscent of the Elbs cyclisation (J. Fu, M.J. Sharp and V. Snieckus, Tetrahedron Letters, 1988, *29*, 5459).

5-(2,4,7-Trimethyl-3-benzofurylidene)-2,2-dimethyl-1,3-dioxan 4,6-dione on flash-vacuum pyrolysis at 650°C and 0.02mm Hg affords a 82% yield of 6,9-dimethylbenzofuran-3-ol, by a condensation/hydrolysis/ condensation pathway (R.F.C. Brown and C.M. Jones, Austral. J. Chem., 1980, *33*, 1817).

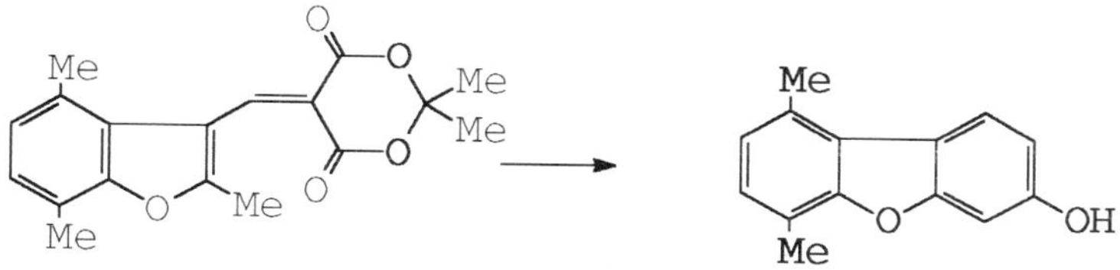

Enamines have been used to obtain hydroxybenzofurans. Thus, 1,4-benzoquinone in benzene upon treatment with 2-(4-morpholino)-1-phenyl-1-propene and finally hydrolysis with ethanolic hydrochloric acid gives in 46% yield 2-benzyl-5-hydroxybenzofuran. This compound is inaccesible by the standard Claisen intramolecular route (V.M. Lyubchanskaya et al., Khim. Geterotsikl. Soedin., 1989, 1031).

Aryldiazoketones afford arylketones upon photochemical irradiation and these are used to synthesise naphthols (A.Padwa et al., Tetrahedron letters, 1991, *32*, 5923).

(b) Reactions of Phenols

(i) Esterification

Aliphatic Acids: Methods for the formation of phenolic esters with higher carboxylic acids may be catalysed with 4A molecular sieve and improved by the addition of B_2O_3 (Proctor and Gamble Co., EP 108540). Phenyl acetate is formed nearly quantitatively from phenol and isopropenyl acetate plus 0.2 equivalents of potassium carbonate (J. Barry, G. Bram and A. Petit, Tetrahedron Letters, 1988, 29, 4567), and from phenol adsorbed on silica gel treated with nitrogen-diluted ketene (T. Chichara, S. Teratini and H. Ogawa, J. Chem. Soc. Chem. Commun., 1981, 1120).

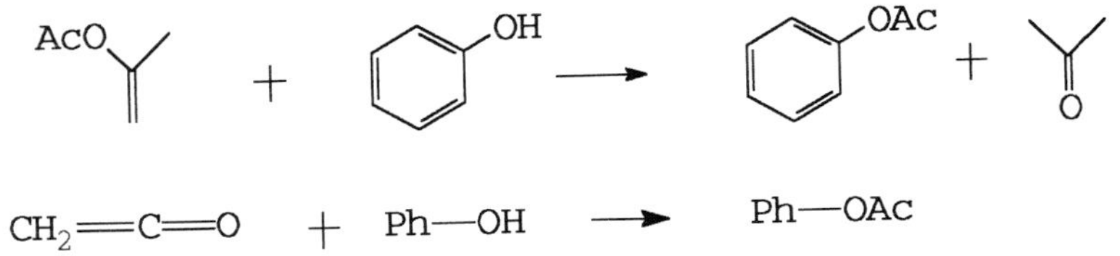

$$AcO-C(=CH_2)-CH_3 + PhOH \longrightarrow PhOAc + CH_3COCH_3$$

$$CH_2=C=O + Ph-OH \longrightarrow Ph-OAc$$

Phenyl 3-phenyl-3-oxopropionate results in 90% yield from the reaction of phenol and 5-phenyl-2,3-dihydrofuran-2,3-dione in boiling benzene (Y.S. Andreichikov and T.N. Tokmakova, Zh. Org. Chem., 1987, 23, 880).

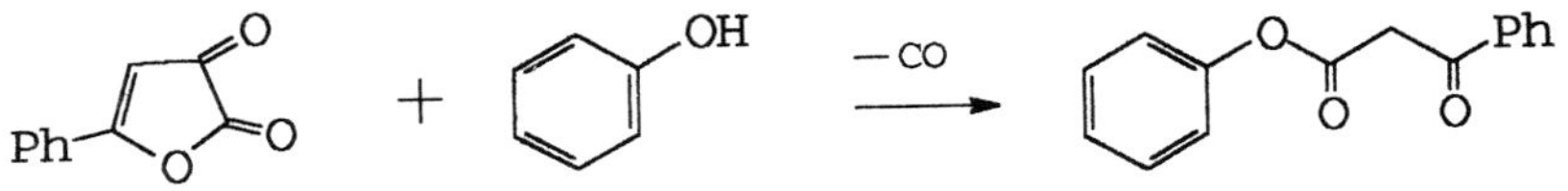

[Reaction: 5-phenyl-2,3-dihydrofuran-2,3-dione + PhOH → (−CO) → PhO-CO-CH$_2$-CO-Ph]

Aromatic Acids: Phenyl benzoate results in 91% yield as shown from the reaction of benzoic acid, phenol, carbon tetrachloride, triethylamine and triphenyl phosphine in acetonitrile at ambient temperature (S. Hashimoto and I. Furukawa, (Bull. Soc. Chim., Japan, 1981, 54, 2227), and in 95% yield from benzoyl chloride and the phenoxide form of Amberlite IRA-400 (D.G. Salunkhe et al., Curr. Sci., 1986, 55, 853).

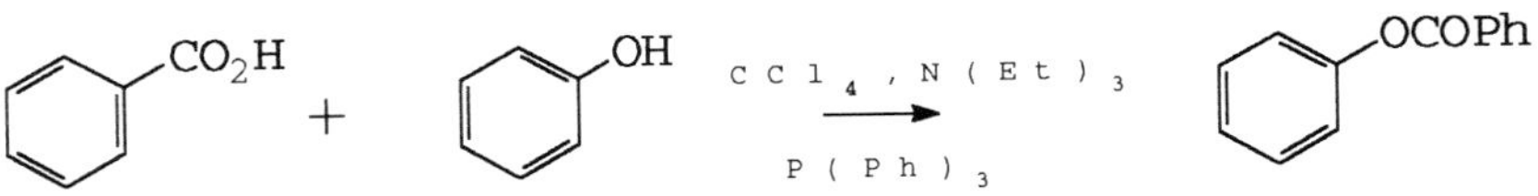

$$PhCO_2H + PhOH \xrightarrow[P(Ph)_3]{CCl_4, N(Et)_3} PhOCOPh$$

Aryl halides have been employed. Thus bromobenzene and phenol afford phenyl benzoate when reacted in pyridine or triethylamine containing a platinum metal catalyst under carbon monoxide at 0.1-10.0 Kg/cm$_2$ (UBE Industries, KK, J59 029641).

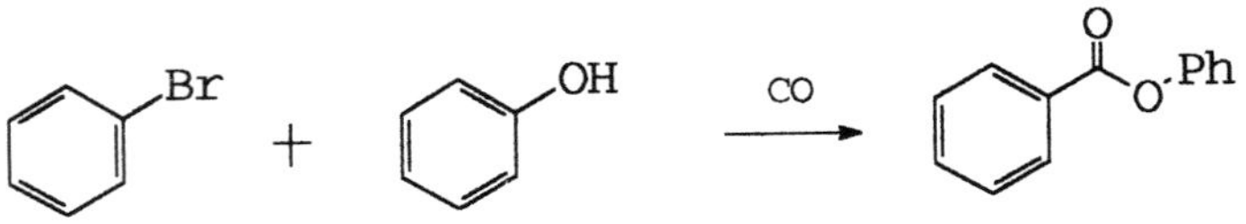

$$PhBr + PhOH \xrightarrow{CO} PhCO-O-Ph$$

Similarly, 4-fluoroiodobenzene and phenol give phenyl 4-fluorobenzoate in 91% yield in triethylamine containing a catalyst Ni(II) acetylacetonate under CO (at 50 Kg/cm$_2$) in a sealed reactor at 200-220°C (Asahi Chemical Ind., KK J62 198646).

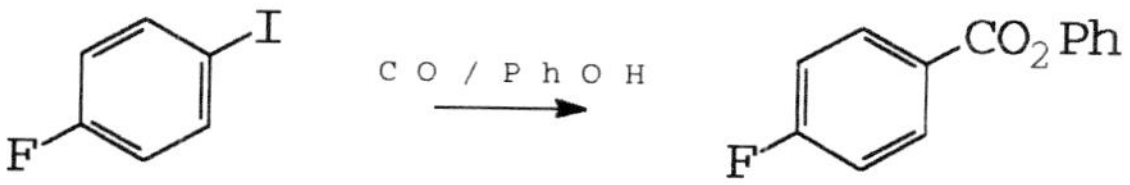

[reaction: 4-fluoro-iodobenzene + CO/PhOH → phenyl 4-fluorobenzoate]

Unsaturated Acids: N-Vinylpyrrolidin-2-one undergoes carbonylation with carbon monoxide and palladium(II) chloride in dichloromethane containing diisopropylamine. If phenol is present esterification occurs at the same time affording the N-acrylate derivative. In the case of 2,6-dichlorophenol (Ar = 2,6-dichlorophenyl) the corresponding ester is obtained in 85% yield (J.H. Hallgren and R.O. Matthews, J. Organometall. Chem., 1980, *192*, C12).

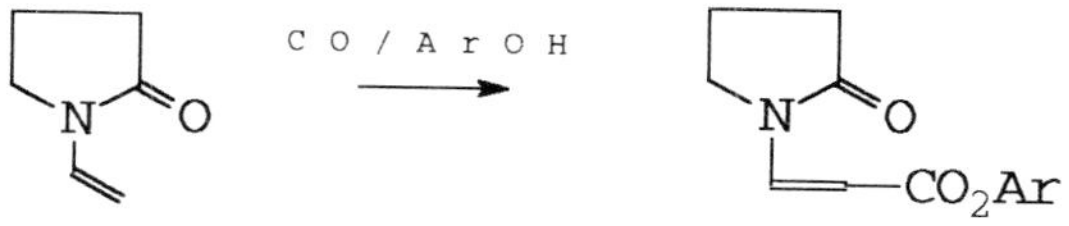

[reaction: N-vinylpyrrolidin-2-one + CO/ArOH → N-(CO$_2$Ar-vinyl)pyrrolidin-2-one]

4-Methylphenyl 2-phenylacrylate is obtained in 83% yield from the reaction of phenyl acetylene and excess 4-methylphenol in toluene with a catalytic amount of palladium(II) acetate at 100°C under CO pressure (1 bar) (K. Itoh, M.M. Miura and M. Nomura, Tetrahedron Letters, 1992, *37*, 5369).

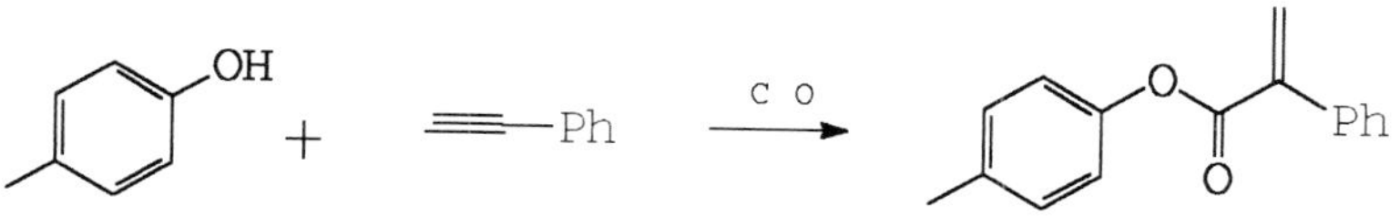

[reaction: 4-methylphenol + HC≡C-Ph + CO → 4-methylphenyl 2-phenylacrylate]

(ii) Diaryl Carbonates

A variety of procedures have been introduced for these compounds which are of interest in polymer chemistry. Diphenyl carbonate results in 85% yield from the transesterification of dimethyl carbonate with phenol catalysed by dioctyl tin oxide and titanium tetra-isopropoxide by passage through a 5A molecular sieve under pressure (203 bar)(General Electric Co., DE 3308921). This compound is also available from di(2,2,2-trifluoroethyl)carbonate and phenol containing sodium methoxide (General Electric Co., NL 7908991/2) and from 2-nitrophenyl carbonate and phenol in dichloromethane in the presence of 4-dimethylaminopyridine (D.J. Brunelle, Tetrahedron letters, 1982, *23*, 1739).

PhOH + MeO-C(=O)-OMe → PhO-C(=O)-OPh

Phenol heated with phenyl chloroformate in 1,2-dichlorobenzene at 150°C containing triphenylphosphine affords an 84% yield of diphenyl carbonate (Dow Chemical Co., USP 513607).
Unsymmetrical diaryl carbonates have been formed. Thus the slow addition of 4-methoxyphenol to N-(dichloromethylene)-dimethylammonium chloride at 0°C in dichloromethane, followed by that of phenol gives a 80% yield of 4-methoxyphenyl phenyl carbonate (C. Copeland and R.V. Stick, Austral. J. Chem., 1984, 37, 1483).

$$ArOH + Cl_2C=N^+Me_2\ Cl^- + PhOH \rightarrow ArOCO_2Ph$$

$$(Ar = 4\text{-MeOC}_6H_4)$$

Mixed aryl alkyl carbonates have also been synthesised (R.C. Larock and N.H. Lee, Tetrahedron Letters, 1991, 32, 6315).
Intramolecular carbonylation leading to a cyclic carbonate has been effected with 67% conversion using 1,1-dichloro-2,2-bis(4-hydroxyphenyl)ethene as the substrate. The reagents are copper(II) bromide, N,N-diisopropylamine, a catalyst, dibromo bis(benzonitrile)-palladium(II) and pressurisation of the mixture with carbon monoxide, oxygen and carbon dioxide (General Electric Co., USP 4329287).

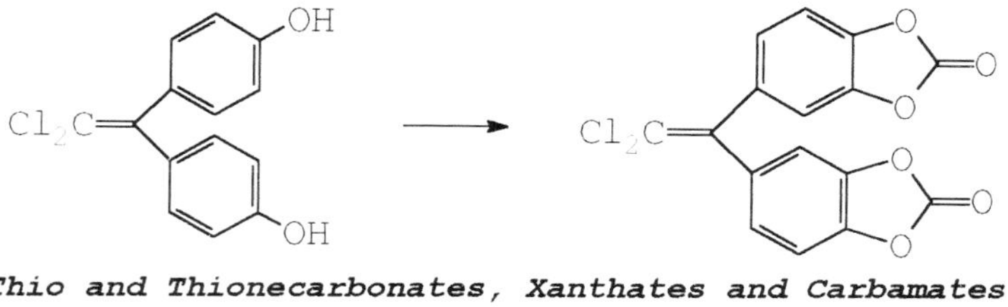

(iii) Thio and Thionecarbonates, Xanthates and Carbamates

Aryl monoalkylthiolcarbonates result from the reactions of potassium phenolates and carbon oxysulphide in dioxan/dimethyl sulphoxide and alkylation with a primary alkyl halide (UFA Aviation Inst., SU 1444335).

$$PhOK + O=C=S + BrPr \rightarrow PhOCOSPr$$

Sodium 4-bromophenoxide in acetonitrile containing copper(I) chloride upon treatment with triethylamine and then with carbon disulphide finally afford di-(4-bromophenyl)thione carbonate (N. Narasimhamurthy and A.G. Samuelson, Tetrahedron Letters, 1982, 29, 827).

$$4\text{-BrC}_6\text{H}_5\text{ONa} \quad + \quad \text{CS}_2 \quad \rightarrow \quad (4\text{-BrC}_6\text{H}_4\text{O})_2\text{C=S}$$

Phase-transfer catalysis has been employed to obtain aryl xanthates in high yield. For example, by adding phenol and iodomethane to 50% aqueous sodium hydroxide and carbon disulphide containing tetra-n-butylammonium bisulphate (A.W.M. Lee, W.H. Chan, H.C. Wong and M.S. Wong, Synth. Commun., 1989, 19, 547).

$$\text{PhOH} \quad + \quad \text{CS}_2 \quad + \quad \text{MeI} \quad \rightarrow \quad \text{EtOCS}_2\text{Me}$$

A general route to vinyl carbamates uses the reactions of phenols and vinyl isocyanate in the presence of triethylamine (BASF AG, DE 3301382).

$$\text{ArOH} \quad + \quad \text{O=C=N-CH=CH}_2 \quad \rightarrow \quad \text{ArOCONHCH=CH}_2$$
$$(\text{Ar} = \text{C}_6\text{H}_4\text{OCH(OMe)CH}_2\text{Cl})$$

Chlorosulphonyl isocyanate reacts with methyl salicylate affording methyl 2-carbamoylbenzoate in 83% yield (A.A. Kamal, M.V. Rao, A.B. Rao, P.V. Diwan and P.B. Sattur, Eur. J. Med. Chem., 1988, 23, 487).

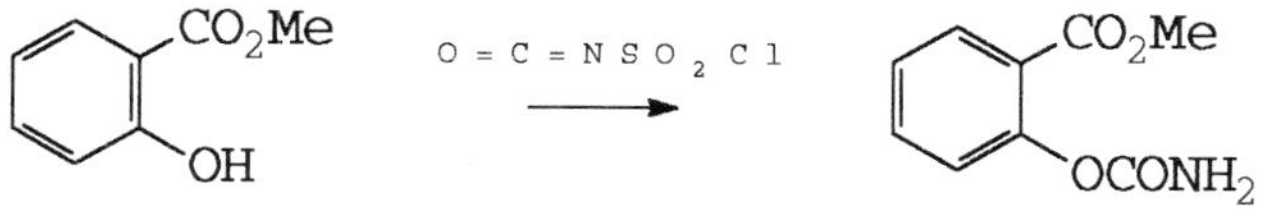

Transcarbamylation is effected with N-(N-methylamido)saccharin; thus 2,3-dihydro-2,2-dimethyl-4-hydroxybenzofuran is O-carbamylated in 57% yield by treatment with this reagent in aqueous acetone containing triethylamine (Chinoin Gyogyszer, DE 3040633).

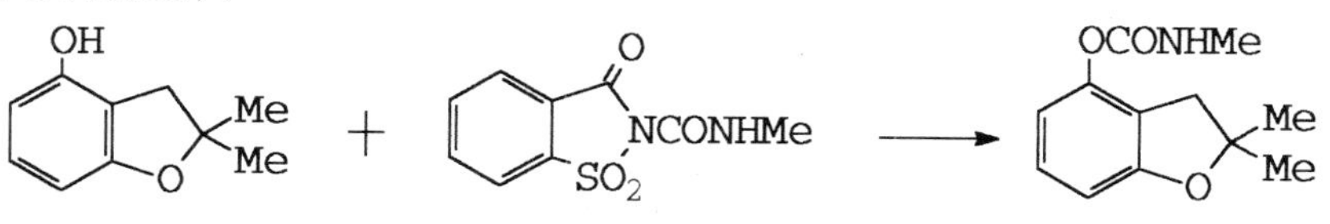

Transesterification of phenol by methyl N-methylcarbamate occurs in 1,2-dichlorobenzene containing a small amount of titanium(IV) ethoxide at 180°C to give phenyl N-methylcarbamate (BASF, DE 4022605).

PhOH + MeNHCO$_2$Me → MeNHCO$_2$Ph + MeOH

1-Naphthol is also esterified by this reagent and phosphorus oxychloride and reaction to afford an 88% yield of 1-naphthyl N-methylcarbamate (G.H. Kulkarni et al., Tetrahedron, 1991, *47*, 1249).

(iv) Phosphites, Phosphates and Aryloxylamines
Triphenyl phosphite has been synthesised in 86% yield by treating phenol in carbon disulphide/benzene containing imidazole with tris(dimethylamino)phosphine (S.D. Stamatov and S.A. Ivanov, Phosphorus-Sulphur, 1989, *45*, 73).

PhOH + (MeN)$_3$P → [PhOP(NMe$_2$)$_2$] → (PhO)$_3$P

Quantitative yields are reported in the formaton o
diaryl phosphites by the simultaneous addition of phosphorus trichloride and 38% hydrochloric acid (fuming HCl) to a two molar proportion of phenol (Phytopathology Res., SU 1421747).

2 PhOH + PCl$_3$ → (PhO)$_2$PO(H)

Triphenyl phosphate is derived in 89% yield by the dropwise addition of phosphorus oxychloride to aqueous sodium hydroxide and chloroform containing the catalyst Aliquat 336 (V.K. Krishnakumar and M.M. Sharma, Synthesis, 1983, 558). A 90% yield is reported when a stirred paste of chloroform, water and sodium phenoxide containing a catalytic amount of polyethylene glycol 400 was treated with phosphorus oxychloride in chloroform (V.K. Krishnakumar, Synth. Commun., 1984, *14*, 189).

3 PhONa + POCl$_3$ → (PhO)$_3$P=O

Tri(4-tert-butylphenyl)thionophosphate is obtained in 93% yield from the reaction of 4-tert-butylphenol with red phosphorus and sulphur heated from 150 to 220°C (F.N. Mazitoza and V.K. Khairullin, Zh. Obsch. Khim., 1980, *50*, 815).

ArOH + P + S → (ArO)$_3$P=S

The quantitative O-phosphorylation of L-tyrosine is effected by addition to a mixture of phosphorus pentoxide and 85% orthophosphoric acid, followed by heating at 80°C (P.F. Alewood et al., Synthesis, 1983, 30).

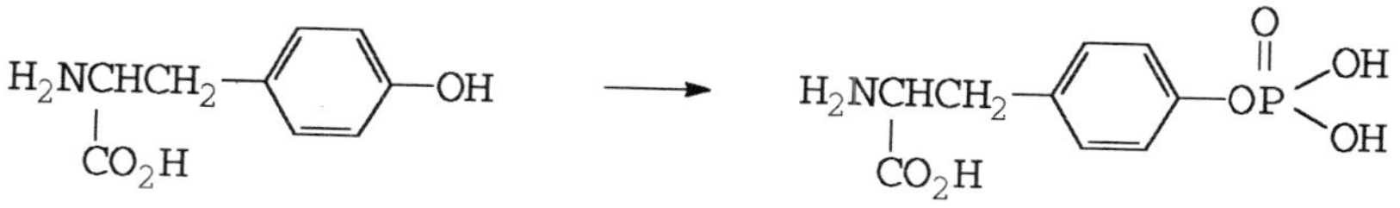

O-Phenylhydroxylamine is formed in 82% yield from the addition of potassium phenoxide in N,N-dimethylformamide to mesitylsulphonylhydroxylamine (Y. Endo, K. Shodo and T. Okamoto, Synthesis, 1980, 461).

$$PhOH \rightarrow PhONH_2$$

(v) Sulphates and sulphonyl esters

O-Sulphonation of 3-hydroxybenzo[a]pyrene by sulphuric acid in dimethylformamide containing dicyclohexylcarbodiimide affords an 85% yield of the corresponding sodium sulphonate after neutralisation (D.B. Johnson and R.M. Thissen, J. Chem. Soc. Chem. Commun., 1980, 598).

$$ArOH \rightarrow ArOSO_2ONa$$

Phase-transfer O-sulphonylation of 3-aminophenol by benzenesulphonyl chloride is observed with tri(C_8-C_{10})alkylmethylammonium chloride (H. Tappe, Synthesis, 1980, 577).

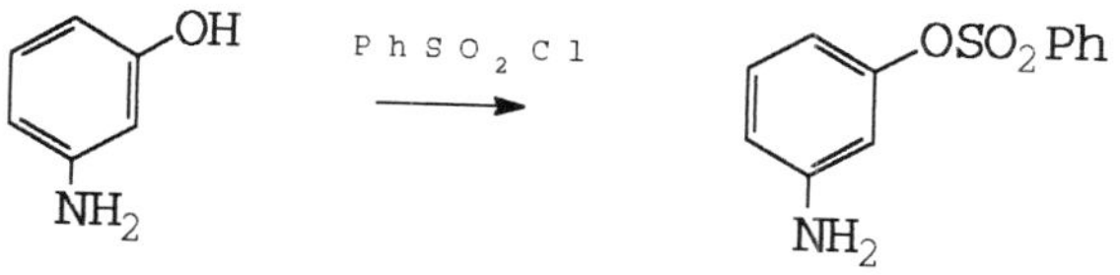

(vi) Replacement of the Hydroxyl Group

Substituents other than Hydrogen: Phenolic hydroxyl groups can be replaced using phosgene in o-xylene. In the case of 4-hydroxybenzophenone, by reaction at 95-120°C in the presence of benzyltrimethylammonium chloride under anhydrous conditions, 4-chlorocarbonylbenzophenone can be obtained in 99% yield (BASF AG, DE 384443).

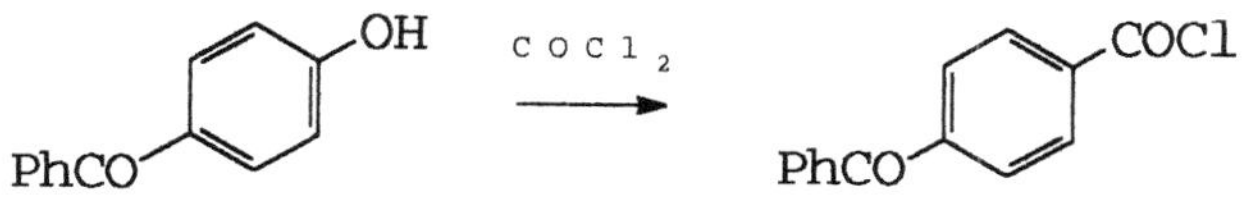

1-Trimethylsilylmethylnaphthalene is formed in 79% yield by the reaction of 1-naphthyl diethylphosphate with trimethylsilyl-methylmagnesium chloride in diethyl ether containing nickel acetylacetonate (T. Hayasi et al., Tetrahedron Letters, 1981,

22, 4449).

[1-naphthyl-OP(O)(OEt)$_2$] $\xrightarrow{Me_3SiCH_2MgCl}$ [1-naphthyl-CH$_2$SiMe$_3$]

Aryl N,N-dimethylsulphonamides can be obtained from the parent phenols through decarbonylation of their aminothiocarbamates via the Newman-Kwart procedure. Thus, 2-methylbenzenesulphonyl chloride (characterised as the dimethylamide) is derived from 2-methylphenol by way of the diethylaminothionecarbamate converted to the diethylaminothiolcarbamate followed by oxidation with chlorine (A. Wagenaar and J.B.F.N. Engberts, Rec. Trav. Chim. Pays-Bas, 1982, 101, 91).

ArOH ---▶ ArO-CS-NEt$_2$ ---▶ ArS-CO-NEt$_2$ → ArSO$_2$Cl → ArSO$_2$NMe$_2$

Treatment of 1-naphthol with ethyl thioglycollate in benzene and trifluoromethanesulphonic acid gives a 88% yield of ethyl 1-thionaphthoxyacetate (T. Nakazawa, N. Hirose and K. Itabashi, Synthesis, 1989, 935).

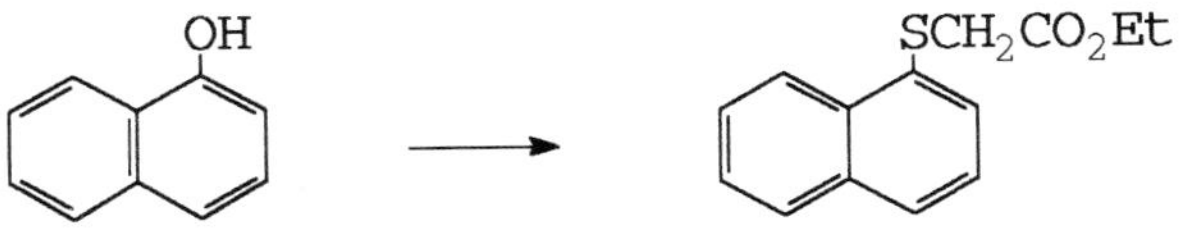

[1-naphthol OH] → [1-naphthyl-SCH$_2$CO$_2$Et]

The conversion of phenol to chlorobenzene in 76% yield occurs when it is added to molten tetrachlorophenylphosphorane (E. Bay et al., J. Org. Chem., 1990, 55, 3415).

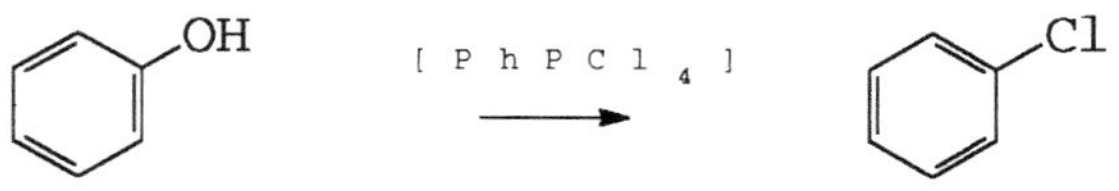

[PhOH] $\xrightarrow{[PhPCl_4]}$ [PhCl]

Phenols can be converted to the corresponding aniline by reaction with 2-bromoisobutyramide to form 2-aryloxybutyramides. These readily undergo the Smiles rearrangement in dimethylformamide containing sodium hydride to form hydroxyisobutyranilides and thence aromatic amines in good overall yields (I.G. C. Coutts and M.R. Southcott, J. Chem. Soc. Perkin I Trans., 1990, 767).

(vii) Replacement by Hydrogen

Several processes are available for this type of transformation. 2-Naphthol as the triflate in dimethylformamide containing triethylamine, Pd(II) acetate and triphenylphosphine upon treatment with 99% formic acid furnished naphthalene in 91% yield (S. Cacchi et al., Tetrahedron Letters, 1986, *27, 5541*). A 'transfer hydrogenation' modification of a known dehydroxylation procedure (W. Musliner and J. W. Gates, J. Am. Chem. Soc., 1966, *68*, 4271) has been reported in which the tetrazolyl ether, obtained from 1-naphthol with 1-chloro-2-phenyltetrazole in benzene/ethanol/water (7:3:2) containing 10% Pd-C, was treated with 64% aqueous hydrazine to afford an 82% yield of naphthalene (J.D. Entwhistle, B.J. Hussey and R.A.W. Johnstone, Tetrahedron Letters, 1980, *21*, 4747).

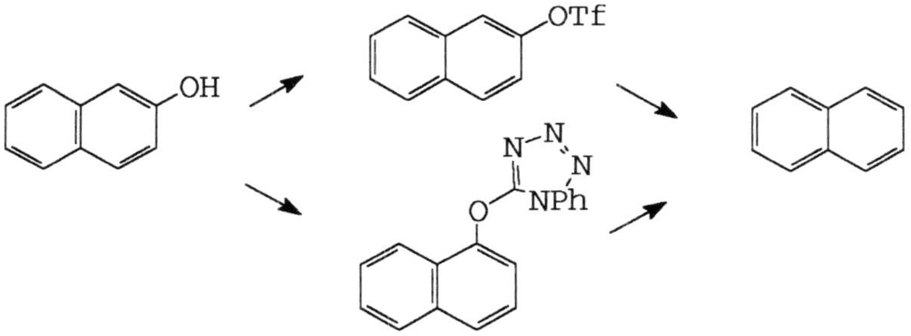

The dehydroxylation of estrone is effected in 96% yield by way of the ether, (obtained quantitatively from reaction with 3-chloro-1,2-benzoisothiazole-1,1-dioxide in toluene, containing triethylamine (A.F. Brigas and R.A.W. Johnstone, Tetrahedron Letters, 1990, *31*, 5789).

An alternative procedure (F.Wang et al., J. Chem. Soc. Perkin Trans, I, 1992, 1897) is exemplified by the formation of naphthalene in 96% yield from 2-naphthyl tosylate by treatment in chloroform/methanol (1:1) containing nickel(II)chloride 6 H_2O with excess sodium borohydride. Isolated alkene and keto groups were also reduced.

Dehydroxylation, accompanied by ring reduction, of 1,1-bis(4-hydroxyphenyl)cyclohexane in ethylcyclohexane containing some ruthenium and Galleon earth occurs when it is heated under hydrogen (75Kg/cm_2). This gives 1,1-dicyclohexylcyclohexane in almost quantitative yield (Idemitsu Kosan, KK, EP 362673).

Biological dehydroxylation of a polyphenolic melanin precursor has been effected and the role of NADPH in this process has been studied (F. Vivian, M. Gaudry and A. Marquet, J. Chem. Soc., Perkin Trans., 1990, 1255).

(viii) Protection of, and by, the Phenolic Group

Esterification and etherification (considered in the next section) remain useful protective techniques and there have been few developments. The definitive text on this topic remains relevant (T.W. Greene, 'Protective Groups in Organic Synthesis', Wiley, New York, 1981).
2-Methyl and 2,6-dimethylphenol afford better protective systems than 4-hydroxydiphenyl for phosphopeptide syntheses in solid phase procedures (M. Tsukamoto et al., Tetrahedron Letters, 1991, *32*, 7083). The technique for selective removal of the 2-bromophenyl group in certain phosphorylnucleotides has been improved (Y. Stabinsky, R.T. Sakata, and M.H. Caruthers, Tetrahedron Letters, 1982, *23*, 25).

(c) Reactions of the Aromatic Ring

(i) Deuteration
Pentadeuteriophenol is obtained by treating phenol in 10% sodium deuteroxide in deuterium oxide with Raney nickel (Nippon Sheet Glass, KK J62 056441).

(ii) Diels-Alder Reaction
It is reported that a 2:1 mixture of phenol and N-phenylmaleide undergoes cycloaddition when heated at 170°C (D. Bryce-Smith et al., J. Chem. Soc. Chem. Commun., 1984, 951).

(iii) Oxyalkyl and Oxycarboxyalkyl compounds by Substitution
Phenols substituted by the 2-hydroxyalkyl group represent the monomeric stage of phenol/aldehyde polymerisation and their regiospecific and asymmetric synthesis has received much attention. Phenol in xylene with excess paraformaldyde and 1,2-dimethoxyethane (1 mole) reacted in a sealed reactor at 135°C to give a 72% yield of salicyl alcohol (2-hydroxymethylphenol) (G. Casiraghi, et al.,). This compound is also obtained in 91.5% yield from phenol with paraformaldehyde, followed by the addition of aluminium phenolate (Isover Saint Gobain, EP 69016). By contrast, the alkyl methyl ether of salicyl alcohol is formed in low yield by passage of phenol and 1,1-dimethoxyethane (1:10) at 250°C through the catalyst H-ZSM-5 zeolite supported on glass wool (Celanese Corporation, USP 4694111).

A reaction of phenol in nitromethane with 1-oxaspiro[4.5]deca-6,9-diene-2,8-dione in the presence of stannic chloride leads to 2,2'-dihydroxy-5-(2-methoxycarbonylethyl)biphenyl in 60% yield (T. Takeya, T. Okubo and S. Tobinaga, Chem. Pharm. Bull., 1987, 35, 1761).

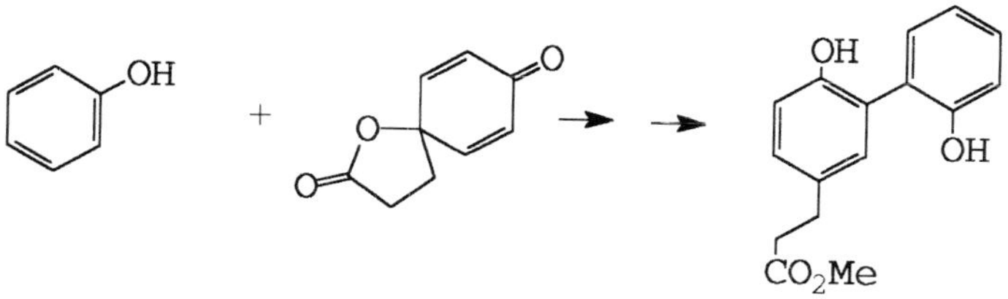

Paraldehyde, phenol and H⁺-Y-Zeolite when heated afford a 86% yield (60% conversion) of 4-(2-hydroxyethyl)phenol (Mitsui Toatsu Chem. Inc., J6 3307834). Phenol, (-)-menthol and diethylaluminium chloride react together to form a phenoxymenthyloxyaluminium chloride intermediate, which on further reaction with chloral affords a 96% yield of (-)-2-(2,2,2-trichloro-1-hydroxyethyl)phenol (36% e.e.) (F. Bigi et al., J. Chem. Soc. Chem. Commun., 1983, 1210).

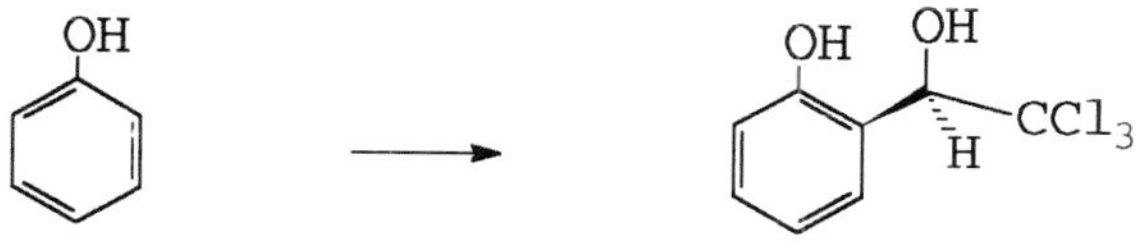

In a procedure which represents the first diastereoselective approach to ephedrine-like compounds, phenol (as its magnesium bromo salt) reacts with N-t-butoxycarbonyl-L-prolinal to give the syn adduct (1) (94-99% d.e.). If the phenate from Ti(O-iPr)$_4$ is used, the anti isomer (2) is obtained (76-84% d.e.) (F. Bigi et al., Tetrahedron, 1989, 45, 1121)

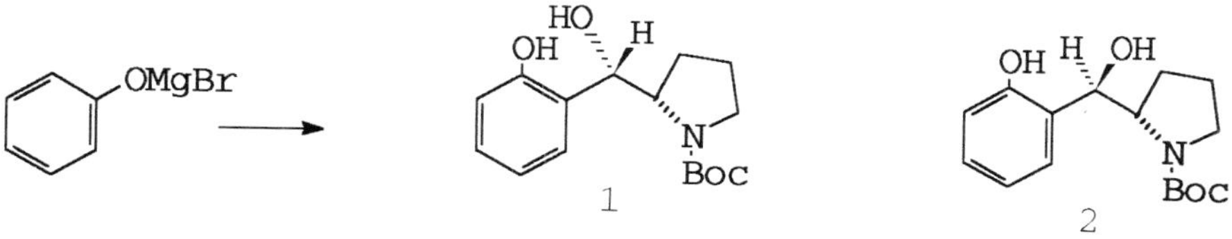

Phenol reacts with ethyl pyruvate in dichloromethane containing titanium(IV) chloride to afford the racemic aldol 2-(1-ethoxycarbonyl-1-hydroxyethyl)phenol (Blaschim. Spa, EP 98012).

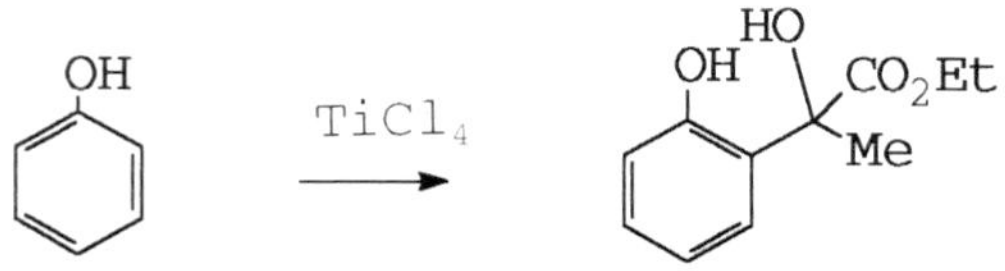

Aldolisation occurs when phenol, as the chromium tricarbonyl complex, is treated with n-butyllithium and tetramethylethylene diamine, and benzaldehyde is added. The product 3-(1-hydroxybenzyl)phenol is isolated in 64% yield (M. Fukui, T. Ikeda and T. Oishi, Tetrahedron Letters, 1982, 23, 1605).
Enhancement of the reactivity of phenol through transition metal complexation is exhibited in the reaction of the osmium compound [Os(NH$_3$)$_5$(η^{2+}-phenol)](OTf)$^{2-}$ with maleic anhydride in acetonitrile to give dimethyl (4-hydroxyphenyl)succinate. This process is effectively one of nucleophilic addition reminiscent of the Michael reaction (M.E. Kopach, J. Gonzalez and W.D. Harman, J. Am. Chem. Soc., 1991, 113, 8972).

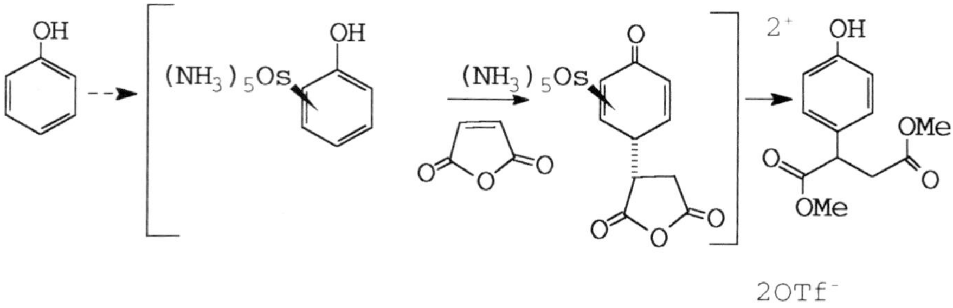

(iv) Alkyl side chains containing S or Si atoms

Phenol in tetrahydropyran treated with tert-butyllithium in hexane followed by the addition of trimethylsilyl chloride affords a 67% yield of 2-trimethylsilylphenyl trimethylsilylether (G.A. Posner and K.A. Canella, J. Am. Chem. Soc., 1985, 107, 2571). 2-Methylthiomethylphenol is formed in 78% yield by the addition of phenol to the reagent [Me$_2$S$^+$OS=O(Cl)]Cl$^-$ (prepared from dimethyl sulphoxide and thionyl chloride) and final treatment of the mixture with triethylamine (K. Sato et al., Perkin Trans I, 1984, 2715).
Variations of the procedure with derivatives such as dimethyl 3,3'-sulphinylpropionate have been listed (ibid., 1987,1384).

(d) Oxidations
(i) Oxidative Coupling
Phenols:
Thymol is oxidised under mild conditions by anhydrous copper(II) chloride and oxygen over 48 hours giving thymylthymoquinone (65%), the corresponding diphenquinone (10%), and thymoquinone (25%).

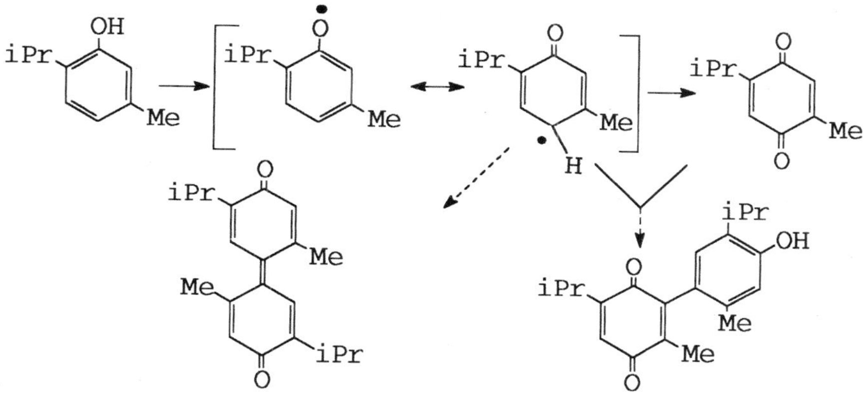

Under similar conditions 2,6-dimethyl and 2,6-di-tert-butylphenol give 3,5,3',5'-diphenoquinones as the major products However, in the first case some 2,6-diethyl-1,4-benzoquinone is also produced (Y. Takizawa et al., J. Org. Chem., 1985, 50, 4383).

By contrast, 2,6-di-tert-butylphenol in 50% aqueous potassium hydroxide containing 50% methyl tri-n-butylammonium chloride in a sealed reactor at 200°C and 300psig. is oxidised by hydrogen peroxide to afford 4,4'-bis(2,6-di-tert-butylphenol) diphenoquinone in 97.5% yield (Ethyl Corporation, USP 4847434).

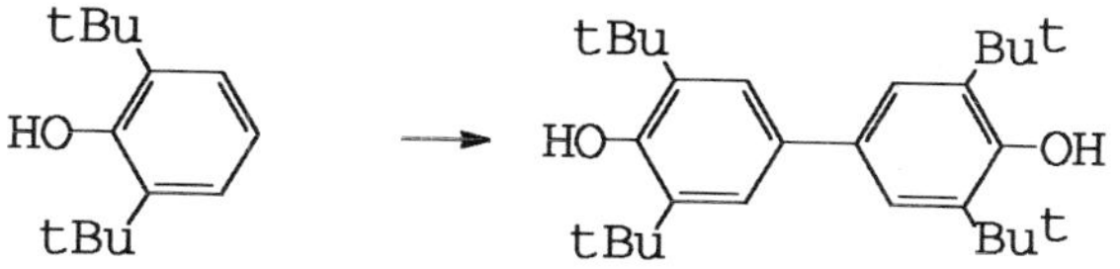

A comparison of the phenolic coupling of 2,6-dimethyl, 2,6-di-isopropyl and 2,6-di-tert-butylphenols in chloroform solution on solid (a)potassium permanaganate and (b)manganate surfaces results in more than 90% of the diphenoquinone with (a) and exclusively poly(2,6-diaryl-1,4-phenylene)oxides with (b) (F.M. Menger and D.W. Carnahan, J. Org. Chem., 1985, 50, 3927). Addition of powdered potassium hydroxide to the permanganate system gives the appropriate polyphenylene oxide (85%) and the diphenoquinone (15%).

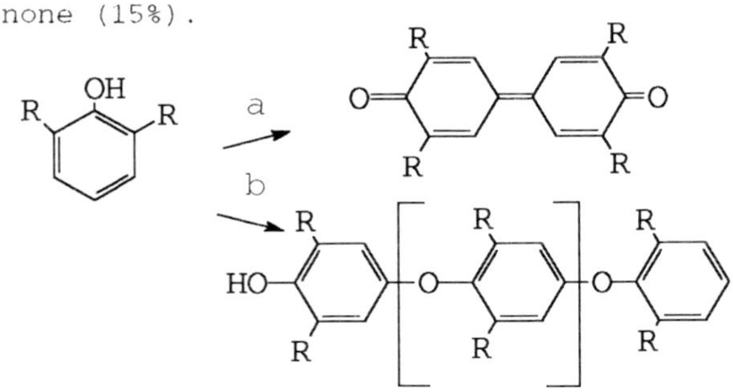

It has been reported that solid iron(III) chloride is more effective for oxidative coupling than the same reagent in solution (F. Toda, K. Tanaka and S. Iwata, J. Org. Chem., 1989, 54, 3007). Cobalt(III) acetate in acetic acid with 2,6-dimethylphenol produces a 75% yield of 3,3',5,5'-tetramethyl-4,4'-diphenoquinone and 23% of 2,6-dimethyl-1,4-benzoquinone (M. Hirano, T. Ishii, and T. Morioto, Bull. Chem. Soc. Japan, 1991, 64, 1434).
Oxidation of 2,6-dimethylphenol in a phosphate buffer at pH 6.8 with mushroom tyrosinase gives the corresponding diphenoquinone in 96% yield. Dissolved oxygen is essential, but 2,6-dichlorophenol proved unreactive and in the co-solvent acetonitrile, biphenols are by-products (G. Pandey, C. Muralikrishna, V.T. Bhalerio, Tetrahedron Letters, 1990, 31, 3771).
The previous examples in general involve the 4-position. Where this is substituted as in 4-methoxyphenol, oxidative coupling occurs at the 2-position probably via an intermediate phenoxide ion, without benzoquinone formation. Thus 4-methoxyphenol introduced into aluminium chloride in nitromethane, after treatment with anhydrous iron(III) chloride affords 2,2'-dihydroxy-5,5'dimethoxydiphenyl in 7% yield (G. Sartori et al., Tetrahedron Letters, 1992, 33, 2207).
Naphthols:
With the copper(II) chloride/oxygen system used for monocyclic phenols, 2-naphthol gives only 4-ethoxy-1,2-naphthoquinone (65%) and 1-chloro-2-naphthol (30%) without any coupled product. By contrast, 2-naphthol with iron(III) chloride hexahydrate gives only 2,2'-dihydroxy-1,1'-dinaphthyl in 95% yield without quinone

formation (F. Toda, K. Tanaka and S. Iwata, J. Org. Chem., 1989, 54, 3007). When the S(+)-amphetamine copper(II) complex is used as oxidant the S(-) form of this compound is obtained in 85% yield (e.e., 95%) with 80% recovery of the amphetamine salt (J. Brussee and A.C.A. Jansen, Tetrahedron Letters, 1983, 24, 3261).

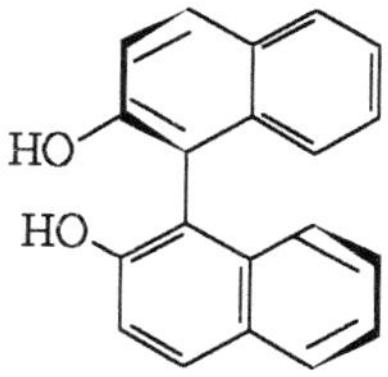

The absolute configuration of the monohydroxy analogue, (+)-2-hydroxy-1,1'-dinaphthyl has been revised to the R form (K. Kabuto, F. Yasuhara and S. Yamaguchi, Bull. Soc. Chem. Japan, 1983, 56, 1263).
Asymmetric oxidative dimerisation of S(+)-3,4,8-trimethyl-5,6,7,8-tetrahydro-2-naphthol in diethyl ether with potassium ferricyanide in aqueous 0.2M sodium hydroxide is effected at ambient temperature to furnish the S,S-(+)-trans-dimer in 62% yield (B. Feringa and H. Wynberg, J. Org. Chem., 1981, 56, 2457).

(ii) Cyclohexadienone Formation

Cyclohexadienones can be considered as hemiquinones and thus essentially the phenolic OH becomes a carbonyl group. A variety of synthetic approaches invoke the tautomeric character of phenols.
2,6-Dimethylphenate is alkylated with chloromethyl methyl ether to give a 85% yield of 2,6-dimethyl-6-methoxymethylcyclohexa-2,4-dienone without any of the 2,5-isomer (R.S. Topgi, J. Org. Chem., 1989, 54, 6125).

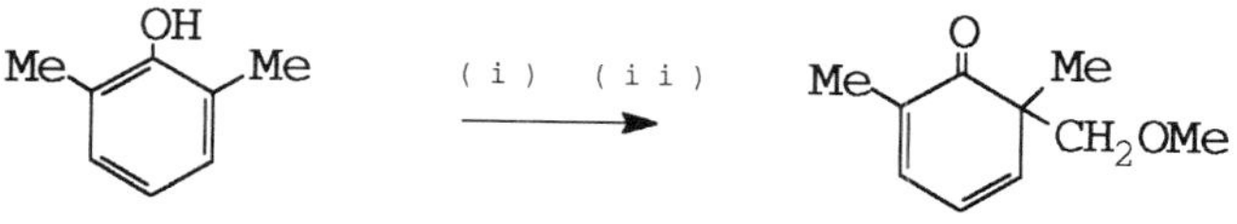

(i) BuLi (ii) ClCH$_2$OMe

A number of different reagents have been used for obtaining 4-hydroxy-2,4,6-trimethylcyclohexadienone from 2,4,6-trimethyl-

phenol. In the simplest approach an 81% yield is obtained from aqueous 2,4,6-trimethylphenol pressured with oxygen during 4 hours (Mitsui Toatsu Chem. Inc., J57 183734). A yield of 70% resulted from the action of chlorine on a carbon tetrachloride solution of the phenol containing sodium bicarbonate (Rhone-Poulenc Sante, Fr.P. 2549044), 66% from the use of manganese dioxide in dilute sulphuric acid with an isopropyl ether solution (Rhone-Poulenc Sante Fr.P. 255225) and 65% with 16% sodium hypochlorite at pH 12.5 (Mitsubishi Gas Chem., KK EP 84158).
By contrast, the electrolytic oxidation of mesitylene at a graphite anode in aqeous sulphuric acid/acetonitrile (4:1) affords the same product in 43% yield (Anic SPA, EP 95206).

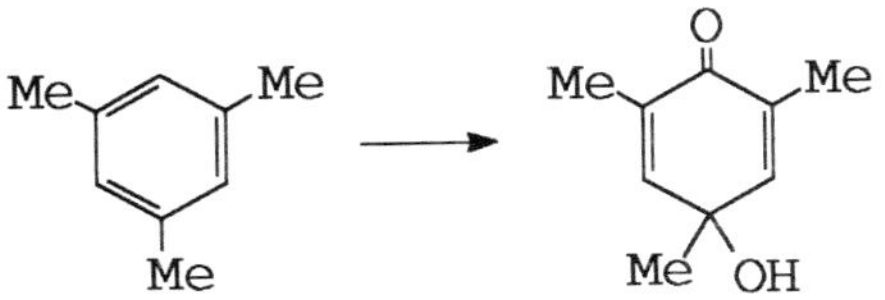

2,4,6-Trimethyl-4-allyl-cyclohexa-2,5-dienone is synthesised from sodium 2,4,6-trimethylphenate in the presence of β-cyclodextrin by treatment with allyl bromide during 24 hours (H. Eiji, J58 194835).

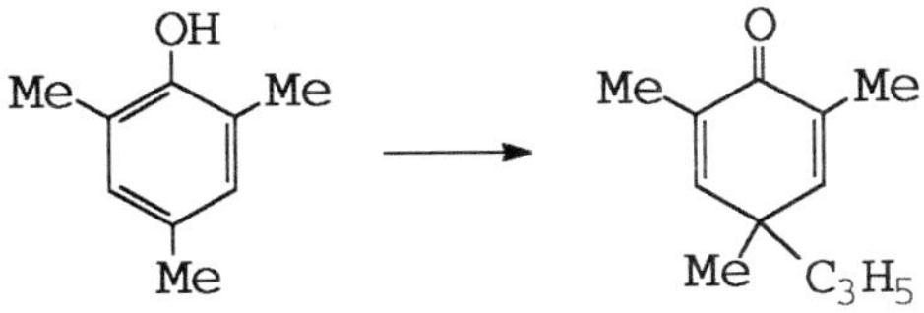

Classical conditions, the use of lead tetraacetate, are employed for the synthesis of 6-acetoxy-6-hexadienyl-5-methylcyclohexa-2,4-dienone from the parent phenol. The yield is 50% (N.K. Bhamare, T. Granger, C.R. John and P. Yates, Tetrahedron Letters, 1991, *32*, 4439).

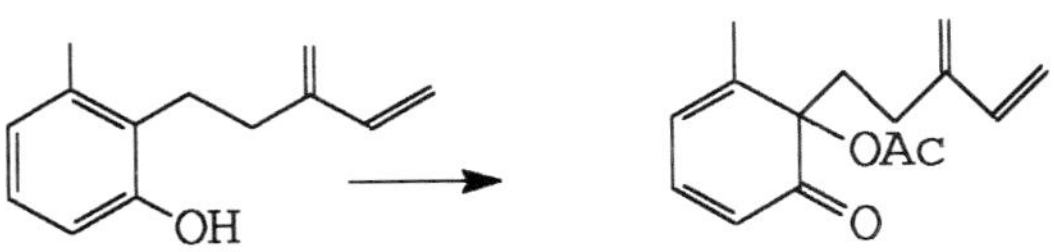

The oxidation of carvacrol, 5-isopropyl-2-methylphenol, with potassium periodate or iodic acid in ethanol affords a cyclohexadienone which can be employed to synthesise 3,10-dihydroxydielmentha-5,11-diene-4,9-dione, a monoterpene dimer from the plant, *Callitris macleayana* (S.K. Parnikar and J. Patel, Chem. and Ind.,(London), 1988, 2639). The oxidation by iodic acid of thymol and isothymol gives similar products.

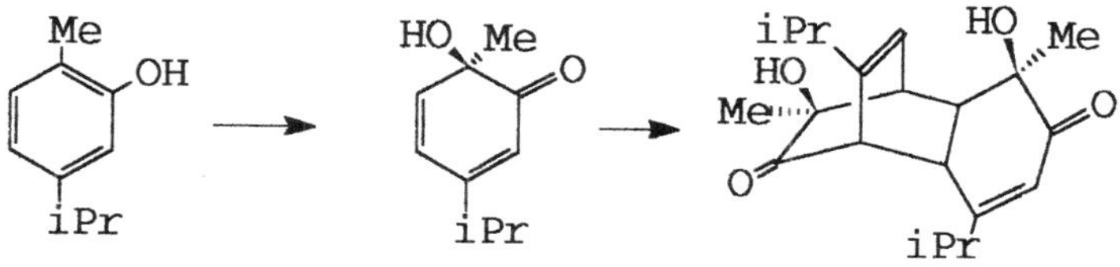

The oxime of 4-(2-oxopropyl)phenol in boiling acetonitrile upon treatment with phenyliodosodi(trifluoroacetate) affords 3-methyl-1-oxa-2-azaspiro[4,5]deca-2,6,9-triene-8-one, in 63% yield (M.Kacan, D. Koyuncu and A.Mckillop, J. Chem. Soc. Perkin Trans., I 1993, 1771).

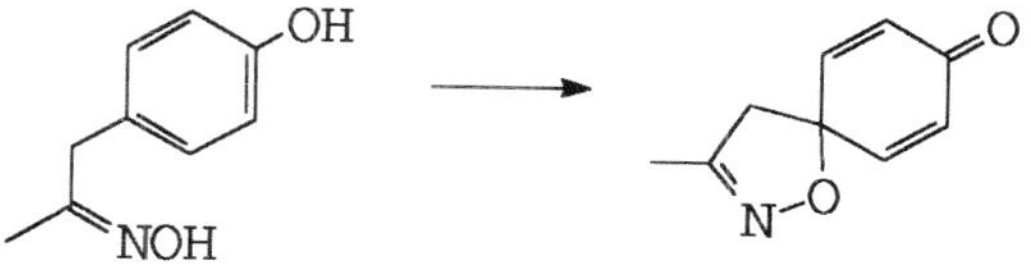

A number of halogenated cyclohexa-2,5-dienones have been synthesised. The addition of phenol and lead dioxide to a solution of 70% hydrogen fluoride in pyridine/dichloromethane leads to a 60% yield of 4,4-difluorocyclohexa-2,5-dienone (Shell Int. Res. Mij BV, EP 188847). 2,6-Dibromo-4-methylphenol in methanolic dichloromethane with phenyliodosodiacetate affords a 63% yield of 2,6-dibromo-4-methoxy-4-methylcyclohexa-2,5-dienone (N. Lewis and P. Wallbank, Synthesis, 1987, 1103).

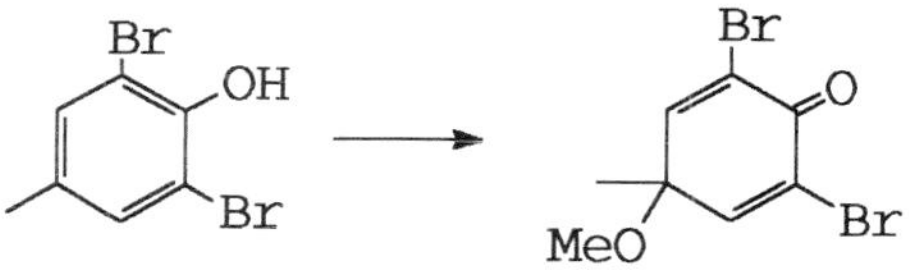

4-Methyl-2,3,5,6-tetrabromophenol is oxidised by nitric acid to 4-methyl-4-nitro-2,3,5,6-tetrabromocyclohexa-2,5-dienone. In turn this product can be used as a regiospecific brominating reagent (M. Lemaire et al., Tetrahedron, 1987, *43*, 835).

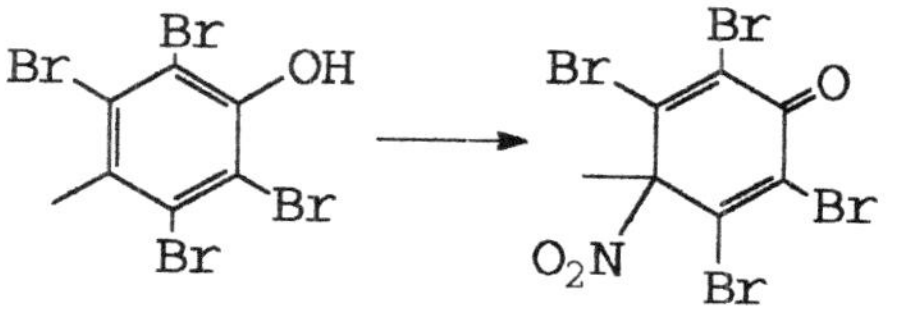

A cyclohexa-2,4-dienone structure incorporating an o-quinonimine residue results from a phenol with an oxidisable group in the o-position. Thus, 4-methoxy-2-{N-[2-(2-methoxyethoxy)-5-nitrophenyl]sulphonamido}-5-methylphenol in acetone with manganese dioxide at ambient temperature affords 4-methoxy-N-{[2-(2-methoxyethoxy)-5-nitrophenyl]sulphonyl}-5-methyl-o-benzoquinonimine in 59% yield (S. Fujita, J. Org. Chem., 1983, *48*, 177).

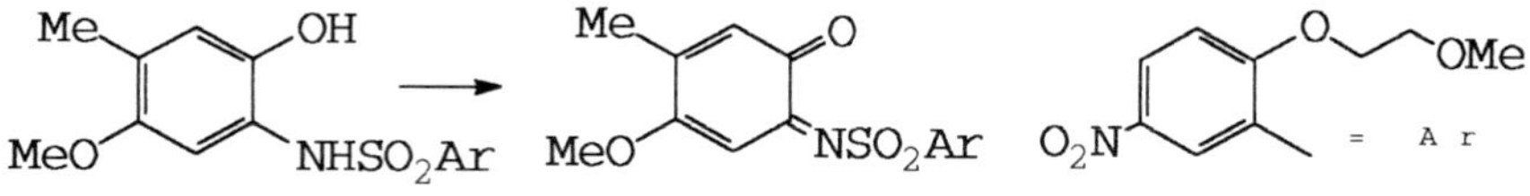

In the bicyclic series oxidative rearrangement and cyclisation has been noted. For example in the formation of 6,7,8,9-tetrahydro-4a,7-methano-4aH-benzocyclohepten-2(5H)-one in 57% yield. Here the bicyclic substrate was simply heated with potassium t-butoxide in t-butanol for several hours (J.D. McChesney and R.A. Swanson, J. Org. Chem., 1982, *47*, 5201).

4,4'-Dimethoxybiphenyl undergoes anodic dimethoxylation when electrolysed in 1% methanolic potassium hydroxide solution. The tetramethoxy cyclohexa-2,5-dienone system is formed in 86% yield (R.E. De Schepper and J.S. Swenton, Tetrahedron Letters, 1985, *26*, 4831).

Spiro-annulated cyclohexa-2,5-dienones result from the oxidation of 4-phenylphenols with phenyliodosodiacetate (K.V.R. Krishna, K. Sujathi and R.S. Kapil, Tetrahedron Letters, 1990, 31, 1351; A. Callinan et al., Tetrahedron Letters, 1990, *31*, 4551). 2-Tert-butyl-1,3-oxazaspiro[5,5]undeca-2,7,10-trien-9-one is obtained in 75% yield by the oxidative cyclisation of N-[2-(4-hydroxyphenyl)ethyl] 2,2-dimethylpropionamide with phenyliodosodi(trifluoroacetate). In ethanol or acetic acid, a 4-hemiquinol ether or an acetate resulted (Y. Kita et al., J. Org. Chem., 1991, *56*, 435).

Cyclohexadienone formation accompanied by oxidative coupling takes place when 4,8-dimethoxy-6-methyl-1-naphthol in chloroform solution containing 0.2% triethylamine is oxidised with siver oxide to produce 4,4',8,8'-tetramethoxy-6,6'-dimethyl-2,2'-dinaphthylidene-1,1'-dione in 95% yield (H. Laatsch, Ann, Chem., 1980, 1321).

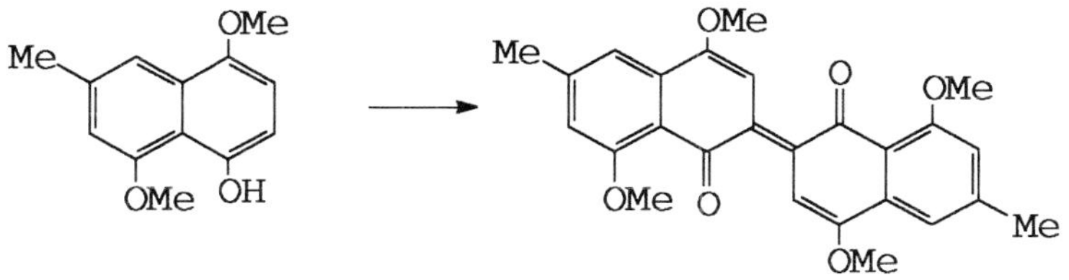

(iii) Quinone Formation

Benzoquinones:

Under more drastic oxidative conditions than those required for the formation of cyclohexadienones, 1,4-quinones result even when the 4-position is occupied. This is illustrated by the case of 2,4,6-trichlorophenol which when oxidised gives 2,6-dichloro-1,4-benzoquinone.

2,6-Dimethyl-1,4-benzoquinone is formed in 84% yield from 2,6-dimethylphenol with Jones reagent (D. Liotta et al., J. Org. Chem., 1983, 48, 2932) and in 85% yield by reaction with bromine and hydrogen peroxide (F. Minisci et al., J. Org. Chem., 1989, 54, 728).

2,6-Di-tert-butyl-1,4-benzoquinone is obtained essentially by the same procedure in 92% yield from 2,6-di-tert-butylphenol (Consiglio Nat. Rierche, EP 249289) and in 68% yield with oxygen in dimethyl sulphoxide containing potassium hydroxide (Showa Denko, KK, J58 121237).

2,3,6-Trimethyl-1,4-benzoquinone results in 90% yield from 2,3,6-trimethylphenol in acetic acid containing ruthenium trichloride trihydrate and 30% hydrogen peroxide (Sagami Chem. Res. Centre, EP 10776; S. Ito, K. Aihara and M Matsumoto, Tetrahedron Letters, 1983, 24, 5249), in 78% yield from the addition of 60% hydrogen peroxide to the phenol in acetic acid containing a little phosphomolybdic acid (M. Shimizu, H. Orita, T. Hayakawa and K. Takehara, Tetrahedron Letters, 1989, 30,

471), and with a 95% yield from the use of tert-butylhydroperoxide in tert-butanol added to an acetic acid solution containing cuprous chloride (British Petroleum plc, EP 220948).
The corresponding quinones were obtained from thymol (2-isopropyl-5-methylphenol,carvacrol,(2-methyl-5-isopropylphenol), 2-methyl and 3-methylphenol in yields of 79%, 88%, 50% and 50% respectively, with oxygen in dimethylformamide containing cobalt salicylaldehyde ethylene diamine (cobalt(II)salen) (E.R. Dockai, Q.B. Cass, T.J. Brocksom, U. Brocksom and A.G. Correa, Synth. Commun., 1985, 15, 1033).
1,4-Benzoquinones are derived from the oxidation of electron-rich methoxyarenes by magnesium monoperoxyphthalate with a water-soluble iron porphyrin as catalyst (I. Artaud et al., J. Chem. Soc. Chem. Commun., 1991, 31). Primin (2-methoxy-6-n-pentyl-1,4-benzoquinone is synthesised from the oxidation of 2-methoxy-6-n-pentylphenol with Fremy's salt (the Teuber reaction) by the route shown (J.H.P. Tyman and S.S. Arogba, unpublished work; S.S. Arogba, Org. Prep. Proc. Int., 1991, 23, 639).

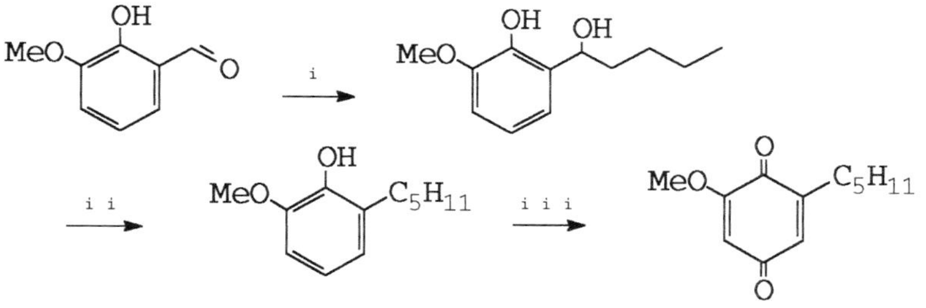

(i)2BuLi, H_3O^+ (ii)Pd-C, H_2, (iii)oxidn.

In the tocopherol series, tocopherylquinone is formed in 93% yield by the oxidation of 2,3,6-trimethyl-5-(3'-hydroxy-3',7',11',15'-tetramethylhexadecanyl)phenol by Fremy's salt, aqueous sodium carbonate and tricaprylmethylammonium chloride (G.L. Olsen et al., J. Org. Chem., 1980, 45, 803).

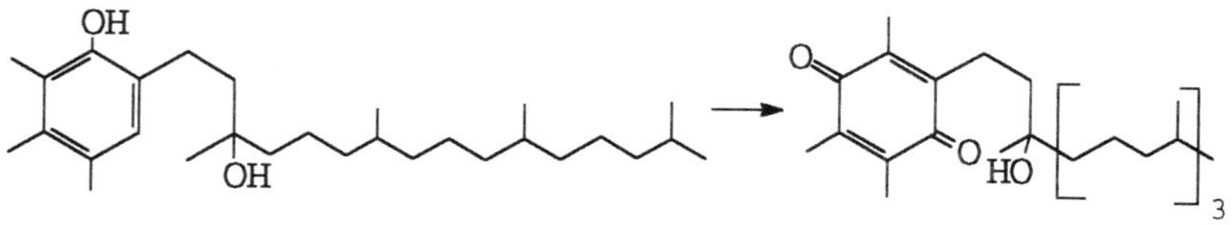

1,2-Benzoquinones result from the oxidation of phenols with

iodoxybenzene in the presence of a small amount of trichloroacetic acid (D.H.R. Barton et al., Tetrahedron Letters, 1982, 23, 957), and from phenols treated with dimethyldioxirane (J.K. Crandall et al., Tetrahedron Letters, 1991, 32, 5441), although catechols are cleaved to (Z,Z)-muconic acids during oxidation (A. Altamura et al., Tetrahedron Letters, 1991, 32, 5445).

Naphthoquinones:

New reagents have been introduced for the formation of 1,4-naphthoquinones. 1-Naphthol in aqueous acetonitrile reacts at 0°C with phenyliodoso di(trifluoroacetate) to afford a 73% yield of the parent quinone (R. Barret and M. Daudon, Tetrahedron Letters, 1990, 31, 4871), and the monooxime is formed in 68% yield by irradiation of a dioxan solution of 1-naphthol and N-nitrosodimethylamine (Y.L. Chow and Z-Z. Wu, J. Am. Chem. Soc., 1985, 107, 3338). Naphthazarin dimethyl ether (5,8-dimethoxy-1,4-naphthoquinone) is produced in 68% yield by treatment of 1-hydroxy-5,8-dimethoxynaphthalene in acetonitrile with aqueous ceric ammonium nitrate (H. Laatsch, Ann. Chem., 1986, 1655). 2-Bromo-5-acetoxy-1,4-naphthoquinone is synthesised in 90% yield from 1,5-diacetoxynaphthalene in warm acetic acid by treatment with an aqueous acetic acid solution of N-bromosuccinimide (S.W. Heinzman and J.R. Grunwell, Tetrahedron Letters, 1990, 21, 4305).

A 1,2-, rather than a 1,4-naphthoquinone, results in 73% yield by reaction of 1,5-dihydroxy-3,7-di-tert-butylnaphthalene with a molybdenum oxide pyridine complex (K. Krohn, H. Rieger and K. Khanabaee, Chem. Ber., 1989, 122, 2323).

(iv) Carbonylic and Other Transformation Products

The versatility of the phenolic system towards oxidation is demonstrated by the occurrence of other oxidation reactions than coupling, quinone and cyclohexadienone formation. For example, 3,5-dimethyl-4-hydroxybenzaldehyde results in 98% yield from 2,4,6-trimethylphenol by aeration in dimethylformamide/methanol solution (1:5) containing cuprous chloride (Mitsui Petrochem. Ind., KK J62 241642).

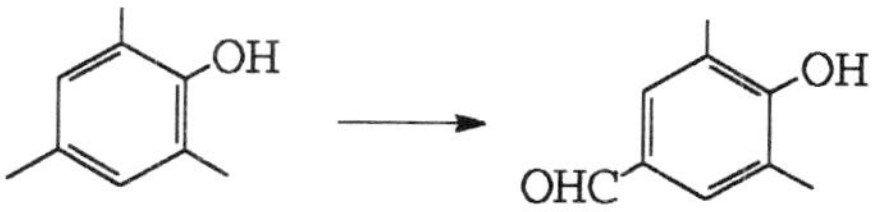

2,4,7-Trimethyltropone is formed in 59% yield by rearrangement and hydrolysis of the intermediate obtained by treating 2,4,6-

trimethylphenol in chloroform with tributyl tin hydride and AIBN, (azoisobutyronitrile), in aqueous alkali under phase-transfer conditions (M. Barbier et al., J. Chem. Soc. Chem. Commun., 1984, 743).

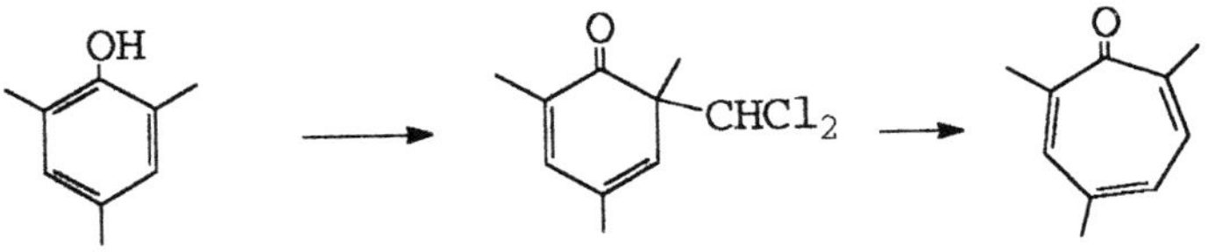

Photochemical rearrangement of thymol in trifluoromethane-sulphonic acid gives ten products of which eight have been characterised. Of these at least two [umbellone (10%) and piperitone (5%)] are interesting in that the hydroxyl group has been convertd to a ketone (P. Baeckstrom et al., J. Org. Chem., 1985, 50, 3728)

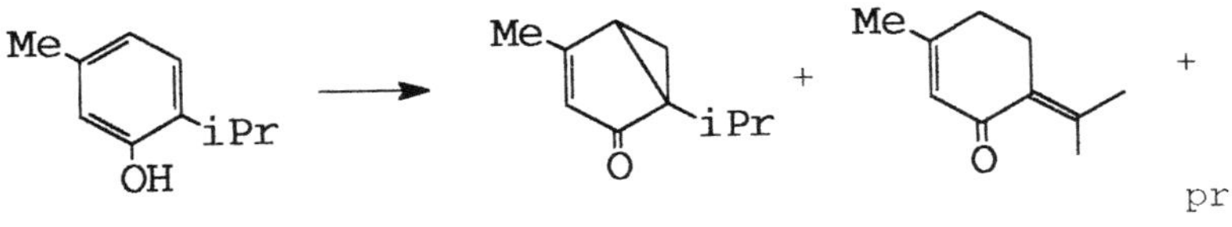

(e) Reactions of Alkylphenols

The reactions of alkylphenols show a great variety partly attributable to the increased activation of the ring and a consequent opportunity for the development of regiospecific electrophilic attack.

(i) Esterification and Hydrogenation
4-Ethylphenol with 2,3-dichloro-5,6-dicyanobenzoquinone in methanol affords 4-(1-methoxyethyl)phenol with a selectivity of 87% and conversion of 89% (M. Bouquet et al., Compt. Rend., Ser.II, 1984, 299, 1389).
Preferential acetylation of the alcoholic hydroxyl group in 4-hydroxymethylphenol is achieved in 60% yield by treatment of the phenolic alcohol with acetic anhydride containing a little boron trifluoride etherate (Y. Nagao et al., Chem. Pharm. Bull., 1981, 29, 3202).

Generally in diols containing both phenolic and alcoholic groups, neutral conditions favour the acylation of the latter, while the former are selectively acylated in the presence of triethylamine (S. Yamada, Tetrahedron Letters, 1992, *33*, 2171). This of course, is the basis of the Chattaway method used many years ago. The selective acetylation of the phenolic hydroxyl in estradiol is an example of the "rediscovery" of the procedure (V. Srivastava, A. Tandon and S. Ray, Synth. Commun., 1992, *22*, 2703).

Phenolic formates are generally troublesome to form. 4-Tert-butylphenol and formic acid in dry dichloromethane when treated with dicyclohexyldiimide (1.4 moles) afford a 80% yield of the formate (J. Huang and H.K. Hall, J. Chem. Res. (S), 1991, 292).

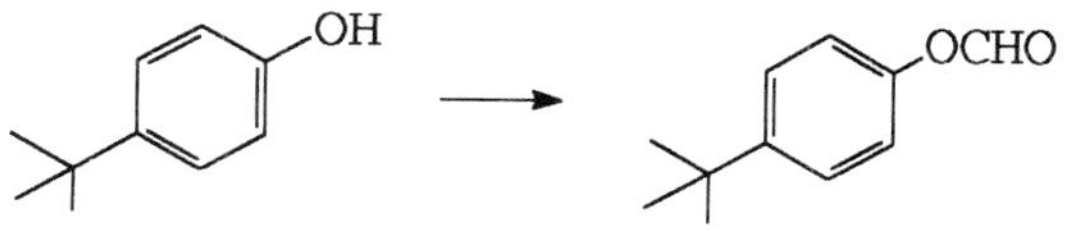

Ring reduction of 4-tert-butylphenol in butyl acetate is effected in 99% yield by warming with 50% aluminium-nickel oxide at 180-190°C (at 1-1.5Pa pressure)(V. Vogtman et al., (Parfum. Kosmet., 1988, *69*, 642,645,648).

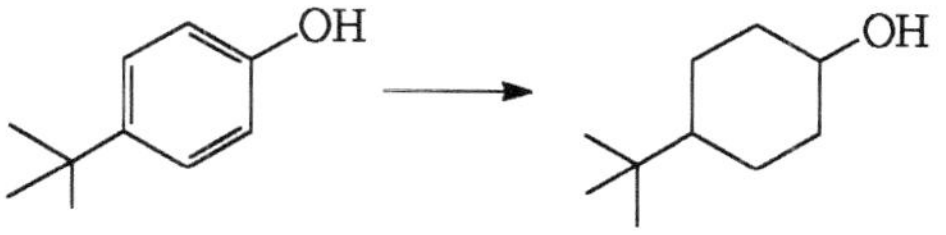

The similar use of a Raney-type catalyst from the interaction of 50% aluminium-nickel alloy with water at 100°C has been described (J.H.P. Tyman, Chem. and Ind., (London), 1964, 404; J. App. Chem. and Biotechnol., 1972, *22*, 465).

(ii)Aldol and Mannich Reactions

Aldol reactions with alkylphenols lead to a number of selective procedures. 3-Methylphenol in dichloromethane treated with titanium tetrachloride and ethyl pyruvate gives a 81% yield of ethyl 2-hydroxy-2-(2-hydroxy-4-methylphenyl)propionate (A. Citterio et al., Synthesis, 1984, 760). 3-Tert-butylphenol in dichloromethane containing (+)-menthol and titanium tetrachloride when reacted with (+)-menthyl pyruvate affords a 67% yield of menthyl (+)(2R)-(2-hydroxy-4-tert-butylphenyl)lactate (d.e. 82%) (F. Bigi et al., Tetrahedron

Letters, 1985, *26*, 2021).

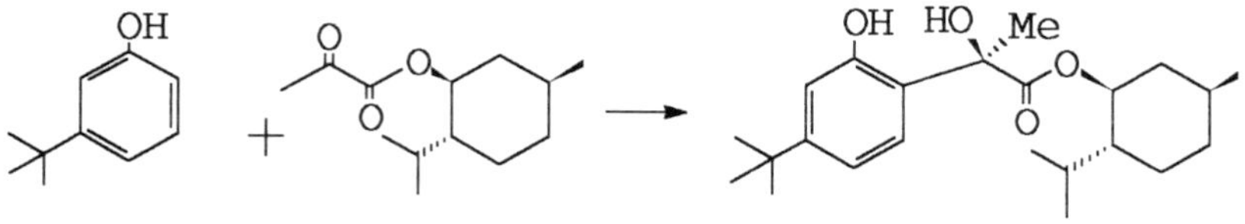

The magnesiobromo derivative of 4-tert-butylphenol upon treatment with (R)-glyceraldehyde acetonide and ultrasonification affords mainly the aldol product (syn:anti,96:4) in 70% yield (G. Casiraghi et al., J. Chem. Soc. Chem. Commun., 1987, 794).

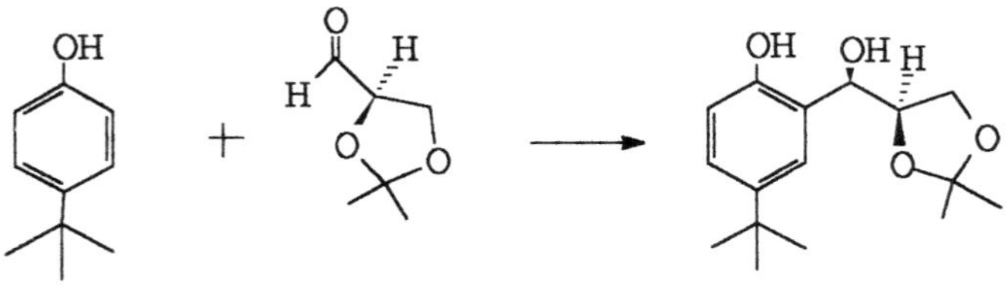

When the same magnesio salt is reacted with 3,4,6-tri-O-acetyl-D-glucal by ultrasonification the C α-glucoside (α:β, 100:1) is formed (G. Casiraghi et al., Carbohyd. Res., 1989, *191*, 243).

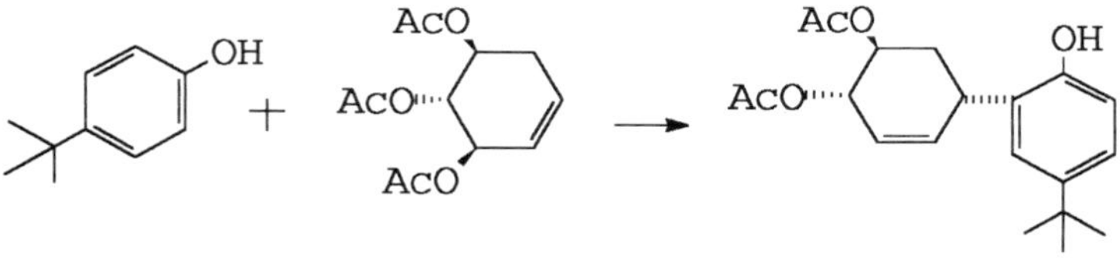

Alkylation at the 4-position occurs when but-3-en-2-one in toluene is reacted with 2-n-propylphenol in toluene containing Amberlyst-15. This gives 2-n-propyl-4-(3-oxobutyl)phenol in 77% yield (R.A. Bunce and H.D. Reeves, Synth. Commun., 1989, *19*, 1109).

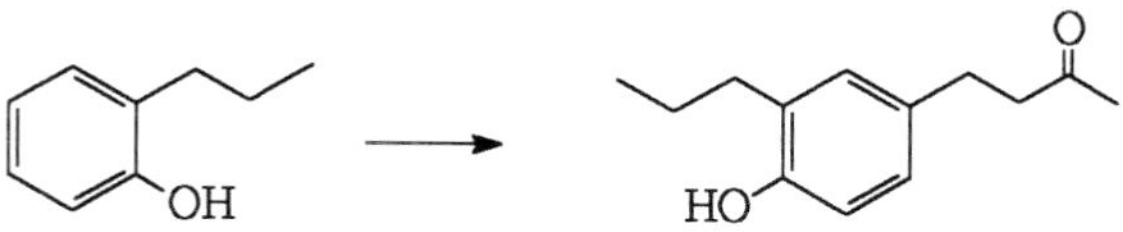

4-Methylphenol reacts with glycals upon treatment with boron trifluoride-etherate. Substitution rather than Michael addition products are formed (C. Booma and K.K. Balasubramaniam, Tetrahedron Letters, 1992, *33*, 3049).

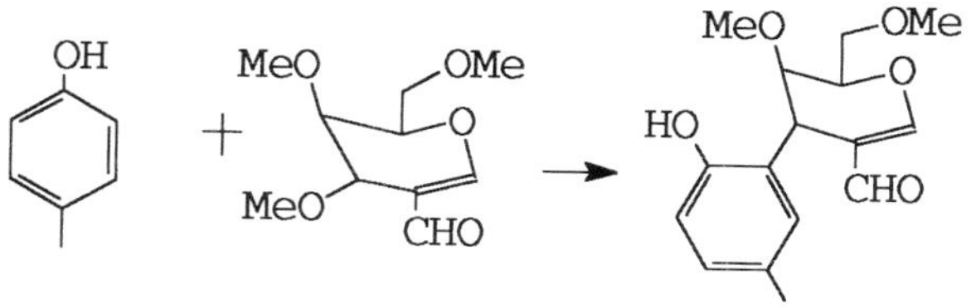

2,4-Dimethylphenol and bis(diethylamino)methane react in acetonitrile containing sulphur dioxide in a modification of the usual Mannich reaction conditions to give a 72% yield of 2-diethylaminomethyl-4,6-dimethylphenol (R.A. Fairhurst et al., Tetrahedron Letters, 1988, *29*, 5801).

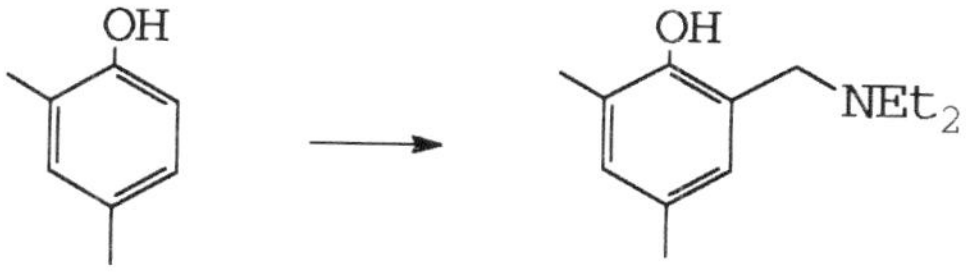

In a study of the Mannich reaction of 3-pentadecylphenol with a number of secondary amines and aqueous formaldehyde the aminal appears a more likely intermediate than the aminol (E.L. Short, J.H.P. Tyman and V. Tychopoulos, J. Chem. Tech. and Biotechnol., 1992, *53*, 389). The rate of the Mannich raection was found to be enhanced in aqueous media (J.H.P. Tyman and V. Tychopoulos, Synth. Commun., 1986, *16*, 1402).

4-Dodecylphenol underwent C-sulphomethylation by heating with sodium sulphite and aqueous formaldehyde during 20 hours (Phillips Petroleum Co., USP 4939293).

(iii) Cyclisation of o-Alkenylphenols

Alkenylphenols, formed as intermediates usually without isolation by alkylation of phenol at the 2-position, cyclise to benzofuran derivatives. Thus phenol and isobutanal (1:10), passed through γ-alumina held in a Pyrex tube at 300°C give 2,3-dihydro-2,2-dimethylbenzofuran in 60% yield (Ube Industries, J5 9184172). 4-Methylphenol and 2-ethylbutanal heated under reflux with a small quantity of stannic chloride with provision for removal of water afford 2,2-diethyl-2,3-dihydro-5-methylbenzofuran with a conversion of 83% and a yield of 44% (Chisso Corpn., J0 3190868).

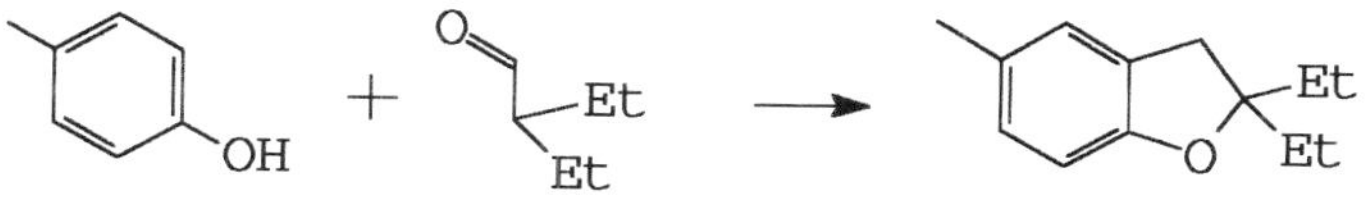

The aldol product from phenol and ethyl pyruvate has been
cyclised and reduced with lithium aluminium hydride to give 3-
methyl-3-hydroxy-2,3-dihydrobenzofuran in 98% yield from the
aldol (F. Bigi et al., Synthesis, 1980, 724). Lithium phenate
when heated with 1,4-dibromobut-2-ene gives a 48% yield of 2-
vinyl-2,3-dihydrobenzofuran (95% based on the recovered phenol
(G. Casiraghi et al., J. Chem. Soc., Perkin Trans I, 1983, 169).

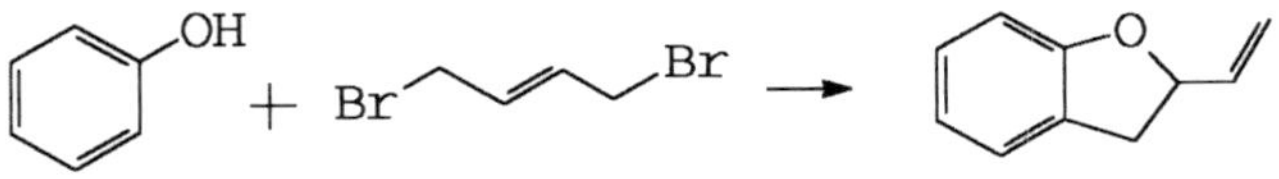

1,3-Cycloalkadienes undergo cycloaddition reactions with phenol
in contact with aluminium and under pressure to afford 2,3-
dihydro-2,3-cycloalkanobenzofurans in yields ranging from 73%
to 86% (Kiev Poly, SU 1180368).

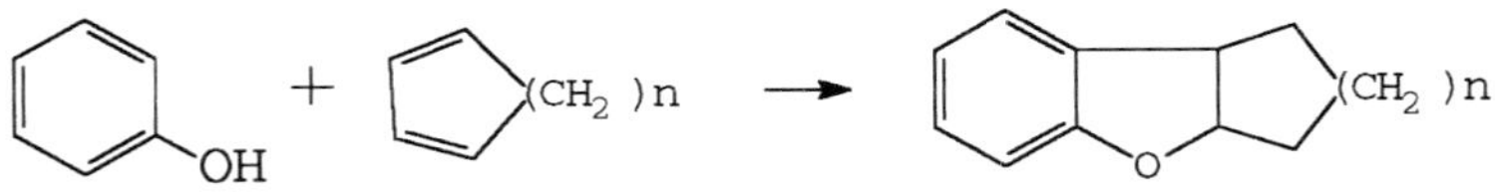

Δ^3-Carene, phenol and aluminium phenolate heated at 150-160°C
produce a tricyclic chroman in yields of 75-80% by preliminary
opening of the cyclopropano ring and ring closure by way of the
double bond (V.N. Rodionov, Y.B. Kozlikovskii and V.A.
Androschenko, Zh. Org. Khim., 1991, 27, 2627).

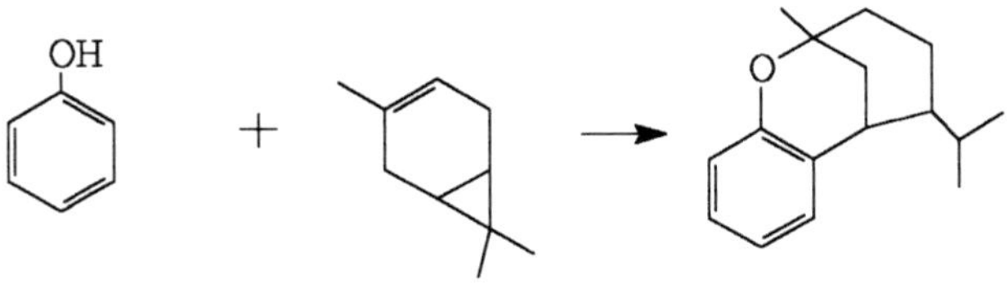

The ability of the phenolate anion to behave as a carbanion
character and hence to participate in Michael addition is
exhibited in the reaction of 2-methylphenol containing a little
lithium hydride with ethyl acrylate to afford a 58% yield of 8-
methyl-2,3-dihydrocoumarin, Crotonates can also be used but
methacrylates or 3,3-dimethylacrylates are much less efective
(J.E. Picket and P.C. Van Dort, Tetrahedron Letters, 1992, 33,
2261).

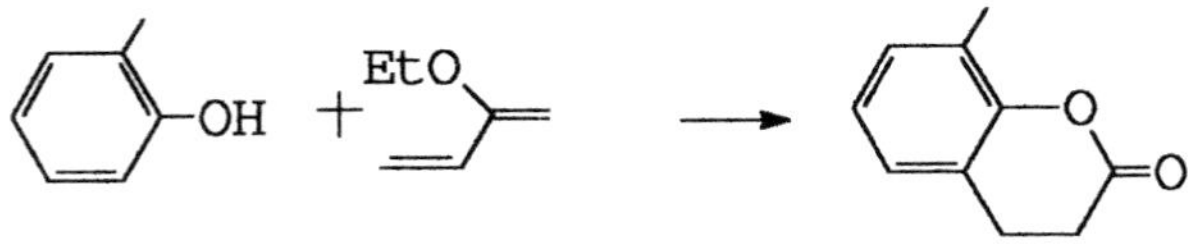

(iv) Dimerisation and Arylation

The possibility of unusual condensation reactions might be anticipated from the increased activation of the alkylphenolic ring compared with phenol itself. 4-Iso-propenylphenol in trifluoroacetic acid at ambient temperature furnishes 6-hydroxy-1-(4-hydroxyphenyl)-1,3,3-trimethylindane in acceptable yields (Upjohn Co., USP 4334106).

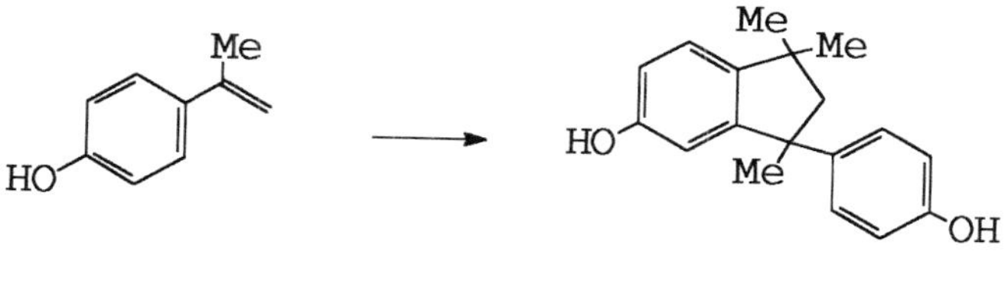

Bacterial degradation of 2,6-xylenol produces a dimeric product (H. Kneifel et al., Angew. Chem. Int. Ed. Engl., 1991, *30*, 202). 4-Methylphenol reacts at the 3-position to the OH group with acetanilide in the presence of a mixture of hydrochloric acid and antimony pentafluoride to give a 87% yield of 3-(4-acetylaminophenyl)-4-methyl-2-cyclohexenone (J.-C. Jacquesy and M.-P. Jouannetaud, Bull. Soc. Chim. Fr., II, 1980, *267*, 295). By contrast, reaction takes place at the 2-position in the reaction of 4-methylphenol with 2-methoxycarbonylbenzo-1,4-quinone in dichloromethane, or benzene, containing a little trifluoroacetic acid to afford 7,10-dihydroxy-2-methyl-6H-dibenzo[b,d]pyran-6-one in 78% yield (P. Mueller, T. Venakis, C.H. Eugster, Helv. Chim. Acta, 1979, *62*, 2833).

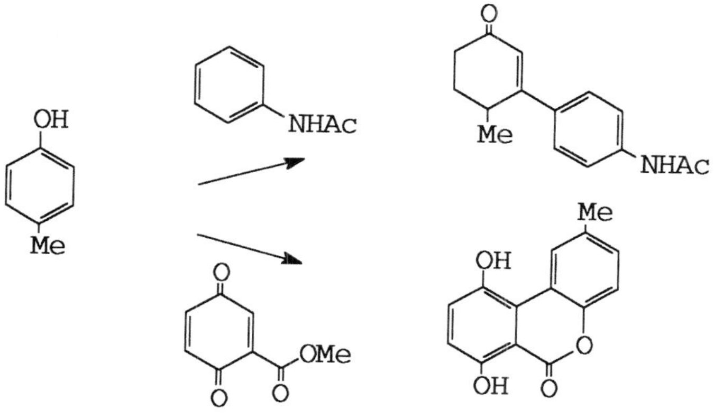

2-Iso-propylphenol and 2-(4-hydroxy-3-isopropylbenzoyl)benzoic acid react together with conc. sulphuric acid to produce 3,3-

bis(4-hydroxy-3-isopropylphenyl)phthalide in 76% yield, presumably by electrophilic attack of the monophthalide carbocation (J.K. Ruminski, and K.D. Przewoska, Chem. Ber., 1982, *115*, 3436).

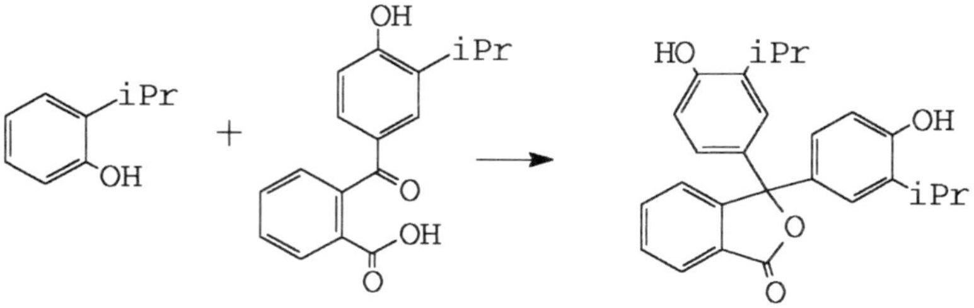

3,5-Dimethylphenol reacts with aromatic esters to form phthalides in the presence of triethyloxonium fluoroborate (T. Kurihara and K. Nasu, Chem. Pharm. Bull., 1982, *30*, 2723).

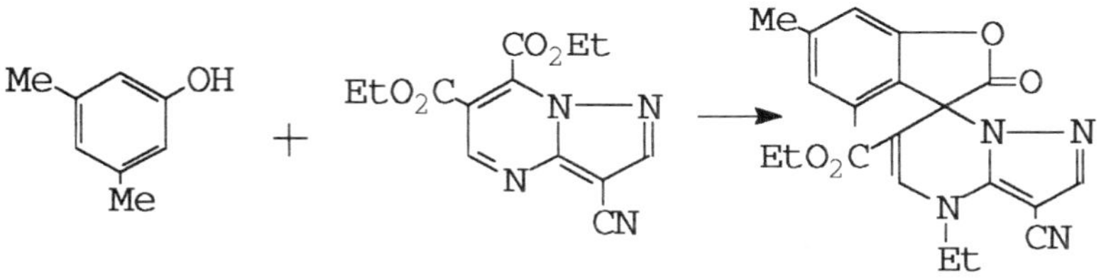

Trans-arylation takes place in the reaction of 2,6-diphenylphenol with 2,2-bis(4-hydroxyphenyl)propane (bis-phenol A) in dry chloroform containing methanesulphonic acid to afford 2,2-bis(4-hydroxy-3,5-diphenyl)propane in 87% yield (Z.Y. Wang and A.S. Hay, Synthesis, 1989, 471).

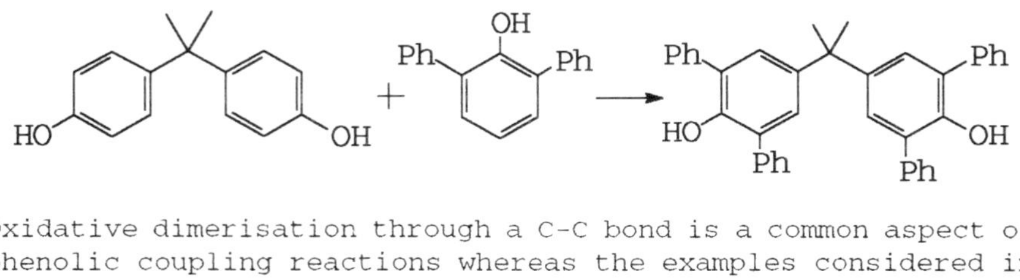

Oxidative dimerisation through a C-C bond is a common aspect of phenolic coupling reactions whereas the examples considered in the previous section generally resulted in both this and carbonyl group formation. 4-Methylphenol with caesium carbonate and a catalytic amount of a rhodium complex when heated, affords a 51% yield of 2,2'-dihydroxy-5,5'-dimethylbiphenyl (A.G.M. Barrett, T. Itoh and E.M. Wallace, Tetrahedron Letters, 1993,

A symmetrically fused naphthopyran has been synthesised, albeit in low yield, from 2-naphthol in ethanol containing Amberlyst-15 and 4-hydroxy-3-hydroxymethylbutan-2-one (N. Talinli, A. Akar and A.C. Aydogan, Tetrahedron, 1987, *43*, 3505).

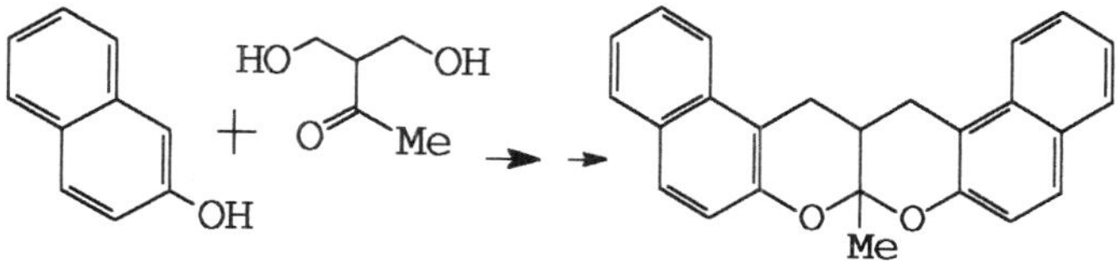

(v) Calixarenes

Calixarenes, a group of compounds first observed many years ago (A. Zinke and E. Ziegler, Chem. Ber.,1944, *77*, 264; R.F. Hunter and V. Vand, J. App. Chem., 1951, *1*, 298) have only been studied intensively in the last few years. Their chemistry has been reviewed (C.D. Gutsche, 'Calixarenes', Royal Society of Chemistry, Cambridge, 1991). Calix[4]arene in dichloromethane is acetylated with acetyl chloride in the presence of aluminium chloride to give a 4,4'-diacetyl derivative in 63% yield without isomer formation (K. No and M. Hong, J. Chem. Soc. Chem. Commun., 1990, 572).

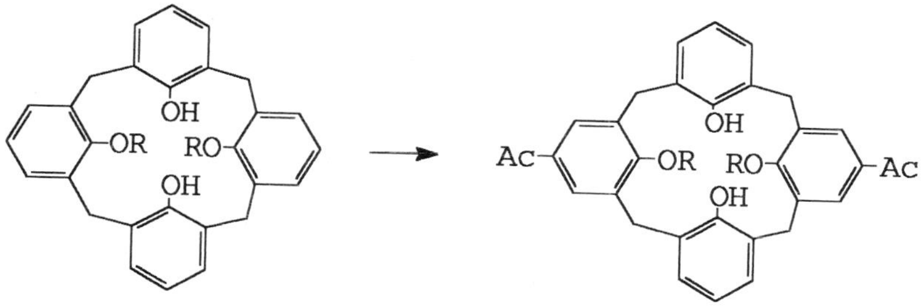

(vi) Crypto- and Hindered Phenols

In this section the reactions of 2-tert-butylphenols and 2,6-di-tert-butylphenols are described. The former group, the cryptophenols are not completely hindered, although their reactions are different from those of 2-methylphenols.

2-Tert-butylphenols:

Potassium 2-tert-butylphenate and aluminium chloride react with trichloroacetaldehyde in toluene to give the 6-trichloromethylol derivative in 58% yield. This affords (E)-[3,3']-dibenzofuranylidene-7,7'-di-tert-butyl-2,2'-dione in 89% yield upon boiling in decalin with basic alumina (F. Bigi, G. Casiraghi and G. Sartori, Tetrahedron, 1983, *39*, 2147).

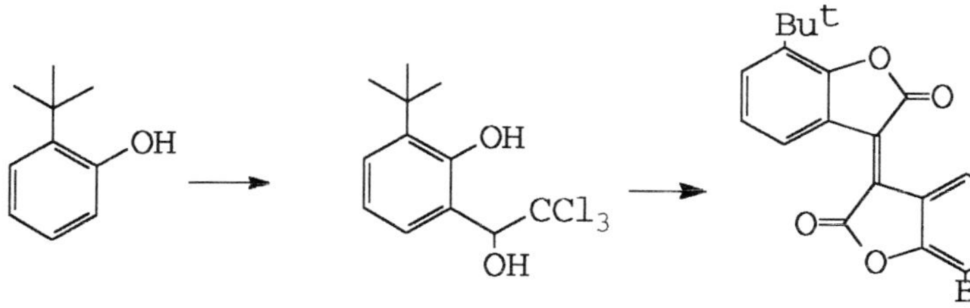

2-Tert-butyl-5-methylanisole is prepared in 94% yield from the phenol in contact with a little polystyrene-bound phosphonium salt (Polymer-$CH_2P^+Bu_3Cl^-$), pre-treated with aqueous sodium hydroxide and chloromethane (Y. Hamada et al., Chem. Pharm. Bull., 1981, *29*, 2246).
4,6-Di-tert-butyl-2,1,3-benzothiadiazole is synthesised in 86% yield from 2,4-di-tert-butyl-6-iodophenol by heating with tetranitrogen tetrasulphide in toluene (S. Mataka et al., J. Chem. Soc. Chem. Commun., 1983, 1136).

In the examples given, the tert-butyl group survives comparatively drastic conditions. Loss of a tert-butyl group occurs, however, upon treatment of 2,4-di-tert-butyl-6-(4-chloropyridazin-1-yl)phenol at ambient temperature in dichloromethane with aluminium chloride to afford a 73% yield of the 4-tert-butyl analogue (N. Lewis and I. Morgan, Synth. Commun., 1988, *18*, 1783).
A similar loss of the tert-butyl group occurs in the ferric/manganese naphthoate promoted 'dimerisation' of 2-methyl-6-tert-butylphenol to give 3,3'-dimethyl-4,4'-dihydroxybiphenyl in 54% yield (Dainippon Ink Chem., KK JO 2196745).
2,6-Di-tert-butylphenols:
The reactions depicted comprise those at the 4-position, substitutions at the hydroxyl group and oxidations. BHT, the commercial antioxidant, has been synthesised in 97% yield by the reaction of 2,6-di-tert-butylphenol with paraformaldehyde in the presence of 10% palladium-carbon, ytterbium chloride and hydrogen (7 Kg/cm^2) at ambient temperature (Idemitsu Petrochem., KK, J63 077831).

In a related process 2,6-di-tert-butylphenol, paraformaldehyde and an alkanol heated under pressure with N,N-diethylbenzylamine affords a high yield of the 4-alkoxymethyl derivative (Idemitsu Petrochem., KK WO 8803129), while with methanol containing 40% aqueous dimethylamine, 2,6-di-tert-butyl-4-methoxymethylphenol specifically results (Ethyl Corporation, WO 8402336).
Di(3,5-di-tert-butyl-4-hydroxybenzyl)ether was formed in 98% yield from the reaction of 2,6-di-tert-butylphenol and paraformaldehde in boiling methanol containing triethylamine (Koppers Co. Inc., USP 4633022).
Ethyl 3,5-di-tert-butyl-4-hydroxybenzoate in acetonitrile containing sodium perchlorate and perchloric acid after pulse-electrolysis affords a 75% yield of ethyl 7-tert-butyl-2-methyl-5-benzoxazolecarboxylate (E.-L. Dreher et al., Chem Ber., 1982, *115*, 288).

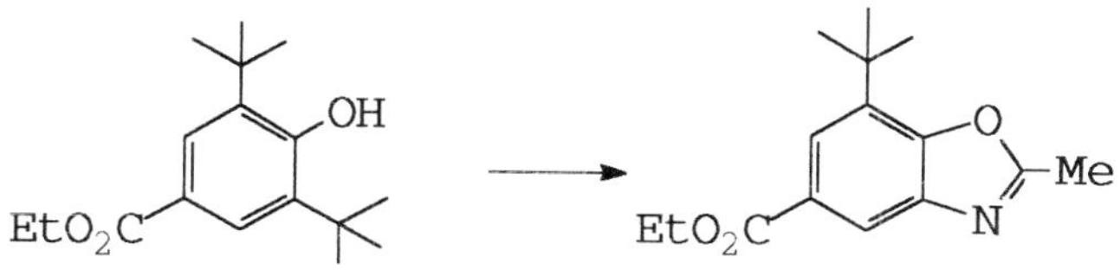

The O-glucoside of 2,6-di-tert-butylphenol is synthesised in 75% yield by reaction with 1-diazirinyl-2,3,4,6-tetra-O-benzyl-1-deoxyglucopyranose in toluene containing powdered molecular sieve 4A (K. Briner and A. Vasena, Helv. Chim. Acta, 1990, *73*, 1764).

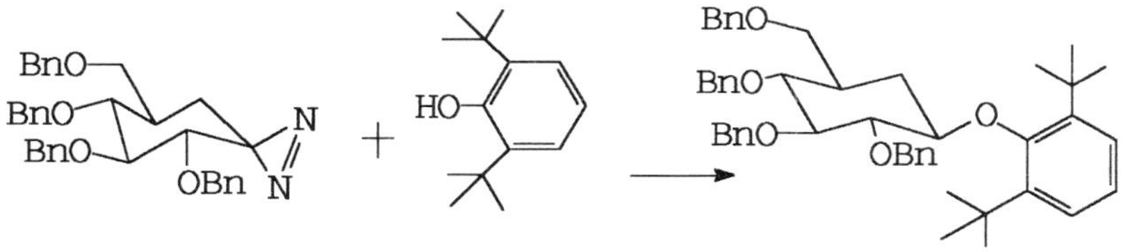

A variety of oxidation products results from hindered phenols. In the oxidation of 2,6-di-tert-butylphenol in benzene with manganese dioxide on alumina at ambient temperature over 60 hours, 3,5,3',5'-tetra-tert-butylstilbenequinone is formed in

72% yield (L.K. Liu, Y-H. Lee and L-S. Leu, J. Chinese Chem. Soc., 1989, *36*, 219). There is an element of rediscovery in that the author of this chapter recalls work of some four decades past on the identification of yellow spots in commercial BHT which proved to be this substance (J.H.P. Tyman and, the late,B.J.F. Hudson, unpublished work).

2,6-Di-tert-butylphenol in dichloromethane with aqueous sodium hydroxide and benzyltrimethylammonium tribromide affords the corresponding diphenoquinone in 61% yield (S. Kajigaeshi et al., Bull. Chem. Soc. Japan, 1991, *64*, 1060).

2,6-Di-tert-butylbenzo-1,4-quinone is formed in 96% yield from 2,6-di-tert-butylphenol in acetonitrile with the chelate [N,N'-bis(2'-pyridinecarboxamido)-1,2-benzene]cobalt(II)-water, by oxygenation (P.A. Ganeshpure, A. Sudalai and S. Satish, Tetrahedron Letters, 1989, *30*, 5929).

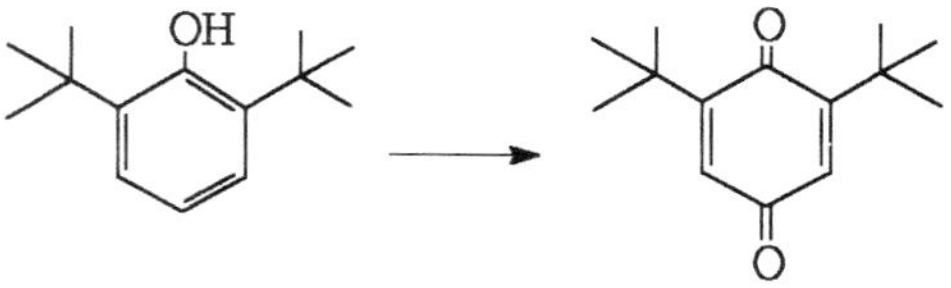

4,6-Di-tert-butylbenzo-1,2-quinone results in 72% yield from 2,4-di-tert-butylphenol in acetone/dichloromethane/water (1:1:2) containing sodium bicarbonate by treatment with ozone (J.K. Crandall et al., Tetrahedron letters, 1991, *32*, 5441).

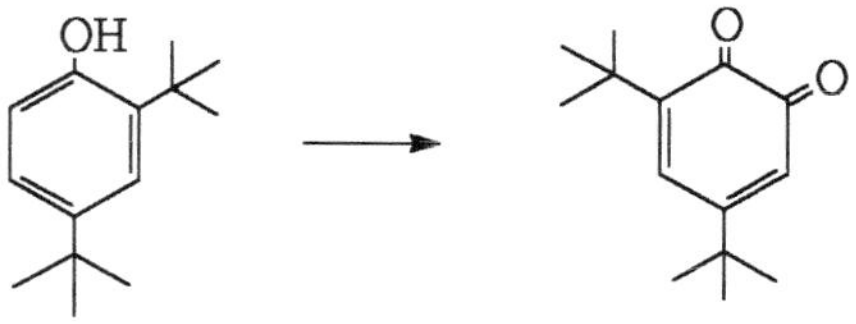

4-Iso-butyroyl-2,6-di-tert-butyl-6-hydroperoxycyclohexa-2,4-dienone is produced from 4-isobutyroyl-2,6-di-tert-butylphenol and bis(salicylidene)ethylenediminecobalt(II) in dichloromethane by gassing with oxygen (A. Nishinaga, T. Shimizu and T. Matsuura, Tetrahedron Letters, 1981, *22*, 5293).

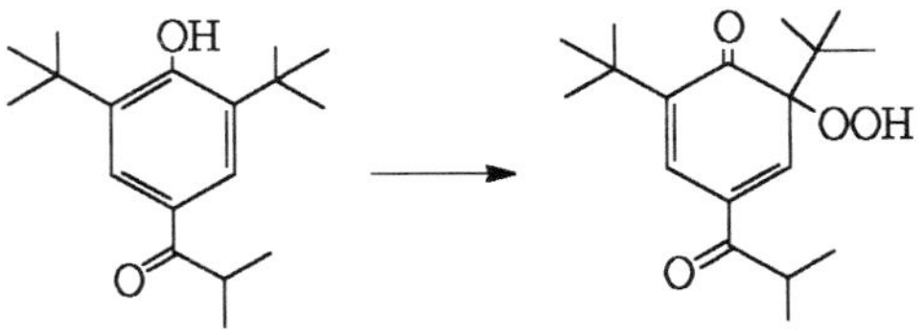

2,6-Di-tert-butyl-4-ethylphenol in ethanol containing potassium hydroxide affords 2,6-di-tert-butyl-4-ethyl-4-hydroxycyclohexa-2,5-dienone in 82% yield by treatment with oxygen followed by reduction of the intermediate hydroperoxide (A.A. Volodkhin, R.D. Malysheva and V.V. Ershov, (Izv. Akad. Nauk SSSR Ser. Khim., 1982, 1594).

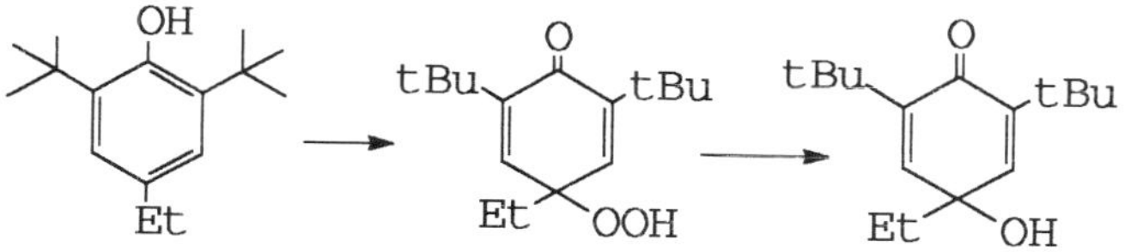

Other Applications of Hindered Phenols:
2,6-Di-tert-butyl-4-methoxyphenyl esters are used for stereospecific aldol reactions. Thus the threo-aldol obtained from isobutanal and 2,6-di-tert-butyl-4-methoxyphenyl propionate, after acetylation, is converted with ceric ammonium nitrate to the acetoxy acid which was then hydrolysed to afford 2,4-dimethyl-3-hydroxypentanoic acid (C.H. Heathcock et al., Tetrahedron, 1981, 37, 4087).
Methyl aluminium bis(2,6-di-tert-butyl-4-methylphenoxide) complexes methyl ethyl ether in preference to diethyl ether and when adsorbed on silanised silica gel can separate chromatographically THF from diethyl ether (K. Maruoka, S. Nagahara and H. Yamamoto, J. Am. Chem. Soc., 1990, 112, 6115). The analogue from 2,6-diphenylphenol complexes formaldehyde and prevents its polymerisation (ibid., 7422). The same reagent(s) exert an influence on certain Claisen rearrangements to give E-isomers whereas those based on 2,6-di-tert-butyl-4-bromophenol afford Z-isomers (ibid., 316). 2,4-Di-tert-butyl-6-formylphenol, as the manganese derivative, together with *trans*-1,2-diaminocyclohexane has been employed as a catalyst in conjunction with sodium hypochlorite for enantioselective epoxidations (E.N. Jacobsen et al., J. Am. Chem. Soc., 1991, 113, 7063).
Highly hindered phenols can be deoxygenated through hydrogenation of their triflates (J.M. Saa et al., J. Org. Chem., 1990, 55, 991).

(f) Phenolic Aldehydes, Ketones, Acids and Esters

Synthesis

(i) 2-Hydroxyaldehydes
In this section, phenolic aldehydes, ketones, acids and esters

are considered in which the carbonyl group is conjugated with the ring. Some mention is made of amides and nitriles. Improvements in the regiospecific synthesis of salicylaldehyde have enabled this compound to be synthesised in 99% yield from the reaction of phenol in toluene containing stannic chloride and tri-n-butylamine with paraformaldehyde (G. Casiraghi et al., J. Chem. Soc. Perkin Trans I, 1980, 1862). Salicylaldehyde is obtained in 97% yield by heating phenol and hexamethylenetetramine in ethanol (G.A. Tolstikov, SU 829613). 2-Methylphenol in methanol containing sodium hydroxide and a catalytic quantity of tetrakis(2,4-dimethoxyphenyl) porphyrinato -iron(III) chloride and copper(II) nitrate trihydrate when oxygenated, affords a 78% yield of salicylaldehyde (Bayer AG, EP 451650).

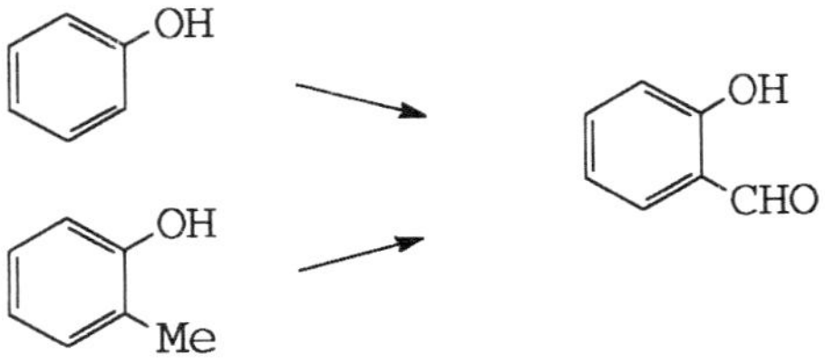

An abnormal Reimer-Tiemann reaction has been observed where, since dichlorocarbene attack at the 2-position to the phenolic hydroxyl was blocked, the incipient formyl group displaced a methoxyl substituent (C.W. Bird, A.L. Brown and C.C. Chan, Tetrahedron, 1985, *41*, 4685).

(ii) 4-Hydroxyaldehydes

4-Methylphenol in methanol has been oxidised by 2,3-dichloro-5,6-dicyanobenzoquinone to afford a 84% yield of 4-hydroxybenzaldehyde (C.W. Bird, and Y-P. S. Chauhan, Org. Prep. Proc. Intern., 1980, 201). In a modification of the Reimer-Tiemann reaction, addition of chloroform to phenol and cyclodextrin in 10% aqueous sodium hydroxide gives a 46% yield of 4-hydroxybenzalehyde (Asahi Kasei Kogyo (Hirai), WP 8203073).

(iii) 2-Hydroxyaryl alkyl ketones

The Fries rearrangement of phenyl acetate in benzene, toluene or xylene with zirconium tetrachloride gives 2-hydroxyacetophenone in high yield (Mitsui Petrochem. Ind., KK, J5 9118730). 2-Propionylphenol is obtained in 62% yield from phenyl propionate with titanium tetrachloride in dichlorobenzene (Mitsui Petrochem. Ind., KK J60 004149).

Photochemical rearrangement of phenyl acetate in hexane containing potassium carbonate furnishes a 74% yield of 2-hydroxyacetophenone although in the absence of the base the yield is only 13% (H. Garcia, J. Primo and M.A. Miranda, Synthesis, 1985, 901).

2-Cinnamoylphenol is derived in 70% yield (with 90% conversion) by a novel photochemical method, said to be higher-yielding than the corresponding Fries rearrangement, consisting of irradiation of a 10^{-4}M solution of phenyl cinnamate as micelles in 10^{-2}M aqueous sodium dodecyl sulphate for 6 hours under nitrogen (A.K. Singh and T.S. Raghuraman, Tetrahedron Letters, 1985, *34*, 4125).

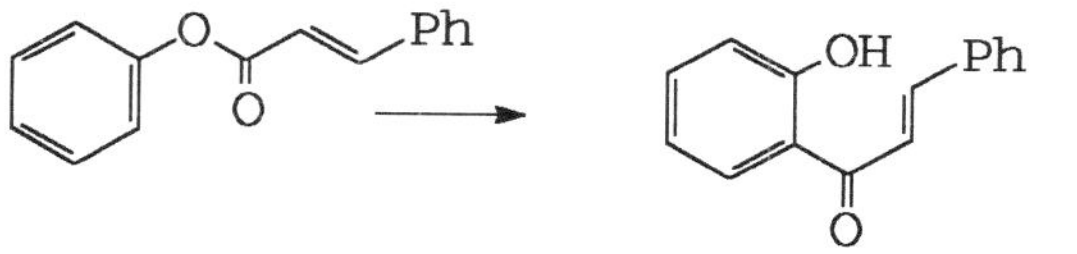

Among the alternative methods to the Fries rearrangement is the ozonolysis of benzofurans. Thus 3-cyclohexylbenzofuran in dichloromethane saturated with ozone affords cyclohexyl 2-(formyloxy)phenyl ketone in 86% yield. This is readily converted by treatment with methanolic sodium hydroxide to cyclohexyl 2-hydroxyphenyl ketone (A.G. Schultz, J.J. Napier and R. Ravichandran, J. Org. Chem., 1983, *48*, 3408).

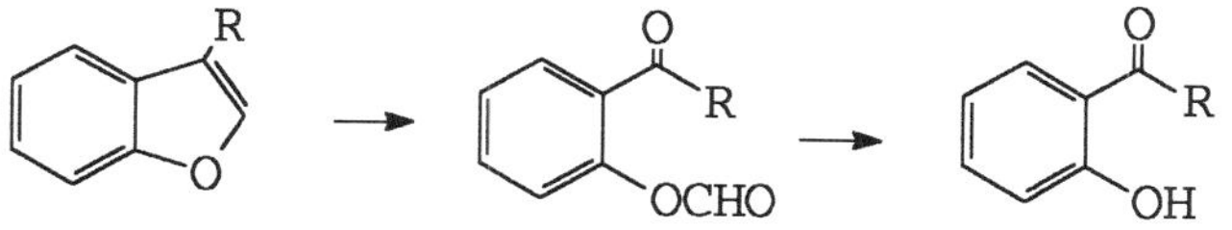

2-Chloroacylphenols form exclusively from the reaction of phenols and chloromethyl cyanide in the presence of both aluminium chloride and boron trifluoride, whereas the Houben-Hoesch procedure results in 4-hydroxyphenyl ketones (S.H. Bertz, Synthesis, 1980, 708).

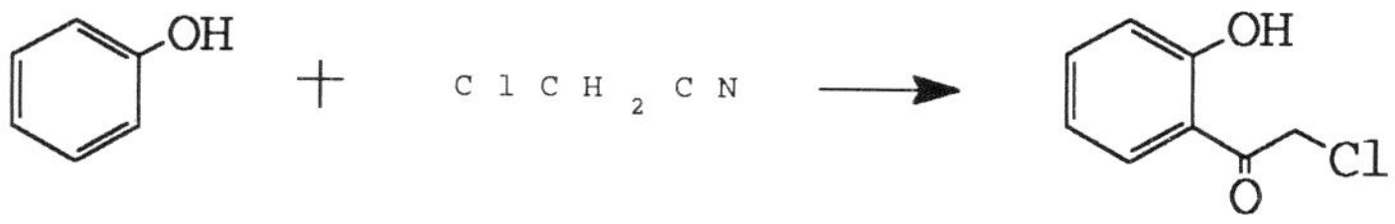

Aromatisation of cyclohexenones by a variety of procedures also gives access to 2-acylphenols (M.A. Elhashhash, A.A. Afify and A. Nagy, Indian J. Chem., Sect.B, 1979, *17*, 581).

(iv) 2- and 4-Hydroxybenzophenones

4'-Methylphenyl 3-chlorobenzoate in boiling nitrobenzene containing Nafion-H affords 3-chloro-2'-hydroxy-5'-methylbenzophenone in 71% yield (G.A. Olah, M. Arvanaghi and V.V. Krishnamurthy, J. Org. Chem., 1983, *48*, 3357).

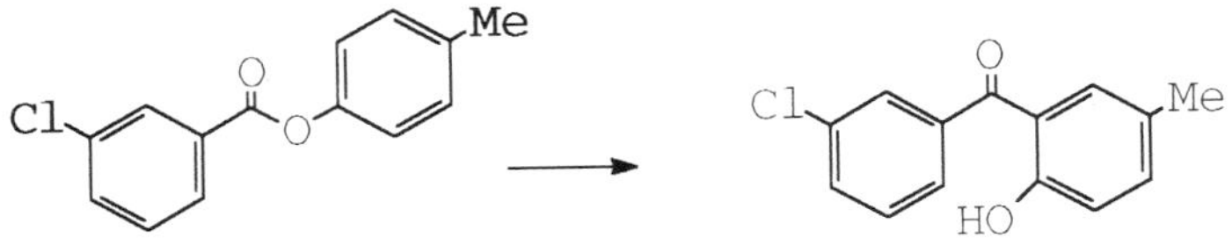

4-Hydroxybenzophenones are derived by various intermolecular reactions. 4,4'-Dihydroxybenzophenone is synthesised in 51% yield by heating 4-hydroxybenzoic acid and phenol with a catalytic amount of trifluoromethanesulphonic acid (Imperial Chem.Ind., EP 57503) and in a similar way, but at ambient temperature, 4-fluorobenzoic acid and phenol afford 4-hydroxy-4'-fluorobenzophenone in 59% yield (Imperial Chem. Ind., EP 75390).

(v) Polycyclic Hydroxyketones

2-Acetyl-4-phenyl-1-naphthols are prepared by the Knoevenagel condensation of diphenylacetaldehyde and acetoacetic esters in the presence of a molecular sieve (G.A. Taylor, J. Chem. Soc., Perkin Trans I, 1981, 2920).
Nucleophilic displacement of the dimethylamino group in 1-dimethylamino-2,4-bis(trifluoroacetyl)naphthylamine by the hydroxide, alkoxide or alkylthioxide ion leads to the corresponding hydroxy, alkoxy or thioalkoxy derivative (M. Hojo et al., Synthesis, 1989, 870).
A synthesis of 1-oxo-1,2,3,4-tetrahydro-9-hydroxyanthracene, giving a 23% yield, involves the thermolysis of a phosphorane and homophthalic anhydride in a sealed tube in toluene containing trimethylsilane (H. Ohmori et al., J. Chem. Soc., Chem. Commun., 1988, 874).

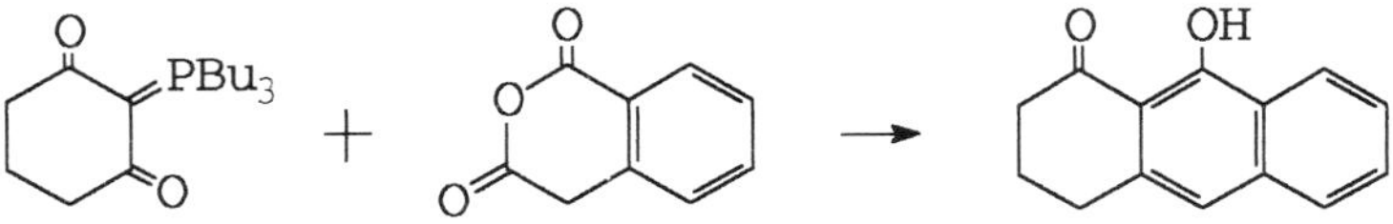

(vi) Phenolic Acids

A range of procedures is available for the synthesis of 2-hydroxy, 4-hydroxybenzoic and hydroxynaphthoic acids.
For example, 3-methylsalicylic acid can be obtained from the magnesiobromide of 2-methylphenol and phosgene (G. Sartori et al., Synthesis, 1988, 763).

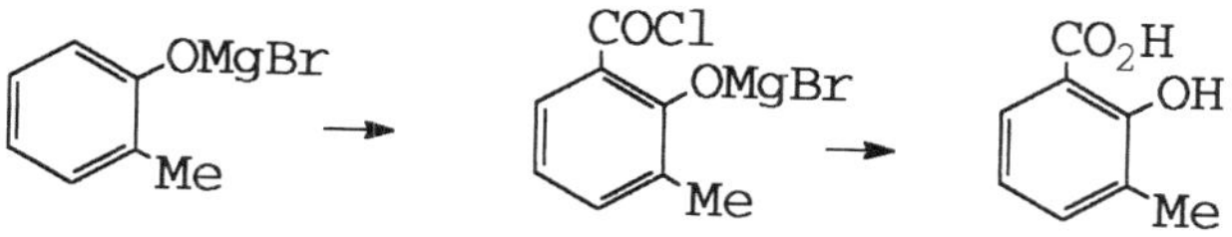

3-Methylsalicylic acid is also synthesised by the reaction of 2-methylphenol with ninhydrin (indane-1,2,3-trione) in boiling acetic acid via the intermediate 2-hydroxy-2-(2-hydroxy-3-methylphenyl)indan-1,3-dione. This when heated with 4-methylaniline in acetic acid affords 3-(2-hydroxy-3-methylbenzoyl)-2-(4-tolyl)-1-oxo-1,3-dihydroisoindole in 70% yield and thence, by hydrolysis with aqueous sodium hydroxide, the required product in a 55% overall yield (G. Schmitt et al., Synthesis, 1984, 385).
Copper(II) salts of N-benzoyl-2-methylalanines upon oxidation and hydrolysis of the intermediate product furnish salicylic acids in quantitative yields (O. Reinaud, P. Capdevielle and M. Maumy, J. Chem. Soc. Chem. Commun., 1990, 566).
In the reworking of an old method, 4-hydroxybenzoic acid has been produced in high yield by the copper-catalysed carboxylation of phenol in hot alkaline solution with carbon tetrachloride (H. Sasson and M. Razintsky, J. Chem. Soc. Chem. Commun., 1985, 1134). The author of this chapter recalls preparing 4-hydroxy-3-methylbenzoic acid by exactly the same route in 1947.
A mixture of 4-hydroxybenzoic acid and benzoic acid (9:1) results in 55% yield from treatment of 4-phenylphenyl triflate in carbon tetrachloride with ruthenium tetroxide, protection of the OH group being necessary to avoid fragmentation of the ring (D.C. Ayres and D.P. Levy, Tetrahedron, 1986, *42*, 4259).

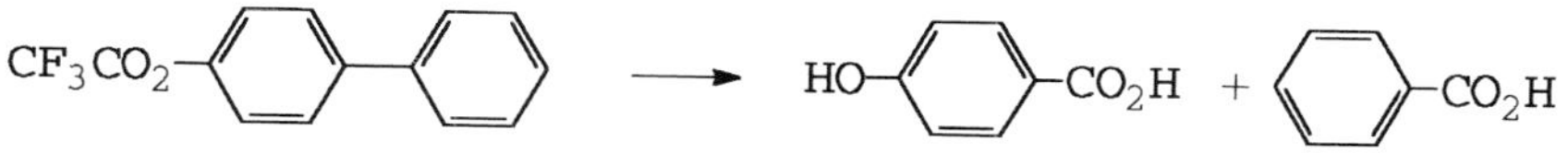

2,4-Dicarboxyphenol can be obtained in 95% yield from its dibenzyl ether by 'transfer hydrogenation' in ethanol solution containing 10% palladium-carbon and aqueous sodium hypophosphite. In related compounds, carbonyl, halogen, and amide groups are unaffected (R. Sala, G. Doria and G. Passarotti, Tetahedron Letters, 1984, *40*, 4565).

1-Hydroxy-2-naphthoic acid is derived in 96% yield from 1-naphthol and methoxymagnesium methyl carbonate in dimethylformamide under heat and pressure; phenol and catechol did not respond to this methodology (L.A. Cate, Synthesis, 1983, 385).

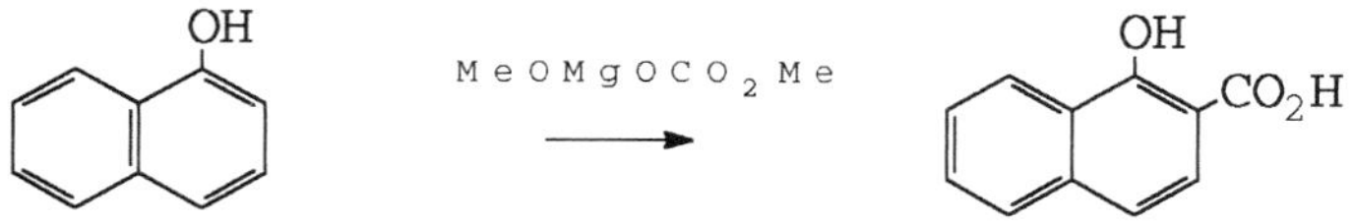

2-Hydroxy-1-naphthoic acid has been synthesised by the oxidative cyclisation of 5-aryl-3-oxopentanoates (A. Citterio et al., Synthesis, 1990, 142).

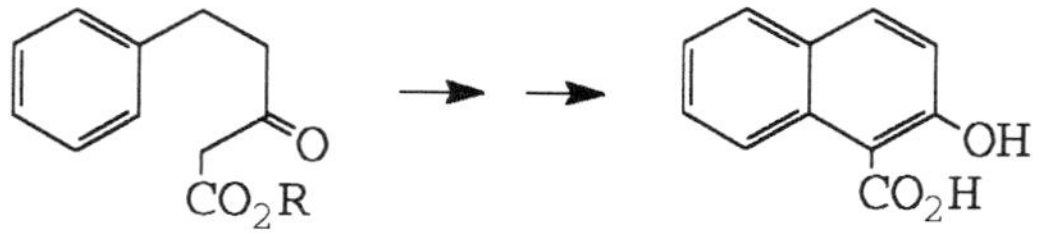

(vii) Phenolic Esters

A number of procedures involving non-aromatic precursors have been developed for the synthesis of salicylates. α,β-Unsaturated aldehydes are reacted with the dianion (DA) of ethyl acetoacetate to afford 4-substituted salicylates in good yield, (D.H.R. Barton et al., J. Chem. Soc., Perkin Trans I, 1983, 665); 6-methylsalicylates result from the addition of the monoanion (MA) at C2 (F.M. Hauser and S.A. Pogany, Synthesis, 1980, 814).

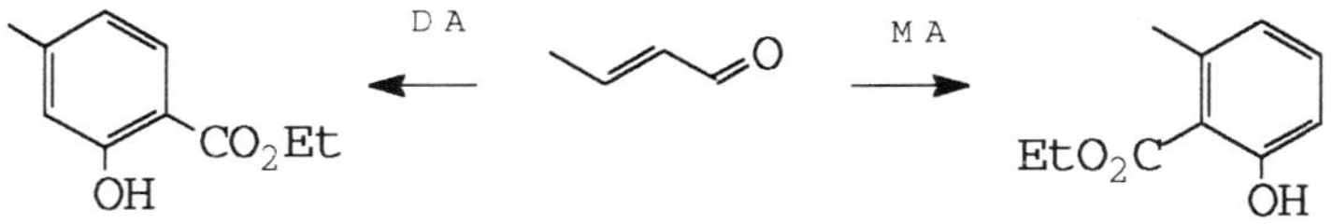

Ethyl 3-cyano-4,6-diphenyl-2-hydroxybenzoate is formed in 52% yield from the Michael addition of ethyl 2-cyano-3-phenyl-2-butenoate and ethyl cinnamate in benzene containing sodium hydride and some ethanol followed by aromatisation of the intermediate (C. Ivanov and T. Tcholakova, Synthesis, 1982, 730).

Methyl 5-methylsalicylate has been obtained in 71% yield by intramolecular cyclisation of an open-chain ketoester in acetic acid containing manganese(III) acetate dihydrate and 4 moles lithium chloride (B.B. Snider and J.J. Patrick, J. Org. Chem., 1989, *54*, 38).

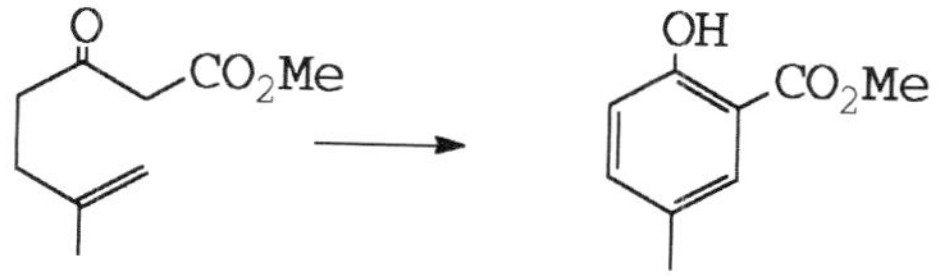

Phenyl salicylate is synthesised in 53% yield from phenol and dichloro bis(benzonitrile)palladium(II) in dichloromethane under carbon monoxide with the addition of diisopropylamine (General Electric, USP 4221920).

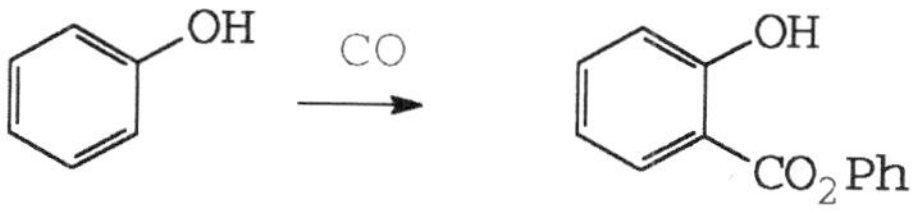

The dihydroaromatic compound, methyl 3-methoxycarbonylcyclohexa-3,5-dienol is derived by the Diels-Alder addition of furan and methyl acrylate in the presence of zinc iodide. The adduct which is obtained in 55% yield is themolysed to the cyclohexadienol in 85% yield (F. Brion, Tetrahedron Letters, 1982, *50*, 5298).

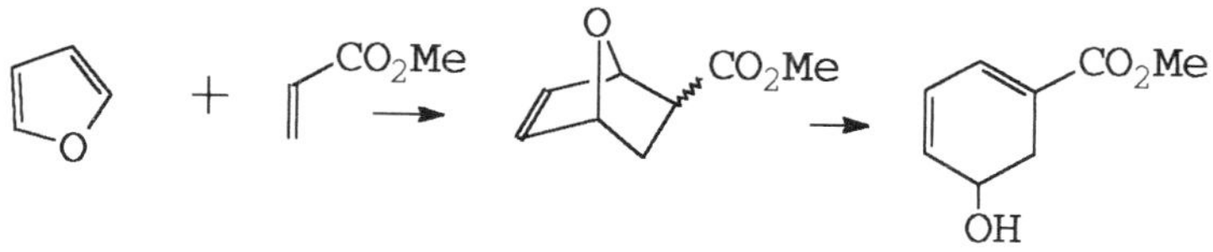

In the naphthalenic series, methyl 3-methoxy-5,6,7,8-tetrahydronaphthalene-2-carboxylate is obtained from the Michael addition of the cyclohexanone anion with dimethylmethoxymethylene malonate in tetrahydrofuran and treatment of the intermediate pyrone with dimethoxyethene (D.L. Boger and M.D. Mullican, Tetrahedron Letters, 1982, *44*, 451).

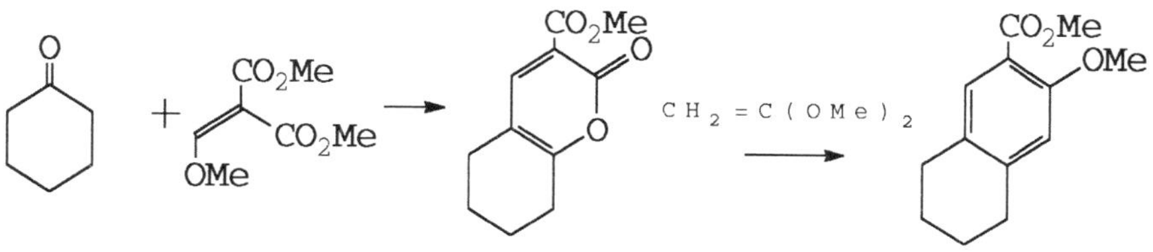

(viii) Alkyl 4-Hydroxybenzoates

Ethyl 4-hydroxy-3-methyl-6-ethylbenzoate is synthesised in 80% yield by the cyclisation of ethyl 4-methyl-5,7-dioxo-2(Z)-nonenoate in ethanol containing sodium ethoxide (A.B. Smith.III and S.N. Kikenyi, Tetrahedron Letters, 1985, 4419).

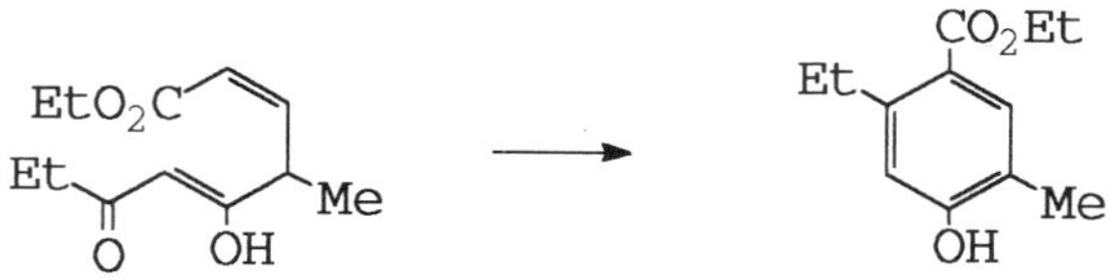

Methyl 4-hydroxy-3,6-dimethylbenzoate is obtained as the O-methyl ether in 84% yield by the aromatisation and dehydration of methyl 3,6-dimethylresorcylate, (2,5-dimethyl-4-methoxycarbonylcyclohexane-1,3-dione) by treatment with thionyl chloride, air and copper oxide (BASF, EP 64605).

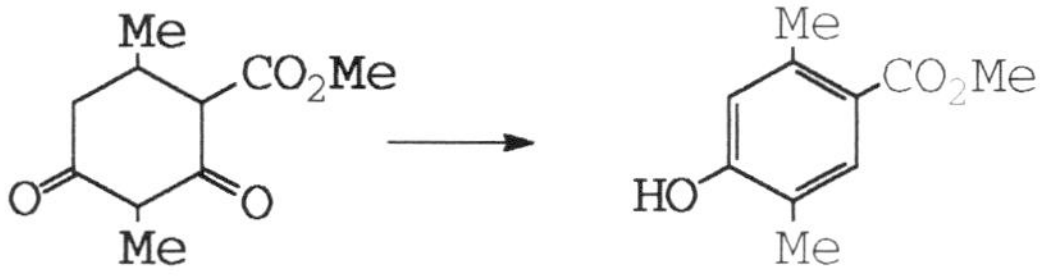

Pyrimidines (as masked malonaldehydes) can act as quasi-aromatic intermediates to afford acceptable yields of 4-phenyl compounds (P. Bamfield and P.F. Gordon, Chem. Soc. Rev., 1984, *13*, 441). Similarly, 5-formyl-1,3-dimethyluracil in ethanolic sodium ethoxide reacts with acetylacetone, to give ethyl 3-acetyl-4-hydroxybenzoate in 55% yield (K. Hirota, Y. Kitade and S. Senda, J. Org. Chem., 1981, *46*, 3949).

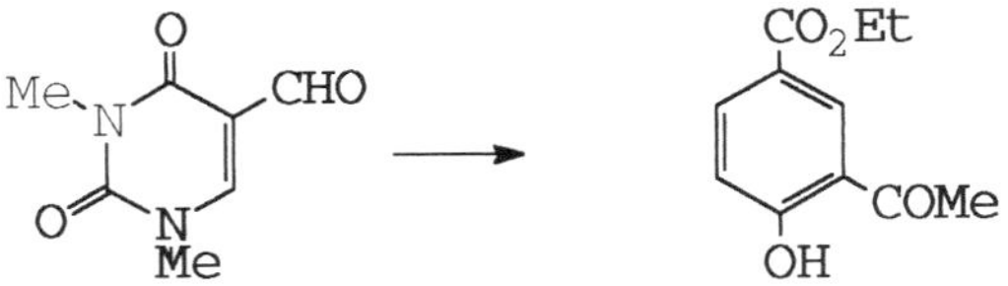

(ix) Dialkyl Hydroxyphenyldicarboxylates

A number of syntheses of hydroxyphthalates, isophthalates and terephthalates have been devised.

The double Claisen condensation of malondialdehyde with diethyl 3-oxoglutarate which results in 2,6-diethoxycarbonylphenol is the simplest example of this approach, although the yield is modest (S.H. Bertz, W.O. Adams and J.V. Silverton, J. Org. Chem., 1981, *46*, 2828).

Ethyl formylacetate and diethyl 3-oxoglutarate also afford moderate yields of compounds otherwise difficult to obtain (G.A. Kraus, J. Org. Chem., 1981, *46*, 201).

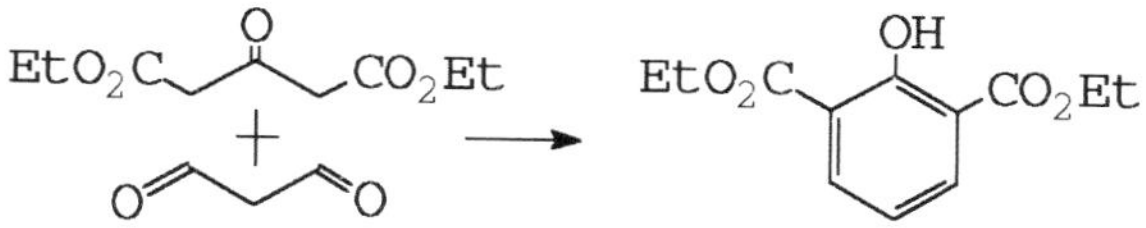

Dimethyl 2-hydroxy-4-(2-phenylethenyl)-1,3-benzenedicarboxylate is synthesised in 63% yield, by heating a mixture of dimethyl 3-oxoglutarate and 5-dimethylamino-1-phenyl-1,4-pentadiene-3-one in dioxan containing acetic acid and potassium fluoride (N. Takeuchi, N. Okada and S. Tobinaga, Chem. Pharm. Bull., 1983, *31*, 4355).

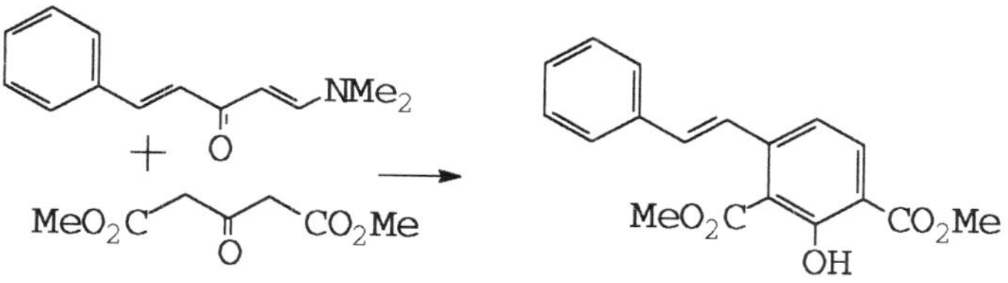

3-Hydroxy-6-(aryloxy)phthalates result from aromatisation of the Diels-Alder addition product of 2-aryloxy-5-hydroxymethylfurans and dimethyl acetylene dicarboxylate (S.D. Carter et al., J. Chem. Soc., Perkin Trans I, 1990, 1231).

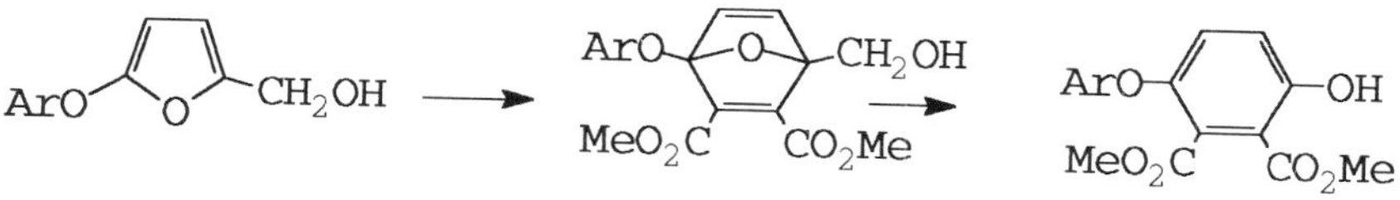

(x) Phenolic nitriles, Amides, Ketoacids and Phthalides

2-Hydroxybenzonitrile is synthesised in nearly quantitative yield by treatment of 6-chloro-6-cyano-2-cyclohexenone with triethylamine (ICI plc, GB 2138419).

2-Hydroxy-N-methylbenzamide is prepared in 73% yield by the reaction of phenol in benzene containing boron trichloride with methyl isocyanate in benzene (O. Piccolo et al., Tetrahedron, 1986, *42*, 885).

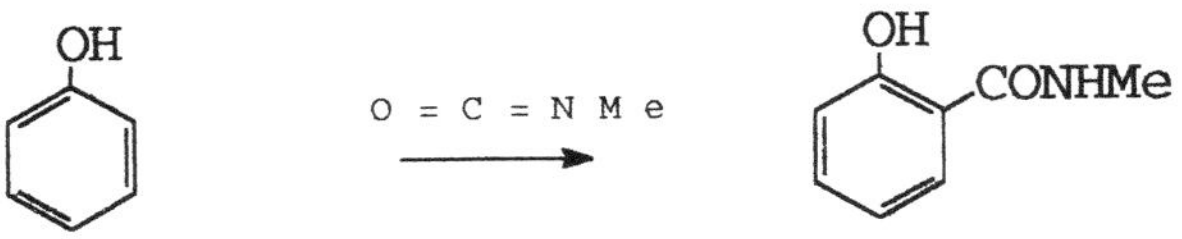

2-Hydroxyphenylglyoxylic acid is synthesised in 44% yield from phenoxymagnesium bromide by treatment with oxalyl chloride (F. Bigi et al., J. Chem. Soc., Perkin Trans I 1984, 2655). An alternative synthesis has been achieved by the use of glyoxylic acid and boric acid leading for example to the important aldehyde, 5-tert-nonylsalicylaldehyde through a final oxidative decarboxylation (ICI, EP 536960).

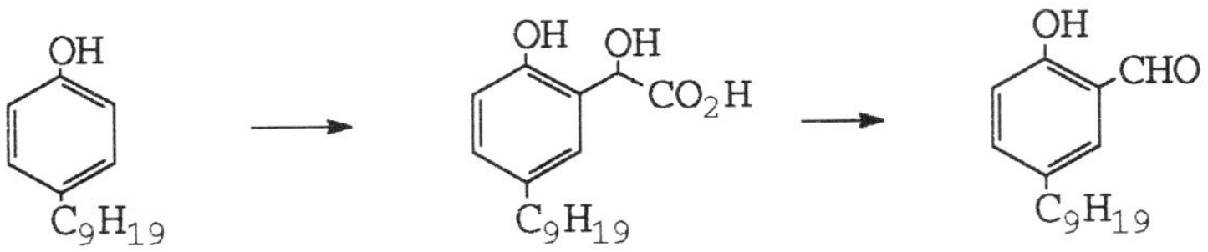

2-Hydroxyphthalic anhydride ring formation by the Diels-Alder addition of 2-pivaloyloxyfuran and maleic anhydride, followed by aromatisation with sulphuric acid is reported (J.H. Nasman, Synthesis, 1985, 788).

Reactions

(i) Phenolic Aldehydes
In this section the reactions of phenolic aldehydes, ketones, phenolic acids and esters which have been listed in the preceding decade are summarised. For the aldehydes typically, these comprise etherification, replacement of the formyl group and condensation reactions.
Etherification:
Metallation occurs ortho to the modified formyl group in N,N'-dimethylethylene adducts of salicylaldehyde (M. Gray and P.J. Parsons, Synlett., 1991, 729).
2-Methoxymethoxybenzaldehyde is produced in 90% yield by treatment of salicylaldehyde in toluene containing methylal and dimethylformamide with phosphorus oxychloride (American Home Products Corpn., USP 4500738).

2-Difluoromethoxybenzaldehyde results in 90% yield by passage of difluorochloromethane through an aqueous alkaline solution of salicylaldehyde (As. Ukr. Org. Chem., GB 2075006). 3-Decyloxybenzaldehyde is synthesised in 96% yield from 3-hydroxybenzaldehyde and 1-bromodecane in dimethylformamide containing caesium carbonate (H. Strzelecka et al., Mol. Cryst. Liq. Cryst., 1988, *156A*, 347). Phase-transfer catalysis with 1-chlorodecane and potassium hydroxide solution has also been used.

Salicylaldehyde upon addition to an aqueous solution of barbituric acid affords a precipitate which in hot acetic anhydride (9:1) gives a chromenopyrimidine (J.D. Figueroa-Villar, E. Rangel Cruz and L. dos Santos, Synth. Commun., 1992, *22*, 1159).

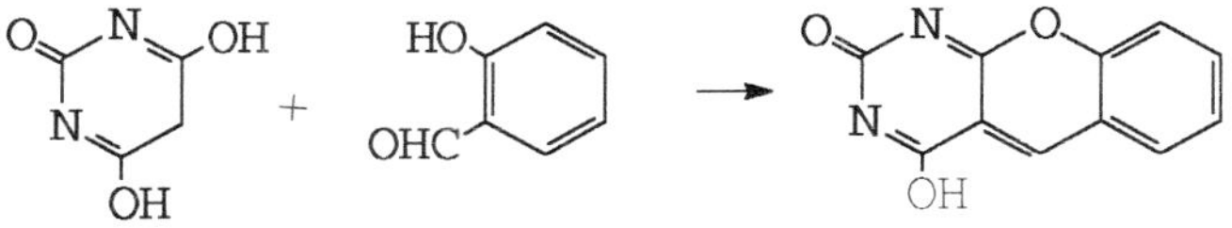

Replacement of the Formyl Group:
Pyrogallol can be obtained in 64% yield by the Dakin oxidation of 2-hydroxyisophthalaldehyde (Fisons Ltd., EP 25659). By contrast, 2-methoxyphenol is obtained in 94% yield from 2-methoxybenzaldehyde in methanol containing 31% hydrogen peroxide and a little sulphuric acid (M. Matsumota, H. Kobayashi and Y. Hotta, J. Org. Chem., 1984, *49*, 4740).

Reactions of the Formyl Group:
These reactions essentially comprise Claisen or Knoevenagel procedures followed by cyclisation.
2-Hydroxybenzaldehyde and ethyl cyanoacetate when reacted with aluminium phosphate/alumina afford 3-ethoxycarbonyl-2-imino-benzopyran, hydrolysable with dilute hydrochloric acid to give ethyl 3-ethoxycarbonylcoumarin in 58% yield (J.A. Carbello et

al., J. Org. Chem., 1984, 49, 5195).

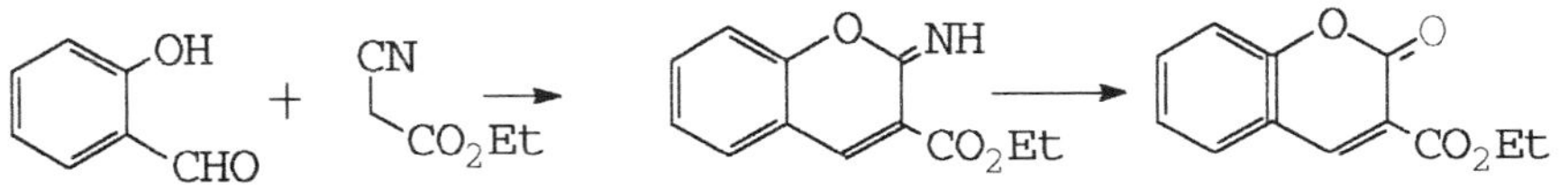

2-Hydroxybenzaldehyde gives coumarin in quantitative yield when it is treated with phosphorus oxychloride and N,N-dimethylacetamide. An intermediate 2-dimethylaminobenzopyrylium salt is produced which can be hydrolysed with hot aqueous sodium carbonate (M.A. Kira, K.Z. Gadalla, Egypt J. Chem., 1989, 21, 395).

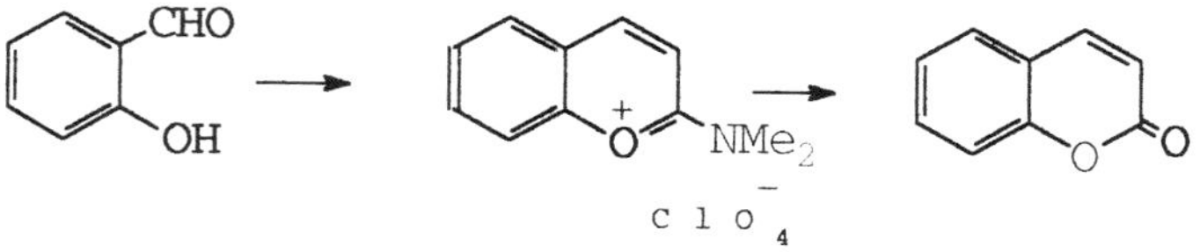

An alternative procedure, avoiding phosphorus oxychloride, uses the dimethyl acetal of N,N-dimethylacetamide. With 6-hydroxy-7-formyl-2,2-dimethyl-2H-chromen, for example, this reagent affords 7,7-dimethyl-2(7H)-benzo[1,2,4,5-b]bipyranone in 64% yield (S. Yamaguchi, R. Mikayama, S. Yonezawa and Y. Kawase, Bull. Chem. Soc. Japan, 1989, 62, 3593).

2-Hydroxybenzaldehyde on treatment with a 2-halogenomethyl-3-methylquinoxaline forms the phenolic ether, 2-(3-methyl-2-quinoxalinyl)benzaldehyde which, in ethanolic potassium hydroxide under reflux affords a 92% yield of 2-(3-methylquinoxalin-2-yl)benzofuran (Veb Fahlberg List Chem., DD 276479). O-Alkylation apparently occurs rather than condensation of the 3-methyl with the formyl group.

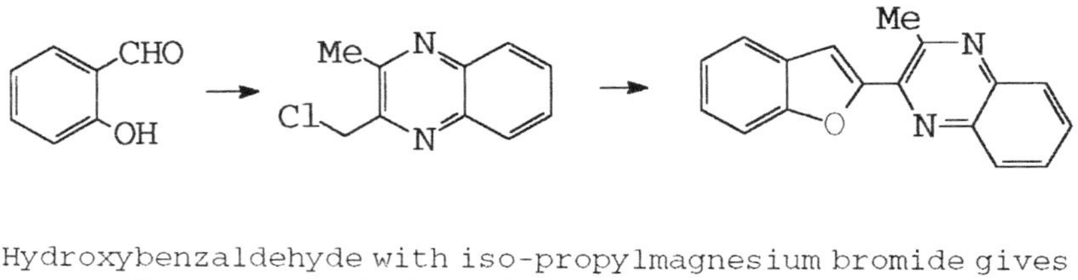

2-Hydroxybenzaldehyde with iso-propylmagnesium bromide gives the coresponding sec-alcohol, which when heated in hexane affords a 85% yield of 2-iso-butenylphenol (A. Arduini, A. Pochini and R. Ungaro, Synthesis, 1984, 950).

(ii) Phenolic Ketones
Under basic conditions 4-hydroxyacetophenone with di-iso-

propylamine reacts at -78°C with fluorosulphonic anhydride to afford 4-acetylphenylfluorosulphonate in 95% yield (G.P. Roth and C.E. Fuller, J. Org. Chem., 1991, 56, 3493).

Ozonolysis of 3,6-dihydroindan-1-one acetal leads to a polyketide which can be recyclised to 1,8-dihydroxy-3-methylnaphthalene (G. Bringmann, Angew. Chem. Int. Ed., 1982, 21, 200).

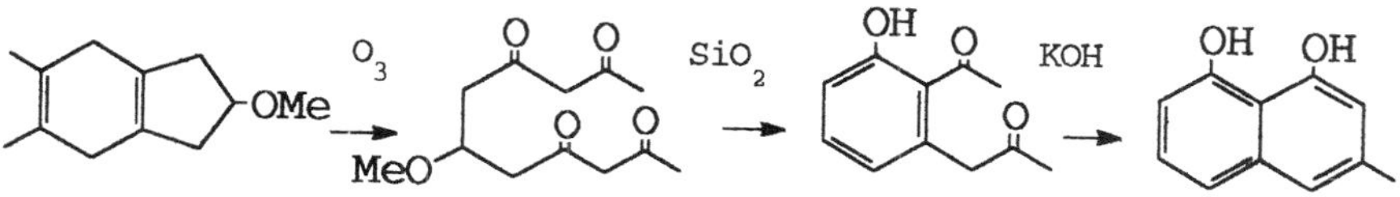

The condensation of 2-hydroxyacetophenone with formaldehyde and methylamine to give 4-chromanone is improved if the Mannich base is isolated and then cyclised with potassium hydroxide (B. Cox and R.D. Waigh, Synthesis, 1989, 709).

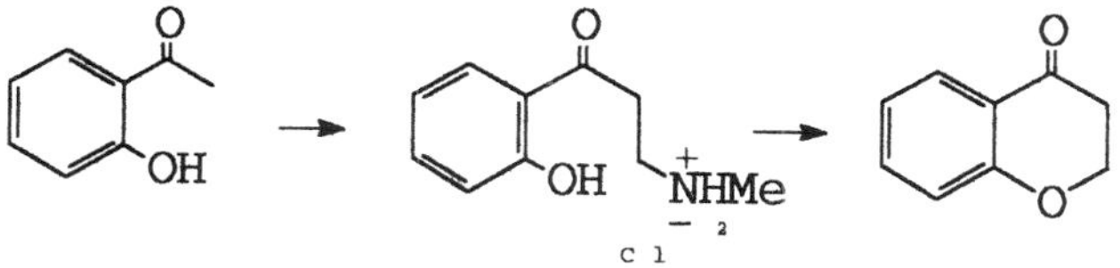

2-Hydroxyphenyl benzyl ketones can be oxidised and cyclised to benzo-3(2H)-furanones with such reagents as thallic nitrate (S. Antus et al., Ann. Chem., 1980, 1271) or phenyliodosodiacetate (M.S. Khanna et al., Synth. Commun., 1992, 22, 2555).

Phenolic polycyclic ketones:
Replacement of the phenolic hydroxyl group occurs in the reaction of 9-hydroxy-1-oxophenalene with 40% aqueous methylamine at 125°C under pressure producing 9-methylamino-1-oxophenalene in 83% yield (R.C. Haddon, S.J. Chichester and S.L. Mayo, Synthesis, 1984, 639).

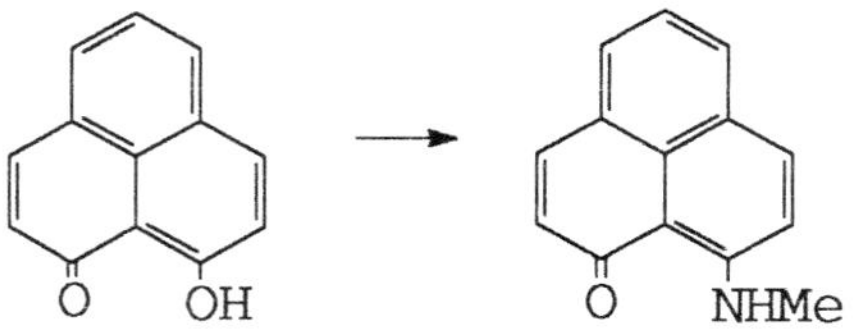

In the anthraquinone series 2-phenyl-4H-anthra[1,2-b]pyran-4,7,12-trione can be synthesised in 82% yield from 1-hydroxy-2(3-phenylprop-2-ynoyl)anthraquinone by cyclisation in piperidine followed by silica gel treatment (M.A. Mzhelskaya, A.A. Moroz and M.S. Shvartsberg, Izv. Akad. Nauk SSSR Ser. Khim., 1991, 7, 1656).

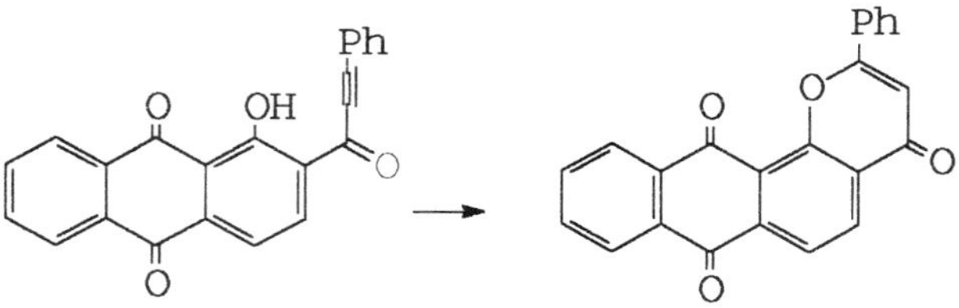

(iii)Phenolic Acids, their Ethers and Phenolic esters
2-Methoxybenzoic acid undergoes Birch reduction to give a product which can be methylated with methyl iodide to furnish 2-methoxy-1-methylcyclohexa-2,5-diene-1-carboxylic acid in 82% yield (J.M. Hook, L.N. Mander and M. Woolias, Tetrahedron Letters, 1982, 1095

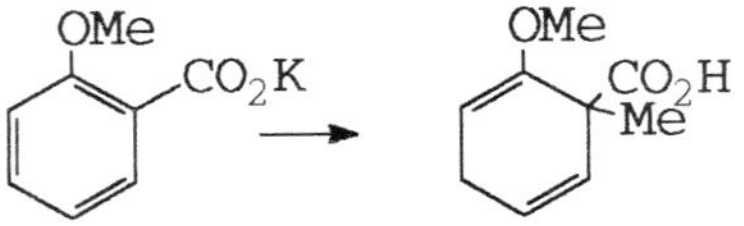

4-Hydroxybenzoic acid, in contact with β-cyclodextrin, is formylated in aqueous alkaline solution in the presence of a little copper powder and carbon tetrachloride, by reaction with chloroform. The product may then be treated with nitromethane to afford 4-hydroxy-3-(2-nitrovinyl)benzoic acid in 87% yield (H. Eiji, J58 194842)

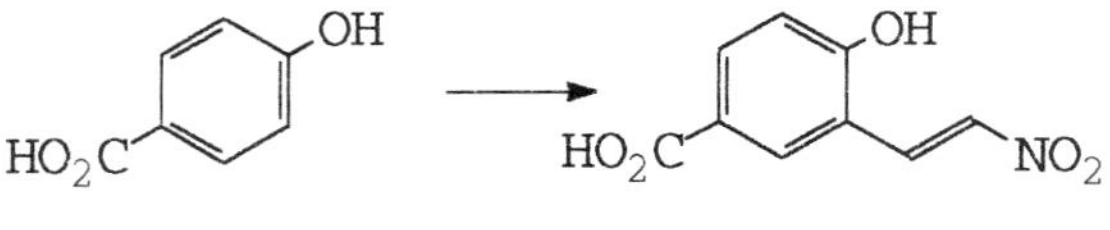

Cyclic acetal formation takes place in the reaction between 4-hydroxybenzoic acid and paraformaldehyde in 1,2-dimethoxyethane containing Nafion-501 (or less effectively with Amberlite IRA 120) which gives 6-carboxy-1,3-benzodioxan in 80% yield (A. Denis, M. Delmas and A. Gaset, J. Heterocyc. Chem., 1984, 517).

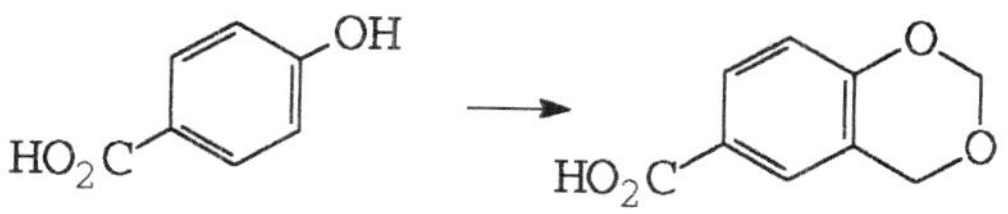

The hydrolysis of alkyl salicylates, often protracted under alkaline conditions, has been effected quantitatively and expeditiously with trifluoroacetic acid. Thus, methyl 2-hydroxybenzoate in excess trifluoroacetic acid after heating for 13 hours at 100°C in a sealed tube gave 2-hydroxybenzoic acid in 100% yield (D.C. Tabor and S.A. Evans Jr., Synth. Commun., 1982, 12, 855).

Methyl 4-hydroxybenzoate in dioxan reacts with aziridine at 50-90°C over affording 2-(4-methoxycarbonylphenoxy)ethylamine in 53% yield (Nippon Shokubai Kagaku, J63 112543).

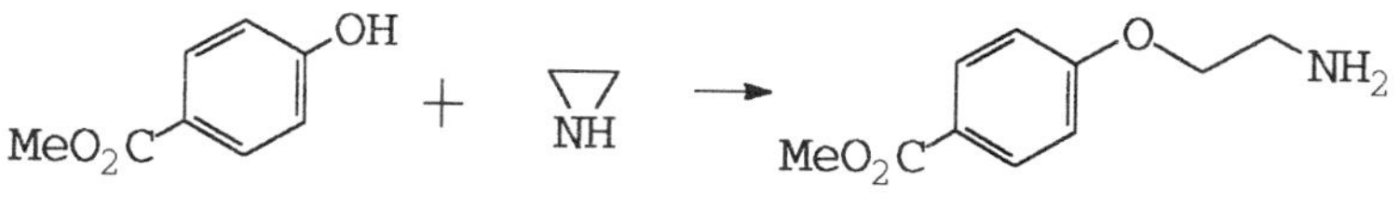

Ethyl salicylate and 1-naphthol in boiling diphenyl ether furnished benzo[b]xanthone in 74% yield (R.J. Patolia and K.N. Trivedi, Indian J. Chem., 1983, 22b, 444).

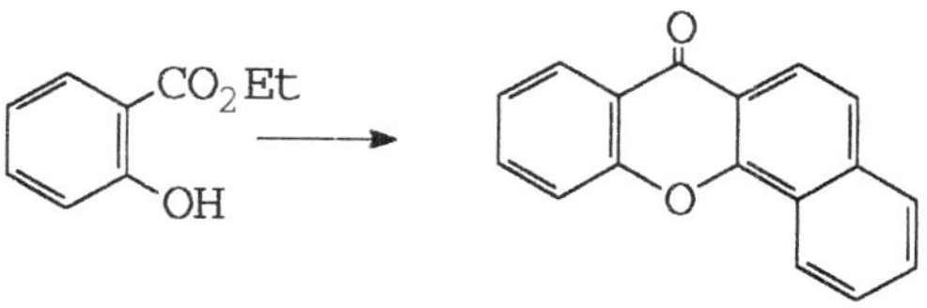

(g) Phenolic Ethers

Synthesis of Phenolic Ethers

During the preceding decade the chemistry of phenolic ethers has expanded considerably and this justifies their separate consideration rather than as an aspect of the reactions of phenols. Their preparation by intermolecular reactions generally involving substitution, by intramolecular procedures and transetherification, by displacement of nitro and cyano groups, and from acyclic and alicyclic precursors is described.

(i) Alkyl Ethers from Phenolic Intermediates by Alkylation under catalytic or acidic conditions

These procedures illustrate a wide range of reagents, reaction environments and 'leaving groups'. A methanolic solution of phenol passed through the catalyst $Ce(SO_4)_2-Al_2O_3$ at 300°C gave a 38% yield of anisole (Amoco Corpn., USP 4675456), while 4-methylphenol with methanol briefly contacted at 450°C through a Zeosorb 13X/K catalyst affords a 94% yield of 4-methylanisole (Wilh-pieck Univ., Rostock, DD 267034). Phenol and benzenesulphonic acid in hot xylene with the addition of methanol and azeotropic removal of water affords a 69% yield of anisole (Dainippon Inc. Chem., KK, J61 103848).
Phenol and polystyrene sulphonic acid treated at 120°C with Me_2O produced anisole (Bayer AG, DT 29030), and phenol with BPO_4 and Me_2O in toluene at 280°C and 40 atm. gives a 53% yield (Eniricerche spa, EP 196687). Other methylating agents such as methyl carbonate, $(MeO)_2CO$, with phenol over alumina at 240°C furnishes a 74% yield (Coalite Group plc, Rhodesia, WO 8603485), and methyl phosphite, $(MeO)_3P$, with phenol in tetrahydrofuran containing diethyl azodicarboxylate in a 2 hour reaction gives a 86% yield (G. Grynkiewicz, Pol. J. Chem., 1982, *56*, 149),

(ii) Alkyl Ethers from Phenoxide Intermediates

Methyl trichloroacetate and phenol with catalytic amounts of K_2CO_3 and 18-crown-6 at 150°C produces a 96% yield of anisole (J.M. Renga and P.-C. Wang, Synth. Commun., 1984, *14*, 69) and a solution of phenol in 1,4-dioxane treated at 65°C with powdered KOH and then Me_2SO_4 afforded a 90% yield (D. Achet et al., Synthesis, 1986, 642). 2-Methylphenol in aqueous sodium hydroxide with methyl fluorosulphonate gives a 70% yield of 2-methylanisole (U. Utebaev, V.S. Sagulenko and V.V. Filippenkova, Izv. Akad. Nauk Kaz., SSR Ser. Khim., 1981, 68).
4-Tert-butylcalixarene is methylated in 94% yield by triethyleneglycol monomethyl ether 4-tosylate in chloroform/dichloromethane/methanol containing benzyltrimethyl-ammonium chloride and potassium hydroxide (Wakayama Prefecture, JO 2053749).
4-Chlorophenol and 2-trichloromethylpropan-2-ol in methanol containing NaOH form methyl 4-chlorophenoxy-2,2-dimethylacetate

in 86% yield (J. Bobowska, J. Sosnowska and Z. Eckstein, Przem. Chem., 1980, *59*, 495.
Preparations of phenolic ethers based essentially on nucleophilic substitution, as in the three preceding examples, are generally favoured. They resemble the phase-transfer approach, utilising unusual bases, and crown ethers or calixarenes to render the phenoxide ion more 'naked'. Thus phenol in $CHCl_3$ treated with EtI, catalytic quantities of dicyclohexyl 18-crown-6 and a copolymer of 2-vinylpyridine and styrene and aqueous KOH affords phenetole in 97% yield (L.P. Ivanova et al., Zh. Prikl. Khim., 1990, *63*, 226). Phenol in ethanol containing 1,8-diazabicyclo[5.4.0]undec-7-ene by treatment with C_3H_7I gives a 76% yield of phenyl n-propyl ether (BP Chemicals Ltd., EP 20667). 4-Tert-butylphenol in aqueous NaOH, $(n-Bu)_3N$ and $CHCl_3$ in benzene furnishes 4-tert-butylphenyl n-butyl ether in 53% yield, together with N,N-dibutylformamide (M. Yonovich and Y. Sasson, Tetrahedron Letters, 1980, *21*, 1845). Di(2-methylphenoxy)methane results in 91% yield by heating Amberlite IRA-400 in the 2-methylphenoxide form in CH_2Cl_2 over 12 hours (M.M. Salunkhe et al., Synth. Commun., 1990, *20*, 1143).

$$P-C_6H_4CH_2N^+(CH_3)_3\ ArO^- + CH_2Cl_2 \rightarrow ArOCH_2OAr + P-C_6H_4CH_2N^+(CH_3)_3Cl^-$$

A variation in the choice of inorganic base for phenoxide ion formation is seen in the synthesis of neopentyl phenyl ether in 94% yield from phenol and neopentyl iodide in diethyleneglycol dimethyl ether containing caesium carbonate (H. Quast and J. Schulze, Ann. Chem., 1990, 509).
Benzyl phenyl ether is formed quantitatively from phenol, benzyl bromide, KOH, and a calixarene in CH_2Cl_2 (H. Taniguchi and E. Nomura, Chem. Letters, 1988, 1773). Diphenyl ether is obtained in 60% yield from phenol with 85% KOH/NaOH and chlorobenzene followed by azeotropic removal of water, and the addition of CuCl (Rhone-Poulenc Industries, FR 2460283). Phenyl 2-pyridyl ether results in 72% yield from heating excess phenol and 2,2'-dipyridyl sulphide in a sealed tube (S. Inoue, Phosphorus-Sulphur, 1985, *22*, 141). 3-Methylphenyl trimethylsilyl ether is formed in 90% yield from 3-methylphenol and trimethylsilyl trichloroacetate with K_2CO_3 and 18-crown-6 (J.M. Renga and P.-C. Wang, Tetrahedron Letters, 1985, *26*, 1175). Methyl 2-(4-hydroxyphenyl)-3-methylbutyrate reacts with excess chlorodifluoromethane in the presence of benzyltriethylammonium chloride and aqueous sodium hydroxide to form the difluoromethyl ether in 79% yield; the phase transfer-type conditions

preserving the ester group (Am. Cyanamid Co., USP 4407760).

(iii) Naphthyl Ethers from Naphthoxide Intermediates

2-Methoxynaphthalene is formed quantitatively from 2-naphthol in acetonitrile/methanol (9:1) by treatment with trimethylsilyl -diazomethane at ambient temperature (T. Aoyama et al., Chem. Pharm. Bull., 1984, 32, 3757). By contrast, 2-ethoxynaphthalene is produced in 50% yield from 2-naphthylchloroacetate in dimethylformamide containing sodium telluride and ethyl iodide. In the O-glucoside series, 1-naphthol and tetramethylguanidine in acetonitrile react with 2,3,4,6-tetra-O-acetyl-α-D-glucopyranosyl fluoride and afford the O-isomer (97%) in 90% yield (M. Yamaguchi et al., J. Chem. Soc., Perkin Trans. I, 1990, 1079).

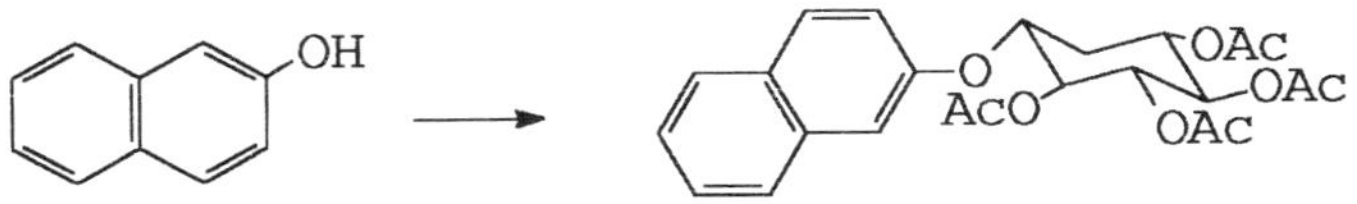

(iv) Substituted Alkyl Ethers from Phenoxide Intermediates

Phenol and epichlorohydrin in contact with piperidine give the intermediate chlorohydrin which by brief reaction with NaOH in tetrahydrofuran affords 2,3-epoxypropyl phenyl ether in 85-90% yield (W.E. Kreightbaum, J. Med. Chem., 1980, 23, 285).

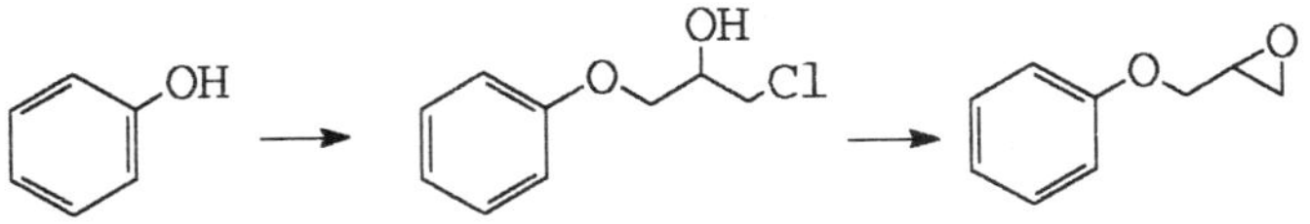

Phenyl 1-propynyl ether is obtained in 85% yield from phenol and propargyl alcohol in benzene containing Ph_3P, followed by addition of diethyl azodicarboxylate (R.S. Subramaniam and K.K. Balasubramanian, Synth. Commun., 1989, 19, 1285).

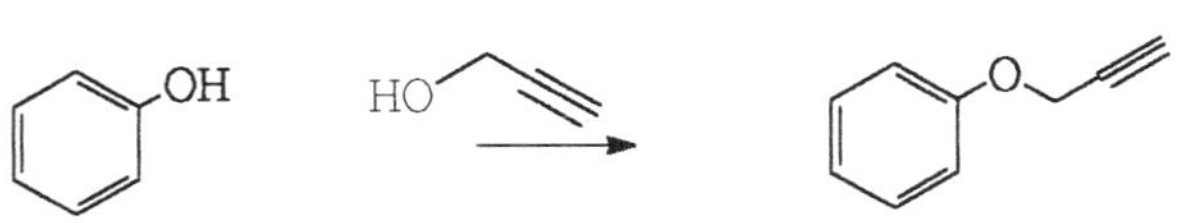

(Z)-1-Hydroxy-4-acetoxybut-2-ene and phenol in THF containing bis[1,2-(diphenylphosphino)ethane]palladium(0) and KF/Al_2O_3 react at ambient temperature to give 2-phenoxybut-3-enol in 65% yield (J. Muzart, J.-P. Genet and A. Denis, J. Organometall.

Chem., 1989, *C23*, 326).

Phenol and dimethyl acetylenedicarboxylate in CH_2Cl_2 containing neutral alumina afford a 91% yield of dimethyl phenoxyfumarate (M. Kodomari, T. Sakamoto and S. Yoshitomi, J. Chem. Soc., Chem. Commun., 1990, 701).

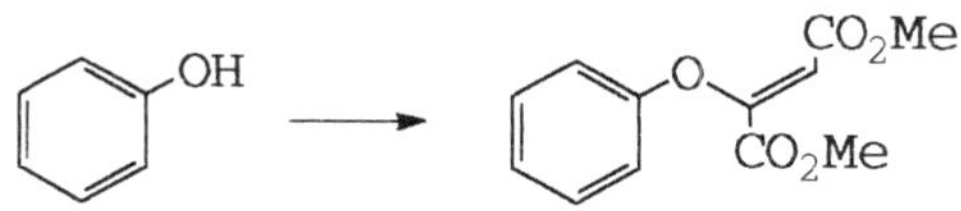

In the reaction of 2,4,6-tribromophenol with epichlorohydrin in the presence of dilute aqueous sodium hydroxide, both alkylation and addition occur to give 1,3-bis(2,4,6-tribromophenyl)propan-2-ol (Y. Liu and H. Chen, Shandong Daxue Xuebao Ziran Kexueban, 1987, *22*, 109).

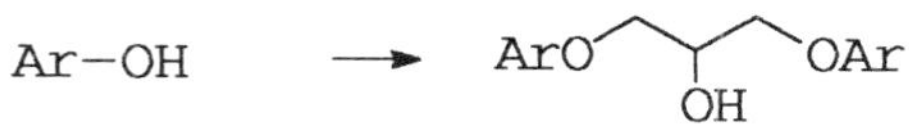

2-Phenoxyethanol resulted in 91% yield in the reaction of phenol with ethylene carbonate and imidazole at 160°C (Dow Chemical Co., USP 4310706).

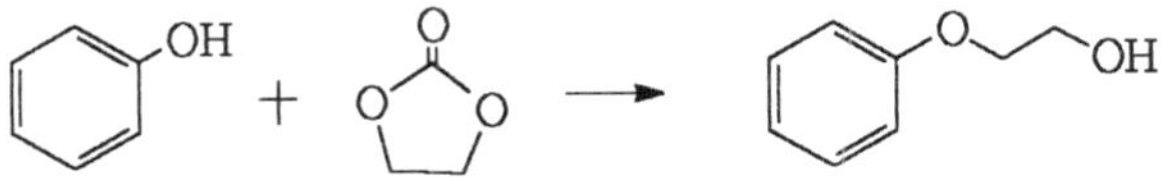

2,4-Dimethyl-1,3-benzodioxane is produced in 70% yield, rather than the expected diphenyl acetal, when phenol and potassium phenoxide in toluene are treated with $TiCl_4$ and then with ethanal in toluene (F. Bigi et al., Synthesis, 1980, 724).

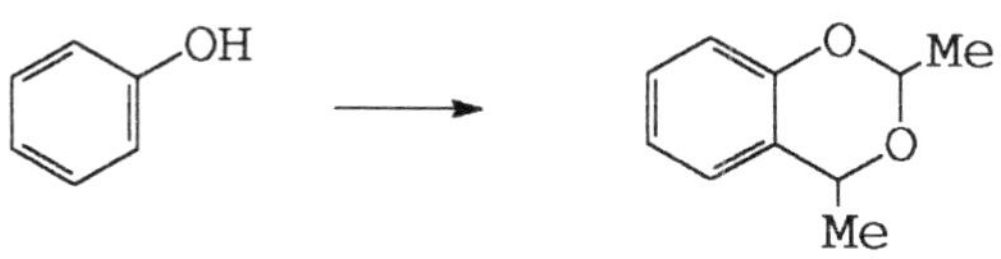

The conversion of cyclohex-2-enyl methyl carbonate to cyclohex-2-enyl phenyl ether in 60% yield, in preference to transesterification, is realised by heating with phenol and small quantities of $Pd_2(dba)_3$ and 1,4-

bis(diphenylphosphino)butane in tetrahydrofuran (G. Goux, P. Lhoste and D. Sinou, Synlett., 1992, 725).

A double alkylation is effected in the reaction of dichloromethyl tert-butyl ketone with 4-chlorophenol and 1H-1,2,4-triazole in benzene/methyl iso-propyl ketone at 100°C. The product is a triazinylalkyl t-butyl ketone isolated in 97% yield (Y. Lin et al., Xiangtan Daxue Ziran Kexue Xuebao, 1986, 6).

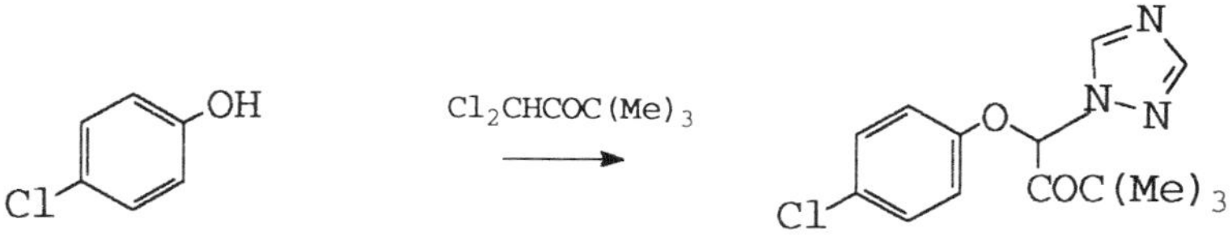

A glycosyl fluoride (from L-rhamnose) when reacted with phenol in dichloromethane containing 1.5 moles each of dicyclopentadienylhafnium dichloride and silver perchlorate in the presence of a 4A molecular sieve gives the corresponding O- and not the C-glycoside (T. Matsumoto, M. Katsui and K. Suzuki, Chem. Letters, 1989, 437).

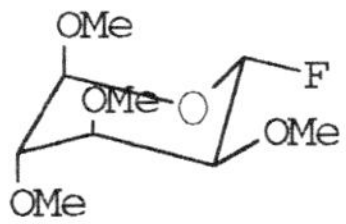

Tert-butyl ethers are usually formed under acidic conditions, although reaction occurs under neutral conditions when 6-O-acetylmorphine is heated in dimethylformamide solution with the di-tert-butyl acetal of formamide (E. Mohacsi, Synth. Commun., 1983, *13*, 827). A by-product is also formed through raction at the acetoxy group .

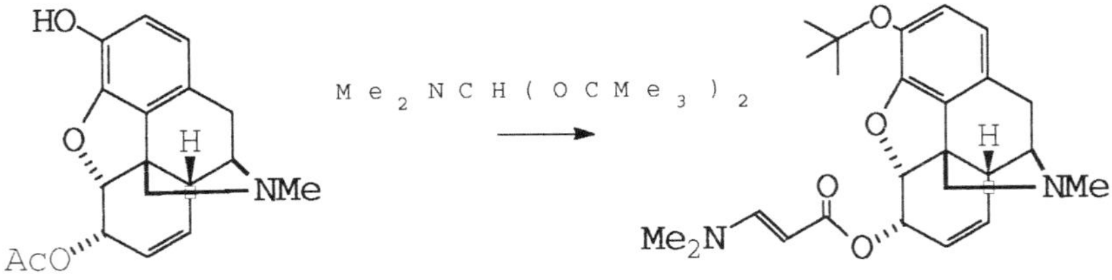

(v) Phenolic Ethers from Transetherification Reactions

Benzyl ethers are formed by transetherification reactions particularly from the silyl ethers, the prepararation of which is straightforward: from trimethylsilyl chloride/triethylamine, from bis(trimethylsilylacetamide) in benzene, or from hexamethyldisilazane. Trimethylsilyl phenyl ether, tris(diethylamino)sulphonium difluoro trimethylsiliconate and benzyl bromide in tetrahydrofuran all react with phenol and give benzyl phenyl ether (R.Noyori, I. Nishida and J. Sakata, Tetrahedron Letters, 1981, 22, 3993). The same type of compound is also obtained from dimethyl-tert-butylsilyl 4-bromophenyl ether and benzyl chloride treated with tetrabutylammonium fluoride trihydrate (D.G. Saunders, Synthesis, 1988, 377). The reaction of methyl phenyl carbonate with benzyl chloride and tetra-n-butylphosphonium bromide is another route (Dow Chemical Co., USP 4642349). The bis(dimethyl-tert-butylsilyl) ether of 3,5-dihydroxy-2-fluorobenzaldehyde and benzyl bromide in dimethylformamide containing anhydrous potassium fluoride afford a 92% yield of the corresponding dibenzyl ether (A.K. Sinhababu, M. Kawase and R.T. Borchardt, Tetrahedron Letters, 1987, 28, 4139).

(vi) Phenolic Ethers formed by the Displacement of Nitro, Chloro and other Groups

Nitro compounds activated by 2- or 4-groups, such as nitro, cyano or alkoxycarbonyl, form ethers by inter- or intramolecular pathways.

Methyl 3-chloro-2-nitrobenzoate and the monopotassium salt of methyl 3,5-dihydroxybenzoate in hot dimethylformamide containing cupric oxide give the corresponding diphenyl ether in 60% yield (C.W. Bird and M. Latif, Chem. and Ind.,London, 1987, 795).

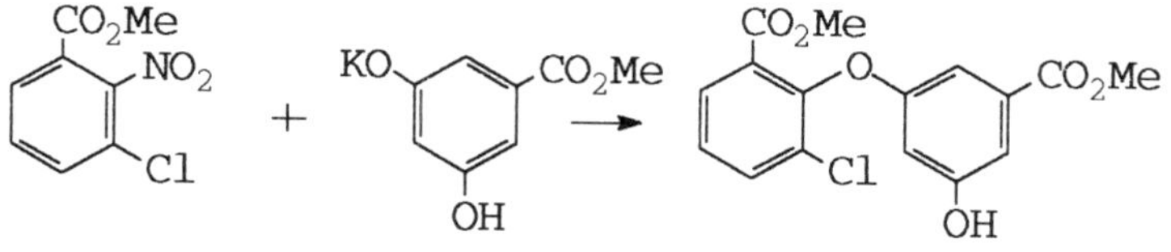

4,4'-Bis(4-nitrobenzoyl)biphenyl ether in dimethyl sulphoxide containing methanolic sodium methoxide affords the corresponding dimethoxy compound in 91% yield (Asahi Chemical Ind., KK JO 104252). The nitro group of 4-nitrobenzonitrile can be displaced by lithium alkoxides as in the following illustration (I.T. Kirillov et al., Zh. Obsch. Khim., 1990, 60, 1140).

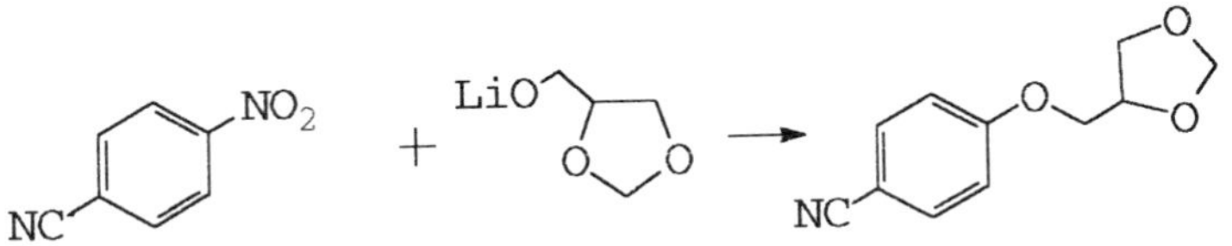

4-Methoxyphthalonitrile is obtained in 98% yield by the reaction of 4-nitrophthalonitrile in methanol with dimethylformamide/1,8-diazobicyclo[5,4,0]undec-7-one (Rikagaku Kenkyusho, J6 1207365). The selective synthesis of ethers by the displacement of the halide ion from nitro halogenobenzenes is commonplace. Thus sodium 2,2,2-trifluoroethoxide upon treatment with 3,4-dichloronitrobenzene forms 3-chloro-4-(2,2,2-trifluoroethoxy)-nitrobenzene in 58% yield (J.J. Gupton, G. DeCrescendo, C. Colon and J.P. Idoux, Synth. Commun., 1984, *14*, 621).

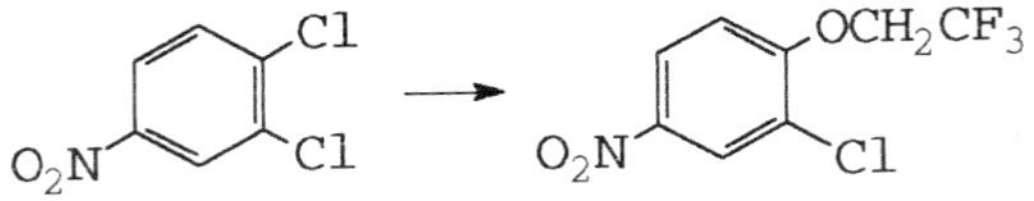

By contrast, sodium n-butoxide in toluene can be reacted with 4-chloro-3-(N-butylamido)nitrobenzene and a little tetrabutyl-ammonium bromide to afford a 89% yield of the 4-n-butoxy compound (D. Nisato et al., Org. Prep. Proc. Int., 1985, *17*, 75).

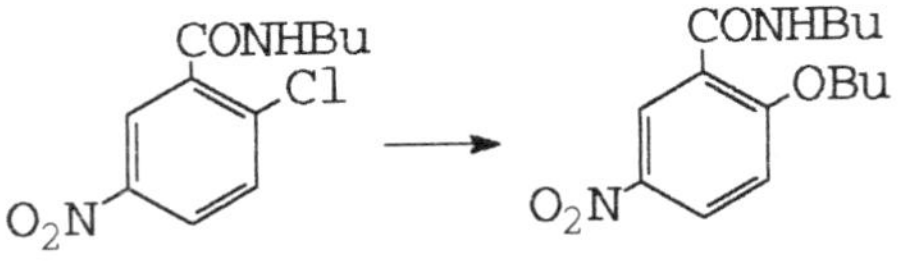

Bromobenzene in methanol containing sodium methoxide plus sodium and cuprous bromides as catalysts furnishes anisole after heating at reflux for 7 hours (Sterwin AG, GB 2089672).

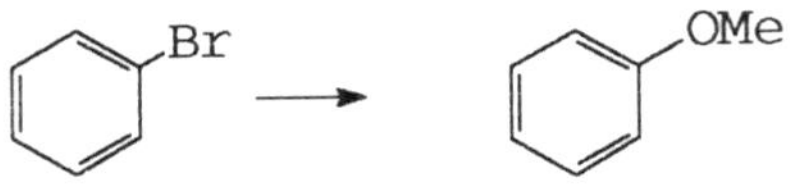

Displacement of the nitrite ion is used for the synthesis of 3-nitroanisole from 1,3-dinitrobenzene, either by its reaction in chlorobenzene containing Aliquat 336 with sodium methoxide at 80°C over 2 hours (F. Montanari, M. Pelosi and F. Rolla, Chem. and Ind., (London), 1982, 412) or more drastically with methanol at 180°C under mild pressure in the presence of carbon dioxide during 20 hours. In this case the yield was only 38%, albeit with a conversion of 40% (Mitsui Toatsu Chem. Inc., J59 044343). In the diphenyl ether series, 4-nitrodiphenyl ether is produced in 70-77% yield from 4-nitrophenol in hot pyridine containing Cu(I)phenyl acetylide and bromobenzene (A. Afzali, H. Fironzabadi and A. Khalafi-Nejad, Synth. Commun., 1983, 13, 335).

The 4,4'-dinitro compound is derived in 82% yield from 4-nitrochlorobenzene and 4-nitrophenol in dimethyl sulphoxide containing sodium tetraborate and nitrite (Mitsui Toatsu Chem. Inc., J56 158740).

Although 1-chloroanthraquinone in THF/hexadecanol containing sodium hydride does not form the corresponding ether, with ethylene glycol oligomers an ether is obtained. A mixture of hexadecanol and these oligomers does however afford the required C16 ether. It is concluded that ethylene glycol alkoxide is more nucleophilic and equally it is a better leaving group than chloride under the experimental conditions (H.Y. Yoo et al., Tetrahedron Letters, 1990, 31, 55).

Cationic compounds and organometallic compounds are useful in the preparation of diaryl ethers. Thus, triphenylbismuth diacetate in ethanol containing copper(II) chloride affords phenyl ethyl ether (B.A. Dodonov, A.V. Guschin and T.G. Brilkina, Zh. Obsch. Khim., 1984, 54, 2157).

$Ph_3Bi(OAc)_3$ + EtOH → EtOPh

An unusual synthesis of anisole gives an excellent yield. This probably involves an organometallic intermediate formed by adding N-phenyl-N'-toluenesulphonylhydrazine to mercuric acetate in methanol (F. Gasparrini, L. Caglioti and D. Misiti, Tetrahedron, 1982, 38, 3609).

$PhNHNHSO_2C_6H_4CH_3$ + $Hg(OAc)_2$ + MeOH → PhOMe

The classical diazonium route to phenolic ethers has been modified. Thus the slow addition of excess methoxytrimethylsilane to diazonium fluoroborate in ice-cold

Freon 115 under nitrogen and ultrasonification gives a 51% yield of anisole (G.A. Olah and A. Wu, Synthesis, 1991, 204).

$$\text{PhN}_2^+ \text{ BF}_4^- \xrightarrow{\text{MeOSiMe}_3} \text{PhOMe}$$

An iodonium salt reacted with the sodium salt of N-protected O-methyl tyrosine in DMF gave a diaryl ether in 51% yield. It is a preferred route to the Ullman reaction (M.J. Crimin and A.G. Brown, Tetrahedron Letters, 1990, *31*, 2017).

(vii) Phenolic Ethers from Alicyclic and Acyclic Precursors

The simplest example of this type of transformation, in effect a reversal of the Birch redution, is the conversion of cyclohex-2-enone to phenyl ethyl ether in 93% yield. This procedure uses VO(OEt)Cl$_2$ as the reagent with oxygen in hot ethanol (T. Hirao, M. Mori and Y. Oshiro, J. Org. Chem., 1990, *55*, 358).

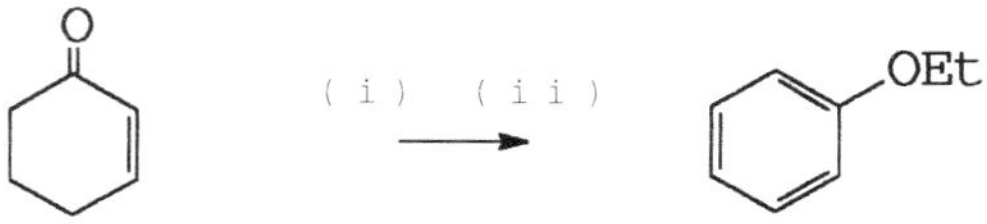

(i) VO(OEt)Cl$_2$ (ii) EtOH, O$_2$, Δ

This methodology has been extended to ethoxycarbonyl derivatives (the so-called Hagermann esters) to give, for example, 3-methyl-4-ethoxycarbonylanisole in 90% yield by heating 4-ethoxycarbonyl-3-methylcyclohex-2-enone in methanol containing iodine (A. Kotnis, Tetrahedron Letters, 1990, *31*, 481). Similarly iso-phorone in hot n-butanol containing iodine and ceric ammonium nitrate is converted into 5-n-butoxy-1,2,3-trimethylbenzene in 90% yield (C.A. Horiuchi et al., Chem. Letters, 1991, 1821).

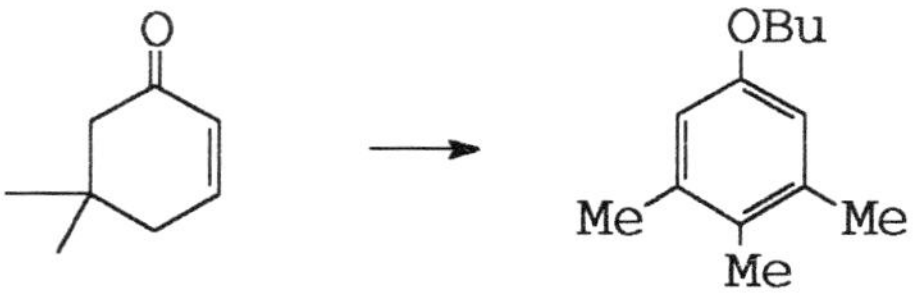

5-(4-Chlorophenyl)-3-methylcyclohex-2-enone in methanol is converted to 4-chloro-3'-methyl-5'-methoxybiphenyl in 86% yield by treatment with hot methanolic iodine (Y. Tamura and Y. Yoshimoto, Chem. and Ind., London, 1980, 888).

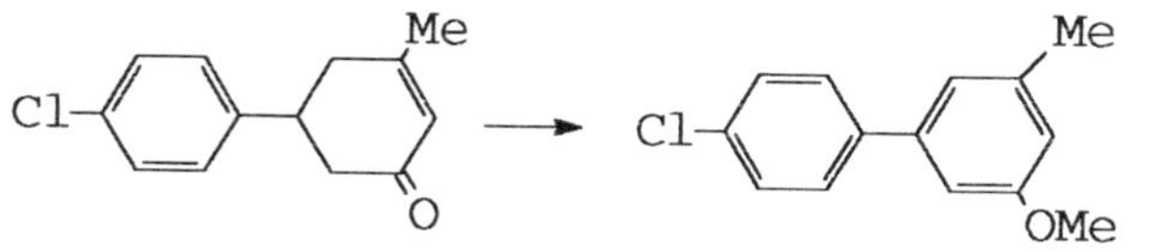

Intramolecular cyclisations are used for the preparation of a variety of complex anisole derivatives. For example, 4-(2'-indolyl)-4-oxobutanal affords 4-methoxycarbazole in 61-83% yield when treated with boron trifluoride/methanol (T. Martin and C.J. Moody, Tetrahedron Letters, 1985, 47, 5841).

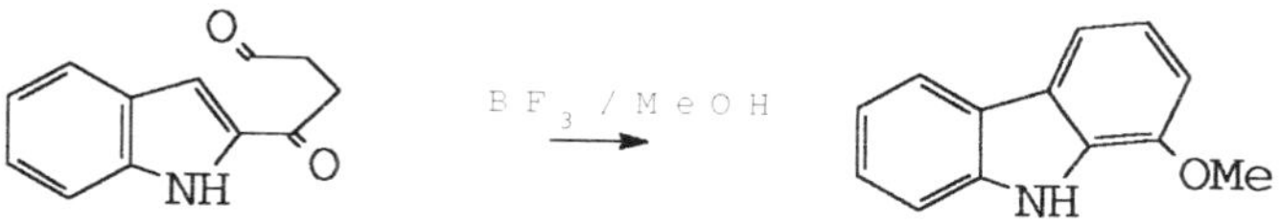

In the polycyclic series, cyclisation of an o-quinomethide derived from a 3-methylanisole furnishes an intermediate for a synthesis of racemic estrone (G. Quinkert et al., Angew. Chem. Int. Edn. Engl., 1980, 19, 1027).

Acyclic intermediates are employed in the synthesis of phenolic ethoxylates. Thus Diels-Alder addition of dimethyl acetylene dicarboxylate and 5,6-dihydro-2-isopropenyl-1,4-dioxine gives 2,3,4a,7-tetrahydro-5,6-di(methoxycarbonyl)-8-methylbenzo-1,4-dioxine which in boiling THF with DBU during 1-3 hours is aromatised and cleaved to the 2-hydroxyethoxyphthalate ester (M. Fetizon, P. Goulaouic and I. Hanna, J. Org. Chem., 1988, 53, 5672).

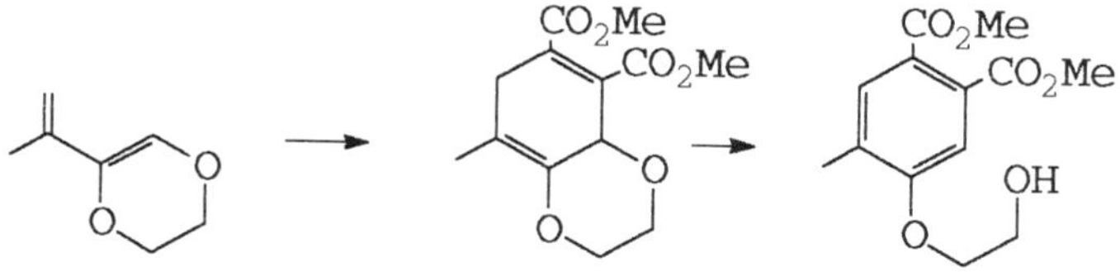

Reactions of Phenolic Ethers

Since the methyl group is widely used for the protection of phenols the reactions described in this section effectively extend the range of syntheses of the phenolic ring system. The general reactions of phenolic ethers comprise dealkylation and dealkoxylation, substitution of the ether group, ring substitution, cyclisation of phenolic ether derivatives, carbonyl compound formation, coupling of phenolic ethers, ring saturation and Diels-Alder reactions.

(i) Dealkylation and Dealkoxylation

A number of examples of the chemical and catalytic dealkylation of aromatic ethers are given in Table 1,
A very general method for the cleavage of phenolic ethers consists of heating with an excess of 37% hydrobromic acid and a small amount of hexadecylammonium bromide. In this way phenyl pentyl ether afforded phenol in 71% yield (B. Jursic, J. Chem. Res., 1989, 284).
Selective cleavage of the dibenzyl ether of estradiol in chloroform containing dimethylaniline was effected with aluminium chloride during 30 minutes at ambient temperature (T. Akiyama, H. Hirofuji and S. Ozaki, Tetrahedron Letters, 1991, *32*, 1321).

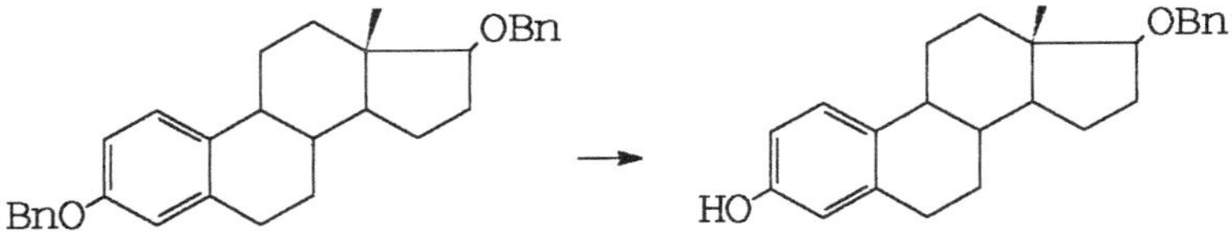

TABLE 1

DEALKYLATION OF AROMATIC ETHERS

Reactant	Reagent	Conditions (R,reflux)	Product	Yield(%)	Ref.
$MeOC_6H_5$	$BBr_3.Me_2S$	$ClCH_2CH_2Cl$ R(12h).	Phenol	86	1
1,2,4-Tri $OMeC_6H_3$	NaH,HMPA	$MeNHC_6H_5, C_6H_4(Me)_2$ R,65C (6h)	2,5-diOMe Phenol	90	2
i-$ProC_6H_5$	$NaI/SiCl_4$	Reactant/NaI, CH_2Cl_2/MeCN. $SiCl_4$,R (12h)	Phenol	–	3
$C_3H_5OC_6H_5$	$(PhCN)_2PdCl_2$	Reactants in Benzene. R	Phenol	89	4
2,4-$diCO_2H$ $C_6H_3OCH_2Ph$	10%Pd-C, $NaH_2PO_2.H2O$	EtOH R(2.5h).	2,4-$diCO_2H$ C_6H_3OH	92	5

1. P.G. Williard and C.B. Fryhle, Tetrahedron Letters, 1980, *39*, 3731.

2. B. Loubinoux, G. Coudert and G. Guillaumet, Synthesis, 1980, 638.
3. M.V. Bhatt and S.S. El-Morey, Synthesis, 1982, 1048.
4. J.M. Bruce and Y. Roshan-Ali, J. Chem. Res., 1981, 193.
5. R. Sala, G. Doria and C. Passaroti, Tetrahedron Letters, 1984, *40*, 4565.

Oxidative cleavage of allyl phenyl ethers in dichloromethane containing chromia-pillared montmorillonite is effected by treatment with an iso-octane solution of tert-butyl hydroperoxide in 80-90% yield (B.M. Choudary et al., Tetrahedron, 1992, *48*, 953). Dealkoxylations are less widely used, but in recent years a number of new approaches have been adopted.

1-Methoxynaphthalene in dichloromethane treated with aluminium chloride and ethanethiol gives naphthalene in 76% yield by reaction through a route believed to involve the thioether as an intermediate (M. Node et al., Tetrahedron Letters, 1982, *23*, 689).

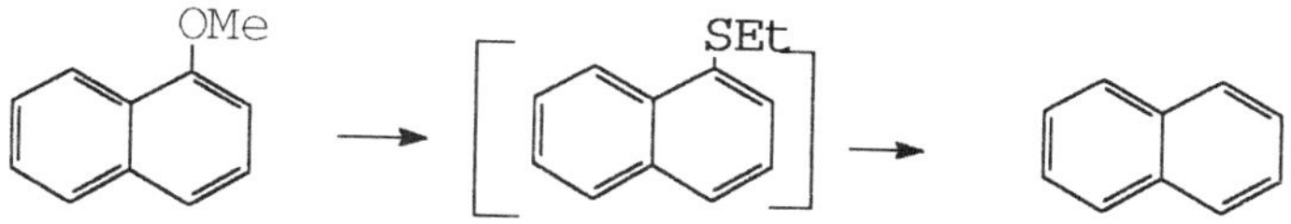

Demethoxylation and intermolecular coupling take place when 2-methoxyacetophenone is treated with hot THF containing Ti(0), from titanium trichloride and lithium. The product is 9,10-dimethylphenanthrene (A. Banerji and S.K. Nayak, J. Chem. Soc. Chem. Commun., 1991, 1432).

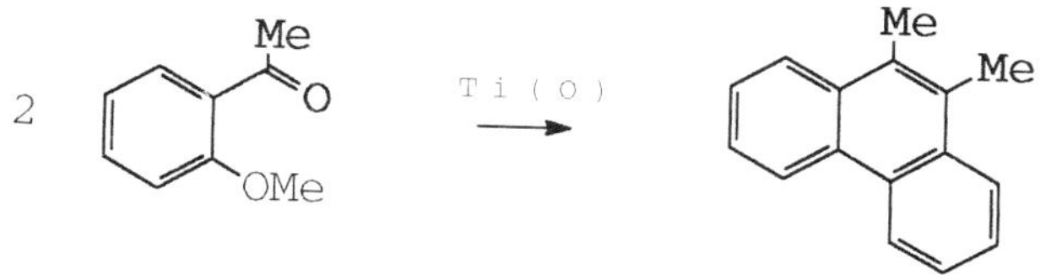

Dealkylation is also related to substitution of the phenolic ether group. Grignard reagents have traditionally been used as demethylating agents. Thus in the reaction of phenyl magnesium bromide with 2,6-di-tert-butyl-4-methylphenyl 2-methoxybenzoate, demethylation took place in preference to attack at the hindered ester position to give the benzoyl derivative in 96% yield (T. Hatiori, T. Suzuki and S. Miyano, J. Chem. Soc. Chem. Commun., 1991, 1375).

(ii) Substitution of the Phenolic Ether Group
Phenyl tetrazolyl ethers are often used in procedures for the removal of an OH group. Thus 1-phenyltetrazol-5-yl ether may be reductively cleaved when heated with sec-butylmagnesium chloride and dichloro[1,3-bis(diphenylphosphino)propane]nickel(II) as catalyst. The product, sec-butylbenzene, is isolated in 77% yield (R.A.W. Johnstone and W.N. McLean, Tetrahedron Letters, 1988, *29*, 5553).

9-Methoxyanthracene in benzene containing a little methanesulphonic acid, reacts with 1-butanethiol to give 9-anthryl butyl sulphide in 85% yield (W.H. Pirkle and J.M. Finn, J. Org. Chem., 1983, *48*, 2779).
The cyanomethyl group can replace the methoxyl group in a estrone methyl ether derivative in the form of its arene chromiumtricarbonyl complex (H. Kunzer and M. Thiel, Tetrahedron Letters, 1988, *29*, 1135).

(iii) Substitution in the Aryl Ring of Phenolic Ethers
Halogenation: Anisole in cold chloroform or in Freon is regiospecifically fluorinated by addition to acetyl hypofluorite, resulting in 2-fluoroanisole in 85% yield plus 10% of the 4-isomer (D. Lerman, Y. Tor and S. Rozen, J. Org. Chem., 1981, *47*, 4629).
A variety of methods for the selective chlorination of anisole have been described. Anisole in chlorobenzene when treated with copper(II) chloride on neutral alumina gives 4-chloroanisole in 98% yield (4-/2-; 32.3/1) (M. Kodomari, S. Takahashi and S. Toshimoto, Chem. Letters, 1987, 1901). In another approach chlorine water on a reversed-phase HPLC column at ambient temperature can be employed (D.A. Jeger and M.W. Clennan, Tetrahedron Letters, 1987, *28*, 4805).

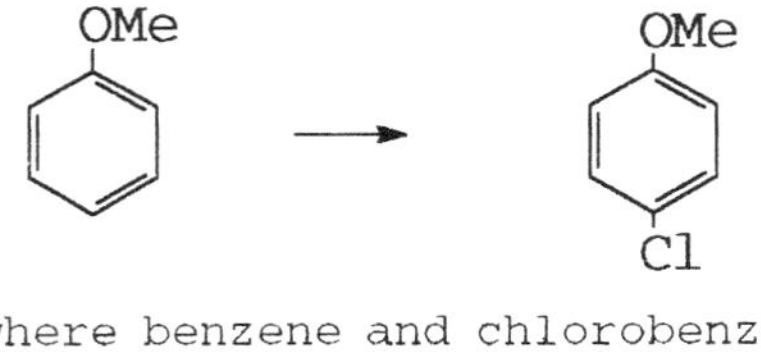

Under conditions where benzene and chlorobenzene fail to react, anisole and aluminium chloride added to stirred phenylselenyl

chloride in dichloromethane at -70°C give a quantitative yield of 4-chloroanisole. (N. Kamiata et al., Bull. Chem. Soc. Japan, 1988, *61*, 2226).
Benzyltrimethylammonium tetrachloroiodate is less effective, but with anisole in acetic acid gives 4- and 2-chloroanisole (6:1) in 73% yield (S. Kajigaeshi et al., Chem. Letters, 1989, 415). Chlorination of the methylene group in diphenoxymethane and not the ring occurs in 72% yield with chlorine in carbon tetrachloride containing azoisobutyronitrile (Bayer AG, EP 347667).
4-Bromoanisole is obtained in 98% yield from the bromination of anisole in dichloromethane/methanol (5:2) with benzyltrimethylammonium bromide (S. Kajigaeshi et al., Chem. Express, 1988, *3*, 219. Anisole in dichloromethane and bromodimethylsulphonium bromide furnish 4-bromoanisole in 94% yield (G.A. Olah, L. Ohannesian, M. Arvanaghi, Synthesis, 1986, 868).

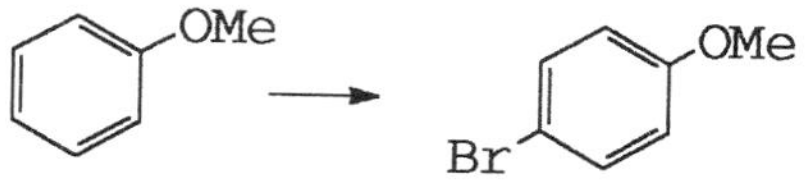

PhOMe → 4-Br-C6H4-OMe

Iodination of anisole in acetic acid containing anhydrous zinc chloride and benzyltrimethylammonium dichloroiodate takes place in 92% yield giving 4-iodoanisole (S. Kajigaeshi et al., Chem. Letters, 1988, 795).

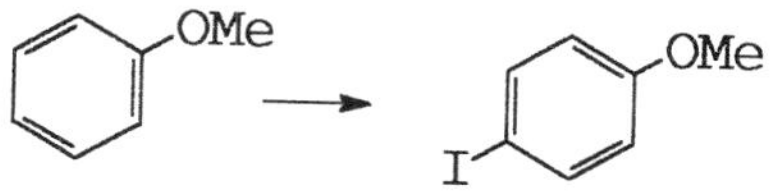

PhOMe → 4-I-C6H4-OMe

Alkoxybenzenes are also iodinated with iodine-silver sulphate (W.W. Sy, B.A. Lodge and A.W. Sy, Synth. Commun., 1990, *20*, 877).

Nitro, Sulphonic, Sulphur, Thioalkyl, Hydroxyalkyl and Acyl Derivatives of Phenolic Ethers:
Although anisole is unreactive in the two-phase system of concentrated nitric acid-diethyl ether, 4-nitroanisole is formed if the ether is replaced with other organic sovents such as alkanes and nitrobenzene (A. Germain, J. Chem. Soc. Chem. Commun., 1990, 1697).

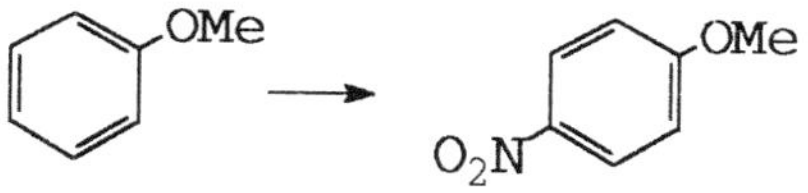

PhOMe → 4-O_2N-C6H4-OMe

Anisole when heated with bis(trimethylsilyl)sulphate in a Dean-Starke apparatus affords the trimethylsilyl sulphonate in 92% yield. This can be converted in aqueous ether to 4-methoxybenzenesulphonate in quantitative yield (P. Bourgeois and N. Duffaut, Bull. Soc. Chim., France, II, 1980, 195).

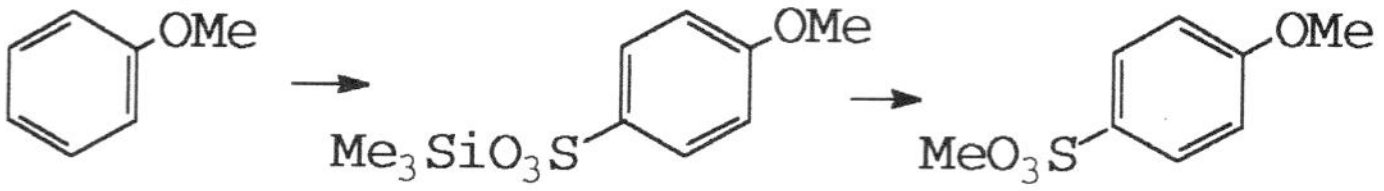

Powdered 2-methoxynaphthalene if treated with thionyl chloride gives the 1-sulphinyl chloride derivative in 91% yield (K.H. Bell, Austral. J. Chem., 1985, 38, 1209).

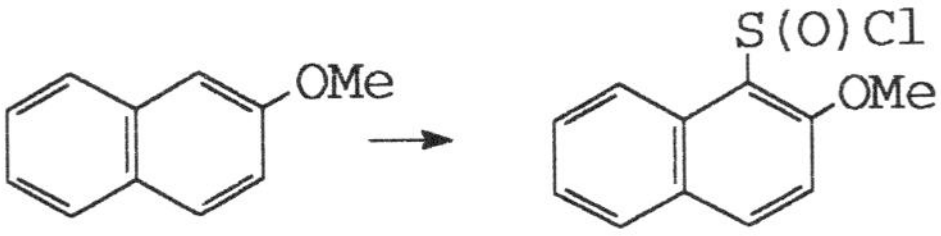

Substitution m- to the methoxy group, established earlier for arene/chromium carbonyl systems (M.F. Semmelhack and G. Clark, J. Am. Chem. Soc., 1977, 99, 1675), has been used to synthesise the 1-(2-dithianyl) derivative of (HO-protected) O-methylestrol by addition of the chromium tricarbonyl complex to 2-lithio-1,3-dithian (H. Kunzer and M. Thiel, Tetrahedron Letters, 1988, 29, 3223).

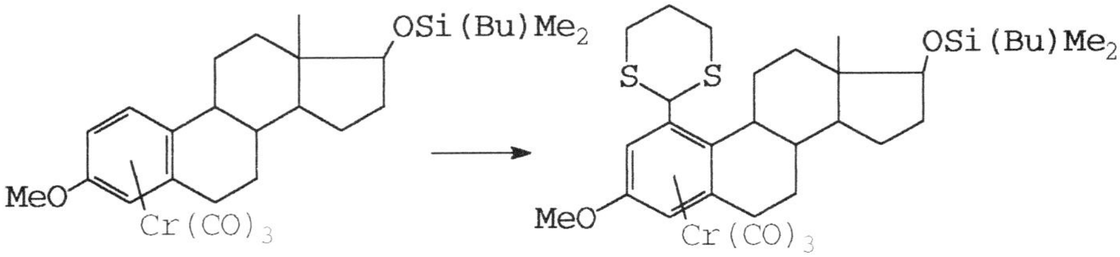

o-Lithiated anisole reacts with acetophenone under ultrasonification to give 1-(2-methoxyphenyl)-1-phenylethanol in 86% yield (G. Banerji and S.K. Nayak, Curr. Science, 1989, 58, 249).

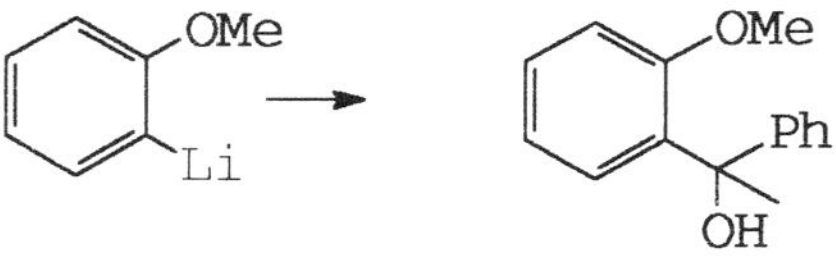

Estrone methyl ether in chloroform solution is 2-acetylated in

89% yield by the addition of acetyl mesylate (G. Mikhail and M. Demuth, Synlett., 1989, 54).

1-(4-Methoxyphenyl) pentyl ketone is formed in 85% yield by the addition of anisole and caproic anhydride to a suspension of lithium perchlorate in dichloromethane containing a small amount of antimony perchlorate (T. Mukaiyama et al., Chem. Letters, 1992, 435).

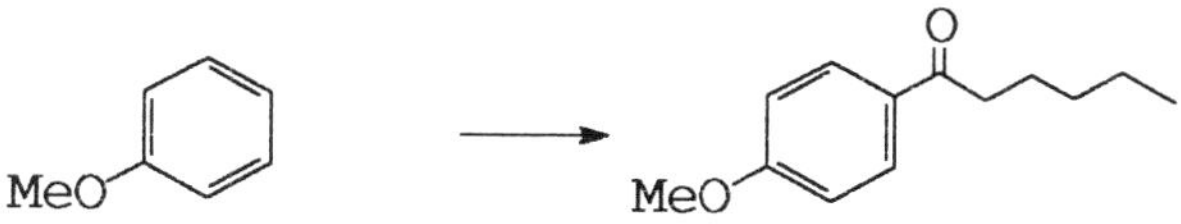

6-Acyl-2-methoxynaphthalenes are synthesised by 'reversible' acylation of 2-methoxynaphthalene (C. Giordano, M. Villa and R. Annunziata, Synth. Commun., 1990, 20, 383).

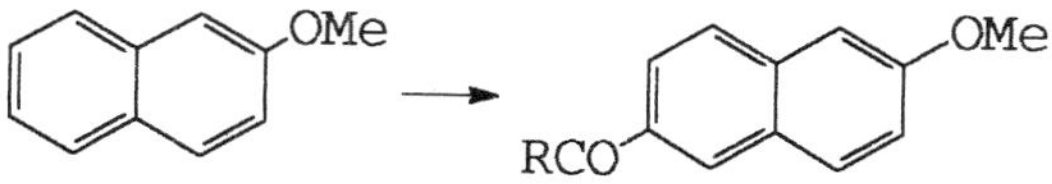

By the addition of anisole and benzoyl chloride to a solution of anhydrous cobalt(II) chloride in acetonitrile, 4-methoxybenzophenone is produced in 72% yield (J. Iqbal, M.A. Khan, N.K. Nayar, Tetrahedron Letters, 1991, 32, 5179).

The 2-carboxy derivative of the ketal of 3-benzyloxyacetophenone is synthesised in 75% yield by lithiation with n-butyllithium and addition of the reaction mixture to excess dry ice (Ihara Chem. Ind. Co. Ltd., JP 4134080).

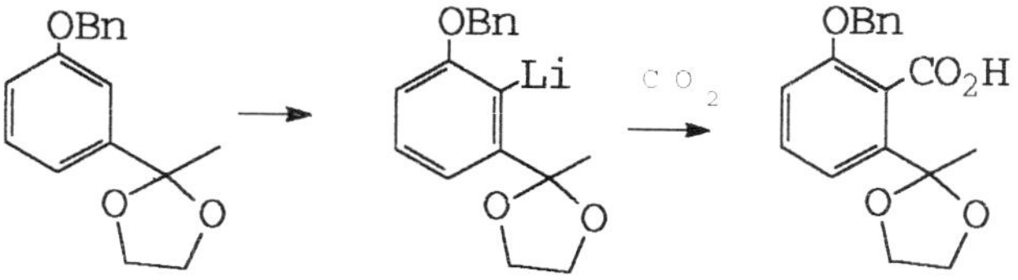

Lithiation of 4-methylphenyl 2-(trimethylsilyl)ethoxymethyl ether with n-butyllithium followed by treatment with benzaldehyde affords α-[5-methyl-2-(2-trimethylsilyl)ethoxymethyl)phenylbenzyl alcohol in 80% yield (S. Sengupta and V. Snieckus, Tetrahedron Letters, 1990, 31, 4267).

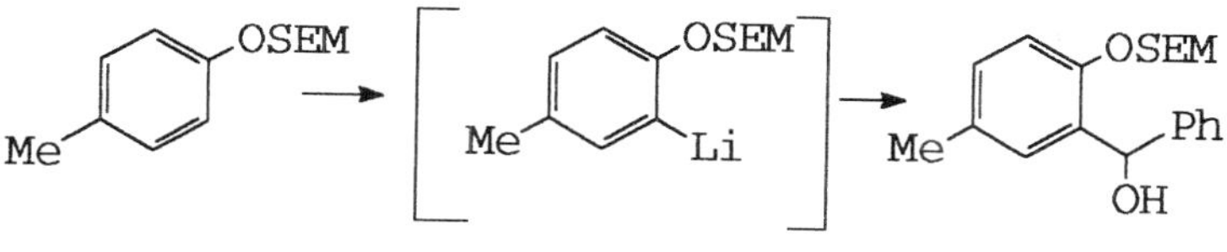

(iv) Reactions in the Alkyl Side Chain of Phenolic Ethers

The benzylic position of 4-methoxy-n-propylbenzene is substituted by the azido group in 97% yield through prolonged reaction with excess trimethylsilylazide in chloroform containing 2,3-dichloro-5,6-dicyanobenzoquinone. Catalytic hydrogenation of the product in ethanol with Pd-BaSO4/ hydrogen readily affords 1-(4-methoxyphenyl)propylamine, isolated as the hydrochloride in 60% yield (A. Guy et al., Synthesis, 1988, 900).

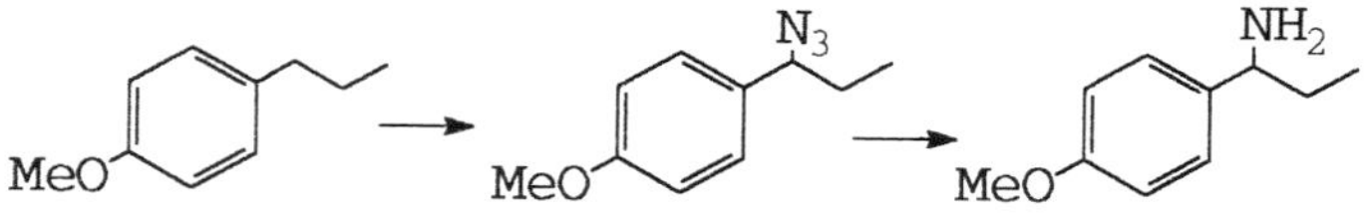

Asymmetric hydroxylation of the side-chain of (Z)-1-(4-methoxyphenyl)-1-(tert-butyldimethylsiloxy)-1-propene to give (R)-1-hydroxyethyl 4-methoxyphenyl ketone in 94% yield (99%e.e.) is effected by addition of the alkene to a stirred mixture of osmium tetroxide, potassium ferricyanide, potassium carbonate, dihydroquinidine, 9-O-(9'-phenanthryl ether and one mole of methanesulphonamide in aqueous tert-butanol (1:1) (T. Hashiyama, K. Morikawa and K.B. Sharpless, J. Org. Chem., 1992, *57*, 5067).

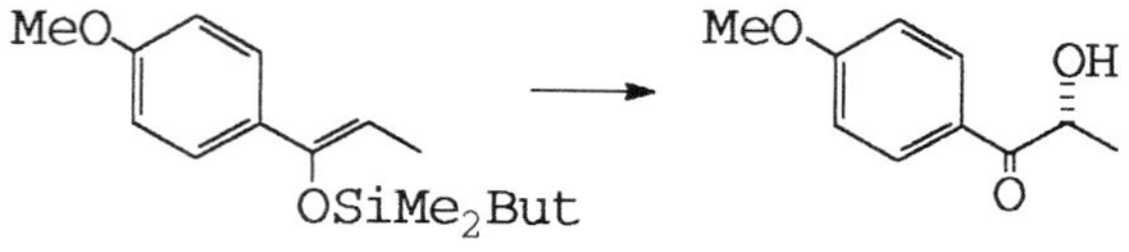

Addition to the side-chain of the aldimine from 4-anisaldehyde and aniline occurs when heated with 5-phenylfuran-2,3-dione affording 2-(4-methoxyphenyl)-3,6-diphenyl-2,3-dihydro-4H-1,3-oxazine-4-one in 93% yield (Y.S. Andreichikov et al., Zh. Org. Khim., 1992, *28*, 779).

Treatment of 4-methyliodobenzene with 3-(4-methoxyphenyl)-1-propyne and carbon monoxide in benzene containing triphenylphosphine and triethylamine with palladium acetate as catalyst affords 3-(4-methoxybenzylidene)-5-(4-methylphenyl)-furan-2(3H)-one in 40% isolable yield (Y. Huang and H. Alper, J. Org. Chem., 1991, *56*, 4534).

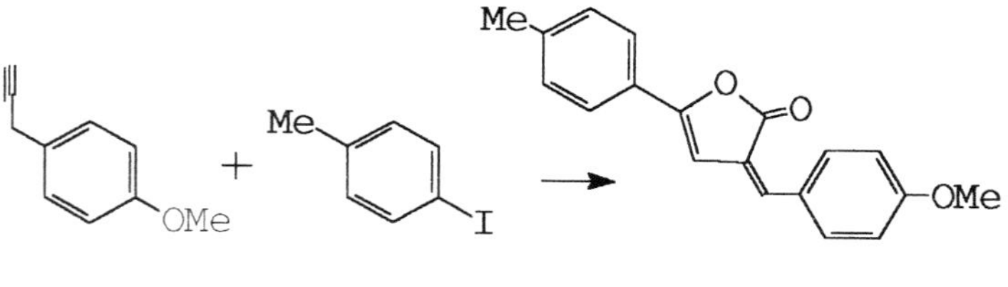

(v) Displacement of Nuclear Chlorine, Bromine or Fluorine from a Phenolic Ether

Conversion of 4-tert-butoxychlorobenzene to the Grignard reagent and its subsequent reaction with vinyl chloride in the presence of a small proportion of dichloro[1,3-bis(diphenylphosphino)-propane]nickel gives 4-tert-butoxystyrene in 93% yield (Hokko Chem. Ind., KK, JO 1106835).

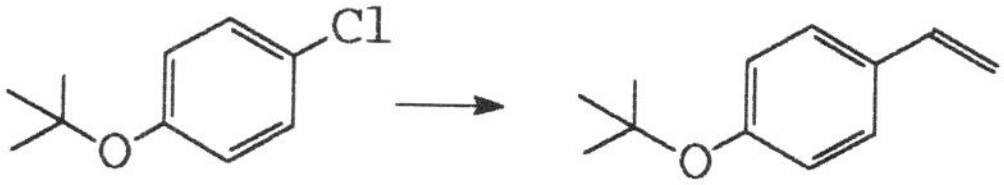

The Grignard reagent prepared from 4-bromophenyl trifluoromethyl ether in diethyl ether combines with trifluoroacetic anhydride, to yield 4-trifluoromethoxytrifluoroacetophenone (Imperial Chemical Industries, plc, GB 2210881).

The formation of 3,5-bis[(2-hydroxy-2-methyl)ethyl]anisole from 2,6-dibromoanisole via a zirconocene intermediate occurs through treatment with excess butyllithium in THF containing methylzirconocene chloride and completion of reaction by the addition of trimethylphosphine and acetone in 66% yield (D.P. Hsu, E.A. Lucas and S.L. Buchwald, Tetrahedron Letters, 1990, 31, 5563).

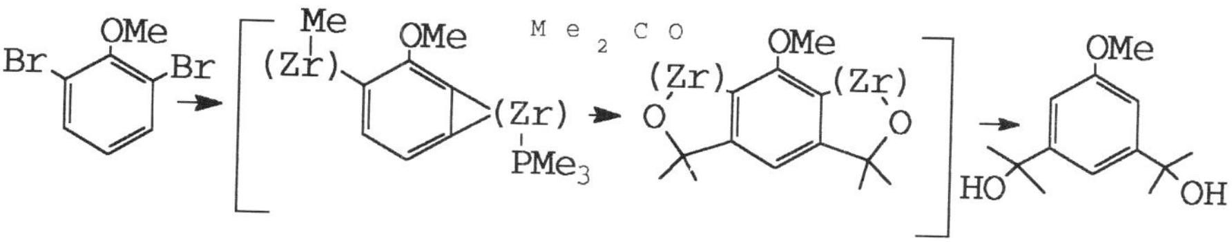

Displacement of bromide ion from 2-bromophenoxy(dimethyl)chloromethylsilane with sodium in boiling toluene, occurs with formation of 2,3-dihydro-1,3-benzoxasilole in 60% yield (Luther University, Halle, DD 296280).

(vi) Cyclisation Reactions of Phenolic Ethers

Intramolecular cyclisations of phenolic ether derivatives often containing an activating group at the 4-position, are used to obtain a variety of heterocycles.

Diphenyl ether is oxidatively convered to dibenzofuran upon heating at 140°C with oxygen (300psig) in the presence of palladium(II) and copper(I) acetates in acetic acid (Halcon S.D. Inc., USP 4362883).

5-Methoxy-1-naphthol reacts with cyclohexylamine and aqueous formalin at ambient temperature to give 3-cyclohexyl-3,4-dihydro-7-methoxy-2H-naphtho[2,1-e][1,3]oxazine in 80% yield (H. Mohrle and H. Folttmann, Arch. Pharm., 1988, *321*, 259).

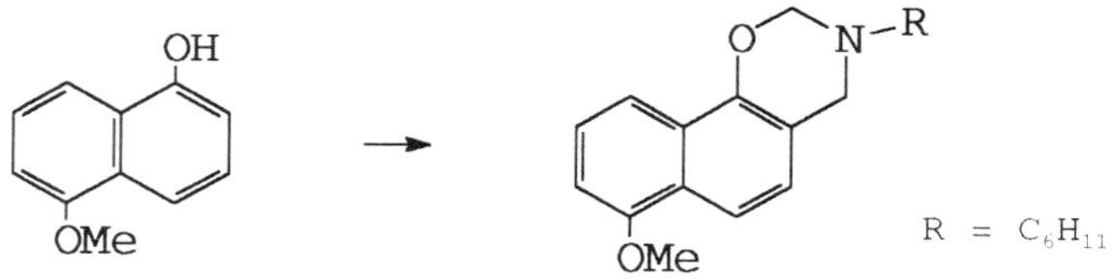

$R = C_6H_{11}$

Regioselective photocyclisation of 2-methoxystilbene to phenanthrene with loss of a methoxyl group as methanol, occurs in a mixture of sulphuric acid and tert-butanol (F.B. Malory, M.J. Rudolph and S.M. Oh, J. Org. Chem., 1989, *54*, 4619).

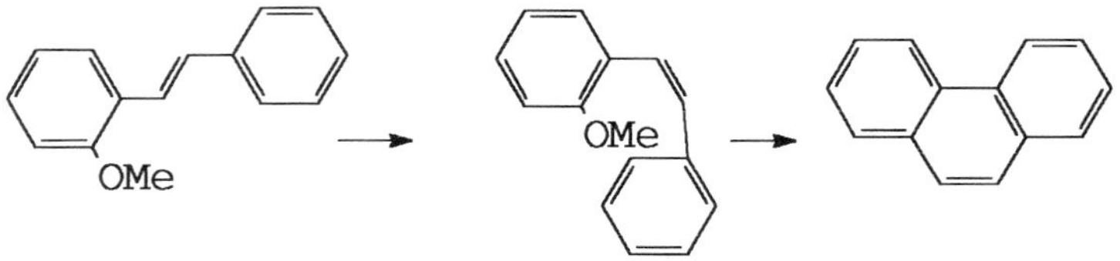

α-Aryloxyacetophenones, undergo cyclodehydration with concentrated sulphuric acid affording 3-arylbenzofurans after initial formylation under Vilsmeier conditions. Thus, α-(3-diethylaminophenoxy)acetophenone gives 2-benzoyl-6-diethylamino-benzofuran in 78% yield (T. Hirota et al., Heterocycles, 1986, *24*, 771).

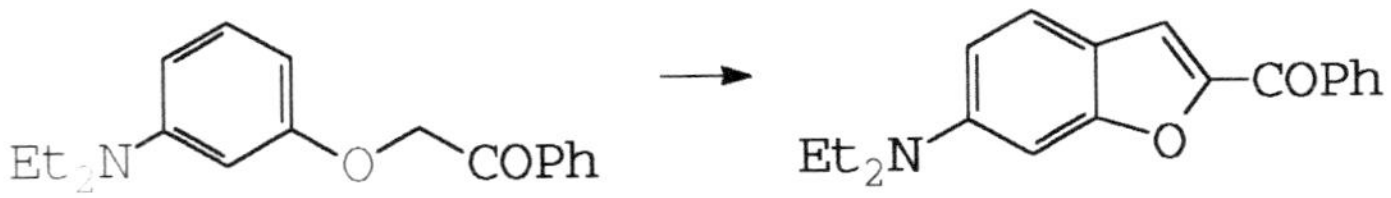

Formylation of N-3-trifluoromethylphenyl-3-methoxyphenyl - acetamide followed by cyclisation of the resulting diformyl compound gives a 60% yield of 4-formyl-6-methoxy-2-(3-trifluoromethylphenyl)isoquinol-3-one (D.J. Le Count, R.J. Pearce and A.P. Marson, Synthesis, 1984, 349).

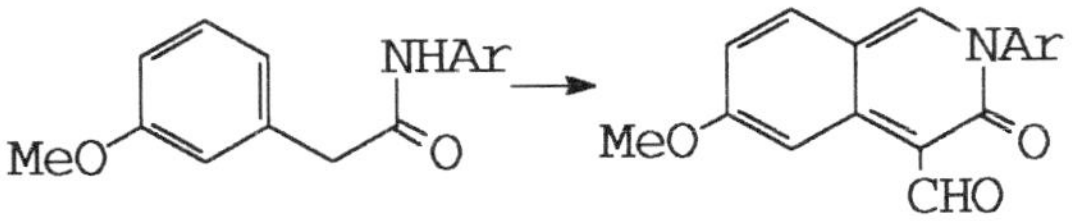

In a similar general way N-trifluoroacetyl derivatives have been cyclised. A Bischler-Napieralski type reaction occurs when N-hydroxy 1-(3-methoxyphenyl)-2-aminopropane is heated in trifluoroacetic acid and anhydride affording 6-methoxy-3-methyl-1-trifluoromethyl-3,4-dihydroisoquinoline in 86% yield (M. Kawase and Y. Kikugawa, Chem. Pharm. Bull. 1981, 29, 1615).

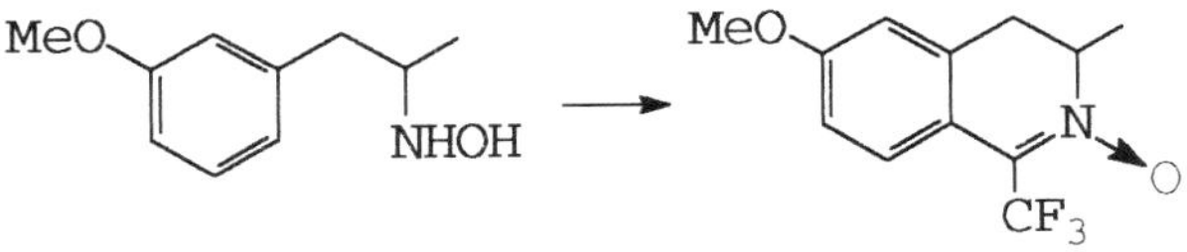

Pyran derivatives are synthesised from propynyl ethers. Two examples are the cyclisation of a coumarinyl ether in the presence of N,N-diethylaniline and caesium fluoride to give a naphthopyran in 88% yield (H. Ishii et al., Chem. Pharm. Bull., 1992, 40, 1148) and a quinolonyl ether in hot dimethylaniline to give a pyranoquinolone (Kodama KK JO 1040485).

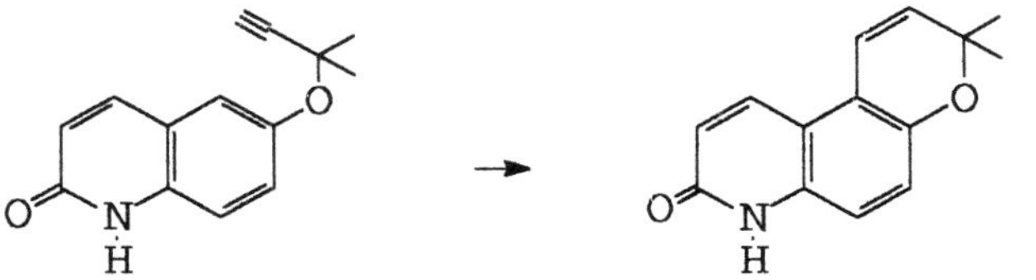

(vii) Reactions of Keto Derivatives of Phenolic Ethers
Certain of these reactions are also examples of side-chain

substitution in phenolic ethers.
4-Methoxypropiophenone in methyl orthoformate containing iodine gives, by a 1,2-migration, methyl 2-(4-methoxyphenyl)propionate in 85% yield (Kyowa Hakko Kogyo, KK, J62 238233).

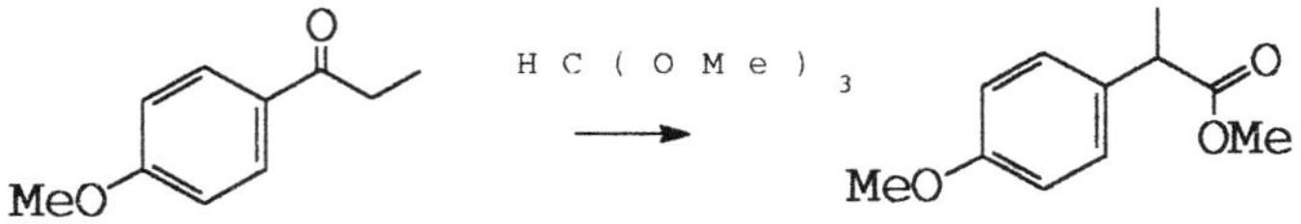

The Mannich base from 4-methoxyacetophenone, formaldehyde and dimethylamine heated in dimethylformamide with the imino ester hydrochloride of ethyl benzoylacetate, initially gives the amidine but this, with ammonium acetate, affords 2-amino-6-(4-methoxyphenyl)pyrid-3-yl phenyl ketone in 66% yield (M. Solhuber-Kretzer, R. Troschutz and H.J. Roth, Arch. Pharm., 1982, *315*, 199).

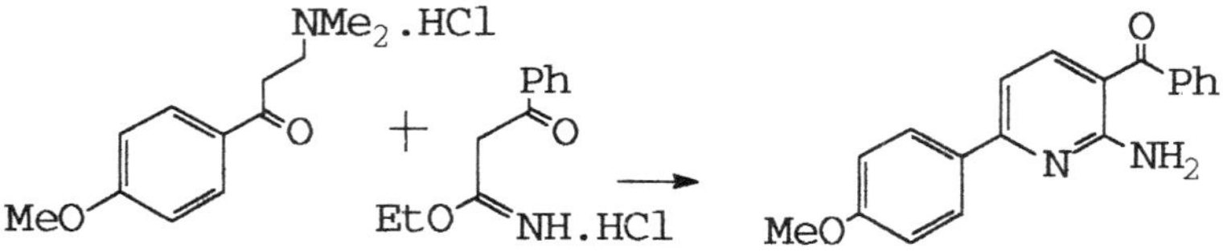

The phenolic ether depicted, reacts with hydrazine hydrate in hot ethanol giving a 93% yield of 4-methylthiomethyl-5-(4-methoxyphenyl)pyrazole (S. Apparao et al., Synthesis, 1982, 792).

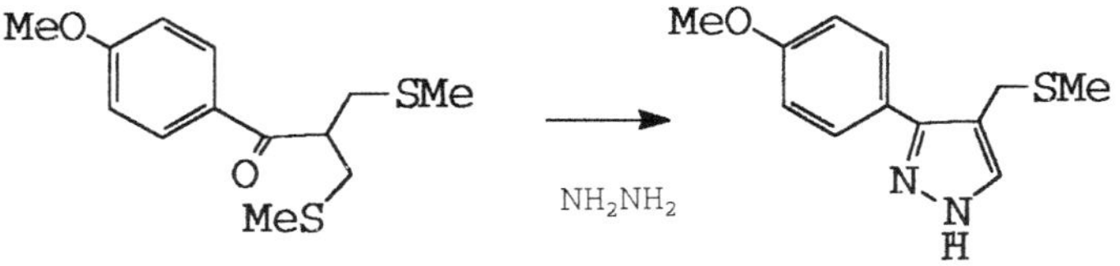

Chloro, bromo, dichloro, tert-butyl and unsubstituted 1-benzoxepine-3,5(2H,5H)-diones are synthesised from the corresponding ethyl 2-(2-acetophenoxy)propionates in good overall yields by the action of ethanolic sodium ethoxide (J.H.P. Tyman, G. Gabriel and R. Pickles, J. Chem. Res., *(S)*, 1989, 348; *(M)*, 2710-18.

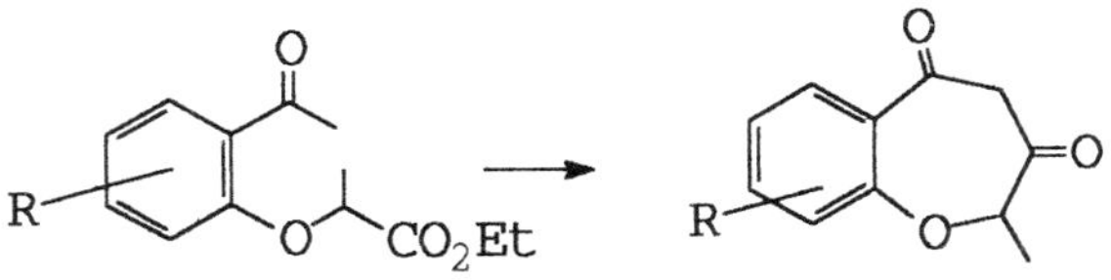

Ozonisation of 2-furfurylidene-6-methoxy-1-tetralone in ethyl acetate at -78°C and treatment of the ozonide with hydrogen peroxide gives 3-(2-carboxy-5-methoxyphenyl)propionic acid in 82% yield which, as the acid chloride, is cyclised in 45% yield in the presence of anhydrous aluminium chloride at ambient temperature to 7-methoxy-1-oxoindan-4-carboxylic acid (T. Girija, P.S. Shanker and G.S.R. Subbs Rao, J. Chem. Soc. Chem. Commun., 1991, 1467).

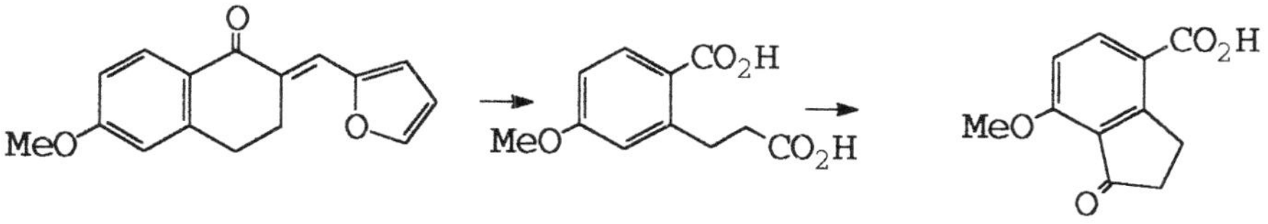

(viii) **Reactions of Carboxy, Amido, Formyl and Nitro Derivatives of Phenolic Ethers**

Some of these reactions also represent examples of side-chain substitution in phenolic ethers.

The demethoxylation of 2-methoxybenzoic acid has been referred to earlier. Its aryl ester derived from 2,6-dialkylphenol reacts with phenyl Grignard reagents to undergo replacement of the methoxy group, rather than attack at the ester function, and furnish 1,1'-biphenyl-2-carboxylates which are of interest for the synthesis of cannabinoids (T. Hatiori, T. Suzuki and S. Miyano, J. Chem. Soc. Chem. Commun., 1991, 1375).

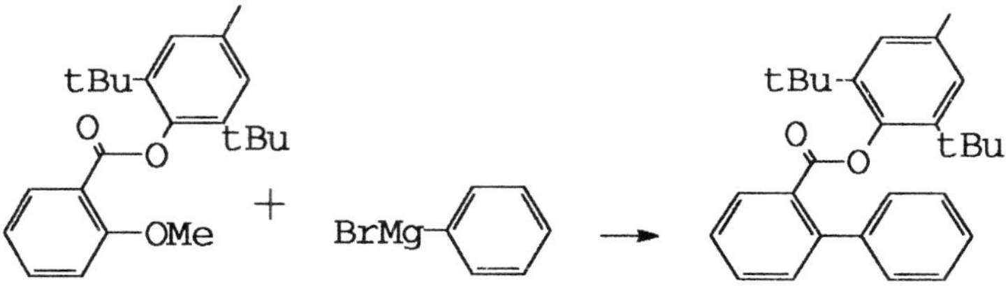

2-Methoxybenzoyl chloride when reacted with the carbanion from 4-phenylsulphonylchroman, furnishes the corresponding benzoyl derivative which after desulphonylation with Raney nickel gives the corresponding α,β-unsaturated ketone. Demethylation and cyclisation of this with ethanolic potassium acetate gives the parent ring system of rotenone (S.M.F. Lai, J.J.A. Orchison and D.A. Whiting, Tetrahedron, 1989, 45, 5895).

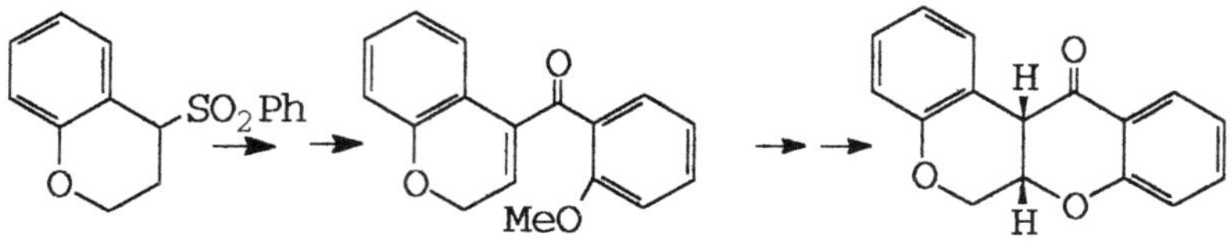

4-Methoxybenzaldehyde is converted to 4-methoxystyrene in 84% yield by reaction with methylene trichloromolybdenum (T. Kaufmann et al., Angew. Chem. Suppl., 1983, 222).
In the photoreaction of 4-nitroanisole with n-hexylamine the methoxyl group is replaced, while with ethyl glycinate the nitro group is substituted (A. Cantos et al., J. Org. Chem., 1990, 55, 3303).

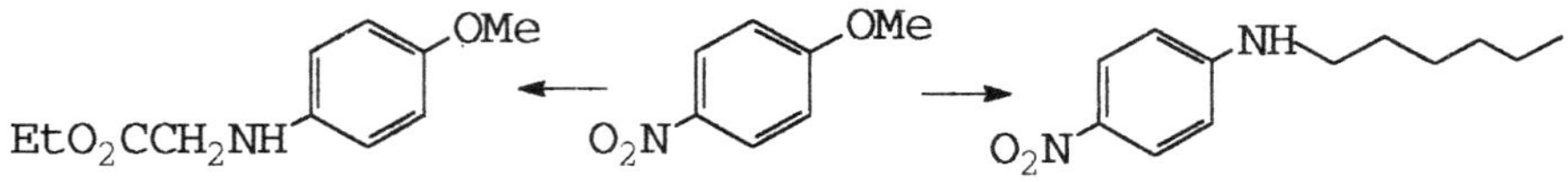

(ix) Phenolic Ether Coupling

1-Methoxy-2-methylnaphthalene affords a 4,4'-binaphthyl derivative in 88% yield upon heating in trifluoroacetic acid containing cobalt trifluoride, or by oxidation with lead tetraacetate, ferric chloride, thallium(III) or mercuric trifluoroacetates (E.C. Taylor et al., J. Am. Chem. Soc., 1980, 102, 6513).
Asymmetric coupling is reported by the use of a chiral leaving group. Thus, 1-(1-menthoxy)-2-oxazolinonaphthalene reacts with 1-lithionaphthalene to furnish 2-(4,5-dihydro-4,4-dimethyloxazol-2-yl)-1,1'-binaphthyl in 80% yield (67% e.e.) (J.M. Wilson and D.J. Cram, J. Am. Chem. Soc., 1982, 104, 881; A.I. Meyers and K.A. Lutomski, ibid., 879).

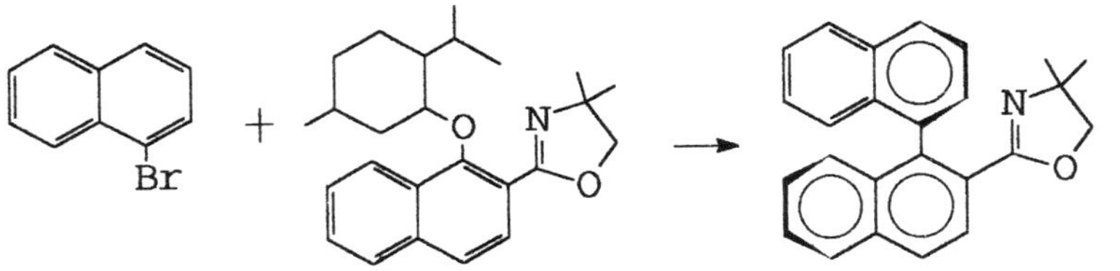

(x) Miscellaneous Reactions of Phenolic Ethers

Ring Saturation of Phenolic Ethers:

Birch reduction of 2-methylanisole and subsequent reaction of the product with trimethylsilylmethyl chloride gives 3,6-dihydro-2-methyl-4-trimethylsilylmethylanisole. This compound upon treatment with concentrated hydrochloric acid affords 4-methylene-2-methylcyclohexanone in 60% yield (R.G. Salomon et al., J. Am. Chem. Soc., 1982, *104*, 1008).

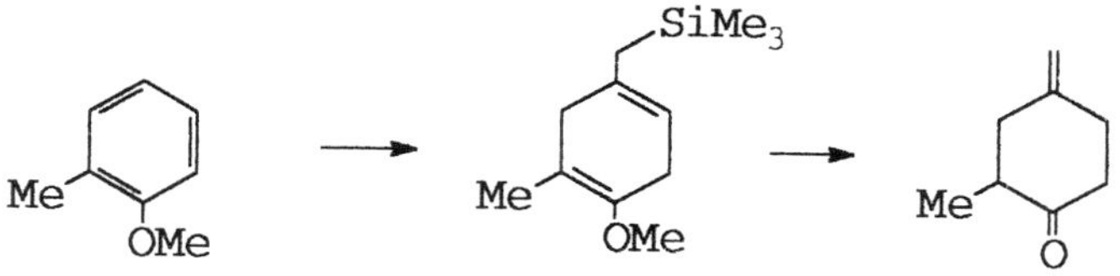

Diels-Alder Reactions of Phenolic Aryl Ethers:

In a one-pot synthesis with simultaneous decarboxylation, 5-(4-methylphenoxy)-2-furoic acid and N-phenylmaleimide in chloroform containing trifluoroacetic acid upon heating in a sealed tube under nitrogen for 1 hour resulted in N-phenyl 3-(4-methylphenoxy)phthalimide in 95% yield (J.A. Cella, J. Org. Chem., 1988, *53*, 2099).

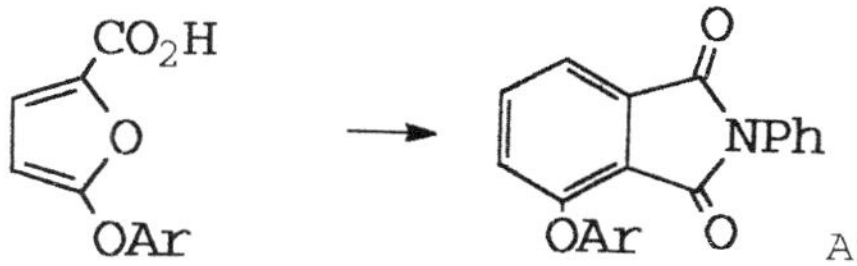

Ar = C_6H_4-Me

(h) Halogeno, Nitro, Amino, Azo, Sulpho and Thio Derivatives of Phenols

Synthesis of Halogeno, Nitro, Amino, azo and Thio derivatives
(i) Fluorophenols

4-Fluorophenol is synthesised from 4-bromofluorobenzene by heating with 28% methanolic sodium methoxide containing a little cuprous iodide to give fluoroanisole in 83% yield and thence demethylation with hot 47% hydrobromic acid to form the product in 92% yield (Central Glass KK, J60 004144).

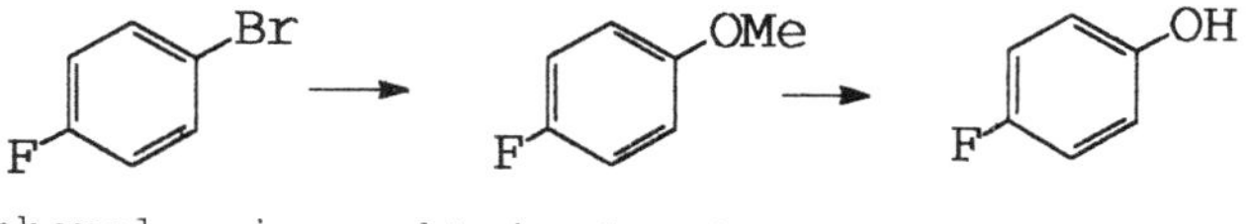

2-Fluorophenol is obtained in 67% yield from 2-chlorofluorobenzene and cupric oxide during 24 hours at 250°C in a sealed Parr reactor with dipotassium hydrogen phosphate (R.G. Pews and J.A. Gall, J. Fluorine Chem., 1990, *50*, 377).

(ii) Chlorophenols

A range of procedures has been introduced for the formation of mono- and dichlorophenols from phenolic intermediates, in the majority of which free chlorine is not used.

Phenol in dichloromethane with a slight excess of phenylselenyl chloride at ambient temperature affords a 50% yield of 4-chlorophenol (F.O. Ayorinde, Tetrahedron Letters, 1983, 2077).

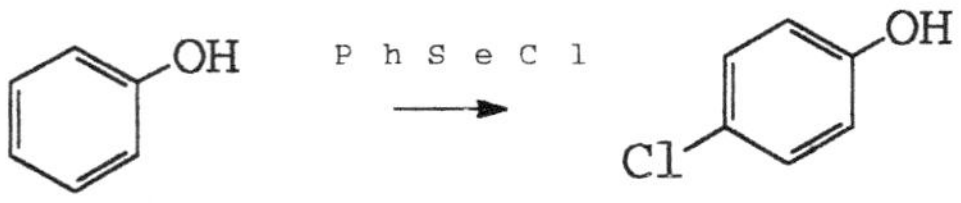

2,3,4,5,6,6-Hexachlorocyclohexa-2,4-diene-1-one in carbon tetrachloride solution containing phenol, upon irradiation at -5°C during 10 hours (100 watt Hg lamp) affords 2-chlorophenol in 92% yield (A. Guy, M. Lemaire and J-P. Guette, J. Chem. Soc. Chem. Commun., 1980, 8), while by contrast, the 2,3,4,4,5,6-isomer in DMF for 72 hours at ambient temperature gives 4-chlorophenol in 72% yield (ibid.).

2-Methylphenol with N-chloro-bis(2-chloroethyl)amine, at 25°C in carbon tetrachloride containing silica gives a (4:1) ratio of o/p monochloro products in 100% yield (K. Smith, M. Rutters and B. Nay, Tetrahedron Letters, 1988, 1319).

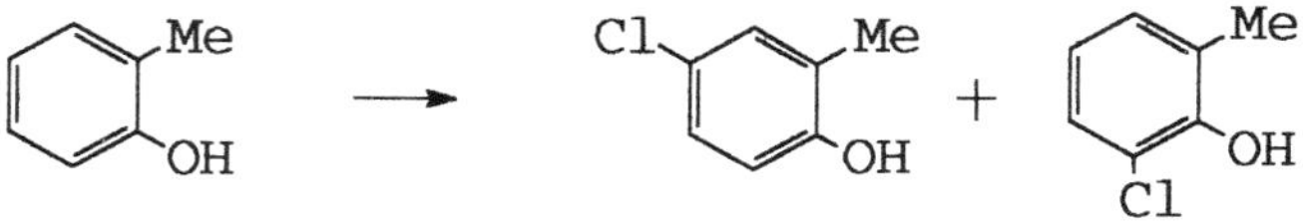

Again by comparison, 3-methylphenol in carbon tetrachloride at ambient temperature with N,N-dichloro-tert-butylamine gives a 62% yield of o/p monochloro products in a (18:1) ratio (Y. Ogata et al., J. Chinese Chem. Soc., Ser.II, 1983, 261).

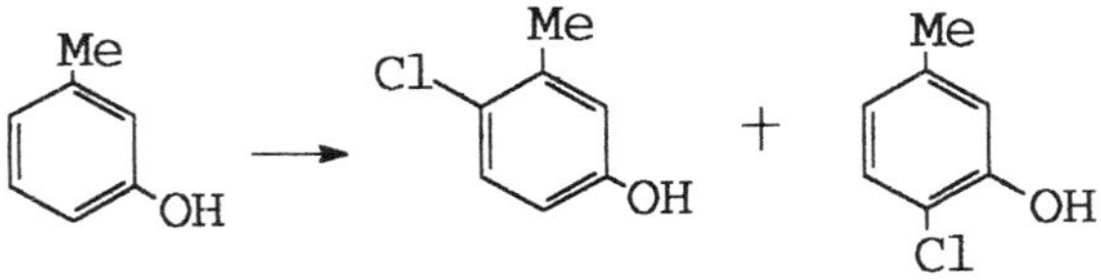

3-Tert-butylphenol with copper(II) chloride dihydrate in boiling acetic acid gives the p-chloro compound in 71% yield (selectivity, 79%), (Sagami Chem. Res. Centre, J62 116532).

2,6-Dimethylphenol upon treatment with benzyltrimethylammonium tetrachloroiodate during 1 minute at ambient temperature affords a 91% yield of the 4-chloro product (S. Kajigaeshi et al., Chem.

Express, 1990, 141).
In a series of patented procedures, 2,4,6-trichlorophenol has been derived in 100% yield from 2,4-dichlorophenol with chlorine at 70°C in the presence of diisopropylamine, containing 2,6-dichlorophenol (Rhone-Poulenc Spec., EP 216714), or in similar yields from 2,6-dichlorophenol by chlorination at 70°C with added tetra n-butylammonium chloride or trifluoromethanesulphonic acid (Rhone-Poulenc Spec., EP 299890, EP 255452). 2,4,5,6-Tetrachlorophenol is obtained from 2,4,6-trichlorophenol by the use of chlorine and a litte added zirconium tetrachloride or less effectively with aluminium chloride (Rhone-Poulenc Spec., FR 2636942).
Other strategies exist for the synthesis of polychlorophenols containing functional groups. Thus 2,5-dichlorothiophene in toluene containing rhodium(II) acetate when treated with ethyl diazoacetoacetate affords a bicyclic intermediate which by prolonged heating in toluene produces ethyl 2,4-dichloro-5-hydroxy-6-methylbenzoate in 86% yield (R.J. Gillespie, J. Murray-Rust and A.E.A. Porter, Tetrahedron, 1981, 37, 743).

(iii)Bromophenols

Although bromination is inherently more selective than chlorination a variety of new procedures have been experimented with.
Phenol in ethyl acetate when added to bromine (or BrCl) in ethylacetate at 0-5°C and excess halogen is removed with sodium bisulphite, gives 4-bromophenol (98%) in 99% yield (Bromine Compounds DE 3742515). When bromine (2 moles) in carbon tetrachloride is added at ambient temperature to phenol in acetonitrile containing 2,2'-azo-(2-phenoxypropane) a 68% yield of 2-bromophenol (98%) is obtained, whereas a 7:1 excess of bromine /phenol affords a 88% yield of 4-bromophenol exclusively (K. Kim and Y-C. Cang, J. Chem. Soc. Chem. Commun., 1986, 1159).

Phenol in chloroform at ambient temperature with tetra n-butylammonium bromide results in a 95% yield of 4-bromophenol (J. Berthelot et al., J. Chem. Res., 1986, 381). The similar use of bromodimethyl sulphonium bromide has been reported (G.A. Olah, L. Ohannesian and M. Arvanghi, Synthesis, 1986, 868).

The novel brominating system, potassium bromide/3-chloroperbenzoic acid has been reported. Potassium chloride can replace potassium bromide leading to a suitable reagent for chlorination (M. Srebnik, R. Mechoulam and I. Yona, J. Chem. Soc. Perkin Trans.I, 1987, 1423).
4-Methylphenol in hydrofluoric acid containing antimony pentafluoride when treated with bromine gives 3-bromo-4-methylphenol in 85% yield, attributable to the intermediate formation of an oxonium ion (J-C. Jacquesy, M-P. Joannetaud and S. Makani, J. Chem. Soc. Chem. Commun., 1980, 110), while benzyl trimethylammonium tribromide added at ambient temperature to 4-methylphenol in dichloromethane/methanol (5:2) gives a 93% yield of 2,6-dibromo-4-methylphenol (S. Kajigaeshi et al., Chem letters, 1987, 627).

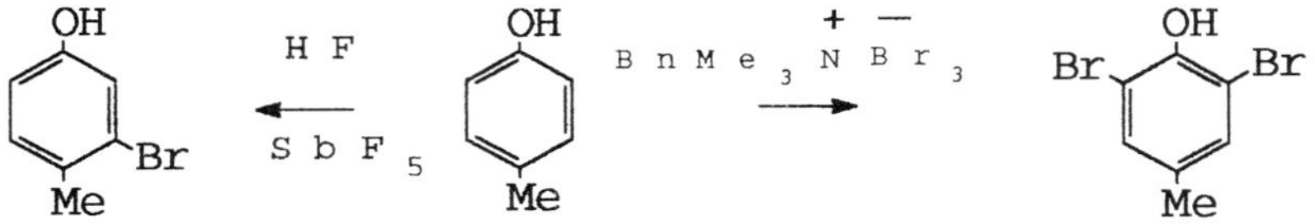

Simultaneous ring and chain substitution take place when bromine is added to 2,4,6-trimethylphenol in bromochloromethane at ambient temerature and the mixture is then heated at 66-68°C. This results in the formation of 4-bromomethyl-3,5-dibromo-2,6-dimethylphenol (Dow Chemical Co., USP 4684752).

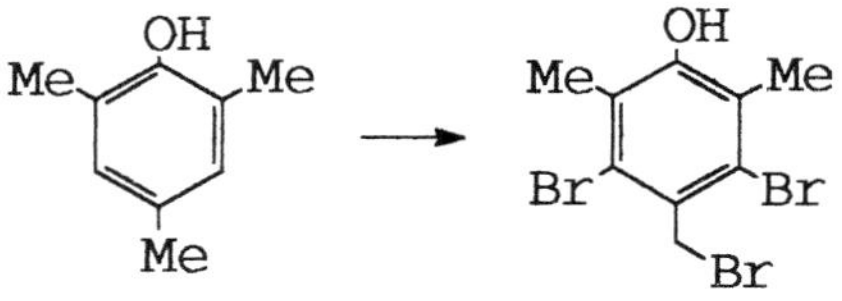

Reductive aromatisation of 2-bromo-4-hydroxycyclohexa-2,5-dienones to 2-bromophenols can be achieved by treatment with sodium hydrosulphite in THF/water (3:1) (K.A. Parker et al., J. Org. Chem., 1992, 57, 5547). (For related reactions see S.R. Ramadas, D. Rav and W. Sucrow, Chem. Ber., 1980, 113, 2579).

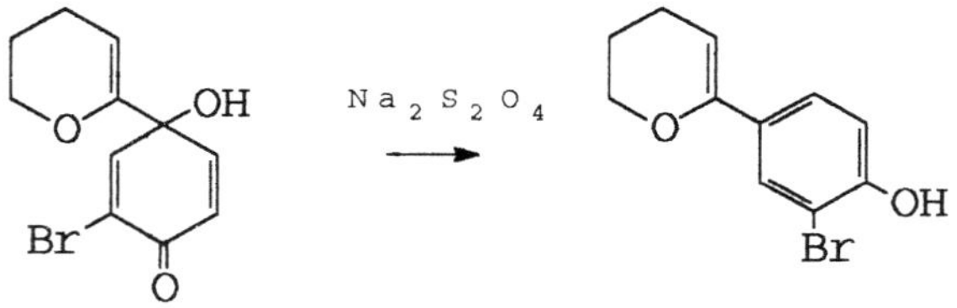

(iv) Iodophenols

Iodoarenes ordinarily difficult to obtain are more accessible in the phenolic series due to the activating influence of the hydroxyl group.

4-Iodophenol is synthesised in 96% yield from the treatment of phenol and cyclodextrin in 20% sodium hydroxide with iodine in aqueous sodium iodide solution (Asahi Chemical Co., KK J63 101342).

2,4,6-Triiodophenol is obtained in 72% yield from phenol, benzyltrimethylammonium dichloroiodate (3 moles) and sodium bicarbonate in dichloromethane/methanol (2.5:1) at ambient temperature (S. Kajigaeshi et al., Chem. Letters, 1987, 2109).

4-Iodo-2-methylphenol results from the treatment at 0°C of a methanolic solution of 2-methylphenol containing equimolar amounts of sodium iodide, sodium hydroxide and sodium hypochlorite (K.J. Edgar and S.N. Falling, J. Org. Chem., 1990, 55, 5287).

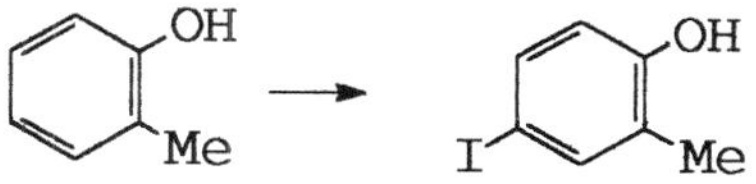

Estradiol heated at 55°C with copper(II) acetate monohydrate and iodine in acetic acid affords 2-iodoestradiol (G.A. Horiuchi and J.Y. Satoh, J. Chem. Soc. Chem. Commun., 1982, 671).

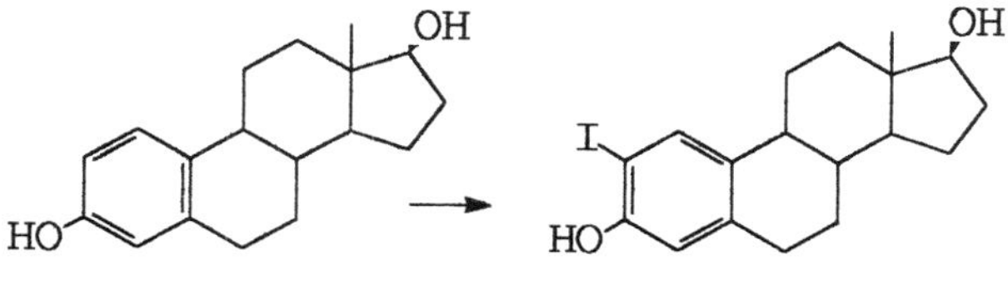

The preceding examples are all based upon chemical oxidation, although anodic oxidation of iodine in trimethyl orthoformate gives the iodonium ion, an effective iodinationg agent for appropriately substituted compounds (T. Shono et al., Tetrahedron Letters, 1989, 30, 1649).

(v) Nitrophenols

The development of regiospecific routes to nitrophenols continues to engage attention. The indirect nitration of phenol in ether/water (1:1), with sodium nitrate and concentrated hydrochloric acid and reaction at ambient temperature during 12 hours produces 2- and 4-nitrophenol (3:2) in 65% yield (D. Gaude, R. Le Goaller and J.L. Pierre, Synth. Commun., 1986, 16,

63).
4-Methylphenol in ethanol treated with ferric nitrate nonahydrate and briefly heated during 30 seconds gives a 83% yield of 4-methyl-2-nitrophenol (J-M. Poirer and C. Vottera, Tetrahedron, 1989, *45*, 1415).
2-Nitroestrone is derived in 79% yield by adding estrone to molar proportions of benzeneselenyl chloride and silver nitrate in acetonitrile and reaction over 3 hours at ambient temperature (H. Ali and J.E. van Liet, J. Chem. Soc. Perkin Trans I, 1991, 269).
The zwitterion, N-dodecylpyridinium-3-carboxylate when converted in acetonitrile suspension to the 'nitro ester', by the addition of an acetonitrile solution of nitronium fluoroborate, reacts with phenol and affords 2-nitrophenol in 95% yield (H. Pervez, L. Rees and C.J. Suckling, J. Chem. Soc. Chem. Commun., 1985, 512).

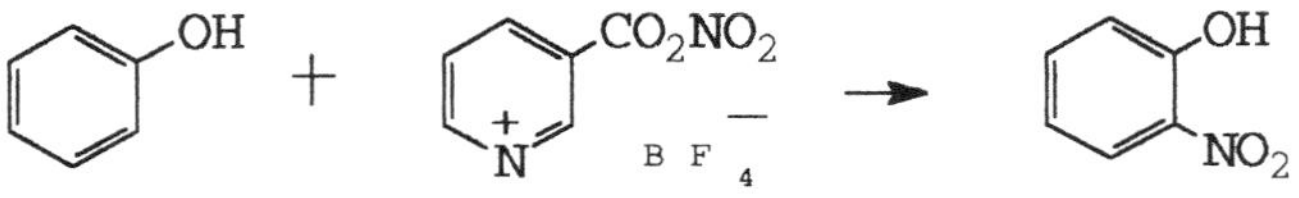

Non-aromatic precursors have been employed mostly for the synthesis of either highly substituted acylphenols (F. Eiden, H.P. Leister and D. Mayer, Arnzeim-Forsch, 1983, *33*, 101), or phenolic esters (E. Matsumura, M. Ariga and Y. Tohda, Bull. Chim. Soc. Japan, 1980, 53, 2891). Simpler 4-nitrosophenols are obtained from propanone by condensation with 3-nitrosopentane-2,4-dione (E. Belyaev et al., Zh. Org. Chim., 1978, *14*, 2189).

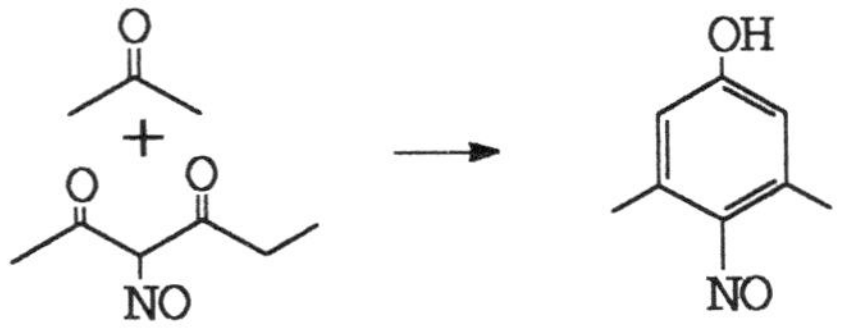

(vi) Aminophenols
Syntheses from benzenoid compounds, cyclohexenones, cyclohexanediones, heterocyclic sources and the use of C2-C4 intermediates have all found a place in the synthesis of aminophenols.
Anilines can been hydroxylated with hydrogen peroxide in superacids (J-C. Jacquesy et al., Bull. Soc. Chim. Fr., 1986, 625).

In a novel approach, N-methyl-N'-phenoxyurea rearranges in contact with trifluoroacetic acid to afford a 85% yield of N-(2-hydroxyphenyl)-N-methylurea from which 2-methylaminophenol is readily obtained (Y. Endo, K. Shudo and T. Okamoto, Synthesis, 1983, 471).

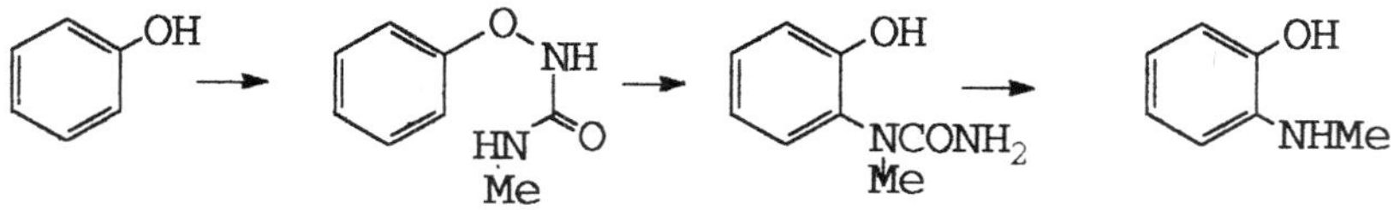

3-Propylaminocyclohex-2-enone in acetonitrile containing mercuric acetate upon heating gives a 78% yield of 3-propylaminophenol (H. Iida, Y. Yuasa and C. Kibayashi, Synthesis, 1982, 471). Cyclohexane-1,3-diones are submitted to this and other techniques to give aminophenols in yields in the range 32-79% (W.H. Mueller, USP 4212823). 3-Dimethylaminocyclohex-2-enone is catalytically dehydrogenated to 3-dimethylaminophenol in 98% yield in the vapour phase at 300°C over a palladium-potassium-charcoal catalyst (Hoechst AG, EP 65733).

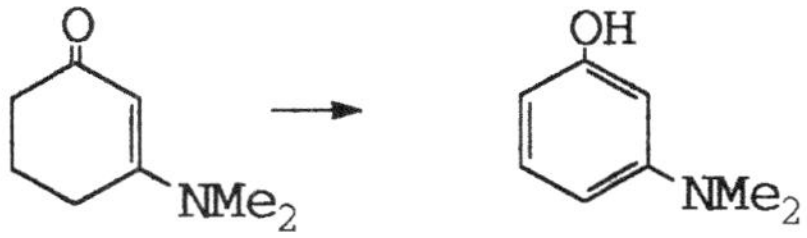

The combination of two acyclic C3 precursors is widely employed for the synthesis of 3-aminophenols. Appropriate ethoxymethylene compounds having cyano or alkoxycarbonyl groups have been condensed with β-ketoesters, malononitrile or cyanoacetic esters to afford substituted aminophenols (R = CO_2R, alkyl; R^2 = CN, CO_2R; R^1 = COMe, CO_2R) in good yield (H.W. Schmidt and H. Junek, Justus Liebig's Ann. Chem., 1979, 2005; S.R. Baker et al., J. Chem. Soc. Perkin Trans I, 1979, 677).

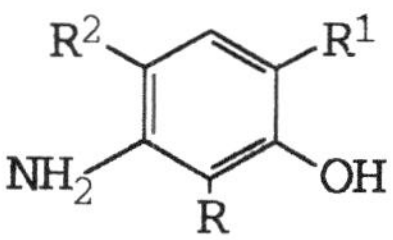

The reactions of enamines has been used as a route to aminophenols (H. Bohme, J.B. von Gratz, F. Martin, R. Matusch and J. Nehne, Justus Liebig's Ann. Chem., 1980, 394) and is improved, with avoidance of dimerisation by the incorporation of a silyl group in one of the participating enamines (T.H. Chan and G.J. Kang, Tetrahedron Letters, 1983, 24, 3051).

The direction of cyclisation, in dichloromethane containing titanium(IV) chloride, is dependent upon the structure of the co-reacting enamines. Those from acyclic ketones afford aminophenols by 3C + 3C addition whereas enamines from cyclic ketones (n = 2-5) give bicyclic products by 4C + 2C addition.

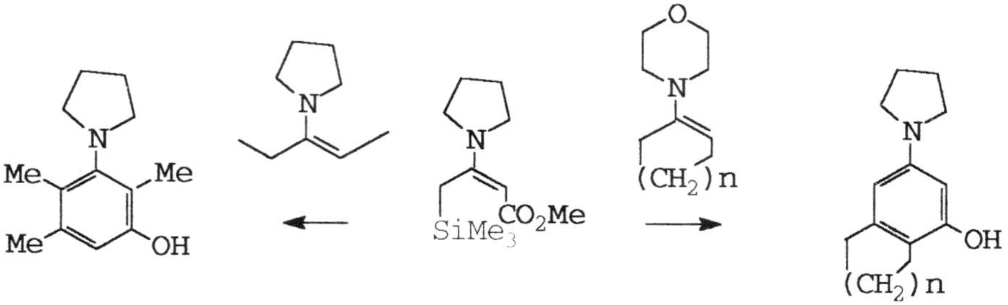

More complex substituted aminophenols are constructed from aldimines and dicarbonyl compounds as in the case shown below where a large excess of the imine reacts with a diketo ester to furnish the carbazole in 63% yield (H.H. Wasserman, J.A. van Dozer and C.B. Vu, Tetrahedron Letters, 1990, *31*, 1609).

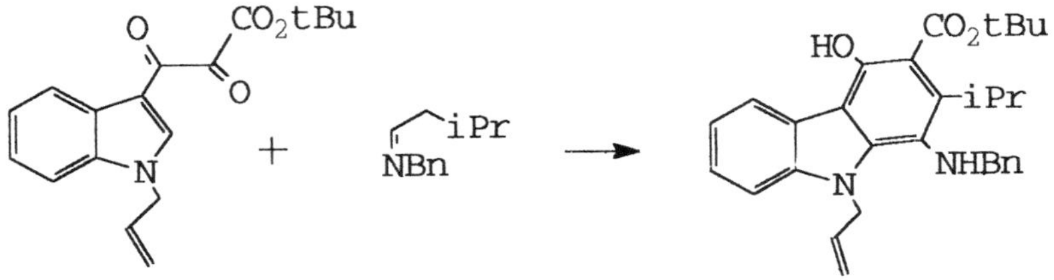

Heterocyclic intermediates can also be used such as bis(isoxazol-5-yl)methanes which rearrange to give 2-amino-4-hydroxyphenyl ketones on heating (S. Auricchio, S. Morrochi and A. Ricca, Tetrahedron Letters, 1974, 793). Reactions of diacylpyrones with amines also afford aminophenols in good yield (F. Eiden, E.G. Teure and H.P. Leister, Arch. Pharm., 1981, *314*, 419).

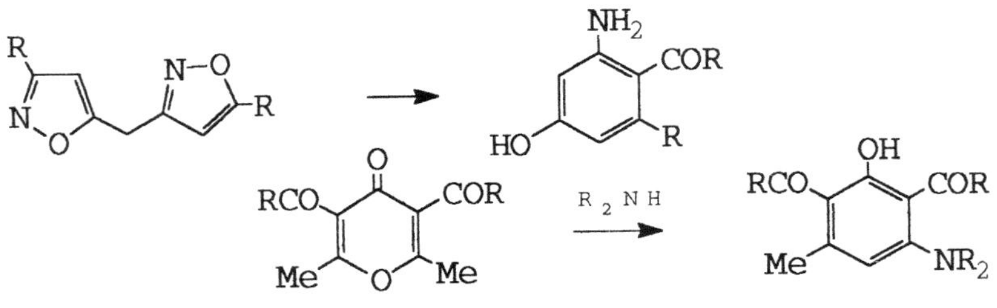

(vii) Hydroxyazo and Hydroxydiaryl Compounds

Diazo aryl ethers are synthesised by the addition of aqueous sodium selenite to an aqueous solution of the diazonium salt at 0 to -5°C, followed by that of the sodium phenate (E.P. Nesynov and T.F. Aldhokhina, Zh. Obsch. Khim., 1981, 52, 432).

$$ArN_2^+ Cl^- + Na_2SeO_3 \rightarrow [ArN=NOSe(O)ONa] \xrightarrow{ArONa} ArN=NOAr$$

Biaryl formation is effected by displacement of the phenylthioazo group. 4-Cyanophenylazo phenyl sulphide with tetra n-butylammonium phenoxide in dimethyl sulphoxide reacts at ambient temperature to afford a mixture of 2-(4-cyanophenyl)phenol and its 4- isomer in 60% yield (G. Petrillo, M. Novi and C. Dell'Erba, Tetrahedron Letters, 1989, 30, 6911).

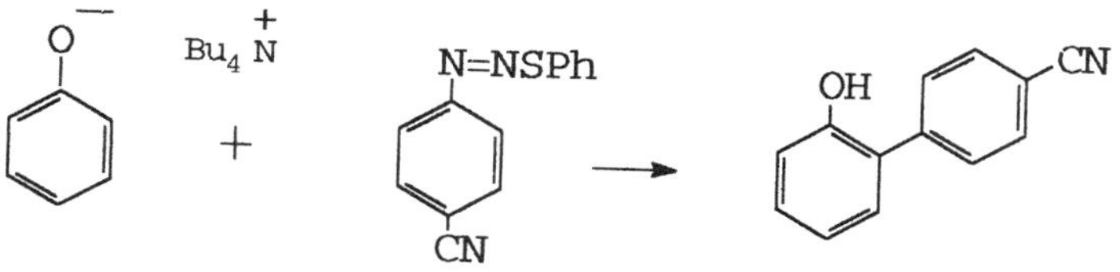

(viii) Thio Derivatives and Thiophenols

Sulphur halides and dialkyl disulphides are employed to obtain a variety of arylthio and alkylthio derivatives of phenols, generally by electrophilic substitution in the benzene series. Phenol and zirconium tetrachloride heated to 156°C, upon addition of diethyl disulphide afford 2-(ethylthio)phenol in 41% yield (du Pont de Nemours Co., USP 4599451). An aluminium salt has also been employed for this type of process and with dimethyldisulphide, 2-(methylthio)phenol is obtained in 41% yield (P.F. Ranken and B.G. McKinnie, Synthesis, 1984, 117).

For the preparation of diphenyl sulphides sulphur dichloride and zinc chloride can be used under milder conditions. For example, 4-methylphenol in hexane containing a small proportion of $ZnCl_2$ is kept at 0-10°C during the addition of SCl_2, and gives di-(2-hydroxy-5-methylphenyl)sulphide in 87% yield (Mitsui Toatsu Chem. Co., NL 7906800).

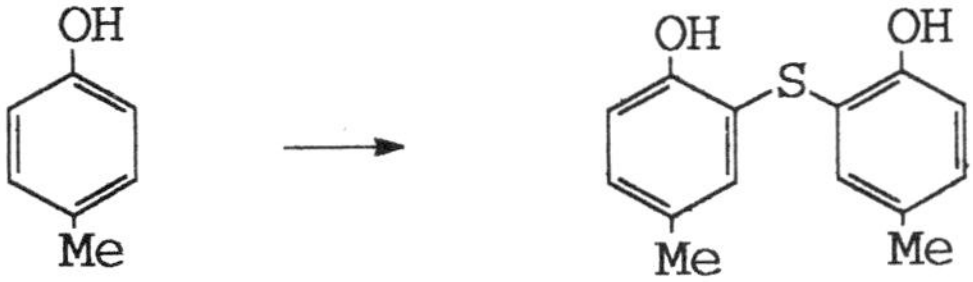

[Scheme: 4-methylphenol (p-cresol) → bis(2-hydroxy-5-methylphenyl) sulfide]

Thiophenols can be obtained from phenolic sources by a number of processes (J.H.P. Tyman, 'Thiophenols', 2nd. Edn. Supplement, Rodd's 'Chemistry of Carbon Compounds', Vol.IIIA, Ch5, p247 Ed. M.F. Ansell, Elsevier, Amsterdam, 1983, Ch.5, p247). They are susceptible to reaction either at the sulphur atom or in the ring.

Thus n-octyl phenyl sulphide is formed in 98% yield at ambient temperature from 1-bromooctane and potasium thiophenoxide in methyl 2-pyridylsulphoxide (N. Furukawa et al., Heterocycles, 1981, 16, 1927).

$$n\text{-}C_8H_{17}Br + PhSK \rightarrow n\text{-}C_8H_{17}SPh$$

Since the phase-transfer method is unsatisfactory for the preparation of n-butyl phenyl sulphide, it is necessary to pass a mixture of thiophenol and 1-bromooctane through a solid bed containing sodium bicarbonate and a little carbowax 6000 at 170°C and 20mm. Hg pressure to furnish a 71% yield of this product (conversion 95%) (E. Angeletti, P. Tundo, and P. Venturello, J. Chem. Soc, Perkin Trans I, 1982, 1137).

Substitution o- to the thiol group is employed for the synthesis of thiochromones. For example, 3-ethoxycarbonyl-4-piperidone and thiophenol in polyphsphoric acid, heated at 100°C, undergo cyclocondensation to afford piperidino[3,4-b]thiochromone in 54% yield (Roussel Uclaf, EP 99303).

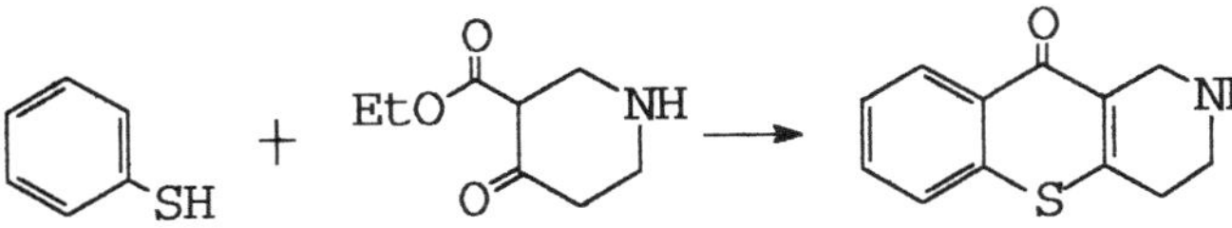

[Scheme: thiophenol + ethyl 4-oxopiperidine-3-carboxylate → piperidino[3,4-b]thiochromone]

The selective dealkylation of an alkylthiophenyl alkyl ether containing both groups can be achieved by the appropriate choice of reagent. Thus, 3-(methylthio)anisole in hexamethylphosphorictriamide treated with sodium iso-propylthiolate at 120°C gives 3-(methylthio)phenol in 88% yield, while sodium alone affords 3-methoxythiophenol (L. Testaferri et al., Tetrahedron, 1982, 38, 2721).

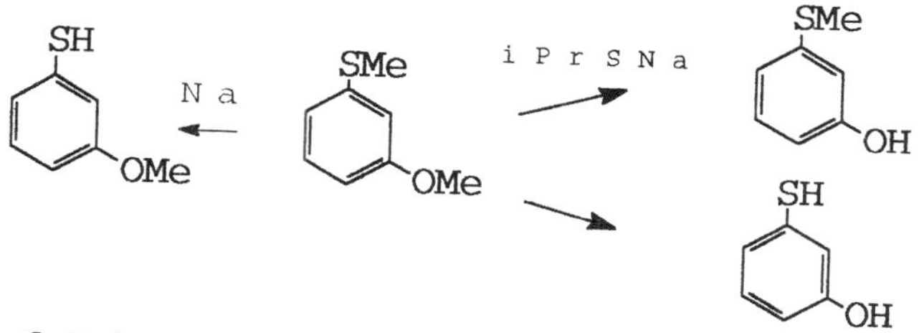

Reactions of Halogeno, Nitro, Amino, and Thio Derivatives of Phenols

(i) Fluorophenols

Methyl 2-(3-fluoro-4-hydroxyphenyl)propionate in DMF containing potassium iodide and carbonate gives the corresponding ether with 3-chloro-3-methyl-1-butyne, cyclisation of which in N,N-diethylaniline by refluxing at 215°C affords methyl 2-(2,2-dimethyl-8-fluoro-1,2-benzopyran-6-yl)propionate in 60% overall yield (Nippon Zoki Pharm., KK, EP 74170).

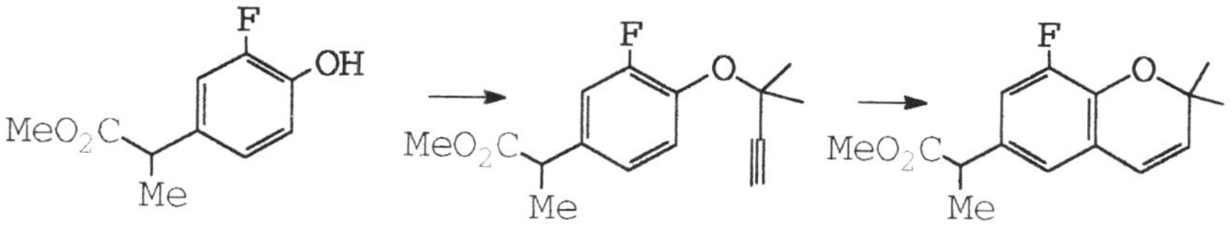

(ii) Chlorophenols

2-Chlorophenol and propargyl methyl ether in DMF containing triethylamine, a fluoroalkane sulphonyl fluoride and a catalytic amount of bis-(triphenylphosphine)palladium(II)dichloride, when heated, give 3-methoxy-1-(2-chlorophenyl)propyne in 63% yield rather than the expected aryl propynyl ether (Q-Y. Chen and Y-B. He, Synthesis, 1988, 896).

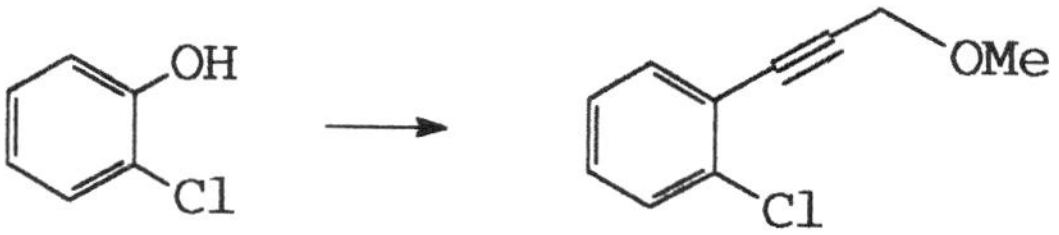

4-Chlorophenol undergoes condensation and cyclisation with 1-(3,3-dimethyltetrahydro-2-furfurylidene)-1,2,4-triazole and para-formaldehyde in trifluoroacetic acid by heating at 100°C giving 6'-chloro-3,3-dimethyl-3'-(1,2,4-triazol-1-yl)-spiro(tetrahydrofuryl)-2,2'-chroman in 62% yield (Bayer AG, EP 315007).

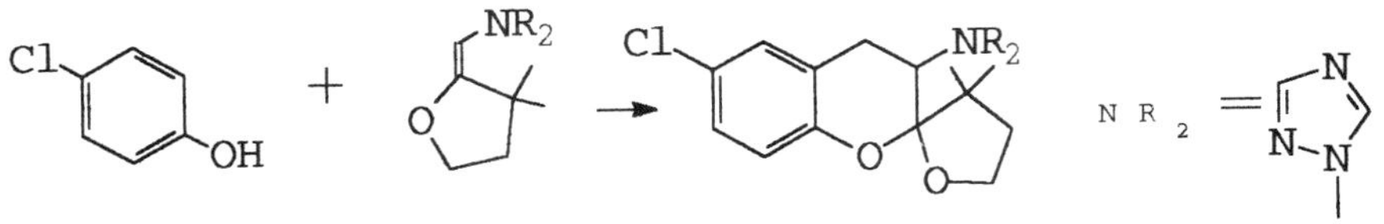

Methyl 5-chlorosalicylate with hexamethylenetetramine in the presence of methanesulphonic acid undergoes a Duff reaction to furnish the 3-formyl derivative in 77% yield (Y. Suzuki and H. Takahashi, Chem. Pharm. Bull., 1983, *31*, 1571).

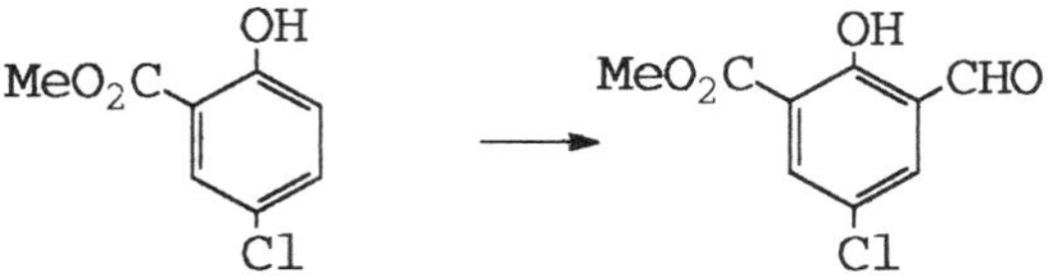

2-Benzyl-4-chlorophenol is formed in 95% yield by the Friedel-Crafts alkylation of 4-chlorophenol with benzyl alcohol at 200°C in the presence of Na-Y-Zeolite (Bayer AG, DE 3700917).

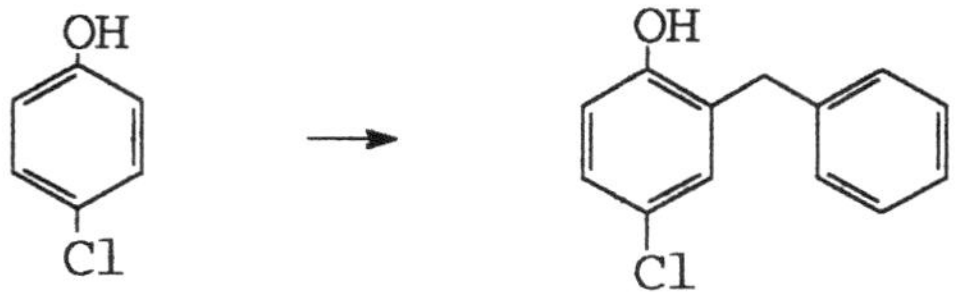

A reaction representing a Beckmann-like rearrangement as the final stage takes place when 2,6-dichlorophenol and N-phenylbenzimidoyl chloride in isopropyl ether containing methanolic sodium methoxide are heated together. The final product is N-phenyl-2,6-dichloroaniline (Fuso Kagaku Kogyo, KK, J58 144350).

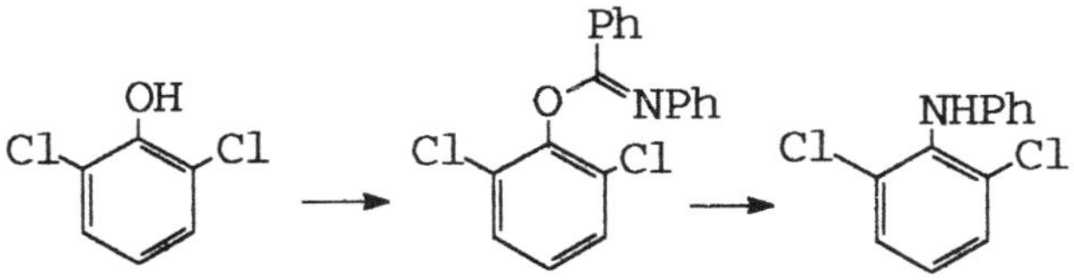

2,4,6-Trichlorophenyl acetate forms in 86% yield in a phase-transfer type procedure in which the chlorophenol in dichloromethane at 0°C containing aqueous sodium hydroxide and tetra n-butylammonium bromide is reacted with acetic anhydride (K-T. Liu and M-Y. Kuo, J. Chinese Chem. Soc., 1981, *28*, 209).

(iii) Bromophenols

The reaction of 4-bromophenol with excess formaldehyde occurs in the presence of fuming (46%) hydrochloric acid and results in a 92% yield of 6-bromo-8-chloromethyl-1,3-dioxan (V.M. Ledovskikh, Y.P. Shapovalova and N.V. Sumliuenko, Ukr. Khim. Zh., 1990, *56*, 406).

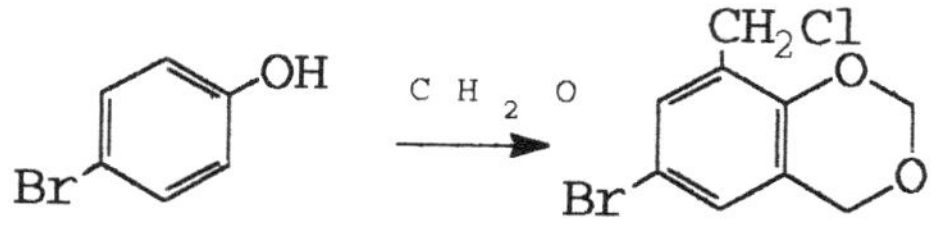

Electrophilic substitution processes effected either in solution or the vapour phase using supported reagents have been studied increasingly in recent years. An 88% yield of 2-nitro-4-bromophenol is realised from the agitation of 4-bromophenol in dichloromethane containing silica gel impregnated with nitric acid at ambient temperature for 3 minutes. Phenol itself gives a 1:1 mixture of 2- and 4-nitrophenols. This reagent enables reduced reaction times to be achieved, compared with similar reactions using ferric nitrate supported on clay (R. Tapia, G. Torres and J.A. Valderrama, Synth. Commun., 1986, *16*, 681).

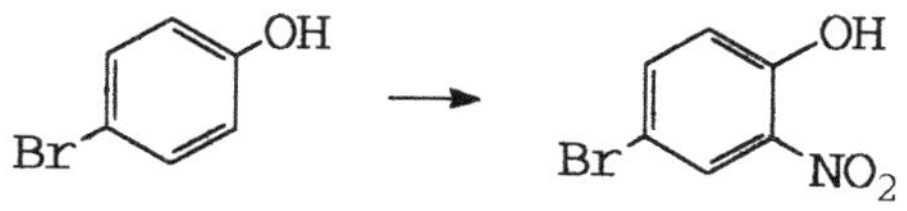

Methods for the removal of bromo groups generally consist of catalytic or chemical reduction by, for example, lithium aluminium hydride and there is interest in alternative selective procedures. Phenol is formed in 84% yield by the reaction of 2,4-dibromophenol in dichloromethane containing ethanethiol and aluminium chloride (M. Node et al., J. Org. Chem., 1984, *49*, 3641).

1-Bromo-2-naphthol is debrominated and esterified by heating with benzoic acid in the presence of triphenylphosphine and triethylamine during 3 hours at 150°C affording a 70% yield of 2-naphthyl benzoate (I. Furijkawa, N. Takenaka, and S. Hashimoto, Chem Express, 1989, *4*, 337).

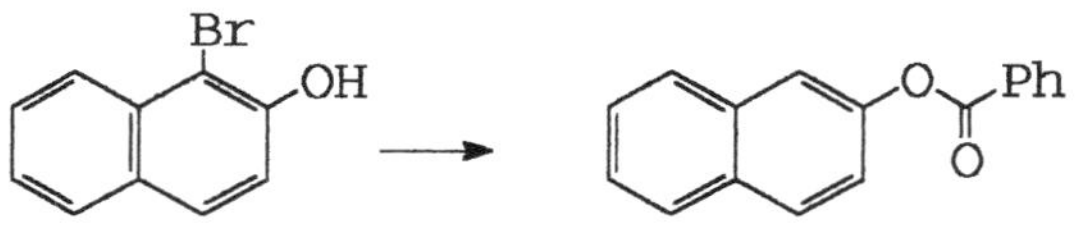

2,4-Dibromo-3,6-dimethylphenol upon nitration treatment in acetic acid gives 4,6-dibromo-2-hydroxy-2,5-dimethyl-3,6-dinitrocyclohexenone while 2,4,5-tribromo-3,6-dimethylphenol affords the 3-bromo analogue. By nitration in aqueous acetic acid ring-contracted cyclopentenes result from both reactants. The tribromo reactant undergoes the Favorsky reaction with aqueous sodium carbonate to give the same cyclopentene as that formed in the previous reaction (P.A. Bates et al., Tetrahedron Letters, 1981, 2325).

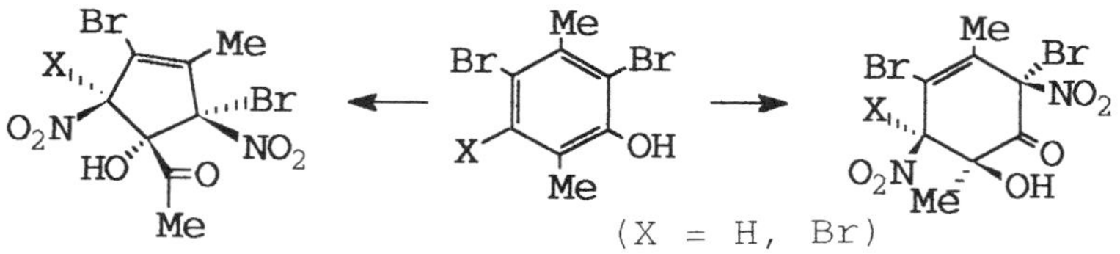

(X = H, Br)

(iv) Iodophenols

Here the main reactions are the displacement either of iodine atoms or of adjoining groups. 2-Iodoestradiol in DMF containing copper(II) chloride and methanolic sodium methoxide give a 95% yield of the 2-methoxy compound when heated under reflux for 5 minutes (M. Numazawaa and Y. Ogura, J. Chem. Soc. Chem. Commun., 1983, 533).

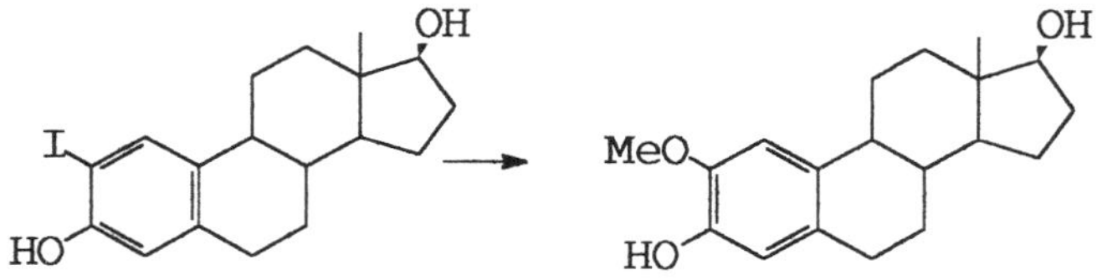

In contrast to the preceding sequence, 4-nitro-2,6-diiodomethoxybenzene in ethanol reacts with excess iso-propylamine giving a 70% yield of 2,6-diiodo-N-iso-propyl-4-nitroaniline by loss of the methoxyl group. The corresponding phenol results, by simple dealkylation, when a secondary amine is employed, a difference perhaps attributable to the degree of steric hindrance (J.F. Pilichowski and J.C. Gramain, Synth. Commun., 1984, 14, 1247).

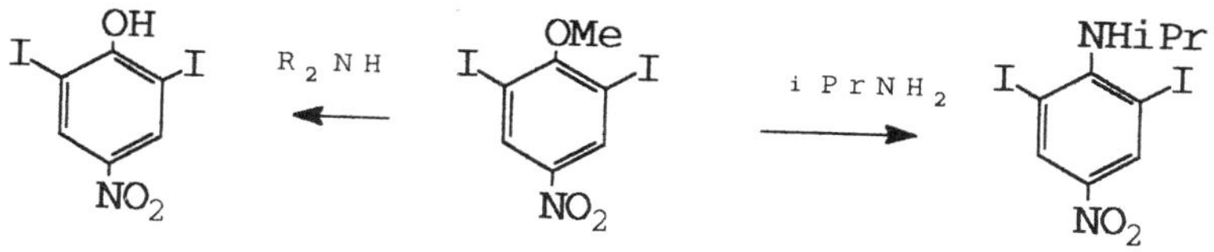

2-Iodophenol in DMF containing vinylcyclopropane, tetra n-butylammonium chloride and a little palladium(II) acetate together with some potassium acetate when heated gave, 2-allyl-2,3-dihydrobenzofuran in 70% yield (R.C. Larock and E.K. Yum, Synlett., 1990, 529).

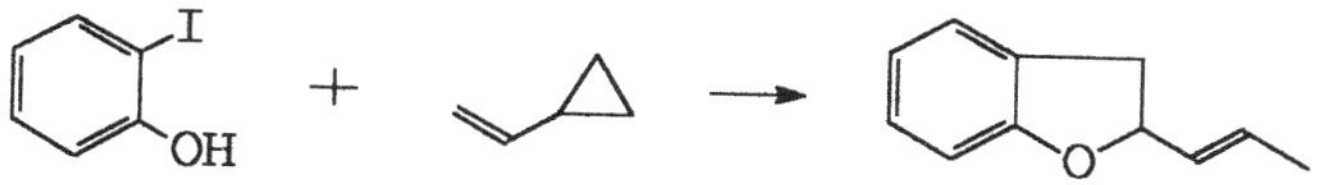

Selective complexation of iodophenols with β-cyclodextrin under aqueous conditions has been investigated (J. Anzai et al., J. Chem. Soc. Chem. Commun., 1985, 1023). The regiospecific deiodination of diiodophenols by n-butyllithium is an intrinsic step in the synthesis of aflatoxin B2 (S. Horne, G. Weeratunga and R. Rodrigo, J. Chem. Soc. Chem. Commun., 1990, 39).

(v) *Nitrophenols*
Reactions in this group comprise substitution of the nitro or the hydroxyl group and etherification of the latter.
2-Hydroxy-3,5-dinitrobenzoic acid reacts in DMF/phosphorus oxychloride (1:1) during 2 hours at ambient temperature giving more than a 90% yield of 2-chloro-3,5-dinitrobenzoic acid (N. Akhtar, M.A. Munawar and M. Siddiq, Indian J. Chem., 1986, *25B*, 328).

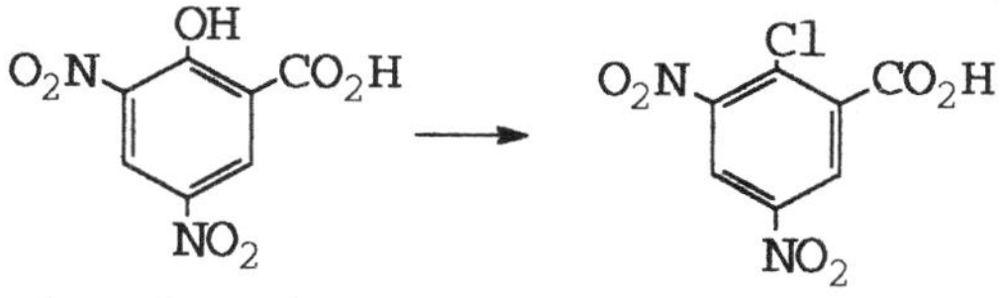

2-Chloro-4-nitrophenol is converted into 2-chloro-6-methoxymethyl-4-nitrophenol in 93% yield by treatment with methanol solution containing paraformaldehyde and concentrated sulphuric acid (Sumitomo Chemical, KK, J59 206336).

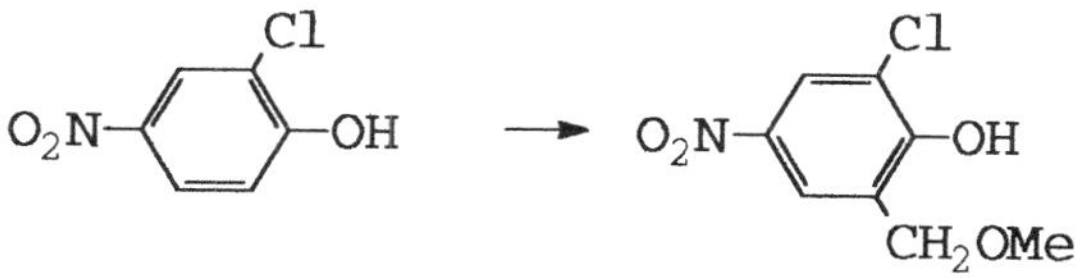

4-Nitrophenol (1 mole), together with 5 mol% palladium tetra(triphenylphosphine) in THF is employed for the ring-opening and rearrangement of an epoxide to an alkylidenecyclopentanocyclopentanone (S. Kim et al., Tetrahedron Letters, 1991, *32*, 3395).

(vi) Aminophenols

A range of reactions are exhibited at the hydroxyl group, the amino group or through substitution in the ring.
3-Diethylaminophenetole (3-diethylaminophenyl ethyl ether) is synthesised in 97.5% yield from a reaction of the parent phenol in methyl iso-butyl ketone containing powdered potassium carbonate and ethyl bromide (Ciba Geigy AG, EP 258190).

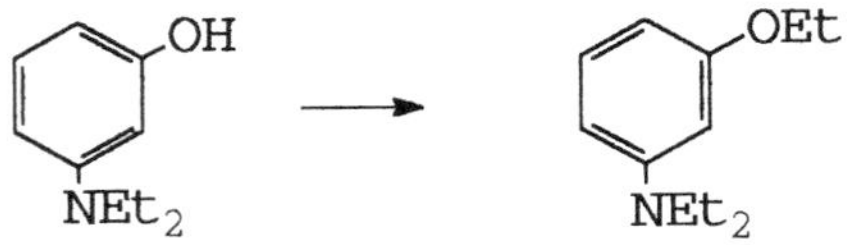

4-Acetylaminophenol and O-acetylsalicyloyl chloride in chloroform containing pyridine react to afford a moderate yield of 2-methyl-2-(4-acetylaminophenoxy)-1,3-benzodioxan-4-one (Medea Res. SRL, EP 258190).

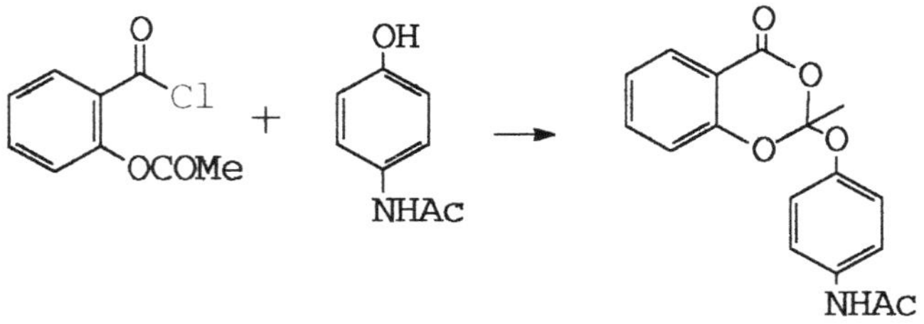

Specific N-acetylation occurs during the sequential addition of magnesium oxide and acetic anhydride to 2,4-diaminoanisole in methanol to give a 80% yield of 3-amino-4-methoxyacetanilide (Toms River Chemical Corpn., EP 11048).

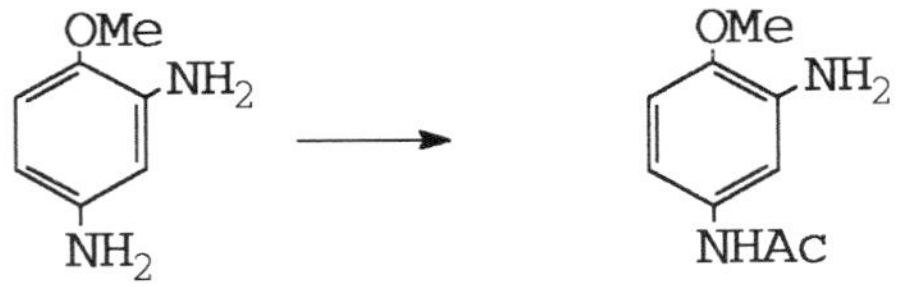

Formylation of 3-aminophenols leading to intermediates for the synthesis of coumarins and other carbonyl compounds has been described. Nuclear formylation of 3-dimethylaminophenol with hot ethyl orthoformate in the presence of 4-nitroaniline results in the formation of 2-hydroxy-4-dimethylaminobenzylidene-4-nitroaniline in 88% yield (P. Czerney, DD 156080).

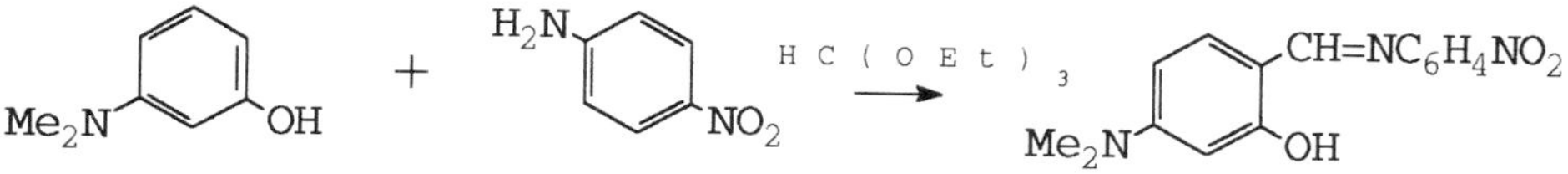

3-Ethylamino-4-methyphenol in DMF reacts with 3-amino-1-oxo-isoindoline hydrochloride at 120°C during 1 hour to afford 3-(4-ethylamino-2-hydroxy-5-methylphenyl)isoindolinone in 94% yield. Hydrolysis with 20% sodium hydroxide produced 2-(4-ethylamino-2-hydroxy-5-methylbenzoyl)benzoic acid in 78% yield (BASF AG, DE 3800577).

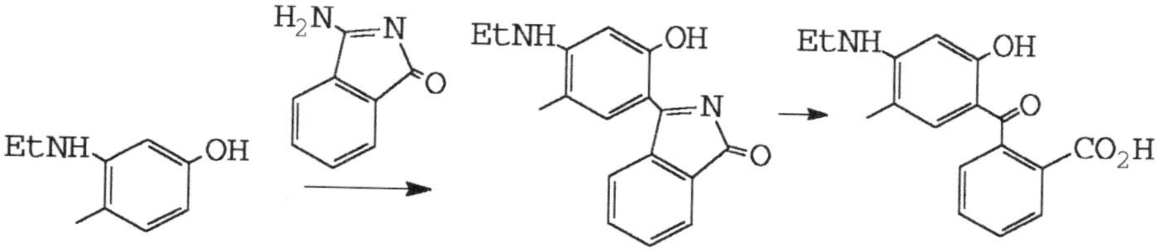

o-Directed metallation, followed by carbonation of 3-aminophenol, after protection as the O-methoxymethoxy and N-pivaloyl derivative, leads to an efficient synthesis of 6-hydroxyanthranilic acid (L.R. Hillis and S.J. Gould, J. Org. Chem., 1985, 50, 718).

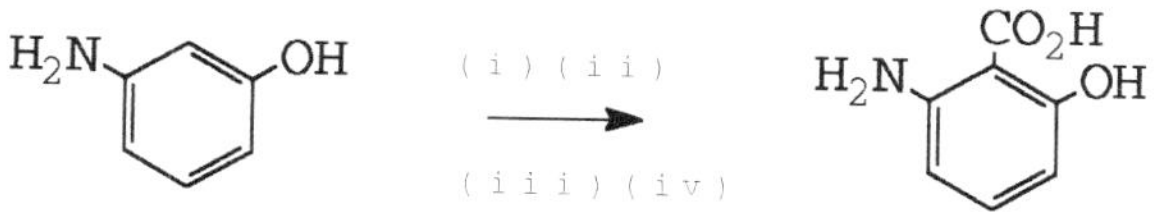

(i) HCO_3^-, (Me)$_3$CCOCl (ii) OH$^-$, ClCH$_2$OMe (iii) BuLi, CO_2, H_3O^+ (iv) HCl, MeOH

Oxidative arylation of 1,4-naphthoquinone takes place with 3-diethylaminophenol by reaction at 50°C in acetic acid solution containing copper(II) acetate monohydrate giving a 71% yield of 2-[2-hydroxy-4-(diethylamino)phenyl]-1,4-naphthoquinone (Ricoh KK, JO 3264575).

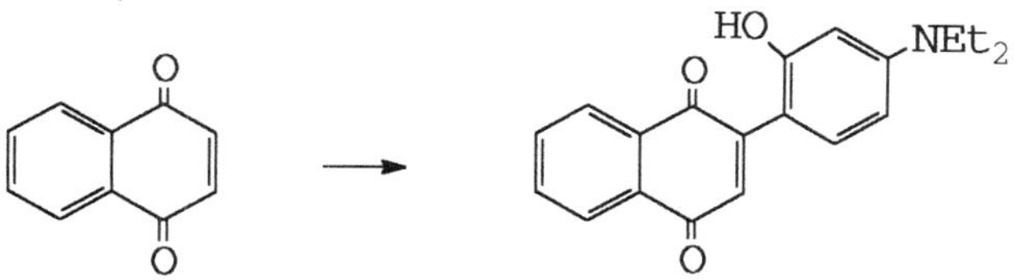

Coumarins from 3-aminophenols:
A variety of methods has been described for the introduction of the 7-substituted amino group into the resulting highly fluorescent substances which have a range of industrial uses. 3-Pyrrolidinophenol is converted into the 6-formyl derivative with ethyl orthoformate and thence to the benzylidene derivative with 4-nitroaniline. This product reacts with acetoacetic esters in the presence of piperidine to afford 7-pyrrolidinyl-3-acetylcoumarin in 94% yield (P. Czerney and H. Hartmann, J. Prakt. Chem., 1982, *324*, 21).

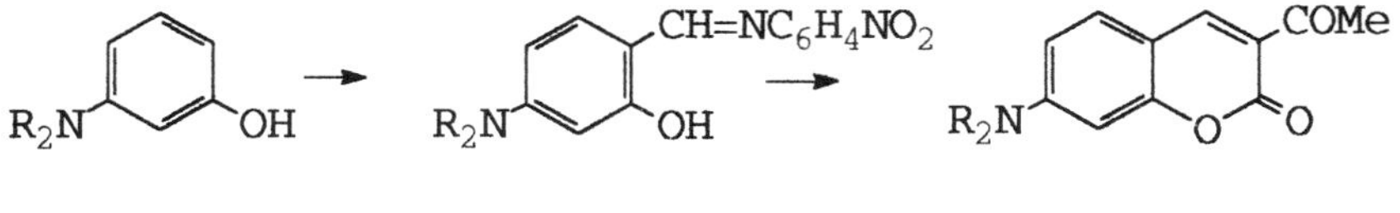

(R_2N = Pyrrolidinyl)

3-Ethylamino-4-methylphenol and diethyl ethoxymethylene malonate in boiling THF containing titanium tetrachloride give a 81% yield of 7-ethylamino-6-methyl-3-ethoxycarbonylcoumarin (US Dept. Energy, USP 6340958).

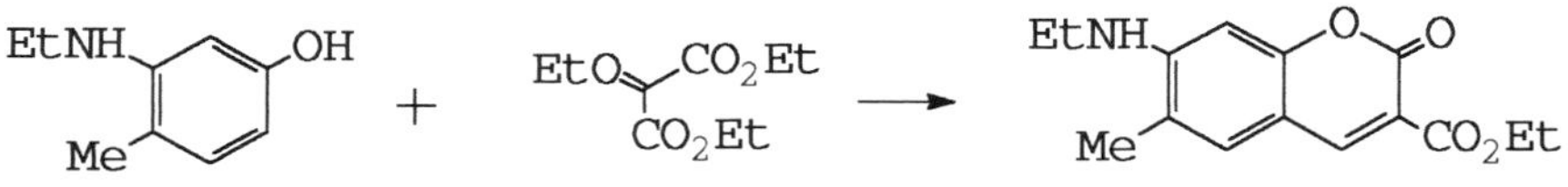

3-Diethylaminophenol with 2-chloro-1-cyanoacrylonitrile in diethyl ether initially, and subsequently in boiling aqueous ethanol forms 7-diethylamino-3-cyanocoumarin in a 56% yield (A. Ya. Ilchenko et al., Zh. Org. Khim., 1981, *17*, 2630).

3-Aminophenol and ethyl 4,4,4-trifluoroacetoacetate upon heating in ethanol containing zinc chloride produce 7-amino-4-trifluoromethylcoumarin in 75% yield (E.R. Bissel, A.R. Mitchell and R.E. Smith, J. Org. Chem., 1980, *45*, 2283).

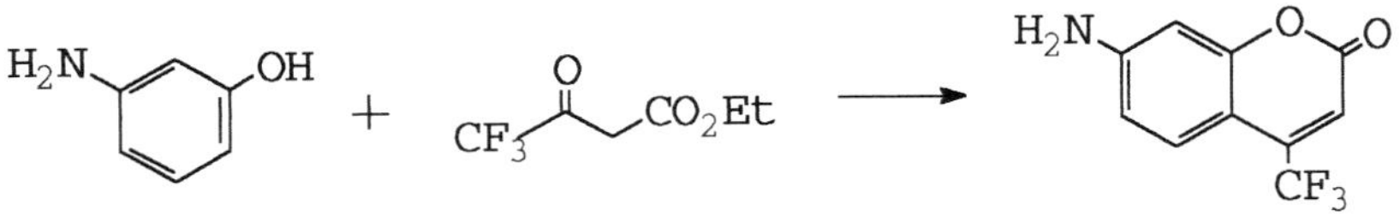

3-Dimethylaminophenol and 4-chloro-3-formylcoumarin together with 70% perchloric acid heated in acetic acid during 1 hour afford 11-dimethylaminocoumarino[4,3-b]-1-benzopyrylium perchlorate in 60% yield (Karl Marx Univ., Leipzig, DD 272852).

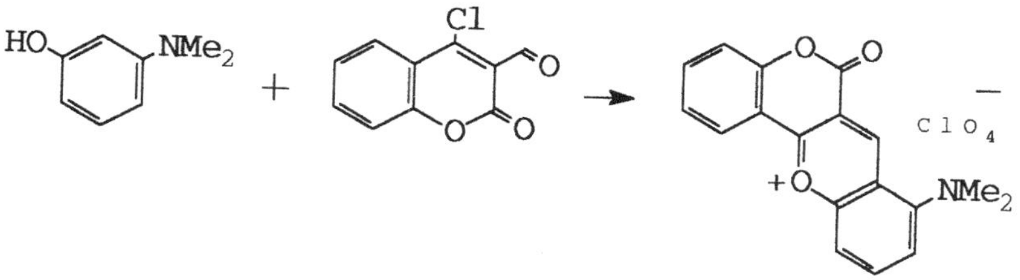

3-Diethylaminophenol and the dimethyliminium perchlorate of 2-aryl-2-chloropropenal when boiled together in acetic acid afford a flavylium perchlorate in moderate yield (J. Liebscher, DL 140353).

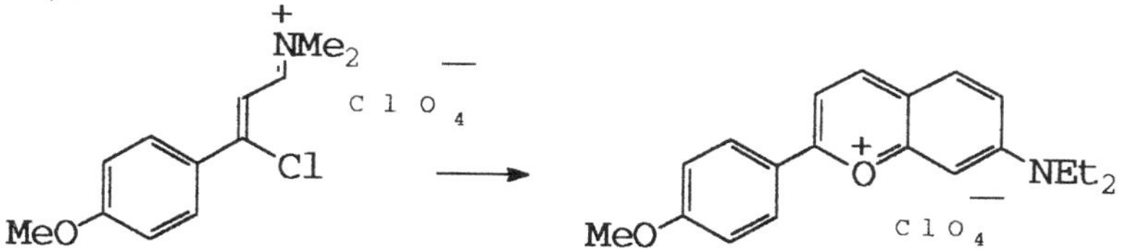

2-Aminophenols are valuable precursors of heterocycles. Thus benzoxazoles are obtained from N-acyl-2-aminophenols. For example, 2-hydroxymaleanilic acid upon heating with triethylamine in methanol affords a nearly quantitative yield of methyl [3,4-dihydro-1,4-benzoxazin-3(2H)-one-2-yl]acetate (T. Teitei, Austral. J. Chem., 1986, *39*, 502).

2-Aminophenol reacts with phenyl chloroformate in methanol/water (1:1) containing sodium bicarbonate, followed by treatment of the intermediate carbamate with aqueous sodium hydoxide, to give benzoxazolin-2(3H)-one in 83% yield (R.J. Maleski, C.F. Osborne and S.M. Cline, J. Heterocyc. Chem.,1991, *28*, 1937).

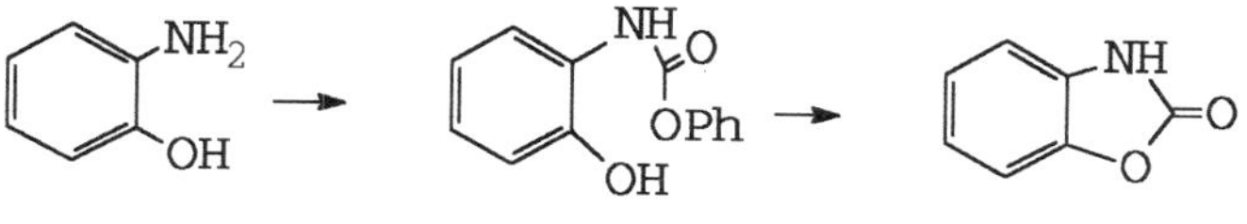

(vii) Phenolsulphonic Acids

Sodium 4-hydroxybenzenesulphonate dihydrate is employed to

transacylate phenyl nonanoate. The sodium sulphonate in octane/hexadecane when heated with phenyl nonanoate affords sodium nonanoyloxybenzene sulphonate in 87% yield (Ethyl Corporation, EP 125641).

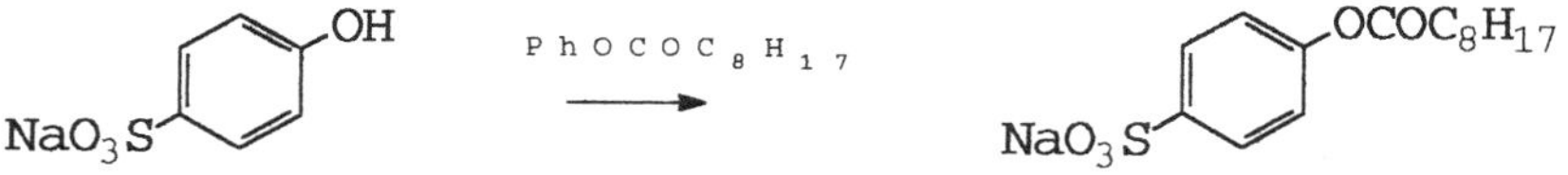

$$\text{NaO}_3\text{S-C}_6\text{H}_4\text{-OH} \xrightarrow{\text{PhOCOC}_8\text{H}_{17}} \text{NaO}_3\text{S-C}_6\text{H}_4\text{-OCOC}_8\text{H}_{17}$$

An alternative method consists of heating nonanoic acid with acetic anhydride, sodium phenolsulphonate and some sulphuric acid. In this case the yield is 96% (Proctor and Gamble Co., EP 1056730).

2 Dihydric Phenols

In this section advances in the chemistry of 1,2-dihydroxybenzenes (catechols), 1,3-dihydroxybenzenes (resorcinols), 1,4-dihydroxybenzenes (hydroquinones) and dihydroxynaphthalenes are discussed.

Synthesis

(i) 1,2-Dihydroxybenzenes (catechols) and their Derivative
There has been considerable interest in the hydroxylation of phenol. The oxygenation of phenol in acetonitrile in the presence of catalytic quantities of copper(I) chloride affords catechol in yields of 70-90% (P. Capdevielle and M. Maimy, Tetrahedron Letters, 1982, *23*, 1577). With hydrogen peroxide a similar transformation is effected (S.H. Dai et al., J. Org. Chem., 1985, *50*, 1722).

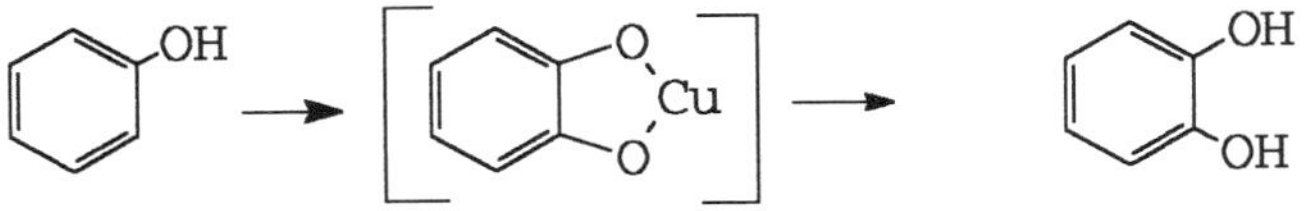

3-Phenylcatechol is obtained by heating 3-phenylcyclohexane-1,2-dione with 4-toluenesulphonyl chloride, potassium carbonate and a small proportion of azoiso-butyronitrile. Catechol is similarly obtained from cyclohexane-1,2-dione (A. Fergenbaum, J-P. Pete and A. Poquet-Dhimane, Tetrahedron Letters, 1988, *29*, 498).

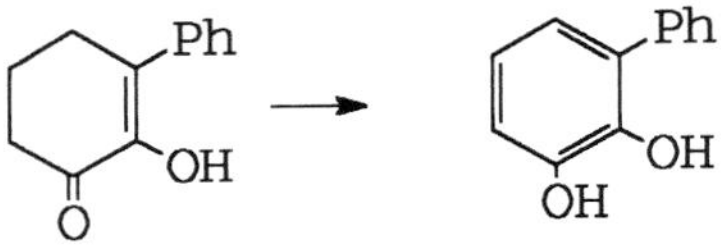

3-Alkylcatechols are synthesised through ring-opening and recyclisation of 2-alkanoyl-2,5-dimethoxytetrahydrofurans (T. Miyakoshi and H. Togashi, Synthesis, 1990, 407) by an adaptation of an earlier method (W.R. Boehme, J. Am. Chem. Soc., 1960, 82, 498). Typically furfural reacts with RCH_2MgBr to afford a carbinol, oxidation of which with manganese dioxide, followed by electrochemical oxidation in methanolic solution and catalytic reduction gives an acyltetrahydrofuran. This opens and ring closes again to give the catechol, although on brief exposure to acid the intermediate dioxoaldehyde can be isolated.

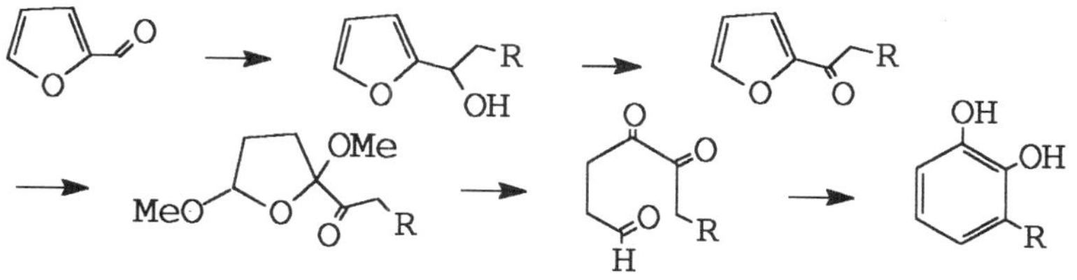

In the alizarin series a Diels-Alder adduct from naphtho-1,4-quinone and 1-methoxy-3-methyl-1-trimethylsilyl-1,3-butadiene can be oxidatively converted with accompanying dehydration to afford 3-methyl-1,2-dihydroxyanthra-9,10-quinone in 92% yield (D.W. Cameron, G.I. Feutrill, G.R. Gamble and J. Stravrakis, Tetrahedron Letters, 1986, 26, 4999).

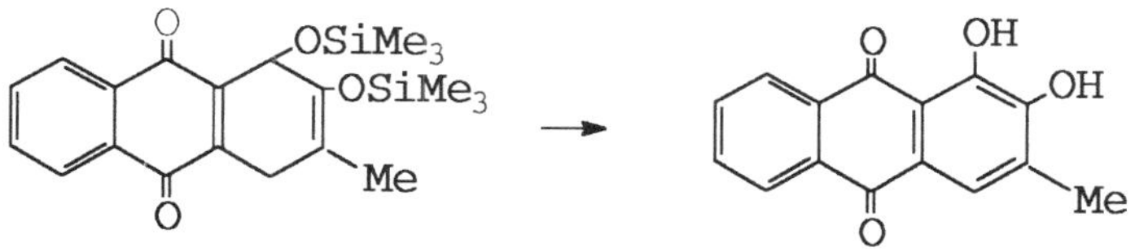

(ii) Mono O-Alkyl and Acyl Derivatives of Catechol

2-Methoxyphenol (guaiacol) has been synthesised in nearly quantitative yield from catechol, sodium acetate and acetic acid by heating an aqueous methanolic solution in a sealed tube at 250°C (Rhone-Poulenc Industries, EP 37353).

The mono 4-methoxybenzoylacetyl derivative of catechol is produced in 92% yield by heating 5-(4-methoxyphenyl-2,3-dihydofuran-2,3-dione with catechol in benzene (Y.S.

Andreichikov et al., Zh. Org. Khim., 1980, 16, 3087).

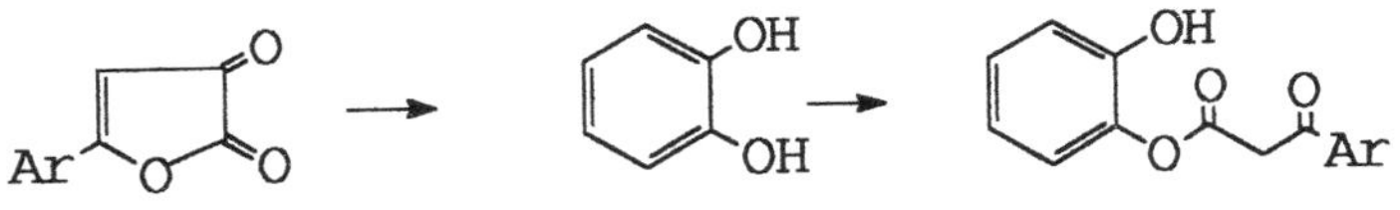

(Ar = 4-MeOC$_6$H$_4$)

Catechol monobenzoate is synthesised by treatment of O-phenyl-N-benzoylhydroxylamine in trifluoroacetic acid with trifluoromethanesulphonic acid (Y. Endo, K. Shudo and T. Okamoto, Synthesis, 1980, 4612).

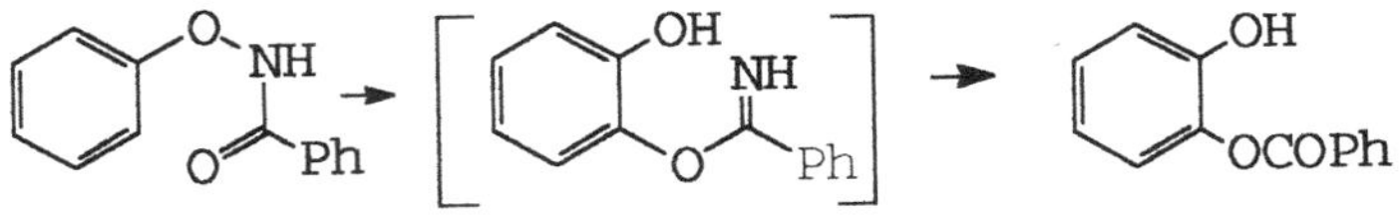

Catechol monoethers are prepared from polychlorocyclohexanes by dehydrochlorination through treatment with bases (van Winckel, USP 4267388). For example, 2,2,6-trichlorocyclohexanone and sodium methoxide afford 2-methoxyphenol. This is also obtained in 85% yield from 2-bromophenol with the same reagent at 125°C under CO_2 in the presence of a little $Cu(OH)_2 \cdot CuCO_3$ (D. Nobel, J. Chem. Soc. Chem. Commun., 1993, 419)

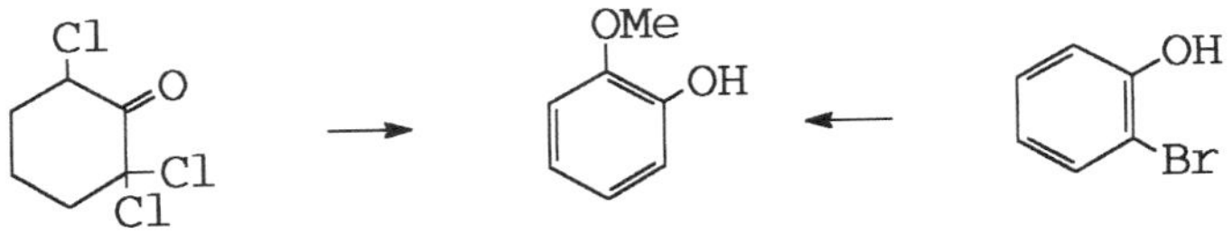

(iii) 1,3-Dihydroxybenzenes (resorcinols) and their Derivatives

A solution of 4-methylphenol in antimony pentafluoride/hydrogen fluoride (2:5) treated with hydrogen peroxide gives 4-methylresorcinol in 78% yield (J-P. Gesson, J-C. Jacquesy and M-P. Jouannetaud, J. Chem. Soc. Chem. Commun., 1980, 1128).

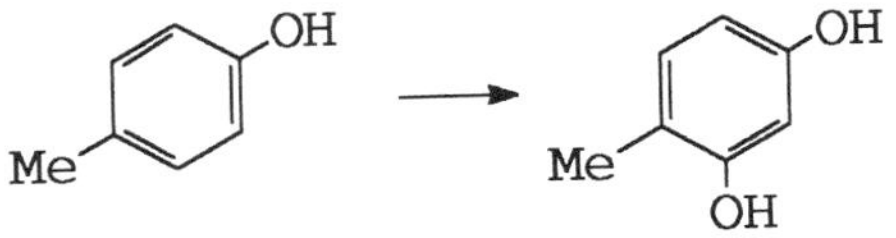

5-Methylresorcinol monomethyl ether is synthesised in 71% yield from excess 3-methylcyclobut-2-enone and methoxyacetylene in the presence of 2,4,6-tri-tert-butylphenol heated in a sealed tube (R.L. Danheiser and S.K. Gee, J. Org. Chem., 1984, 49, 1672).

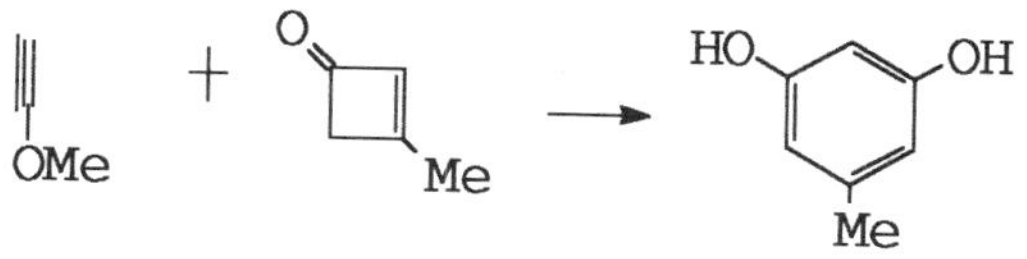

5-Phenylresorcinol is produced in 88% yield by a type of Horner-Emmons reaction in which 4-methoxy-6-phenyl-2-pyrone is added to the anion of dimethyl phosphonate (C. Tanyeli and O. Tarhan, Synth. Commun., 1989, *19*, 2749).

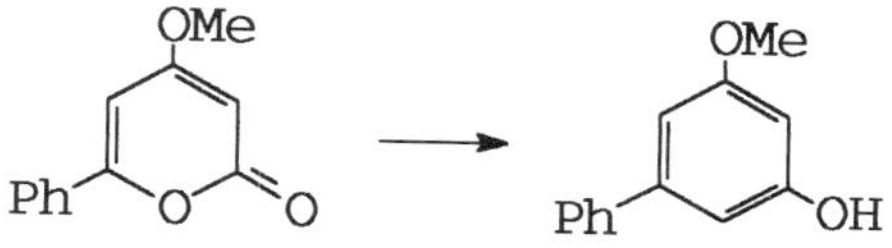

4-Alkylresorcinol derivatives are obtained by the Michael addition of the carbanion of α-phenylsulphinyl-γ-butyrolactone with methyl vinyl ketone, followed by heating (Y. Ozaki and Sang-Won Kim, Synth. Commun., 1985, *15*, 1171).

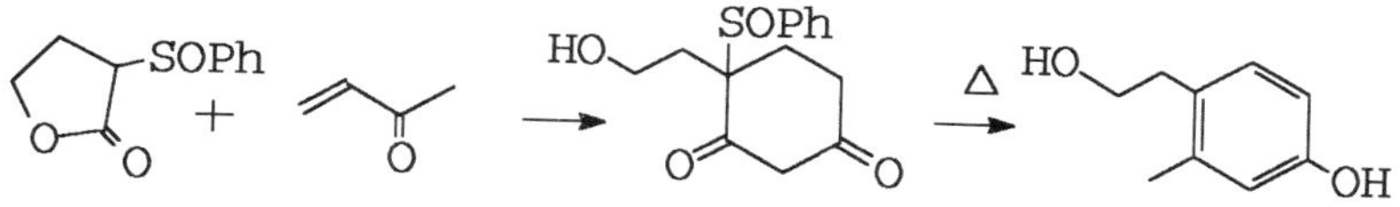

This method is related to an earlier synthesis from $PhSOCH_2COCH_2R$ and $PhCH=CHCOMe$ (A.A. Jaxa-Chamiec, P.G. Sammes and P.D. Kennewell, J. Chem. Soc. Perkin Trans I, 1980, 1701). 2-Substituted resorcinol dimethyl ethers are synthesised by the 2-demethoxylation of 1,2,3-trimethoxybenzene by electron-transfer from alkali metals and reactions with electrophiles (U. Azzena et al., J. Org. Chem., 1990, *55*, 5386).
5-Methylresorcinol dimethyl ether is produced from 5-methylcyclohexane-1,3-dione in methanol by the addition of iodine and heating (A.S. Kotnis, Tetrahedron Letters, 1991, *32*, 3441).

Resorcinol Carboxylic Acids and other Carbonyl Derivatives:
The facile decarboxylation of resorcylic acids particularly of

orsellinic and its homologous acids (5-alkylresorcinol-4-carboxylic acids) notably in alkaline solution provides a route to 5-alkylresorcinols. For the acids the synthesis of alkyl orsellinates is adopted and the free acids then obtained by acidic hydrolysis. Various methodologies for obtaining orsellinic acids are summarised in the scheme shown. Homologous alkyl acetoacetates with thallium ethoxide or sodium hydride, followed by reaction with diketene (route a) afford the corresponding alkyl orsellinates (T. Kato, M. Sato and T. Hozumi, J. Chem. Soc. Perkin Trans I, 1979, 529). In a related method, methyl orsellinate (R = Me) is formed from the monoanion with the dianion of methyl acetoacetate(R = R'= Me, route b) (P.E. Sum and L. Weiler, J. Chem. Soc. Chem. Commun., 1977, 91). A similar methodology (route c) uses the dianion from a diketone with the anion of dimethyl malonate (J.E. Hill and T.M. Harris, Synth. Commun., 1982, *12*, 621). The trimethylsilyl ether from methyl acetoacetate with a ketalised acid chloride affords the methoxycarbonyl derivative of olivetol (R = C_5H_{11}, route d) (A.G.M. Barrett, T.M. Morris and D.H.R. Barton, J. Chem. Soc. Perkin Trans I, 1980, 2272) and pentane-2,4-dione with dimethyl carbonate furnishes methyl orsellinate (ibid.) In a biomimetic approach a tetraketone (R = Me, route e) has been enzymically cyclised to give the corresponding orsellinic acid (R' = H) (T.M. Harris and C.M. Harris, Tetrahedron, 1977, *33*, 2159).

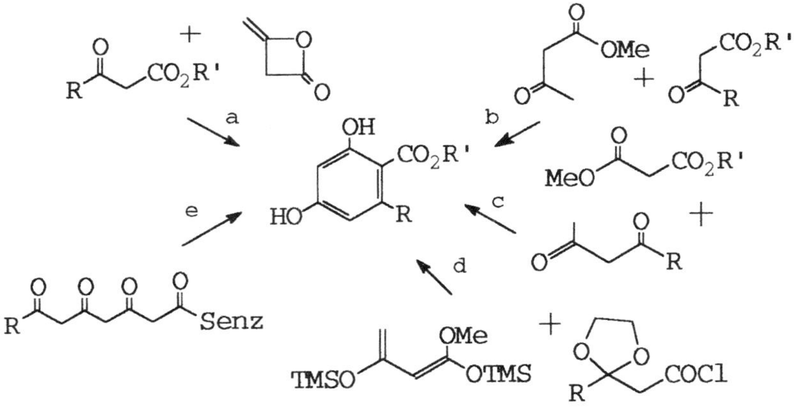

In contrast to the preceding strategies 2-methoxycarbonyl resorcinol compounds can be synthesised from the trimethylsilyl derivative of methyl acetoacetate and the dimethyl acetal of methyl 3-oxooctanoate (T.H. Chan and T. Chaly, Tetrahedron Letters, 1982, *23*, 2935).

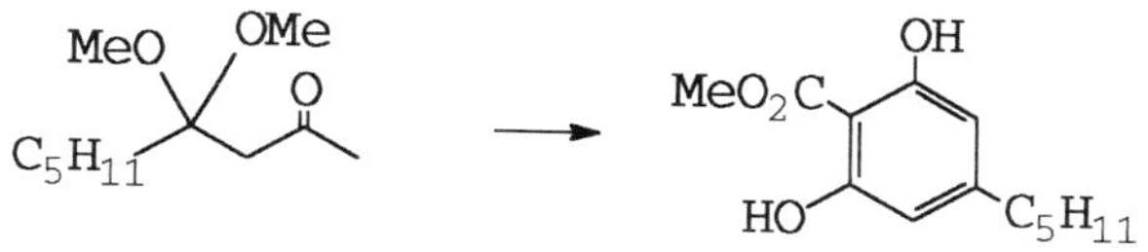

2,5-Dimethyl derivatives are synthesised by the Michael addition of the enol acetate of hexane-3,4-dione, with the anion of dimethyl malonate (K.K. Light, USP 3928419).

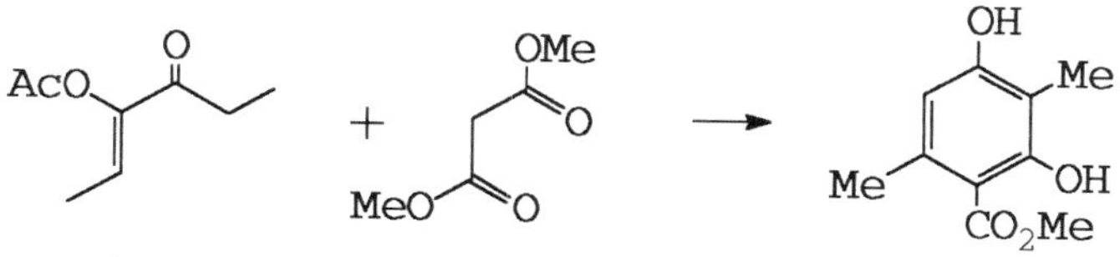

In the previously described procedure for orsellinic acid (G.M. Gaucher and M.G. Shepherd, Biochem. Prep., 1971, *13*, 70), the dihydro intermediate was aromatised by a bromination/dehydrobromination step. This is obviated in the preceding method by the presence of an acetoxy group. Aromatisation can also be achieved with an acetylenic intermediate as in the case of the synthesis of 2,4-diethoxycarbonyl-3,5-dihydroxybiphenyl (D.O.S. Patent 2359410, 1973).

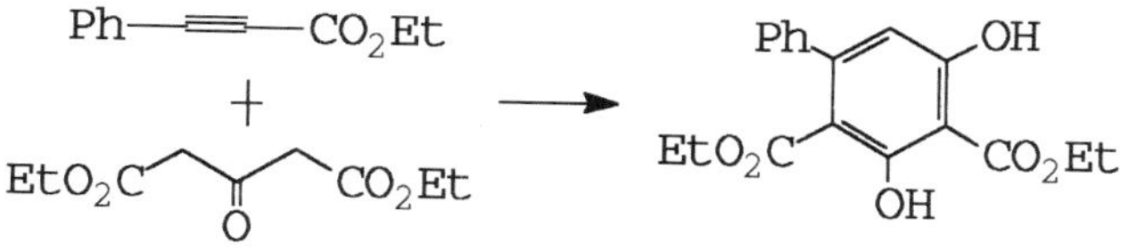

The probable biogenetic precursor of the C15 phenolic lipids from *Anacardium occidentale*, 2,4-dihydroxy-6-n-pentadecylbenzoic acid has been prepared from octadecen-2-enoic acid and the anion of ethyl acetoacetate (J.H.P. Tyman, I.E. Bruce and R.N. Spencer, J. Chem. Res., *(S)*, 1992,224; *(M)*, 1773-95). The aromatisation step can be effected with 2,3-dichloro-5.6-dicyanobenzoquinone in very moderate yield.

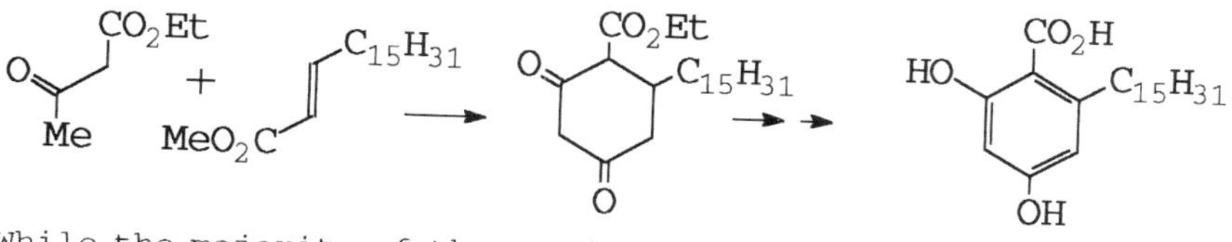

While the majority of the syntheses discussed have been directed to orsellinic acids and their homologues, keto and cyano analogues have also been obtained. The ketoester shown below affords the corresponding salicylate upon cyclisation with sodium ethoxide or magnesium methoxide and none of the resorcinol; although an excess of magnesiumethoxide gives a mixture of both products (L. Crombie et al., J. Chem. Soc. Perkin Trans I, 1979, 464; L. Crombie et al., ibid., 1979, 478; S.R. Baker and L. Crombie, J. Chem. Soc. Chem. Commun., 1979, 666).

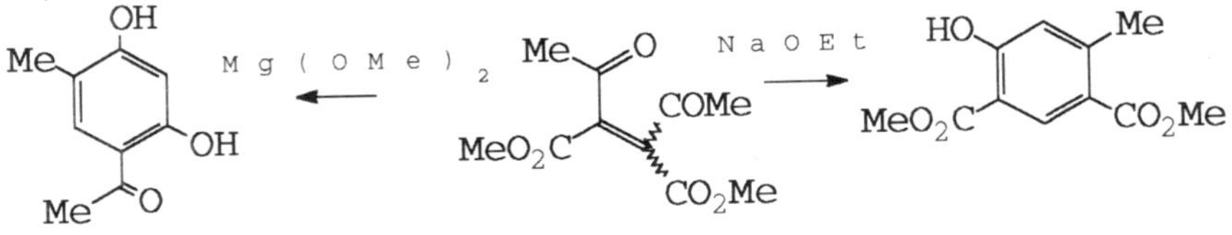

A related synthesis has been recorded (L. Crombie, D.E. Games and A.W. G. James, J. Chem. Soc. Perkin Trans I, 1979, 472).

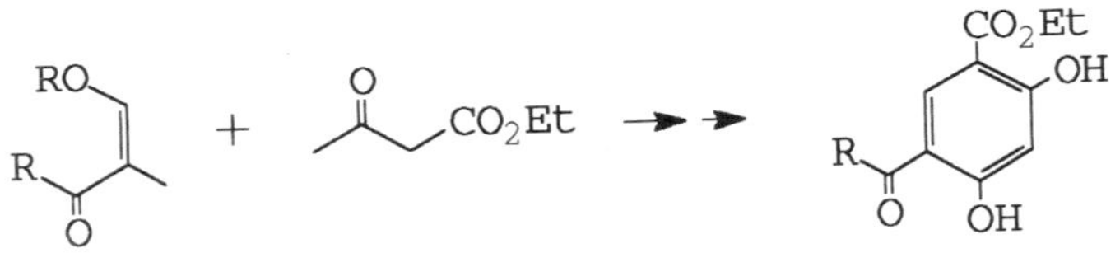

2-Acylresorcinols are derived by the aromatisation of 2-acyl-3-hydroxycyclohex-2-enones (J.E. Oliver, K.R. Wilzer and R.M. Waters, Synthesis, 1990, 1117).

Preferential O-deacylation occurs when methyl 2,6-diacetoxy benzoate is heated with 4-toluenesulphonic acid and silica suspended in toluene and water. The product is methyl 2,6-dihydroxybenzoate (G. Blay, M.L. Cardona, M.B. Garcia and J.A. Pedro, Synthesis, 1989, 438).

In the bicyclic series, 2,4-dihydroxy-5-methylphthalide, a

synthetic precursor of mycophenolic acid is produced in an overall yield of 20% from an isoxazoline in three stages (S. Aurricchio, A. Ricca and O.V. de Pava, J. Org. Chem., 1983, 48, 602).

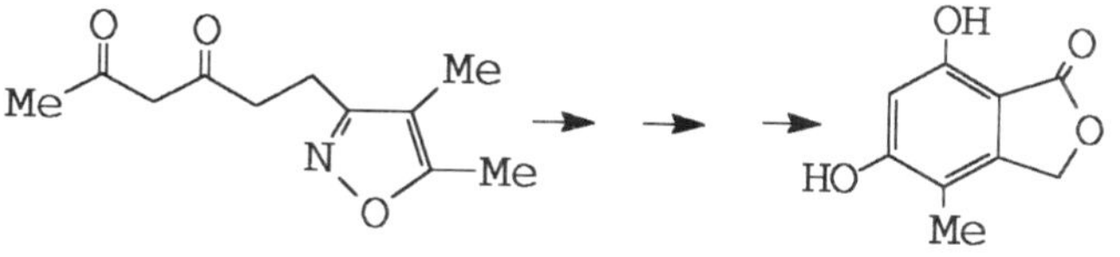

(iv) 1,4-Dihydroxybenzenes (hydroquinones) and their Derivatives

Classical diazotisation procedures have been improved. 2,3-Dimethyl-4-aminophenol in 50% aqueous sulphuric acid containing acetone is diazotised with aqueous sodium nitrite and the isolated diazonium salt after being washed with acetone is suspended in 50% sulphuric acid and added to 50% boiling sulphuric acid and toluene until nitrogen evolution ceases. Cooling affords 2,3-dimethyl-1,4-hydroxybenzene in 77% yield (EMS-Inventa AG, DE 3818696).

An autoxidative technique involves the alkylation of phenol with cyclopentadiene in the presence of phosphoric acid to produce a 80% yield of 4-(cyclopenten-2-yl)phenol. Isomerisation of this product is effected by heating with dichloro bis-(benzonitrile)palladium(II) to give the conjugated isomer in 91% yield. This with conc. hydrochloric acid upon stirring for 3 hours at 50°C gives 1,4-dihydroxybenzene in 92% yield, accompanied by cyclopentanone. These products suggest that the reaction proceeds via a hydroperoxide intermediate (D.V. Rao and F.A. Stuber, Tetrahedron Letters, 1981, 22, 1337).

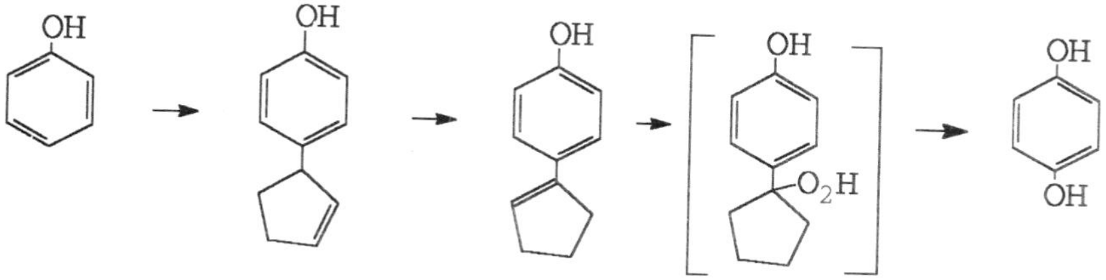

In an alternative procedure 2-dimethylaminoethyl 2,5-dimethylphenyl ether is oxidatively transformed by reaction with mesitylene iodosoacetate to 2-dimethylaminoethyl 4-hydroxy-2,5-dimethylphenyl ether (Nippon Chemifar, J63 063644).

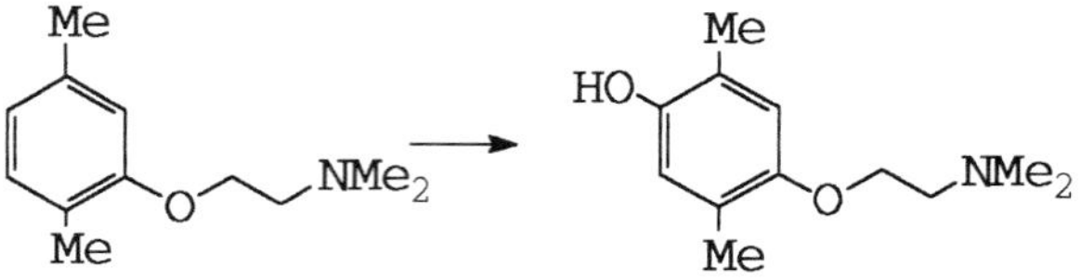

As with catechol and resorcinol which are obtainable from cyclohexane diones, cyclohexane-1,4-dione can be converted eventually to 1,4-dihydroxybenzene in high yield by treatment with acetic anhydride and inorganic acids (M.S. Kablaou, J. Org. Chem., 1974, *39*, 3696).

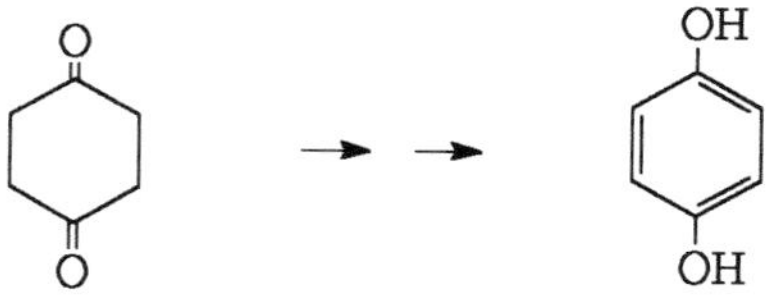

The Diels-Alder adduct of 2-acetyl-1,4-benzoquinone and buta-1,4-diene in xylene solution containing a little pyridine isomerises upon heating at 140°C to 2-acetyl-1,4-dihydroxy-5,8-dihydronaphthalene in 85% yield (S.C. Cooper and P.G. Sammes, J. Chem. Soc. Chem. Commun., 1980, 633)

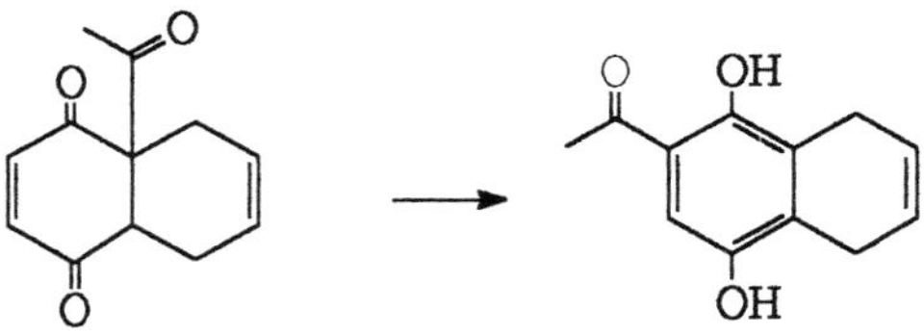

The 1,4-dihydroxy system is prevalent in a wide range of bicyclic amd polycyclic natural products.
Polyfunctional naphthalenes are available from Diels-Alder reactions between vinylcyclohexenes and acetylene dicarboxylates. An example is shown below (K.A. Parker and S.M. Ruder, J. Am. Chem. Soc., 1989, *111*, 5948).

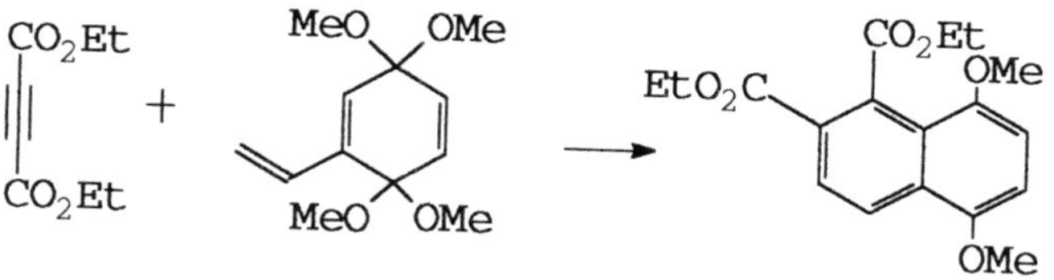

Similarly a number of anthraquinones having 1,4-dihydroxy systems have been synthesised by Diels-Alder reactions. 2-Chloronaphthazarin with di-trimethylsilyldienes shown afford alkyl 1-methyl-3,5,8-trihydroxyanthraquinone-2-carboxylates (alkyl 6-deoxykermesates) in high yield (S.J. Bingham, Ph.D. Thesis, Brunel University, 1992; P.Allevi et al., J .Chem. Soc. Chem. Commun., 1991, 1319).

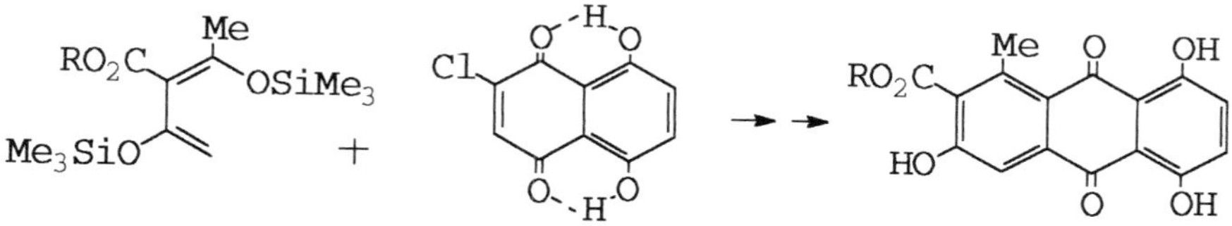

Other 5,8-dihydroxy compounds in the leuco form undergo aldol (Marschalk) reaction with alkanals followed by dehydration and isomerisation (L.M. Harwood, L.C. Hodgkinson and J.K. Sutherland, J. Chem. Soc. Chem. Commun., 1978, 712; J.H.P. Tyman and S.J. Bingham, unpublished work).

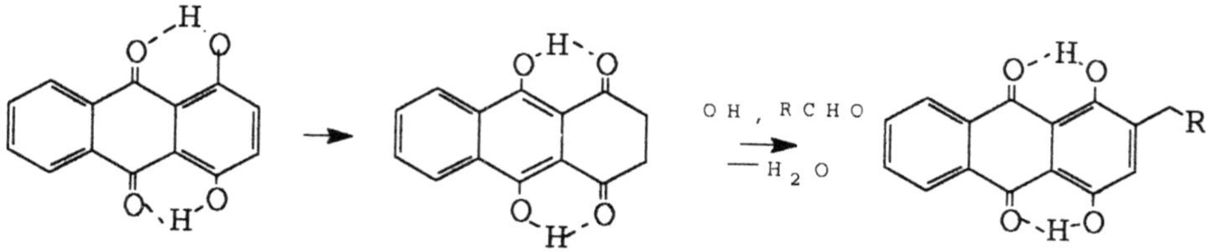

Anthracyclines are available by the reaction of 3-ethinylphthalide and the pentacarbonylchromium carbene complex, followed by treatment of the product with ferric chloride-DMF (W.D. Wulff and and P.-C. Tang, J. Am. Chem. Soc., 1984, 106, 434).

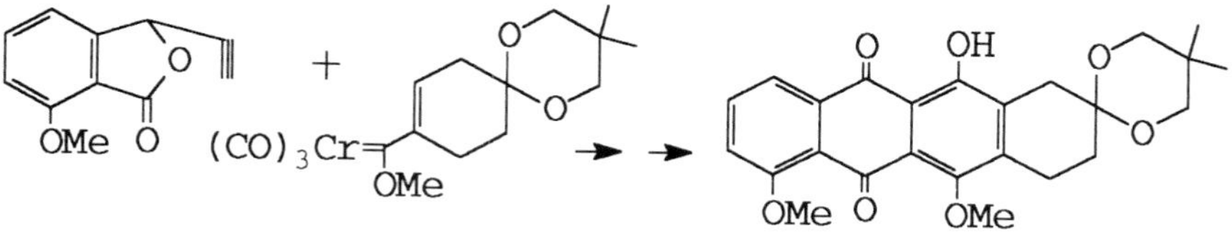

Related tetracyclic structures have been derived, for example, by the reaction of phthalic acid chloride with L-2-acetyl-2-amino-1,2,3,4-tetrahydro-5,8-dimethoxynaphthalene in dichloromethane containing aluminium chloride to afford, L-9-acetyl-acetylamino-6,11-dihydroxy-7,8,9,10-tetrahydro-5,12-naphthacene-dione in 90% yield (Sumitomo Seiyaku, KK, J62 132838).

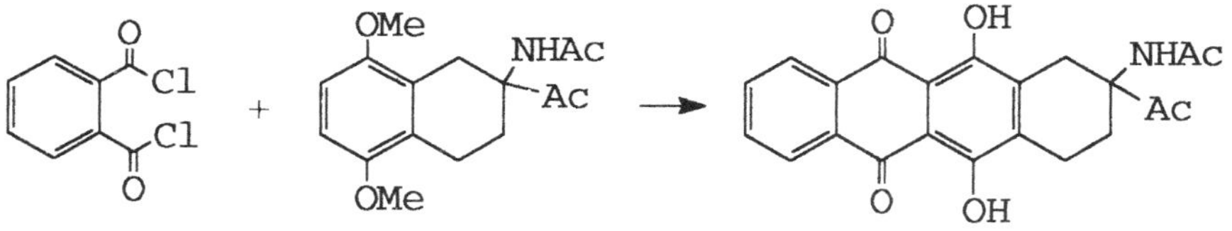

Reactions of Dihydric Phenols

(i) 1,2-Dihydroxybenzenes (Catechols) and their Derivatives

5,5',6,6'-Tetrahydroxy-3,3,3',3'-tetramethyl-1,1'-spirobisindan has been produced in 87% yield from catechol and acetone in a remarkably simple way by reaction in glacial acetic acid containing hydrogen bromide (Fuji Photo Film, KK, JO 2286642).

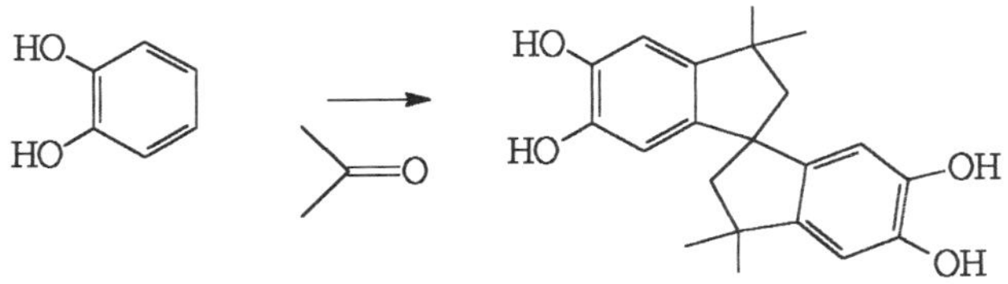

The majority of the reactions of catechols relate to their mono and dialkyl ethers (guaiacol and veratryl systems) and remarkably few to the parent compounds which are prone to oxidation. For example, 3,5-di-tert-butylcatechol in an oxygen atmosphere in 1,1,2,2-tetrachloroethane containing a catalytic quantity of dichloro tris(triphenylphosphine)ruthenium(II) affords the corresponding 2H-pyran-1-one (M. Matsumoto and K. Kuroda, J. Am. Chem. Soc., 1982, *104*, 1433).

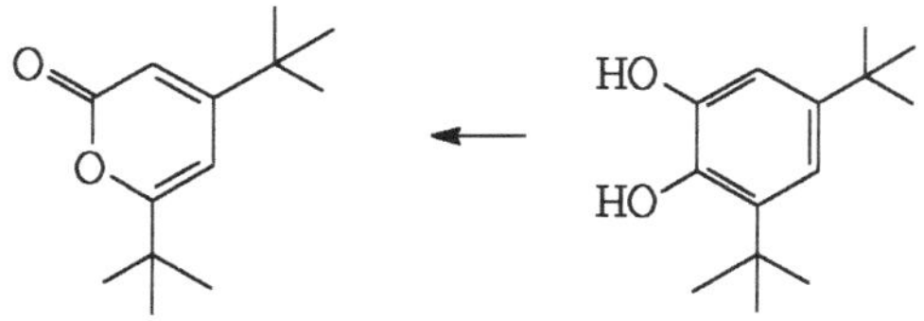

2,3-Dihydroxy-1-naphthaldehyde with iodomethane and potassium carbonate, gives the 2-methyl ether (F.M. Dean, G. El-Kass, L. Prakash, J. Chem. Soc. Chem. Comun., 1985, 502).

5,6-Methylenedioxy-1-tetralone is obtained in 92% yield from 5,6-dihydroxy-1-tetralone in DMF containing caesium carbonate by the addition of chlorobromomethane (R.E. Zelle and W.J. McClellan, Tetrahedron Letters, 1991, *32*, 2461).

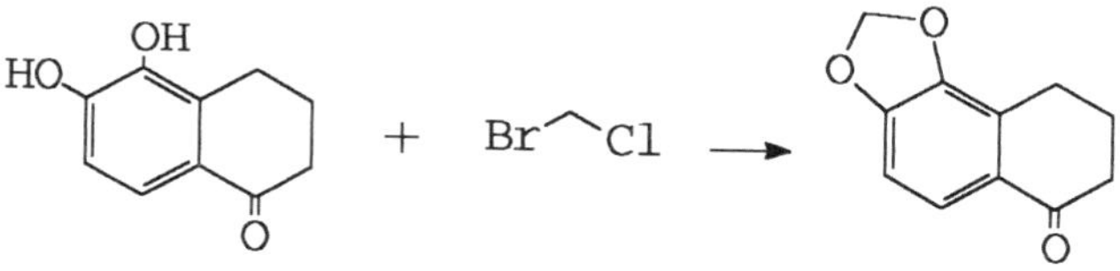

(ii) Mono O-Alkyl Derivatives of 1,2-Dihydroxybenzenes

Reactions of the single OH group in this class of compounds comprise chlorocarbonylation, reduction, substitution by an alkyl group and methylation.

2-Iso-propoxyphenol containing a small amount of triphenylphosphonium chloride on treatment with phosgene at 118-150°C gives 2-isopropoxyphenyl chloroformate in 97% yield (PPG Industries Inc., FrP 2510989; Bayer AG, DE 3135947).

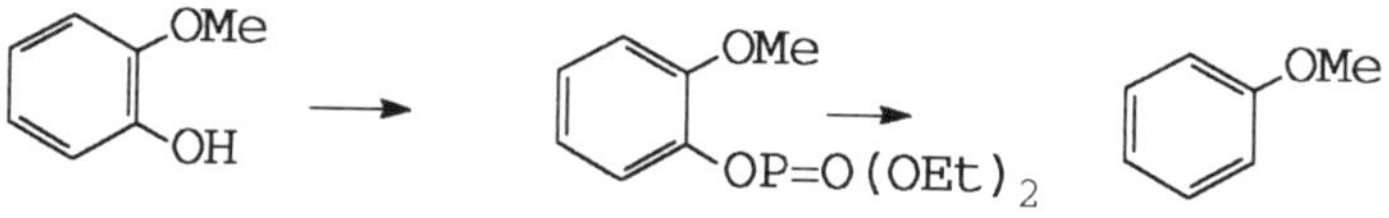

Electrochemical dehydroxylation of 2-methoxyphenol as its diethyl phosphate is achieved in a divided cell with DMF as the solvent and tetraethylammonium 4-tosylate as the supporting electrolyte (T. Shono, Y. Matsumura, K. Tsubata and Y. Sugihara, J. Org. Chem., 1979, *44*, 4508).

Replacement of the OH group in 2-methoxyphenol is effected by adding 2-methoxyphenyl trifluoromethane sulphonate in THF at -70°C to Li$_2$Cu(CN)Bu$_2$ prepared from n-butyllithium and cuprous cyanide in THF, and affords 2-n-butylanisole in 50% yield (J.E. McMurry and S. Mohanraj, Tetrahedron Letters, 1983, *24*, 2732). A catechol mono ether derivative, morphine, in toluene containing potassium carbonate and trimethylphenylammonium chloride can be reacted at 45-120°C to afford codeine in 99% yield (Council Sci. Ind. Res., EP 268710). There is an element of rediscovery in this since the author of this chapter recalls an identical process being investigated in 1945 in the laboratories of May and Baker Ltd. (now Rhone-Poulenc).

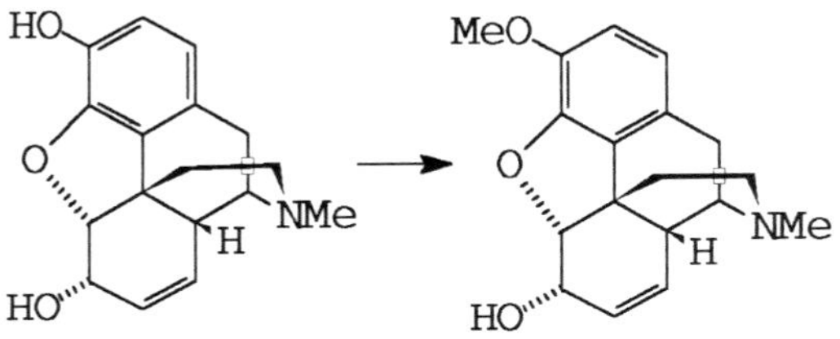

Reactions of the aromatic ring such as halogenation, cyclisation and oxidative dimerisation are shown in the following examples. Thus, 2-methoxyphenol in di-isopropyl ether on treatment with N,N-dibromo-tert-butylamine gives 6-bromoguaiacol in 75% yield (Rhone-Poulenc Chim., EP 338898).

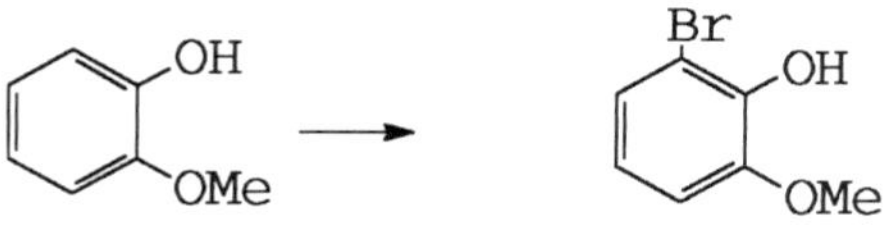

Iodination of vanillin (3-methoxy-4-hydroxybenzaldehyde) in DMF containing sodium iodide and chloroamine T at 25°C gives 5-iodovanillin in 94% yield (T. Kometani, D.S. Watt and T. Ji, Tetrahedron Letters, 1985, *26*, 2043).

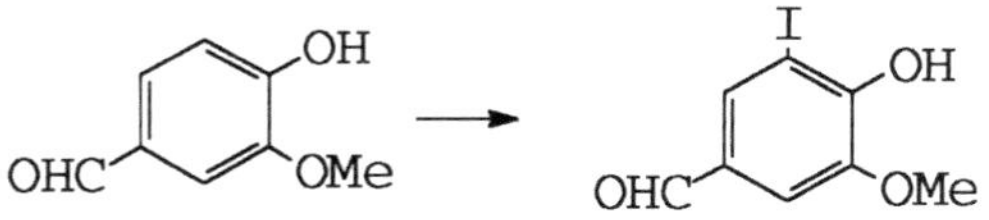

Oxidative dimerisation of prestegane B in dichloromethane containing boron trifluoride etherate is achieved in 80% yield by addition of the substrate to ruthenium dioxide in dichloromethane containing triflic acid/triflic anhydride (8:2:1) (J-P. Robin and Y. Landais, J. Org. Chem., 1988, *53*,

224).

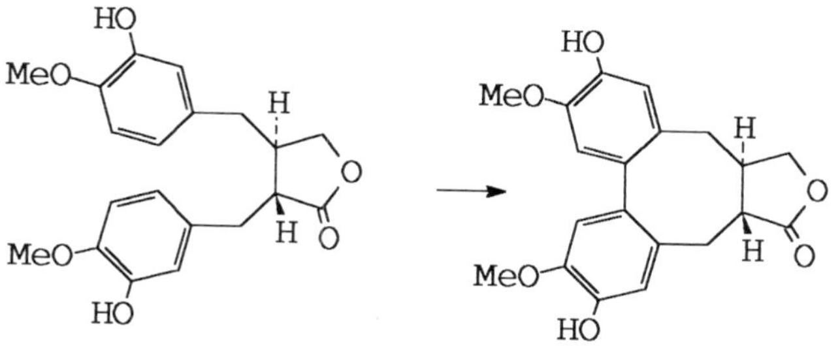

(iii) Di-O-Alkyl derivatives of 1,2-Dihydroxybenzenes

Because of their interest as natural products 1,2-dimethoxy derivatives from benzenoid and polycyclic systems have been submitted to a variety of reactions including oxidative dimerisations, other intramolecular oxidative reactions, carboxylations, acylations and alkylation, miscellaneous cleavage reactions and asymmetric syntheses.

3,4-Dimethoxytoluene in dichloromethane added to silica gel impregnated with ferric chloride gives, after 1 hour 2,2'-dimethyl-4,4',5,5'-tetramethoxybiphenyl in 95% yield (T.C. Jempty, L.L. Miller and Y. Mazur, J. Org. Chem., 1980, 45, 749).

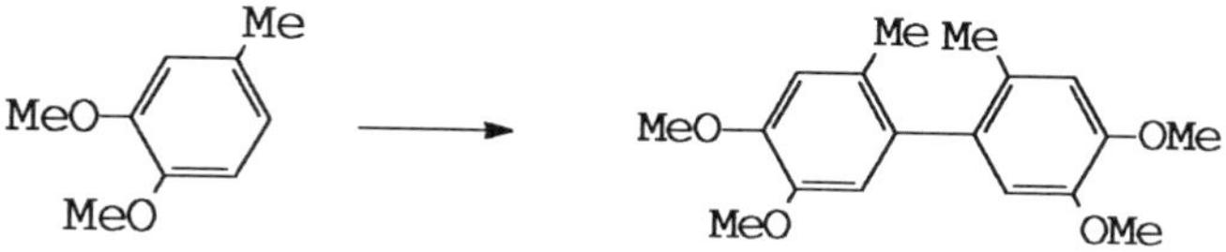

1,2-Dialkoxybenzenes and benzenoid crown ethers when submitted to mixed anodic trimerisation afford triphenylenes which possess one or two complexing sites (J.-M. Chapuzet, N. Simonet-Gueguen, I. Taillepied and J. Simonet, Tetrahedron Letters, 1991, 32, 7405).

The tert-butyl carbamate ester of 4-amino-1,2-dimethoxybenzene (veratrylamine) is lithiated with n-butyllithium and regiospecifically carboxylated to afford a combined 70% yield of the corresponding 3- and 5-carboxylic acids, (20:1). With tert-butyllithium a mixture of the same isomers (7:3) is formed in 66% yield (S. Bengtsson, and T. Hogberg, J. Org. Chem., 1989, 54, 4549).

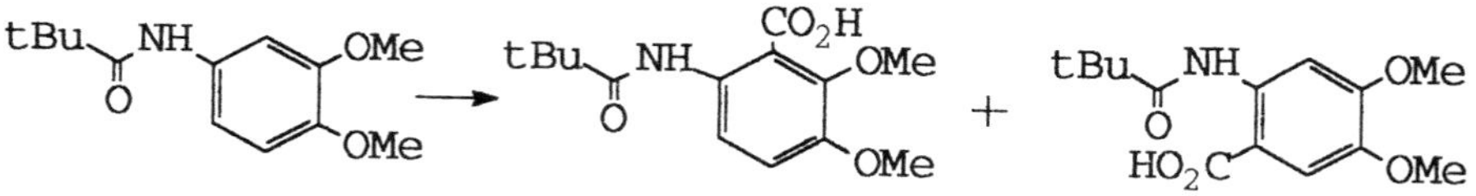

C-4 formylation of 2-methoxy-4-methylphenyl methyl carbonate occurs at 0°C in dichloromethane containing titanium tetrachloride and reaction with dichloromethyl methyl ether. This product can be cleaved to give 3-hydroxy-4-methoxy-6-methylbenzaldehyde (M.L. Scarpati et al. Synth. Commun., 1990, *20*, 2565).

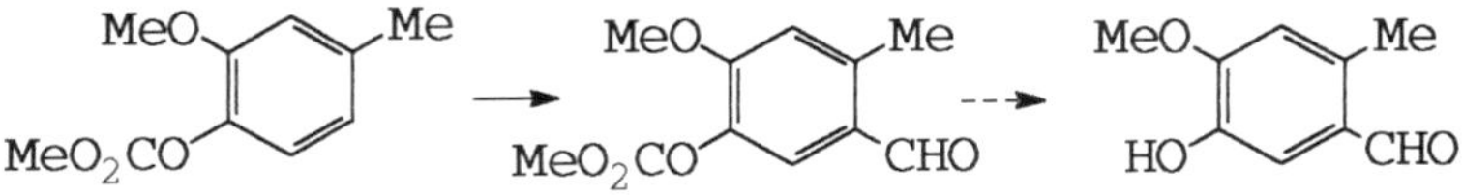

Two modes of cyclisation of 2-(3,4-dimethoxybenzyl)-3-phenylpropionyl chloride in dichloromethane have been described. Aluminium bromide in 100 mole% proportion in dichloromethane at -15°C gives 2-benzyl-5,6-dimethoxyindan-1-one in 78% yield while 300 mole% over a longer reaction period affords an 86% yield of 2-(3,4-dimethoxybenzyl)indan-1-one with some demethylated product (T.F. Buckley (III) and H. Rapoport, J. Am. Chem. Soc., 1980, *102*, 3056).

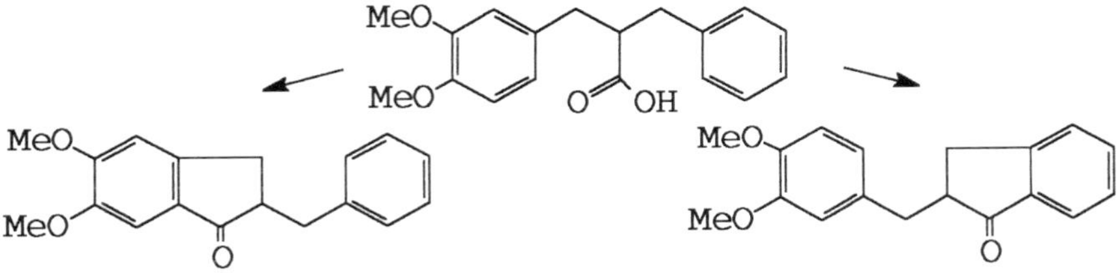

2-(3,4-Dimethoxyphenyl)-N,N-dimethylacetamide in acetonitrile is cyclised by the addition of excess phosphorus oxychloride to 3-dimethylamino-6,7-dimethoxy-1-methylisoquinoline in 56% yield (A.J. Liepa, Austral. J. Chem., 1982, *35*, 1391).

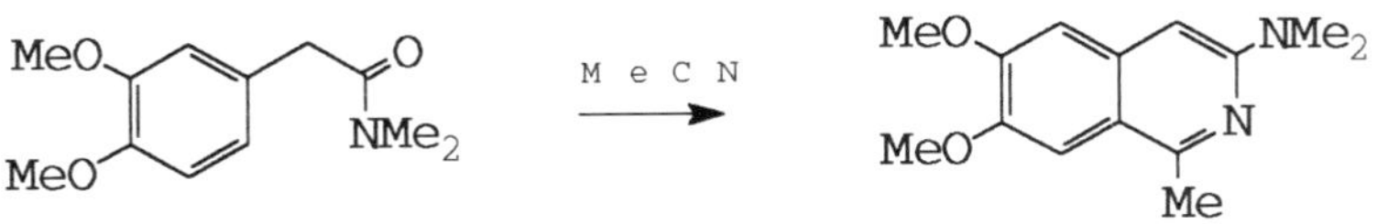

Amongst the miscellaneous reactions of dimethoxy systems, 3,4-dimethoxybenzoylcyanide in ether, benzene, toluene or dichloromethane can be reacted with phenylhydroxylamine at ambient temperature to give the O-acylhydroxylamine in 90% yield (S. Prabhakar, A.M. Lobo and M.M. Marques, Tetrahedron Letters,

1982, *23*, 1391).

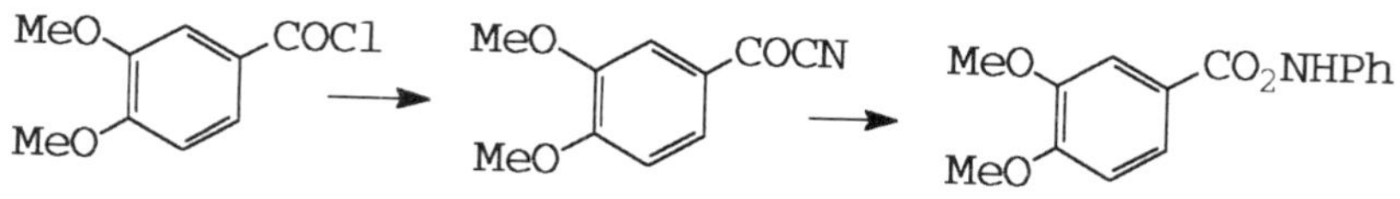

Some selective cleavage reactions are listed. Preferential O-deacetylation of a phenyl acetate in the presence of an alkyl acetate is effected by treatment of 3-(4-acetoxy-3-methoxyphenyl)propyl acetate with pyrrolidine giving 3-(3-methoxy-4-hydroxy)propyl acetate in more than 80% yield (P. Mansson, Tetrahedron Letters, 1982, *23*, 1845).

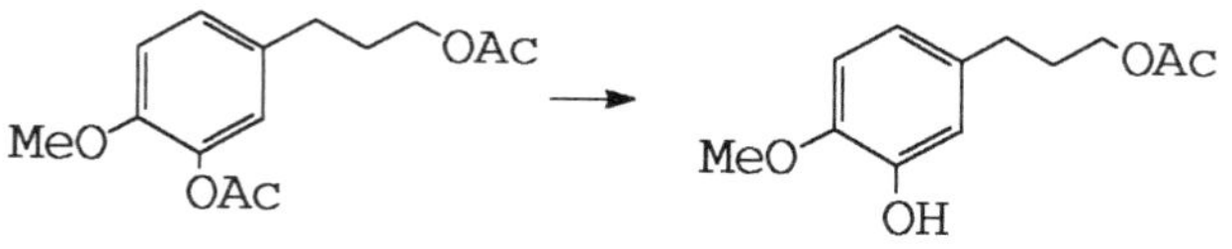

2,3-Dimethoxybenzaldehyde in toluene is regiospecifically demethylated by treatment with lithium N-methylpiperazide, followed by butyllithium and warming to 65°C, affording 3-methoxy-2-hydroxybenzaldehyde (o-vanillin) in 55% yield. Methoxy and ethoxy, but not tert-butoxy, groups can be similarly cleaved (L. Gillies and M.S. Loft, Synth. Commun., 1988, *18*, 191).

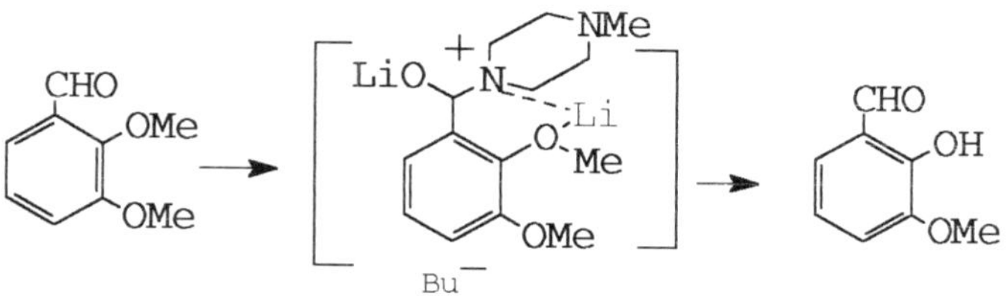

O-Ethylisoeugenol is demethylated preferentially upon addition to 1,2-dihydroxyethane containing potassium fluoride-alumina at 150°C and then heating at 210°C to give 4-ethoxy-3-hydroxyprop-2-enylbenzene (A.S. Radhakrishna et al., Synth. Commun., 1991, *21*, 379).

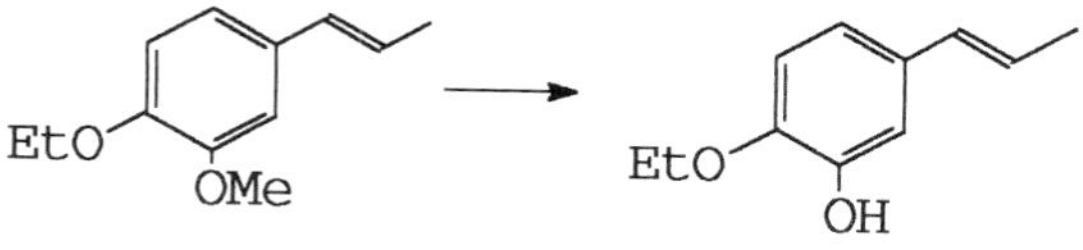

(iv) Methylenedioxy Derivatives of 1,2-dihydroxybenzenes

3,4-Methylenedioxybenzaldehyde is synthesised in 98% yield from the oxidation of 3,4-methylenedioxybenzyl alcohol in dichloromethane solution with excess 4-(dimethylamino)pyridinium chromate in dichloromethane (F.S. Guziec Jr. and F.A. Luzzio, J. Org. Chem., 1982, *47*, 1787).

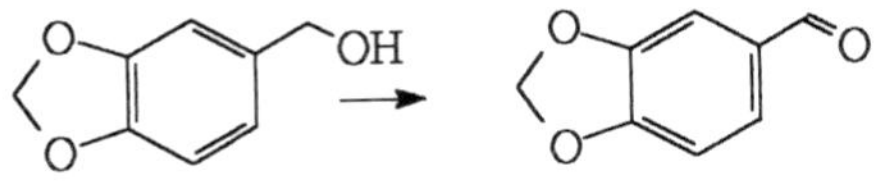

Etherification followed by ring closure is noted when the anion of 4-hydroxymethylenedioxybenzene (sesamol) in THF containing tetra-n-butyl ammonium iodide is heated with 2-(2-chloroethyl)-1,3-dioxolane. The resultant ether-acetal, formed in 81% yield, may be cyclised in benzene containing 4-toluenesulphonic acid to afford 6,7-methylenedioxy-2H-benzopyran in 60% yield (P.F. Schuda and J.L. Phillips, J. Heterocyc. Chem., 1984, *21*, 669).

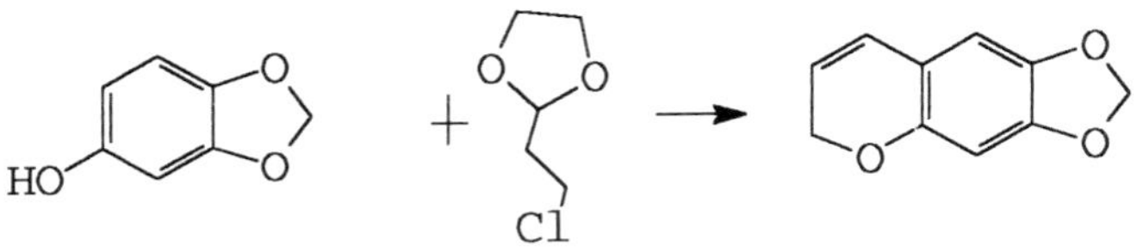

Methylenedioxybenzene in acetonitrile containing sodium iodide gives the crown ester 8,8,19,20-tetrahydro-7H,18H-dibenzo[b,k]-1,7,10,16-tetraoxacyclooctadecane-6,10,17,21-tetrone in 51% yield by gradual treatment with glutaroyl chloride initially at 0°C and subsequently at ambient temperature during 24 hours (A.M. Fadda, A.M. Maccioni and G. Podda, Gazz. Ital. Chim., 1987, *117*, 555). If 1,3-benz-oxathioles are used with magnesium bromide as catalyst, dithia analogues are obtained.

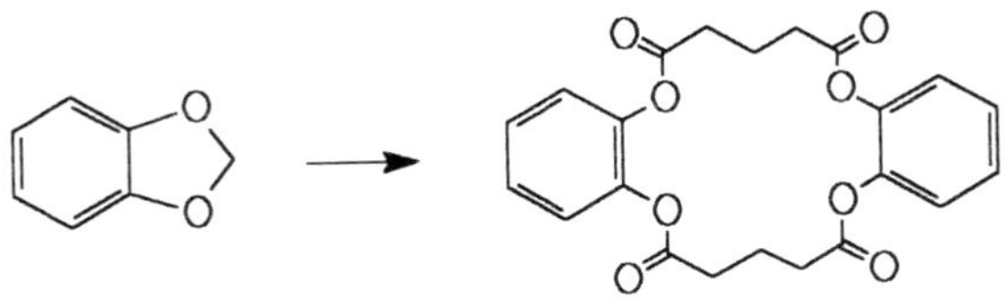

Lithiation of the methylenedioxy benzylamine indicated followed by quenching with an electrophile furnishes a derivative which can be cyclised to an 8-substituted isoquinoline (G. Simig and M. Schlosser, Tetrahedron Letters, 1990, *31*, 3125).

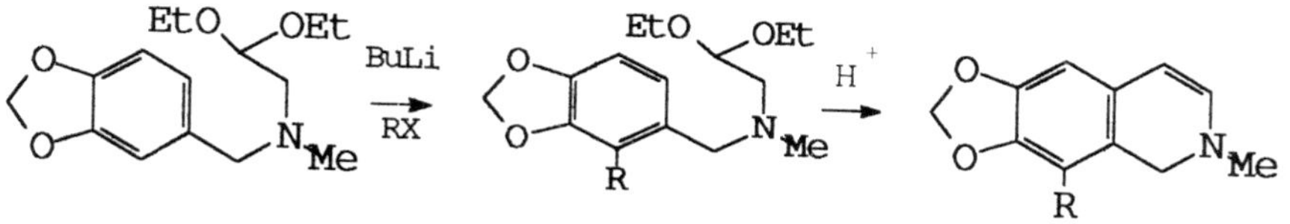

(v) Dioxin Derivatives

The dioxin ring system (effectively a 1,2-ethanodioxy structure) is derivable from catechol. The Diels-Alder reaction between 2,3-dimethylene-2,3-dihydro-1,4-benzodioxin and methyl vinyl ketone at 70°C enables a tricyclic compound in the series to be synthesised in 84% yield (N. Ruiz, M.D. Pujol, G. Guillaumet and G. Caudert, Tetrahedron Letters, 1992, *33*, 2965).

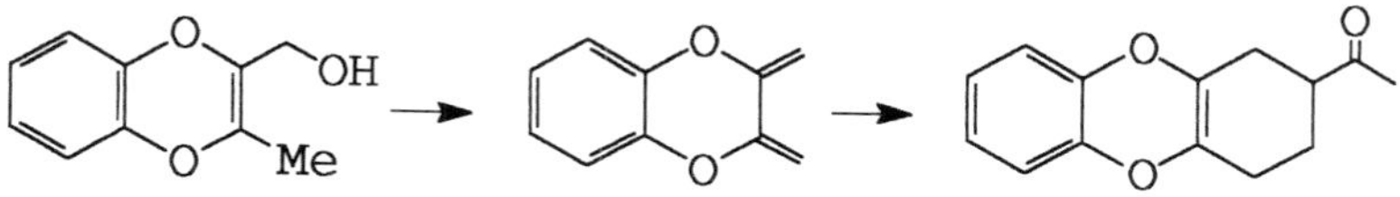

(vi) 1,3-Dihydroxybenzenes (Resorcinols) and their Derivatives

In this section, reactions resulting in O-alkyl, C-acyl derivatives, C-alkylation leading to bicyclic structures, and substitution of the hydroxyl group and of the ring are described.

Resorcinol reacts with ethylene carbonate (98% purity) in the presence of triphenylphosphine on heating at 150-170°C to give a quantitative yield of the bis(2-hydroxyethyl) ether (Indspec Chemical Co., USP 5059723).

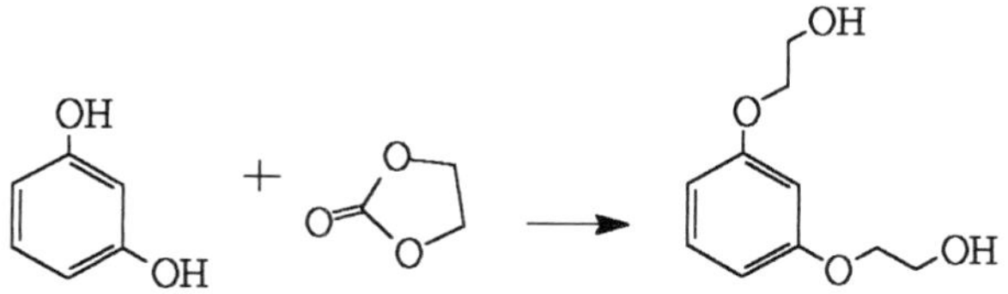

Selective C-4 methylation of 2,4-dihydroxy acetophenone occurs when the compound is heated in DMF containing lithium carbonate and methyl iodide. Generally this method is effective for the selective methylation of phenolic hydroxyl groups having pKa ≤8, or in chelated phenolic/enolic instances where the pKa is approx. 10 (W.E. Wymann et al., Synth. Commun., 1988, *18*, 1379).

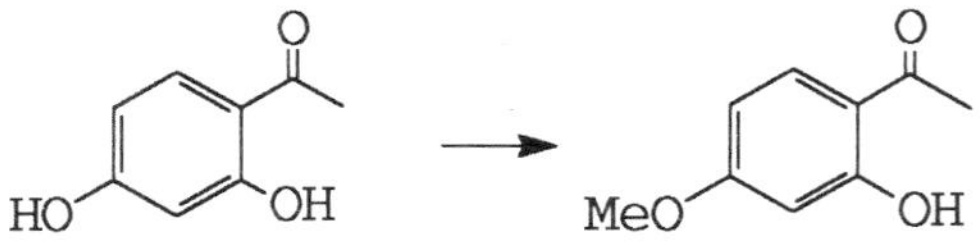

In a similar way 2,4-dihydroxy acetophenone with benzyl tosylate and potassium carbonate in hot acetone gives the 4-benzyl ether in 62% yield (P.M. Dewick, Synth. Commun., 1981, *11*, 853). Chelation of one hydroxyl group in each ring of 2,4,6-tris(2,4-dihydroxyphenyl)-1,3,5-triazine with a nitrogen atom enables the benzylation of the other to be effected when the compound is reacted with benzyl bromide and sodium carbonate (W. Pei, Y. Gao and R. Rishan, Huaxue Shiji, 1991, *13*, 61; Chem. Abs., *115*, 92227).

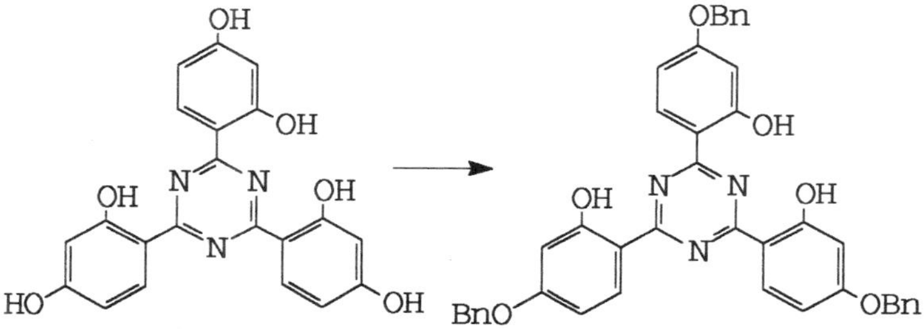

Resorcinol with glacial acetic acid containing 5% hydrobromic acid affords 2,4-dihydroxyacetophenone, whereas traditional reagents and catalysts result in over-reaction. When applied to resorcinol ethers these conditions cause simultaneous ether cleavage (Eli Lilly and Co., USP 4777298).

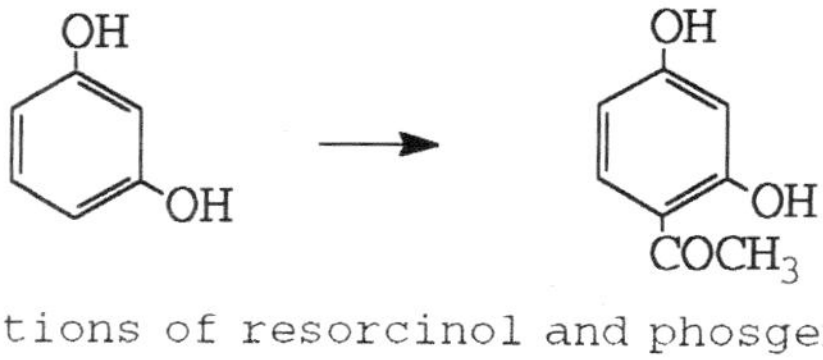

Equimolar proportions of resorcinol and phosgene in nitrobenzene at 100°C give a 73% yield of 3-hydroxyphenyl 2,4-dihydroxybenzoate (Sumitomo Chem. Ind., KK J62 103085).

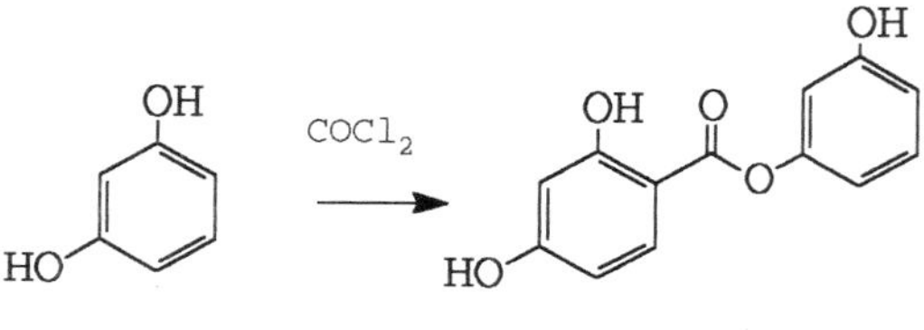

Mono-iodination of 5-acetyl-2,4-dihydroxyacetophenone is effected in methanol by the addition of phenyl iodosodiacetate and potassium hydroxide (O. Prakash et al., Tetrahedron Letters, 1992, *33*, 6519).

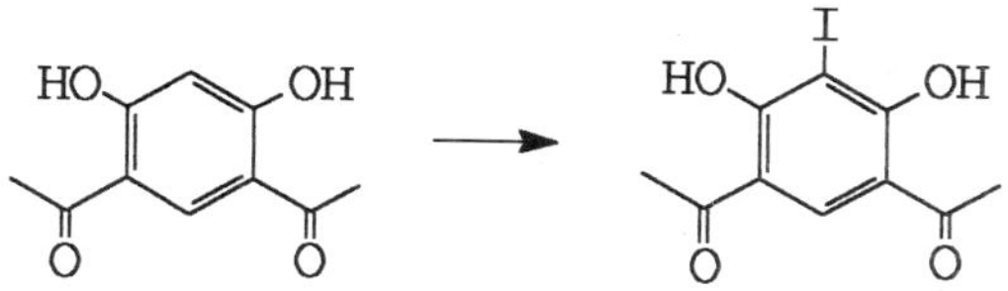

C-Alkylation leading to bicyclic structures

This useful property is shown in the following syntheses. Resorcinol in dichloromethane containing aqueous sulphuric acid upon dropwise treatment with acetone and heating gives a spirochroman derivative in 30% yield (Sumitomo Chem. Ind., KK J62 103085). The mechanism of the reaction probably involves ketal formation, followed by 2,2'-iso-propenylation and ring closure.

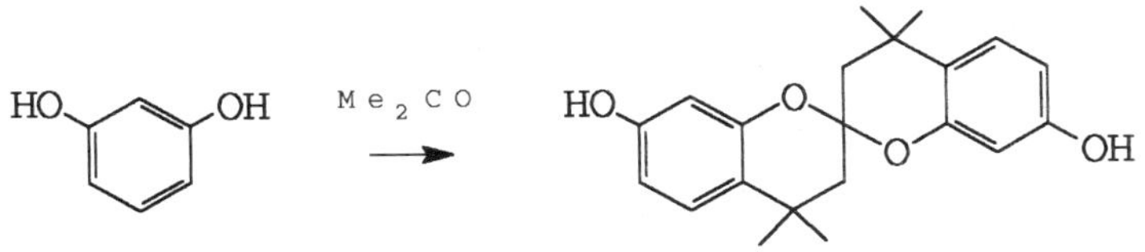

7-Hydroxy-2-methoxy-2,4,4-trimethylchroman is formed in 17% yield through heating resorcinol with 4 moles of acetone in toluene containing methanol and a litle concentrated sulphuric acid at 70°C in an autoclave (Mitsui Petrochem. Ind., KK EP 405748).

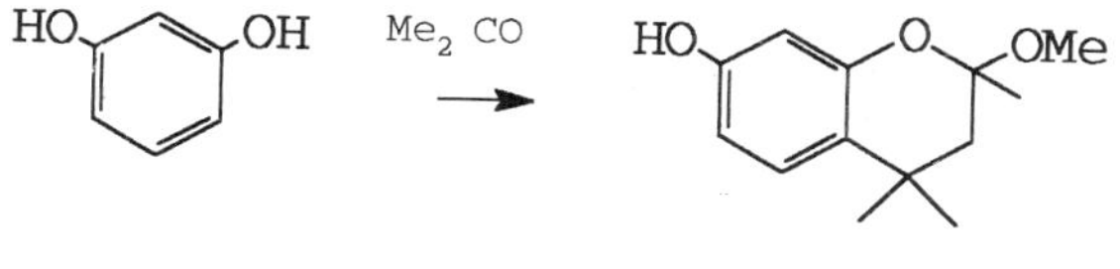

Resorcinol and 3-methyl-2-butenoic acid when added to methanesulphonic acid and phosphorus pentoxide at 70°C furnish a 94% yield of 2,2-dimethyl-7-hydroxy-4-chromanone (F. Camps et al., Synthesis, 1980, 725).

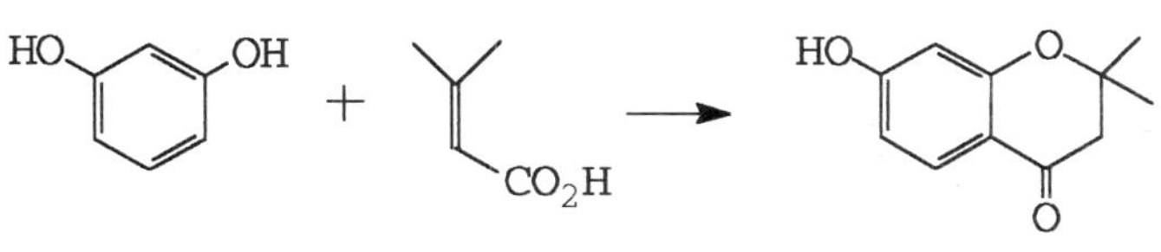

However, resorcinol with mesityl oxide (2-methylpent-2-ene-4-one) in the presence of barium hydroxide at 150-170°C in vacuo (165-170mm. Hg) with azeotropic removal of water affords a 34% yield of 2,2,4-trimethyl-5-hydroxy-2H-chromene rather the 7-hydroxy isomer (Mitsui Petrochem. Ind., KK J57 109779). Under acidic conditions a stirred mixture of resorcinol and mesityl oxide in carbon disulphide containing boron trifluoride etherate, after heating results in two electrophilic substitutions and a ring closure to furnish 2,4,4-trimethyl-2-(2,4-dihydroxyphenyl)-7-hydroxychroman in 77% yield (Mitsui Petrochem. Ind., KK J57 114585).

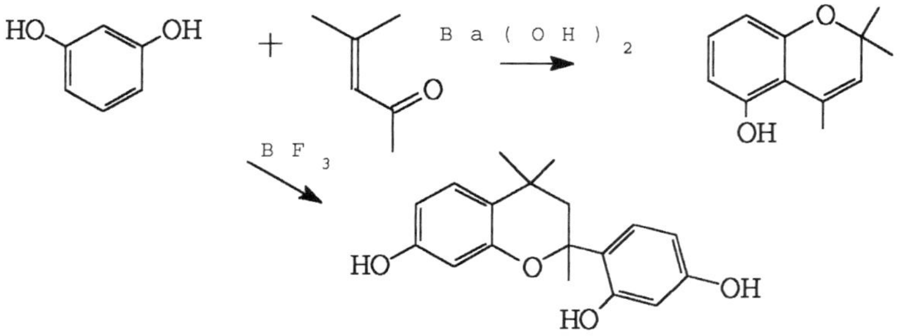

A Michael reaction between resorcinol and 4-bromobenzylidene-malononitrile in ethanol containing morpholine, is succeeded by cyclisation to produce a 70% yield of 2-amino-4-(4-bromophenyl)-3-cyano-7-hydroxy-4H-chromene (Y.A. Sharanin and G.U. Klokol, Zh. Org. Khim., 1983, 19, 1782).

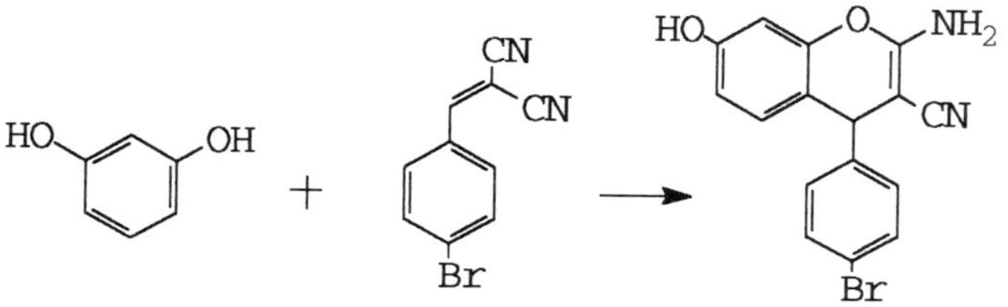

Resorcinol with methyl acetoacetate in octane containing a small amount of sulphuric acid, in a variation of the usual conditions for the Pechmann reaction, upon boiling for 5 minutes furnishes 7-hydroxy-4-methylcoumarin, which without isolation, is transformed to the 3-chloro derivative in 98% yield by the addition of acetic acid/thionyl chloride at 80°C (Eastman Kodak Co., USP 4788298).

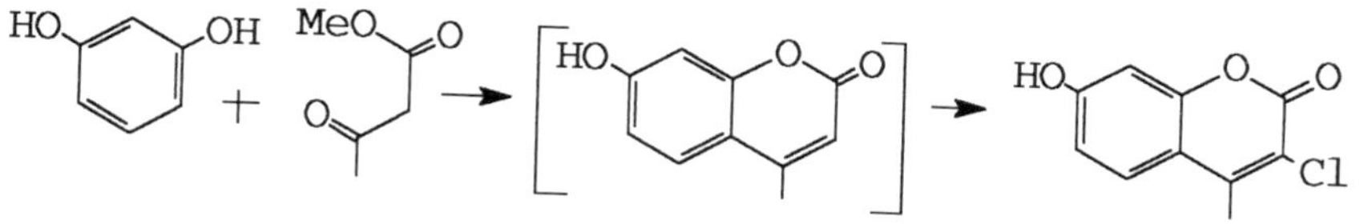

With Nafion-H in place of sulphuric acid the reaction of resorcinol and ethyl acetoacetate, gives a 90% yield of 7-hydroxy-4-methylcoumarin (D.D. Chaudhari, Chem. and Ind., London, 1983, 568).

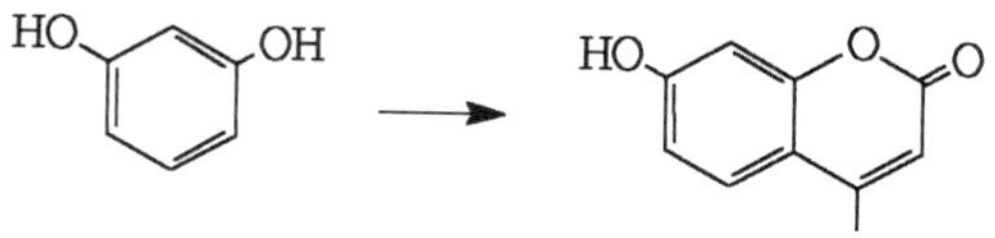

With the cation exchange resin Amberlyst 15, resorcinol and isopropyl methyl ketone at 100°C afford a 66% yield of 6-hydroxy-2,2,3-trimethyl-2,3-dihydrobenzofuran (Mitsui Petrochem. Ind., KK J62 138487).

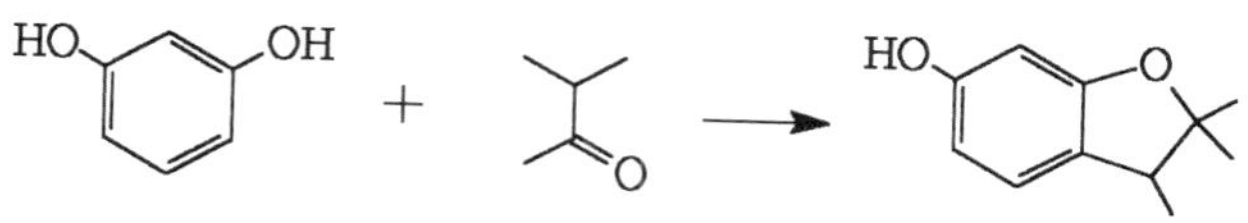

With benzil, resorcinol in xylene saturated with hydrogen chloride reacts at 50-60°C to give a 81% yield of 3,8-dihydroxy-5a,10a-diphenyl-5a,10a-dihydrobenzofurano[2,3-b]benzofuran (Bayer AG, DE 3938282).

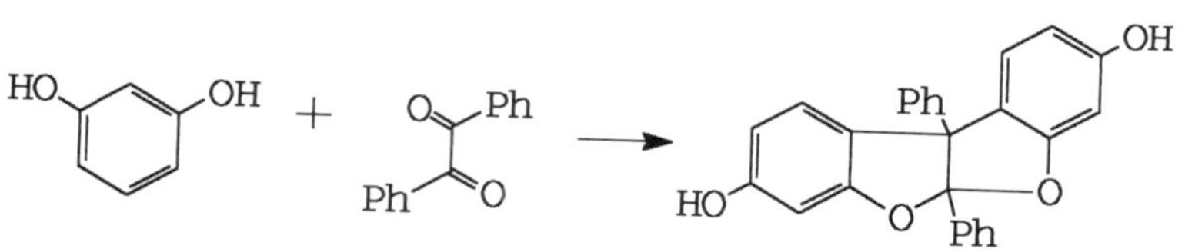

Equimolar proportions of resorcinol and salicylic acid when heated at 130°C in an equimolar mixture of 85% orthophosphoric acid and phosphorus oxychloride, previously heated at 50°C with fused zinc chloride, give a low yield of 1-hydroxyxanthone (R = H) (N.B. Nevrekar et al., Chem. and Ind., London, 1983, 479). 1-Hydroxy-3-methylxanthone (R = Me), has also been obtained in low yield from salicyloylurea and orcinol in the same reagent (M.V.R. Mucheli and N.A. Kuda, Chem. and Ind., London, 1985, 31).

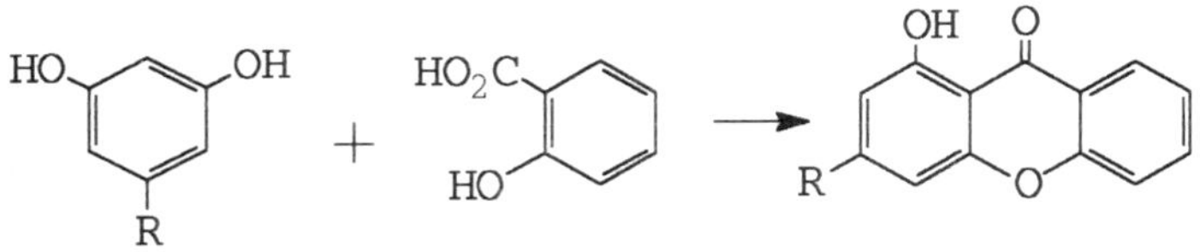

Substitution of the 3-hydroxy group in 3,5-dihydroxybenzoic acid, can be achieved by heating wih ammonium chloride and 28% ammonia in an autoclave. After esterification with methanol containing sulphuric acid, a 75% yield of methyl 3-amino-5-hydroxybenzoate is obtained. (P.R. Iyer, C.S.R. Iyer and K.J.R. Prasad, Indian J. Chem., 1983, 22b, 1055).

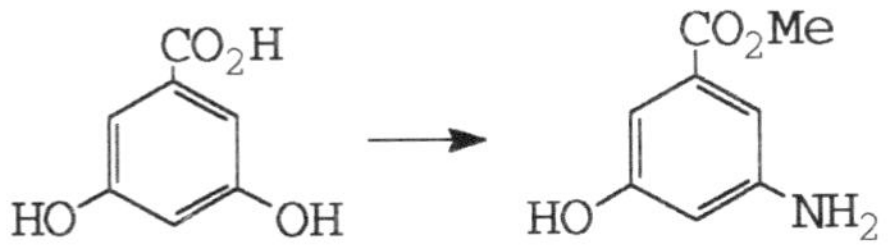

3,5-Dihydroxybenzyl alcohol is employed as starting material for the synthesis of dendritic macromolecules or molecules with controlled 'fan' molecular architecture. It is normally prepared by the reduction of the corresponding acid (K.L. Wooley, C.J. Hawker and J.M. Frechet, J. Chem. Soc. Perkin Trans I, 1991, 1059).

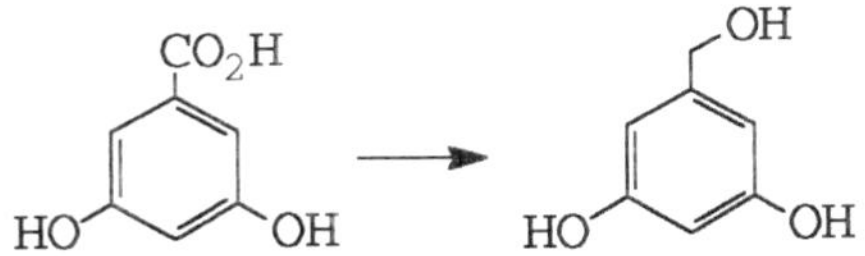

In a range of studies on carcerands and hemicarcerands involving resorcinol (and pyrogallol) structures some remarkable reactions have been effected on trapped molecules (D.J. Cram, M.E. Tanner and R. Thomas, Angew. Chem. Int, Ed. Engl., 1991, 113, 1024). Monosubstitution of dihydric amd polyhydric phenols has never proved easy. In this connection the regiospecific monoiodination of resorcinol and of phloroglucinol has been described (I. Thomsen and K.J.B. Torsell, Acta Chem. Scand., 1991, 45, 539).

(vii) Mono O-Alkoxy Derivatives of 1,3-Dihydroxybenzenes

3-Alkoxyphenols are generally more easily substituted than the parent phenols.
Resorcinol monomethyl ether in benzene containing boron trichloride reacts with benzoyl chloride to give 2-hydroxy-4-methoxybenzophenone in 85% yield (G. Piccolo, L. Filippini and

L. Tinucci, Tetrahedron, 1986, 42, 885).

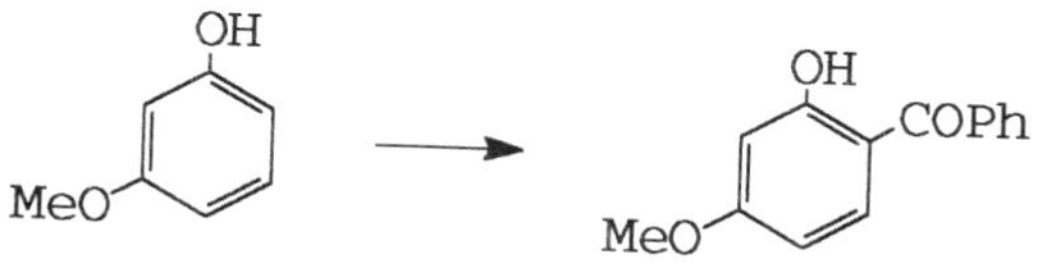

Resorcinol monomethyl ether, chloroacetonitrile and aluminium chloride plus boron trichloride in dichloroethane afford an 81% yield of 2-hydroxy-4-methoxy chloroacetophenone (T. Toyoda, K. Sasakura and T. Sugasawa, J. Org. Chem., 1981, 46, 189).

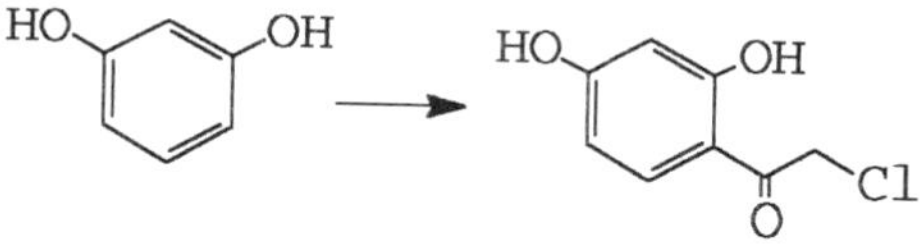

Bicyclic structures are readily assembled from resorcinol mono methyl ether as in the synthesis of benzofuranones. For example, cinnamic acid and the monomethyl ether in trifluoroacetic acid give 6-methoxy-3-(4-methoxybenzyl)benzofuran-2-one (R. Chaturvedi and N.B. Mulchandani, Synth. Commun., 1990, 21, 3317).

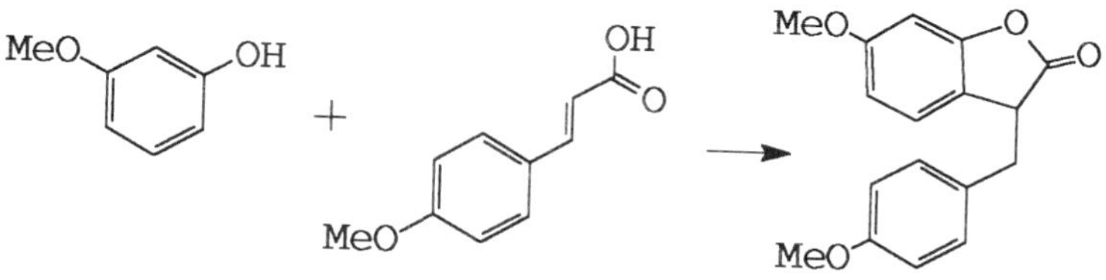

6-Methoxybenzoxazolin-2-one is produced from 5-methoxy-2-nitrophenol in 75% yield by hydrogenation in methanol containing palladium-carbon at normal pressure and temperature, followed by treatment of the intermediate aniline with triethylamine and bis(trichloromethyl)carbonate (Karl Marx Univ., Leipzig, DD 280531).

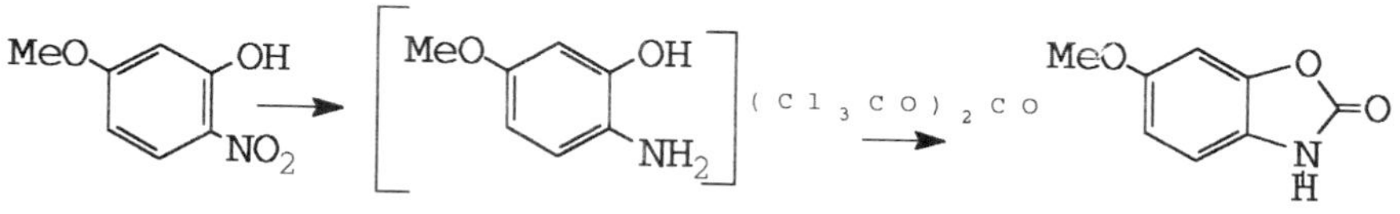

3-Tosylphenol in the form of its phenoxide is O-aminated by 2,4-dinitrophenoxylamine in DMF giving 3-tosyloxyphenoxylamine in 83% yield (A.J. Castellino and H. Rapoport, J. Org. Chem., 1984, *49*, 1348).

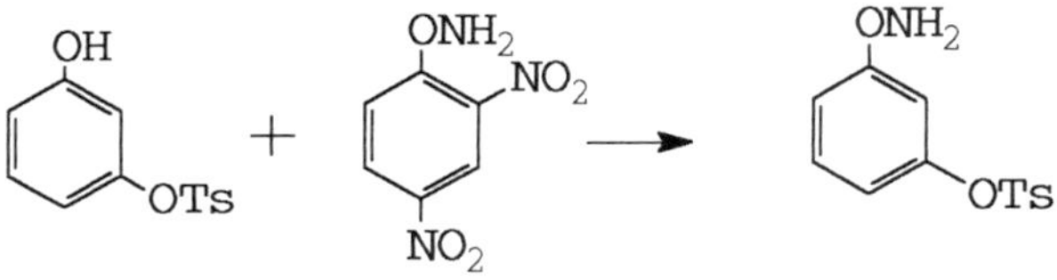

(viii) Mixed Unsymmetrical O-Substituted Derivatives of 1,3-dihydroxybenzenes

Synthetic uses are made of resorcinol monomethyl ethers which differentiate between the two functional groups. Thus desulphonylylation of 3-methoxyphenyl 1,1,2,2-tetrafluoroethoxy-1,1,2,2-tetrafluoroethyl sulphonate is effected with tri-n-butylamine and a small proportion of dichloro bis(triphenyl-phosphine)palladium(II) in dimethyl sulphoxide at 110°C furnishing a 89% yield of anisole (Q-Y. Chen, Y-B. He and Z-Y. Yang, J. Chem. Soc. Chem. Commun., 1986, 1452).

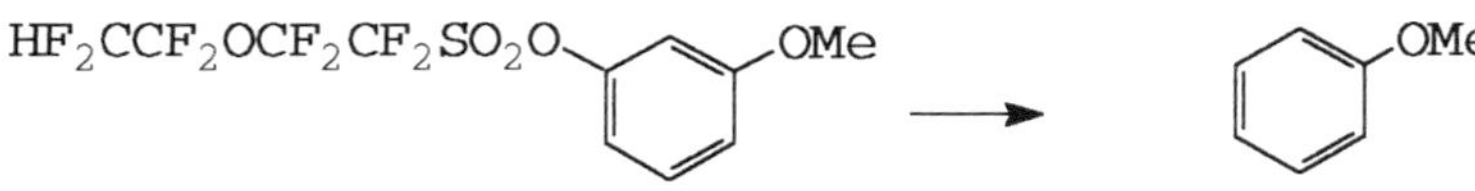

1-Acetoxy-3-methoxy-5-methylbenzene when heated in carbon tetrachloride containing some benzoyl peroxide and N-bromosuccinimide, whilst being irradiated, gives a tribromide intermediate in 90% yield. This with sodium thiophenoxide in DMF gives 3-acetoxy-2-bromo-5-methoxy-α,α-di(phenylthio)toluene in 85% yield (K. Wahala et al., Synth. Commun., 1987, *17*, 137).

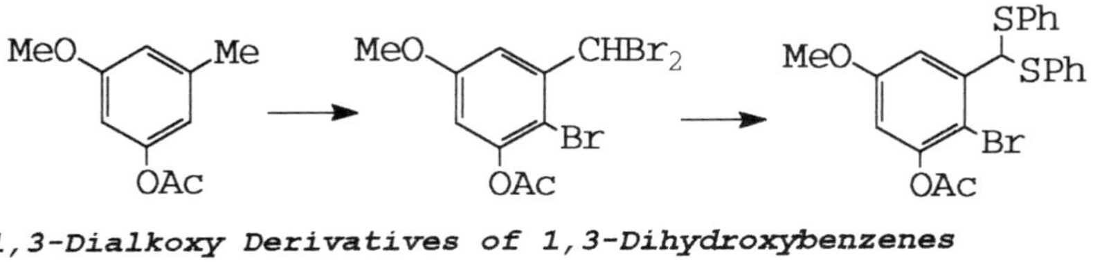

(ix) 1,3-Dialkoxy Derivatives of 1,3-Dihydroxybenzenes

The reactions in this group comprise ring substiution and dealkylation.

Monobromination of 1,3-dimethoxybenzene in carbon tetrachloride is effected at 30°C during 2 hours with N-bromosccinimide in the presence of acidic silica gel (type microbead 3A) giving 2,4-dimethoxybromobenzene in 99% (H. Konishi et al., Bull. Chem.

Soc. Japan, 1989, 62, 591).

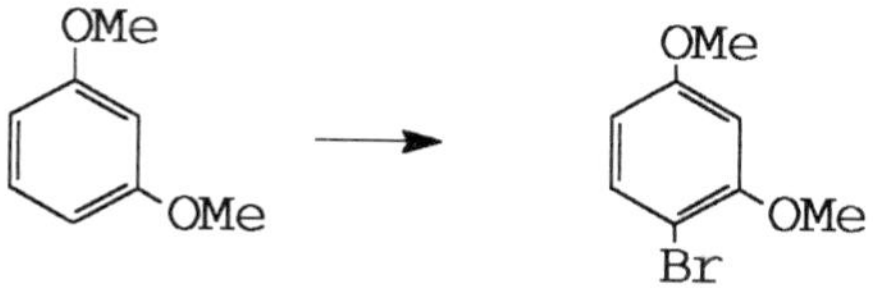

Mild monoformylation of 1,3-dimethoxybenzene takes place in dichloromethane containing tris(phenylthio)methane and subsequent treatment with dimethyl methylthiosulphonium fluoroborate affording 2,4-dimethoxybenzaldehyde in 70% yield (R.A.J. Smith, and A.R. Bin Manas, Synthesis, 1984, 166).

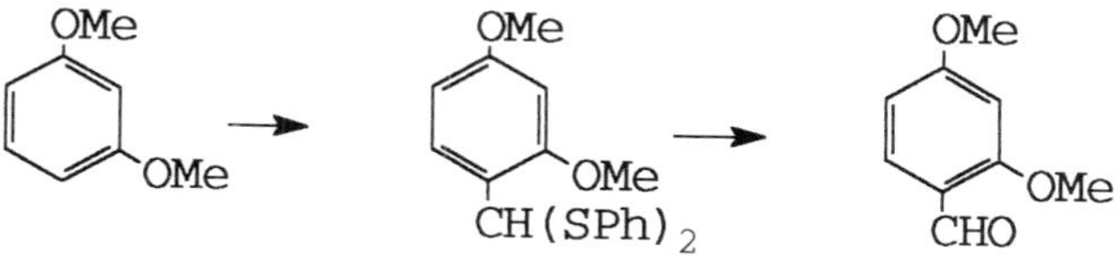

1,3-Dimethoxybenzene, magnesium monoperoxyphthalate and iron(III) meso-tetrakis(2,3,5,6-tetrafluorophenyl)tetrasulphonatoporphyrin in the proportions (1:3:0.01) in acetonitrile/tartrate buffer (pH 3) at 20°C affords 2-methoxy-1,4-benzoquinone in 95% yield (I. Artaud et al., J. Chem. Soc. Chem. Commun., 1991, 31).

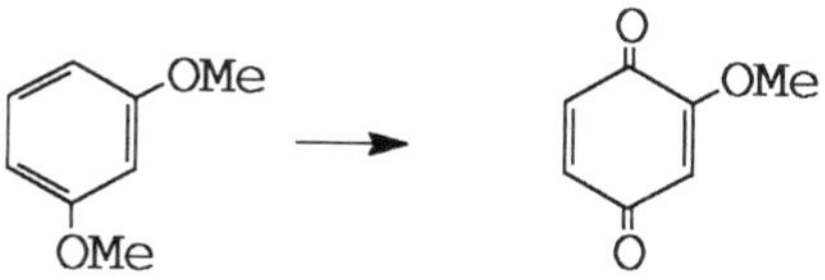

Although O-demethylation can be achieved readily by reagents such as lithium iodide in hot collidine (C.J. Baylis, S.W.D. Odle and J.H.P. Tyman, J. Chem. Soc. Perkin Trans I, 1981, 132), milder conditions are frequently desirable. Thus, 1-(5-chloro-3-formyl-2,6-dimethoxy-4-methylphenyl)-3-methylbut-2-ene on treatment with bromomagnesium ethyl mercaptide is converted in 53% yield to the natural resorcinol, colletochlorin, 2-(3-methylbut-3-enyl)-3-chloro-5-formyl-4-methylresorcinol (H. Saimoto and T. Hiyama, Tetrahedron Letters, 1986, 27, 597).

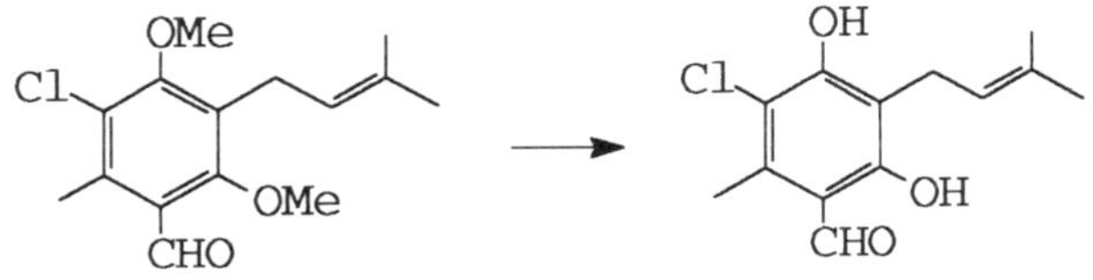

Other 1,3-dimethoxyarenes can be O-demethylated by reaction with methyl fluorosulphonate under mild conditions. In the example depicted a yield of 92% of the dihydroxy ketone was obtained (L. Colombo et al., J. Chem. Soc. Perkin Trans I, 1980, 2549).

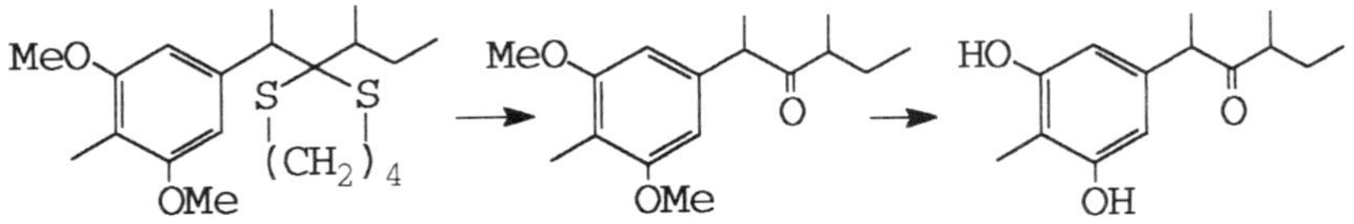

3-Cyano-5-hydroxy-7-pentylcoumarin is obtained through simultaneous demethylation and cyclisation of 4-(2,2-dicyanoethenyl)-3,5-dimethylpentylbenzene by heating with aluminium chloride in chlorobenzene. The yield is 96% (J.W. ApSimon, A.M. Holmes and I. Johnson, Can. J. Chem., 1982, *60*, 308).

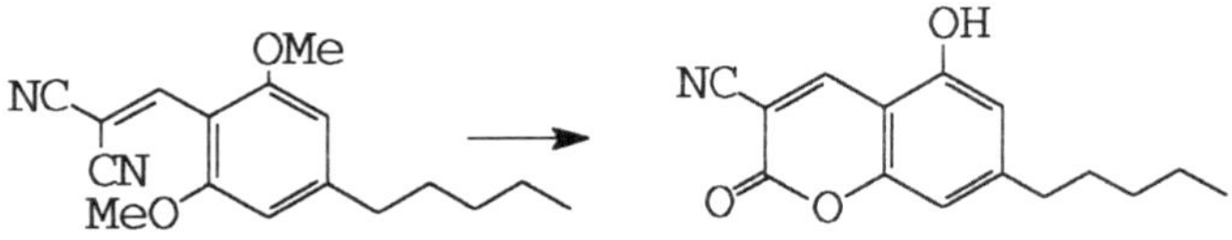

In the case of certain spiro cannabinoids treatment with the lithium salt of 2-methylpropane-2-thiol in HMPT at 70°C results in partial O-demethylation, giving the corresponding (+)-dehydrocannabispiran. (F.S. El-Feraly and Y.-M. Chan, J. Natur. Prod. (Lloydia), 1981, *44*, 557).

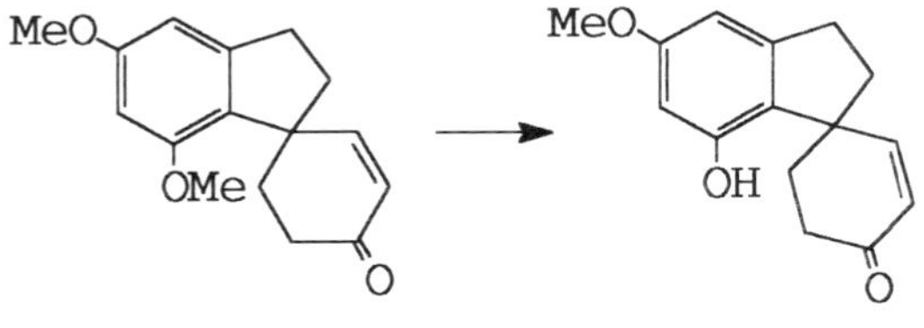

O-Methyl ethers of methyl 6-alkylsalicylates are demethylated by treatment with potassium 2-methylpropane-2-thiolate in DMF.

(J.H.P. Tyman and N. Visani, (unpublished work); PhD thesis, N. Visani, Brunel University, 1988).

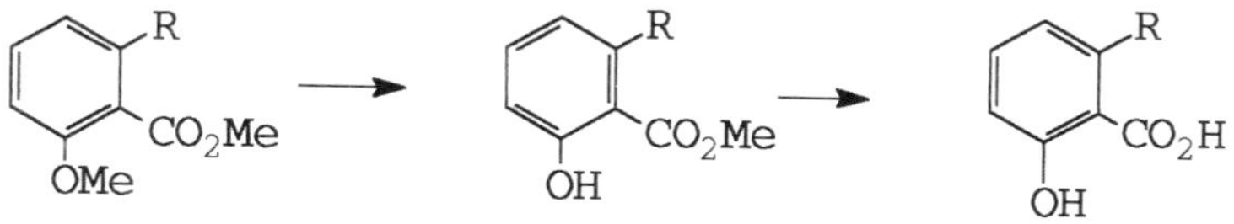

(x) 1,4-Dihydroxybenzenes (Hydroquinones) and their Derivatives
The reactions reviewed in this section consist of oxidation and substitution in the ring for the benzenoid series and a number of sequences for polycyclic compounds having 1,4-dihydroxy systems.
Hydroquinone in isopropanol is oxidised by a small amount of iodine and 35% hydrogen peroxide affording 1,4-benzoquinone in 86% yield (Sanko Kaihatsu KAG, DE 3834239).

2,3,6-Trimethylhydroquinone and phenyl iodosoacetate in methanol at ambient temperature gives a quantitative yield of 2,3,6-trimethyl-1,4-benzoquinone. The reagent is also applicable to the oxidation of 4-alkylphenols and 4-unsubstituted phenols (A. Pelter and S. Elgendy, Tetrahedron Letters, 1988, *29*, 677).

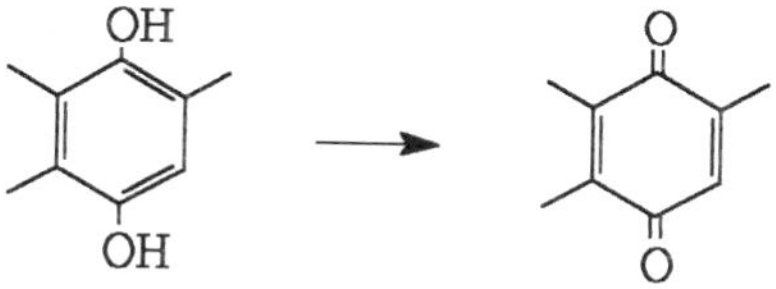

2,5-Dihydroxyacetophenone is formed in 76% yield by the reaction of hydroquinone with acetic anhydride and zinc chloride at 145-150°C. The same method applied to either resorcinol or phloroglucinol gives mixtures of mono and diacetyl derivatives while phenol, catechol and pyrogallol produce monoacyl compounds (A.S.R. Anjanryulu et al., Indian J. Chem., 1981, *26b*, 823).

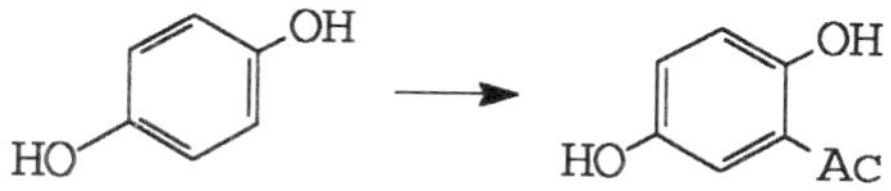

Although lithiation of hydroquinone has no useful outcome the 2-bromo derivative gives the corresponding lithio intermediate by halogen-metal exchange and thence 2-methylthio-1,4-dihydroxybenzene by reaction with dimethyldisulphide, together with some hydroquinone by reduction (H.J. Reich, D.P. Green and N.H. Phillips, J. Am. Chem. Soc., 1991, *113*, 1414; J.M. Saa, J. Morey, G. Suner, A. Frontera and A. Costa, Tetrahedron Letters, 1991, *32*, 7313).

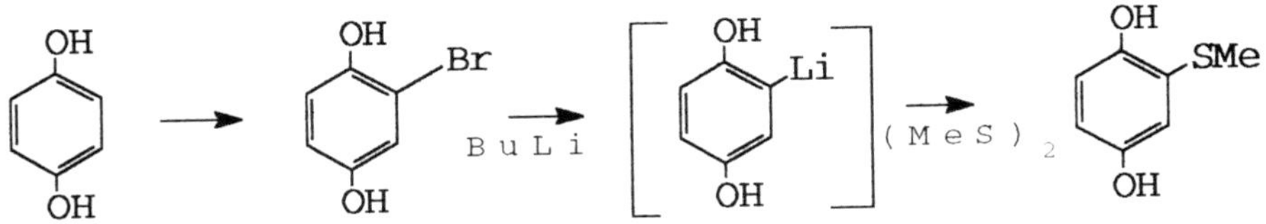

Addition of hydroquinone to cold acetic acid containing acetaldehyde and concentrated hydrochloric acid gives 6-hydroxy-2,4-dimethyl-1,3-benzodioxane in 72% yield, presumably by electrophilic substitution and cyclic acetal formation (V.Y. Denisov, T.N. Grishchenkova and E.P. Fokin, Khim. Geterosikl. Soedin., 1983, *17*, 1195).

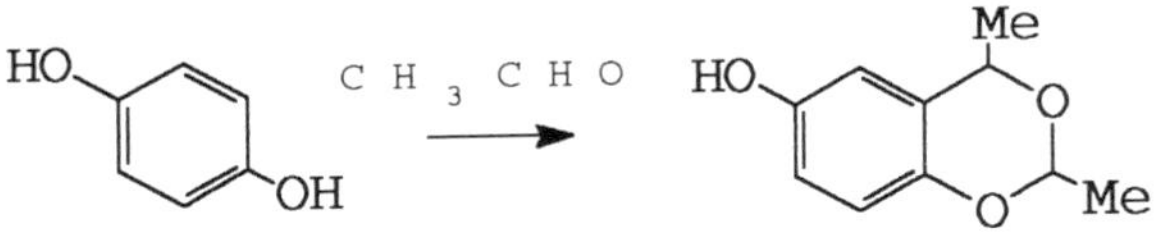

A reaction of 2-(2,2,4,4-tetramethyl)butylhydroquinone with 2-methyl-2,4-pentanediol in heptane containing Amberlyst 15 gives 3,4-dihydro-6-hydroxy-2,2,4-trimethyl-7-(1,1,3,3-tetramethyl-butyl)chroman in 71% yield (J.R.I. Eubanks, and J.G. Pacifici, Synth. Commun., 1987, *17*, 829).

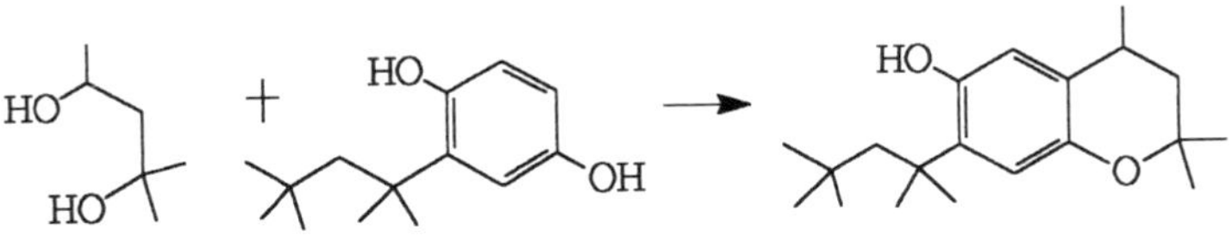

The reaction of 2,3,6-trimethylhydroquinone and boron trifluoride in toluene with methyl vinyl ketone cyanohydrin in toluene/nitromethane containing phosphoric acid at 0-5°C and then at ambient temperature gives 2-cyano-2,5,7,8-tetramethyl-6-

hydroxychroman in 81% yield (BASF AG, EP 15562).

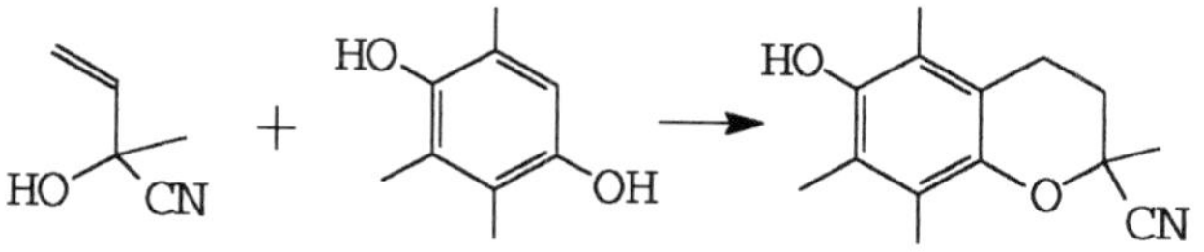

The same phenol, together with 1,1-dimethylprop-2-enol in a small volume of trifluoroacetic acid affords a 63% yield of 6-hydroxy-2,2,5,7,8-pentamethylchroman (F.M.D. Ismail, M.J. Hitton and M. Stefinovic, Tetrahedron Letters, 1992, 33, 3795).

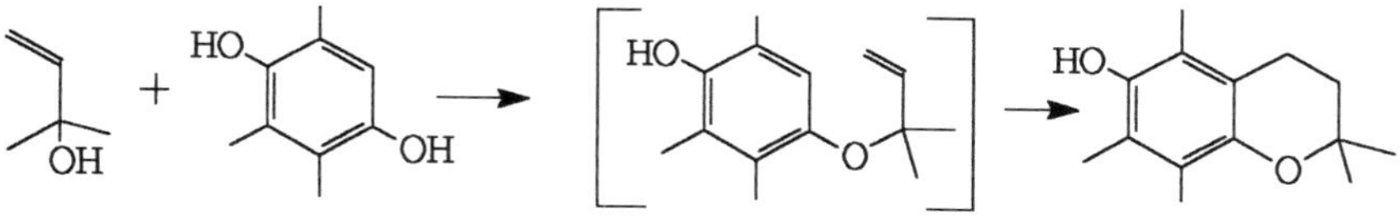

Bicyclic and polycyclic compounds containing 1,4-dihydroxy systems undergo reactions not exhibited by the parent compound. For example, naphthazarin after treatment first with boron trifluoride etherate and then with tetrahydro-5,5-dimethyl-2-furyl acetate gives, at ambient temperature, 5,8-dihydroxy-2-(2,2-dimethyltetrahydrofuran-5-yl)-1,4-naphthoquinone in 56% yield (Nippon Oil, KK, EP 200219).

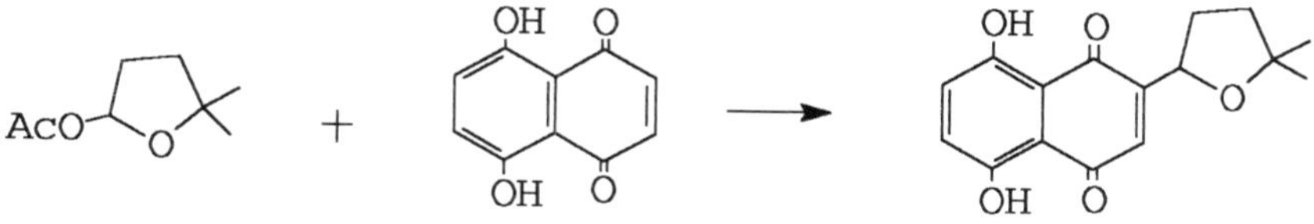

The preceding electrophilic substitution compares with the following nucleophilic reaction. 4-Hydroxyquinizarin (4,5,8-trihydroxyanthra-9,10-quinone in methanolic solution containing sodium methoxide upon refluxing with a nitroalkane produces the 7-alkylsubstituted compound regiospecifically in 65% yield (A.E. Ashcroft and J.K. Sutherland, J. Chem. Soc. Chem. Commun., 1981, 175).
Alkyl 3,5,8-trihydroxy-9,10-anthraquinone-2-carboxylates after conversion to the more activated alkyl 3,5,8,9,10-pentamethoxyanthracenes undergo selective 7-substitution under acidic conditions with 1-trifluoroacetoxy-2,3,4,6-tetra-O-benzyl-β-D-glucopyranose, leading after several further steps,

to a synthesis of carminic acid, the colourant principle of cochineal (P. Allevi et al. J. Chem. Soc. Chem. Commun., 1991, 1319).
Quinizarin reacts with 1,3-diaminoethane in pyridine containing copper(II) chloride giving a 98% yield of 6-hydroxy-1,2,3,4-tetrahydronaphtho[2,3-f]quinoxaline-7,12-dione (M. Matsuoka et al., Chem. Letters, 1980, 743).

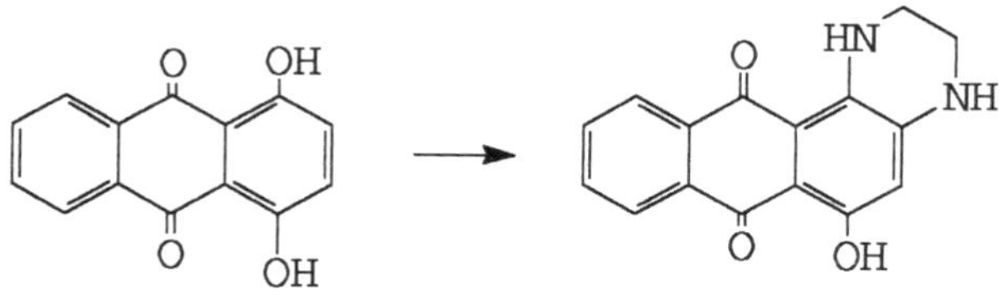

(xi) Mono O-Alkyl Derivatives of 1,4-Dihydroxybenzenes
Reactions in this group comprise those at the hydroxyl group and substitution in the ring.
The monoester, butyl 2-(4-hydroxyphenoxy)propionate with 2-chloro-5-trifluoromethylpyridine in DMF containing potassium fluoride on alumina upon heating gives the corresponding ether in 88% yield. The 5-nitro analogue behaves similarly (K-J. Hwang and S.K. Park, Synth. Commun., 1990, *20*, 949).

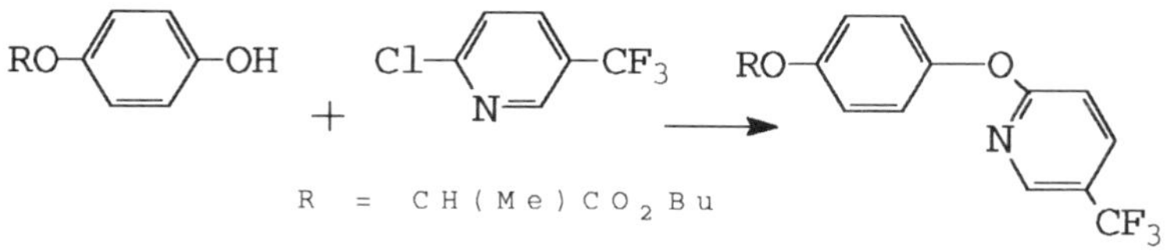

$R = CH(Me)CO_2Bu$

4-Methoxyphenol as the allyl, or propargyl, ether is oxidised anodically to the corresponding quinonemonoketal (S. Dhanalekshmi, K.K. Balsubramaniam, C.S. Venkatachalam, Tetrahedron Letters, 1991, *32*, 7591).

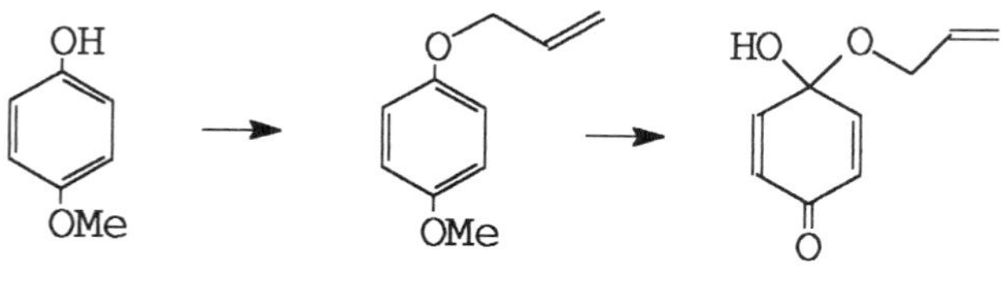

Etherification of 4-methoxyphenol occurs with (R)-4-phenyl-1,1,1-trichloro-2-butanol in dimethoxyethane/water (3:2) by the action of concentrated aqueous sodium hydroxide. This affords an 84% yield of (S)-2-(4-methoxyphenyl)-4-phenylbutanoic acid (E.J. Corey and J.O. Link, Tetrahedron Letters, 1992, 33, 3431).

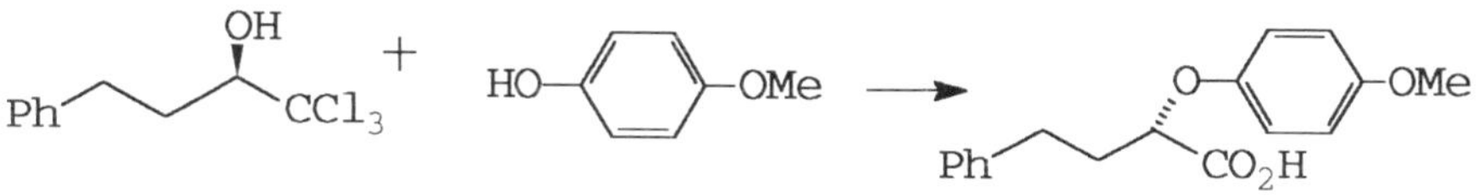

Substitution in the ring of 4-methoxyphenol occurs, when it is treated with glyoxylic acid, alumina and aqueous sodium hydroxide. This furnishes a 75% yield of 2-hydroxy-5-methoxymandelic acid, which upon oxidation gives 2-hydroxy-5-methoxybenzaldehyde (Ube Industries, KK, J54 044634).

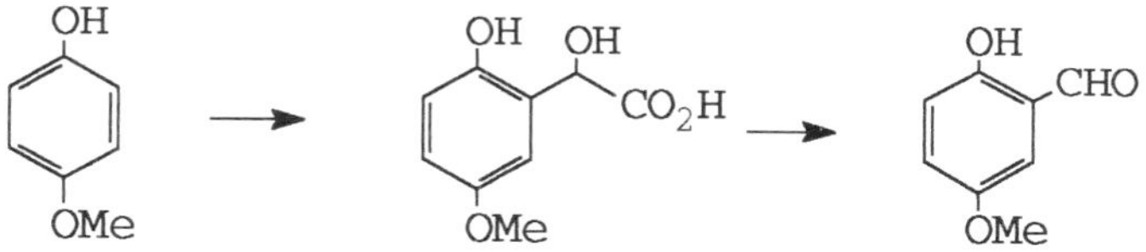

2-Tert-butyl-4-methoxyphenol in acetic acid containing copper(II) acetate when heated under argon gives 2-acetoxy-4-methoxy-6-tert-butylphenol, although the corresponding 4-methyl and 4-phenyl analogues fail to react (Y. Takizawa et al., J. Chem. Soc. Chem. Commun., 1991, 104).

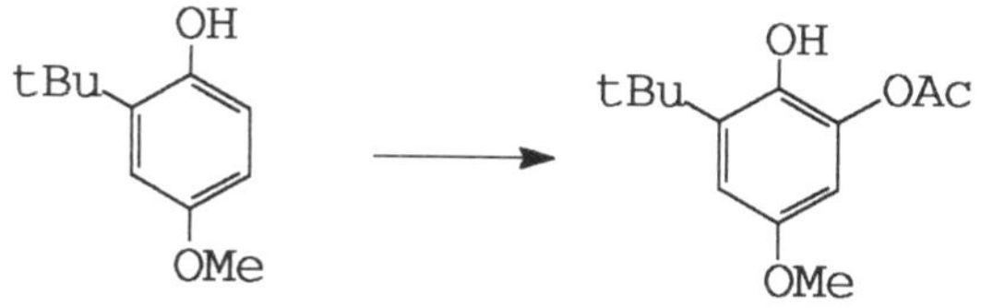

4-Methoxyphenol, the aglycone of arbutin, in dichloromethane containing cyclopentadienylhafnium dichloride, silver perchlorate and a powdered molecular sieve (type 4a) reacts with fluorofuranosides in dichloromethane to afford α-glycosides (T. Matsumoto, M. Katsuki and K. Suzuki, Tetrahedron Letters, 1988, 296935).

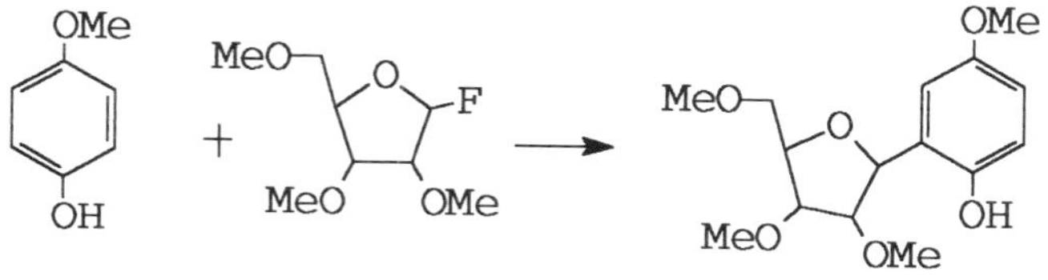

(xii) Mono O-Acyl Derivatives of 1,4-Dihydroxybenzenes

Most of the reactions in this section involve formation of oxygen heterocyclic bicyclic compounds.

Highly substituted hydroquinone monoacetates react with 3-chloro-2-methyl-1-propene in dichloromethane/acetic acid (10:1) containing zinc chloride to give dihydrobenzofurans. 2,2,7-Trimethyl-4,6-diisopropyl-5-acetoxy-2,3-dihydrobenzofuran, for example can be obtained in 72% yield (Eisai, KK JO 7145871).

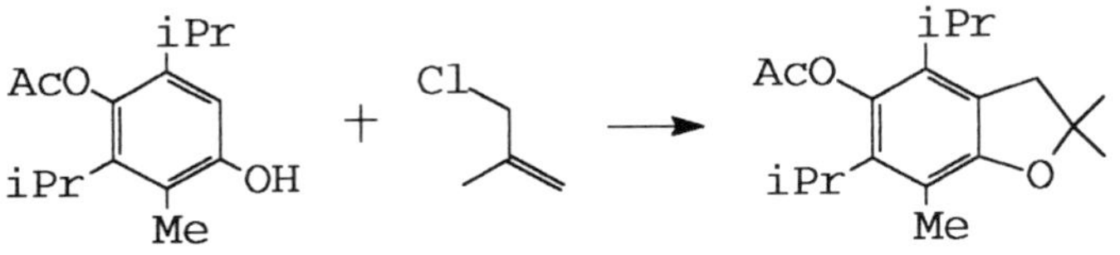

Cleavage of 3,4-epoxy-3-methyl-1-butene with 2,3,5-trimethyl-4-acetoxyphenol at ambient temperature with a catalytic amount of tetrakis(triphenylphophine)palladium(0) occurs to give 2,3,6-trimethyl-4-(1-hydroxymethyl-1-methylallyloxy)phenyl acetate without chroman formation (Hoffman La Roche AG, EP 183042).

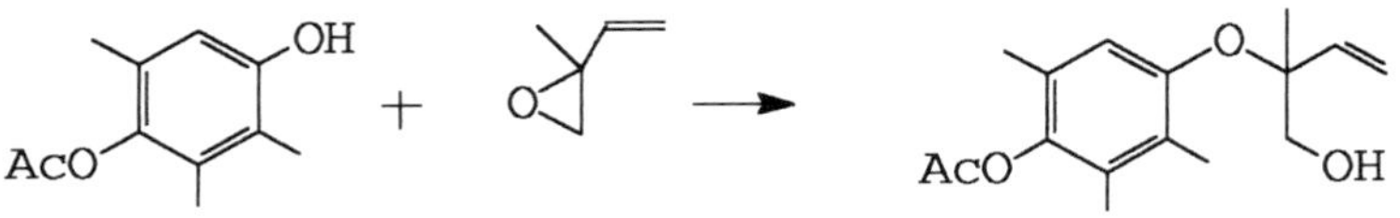

By contrast, in the reaction of the 5-methyl-5-vinylbutyrolactone with 2-methyl-4-benzoyloxyphenol in boiling dioxan in the presence of boron trifluoride etherate, 2,8-dimethyl-2-(2-carboxyethyl)-6-benzoyloxychroman is obtained (Eisai, KK, J57 145871).

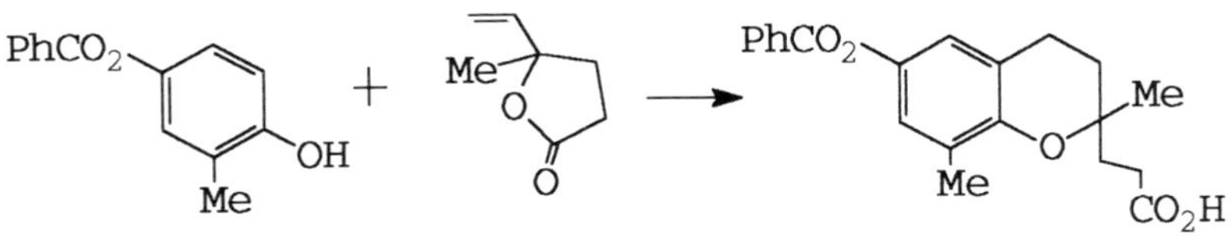

3-Phenyl-5-hydroxy-2-oxo-2,3-dihydrobenzofuran when stirred with methyl phenyl glyoxylate in the presence of 73% sulphuric acid at 120°C affords the symmetrical derivative 3,7-diphenyl-2,6-dioxo-2,6-dihydrobenzo[1,2-b:4,5-b']difuran (Bayer AG EP

AG EP 252406).

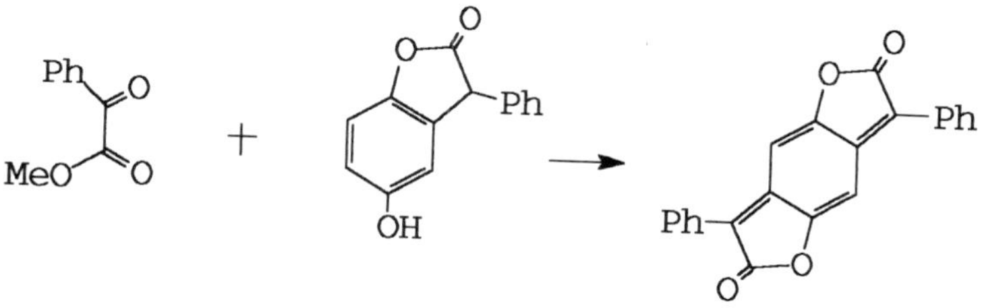

(xiii) 1,4-Dialkoxy Derivatives of 1,4-Dihydroxybenzenes

1,4-Dimethoxybenzene in benzene at 0°C, on treatment with a mixture of nitric acid-impregnated manganese dioxide and celite, followed by ultrasonification gives a 87% yield of benzoquinone (R. Araya, R. Tapia and J.A. Valderrama, J. Chem. Res., (S), 1987, 84).

The activity of 1,4-dimethoxybenzene towards electrophilic substitution is shown by the formation of 1,4-dimethoxythioxanthone in 87% yield from its addition to 2-chlorosulphenylbenzoyl chloride in 1,2-dichloroethane containing stannic chloride (J.F. Honek, M.L. Mancici and B. Belleau, Synth. Commun., 1983,

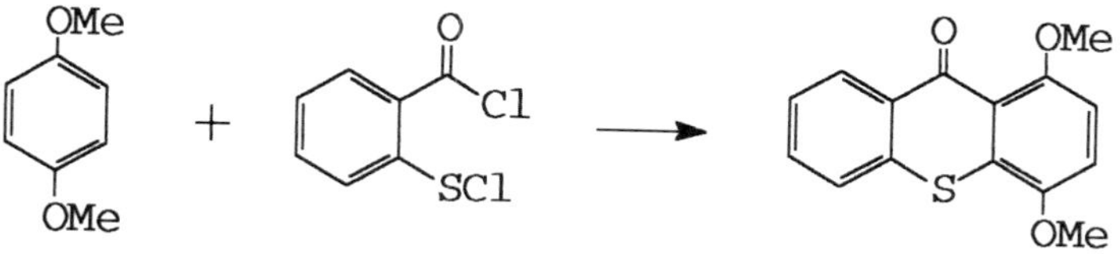

Preferential and selective O-deacetylation of a variety of structures in THF is achieved by reactions with N-methyl 2-dimethylaminoacetohydroxamic acid in a phoshate buffer at ambient temperature. For example, ethyl 6-acetoxy-3-chlorocoumarin-4-carboxylate is cleaved to give the 6-hydroxy analogue in 95% yield (M. Ono and I. Itoh, Tetrahedron Letters, 1989, 30, 207).

Selective debenzylation of 4-methoxyphenyl benzyl ether in HMPT is effected by the addition of sodium metal and heating at 100°C. This furnishes a 93% yield of 4-methoxyphenol (L. Testaferi et al., Tetrahedron Letters, 1989, *38*, 207).

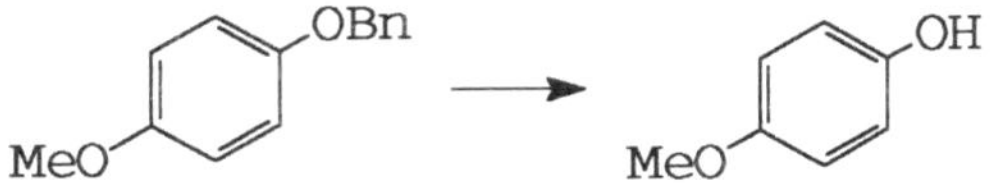

Selective demethylation can be effected with the boron trifluoride. Thus, with 1,4,5-trimethoxyanthraquinone in boiling benzene containing boron trifluoride etherate an intermediate 4,10-chelated compound is formed, which is transformed with hot methanol to 4-hydroxy-1,5-dimethoxyanthraquinone in 94% yield (P.N. Preston, T. Winnick and J.O. Morley, J. Chem. Soc. Perkin Trans. I, 1983, 1439).

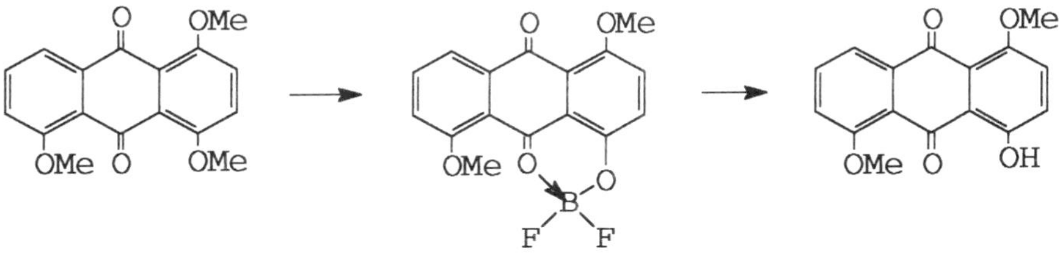

(xiv) Diacyl Derivatives of 1,4-Dihydroxybenzenes
Often the partial hydrolysis of symmetrical aromatic derivatives proceeds poorly. However, the di-pivalyl ester of hydroquinone when added to methanolic potassium hydroxide and reacted at ambient temperature gives the half-hydrolysed ester in 94% yield (Fuji Photo Film, KK, J62 249945).

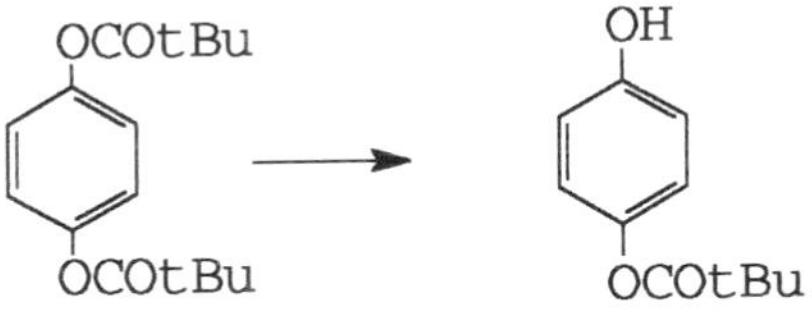

Partial deacetylation of 9,10-diacetoxy-1,2,3,4-tetrahydro -

anthracene in dimethoxyethane is caused by stirring it with a 10-fold excess of sodium borohydride for 30 hours at 60°C. A 90% yield of 10-acetoxy-9-hydroxy-1,2,3,4-tetrahydroanthracene is obtained (S.R. Angle and W. Yang, J. Am. Chem. Soc., 1990, *112*, 4524).

(xv) Naphthalenic and Polycyclic Dihydric Systems

Naphthalene-1,3-diol in pyridine added to a toluene suspension of potasium superoxide under argon furnishes 2-hydroxy-1,4-naphthaquinone in 69% yield (D. Vidril-Robert and M. Hocquax, Tetrahedron Letters, 1984, *25*, 533).

Naphthalene-1,5-diol and 4-dimethylaminoaniline in ethanol/water (1:1) with excess potassium ferricyanide at ambient temperature during 1 hour afford the corresponding quinonimine in 85% yield (L.V. Ektova and R.O. Shishkina, Izv. Akad. Nauk. SSSR Ser. Khim., 1990, 2851).

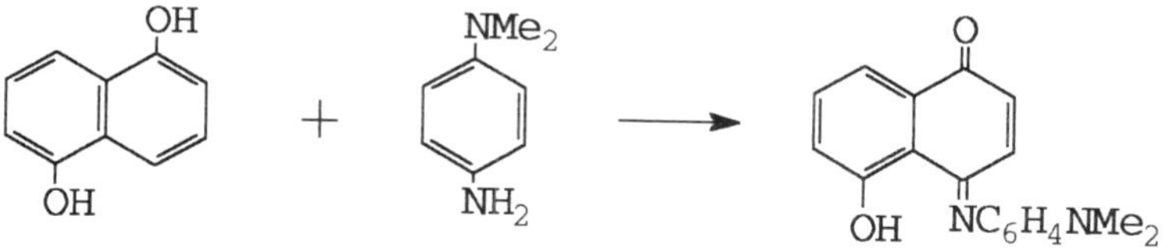

The Mannich reaction of 1-hydroxy-5-methoxynaphthalene with primary amines, such as cyclohexylamine, and 35% formalin in methanol leading to a 80% yield of 3-cyclohexyl-2,4-dihydro-7-methoxy-2H-naphth[2,1-e][1,3]oxazine has been referred to earlier; secondary amines give 2-dialkylaminomethylation without oxazine ring formation (H. Mohrle and H. Folttmann, Arch. Pharm., 1988, *321*, 259).

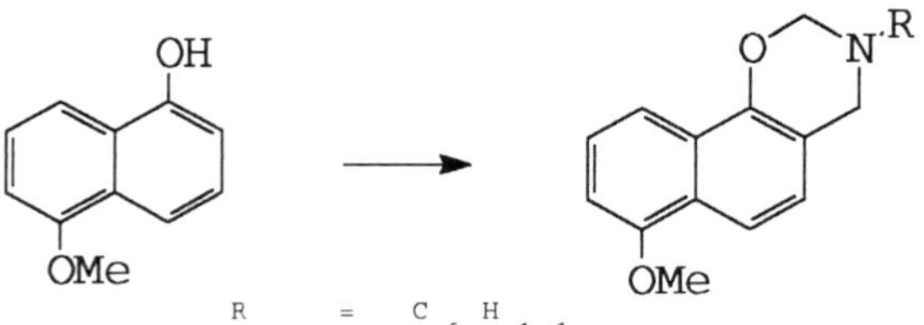

2,7-Dihydroxynaphthalene combines with 2,4-dioxophenylbutanoic acid in the presence of phosphorus oxychloride to give 6-hydroxy-1-oxo-9-phenyl-1H-phenalene-7-carboxylic acid (R.G. Cooke et al., Austral. J. Chem., 1980, *33*, 237).

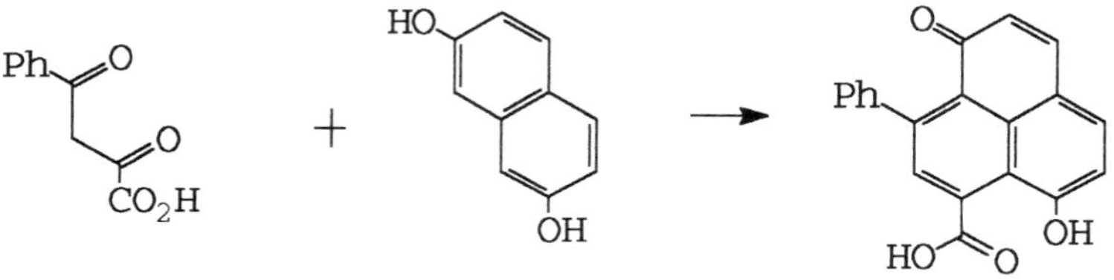

3 Polyhydric Phenols

Synthesis of Trihydroxybenzenes and Derivatives of Tetrahydroxybenzenes

This section recounts progress in the chemistry of 1,3,5-trihydroxybenzene (phloroglucinol), the 1,2,4-trihydroxy (pyrogallol) and the 1,2,4-isomer (hydroxyquinol). This series continues to attract interest mainly because numerous complex derivatives possess these substitution patterns.

While no new synthetic methods have appeared, phloroglucinol attracts interest because of its industrial usage. The classic route based on trinitrotoluene, trinitrobenzoic acid and triaminobenzoic acid has been challenged by other approaches using acyclic precusors.

More recently enol ethers of cyclic alkenes have been reacted with malonyl chloride to give 4,6-cycloalkano-1,3,5-trihydroxybenzene systems with yields ranging from 18% to 70%. For example, the methyl enolic ether of cyclooctene affords the phloroglucinol compound depicted (F. Effenburger, K-H. Schonwalder and J.J. Stezowski, Angew. Chem. Int. Ed., 1982, *21*, 871).

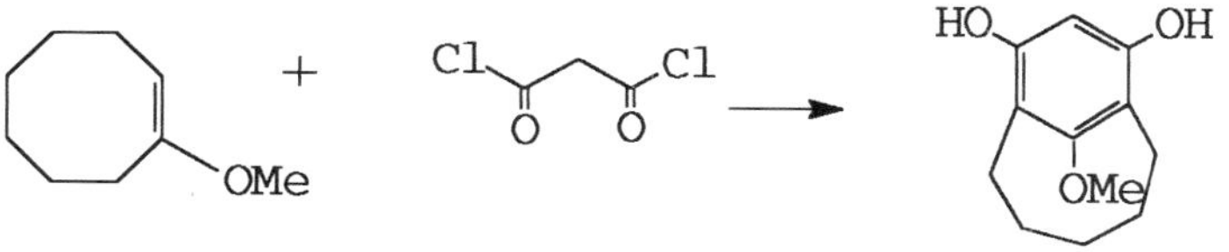

The traditional synthesis of 1,2,3-trihydroxybenzenes by decarboxylation of gallic acid has been superseded by the use of acyclic precursors. Thus, dimethyl oxomalonate and dimethyl glutarate undergo cyclodehydration and tautomerism to a product which when hydrolysed decarboxylates to pyrogallol (M.T. Shipchandler, C.A. Peters and C.D. Hurd, J. Chem. Soc. Perkin Trans I, 1975, 1400).

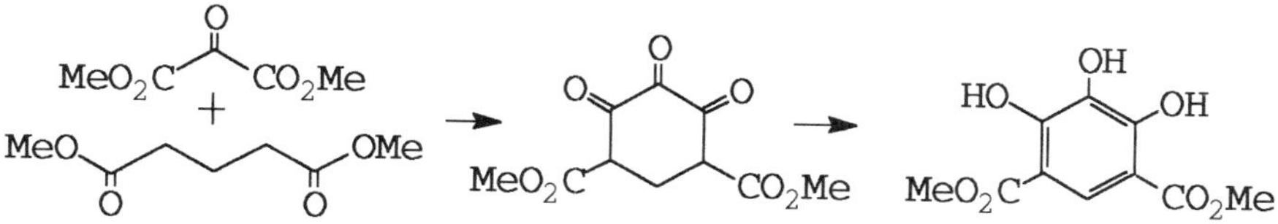

2,2',6,6'-Tetrahalogenocyclohexanones can be hydrolysed to pyrogallol derivatives, although the yields are only in the region of 20% (UK Patent 1574713).

Acyclic precursors can be employed as in the synthesis of 2,6-diphenyl-5-benzyl-1,3,4-trihydroxybenzene from 1,3-diphenyl acetone and 3-phenylcyclobutenedione (W.Ried and W. Kunkel, Justus Liebig's Ann. Chem., 1968, 717, 54).

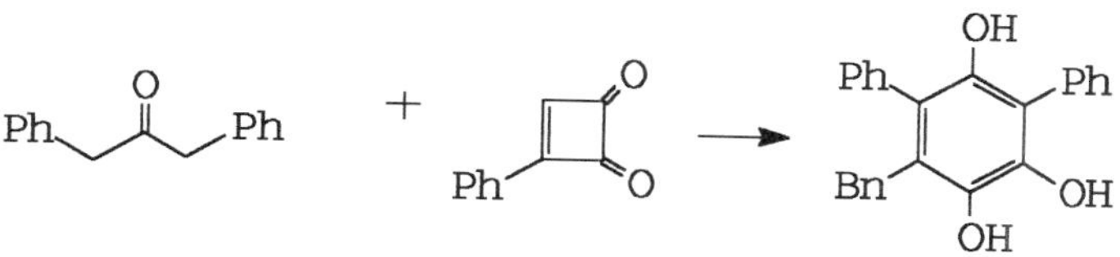

3,4,5,6-Tetraacetoxy-2,7-diphenylisobenzofuran is prepared from 3,4-diformyl-2,5-diphenylfuran and glyoxal through cycloaddition and reductive acetylation (D. Passerieux et al., Bull. Chim. Soc. Fr., 1989, 441).

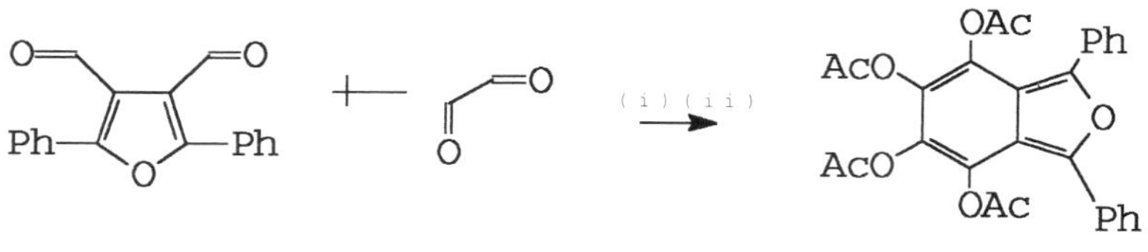

(i) Δ (ii) Ac$_2$O

Reactions of Polyhydroxybenzenes

(i) 1,3,5-Trihydroxybenzene and its Alkoxy Derivatives
Acylation reactions leading to coumarins, chromones and flavenols might be anticipated with this highly activated system, and represent the main development. Phloroglucinol and phenylacetonitrile plus a catalytic amount of boron trifluoride etherate upon saturation wih hydrogen chloride afford a ketimine hydrochloride which by acidic hydrolysis gives a 82% yield of benzyl 2,4,6-trihydroxyphenyl ketone. The use of zinc chloride and acid anhydrides is much inferior (S. Mohanty and S.K. Grover, Current Science, 1988, 57, 537).

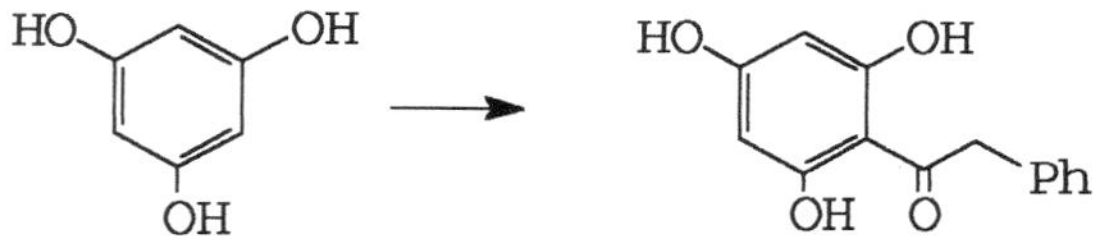

Phloroglucinol and Meldrum's acid in pyridine with benzaldehyde form 4-phenyl-5,7-dihydroxy-3,4-dihydrocoumarin in 87% yield (V. Nair, Synth. Commun.,1987, 17, 723).

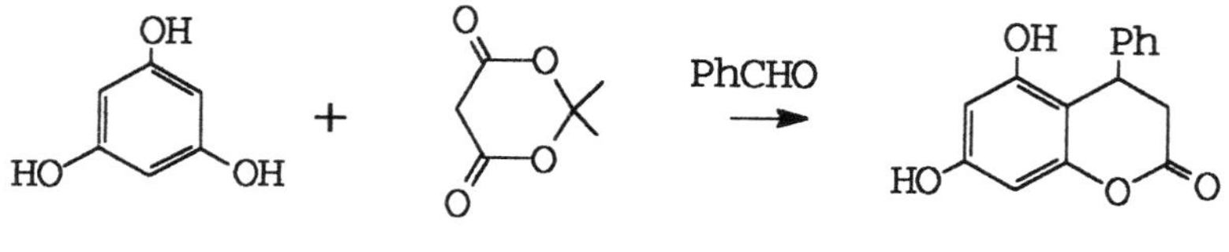

By contrast, reaction of phloroglucinol with 3,3-dimethylacrylic acid in the presence of anhydrous aluminium chloride and phosphorus oxychloride results in a 90% yield of 2,2-dimethyl-5,7-dihydroxychroman-4-one (D. Somithran and K.J. Rajendraprasad, Synthesis, 1984, 545).

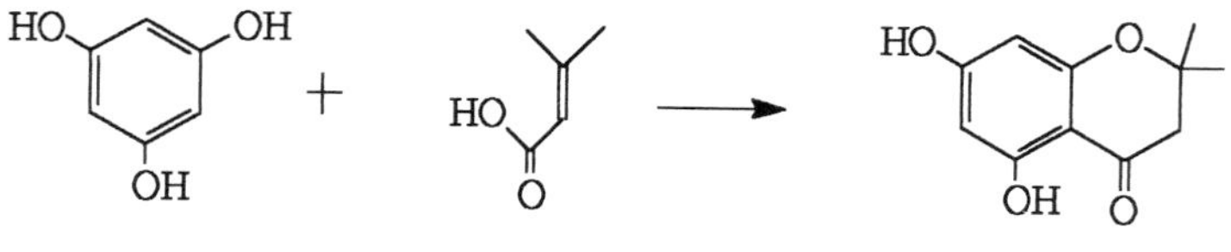

Phloroglucinol reacts with the quinonemethide, 2,6-dimethoxy-4-propenylidenecyclohexa-2,5-dienone in benzene/DMF solution (25:1) containing 4-toluenenesulphonic acid to give a C-alkylated intermediate which when stirred at ambient temperature in benzene/acetone with silver oxide and then acetylated cyclises to 2-(4-acetoxy-3,5-dimethoxyphenyl)-5,7-diacetyl-3-acetoxyflav-3-ene (A. Zanarotti, Tetrahedron Letters, 1982, 23, 3963).

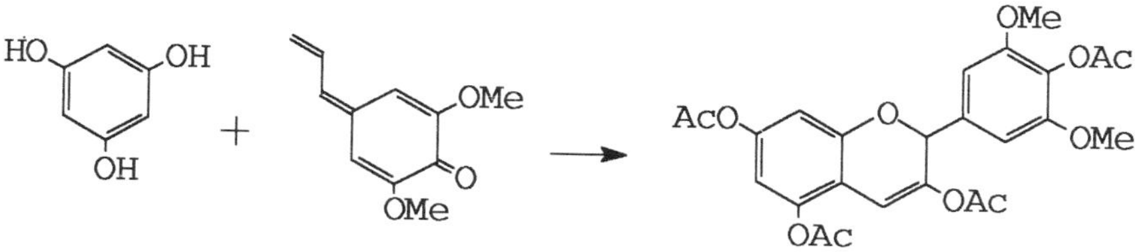

Mono and dialkoxy as well as diacyloxy derivatives of phloroglucinol serve as protected compounds in various syntheses. Rearrangement of diethyl 3,5-dimethoxyphenyl phosphate (derived from 3,5-dimethoxyphenol) in THF at -78°C with lithium diisopropylamide and reaction at 0°C for 30 minutes furnishes a 93% yield of 2,4-dimethoxy-6-hydroxyphenyl phosphonate (L.S. Melvin, Tetrahedron Letters, 1981, 22, 3375). The phosphorodiamidate group (prepared by a standard procedure on the phenol) has been used to direct and effect the ortho methylation of 3,5-dimethoxyphenol. Thus the O-phosphorodiamidate is treated with sec-butyllithium and then quenched with methyl iodide to give a 93% yield of 2-methyl-3,5-dimethoxyphenol, after acidic hydrolysis of the bis(dimethylamino)phosphoryl group with hot formic acid (M. Watanabe et al., Chem. Pharm. Bull., 1981, 37, 2564).

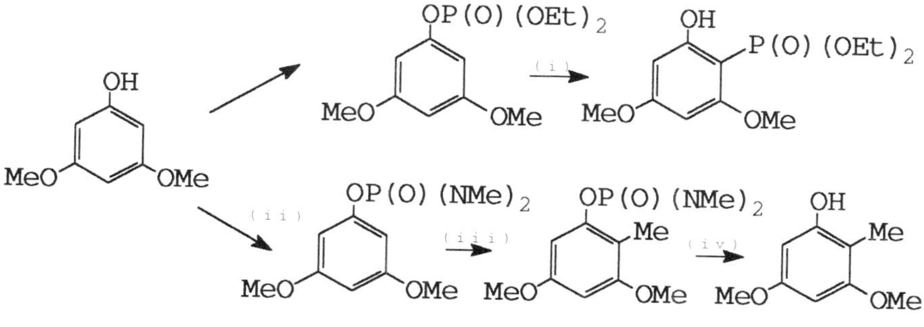

(i) LiN(iP)$_2$ (ii) ClP(O)(NMe)$_2$ (iii) s-BuLi, MeI (iv) H$_3$O$^+$

Trimethylphloroglucinol and hexamethylenetetramine after boiling in methanol during several hours afford a aminoalkylated product in 89% yield (N. Risch, J. Chem. Soc. Chem. Commun., 1983, 532).

Dehydroxylation of 4-tert-butyl-2,6-dimethoxyphenol is achieved from the diethylphosphate in THF/t-butanol by addition to 100% ammonia containing lithium. This affords 3-tert-butyl-1,5-dimethoxy-1,4-cyclohexadiene (R = H)(F.J. Sandina et al., J. Org. Chem., 1982, 47, 1576).

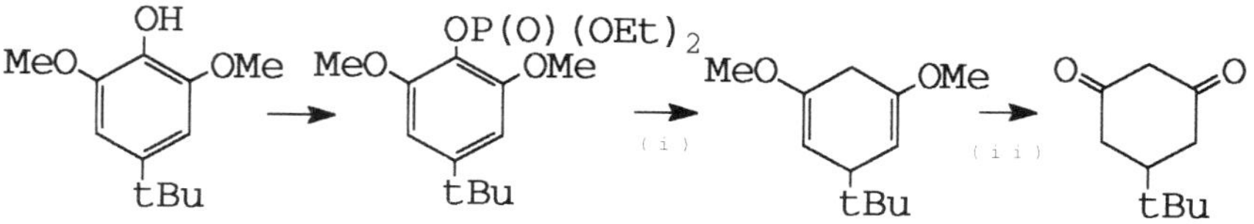

(i) Li/NH$_3$ (ii) H$_3$O$^+$

3,5-Dibenzyloxyphenol in hydrogen bromide-saturated dichloromethane when treated with 4-acetoxy-2-bromotetrahydrofuran-3-one affords a 42% yield of 3-acetoxy-3a-bromo-3[4,6-bis(dibenzoyloxy)]-2,3,3a,8a-tetrahydrofuro[2,3b]benzofuran(G. Buchi et al., J. Am. Chem. Soc., 1981, 103, 3497).

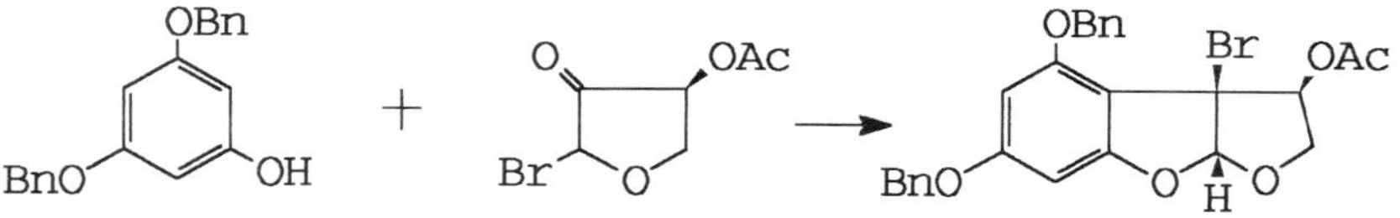

5-Hydroxy-7,4'-dimethoxyflavone (Ar = p-OMeC$_6$H$_4$) is formed in 80% yield from 2'-hydroxy-3'-iodo-4,4',6'-trimethoxychalcone in DMF by treatment with nickel(II) chloride, zinc powder and potassium iodide and heating (S.M. Ali, J. Iqbal and M. Ilyas, J. Chem. Res., (S), 1984, 236).

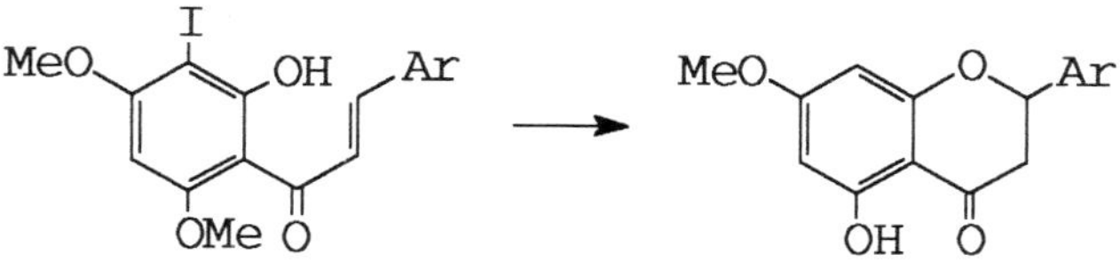

6-Cinnamoyl-5,7-dihydroxychroman is derived in 80% yield from 5,7-dihydroxychroman and cinnamic acid by treatment of the mixture with freshly-fused zinc chloride and phosphorus oxychloride and reaction at ambient temperature (R.V. Suresh, C.S.R. Iyer and P.R. Iyer, Heterocycles, 1986, *24*, 1925).

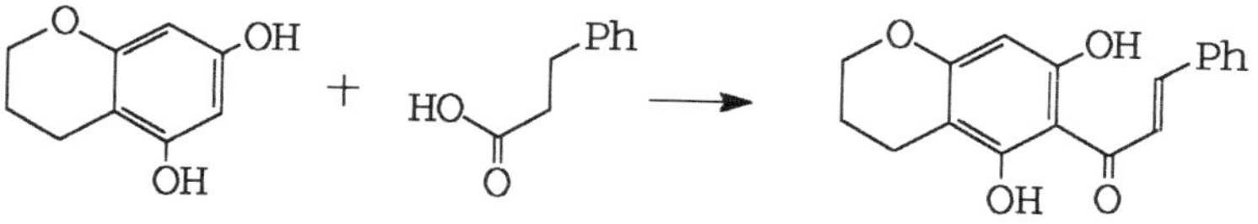

The benzylic position reacts, rather than the ring, when (+)-penta-O-acetylcatechin is brominated with N-bromsuccinimide and a small amouunt of benzoyl peroxide in carbon tetrachloride giving a 83% conversion (J.A. Steenkamp, J.C.S. Malan and D. Ferreira, J. Chem. Soc. Perkin Trans. I, 1988, 2179).

(ii) *1,2,3-Trihydroxybenzenes and Alkoxy Derivatives*

2,3,4-Trihydroxyacetophenone in xylene containing orthophosphoric acid on treatment with buta-1,3-diene at 30-35°C gives 5-(but-2-enyl)-2,3,4-trihydroxyacetophenone and a small quantity of 6-acetyl-7,8-dihydroxy-2-methyl-3,4-dihydro-2H-benzpyran which accumulates on heating the reaction mixture (V.K. Aluwalia, M. Singh and R.P. Singh, Synth. Commun., 1984, *14*, 127).

The phenyl analogue of this compound, 6-acetyl-7,8-dihydroxy-2-phenyl-3,4-dihydro-2H-benzopyran is synthesised in 35% yield by the addition of cinnamyl alcohol in benzene to a suspension at 60°C of 2,3,4-trihydroxyacetophenone in orthophosphoric acid in benzene (2:5). Some uncyclised 5-cinnamyl-2,3,4-trihydroxyacetophenone (25%) can be cyclised by heating with phosphoric acid at 80°C at which temperature the starting material gave the cyclised product directly in 55% yield (V.K. Aluwalia, K.K. Arora and K. Mukherjee, Synthesis, 1984, 127).

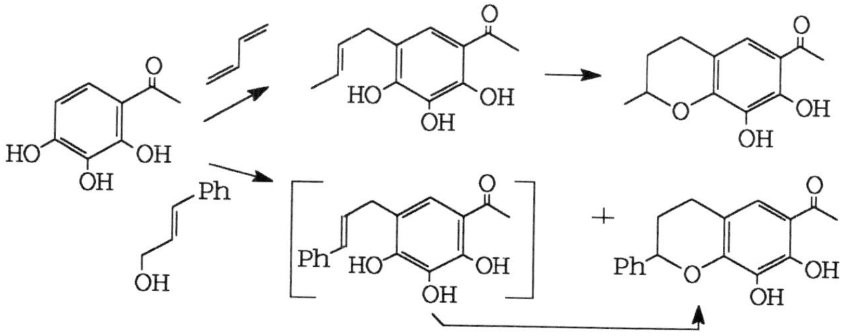

The corresponding aldehyde, 2,3,4-trihydroxybenzaldehyde in light petroleum containing 85% orthophosphoric acid upon treatment with 2-methylbut-3-en-2-ol in light petroleum affords 6-formyl-7,8-dihydroxy-2,2-dimethyl-3,4-dihydro-2H-benzopyran in 75% yield (V.K. Aluwalia, K.K. Arora and R.S. Jolly, Synthesis, 1981, 527).

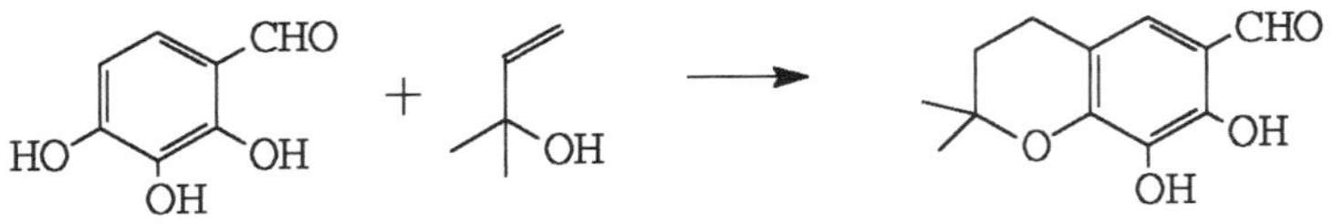

Pyrogallol trimethyl ether in dichloromethane is selectively demethylated wih boron trichloride by reaction at -10°C initially and during 3 hours at ambient temperature to give a 90% yield of 2,6-dimethoxyphenol (C.F. Carvalho and M.V. Sargent, J. Chem. Soc. Perkin Trans I, 1984, 27).

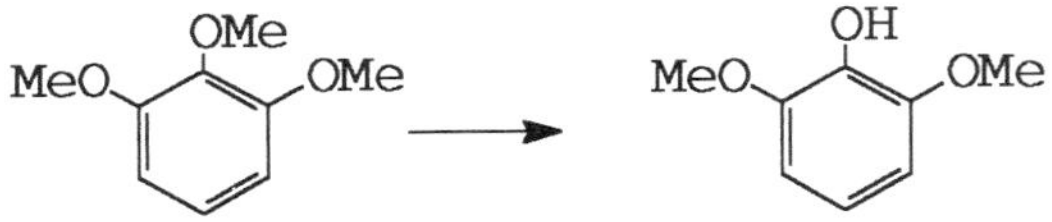

By contrast, 2,3,4-tris(2-ethylhexyloxy)benzaldehyde in 1,2-dichloroethane is monodealkylated selectively by treatment with titanium tetrachloride at 8-12°C with continued reaction at ambient temperature to afford 2-hydroxy-3,4-bis(2-ethylhexyloxy)- benzaldehyde in 73% yield (Fuji Photo Film, KK, J59 082333).

[R = CH_2 CH(Et)Bu]

3,4,5,4'-Tetramethoxystilbene is selectively demethylated in 80% yield giving 3,4-dihydroxy-4',5-dimethoxystilbene by heating for 6 hours in benzene containing lithium aluminium hydride (J.F. Carroll, S. Kulkowit and M.A. Mckervey, J. Chem. Soc. Chem. Commun., 1980, 507).

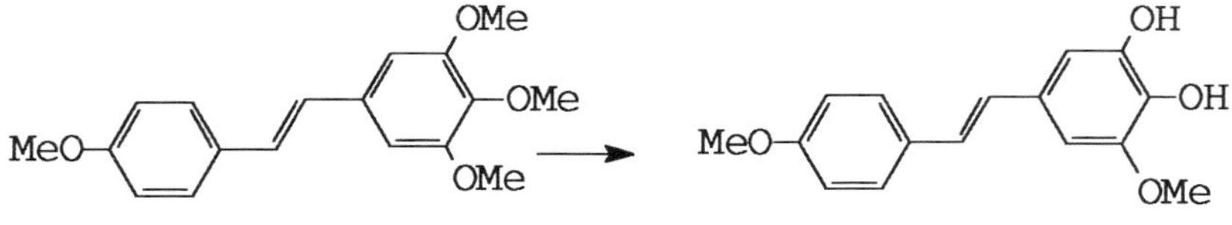

The pyrogallol ring system is obtained through a Dakin reaction on 2-methoxy-4,6-diformylphenol in aqueous sodium hydroxide by treatment at 5°C with a small quantity of copper sulphate pentahydrate and 10% hydrogen peroxide. This gives a 91% yield of 3-methoxy-5-formylcatechol (Ube Industries KK, J57 072932).

(iii)1,2,4-Trihydroxybenzene and Alkoxy Derivatives
3,4-Dimethoxyphenol and (E)-3,4-dimethoxypropenylbenzene in acetonitrile give, in 65% yield, trans-2,3-dihydro-2-(3,4-dimethoxyphenyl)-5,6-dimethoxy-3-methylbenzofuran upon treatment with phenyl iodosodi(trifluoroacetate), although the triacetate gives lower yields (S. Wang, B.D. Gates and J.S. Swenton, J. Org. Chem., 1991, 56, 1979).

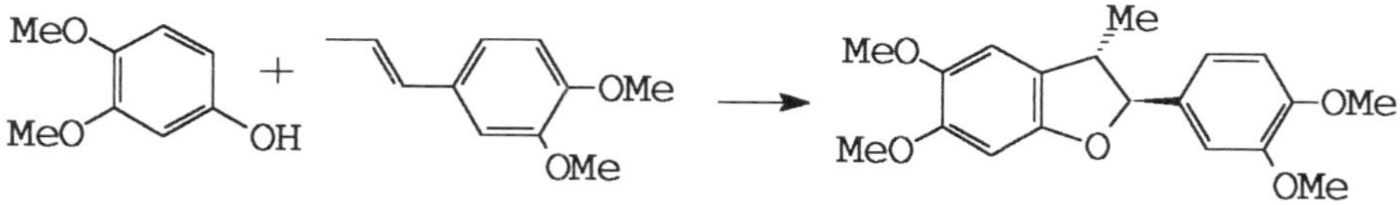

3,4-Dimethoxy-6-methylphenol and furan in acetic anhydride containing tetra-n-butylammonim fluoroborate, on anodic oxidation, couple together to form 2-(furan-3-yl)-4,5-dimethoxy-2-methylcyclohexa-3,5-dienone (Y.Shizuri, K. Nakamura amd S. Yanamura, J. Chem. Soc. Chem. Commun., 1985, 530).

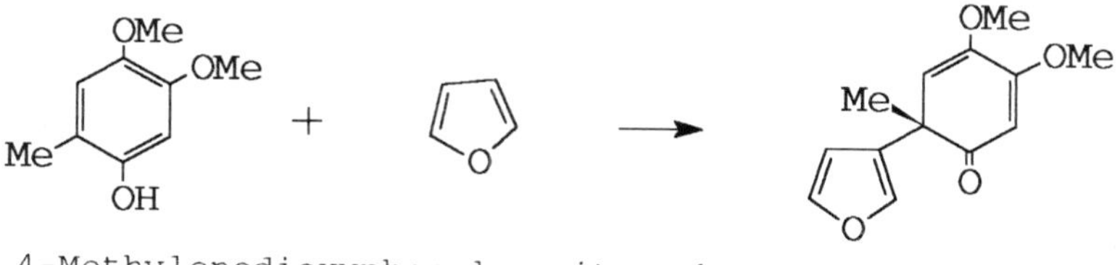

3,4-Methylenedioxyphenol as its ethoxymagnesio derivative is acylated with cinnamoyl chloride in toluene affording 2-hydroxy-4,5-methylenedioxychalcone in 70% yield (F. Bigi et al., Tetrahedron 1984, 40, 4081).

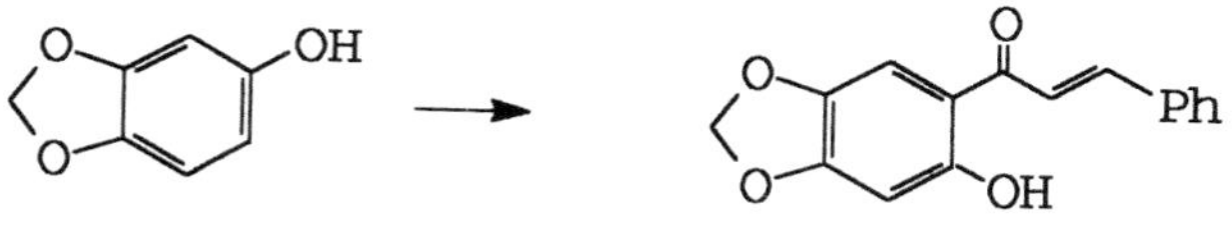

2-Hydroxymethyl-8-methoxy-1,4-benzodioxane in benzene containing acetic anhydride upon treatment at ambient temperature with the enzyme, Amano P-30, produces a 44% yield of (S)-2-hydroxymethyl-8-methoxy-1,4-benzodioxane (e.e. 83%) with a 50% conversion (M.D. Ennis and D.W. Old, Tetrahedron Letters, 1992, *33*, 6283).

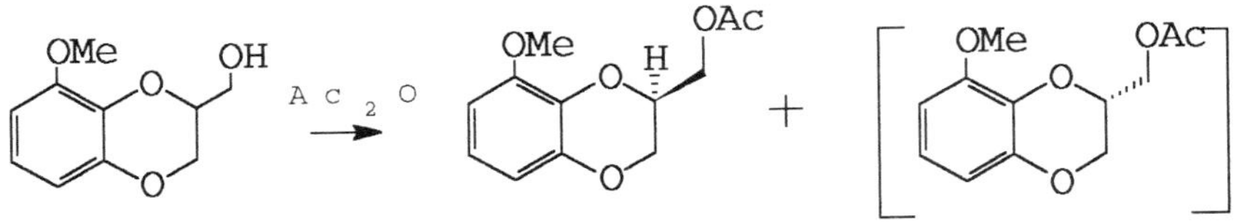

Alkylation of 3,4-methylenedioxyphenol with 1-(4-methoxyphenyl)ethanol in glacial acetic acid/water (6:1) is effected by heating to give a 97% yield of 3,4-methylenedioxy-6-[α-(4-methoxyphenyl)ethyl]phenol (US Sec. Agric. USP 4342777).

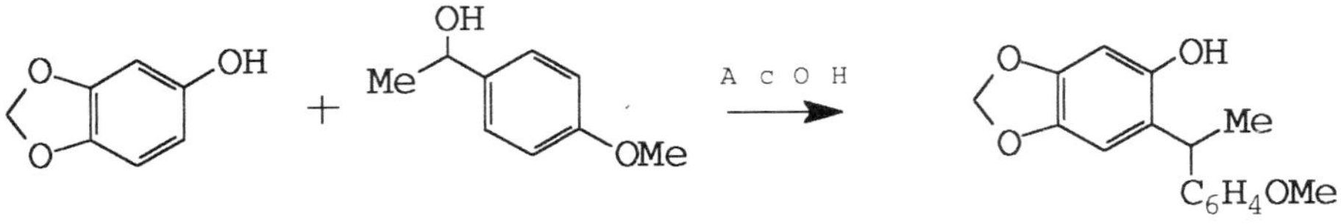

The 2,5-dimethyl analogue of the preceding phenol in benzene containing anhydrous zinc chloride is alkylated with 3-methylbut-3-enyl bromide at 60°C to afford 2,2,5,8-tetramethyl[1,3]dioxolo[6,7-d]chroman in 72% yield (F. Dallacker and W. Coerver, Z. Naturforsch., B, 1981, *36*, 1037).

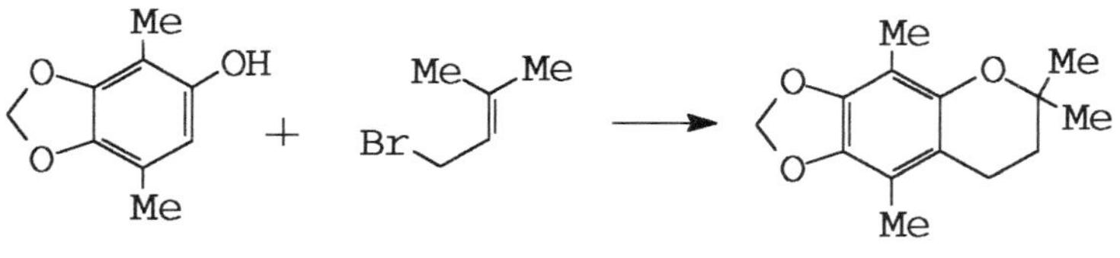

2-Methoxyhydroquinone with benzofurazan oxide in methanolic suspension upon treatment with powdered sodium hydroxide and reaction over 6 hours furnishes a 78% yield of 2-hydroxy-3-methoxyphenazinedi-N-oxide (A. Romer et al., Z. Naturforsch., B, 1981, *36*, 1037).

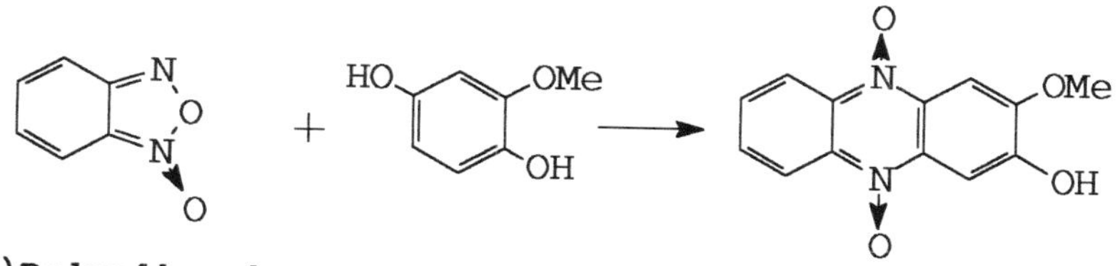

(iv) Polyalkoxybenzenes, Tetraacylbenzenes and Polyalkoxy naphthalenes

1,3,5,6,6-Pentamethoxy-3-methyl-1,4-cyclohexadiene in methanol by treatment with a small proportion of concentrated sulphuric acid and reaction at ambient temperature affords a 97% yield of 3,4,5-trimethoxybenzyl methyl ether (Otsuka Kagaku Yakunin, J57 075942).

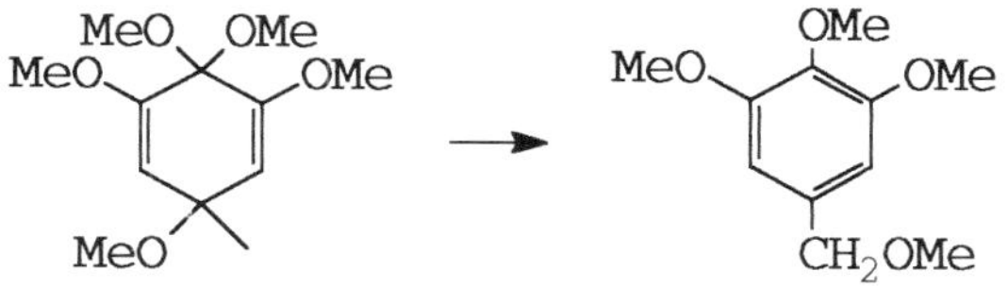

The tetraacetate of 1,2,3,5-tetrahydroxybenzene is synthesised from the cyclohex-2-enone (T. Posternak and J. Deshusses, Helv. Chim. Acta, 1961, 44, 2089).

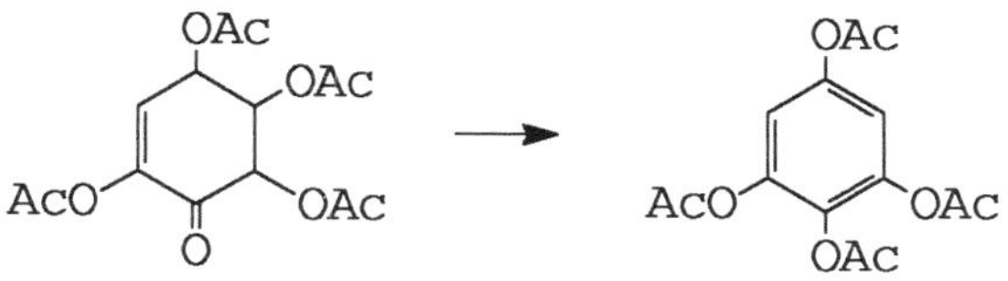

2,3,4,5-Tetramethoxytoluene in methanol containing sodium hydroxide is anodically oxidised in a divided cell affording a 90% yield of 2,3-dimethoxy-5-methyl-1,4-benzoquinone (Otsuka Kagaku Yakunin, J56 147741).

Vanillin in acetonitrile containing potassium bicarbonate when treated with 1-oxo-2,2,6,6-tetramethylpiperidinium fluoroborate affords dehydrovanillin, 2,2'-dihydroxy-3,3'-dimethoxy-5,5'diformylbiphenyl in 85% yield (J.M. Bobbit and Z. Ma, Heterocycles, 1992, 33, 641).

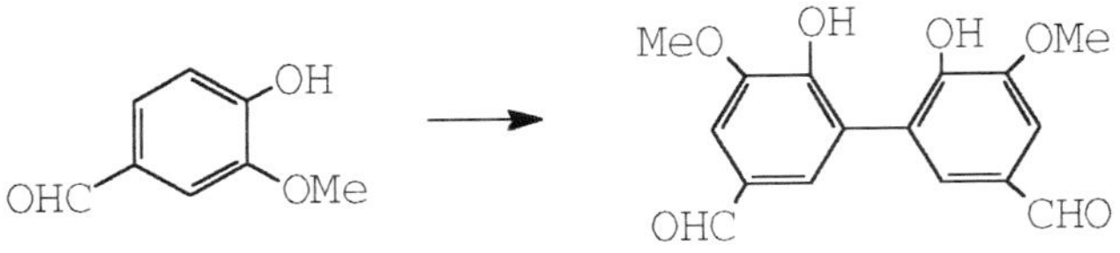

Diaryl oxidative coupling by means of ruthenium dioxide in THF is used to obtain neoisostegane (Y. Landais, J.-P. Robin and A. Leburn, Tetrahedron, 1991, 47, 3787).

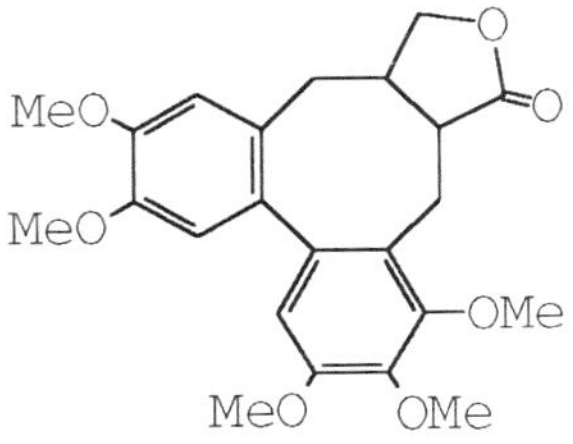

The Grignard reagent from the 2,3,4-trimethoxybromobenzene compound shown (R = CH$_2$OSiMe$_2$t-Bu) by reaction with the chiral oxazinyl-2-yl tetramethoxy compound depicted, affords by methoxyl group displacement a pentamethoxydiphenyl derivative in 60% yield, ee 90%) (A.I. Meyers, A. Meier, D.J. Rawson, Tetrahedron Letters, 1992, 33, 853).

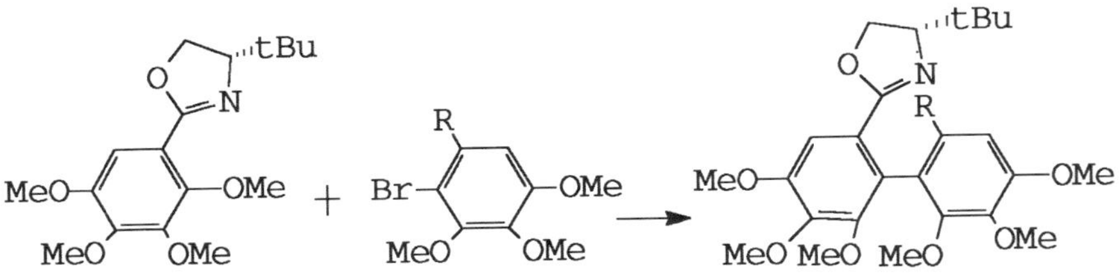

Chapter 5

HYDROCARBONS CARRYING SUBSTITUENTS ATTACHED THROUGH DIVALENT SULPHUR, SELENIUM, OR TELLURIUM: THIOPHENOLS, SULPHIDES ETC

G.C.BARRETT

1. Introduction

Compounds with divalent sulphur functional groups attached to a benzene ring have importance in a range of contexts: in organic chemistry, where the unique reaction characteristics of divalent sulphur are exploited in organic synthesis; in inorganic chemistry where the properties of the sulphur atom as ligand are of continuing interest and are attracting an expanding number of laboratories, and in physico-chemical studies, for example in the synthesis of complex forming host molecules for molecular recognition studies.

For all these major areas, and to meet the needs for exploratory organosulphur chemistry, preparative methods for thiols and sulphides have been undergoing continuous development. In particular, the properties and reactions of these compounds have been investigated within the contexts of the new instrumental techniques and new synthetic methodology of the past 15 years.

2. Organization of this Chapter

The Contents list at the start of this Book should be consulted for the sequence of topics covered. Overlap between sections has been avoided; for example, the preparation of a vinyl sulphide starting from a thiol, is covered as such and is not repeated in the "Reactions of Thiols" section.

3. Reference Sources and Textbooks

Prime amongst reference sources covering this topic, "Houben-Weyl" (Methoden der Organischen Chemie, Volume E11, Georg Thieme Verlag, Stuttgart, 1985; e.g. thiols, p.32) offers valuable support through thorough referencing of the early literature (but with little coverage of the literature in the time window on which this "Rodd" Supplement is based). Volume 3 (Ed.D.N.Jones) of "Comprehensive Organic Chemistry", Eds.Sir Derek Barton and W.D.Ollis, Pergamon Press, Oxford, 1979, includes Chapters by G.C.Barrett on Thiols (p.3), Polyfunctional Thiols (p.21), Sulphides, (p.33), Polyfunctional Sulphides (p.55), and Sulphonium Salts (p.105).

Textbooks include "Chemistry of Organosulphur Compounds: Biochemical Aspects", S.Oae and T.Okuyama, CRC Press, Boca Raton, FA, 1992; "Chemistry of Organosulphur Compounds: Structure and Mechanism", S.Oae and J.T.Doi, CRC Press, Boca Raton, FA, 1991; R.J.Huxtable, "Biochemistry of Sulphur", Plenum Press, New York, 1986; "Chemistry of Organosulphur Compounds: General Problems", Ed. L.I.Belen'kii, Ellis Horwood, Chichester, 1990; "Reactions of Sulphur with Organic Compounds", M.G.Voronkov, N.S.Vyazankin, E.N.Deryagina, A.S.Nakhmanovich, and V.A.Usov, Ed. J.S.Pizey for Consultants Bureau, New York, 1987; "Organic Sulphur Chemistry: Theoretical and Experimental Advances", Eds.F.Bernardi, I.G.Csizmadia, and A.Mangini, Elsevier, Amsterdam, 1985; "Organoselenium Chemistry", Ed. D.Liotta, Wiley-Interscience, New York, 1987; and "Chemistry of Organoselenium and Tellurium Compounds"(Vol.1, Eds. S.Patai and R.Rappoport, 1986; Vol.2, Ed. S.Patai, 1987), Wiley, New York.

4. Properties of Thiols and Sulphides

The relatively high acidity of the thiol functional group and the high nucleophilicity of the thiolate anion are well known, and ^{1}H-n.m.r. studies (M.F.Corrigan, I.D.Rae, and B.O West, Aust.J.Chem., 1978, 31, 587) showing a zwitterionic structure for the Schiff bases of o-formylbenzenethiol are a clear demonstration of the former property. The Schiff base nitrogen is considered to have a low basicity, so the driving force in the proton transfer must be the result of thiolate stability.

The acidity of a given thiol is particularly dependent upon the composition of the solvent and on temperature (G.P.Sharnin, V.V.Nurgatiu, and B.M.Ginzburg, Zh.Org.Khim., 1979, 15, 1638), and may be measured by the straightforward exchange reaction, (J.J.Khurma and D.V.Fenby, Aust.J.Chem., 1979, 32, 755):-

$$RS^1H + {}^2H_2O \rightleftharpoons RS^2H + {}^2HO^1H$$

Direct and indirect consequences of this property account for the differences shown between thiols and alkanols, and the differences are shown in new results from a number of recent studies. The nucleophilicity of the thiol group is so much larger than that of other nucleophiles, that attempts to compare photostimulated and electrostimulated substitution reactions (radical intermediates; $S_{RN}1$ and $S_{RN}2$ reactions) with normal nucleophilic substitution reactions (e.g., pentafluoronitrobenzene + PhSH $\longrightarrow$ C_6F_5SPh) had to exclude thiophenol because the reaction rate was too fast (J.Marquet, Z.Jiang, I.Gallardo, A.Batlle, and E.Cayon, Tetrahedron Lett., 1993, 34, 2801).

Also, the effectiveness compared with analogous catalysts carrying two β-positioned hetero-atoms, of homochiral copper(I) o-(dimethylaminomethyl)thiophenolates as catalysts, e.g. for the asymmetric synthesis of homochiral secondary alcohols in up to 99% enantiomeric excess through addition of diorganozinc reagents to aldehydes, is notable (E.Rijnberg,

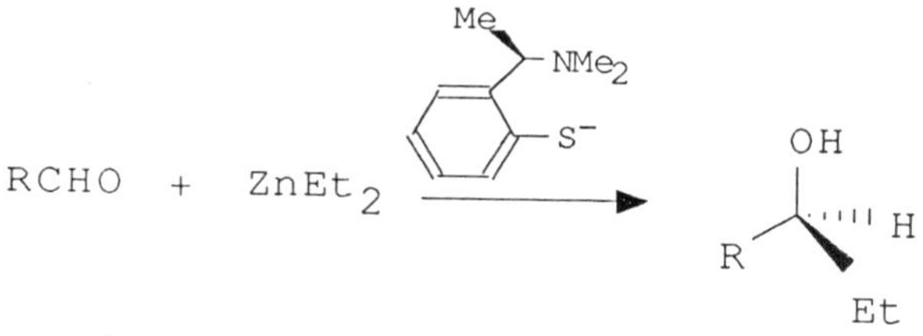

J.T.B.H.Jastrzebski, M.D.Janssen, J.Boersma, and G.van Koten, Tetrahedron Lett., 1994, 35, 6521). This paper describes the structural chemistry of the putative zinc thiolate intermediate involved in this process, and reflects the high current interest in transition metal thiolates (I.G.Dance, Polyhedron, 1986, 5, 1037; P.J.Blower and D.Dilworth, Co-ord.Chem.Rev., 1987, 76, 121). This interest stems partly from the structure and function of the active sites of metalloproteins, and for other reasons that are briefly mentioned in later sections of this Chapter.

The i.r. stretching frequency of the -S-H group in the bulky 2,4,6-triphenylbenzenethiol is seen at 2562 cm^{-1} (K.Ruhlandt-Senge and P.P.Power, Bull.Soc.Chim.Fr., 1992, 12, 594). The "normal" value for thiols is within a range covering somewhat higher wavenumbers (2560 - 2590 cm^{-1}), and the i.r. spectrum of 2,6-diphenylbenzenethiol shows a peak at 2580 cm^{-1} (P.T.Bishop, J.R.Dilworth, T.Nicholson, and J.Zubieta, J.Chem.Soc., Dalton Trans., 1991, 385).

The preparation of sulphides for use in molecular recognition studies is a key feature in some current studies, e.g. the trisulphide moiety at the base

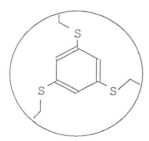

of a cup-shaped receptor, the rim of which consists of a tris[L-phenylalanyl N-(p-carboxy-benzyl)amide] ring. The whole entity shows highly selective binding of the L-enantiomer when offered DL-amino acid methylamides in CHCl$_3$ solution, and shows comparable selectivity towards certain glycosides (R.Liu and W.C.Still, Tetrahedron Lett., 1993, 34, 2573; S.D.Erickson, J.A.Simon, and W.C.Still, J.Org.Chem., 1993, 58, 1305). Benzene-1,3,5-tri-thiol (V.Bellavita, Gazz.chim.Ital., 1932, 62, 655) was used as starting material and reacted readily with the tris(benzylic bromide) in the required 1:1-fashion.

5. Preparation of Arenethiols

Many of the preparations covered in this section, like most standard methods for this functional group, lead to symmetrical disulphides rather than to thiols, but the conversion of disulphides into thiols is routine [see Section 5(f)].

(a) Preparation by Reduction of Arenesulphonyl Halides

An arenesulphonyl chloride reacts with 3 equivalents of an arenethiol during 2 hours in refluxing ethanol, to give the corresponding disulphide (D.Cipris and D.Pouli, Synth.Commun., 1979, 9, 207).

Trifluoroacetic anhydride and NaI reduce toluene-p-sulphonyl chloride to the corresponding disulphide under mild conditions (T.Numata, H.Awano, and S.Oae, Tetrahedron Lett., 1980, 21, 1235).

A long-standing method, reduction of arenesulphonyl chlorides with LiAlH$_4$, has been used for the preparation of 1,3,5-trialkylbenzenethiols (P.J.Blower, J.R.Dilworth, and J.Zubieta, Inorg.Chem., 1985, 24, 2866).

(b) Preparation by Reduction of Arenesulphonic acids

Reduction of sulphonic acids using trifluoroacetic anhydride and Bu$^n_4$N$^+$ I$^-$ (T.Numata, H.Awano, and S.Oae, Tetrahedron Lett., 1980, 21, 1235) involves the formation of the thiol and its trifluorothiolacetate, which is easily cleaved by NaOH in MeOH and the product is easily separated from impurities so as to give pure thiols:-

$$ArSO_3H + (CF_3CO)_2O + Bu^n_4N^+ \, I^-$$

$$\xrightarrow[CH_2Cl_2]{room\ temp} ArSCOCF_3 + ArSH + I_2 + Bu^n_4N^+ \, CF_3CO_2^-$$

Yields are modest (p-MeC$_6$H$_4$SO$_3$H $\longrightarrow$ ArSCOCF$_3$ + ArSH in 43% yield) but improve to excellent figures (90%) if excess trifluoroacetic anhydride is used. The authors do not indicate the amounts recovered, but imply that there is some unreacted starting material, so the method should be regarded as an excellent new route. Similar results are obtained with alkanesulphonic acids.

(c) Reduction of Thiolsulphonates

Arenethiolsulphonates ArSO$_2$SAr can be reduced to disulphides (T.Numata, H.Awano, and S.Oae, Tetrahedron Lett., 1980, 21, 1235) using the trifluoroacetic anhydride - NaI system, in moderate yield.

(d) Preparation by Reduction of Arenesulphinic acids
Benzenesulphinic acid gives diphenyl disulphide on reduction with the trifluoroacetic anhydride - NaI system (T.Numata, H.Awano, and S.Oae, Tetrahedron Lett., 1980, 21, 1235).

(e) Preparation by Reduction of Arenesulphonates
A dialkoxyaluminium hydride [NaAlH(OCH$_2$CH$_2$OMe)$_2$] is more effective than the established reagent (lithium aluminium hydride) for reducing arenesulphonates of monosaccharides (A.Zobacova, V.Hermankova, and J.Jary, Coll.Czech.Chem.Commun., 1977, 42, 2540).

(f) Preparation by Reduction of Disulphides
Standard methods favouring mild inorganic reducing agents continue to be represented, e.g. sodium borohydride in methanol for the cleavage of bis(2-phenylthiazol-5-yl) disulphides (G.C.Barrett and R.Walker, Tetrahedron, 1976, 32, 583).

(g) Preparation from Phenols
Arenethionocarbamates, prepared from a phenol and an alkanethiocarbamoyl chloride, rearrange at elevated temperatures into thiolcarbamates. These give the corresponding thiols on hydrolysis, and the method has been used to make p-cyano- and p-alkylthiophenols (C.F.Shirley, Mol.Cryst.Liq.Cryst., 1978, 44, 193).

2,6-Diphenylphenol, converted into its bis(dimethylthiocarbamate) through reaction with NaH and dimethylthiocarbamoyl chloride, may be rearranged into the thioncarbamate by heating at 250°C (P.T.Bishop, J.R.Dilworth, T.Nicholson, and J.Zubieta, J.Chem.Soc., Dalton Trans., 1991, 385). LiAlH$_4$ Reduction gives 2,6-diphenylbenzenethiol. The overall process represents a standard route for the preparation of those benzenethiols that can withstand the relatively drastic conditions of the rearrangement. The low yield is indicative of limitations imposed by steric hindrance since the route is usually quite efficient.

Flexibility of otherwise crowded structures in which this transformation is to be carried out allows standard reaction conditions to be used, a spectacular example being the preparation in 98% yield by the thioncarbamate method, of the "conformationally-immobile" tetramercapto[1.1.1.1]metacyclophane (X.Delaigue and M.W.Hosseini, Tetrahedron Lett., 1993, 34, 8111):-

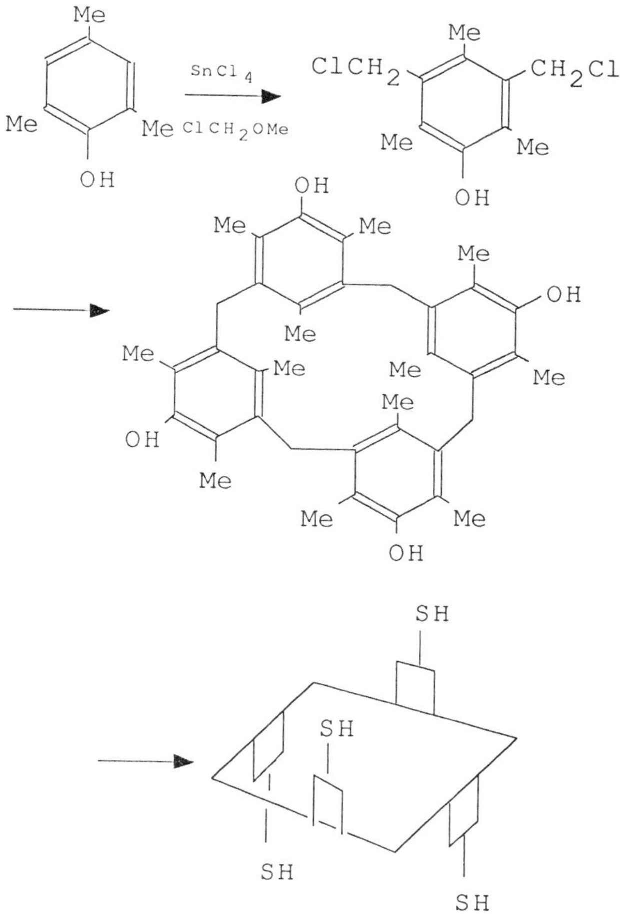

(The preparation of the starting phenol is described by S.Pappalardo, G.Ferguson, and J.F.Gallagher, J.Org.Chem., 1992, 57, 7102)

Stereochemically-interesting products such as these, especially the following atropisomeric 1,1'-binaphthalene-2,2'-dithiol used as homochiral catalyst for asymmetric synthesis, (O.De Lucchi, Phosphorus Sulphur Silicon Relat.Elem., 1993, 74, 195) has motivated the synthesis of quite complex arenethiols:-

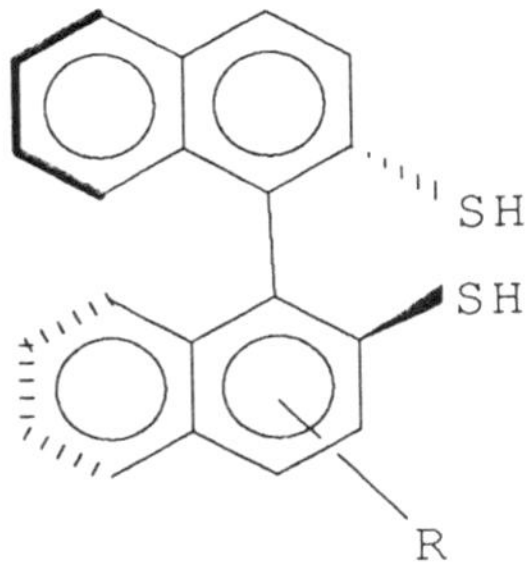

1-Naphthol with S_2Cl_2 followed by Zn/HCl reduction gives 1-hydroxy-naphthalene-2-thiol (A.M.Zeinalov, F.N.Mamedov, M.Morsum-Zade, and A.K.Ibad-Zade, Zh.Org.Khim., 1979, 15, 816).

(h) Preparation from Arenes

A typical preparation involving attack by elementary sulphur (S_8) on a lithiated arene has been described for 4-hydroxy-6-mercaptodibenzofuran (N.Fotouhi, N.G.Galakatos, and D.S.Kemp, J.Org.Chem., 1989, 54, 2803; idem, J.Org.Chem., 1986, 51, 1821). Protection of the the OH group is necessary, using MeI and K_2CO_3, and the methoxy-analogue is refluxed with S_8 after lithiation with Bu^nLi. Demethylation is achieved without side-reactions, using Me_3SiI.

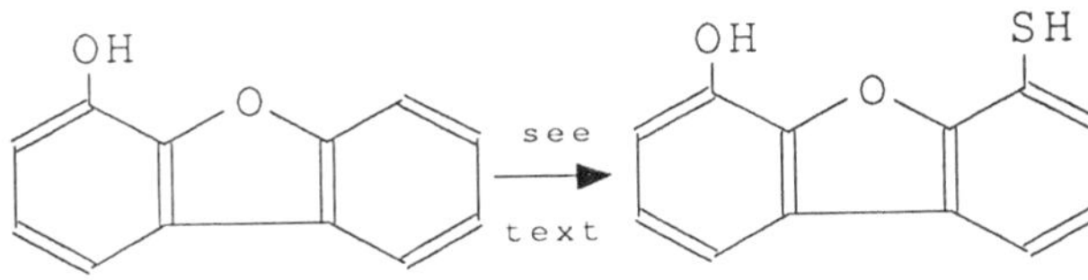

The directing effect of the furan oxygen atom of 4-hydroxydibenzofuran, on the position of lithiation of the adjacent benzene ring, accounts for the position of the mercapto group in the product; a result known from Gilman's early pioneering work on the lithiation of substituted arenes. Dibenzofuran itself gives 4-mercaptodibenzofuran rather than the 4,6-dimercapto-compound through lithiation and reaction with S_8 (M.Janczewski and H.Maziarzyk, Roczniki Chem., 1977, 51, 891).

An analogous preparation of 4-hydroxy-6-mercaptophenoxythiin has been described (D.S.Kemp, N.G.Galakatos, B.Bowen, and K.Tan, J.Org.Chem., 1986, 51, 1829); a procedure that provides the evidence for the position of lithiation and thiation in these reactions, as well as illustrating modern methods of structure assignment to thiols.

The beneficial role of N,N,N',N'-tetramethyl-1,2-ethanediamine as complexing agent for BuLi in this reaction is illustrated in the preparation of 2,6-dimethoxybenzenethiol from dimethoxybenzene and S_8 in benzene under mild conditions (M.Wada, K.Kajihara, K.Nishimura, K.Tenma, and T.Erabi, Kenkyu-Hokoku-Asahi Garasu Zaidan, 1991, 58, 287; Chem.Abs., 1993, 119, 138808).

This simple preparative method is widely useful, though with particular targets in the general family of substituted arenes, its relevance is dependent upon a fortunate coincidence between the intended position of the SH group and the position of lithiation. For this reason it is suitable for the preparation of either highly hindered, or symmetrical, arenethiols, where by definition there is a unique site for lithiation:-

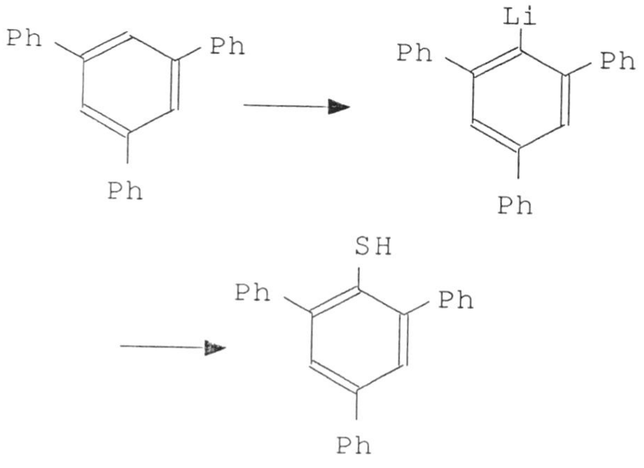

The triphenyl compound above, m.p. 142-4°, has been prepared by the reaction of the Li(Et$_2$O) - 1,3,5-triphenylbenzene complex with S_8, followed by cautious treatment with dilute sulphuric acid (K.Ruhlandt-Senge and P.P.Power, Bull.Soc.Chim.Fr., 1992, 12, 594) following methods that were established for the preparation of 2,4,6-tri-isopropyl-benzenethiol and its tri-t-butyl analogue (G.A.Sigel and P.P.Power, Inorg.Chem., 1987, 26, 2819; P.P.Power and S.C.Shoner, Angew. Chem.Int.Ed., 1991, 30, 330). These new "bulky thiols" are attractive to inorganic chemists searching for monomeric organolithium species, though they present some practical difficulties since they tend to show low solubility in non-polar solvents.

The product from treatment of the triphenylbenzenethiol with Bu^nLi in tetrahydrofuran is a rare example of a monomeric species, $Li(THF)_3SC_6H_2Ph_3$, while $Fe[N(SiMe)_3]_2$ reacts to give the dimeric species $[Fe(SC_6H_2Ph_3)_2]_2$. Other monomeric species are known, e.g. $Li(THF)_3SC_6H_2Pr^i_3$, and the titanium salt $[Li(Et_2O)_3][Ti(SC_6H_2Pr^i_3)_4]$ and the dimers $[M(SC_6H_2Bu^t_3)]_2$. These are interesting examples representing numerous compounds of these types, described in the literature for the preparation of the thiols in these ways (e.g., P.B.Hitchcock, M.F.Lappert, B.J.Samways, and E.Weinberg, J.Chem.Soc., Chem.Commun., 1983, 1492).

Preparation of close relatives of these substituted arenethiols from phenols has been covered earlier in this Chapter [Section 5(a)].

A selenium analogue of these "bulky thiols" synthesized using corresponding methods is represented by the following 2,4,6-tris(trifluoromethyl) compound (N.Bertel, H.W.Roesky, F.J.Edelmann, M.Noltemeyer, and H.G.Schmidt, Z.Anorg.Allg.Chem., 1990, 586, 7):-

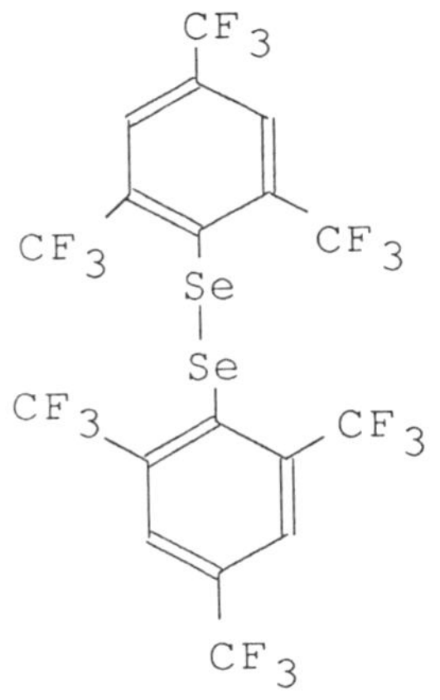

(i) Preparation from Aryl Halides using H_2S, NaSH or MSR; or using Other Simple Inorganic Reagents

High temperature reaction of H_2S with ArBr gives the arenethiol and the symmetrical disulphide (M.G.Voronkov and E.N.Deryagina, Phosphorus Sulphur, 1979, 7, 123). This illustrates standard methodology, but one that is only for the brave! And an efficient, simple alternative method for laboratory use is the reaction of a primary, secondary, or "activated tertiary" halide, or an aryl halide, with tribasic sodium thiophosphate dodecahydrate, $Na_3SPO_3,12H_2O$, in methanol or DMF (C.Bienarz and M.J.Cornwell, Tetrahedron Lett., 1993, 34, 939). This paper gives a list of references to standard methods conforming to the title of this Section.

Another modern reagent that takes the place of H_2S as a thiolating species, is triphenylthiasilanol, Ph_3SiSH. It is a colourless, crystalline solid, essentially odourless and easy to handle. Its preparation from S_8 and Ph_3SiH was disclosed some time ago (L.Birkofer, A.Ritter, and H.Galler, Chem.Ber., 1963, 96, 3289) but the reagent was not put to use in organic synthesis until recently (J.Brittain and Y.Gareau, Tetrahedron Lett.,. 1993, 34, 3363). For example, it has been used for the preparation of β-hydroxyalkanethiols by ring-opening of epoxides in MeOH catalyzed by NEt_3. "Sulphurated borohydride", $NaBH_2S_3$ (J.Lalancette and A.Freche, Can.J.Chem., 1971, 49, 4047), is another example out of the same stable of metal-thiol reagents, used for ring-opening of oxiranes to give β-hydroxyalkanethiols and corresponding sulphides.

(j) Preparation from Arenesulphenyl halides

Benzenesulphenyl chloride and bromide, and their selenium and tellurium analogues, react readily with terminal alkynes in the presence of 2 equivalents of anhydrous copper(I) iodide in DMF, to give the corresponding sulphides, selenides, and tellurides. (A.L.Braga, C.C.Silveira, A.Reckziegel, and P.H.Menezes, Tetrahedron Lett., 1993, 34, 8041). Extension of the study to polyfunctional analogues shows that other functional groups (OH, NHR, C=C, or CO_2R) are not affected by the reaction conditions.

(k) Ring Opening of Sulphur Heterocycles [see also Section 9(a)(1)]

Photolysis of the enol acetate of 4-thiochromanone (I.W.J.Still and T.S.Leong, Tetrahedron Lett., 1979, 3613) gives the monocyclic thiolacetate without loss of carbon, through an electrocyclic mechanism. Although the vinyl ketone moiety is a little sensitive to the conditions needed to release the thiol from its thiolester, the route might be quite versatile in accessing o-substituted thiophenols of a wide variety of types.

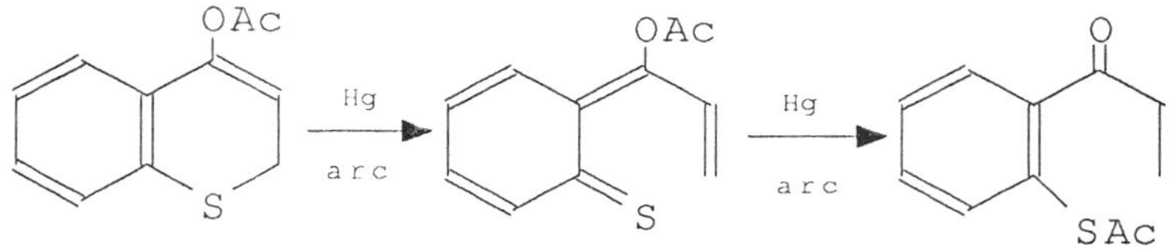

(l) Preparation by Cleavage of Sulphides

An alkyl benzyl sulphide can be cleaved to leave an alkanethiol, as in the Scheme showing the conversion of (+)-pulegone into an oxathian (E.L.Eliel, J.E.Lynch, F.Kume, and S.V.Frye, Org.Synth., 1987, 65, 215):

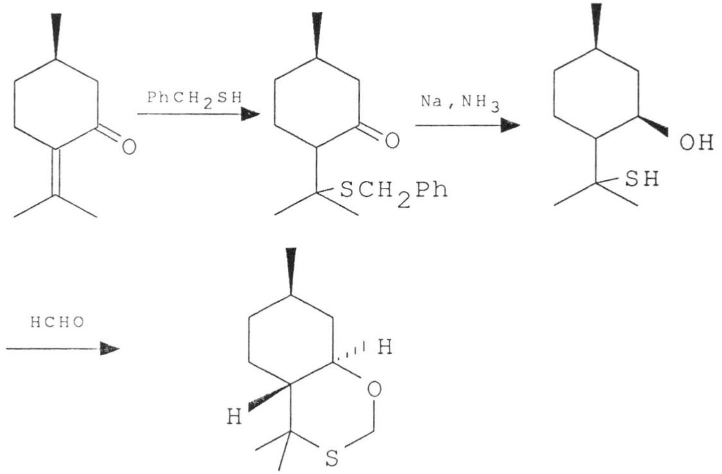

Cleavage of the sulphide is brought about with sodium in liquid ammonia; this also causes the reduction of the carbonyl group. Thioacetal formation with formaldehyde completes the synthesis and the overall Scheme is a good example of standard reactions of thiols and sulphides as applied in synthesis.

Photolysis of sulphides carrying the R-CH_2-S- grouping causes α-cleavage through a Norrish Type-II route, to generate a highly reactive thioaldehyde (E.Vedejs, T.H.Eberlein, and D.L.Varie, J.Am.Chem.Soc., 1982, 104, 1445). This suffers electrophilic attack through its thiocarbonyl carbon atom when generated in the presence of aromatic species, delivering thiols (G.A.Krafft and P.T.Meinke, Tetrahedron Lett., 1985, 26, 135):-

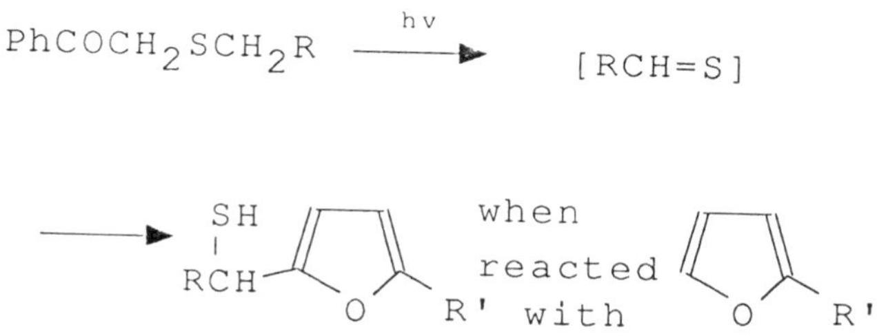

$$\text{PhCOCH}_2\text{SCH}_2\text{R} \xrightarrow{h\nu} [\text{RCH}=\text{S}]$$

$$\longrightarrow \text{RCH}(\text{SH})\text{-furan-R'} \xleftarrow{\text{when reacted with}} \text{furan-R'}$$

The same intermediate can be generated from α-silylated sulphides (G.A.Krafft and P.T.Meinke, unpublished work mentioned in G.A.Krafft and P.T.Meinke, Tetrahedron Lett., 1985, 26, 135).

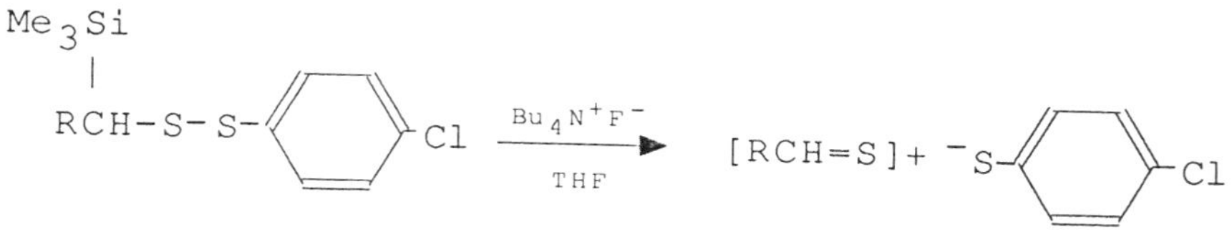

$$\text{RCH(SiMe}_3\text{)-S-S-C}_6\text{H}_4\text{-Cl} \xrightarrow[\text{THF}]{\text{Bu}_4\text{N}^+\text{F}^-} [\text{RCH}=\text{S}] + {}^-\text{S-C}_6\text{H}_4\text{-Cl}$$

Electrochemical oxidation of selenides ArSeCH_2R gives areneselenols ArSeH (in the form of their diselenides) as ultimate products (V.Jouikov, V.Ivkov, and D.Fattahova, Tetrahedron Lett., 1993, 34, 6045), but the process can be directed towards products of α-attack by nucleophilic species derived from solvents or from salts added to support the passage of electric current (e.g. acetoxylation by added AcO^-). Electrochemical cleavage of a sulphur - sp^3-carbon bond is much more difficult, and since there is only one example of this, the rather untypical trityl phenyl sulphide, $\text{Ph}_3\text{C-S-Ph}$ (S.Torii and K.Uneyama, Tetrahedron Lett., 1971, 329), these results can be concluded to be an example of an unambiguous difference between the behaviour of divalent organosulphur compounds and their organoselenium analogues.

6. Thiolesters

Although routinely prepared by acylation of thiols (see following section), useful synthesis opportunities are offered by thiolesters prepared as intermediates by the use of ketenes and tin(II) thiolates (N.Yamasaki, M.Murakami, and T.Mukaiyama, Chem.Lett., 1986, 1013):-

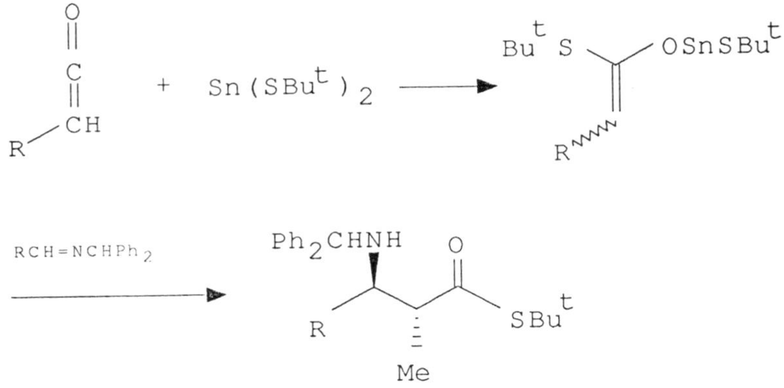

$$R-\underset{\underset{CH}{\|}}{\overset{\overset{O}{\|}}{C}} + Sn(SBu^t)_2 \longrightarrow \underset{R}{\overset{Bu^tS}{\nwarrow}}C=C\overset{OSnSBu^t}{\nearrow}$$

$$RCH=NCHPh_2 \longrightarrow \underset{R}{\overset{Ph_2CHNH}{\wedge}}\underset{\overset{\vdots}{Me}}{\hspace{-0.3cm}}\underset{}{\overset{O}{\|}}C-SBu^t$$

The formation of thiolesters from phenyl chlorothiolformate is also of some value in synthesis, because the method widens the use of Grignard reagents which routinely add to the carbonyl group in other carbonyl compounds (C.Cardellicchio, V.Fiandanese, G.Marchese, and L.Ronzini, Tetrahedron Lett., 1985, 26, 3595):-

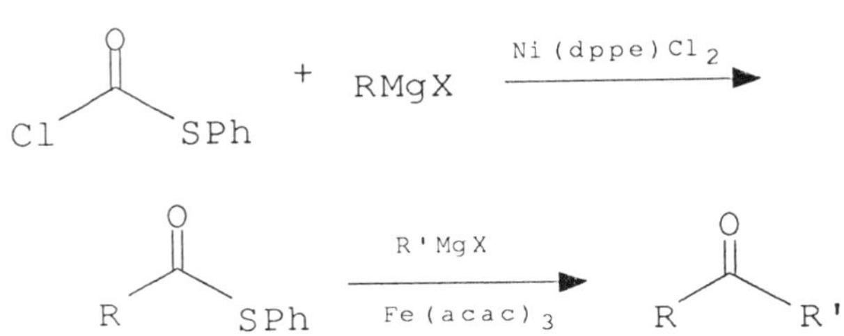

$$Cl\overset{O}{\underset{}{\wedge}}SPh + RMgX \xrightarrow{Ni(dppe)Cl_2}$$

$$R\overset{O}{\underset{}{\wedge}}SPh \xrightarrow[Fe(acac)_3]{R'MgX} R\overset{O}{\underset{}{\wedge}}R'$$

Among numerous applications of thiolesters in synthesis, a stereospecific route to alkenes is notable (R.L.Danheiser and J.S.Nowick, J.Org.Chem., 1991, 56, 1176):-

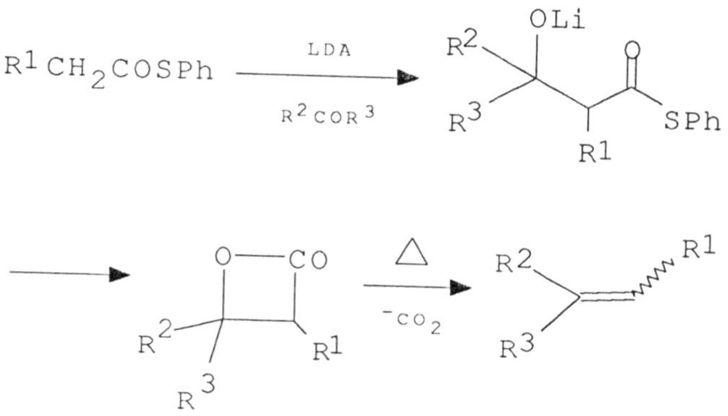

$$R^1CH_2COSPh \xrightarrow[R^2COR^3]{LDA} \underset{R^1}{\overset{OLi}{\underset{R^3}{R^2}}}\! \overset{O}{\underset{SPh}{}}$$

$$\longrightarrow \underset{R^3}{\overset{O-CO}{\underset{R^1}{R^2}}} \xrightarrow[-CO_2]{\Delta} \underset{R^3}{R^2}\!\!\!=\!\!\!\underset{}{R^1}$$

7. Reactions of Thiols

(a) Replacement of SH by H

Thiols dissolved in benzene at 80°C give the corresponding hydrocarbons when treated with Bu^n_3SnH and AIBN (E.Vedejs and D.W.Powell, J.Am.Chem.Soc., 1982, 104, 2046).

(b) Oxidation of Thiols

(1) Oxidation to give Disulphides. Although this is the simplest oxidative change that a thiol can undergo, many mild oxidants cause over-oxidation (as discussed in some following sections). Halogenosilanes in conjunction with CrO_3, however, offer a very effective combination for the simple oxidation of thiols to disulphides, though its mechanism of action is unknown (J.M.Aizpurua, M.Juaristi, B.Lecea, and C.Palomo, Tetrahedron, 1985, 41, 2903). Possible intermediates include chlorochromate esters [$Me_3SiCl + CrO_3 \longrightarrow ClCrO_2OSiMe_3$].

Dimethyldioxirane oxidizes alkanethiols to corresponding alkanesulphinic acids in very good yields (D.Gu and D.N.Harpp, Tetrahedron Lett., 1993, 34, 67). The corresponding reaction with $PhCH_2SH$ leads to lower yields because of alternative oxidation pathways, giving benzaldehyde, for example, and p-thiocresol gives 18% $ArSO_2H$, 29% $ArSO_3H$, 33% ArSSAr, and 20% $ArSO_2SAr$ (Ar = p-Me-C_6H_4-).

(2). Oxidation with oxides of nitrogen. Reaction of thiols with N_2O_4 at -25°C is an effective method of preparing disulphides (S.Oae, K.Shinhama, K.Fujimori, and Y.H.Kim, Bull.Chem.Soc.Jpn., 1980, 53,

775) and probably proceeds by way of the thionitrite (S.Oae, K.Shinhama, and Y.H.Kim, Tetrahedron Lett., 1979, 3307).

$$2\,RSH + N_2O_4 \longrightarrow 2\,RSNO + HNO_3 \longrightarrow RSSR + 2\,NO$$

(3.) Oxidation by Di-oxygen. This topic is either very straightforward in chemical laboratory terms as far as disulphide formation from thiols is concerned, and has not been investigated recently to any considerable extent; or it has earned detailed attention in the biochemical context because of continuing interests in catalyzed oxidation.

In the former category, investigations have been directed at the further oxidative possibilities for thiols and disulphides, and in the presence of base, O_2 has been shown to lead to a mixture of peroxysulphenates $RSOO^-$, and peroxysulphinates $RS(O)OO^-$. These are therefore valid intermediates to be considered as alternative to thiyl radicals (the standard textbook explanation of mild thiol autoxidation). The intermediates are transient species, the presence of which is revealed by the conversion of phosphines to phosphine oxides ($R_3P \longrightarrow R_3PO$) and of sulphoxides to sulphones in oxygenation reaction mixtures (S.Oae and T.Takata, Tetrahedron Lett., 1980, 3689). These intermediates explain the fact that the usual final base-catalyzed oxidation product of a thiol is an equimolar mixture of sulphonic and sulphinic acid anions.

(4). Oxidation by Other Species. This Section is a means of noting interesting recent literature on the uses of thiols as mild reducing agents. A use of the azide ion as the first step in introducing a nitrogen function into an organic compound by nucleophilic substitution is of increasing interest, and as a second step, a combination of a thiol with a tertiary amine offers a valid method of reducing the covalently-bound azide grouping to primary amine (P.Aufranc, J.Ollivier, A.Stolle, C.Bremer, M.Es-Sayed, A.de Meijere, and J.Salaun, Tetrahedron Lett., 1993, 34, 4193; see also H.Bayley, D.N.Standring, and J.R.Knowles, Tetrahedron Lett., 1978, 3633).

(c) Substitution Reactions
m-Bromothiophenols undergo lithiation so as to generate the bromine-substitution product, from which a variety of m-substituted thiophenols can be obtained by electrophilic substitution (H.G.Selnick,

M.L.Bourgeois, J.W.Butcher, and E.M.Radzilowski, Tetrahedron Lett., 1993, 34, 2043).

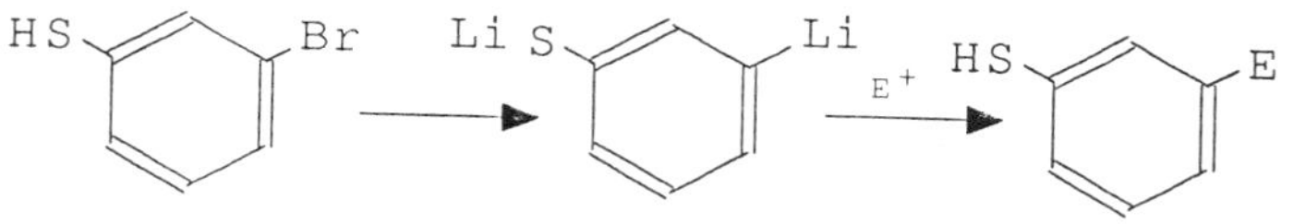

$$HS\text{-}C_6H_4\text{-}Br \xrightarrow{} LiS\text{-}C_6H_4\text{-}Li \xrightarrow{E^+} HS\text{-}C_6H_4\text{-}E$$

The m-lithiation of a thiophenolate seen in this process is the first example of its type. However, the directed o-lithiation of thiophenolates has been thoroughly studied (E.Block, G.Ofori-Okai, and J.Zubieta, J.Am.Chem.Soc., 1989, 111, 2327; G.D.Figuly, C.K.Loop, and J.C.Martin, J.Am.Chem.Soc., 1989, 111, 654; E.Block, V.Edwarakrishnan, M.Gernon, G.Ofori-Okai, C.Saha, K.Tang, and J.Zubieta, J.Am.Chem.Soc., 1989, 111, 658; K.Smith, C.M.Lindsay, and G.J.Pritchard, J.Am.Chem.Soc., 1989, 111, 665).

(d) Synthetic Uses: Heterocyclic Ring Formation

The general possibilities have been represented in the recent literature in many different examples. Tyupically, o-aminothiophenol gives the potent vasodilator (+)-(2S,3S)-diltiazem, marketed by the Tanabe Seiyaku Company, by the addition of o-aminophenol through its sulphur function to the (Z)-enolate ArCH=C(OMEM)CO$_2$R, and cyclization and routine elaboration of the resulting threo-adduct (O.Miyata, T.Shinada, T.Naito, I.Ninomiya, T.Date, and K.Okamura, Tetrahedron, 1993, 49, 8119):-

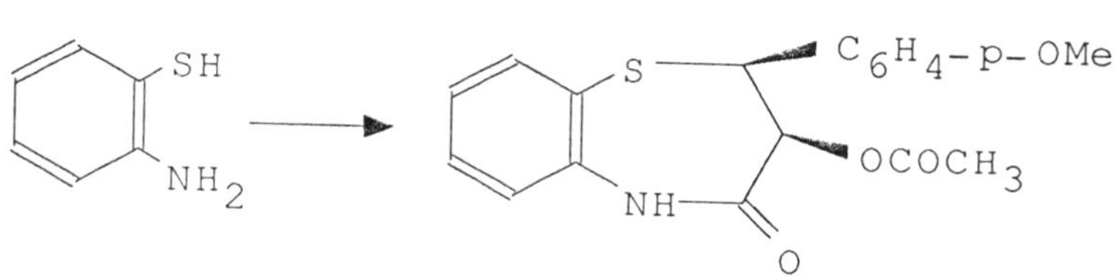

$$\text{o-H}_2\text{N-C}_6\text{H}_4\text{-SH} \xrightarrow{} \text{benzothiazepinone with C}_6\text{H}_4\text{-p-OMe and OCOCH}_3$$

(e) Acylation

Acylation of a thiol, involving condensation with an alkanoic acid, can be readily accomplished using N,N'-dicyclohexylcarbodi-imide if the process is catalyzed with 4-(dimethylamino)pyridine (B.Nieses and W.Steglich, Angew.Chem., 1978, 90, 556).

4-Hydroxy-6-mercaptodibenzofuran has a key role as a template, in an intriguing peptide synthesis route, where intramolecular transfer of the incoming amino acid residue to the growing peptide chain is involved:-

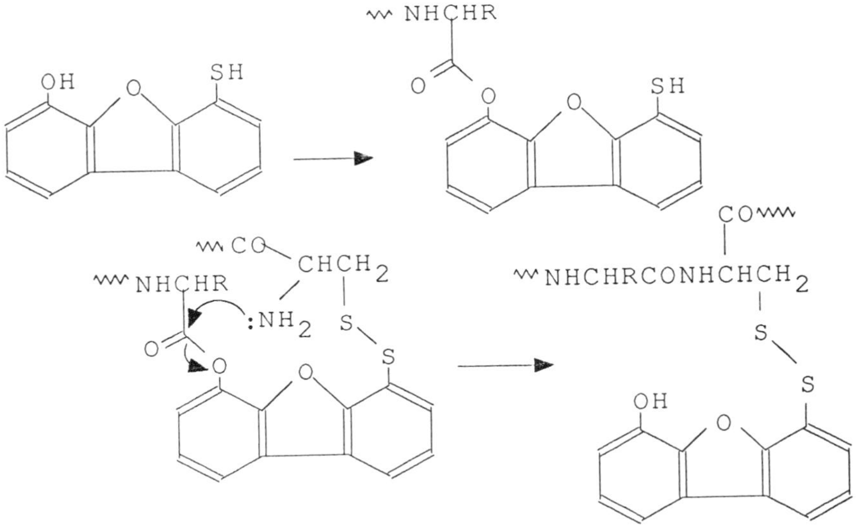

The SH group is used to form a disulphide bond with a cysteine ester, whose amino group is then at an ideal distance to aminolyse an ester group involving the OH group (N.Fotouhi, N.G.Galakatos, and D.S.Kemp, J.Org.Chem., 1989, 54, 2803, idem, J.Org.Chem., 1986, 51, 1821; 1829, for the preparation of the thiol).

8. Protection of the SH Group in Synthesis

There are several strategies available, depending on the particular stages of the synthesis through which the SH group is intended to survive. The simplest is to make use of the easily-reversible disulphide-forming tendency of thiols [Section 7(b)]. This is not always sufficiently secure protection, and S-tritylation is more reliable, and easily achieved using triphenylmethanol with $BF_3.Et_2O$ in AcOH at 60°C (N.Fujui, T.Watanabe, T.Aotake, A.Otaka, I.Yamamoto, J.Konishi, and H.Yajima,

Chem.Pharm.Bull., 1988, 36, 3304). Removal can be accomplished by dissolution in trifluoroacetic acid, or by reaction with I_2 (which leads to the formation of the disulphide:-

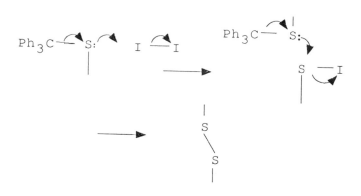

Hg(OAc)$_2$ followed by H$_2$S can be used to free the thiol group from its trityl protecting group (N.Fuju, T.Watanabe, T.Aotake, A.Otaka, I.Yamamoto, J.Konshi, and H.Yajima, Chem.Pharm.Bull., 1988, 36, 3304), as an alternative to the original method (L.Zervas and I.Photaki, J.Am.Chem.Soc., 1962, 84, 3887) in which AgNO$_3$/py in MeOH was recommended. Cleavage of the S-trityl group by trifluoroacetic acid is currently favoured in conventional peptide synthesis protocols.

Apart from the long-known S-benzyl protection strategy [removal by Na/NH$_3$; see Section 5(k)], most of the other SH protecting groups are thioacetals of one form or another, and owe their effectiveness to the ease of cleavage of groupings -S-CHR-X-, where X is a heteroatom, e.g. -NH- in the case of the acetamidomethyl grouping NHCOMe (R = H). This grouping can be introduced using MeCONHCH$_2$OH/TFA and cleaved by I$_2$ in AcOH or by a mercuric salt (B.Kamber, A.Hartmann, K.Eisler, B.Riniker, H.Rink, P.Sieber, and W.Rittel, Helv.Chim.Acta, 1980, 63, 899].

9. Saturated Alkyl, Aryl, and Diaryl Sulphides

(a) Preparation of Sulphides

<u>(1). Preparation from Thiols or Sulphenyl Halides, or their Selenium Analogues.</u> Addition to unsaturated systems is a standard route, illustrative examples including asymmetric synthesis of α-phenylthiosuccinate esters (H.Yamashita and T.Mukaiyama, <u>Chem.Lett.</u>, 1985, 363):-

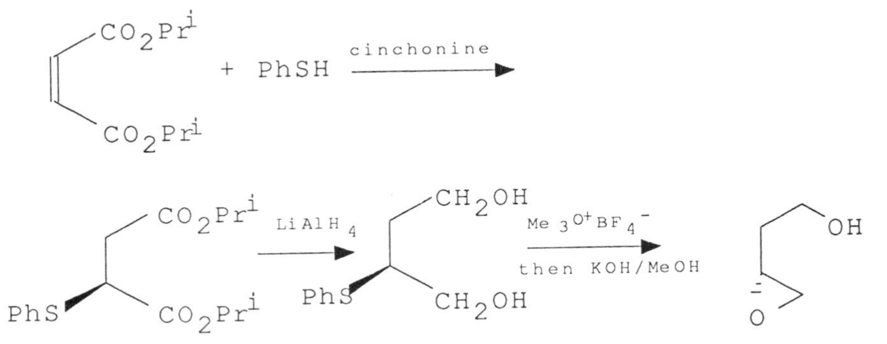

$$\text{(CO}_2\text{Pr}^i\text{)CH=CH(CO}_2\text{Pr}^i\text{)} + \text{PhSH} \xrightarrow{\text{cinchonine}}$$

$$\text{PhSCH(CO}_2\text{Pr}^i\text{)CH}_2\text{(CO}_2\text{Pr}^i\text{)} \xrightarrow{\text{LiAlH}_4} \text{PhSCH(CH}_2\text{OH)CH}_2\text{(CH}_2\text{OH)} \xrightarrow[\text{then KOH/MeOH}]{\text{Me}_3\text{O}^+\text{BF}_4^-}$$

(R)-(+)-3,4-epoxybutan-1-ol
81% enantiomeric excess

and 1,4-addition of PhSMgI to enones, followed by the addition of the resulting carbanion to a ketone R^4COR^5, through efficient carbon-carbon bond-forming steps (T.Shono, Y.Matsumura, S.Kashimura, and K.Hatanaka, <u>J.Am.Chem.Soc.</u>, 1979, 101, 4752):-

$$R^1CH=CR^2COR^3 + PhSMgI \longrightarrow PhS\overline{C}HR^1CR^2COR^3$$

$$\longrightarrow PhSCHR^1CR^2(COR^3)CR^4R^5OH$$

Photoaddition of thiophenols to alkynes gives cis-vinyl sulphides, the expected result (based on the products obtained from other radical addition reactions of alkynes), but the end-products are usually cis/trans mixtures because of isomerization by arenethiyl radical equilibration of the initial stereochemically-pure adducts (M.G.Voronkov, N.N.Vlasova, and G.Yu.Zhila, Zh.Org.Khim., 1984, 20, 211):-

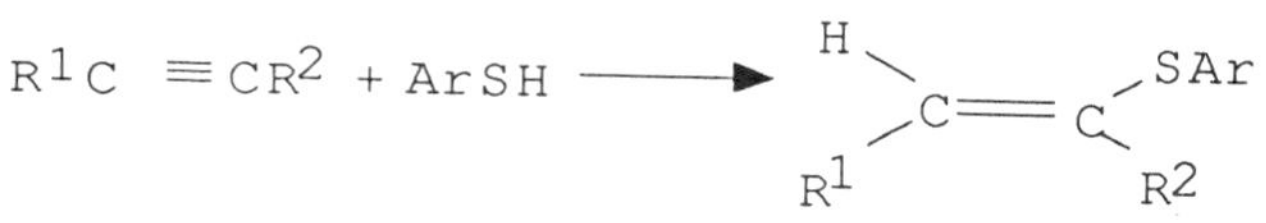

$$R^1C \equiv CR^2 + ArSH \longrightarrow \underset{R^1}{\overset{H}{\diagdown}}C=C\underset{R^2}{\overset{SAr}{\diagup}}$$

Arenesulphenyl and selenenyl halides add to alkenes to give β-halogenoalkyl aryl sulphides offering the possibility of further elaboration through the nucleophilic substituion of the halogen atom. Although this can be useful in synthesis, the β-halogen is also the source of side-reactions that detract from the value of the method, and may call for tedious purification procedures. In this context, some effort has been put in to finding "mild" reaction conditions avoiding the ensuing Cl-substitution in the addition of PhSeCl to alkenes (D.Liotta and G.Zima, Tetrahedron Lett., 1978, 4977).

This "mild" approach has been developed more recently with alkynes so as to prevent "over-addition" (A.L.Braga, C.C.Silveira, A.Reckziegel, and P.H.Menzies, Tetrahedron Lett., 1993, 34, 8041):-

$$R^1C \equiv CH + ArYX \xrightarrow[DMF]{CuI} R^1C \equiv CYAr$$

(Y = S, Se, or Te; X = Cl, Br)

Two equivalents anhydrous CuI are involved in this process, and treatment in DMF at room temperature under nitrogen is effective within 2 hours.

Sodium arenethiolates react with bismuthonium salts, providing the first example of alkyl transfer from such compounds (Y.Matano, N.Azuma, and H.Suzuki, Tetrahedron Lett., 1993, 34, 8457):-

$$RCOCH_2Bi^+Ar_3\ BF_4^- + ArS^-Na^+$$

$$\longrightarrow RCOCH_2SAr + Ar_3Bi$$

Thiolates react with $BrCF_3$ under 2-3 atm pressure in DMF, to give products of radical oxidation through an $S_{RN}1$ pathway (C.Wakselman and M.Tordeux, J.Org.Chem., 1985, 50, 4047):-

$$ArS^-K^+ + CF_3Br \longrightarrow ArS^\cdot + CF_3Br^{-\cdot} \longrightarrow$$

$$CF_3^\cdot + Br^- \xrightarrow{ArS^-} ArSCF_3^{-\cdot} \xrightarrow{CF_3Br} ArSCF_3 + CF_3Br^{-\cdot}$$

Thiols react readily with acetals and their "hidden" analogues to give sulphides (or, more correctly, to give thioacetals). In the following example, formation of 5-phenylthiopyrrolidin-2-ones and the substitution of the phenylthio group illustrates the value of divalent sulphur functions in synthesis (M.Eguchi, Q.Zeng, A.Korda, and I.Ojima, Tetrahedron Lett., 1993, 34, 915):-

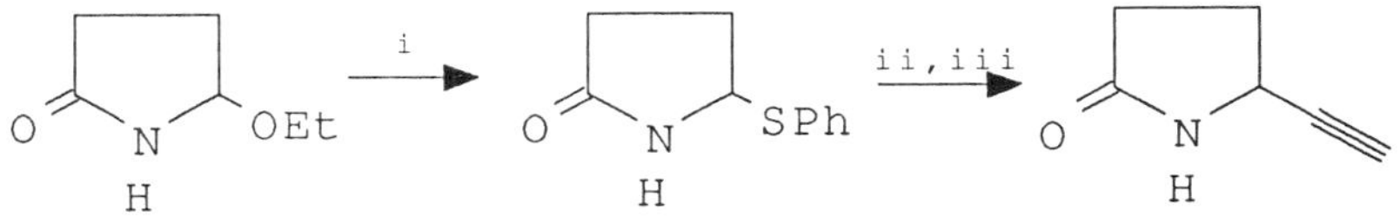

i, $PhSH/H^+$ at r.t., 5h in EtOH

ii, $ClZnC{\equiv}CSiMe_3 / Zn(C{\equiv}CSiMe_3)_2$ in toluene

iii, $Bu_4N^+\ F^-$ in MeCN

Participation by the neighbouring carbonyl group in hex-4-enal allows the formation of good yields of alternative products as well as the normal sulphenyl halide - alkene adducts, and this property opens up a useful synthesis of deoxyglycosides (S.Current and K.B.Sharpless, Tetrahedron Lett., 1978, 5075) using an areneselenenyl bromide:-

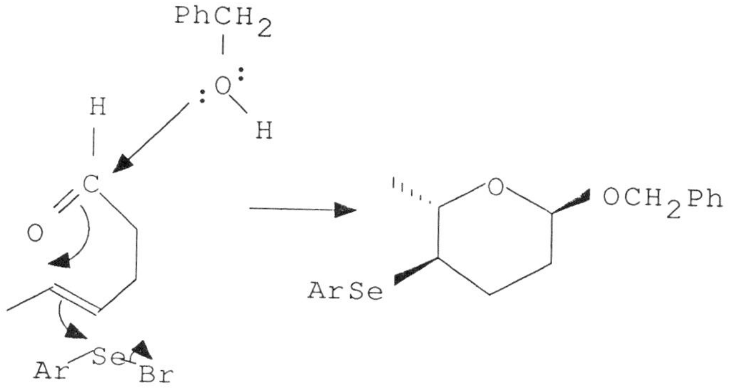

The simple arenethiol + bromoalkane reaction leading to sulphides has been carried out with the aid of Al_2O_3 impregnated with NaOH, to give good yields under mild conditions (B.Czech, S.Quicic, and S.L.Regen, Synthesis, 1980, 113). In some cases, thallium thiolates and selenolates, e.g. PhSTl and PhSeTl,

$$R^1CH_2CHCOR^3 \longrightarrow R^1CH_2CHCOR^3$$
$$| |$$
$$Cl SPh$$

have been used to advantage in substitution processes of α-halogenocarbonyl compounds (M.R.Detty and G.P.Wood, J.Org.Chem.,1980, 45, 80) and for the ring-opening of epoxides (H.A.Klein, Chem.Ber., 1979, 112, 3037).

Related arene substitution reactions leading to sulphides include the reaction of an alkali metal alkanethiolate with an unactivated halobenzene (J.R.Beck and J.A.Yahner, J.Org.Chem., 1978, 43, 2048, 2052), and of nitrobenzenes carrying an o- or p-alkylthio grouping that exerts an activating role assisting the substitution (P.Gogolli, L.Testaferri, M.Tingoli, and M.Tiecco, J.Org.Chem., 1979, 44, 2636; P.Gogolli, F.Maiolo, L.Testaferri, M.Tingoli, and M.Tiecco, J.Org.Chem., 1979, 44, 2642). It should be noted however, that the success of these examples depends upon the use of the highly toxic hexamethylphosphoric triamide as solvent.

The reaction of an arenethiolate with 5-arylthianthrenium salts (S.S.Shin, M.N.Kim, H.O.Kim, and K.Kim, Tetrahedron Lett., 1993, 34, 8469; see also K.Kim and H.J.Rim, Tetrahedron Lett., 1990, 31, 5631) illustrates ring-opening reactions of sulphur heterocycles to give sulphides (see also G.C.Barrett, in "Comprehensive Organic Chemistry", Eds. Sir Derek Barton and W.D.Ollis, Vol.3, Ed. D.N.Jones, Pergamon Press, Oxford, 1979, p.10), and this particular Scheme reveals a number of additional points of interest:-

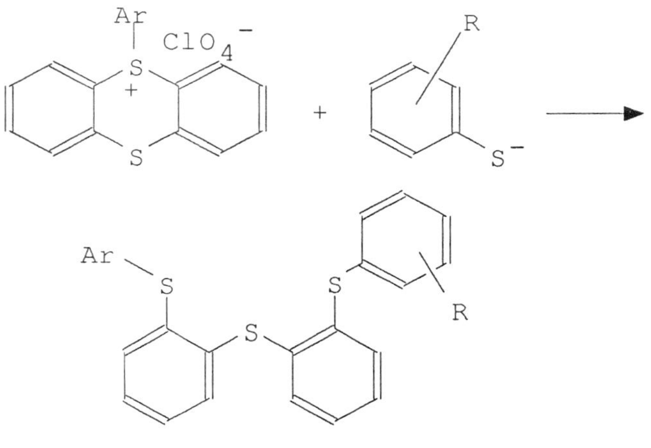

(2). From Disulphides. Disulphides can show properties more usually associated with sulphenyl halides, i.e., as sources of sulphenyl cations or their equivalent. Illustrated by an unusual reaction, in which cyclohexanone is converted into o-phenylthiophenol with diphenyl disulphide, the dual nature of a disulphide can be discerned; presumably the disulphide acts as a mild oxidizing agent as well as sulphenylation reagent, but the stoichiometry has not been worked out (B.M.Trost and J.H.Rigby, Tetrahedron Lett., 1978, 1667):-

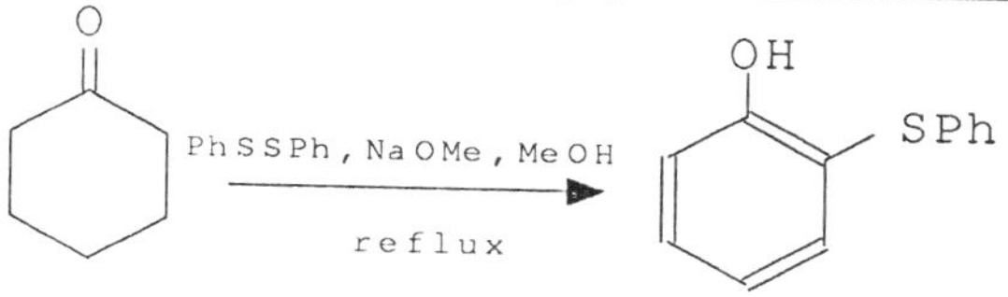

[cyclohexanone + PhSSPh, NaOMe, MeOH, reflux → 2-(phenylthio)phenol]

(3). By the Use of Other Sulphenylation and Selenenylation Reagents.
Benzeneseleninic acid and its anhydride have entered into organic synthesis methodology as mild oxidants that differ in their regioselectivity towards phenols. The former reagent brings about specific p-oxidation of phenols to give p-quinones, but the latter brings about o-substitution leading to diaryl selenides through oxidation involving benzeneselenenic acid, PhSeOH, formed from benzeneselenenic acid, PhSeO$_2$H, by disproportionation (D.H.R.Barton, J.-P.Finet, and M.Thomas, Tetrahedron, 1988, 44, 6397; P.G.Gassman, A.Miura, and T.Miune, J.Org.Chem., 1982, 47, 951).

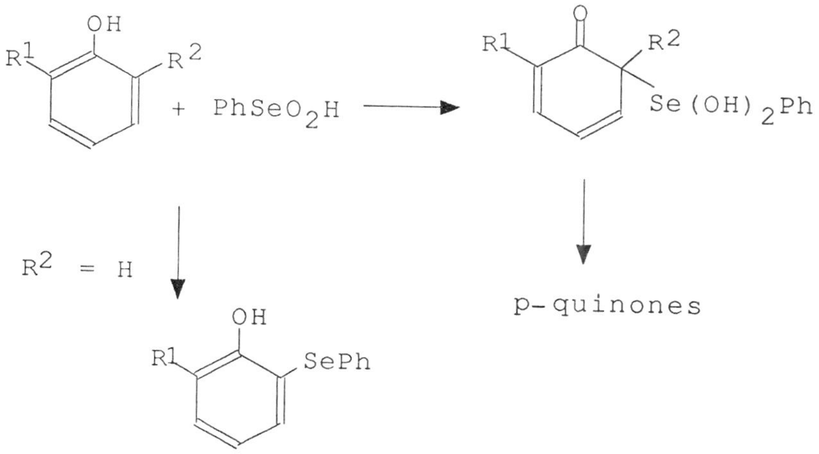

[Scheme: R^1-substituted phenol with R^2 + PhSeO$_2$H → cyclohexadienone intermediate with Se(OH)$_2$Ph; when R^2 = H → o-SePh phenol; otherwise → p-quinones]

Further examples of selenide formation using these reagents can be found in "Organoselenium Chemistry", Ed.D.Liotta, Wiley-Interscience, New York, 1987, and an interesting rationalization of the regiospecificity of phenol oxidation through further selenide-forming examples has been provided (L.Henricksen, Tetrahedron Lett., 1994, 35, 7057).

α-Phenylseleno-β-trimethylsilyloxy-aldehydes are formed using phenylselenenic anhydride with an allyl silyl ether (M.Shimizu, R.Takeda, and I.Kuwajima, Tetrahedron Lett., 1979, 3461):-

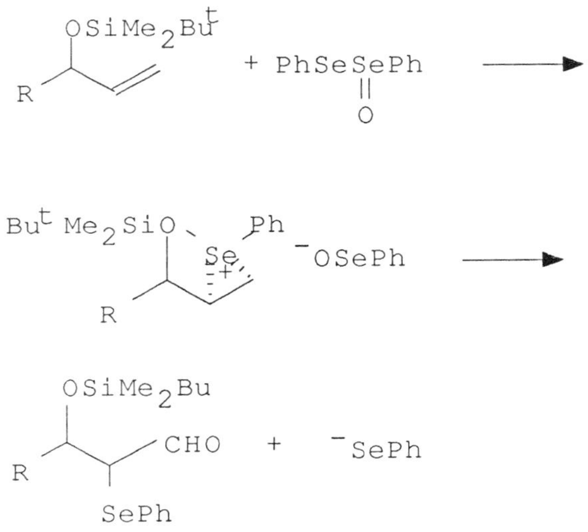

$$R\text{-CH(OSiMe}_2\text{Bu}^t\text{)-CH=CH}_2 + \text{PhSe(O)SePh} \longrightarrow$$

$$R\text{-CH(OSiMe}_2\text{Bu}^t\text{)-CH(Se}^+\text{Ph)-CH}_2\text{-OSePh}^- \longrightarrow$$

$$R\text{-CH(OSiMe}_2\text{Bu)-CH(SePh)-CHO} + {}^-\text{SePh}$$

__(4). By the Use of Simple Sulphur and Selenium Reagents__. Symmetrical sulphides can be obtained in low or moderate yields from an alkyl or activated aryl halide, with NaOH and S_8, in dimethyl sulphoxide. Unsymmetrical sulphides $PhC(CN)R^1SR^2$ are formed in this reaction in the special case of $PhCHR^1CN$ as an additional reactant (A.Jonczyk, Angew.Chem., 1979, 91, 228). $LiEt_3BH$ causes S_8 to dissolve in THF, to give a mixture of lithium sulphide and lithium persulphide that yields various proportions of symmetrical organic sulphide and disulphide through reaction with a halogen compound

(J.A.Gladysz, V.K.Wong, and B.S.Jick, J.Chem.Soc., Chem.Commun., 1978, 838).
Grignard reagents react with Se_8 to give areneselenenylmagnesium halides, which can give selenols by hydrolysis or selenides by interception by an aryldiazonium salt (V.P.Krasnov, V.I.Naddaka, V.P.Garkin, and V.I.Minkin, Zh.Org.Khim., 1978, 14, 2620). Thiophens react with Te_8 and CH_3I to give 2-thienyl methyl tellurides (N.Dereu and J.L.Piette, Bull.Soc.Chim.Fr., 1979, 623).

(5). From Thioacetals. O-Silyl thioacetals give corresponding sulphides through a novel alkylation procedure illustrated in the following examples (T.Mukaiyama, T.Ohno, T.Nishimura, J.S.Han, and S.Kobayashi, Chem.Lett., 1990, 2239):-

(a)

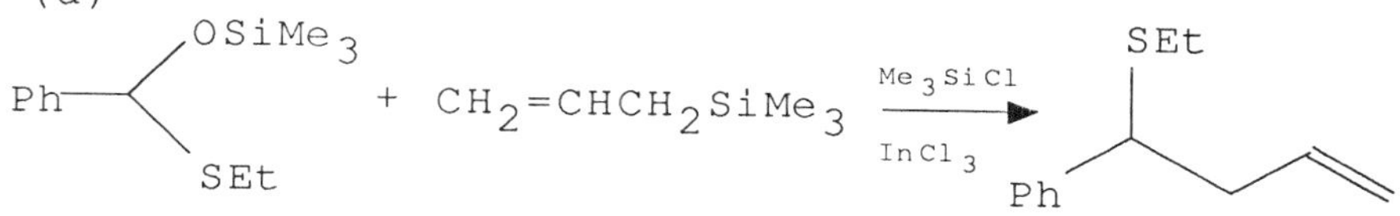

$$Ph\text{-}C(OSiMe_3)(SEt) + CH_2=CHCH_2SiMe_3 \xrightarrow[InCl_3]{Me_3SiCl} Ph\text{-}CH(SEt)\text{-}CH_2\text{-}CH=CH_2$$

(b)

$$R^1R^2C(OSiMe_3)(SEt) + R^3C(OSiMe_3)=CHR^4 \xrightarrow[InCl_3]{Me_3SiCl} R^3\text{-}C(O)\text{-}CHR^4\text{-}CR^1R^2(SEt)$$

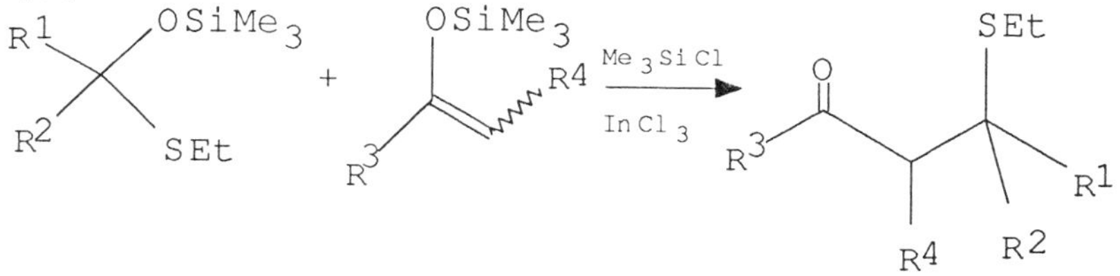

(6). From Thiocarbonyl Compounds. There are special circumstances in which thiocarbonyl compounds yield structurally simple sulphides, e.g. the formation of methyl sulphides from MeSH and 1,2- and 1,3-bis(dimethylamino)cyclobutene-3,4- and -2,4-dithiones (G.Seitz, K.Mann, R.Schmiedel, and R.Matusch, Chem.Ber., 1979, 112, 990; R.Matusch, R.Schmiedel, and G.Seitz, Liebigs Ann.Chem., 1979, 595). However, a wide range of thiocarbonyl compounds carrying adjacent heteroatoms are readily alkylated on S (Se analogues are even more easily alkylated). For example, radical S-phenylation of thiobenzanilides can be accomplished in high yield using phenylazotriphenylmethane (R.Kh.Friedlina, I.I.Kandror, and I.O.Bragina, Izv.Akad.Nauk S.S.S.R., Ser.Khim., 1979, 1165; I.I.Kandror, I.O.Bragina, and R.Kh.Friedlina, Dokl.Akad.Nauk S.S.S.R., 1979, 249, 867). Such processes have been thoroughly studied over many years and little attention has been given recently to their non-routine details.

10. Reactions of Sulphides and Selenides and Their Analogues

(a) Oxidation

Numerous simple oxidants have been uncovered recently for the formation of sulphoxides from sulphides, but any aggressive oxidant will bring about the formation of sulphones from sulphides, so there has been little new in that particular context.

Sulphides $ArSCHR^1R^2$ give sulphoxides with a "physiological oxidant" [O_2 + $R_2Co(II)X$; a cytochrome; or $Fe(ClO_4)_2$ + ascorbic acid] (T.Numata, Y.Watanabe, and S.Oae, Tetrahedron Lett., 1979, 1411; idem, 1978, 4933) though a considerable amount of sulphide cleavage accompanies the process and generates carbonyl compounds:-

$$ArSCHR^1R^2 \longrightarrow ArSSAr + R^1COR^2$$

In the biochemical context, the microsome-induced oxygenation of sulphides (cytochrome P as co-enzyme) shows enantiotopic differentiation, though interestingly, the stereochemical preference established for a model di-aryl sulphide is not in any way total, or in favour of one enantiomer:-

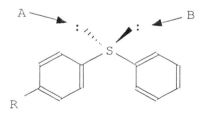

Thus, for the placing of an oxygen atom by the preferred pathway A [leading to (S)-stereochemistry at sulphur] when a cytochrome P450/O_2/NADPH system is used, the enantiomeric excess is only slight (S:R = 55:45), but the opposite preference is shown using an FAD-mono-oxygenase/O_2/NADPH/pig liver microsome system, with an R:S ratio = 82:18(T.Takata, M.Yamazaki, K.Fujimori, Y.H.Kim, S.Oae, and T.Iyanagi, Bull.Chem.Soc.Jpn., 1983, 56, 2300).

11. Synthetic Uses of sulphides

Activation of an adjacent methyl or methylene group by a divalent sulphur atom underlies some applications of sulphides in synthesis and must be the explanation for a novel acetoxy-group transfer reaction (R.R.King, J.Org.Chem., 1979, 44, 4194):-

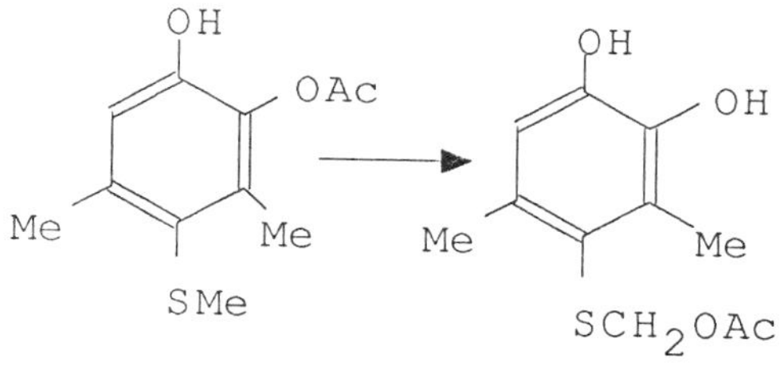

Arenesulphenyl-stabilized carbanions (N.Wang, S.Su, and L.Tsai, Tetrahedron Lett., 1979, 1121) and carbenes (M.Franck-Neumann and J.J.Lohmann, Tetrahedron Lett., 1979, 2075) are also prominent representatives of reaction intermediates in modern organic synthesis methodology, in the former case illustrated by Michael addition of PhSCHLiCN, and in the latter case providing cyclopropanes through reaction with alkenes and alkynes.

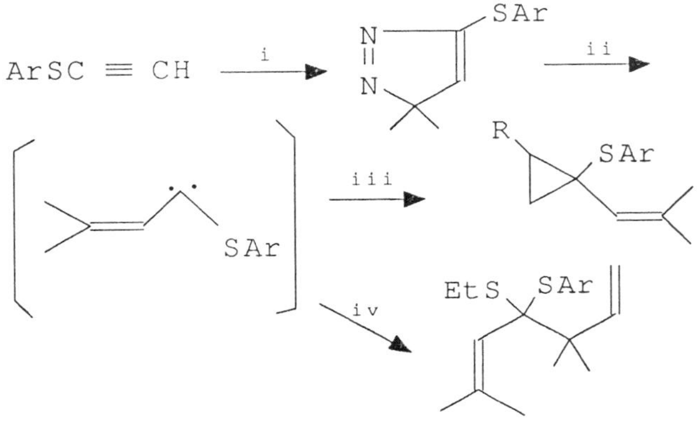

i, Me_2CN_2; ii, $h\nu$; iii, $RCH=CH_2$; iv, $EtSCH_2CH=CH_2$

Propargyl sulphides are valuable synthetic intermediates, as sources of carbanions. This is illustrated clearly in a 1,5-enyne synthesis (E.Negishi, C.L.Rand, and K.P.Jadhav, unpublished results cited in E.Negishi, Pure Appl.Chem., 1981, 53, 2333). Previous routes to these compounds involving allyl-propargyl coupling give disappointing yields.

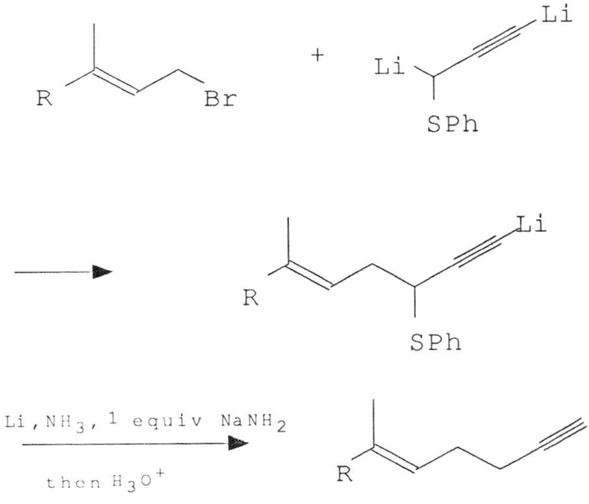

Alkylation adjacent to a phenylthio-grouping initiates a novel β-lactam synthesis (K.Hirai and Y.Iwano, Tetrahedron Lett., 1979, 2031):-

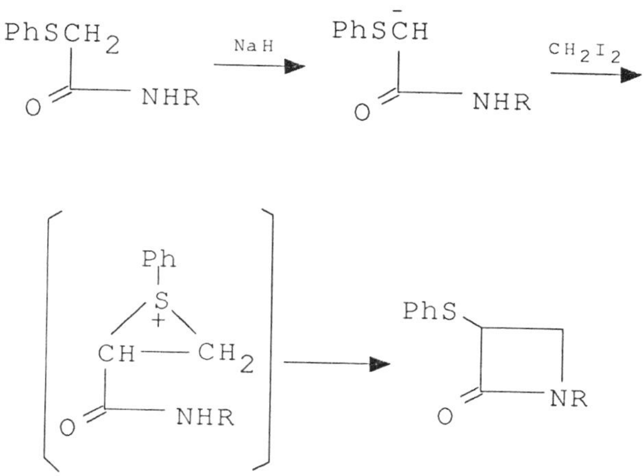

A valuable feature of such species is the delivery of products that carry a phenylthio-substituent whose presence can be exploited in further synthetic operations; or of course, the sulphide group can be replaced by H as in the enyne synthesis above (E.Negishi, C.L.Rand, and K.P.Jadhav, unpublished results cited in E.Negishi, Pure Appl.Chem., 1981, 53, 2333).

α-Acyl sulphides react with $PhI(OCOCF_3)_2$ to generate the α-cation, accounting for the easy cyclization of anilides (Y.Tamura, Y.Yakura, Y.Shirouchi, and J.Haruta, Chem.Pharm.Bull., 1986, 34, 1061):-

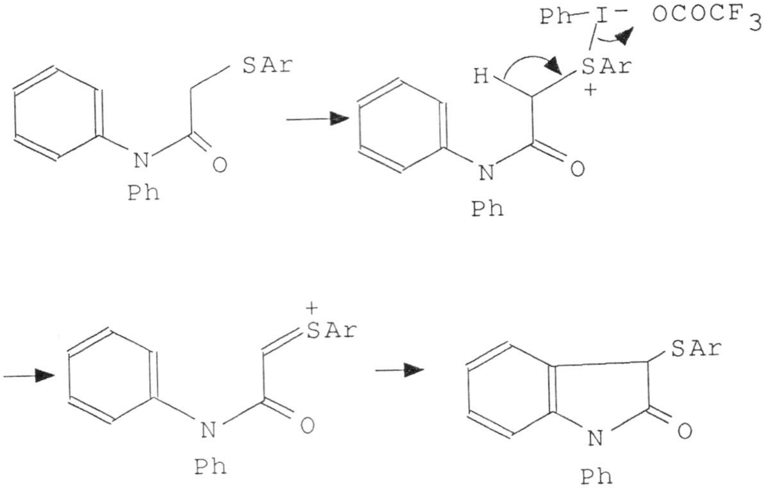

[Scheme showing conversion of PhNHC(O)CH₂SAr through PhI⁺(OCOCF₃) oxidation to a thionium intermediate, then cyclization to 3-(arylthio)oxindole (N-Ph)]

The scope of Petersen olefination has been extended by the use of α-silylated sulphides (D.J.Ager, *J.Chem.Soc., Perkin Trans I*, 1986, 183, 195; S.Hackett and T.Livinghouse, *J.Org.Chem.*, 1986, 51, 879):-

(a) PhS–CHBu–SiMe₃ $\xrightarrow{\text{Li Np}}$ Bu(Li)C(SiMe₃) $\xrightarrow{\text{MeCHO}}$ BuCH=CHMe

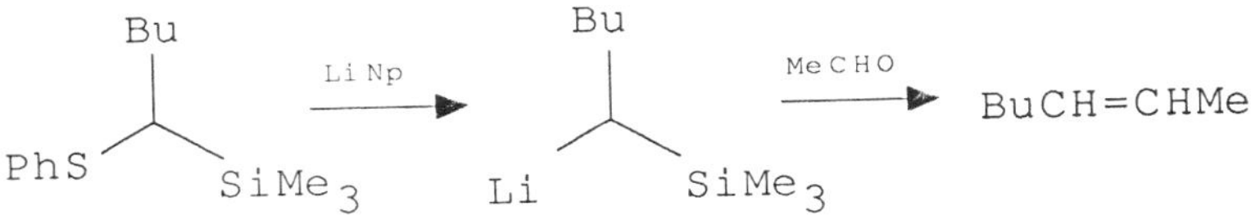

(b) PhS–C(Li)(OMe)(SiMe₃) + RCOR' ⟶ R'R C=C(SPh)(OMe)

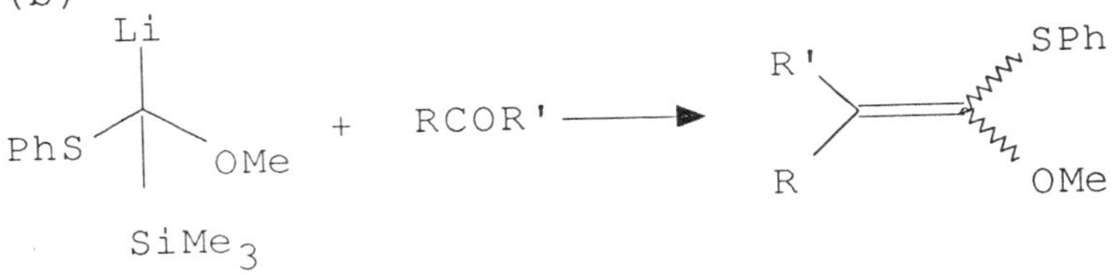

Selenium analogues are also effective in similar ways but their higher reactivity and the lower C-Se bond energy compared with that of the C-S bond provides greater flexibility (M.Sevrin and A.Krief, Tetrahedron Lett., 1980, 21, 585):-

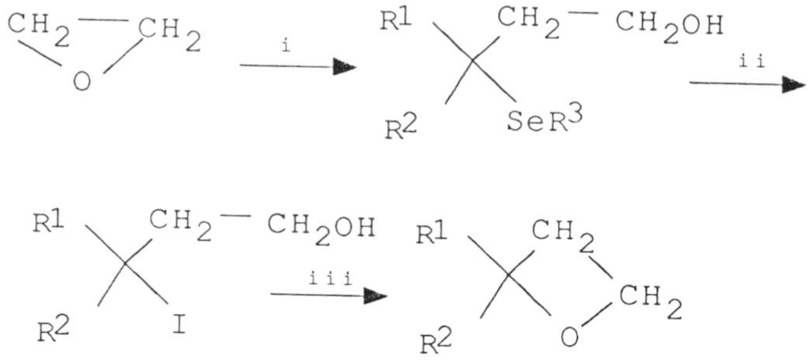

$$\text{CH}_2\text{—CH}_2\text{—O} \xrightarrow{i} \underset{R^2}{\overset{R^1}{\diagdown}}\!\!\!\!\!\bigtimes\!\!\!\!\!\underset{SeR^3}{\overset{CH_2-CH_2OH}{\diagup}} \xrightarrow{ii}$$

$$\underset{R^2}{\overset{R^1}{\diagdown}}\!\!\!\!\!\bigtimes\!\!\!\!\!\underset{I}{\overset{CH_2-CH_2OH}{\diagup}} \xrightarrow{iii} \underset{R^2}{\overset{R^1}{\diagdown}}\!\!\!\!\!\bigtimes\!\!\!\!\!\underset{O}{\overset{CH_2}{\diagup}}\!\!\diagdown\!CH_2$$

i, $R^1R^2C(SeR^3)Li$; ii, MeI, NaI, DMF; iii, Bu^tOK

12 Thioacetals and related compounds

There are numerous examples in the current literature reinforcing the established uses of thioacetals in synthesis. These examples also illustrate their compatibility with valuable organometallic synthons:-

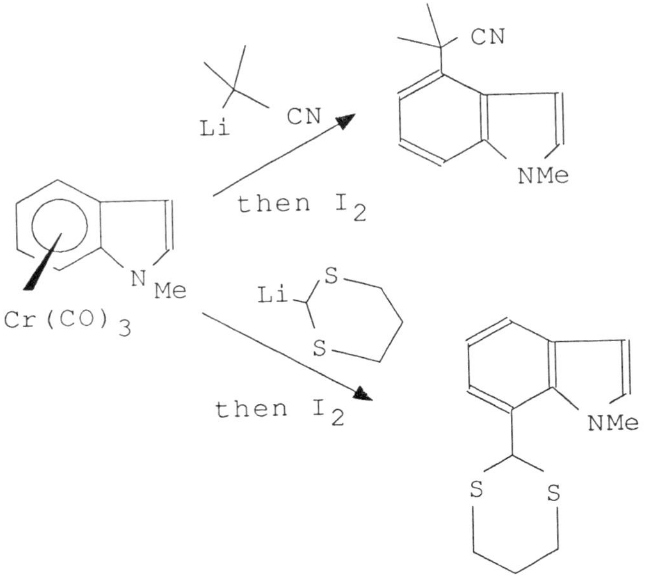

The η-(indole)-Cr(CO)₃ reactions in the accompanying Scheme (M.F.Semmelhack, <u>Pure Appl.Chem.</u>, 1981, 53, 2379) are interesting for a number of reasons. Surprisingly, the Cr(CO)₃ moiety selectively complexes with the six-membered ring, and promotes nucleophilic addition at that location. It again exerts regiocontrol in the reactions at that location, since the thioacetal carbanion enters at a different site to a cyano-stabilized carbanion (T.Albright and B.K.Carpenter, <u>Inorg.Chem.</u>, 1980, 19, 3092).

As is standard practice, the thioacetal moiety can be converted into a formyl group by hydrolysis, and the Scheme represents an example of the combinations that are feasible, of the chemistry of organosulphur compounds with other modern transition metal synthons.

13. Unsaturated sulphides

(a) Preparation

Direct synthesis from alkenes using a sulphenyl halide proceeds by way of the β-halogenosulphide which is subjected to easy dehydrohalogenation using Bu^{t}OK (W.Verboom, J.Meijer, and L.Brandsma, Synthesis, 1978, 577; B.Geise and S.Lachheim, Chem.Ber., 1979, 112, 2503). The geometry of the products can be controlled [cis-vinyl sulphides are formed under kinetic control at -78°C, trans at +77°C] (B.Geise and S.Lachheim, Chem.Ber., 1979, 112, 2503).

(b) Synthetic uses of unsaturated sulphides

Vinyl sulphides are carbonyl group equivalents and have generated numerous synthetic uses, e.g. in the transposition of a carbonyl group from one position in an aliphatic system to the neighbouring carbon atom (T.Nakai and T.Mimura, Tetrahedron Lett., 1979, 531):-

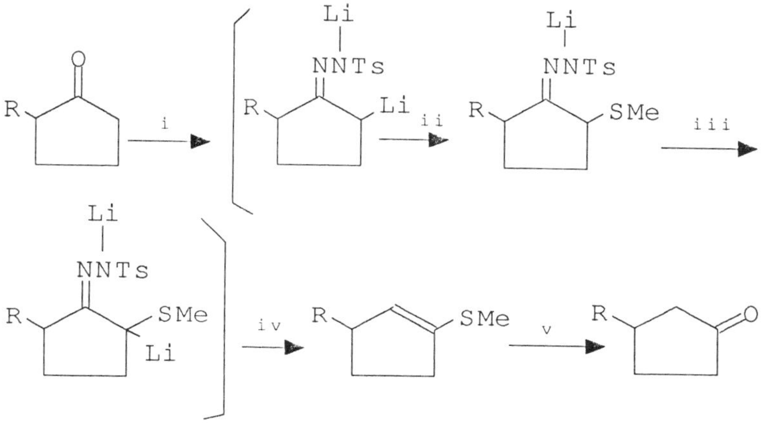

i, TsNHNH$_2$, then BuLi/TMEDA/THF; ii, MeSSMe;

iii, BuLi; iv, -N$_2$ at r.t., then NH$_4$Cl$^-$H$_2$O;

v, HgCl$_2$ in hot MeCN

An inherent feature in this process is the ready cleavage of the C-S bond to complete the synthesis, and ketovinyl sulphides $R^1COCH=CR^2SPh$ undergo desulphurization with $NaBH_4$ and $CoCl_2$ (or $NiCl_2$), or with $LiAlH_4$; however, reduction of the C=C grouping also occurs with $NaBH_4$ and $CoCl_2$ (or $NiCl_2$), but with $LiAlH_4$ a mixture of reduction product and $\alpha\beta$-unsaturated ketone is formed (T.Nishio and Y.Omote, Chem.Lett., 1979, 365).

Cleavage in another way is the result of reduction of allyl sulphides with Bu_3SnH/AIBN in toluene at 0°C to give thiols (Y.Ueno and M.Okawara, J.Am.Chem.Soc., 1979, 101, 1893), illustrated further with complex allyl aryl sulphides (see Section 9.1; S.S.Shin, M.N.Kim, H.O.Kim, and K.Kim, Tetrahedron Lett., 1993, 34, 8469).

Elimination of sulphur, in the form of a thiol, from vinyl sulphides that occurs through reaction with $H_2N(CH_2)_3NH^-K^+$ leads to the corresponding alkyne (C.A.Brown, J.Org.Chem., 1978, 43, 3083).

Chapter 6

BENZENES CARRYING SULFONYL, SULFINYL, OR SULFENYL FUNCTIONAL GROUPS, BUT EXCLUDING THOSE COMPOUNDS BEARING FUNCTIONALIZED SIDE CHAINS

Richard W. Brown

1. Benzenesulfonic Acids and Their Derivatives

Recent reviews on the chemistry of sulfonic acids and derivatives: Andersen, K. K. in *Comprehensive Organic Chemistry;* Barton, D. H. R.; Ollis, W. D., Eds.; Pergamon: Oxford, 1979; Vol. 3, Section 11.19; *The Chemistry of Sulphonic Acids, Esters and their Derivatives*; Patai, S., Ed.; Wiley: Chichester, 1991.

(a) Benzenesulfonic Acids

(i) Preparation

Sulfonation, the introduction of a SO_3H group into a benzene nucleus, or oxidation of an existing sulfur function group are the main means of preparing benzenesulfonic acids. Industrially, the former method is more common. Sulfonic acids and sulfonation have been reviewed from an industrial perspective (Knaggs, E. A.; Nussbaum, M. L.; Schultz, A. in *Encyclopedia of Chemical Technology (Kirk-Othmer)*, 3rd Ed.; Grayson, M., Ed.; Wiley: New York, 1983; Vol. 22, p. 1-45; Schultz, A. in *Encyclopedia of Chemical Technology (Kirk-Othmer)*, 3rd Ed.; Grayson, M., Ed.; Wiley: New York, 1983; Vol. 22, p. 45-63).

Bis(trimethylsilyl)sulfate ($(TMS)_2SO_4$), prepared from sulfuric acid with TMSCl, provides excellent yields of benzenesulfonic acids under mild conditions (Bourgeois, P.; Duffaut, N. *Bull. Soc. Chim. Fr.* **1980**, 195). A related procedure is the reaction of trimethylsilyl chlorosulfonate with substituted benzenes. Hydrohalogenolysis of the resulting trimethylsilyl benzenesulfonates affords anhydrous benzenesulfonic acids (Hofmann, K.; Simchen, G. *Liebigs Ann. Chem.* **1982**, 282).

In addition to activated halogens or nitro groups, other functionalities can be displaced by sulfite ion to give sulfonic acids. Bisulfite ion reacts with resorcinol, replacing a hydroxy group, to give 3-hydroxybenzenesulfonic acid (Lauer, W. M.; Langkammerer, C. M. *J. Am. Chem. Soc.* **1934**, *56*, 1628) and with 2,4-diaminotoluene, replacing an amino group, to give 3-amino-4-methylbenzenesulfonic acid (Allan, Z. J.; Podstata, J. *Coll. Czech. Chem. Commun.* **1965**, *30*, 1024; *Chem. Abstr.* **1965**, *63*, 2918b).

Sulfite ion adds to quinones and other quinoid-type structures to give sulfonic acids. The irreversible Michael addition of sulfite to *p*-benzoquinone has long been a reaction of practical importance in photographic science (Lee, W. E.; Brown, E. R. in *The Theory of the Photographic Process*; James, T. H., Ed.; Macmillan: New York, 1977; p. 309-10). The kinetics and mechanism of this reaction have been examined (Youngblood, M. P. *J. Org. Chem.* **1986**, *51*, 1981).

Sulfur trioxide in carbon tetrachloride reacts with phenyl trialkylstannanes at 70°C to give the corresponding trialkylstannyl benzenesulfonates (Hillgärtner, H.; Baines, K. M.; Dicke, R.; Vorspohl, K.; Kobs, U., Nussbeutel, U. *Tetrahedron* **1989**, *45*, 951).

Many sulfur-substituted benzenes can be oxidized to the corresponding sulfonic acids. These include thiols, thiocyanates, xanthates, disulfides, thiolesters, thiocarbamates, thiolsulfonates, thiolsulfinates, sulfides, sulfoxides, sulfones, and sulfinates. Thiols and disulfides can be oxidized by heating in DMSO containing water and a catalytic amount of I_2 or HBr. Water minimizes the decomposition of DMSO (Lowe, O. G. *J. Org. Chem.* **1976**, *41*, 2061). A quantitative method of converting benzenethiol to benzenesulfonic acid through the reaction with six equivalents of N_2O_4 at ambient temperature has been reported. At lower temperatures, the corresponding disulfide and thiolsulfonate can be isolated (Kim, Y. H.; Shinhama, K.; Fukushima, D.; Oae, S. *Tetrahedron Lett.* **1978**, 1211). Superoxide, generated from KO_2 and a crown ether, in pyridine at 25°C can be used to oxidize thiols, disulfides, thiolsulfinates, and thiolsulfonates (Takata, T.; Kim, Y. H.; Oae, S. *Tetrahedron Lett.* **1979**, 821; Oae, S.; Takata, T.; Kim, Y. H. *Tetrahedron* **1981**, *37*, 37).

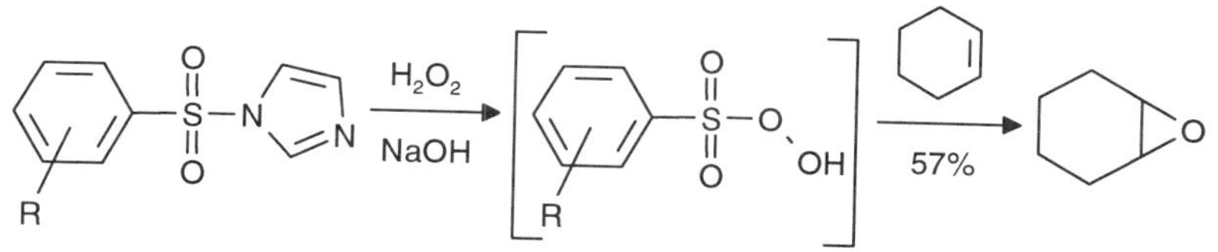

Benzenesulfonic peracids can be prepared *in situ* by reaction of the corresponding benzenesulfonyl imidazolides with hydrogen peroxide under alkaline conditions. This reaction system epoxidizes alkenes in moderate yields (Schulz, M.; Kluge, R.; Lipke, M. *Synlett* **1993**, 915).

(ii) Reactions

Benzenesulfonic acids undergo deprotonation at the acid group and *ortho*-lithiation with *n*-butyllithium in THF at 0°C. Trapping of the dilithiated species with various electrophiles affords a variety of *ortho*-substituted benzenesulfonic acids. Desulfonation of these products with $HgSO_4$ in aqueous sulfuric acid provides a convenient route to *meta*-substituted benzenes that might otherwise be difficult to prepare (Figuly, G. D.; Martin, J. C. *J. Org. Chem.* **1980**, *45*, 3728).

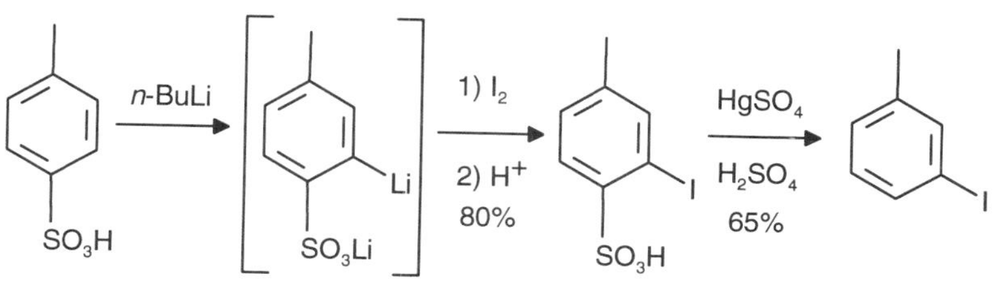

The reduction of benzenesulfonic acids with a catalytic amount of iodine and excess triphenylphosphine in benzene at reflux gives the corresponding benzenethiols in 88-98% yields. In the presence of a catalytic amount of 18-crown-6, sodium benzenesulfonates are similarly reduced to benzenethiols *via* the disulfides (Fujimori, K.; Togo, H.; Oae, S. *Tetrahedron Lett.* **1980**, *21*, 4921). Benzenesulfonic acids may also be reduced to benzenethiols by treatment with activated carbon in technical white oil at 350°C (Dockner, T. *Angew. Chem. Int. Ed. Engl.* **1988**, *27*, 679; *Angew. Chem.* **1988**, *100*, 699). Industrially, benzenesulfonic acids are reduced to benzenethiols by heating with white phosphorus and iodine followed by hydrolysis (Pitt, H., U.S. Patent 3 734 969, 1973). The reduction of benzenesulfonic acids to the corresponding disulfides with boron tribromide or -iodide and potassium iodide in the presence of a phase transfer catalyst has been reported (Olah, G. A.; Narang, S. C.; Field, L. D.; Karpeles, R. *J. Org. Chem.* **1981**, *46*, 2408).

(b) Benzenesulfonyl Halides

(i) Preparation

A large variety of preparative procedures continue to be used for the synthesis of benzenesulfonyl halides. Recently, an improved method of preparing benzenesulfonyl chlorides from aryl halides has been reported. Aryl chlorides or bromides are treated with *n*-butyllithium and sulfur dioxide at low temperature to form the corresponding lithium sulfinate. Further treatment with sulfuryl chloride in hexane provides the benzenesulfonyl chlorides in 82-99% overall yield (Hamada, T.; Yonemitsu, O. *Synthesis* **1986**, 852). Phosgene is used in the industrial preparation of certain benzenesulfonyl chlorides (Blank, H. U.; Pfister, T., U. S. Patent 4 215 071, 1980). Sterically hindered benzenesulfonyl chlorides can be prepared from the corresponding phenols *via* a three-step process. Acylation of the phenols with dialkyl thiocarbamoyl chlorides followed by Newman-Kwart rearrangement at 240-300°C affords the *S*-aryl *N,N*-dialkylthiocarbamates. Oxidation with chlorine in acetic acid gives high yields of the desired sulfonyl chlorides (Wagenaar, A.; Engberts, J. B. F. N. *Recl. Trav. Chim. Pays-Bas* **1982**, *101*, 91). The related benzenesulfonimidoyl chlorides can be prepared in good yields by oxidation of sulfinamides with chlorine (Johnson, C. R.; Jonsson, E. U.; Bacon, C. C. *J. Org. Chem.* **1979**, *44*, 2055).

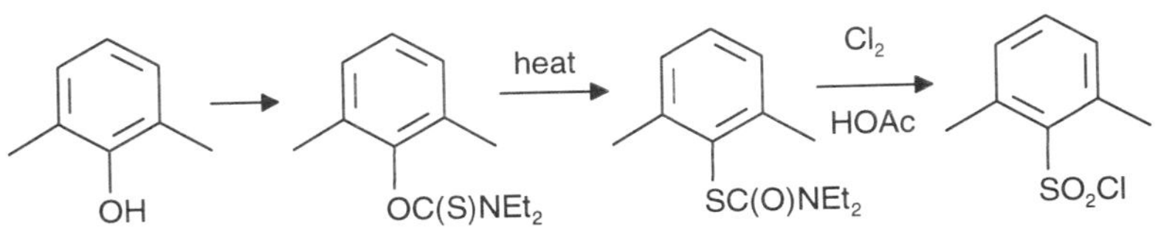

(ii) Reactions

Benzenesulfonyl fluorides are converted into the corresponding chlorides by treatment with anhydrous aluminum chloride in boiling 1,2-dichloroethane. In conjunction with the reverse conversion using potassium fluoride in aqueous dioxane, this reaction constitutes a means of protecting benzenesulfonyl chlorides as their less-reactive sulfonyl fluorides (Norris, T. J. Chem. Soc., Perkin Trans. I **1978**, 1378). The product of the reaction of the enolate of *tert*-butyl methyl ketone with benzenesulfonyl fluoride is strongly dependent on the nature of the gegenion. The β-oxosulfone, the product of *C*-acylation, is formed exclusively with the lithium cation while only vinyl sulfonates are observed in the presence of cesium or quaternary ammonium ions (Hirsch, E.; Hünig, S.; Reißig, H.-U. *Chem. Ber.* **1982**, *115*, 3687).

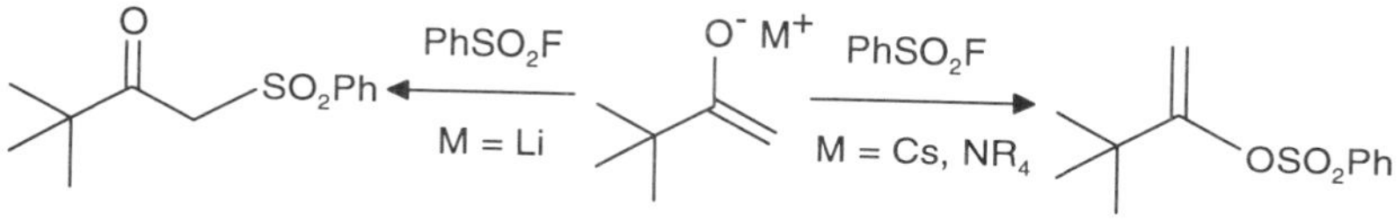

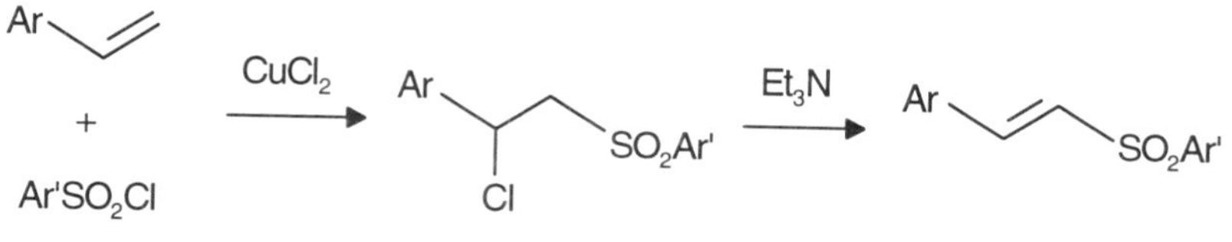

Several synthetically-useful methods for the addition of benzenesulfonyl chlorides to alkenes have been reported. Styrenes react with benzenesulfonyl chlorides in the presence of $CuCl_2$ to afford 1-chloro-1-aryl-2-(benzenesulfonyl)ethanes. Further treatment with triethylamine converts these adducts to the corresponding *trans*-1-aryl-2-(benzene-sulfonyl)ethenes in excellent yields (Truce, W. E.; Goralski, C. T. *J. Org. Chem.* **1971**, *36*, 2536). This transformation can also be affected with dichlorotris(triphenylphosphine)ruthenium(II) as catalyst (Kamigata, N.; Sawada, H.; Kobayashi, M. *Chem. Lett.* **1979**, 159). If allyl(trimethyl)silane is used in place of styrene, allyl aryl sulfones can be prepared (Pillot, J.-P.; Dunogués, J.; Calas, R. *Synthesis*, **1977**, 469). Allyl phenyl sulfides and selenides can also be used to prepare allyl aryl sulfones by treatment with benzenesulfonyl chlorides in the presence of a catalytic amount of dichlorotris(triphenylphosphine)ruthenium(II) (Kamigata, N.; Ishii, K.; Ohtsuka, T.; Matsuyama, H. *Bull. Soc. Chem. Jpn.* **1991**, *64*, 3479). In the presence of a catalytic amount of a ruthenium(II) complex with the chiral phosphine ligands (+)- or (-)-DIOP, benzenesulfonyl chlorides add to styrenes to give optically active adducts in 24-40% enantiomeric excess (Kameyama, M.; Kamigata, N.; Kobayashi, M. *J. Org. Chem.* **1987**, *52*, 3312). The related selenosulfonates ($ArSO_2SeAr'$) also react with alkenes and alkynes to give vinyl and acetylenic sulfones, respectively, after oxidative elimination of the areneseleno group (Back, T. G.; Collins, S. *Tetrahedron Lett.* **1980**, *21*, 2215; Miura, T.; Kobayashi, M. *J. Chem. Soc., Chem. Commun.* **1982**, 438). Benzeneselenosulfonates can be conveniently prepared from benzenesulfonyl hydrazides by oxidation with benzeneselenic acid (Back, T. G.; Collins, S. *Tetrahedron Lett.* **1980**, *21*, 2213) or benzenesulfinates by treatment with either benzeneselenic acid (Gancarz, R. A.; Kice, J. L. *Tetrahedron Lett.* **1980**, *21*, 1697) or benzenesulfenium cation (Wang, L.; Huang, X. *Synth. Commun.* **1993**, *23*, 2817).

Palladium-catalyzed carbonylation of benzenesulfonyl chlorides in the presence of metal alkoxides $M(OR)_x$ (M = B, Al, Ti) gives the corresponding carboxylic esters along with diaryl disulfides. With metal carboxylates $M(OCOR)_x$ (M = Na, K, Ca, Mg, Zn), the free acids can be obtained (Miura,

p-Cl-C6H4-COOH ← [CO, $PdCl_2(AsPh_3)_2$, $Zn(OAc)_2$, 70%] ← p-Cl-C6H4-SO2Cl → [CO, $PdCl_2(PPh_3)_2$, $Ti(O\text{-}i\text{-}Pr)_4$, 59%] → p-Cl-C6H4-COO-i-Pr

M.; Itoh, K.; Nomura, M. *Chem. Lett.* **1989**, 77; Itoh, K.; Hashimoto, H.; Miura, M.; Nomura, M. *J. Mol. Catal.* **1990**, *59*, 325). Similar palladium-catalyzed desulfonylations of benzenesulfonyl chlorides have been used to prepare substituted cinnamates and biaryls (Miura, M.; Hashimoto, H.; Itoh, K.; Nomura, M. *J. Chem. Soc., Perkin Trans. I* **1990**, 2207; Miura, M.; Hashimoto, H.; Itoh, K.; Nomura, M. *Chem. Lett.* **1990**, 459). The palladium-catalyzed cross-coupling of benzenesulfonyl chlorides and substituted vinyl- and allyl-stannanes provides sulfones in 57-90% yields under mild conditions (Labadie, S. S. *J. Org. Chem.* **1989**, *54*, 2496).

(c) Benzenesulfonic Anhydrides

p-Toluenesulfonic acid can be converted to the corresponding anhydride under mild conditions by treatment with methoxyacetylene (Eglinton, G.;

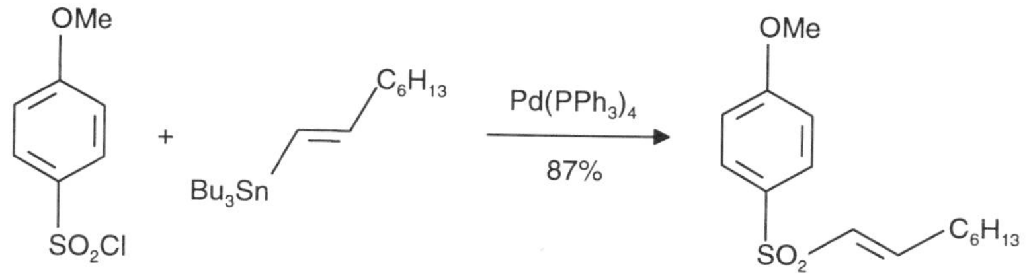

[Scheme: 4-methoxybenzenesulfonyl chloride + Bu₃Sn-CH=CH-CH₂-C₆H₁₃, Pd(PPh₃)₄, 87% → 4-methoxyphenyl-SO₂-CH₂-CH=CH-C₆H₁₃]

Jones, E. R. H.; Shaw, B. L.; Whiting, M. C. *J. Chem. Soc.* **1954**, 1860) or phenyl cyanate (Martin, D. *Chem. Ber.* **1965**, *98*, 3286). Oxidation of aryl disulfides with excess nitrogen tetroxide in CCl$_4$ at 0-50°C gives the corresponding benzenesulfonic anhydrides in 81-92% yields. As no intermediate products are detected, the oxidation does not appear to proceed stepwise (Kunieda, N.; Oae, S. *Bull. Chem. Soc. Jpn.* **1968**, *41*, 233).

The reactions of benzenesulfonic anhydrides closely parallel those of the corresponding benzenesulfonyl chlorides. A notable exception is the preparation of enol tosylates of β-ketoesters. Treatment of the potassium enolates of β-ketoesters with *p*-toluenesulfonic anhydride (Ts$_2$O) affords the enol tosylates in 52-83% yields; if *p*-toluenesulfonyl chloride is used, the yields fall to 20% (Jalander, L.; Mattinen, J.; Oksanen, L.; Rosling, A. *Synth. Commun.* **1990**, *20*, 881).

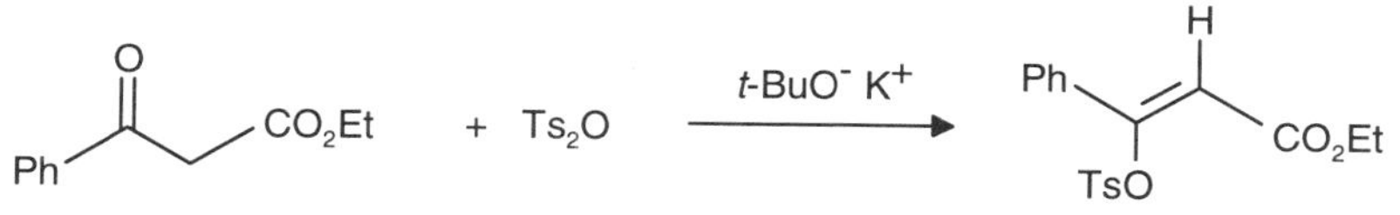

[Scheme: PhC(O)CH₂CO₂Et + Ts₂O, *t*-BuO⁻ K⁺ → Ph(TsO)C=CH(H)CO₂Et]

(d) Benzenesulfonyl Peroxides

Enol derivatives of ketones and esters (enol acetates, silyl enol ethers, enamines, ketene acetals) react with benzenesulfonyl peroxides to give the corresponding α-(benzenesulfonyloxy)ketones and esters, respectively, in high yields (Hoffman, R. V.; Carr, C. S.; Jankowski, B. C. *J. Org. Chem.* **1985**, *50*, 5148; Hoffman, R. V. *Synthesis* **1985**, 760; Hoffman, R. V.; Carr, C. S. *Tetrahedron Lett.* **1986**, *27*, 5811; Hoffman, R. V.; Kim, H.-O. *J. Org. Chem.* **1988**, *53*, 3855). Similar results have been reported with thallium(III) *p*-toluenesulfonate (Khanna, M. S.; Garg, C. P.; Kapoor, R. P. *Tetrahedron Lett.* **1992**, *33*, 1495) and [hydroxy(tosyloxy)iodo]benzene

[Reaction scheme: 1-(trimethylsilyloxy)-3,4-dihydronaphthalene + (ArSO₃)₂ → 2-(arylsulfonyloxy)-1-tetralone, 95%; Ar = p-nitrophenyl]

(Koser, G. F.; Relenyi, A. G.; Kalos, A. N.; Rebrovic, L.; Wettach, R. H. *J. Org. Chem.* **1982**, *47*, 2487; Moriarty, R. M.; Penmasta, R.; Awasthi, A. K.; Epa, W. R.; Prakash, I. *J. Org. Chem.* **1989**, *54*, 1101). A readily available, crystalline oxidizing agent, [hydroxy(tosyloxy)iodo]benzene has also been used to convert alkenes to vicinal bis(tosyloxy)alkanes, terminal alkynes to acetylenic tosylates, and alkenoic acids to lactones (Koser, G. F.; Rebrovic, L. *J. Org. Chem.* **1984**, *49*, 2462; Stang, P. J.; Surber, B. W. *J. Am. Chem. Soc.* **1985**, *107*, 1452; Shah, M.; Taschner, M. J.; Koser, G. F.; Rach, N. L. *Tetrahedron Lett.* **1986**, *27*, 4557).

Oxidation of amines with benzenesulfonyl peroxides at -78°C affords derivatized hydroxylamines (Hoffman, R. V.; Belfoure, E. L. *Synthesis* **1983**, 34). Further treatment with base affords the corresponding aldehydes, constituting a novel method for the oxidative deamination of amines. Similarly, hydrazines can be converted to imines (Hoffman, R. V.; Kumar, A. *J. Org. Chem.* **1984**, *49*, 4011, 4014).

(e) Benzenesulfonates

(i) Preparation

The principal method of preparing esters of benzenesulfonic acids continues to be the reaction of a sulfonyl chloride with a hydroxyl-containing compound in the presence of a base. Improved yields of benzenesulfonates, particularly benzyl benzenesulfonates, has been reported using phase transfer conditions (Szeja, W. *Synthesis* **1979**, 822). The effects of variations in the reactant structure and the base on the acylation of phenols by benzenesulfonyl chlorides has been studied (Vizgert, R. V.; Maksimenko, N. N. *J. Gen. Chem. USSR (Engl. Transl.)* **1978**, *14*, 963; *Zh. Org. Khim.* **1978**, *14*, 1031). An alternative to the silver-mediated reaction of benzenesulfonates with alkyl halides has been reported. At high temperatures and in the presence of a phase transfer catalyst, methyl benzenesulfonates react with alkyl halides to afford alkyl benzenesulfonates. Removal of the byproduct methyl halide is the key to driving this reaction to completion (Hahn, R. C.; Tompkins, J. *J. Org. Chem.* **1988**, *53*, 5783).

A variation of the Mitsunobu inversion has been used to prepare tosylates from chiral alcohols with inversion of stereochemistry. Dihydrocholesterol was converted to 3-α-tosyloxycholestane in 86% yield upon treatment with zinc tosylate, triphenylphosphine, and diethyl azodicarboxylate in benzene at ambient temperature (Galynker, I.; Still, W. C. *Tetrahedron Lett.* **1982**, *43*, 4461; Loibner, H.; Zbiral, E. *Helv. Chim. Acta* **1976**, *59*, 2100).

Benzenesulfonic acids can be converted quantitatively to their simple alkyl esters by reaction with trialkyl orthoformates (Padmapriya, A. A.; Just, G.; Lewis, N. G. *Synth. Commun.* **1985**, *15*, 1057) or -acetates (Trujillo, J. I.; Gopalan, A. S. *Tetrahedron Lett.* **1993**, *34*, 7355). The latter reagent gives superior results. This transformation has been used as the basis for a gas chromatographic method of determining isomers of toluenesulfonic acid (Baker, K. M.; Boyce, G. E. *J. Chromatogr.* **1976**, *117*, 471). Alkyl tosylates can be prepared by reaction of *p*-toluenesulfonic acid with the alkyl esters of a variety of carboxylic acids (Yoshihiro, N.; Arakawa, Y. *Chem. Pharm. Bull.* **1985**, *33*, 1380). Good to excellent yields have been reported for the alkylation of benzenesulfonic acids with trialkyl phosphites, trialkyl phosphates, and dialkyl phenylphosphonates (Karaman, R.; Leader, H.; Goldblum, A.; Breuer, E. *Chem. Ind. (London)* **1987**, 857; Yoshihiro, N.; Arakawa, Y.; Ueyama, N. *Chem. Pharm. Bull.* **1986**, *34*, 2710).

(ii) Reactions

Benzenesulfonates undergo *ortho*-lithiation upon treatment with *n*-butyllithium in THF at -78°C. The lithiated species can be trapped with a variety of electrophiles to yield substituted benzenesulfonates in 63-83% yields. In comparison with the analogous reaction with benzenesulfonic acids (*vide supra*), the products are easily isolated (Bonfiglio, J. N. *J. Org. Chem.* **1986**, *51*, 2833). Upon treatment with *n*-butyllithium, 2-aminophenyl benzenesulfonates rearrange to give *N*-(2-hydroxyphenyl) benzenesulfonamides (Andersen, K. K.; Gowda, G.; Jewell, L.; McGraw, P.; Phillips, B. T. *J. Org. Chem.* **1982**, *47*, 1884).

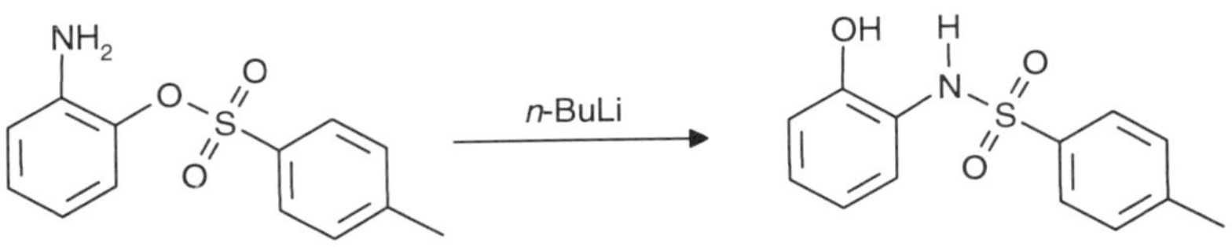

Benzeneselenenyl *p*-toluenesulfonate, prepared *in situ* from benzeneselenenyl chloride and silver *p*-toluenesulfonate, reacts with alkynes in an *anti* stereospecific manner to form β-phenylselenenyl enol tosylates (Back, T. G.; Muralidharan, K. R. *Tetrahedron Lett.* **1990**, *31*, 1957).

Treatment of α-(*p*-nitrophenyl)sulfonyloxy ketones with DBU at ambient temperature results in loss of sulfur dioxide and formation of arylated ketols. Other sulfonyloxy ketones do not react in this way (Hoffman, R. V.;

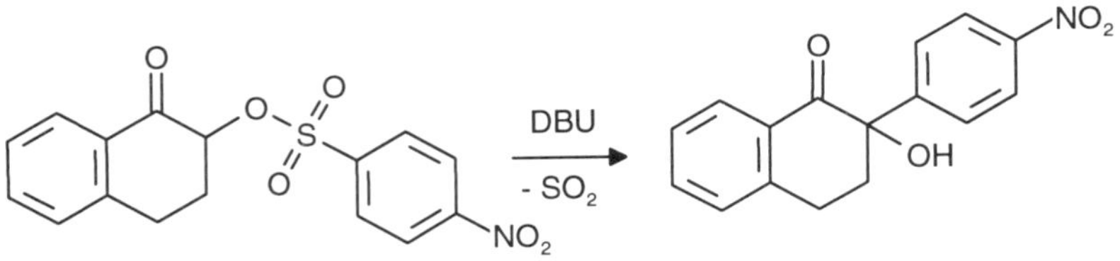

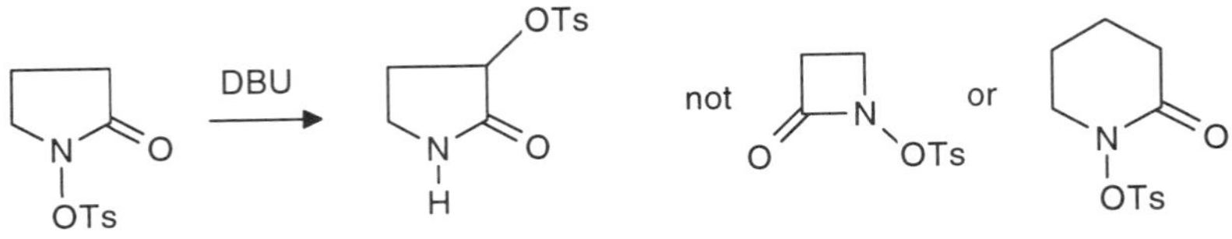

Jankowski, B. C.; Carr, C. S.; Dueslar, E. N. *J. Org. Chem.* **1986**, *51*, 130). A novel rearrangement of N-(*p*-toluenesulfonyloxy)-2-pyrrolidinone to 3-(*p*-toluenesulfonyloxy)-2-pyrrolidinone upon treatment with DBU has been reported. Interestingly, similar treatment of the analogous N-(*p*-toluenesulfonyloxy)-2-piperidinone or N-(p-toluenesulfonyloxy)-2-azetidinone induces no rearrangement (Biswas, A.; Miller, M. J. *Heterocycles* **1987**, *26*, 2849). Photolysis of 2-benzenesulfonyloxy-2-cyclohexenones results in loss of sulfur dioxide and rearrangement to 3-aryl-1,2-cyclohexanediones

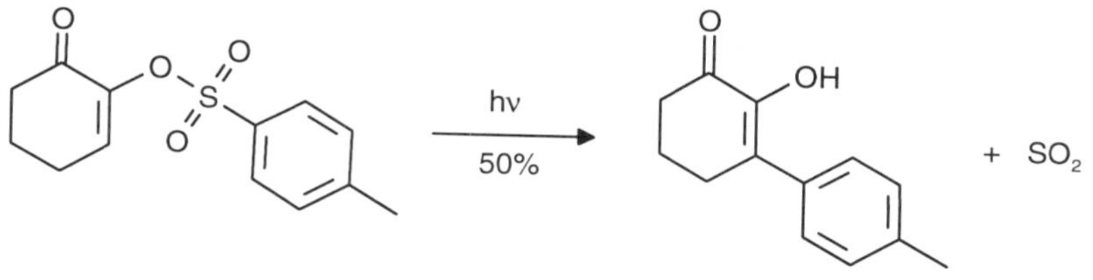

(Feigenbaum, A.; Pete, J.-P.; Scholler, D. *Tetrahedron Lett.* **1979**, 537). 3-Substituted 1,2-cyclohexanediones have also been prepared via conjugate addition of lithium diorganocuprate reagents to the enol tosylates of 1,2-cyclohexanedione (Charonnat, J. A.; Mitchell, A. L.; Keogh, B. P. *Tetrahedron Lett.* **1990**, *31*, 315).

A convenient method of synthesizing geminal difluoro alkenes and ketones has been reported. Sequential treatment of 2,2,2-trifluoroethyl p-toluenesulfonate with LDA and a trialkylborane affords a geminal difluoroalkenylborane. Protonolysis with acetic acid or oxidation with hydrogen peroxide gives 1,1-difluoro-alkenes or -ketones, respectively (Ichikawa, J.; Sonoda, T.; Kobayashi, H. *Tetrahedron Lett.* **1989**, *30*, 1641, 5437).

The regioselective electrochemical cleavage of aryl tosylates has been studied. An electron-withdrawing substituent (i.e., methoxycarbonyl) favors cleavage of tosyl groups that are ortho and para to the directing substituent, whereas an electron-donating substituent (methoxy) favors cleavage of tosylates in the meta position (Civitello, E. R.; Rapoport, H. *J. Org. Chem.* **1992**, *57*, 834).

Primary tosylates have been reduced to the corresponding hydrocarbons in good yields by treatment with $NaBH_4$ and polyethylene glycols in THF at 70°C. Secondary and tertiary tosylates are unreactive under these conditions (Santaniello, E.; Fiecchi, A.; Manzocchi, A.; Ferraboschi, P. *J. Org. Chem.* **1983**, *48*, 3074).

(f) Benzenesulfonamides

(i) Preparation

The reaction of a benzenesulfonyl halide with either ammonia or an amine remains the primary method of preparing benzenesulfonamides. Recently, a mild procedure for the sulfonylation of benzimidazole and benzotriazole has been reported. The amines are treated sequentially with bis(tri-n-butyltin)oxide and a benzenesulfonyl chloride to afford the sulfonamides in high yields (Soundararajan, R.; Balasubramanian, T. R. *Chem. Ind. (London)* **1985**, 92). Primary benzenesulfonamides can be synthesized in excellent yields from the corresponding benzenesulfinates by treatment with hydroxylamine-O-sulfonic acid (Graham, S. L.; Scholz. T. H. *Synthesis* **1986**, 1031). An improved method for the preparation of benzenesulfonamides from benzenesulfinates and chloramines has been reported (Scully, F. E., Jr.; Bowdring, K. *J. Org. Chem.* **1981**, 46, 5077). Sulfinates also react with chloroamides to form N-(phenylsulfonyl) benzenesulfonamides (Kremlev, M. M.; Kharchenko, A. V.; Zyabrev, V. S.; Rudenko, E. A. *J. Org. Chem. USSR (Engl. Transl.)* **1990**, 26, 1182; *Zh. Org. Khim.* **1990**, 26, 1368). Treatment of benzenesulfonimidoyl chlorides with silver nitrite gives the corresponding N-nitroso benzenesulfonamides (De Carvalho, E.; Norberto, F.; Rosa, E.; Iley, J.; Patel, P. *J. Chem. Res. (S)* **1985**, 132). A cyclic benzenesulfonamide has been prepared via the photochemical rearrangement of an α-azidosulfone (Still, I. W. J.; Leong, T. S. *Can. J. Chem.* **1980**, 58, 369).

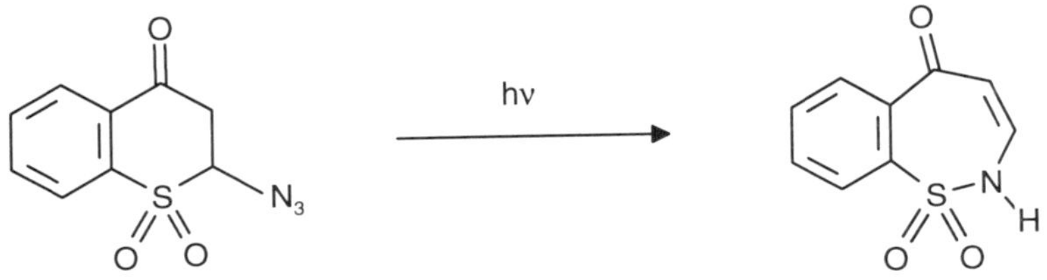

(ii) Reactions

ortho-Lithiated N,N-diphenyl benzenesulfonamides react with aromatic nitriles or N,N-dimethylcarbamonitrile to give directly the corresponding 3-substituted 1,2-benzisothiazole-1,1-dioxides (Hellwinkel, D.; Karle, R. *Synthesis* **1989**, 394).

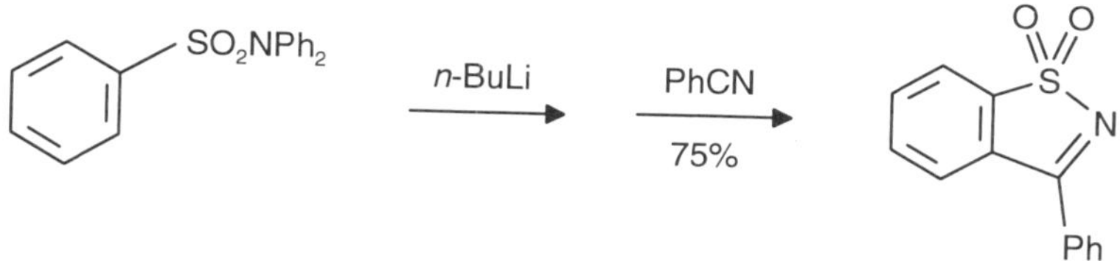

Reaction of N,N-dialkyl o-(carboxy)benzenesulfonamides with excess thionyl chloride results in the formation of N,N-dialkyl o-(chlorosulfonyl)-benzamides rather than the expected benzoyl chlorides. The reaction involves an intramolecular nucleophilic attack of the sulfonamide nitrogen atom on the intermediate benzoyl chloride. The rate of rearrangement depends on the pK_a of the parent amine; if less than 9, no rearrangement occurs (Hovius, K.; Wagenaar, A.; Engberts, J. B. F. N. *Tetrahedron Lett.* **1983**, *24*, 3137).

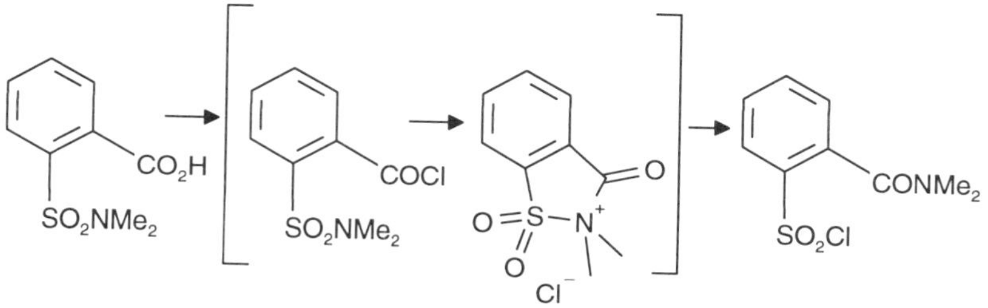

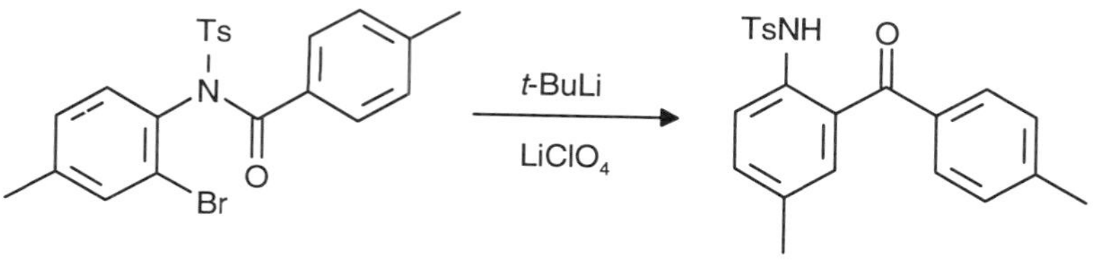

Although the base-catalyzed rearrangement of N-aryl sulfonamides to o-aminosulfones is well-known, the benzoyl group of N-(2-bromo- or iodo-phenyl)-N-benzoyl benzenesulfonamides preferentially rearranges to afford o-(arylsulfonamido)benzophenones under similar conditions (Hellwinkel, D.; Lämmerzahl, F.; Hofmann, G. *Chem. Ber.* **1983**, *116*, 3375; Horne, S.; Rodrigo, R. *J. Chem. Soc., Chem. Commun.* **1991**, 1046). Under radical conditions (tri-n-butylstannane, AIBN), N-(2-iodophenyl) benzenesulfonamides form biaryls *via* an *ipso* substitution reaction (Motherwell, W. B.; Pennell, A. M. K. *J. Chem. Soc., Chem. Commun.* **1991**, 877).

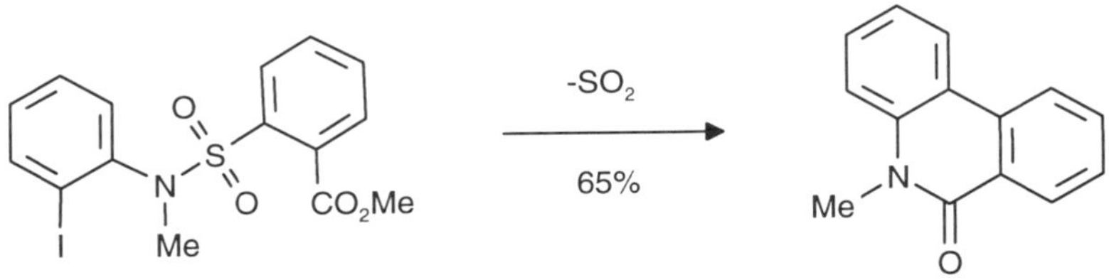

An interesting 1,3-migration of the sulfonyl group in a *p*-toluenesulfonamide has been reported (Fryer, R. I.; Blount, J.; Reeder, E.; Tyrbulski, E. J.; Walser, A. *J. Org. Chem.* **1978**, *43*, 4480).

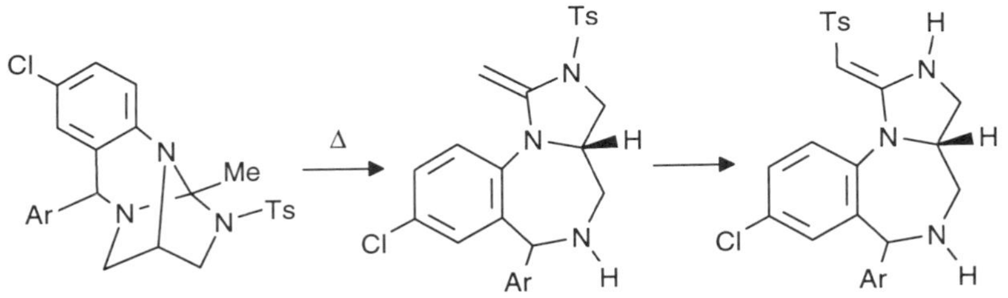

A novel amine protecting group, 4-(4,8-dimethoxynaphthylmethyl)benzenesulfonyl (DNMBS), has been reported. DNMBS groups are cleaved photochemically at wavelengths greater than 300 nm in the presence of ammonia borane (Hamada, T.; Nishida, A.; Yonemitsu, O. *Tetrahedron Lett.* **1989**, *30*, 4241).

Finally, β-acylated pyrroles have been prepared by Friedel-Crafts acylation of N-phenylsulfonylpyrrole in quantitative yields. The phenylsulfonyl group serves two purposes: it deactivates the α-position of the pyrrole ring and suppresses the formation of diacylated products. The sulfonamide can be cleaved with mild base (Rokach, J.; Hamel, P.; Kakushima, M.; Smith, G. M. *Tetrahedron Lett.* **1981**, *22*, 4901).

(g) Benzenesulfonyl Hydrazides and Hydrazones

Benzenesulfonyl hydrazides undergo a variety of reactions resulting from cleavage of the S-N bond. Tosyl hydrazide (p-toluenesulfonyl hydrazide) reacts with two equivalents of alkyl or activated aryl halides in the presence of sodium acetate to afford sulfones in high yields. An equivalent amount of the dehalogenated alkane or arene is also formed (Ballini, R.; Marcantoni, E.; Petrini, M. *Tetrahedron* **1989**, *45*, 6791). Pyrolysis of the lithium salt of trimethylsilyl tosyl hydrazide affords lithium p-toluenesulfinate and a gas, presumably (trimethylsilyl)diazene. A different reaction pathway is followed if the parent compound is pyrolyzed; trimethylsilyl p-toluenesulfinate and diazene are formed (Wuest, J. D. *J. Org. Chem.* **1980**, *45*,

$$\text{tolyl-SO}_2\text{TMS} + \text{HN=NH} \xleftarrow{\Delta} \text{TsN(TMS)(NH}_2) \xrightarrow[\Delta]{n\text{-BuLi}} \text{tolyl-SO}_2^- \text{Li}^+ + \text{TMS-N=NH}$$

3120). Treatment of *N*-alkyl-*N'*-tosyl hydrazides with mercuric acetate in the presence of alcohols or phenols gives high yields of the corresponding ethers (Gasparrini, F.; Cagliotti, L.; Misiti, D.; Palmieri, G.; Ballini, R. *Tetrahedron* **1982**, *38*, 3609). Oxidation with bromine affords alkyl bromides and vicinal alkyl dibromides. The main products of primary hydrazides are monobromides whereas secondary hydrazides normally produce dibromides (Palmieri, G. *Tetrahedron* **1983**, *39*, 4097). N-Substituted *N*-benzenesulfonyl hydrazides, obtained by amination of primary sulfonamides, can be readily oxygenated by atmospheric oxygen under basic conditions to cleanly afford alkyl hydroperoxides in good yields (Collazo, L.; Guziec, F. S., Jr.; Hu, W.-X.; Muñoz, A.; Wei, D.; Alvarado, M. *J. Org. Chem.* **1993**, *58*, 6169). Unsymmetrical azines can be prepared via treatment of *erythro*-1,2-diaryl-2-(2-tosylhydrazido)ethan-1-ols with formic acid (Rosini, G.; Soverini, M.; Ballini, R. *Synthesis* **1983**, 909). The starting sulfonylhydrazides are prepared by stereoselective reduction of the tosylhydrazones of benzoin derivatives with sodium cyanoborohydride (Rosini, G.; Medici, M. *Synthesis* **1979**, 789).

Finally, an interesting N-N bond cleavage of a substituted tosyl hydrazide has been reported. Treatment of 1-acyl-2-*p*-toluenesulfonylhydrazides with either sulfur mono- or di-chloride results in cleavage to the corresponding carboxamides and p-toluenesulfonamide (Bellesia, F.; Pagnoni, U. M.; Pinetti, A. *J. Chem. Res. (S)* **1982**, 222).

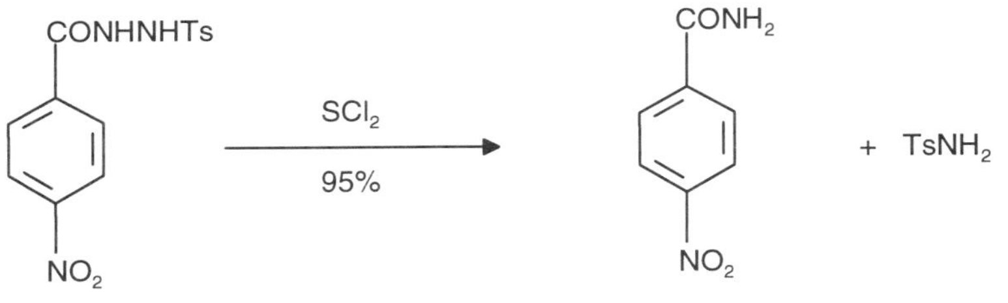

The usual method of preparing benzenesulfonyl hydrazones is the reaction of a sulfonyl hydrazide with an aldehyde or ketone. N-Alkyl sulfonyl hydrazones can be synthesized by treatment of tosyl hydrazones with alkyl halides under phase transfer conditions (Cacchi, S.; La Torre, F.; Misiti, D. *Synthesis* **1977**, 301). Treatment of phenylhydrazones with benzenesulfonyl chlorides affords N-phenyl sulfonyl hydrazones. If the resulting sulfonyl hydrazones have an α-hydrogen, treatment with sodium hydroxide causes 1,4-elimination of the benzenesulfinic acid, forming phenylazoalkenes. In the absence of an abstractable α-hydrogen, the sulfonyl hydrazone undergoes N-N bond cleavage, giving a nitrile and N-phenyl p-toluenesulfonamide (Schantl, J. G.; Hebieson, P.; Karpellus, P. *Synth. Commun.* **1989**, *19*, 39). The analogous N-methyl sulfonyl hydrazones combine with primary and secondary amines to give α-aminodialkyldiazenes and with triethylamine and alcoholic sodium hydroxide to form symmetrical azines (Makhova, N. N.; Mikhajluk, A. N.; Karpov, G. A.; Protopopova, N. V.; Khasapov, B. N.; Khmelnitski, L. I.; Novikov, S. S. *Tetrahedron* **1978**, *34*, 413).

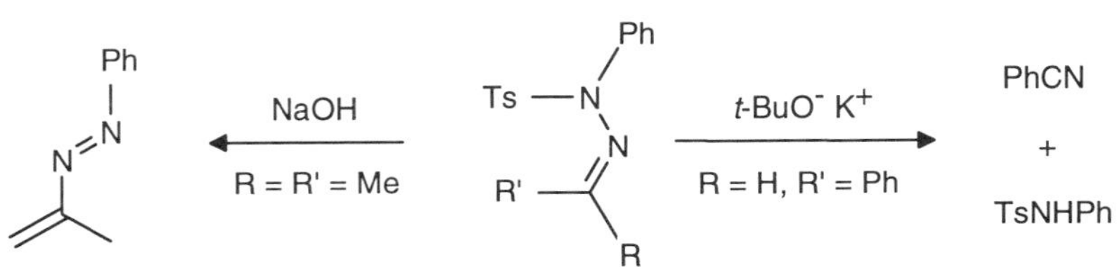

Benzenesulfonyl hydrazones undergo addition of a variety of reagents to the C=N bond of the hydrazone. A new high-yielding method of reducing sulfonyl hydrazones to sulfonyl hydrazides (addition of hydrogen) using pyridine-borane has been reported (Kikugawa, Y.; Kawase, M. *Synth. Commun.* **1979**, *9*, 49). Dimethyl phosphite adds to sulfonyl hydrazides to give substituted sulfonyl hydrazines. Reduction with sodium borohydrides converts the products into *sec*-alkyl phosphonates (Inokawa, S.; Nakatsukasa, Y.; Horisaki, M.; Yamashita, M.; Yoshida, H.; Ogata, T. *Synthesis* **1977**, 179). Benzenesulfonyl hydrazones treated with bromine and sodium acetate in acetic acid are converted into 1-acetyl-1-arylsulfonyl-2-aroylhydrazines in 77-97% yields (Buzykin, B. I.; Izmailova, F. K.; Vasil'eva, R. K.; Kitaev, Y. P. *J. Org. Chem. USSR (Engl. Transl.)* **1977**, *13*, 822; *Zh. Org. Khim.* **1977**, *13*, 898).

One of the most important properties of ketone benzenesulfonyl hydrazones is their reaction with organolithium reagents. Treatment of the hydrazone with two equivalents of alkyllithium reagent at -78°C generates a dianion. Regioselective abstraction of a proton from the less hindered side of unsymmetrical ketone hydrazones is observed (Shapiro, R. H.; Lipton, M. F.; Kolonko, K. J.; Buswell, R. L.; Capuano, L. A. *Tetrahedron Lett.* **1975**, 1811). The dianion sequentially eliminates lithium benzenesulfinate and nitrogen to form a vinyl anion. This vinyl anion can be trapped with a variety of electrophiles (Scheme 1) (Chamberlin, A. R.; Stemke, J. E.; Bond, F. T. *J. Org. Chem.* **1978**, *43*, 147). The initial dianion can also be trapped with ketones or aldehydes to yield β-hydroxytosylhydrazones; further reaction with butyllithium affords homoallylic alcohols in a regiospecific manner (Lipton, M. F.; Shapiro, R. H. *J. Org. Chem.* **1978**, *43*, 1409).

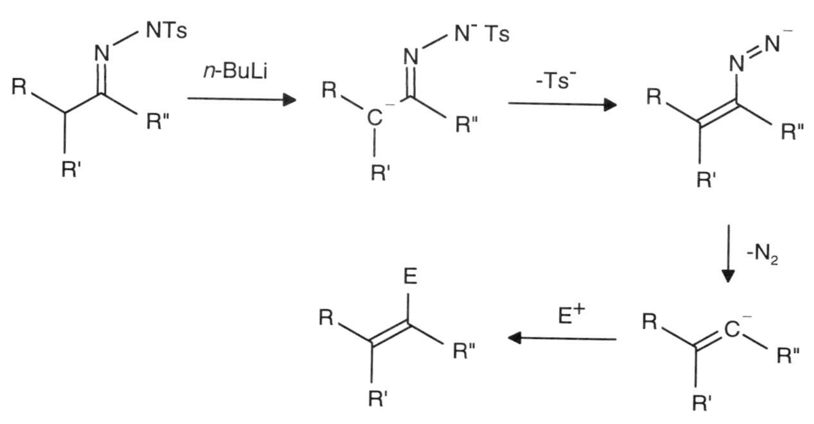

Scheme 1

Reaction of Vinyl Anion with Electrophiles

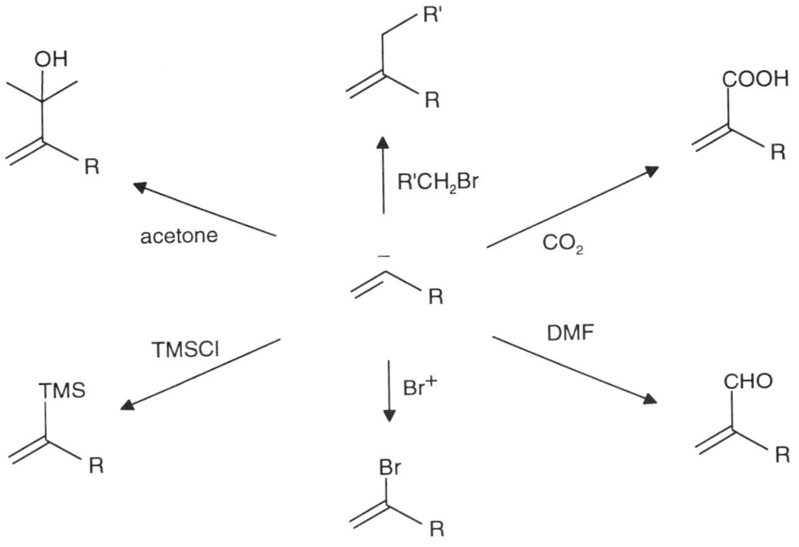

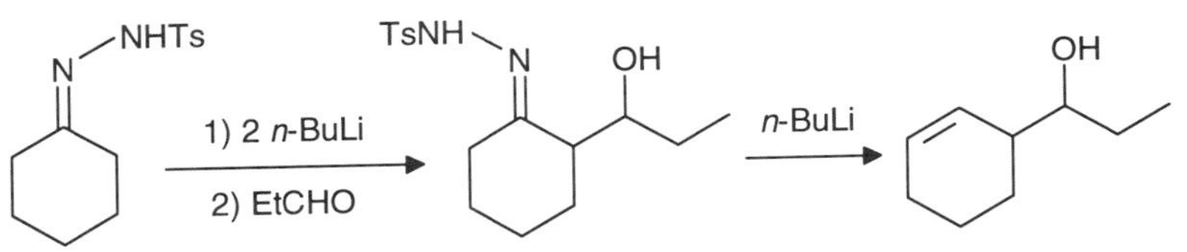

This reaction with alkyllithiums is the basis for a 1,2-carbonyl transformation. The dianion is trapped with dimethyl disulfide to give the corresponding sulfide. Treatment with butyllithium reforms a dianion which, on aqueous workup, regioselectively forms the vinyl sulfide. The desired ketone is obtained by treatment with mercuric chloride in hot acetonitrile. The conversion from tosylhydrazone to vinyl sulfide can be performed in a one-pot operation (Scheme 2) (Nakai, T.; Mimura, T. *Tetrahedron Lett.* **1979**, 531).

Tosylhydrazones of aldehydes and ketones lacking α-hydrogens follow a different pathway. Organolithium reagents add to the C=N bond of the

Scheme 2

1,2-Carbonyl Transformation

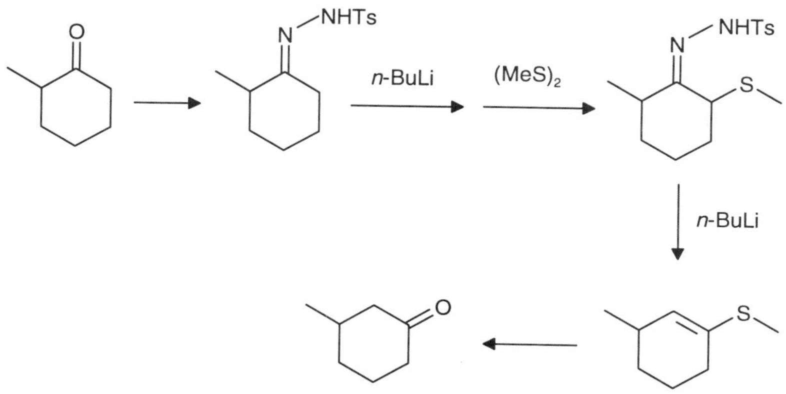

tosylhydrazone, giving a dianion that eliminates *p*-toluenesulfinate and nitrogen. Aqueous workup affords the product of reductive alkylation (Vedejs, E.; Stolle, W. T. *Tetrahedron Lett.* **1977**, 135).

Finally, an easy and mild method of regenerating carbonyl compounds from tosyl hydrazones has been reported. Treatment of tosylhydrazones with sodium nitrite in trifluoroacetic acid at 0°C affected conversion to the corresponding ketone or aldehyde in 88-97% yields (Caglioti, L.; Gasparrini, F.; Misiti, D.; Palmieri, G. *Synthesis* **1979**, 207).

(h) Benzenesulfonyl Azides

The most common method of preparation of benzenesulfonyl azides is the reaction of a sulfonyl halide with azide ion. A high-yielding method using a polymer-supported phase-transfer catalyst has been reported (Kumar, S. M. *Synth. Commun.* **1987**, *17*, 1015). Tosyl azide has been prepared from tosyl hydrazide in 83% yield using clay-supported ferric nitrate ("clayfen") (Laszlo, P.; Polla, E. *Tetrahedron Lett.* **1984**, *25*, 3701).

The mechanism of the reaction of benzenesulfonyl nitrenes, generated by the thermal decomposition of benzenesulfonyl azides, with aromatic substrates has been studied (Krause, J. G. *Chem. Ind. (London)* **1978**,

271). Electron-withdrawing groups (methoxycarbonyl, acetyl, nitro) on the aromatic ring of the substrate lead to the formation of N-sulfonyl-1H-azepines while substituted N-aryl sulfonamides are formed if electron-donating groups are present (Ayyangar, N. R.; Phatak, M. V.; Purchit, A. K.; Tilak, B. D. *Chem. Ind. (London)* **1979**, 853). An improved synthesis of N-sulfonyl-1H-azepines under nitrogen pressure and phase-transfer catalysis conditions has been reported (Ayyangar, N. R.; Kumar, S. M.; Srinivasan, K. V. *Synthesis* **1992**, 499).

A novel synthesis of iminophosphoranes involves the condensation of sulfonyl azides with phosphines and phosphites. This preparation can be combined with the synthesis of sulfonyl azides from sulfonyl hydrazides (*vide supra*) in a one-pot reaction (Laszlo, P.; Polla, E. *Tetrahedron Lett.* **1984**, *25*, 4651).

Benzenesulfonyl azides have long been used as diazo transfer reagents to doubly-activated methylene groups; however, the reaction fails with less acidic substrates such as simple ketones. Treatment of cyclic ketones with 2,4,6-triisopropylphenylsulfonyl azide in a biphasic system containing catalytic amounts of *both* tetra-n-butylammonium bromide and 18-crown-6 ether affords the corresponding α-diazoketones in 48-84% yields. Other benzenesulfonyl azides were found to be unsatisfactory. This method is particularly useful for hindered ketones and substrates containing acid- or base-labile functionalities (Lombardo, L.; Mander, L. N. *Synthesis* **1980**, 368). Certain benzenesulfonyl azides have been promoted as alternatives to the commonly-used tosyl azide. *p*-Dodecylbenzenesulfonyl azide exhibits superior stability while *p*-acetamidobenzenesulfonyl azide is a more practical and cost effective reagent (Hazen, G. G.; Weinstock, L. M.; Connell, R.; Bollinger, F. W. *Synth. Commun.* **1981**, 947; Baum, J. S.; Shook, D. A.; Davies, H. M. L.; Smith, H. D. *Synth. Commun.* **1987**, 1709).

(i) Benzenesulfonyl Isocyanates

The reaction of benzenesulfonamides with phosgene is the most common means of preparing benzenesulfonyl isocyanates. A new method for converting sulfonamides to sulfonyl isocyanates involves treatment with thionyl chloride and chlorocarbonyl sulfenyl chloride (James, D. R., U. S. Patent 4 485 053, 1984). A simplified process for preparing benzenesulfonyl isocyanates is the reaction of sulfonyl chlorides with trimethylsilyl isocyanate in the presence of a catalytic amount of an trialkylstannyl chloride (Mironov, V. F.; Kozyukov, V. P.; Tkachev, A. S.; Dobrovinskaya,

E. K., U.S.S.R. Patent 477 620, 1984; *Chem. Abstr.* **1984**, *101*, 72425u). The reaction of chlorosulfonyl isocyanate with aryltrialkylstannanes affords benzenesulfonyl isocyanates under mild conditions. The products are not isolated, but reacted further to give derivatives of sulfonamides (Arnswald, M.; Neumann, W. P. *Chem. Ber.* **1991**, *124*, 1997). Benzenesulfonyl isocyanates can be prepared by the palladium-catalyzed carbonylation of N-chloro sulfonamidates and (sulfonyliminoiodo)benzenes (Besenyei, G.; Németh, S.; Simándi, L. I. *Angew. Chem. Int. Ed. Engl.* **1990**, *29*, 1147; *Angew. Chem.* **1990**, *102*, 1168; Besenyei, G.; Simándi, L. I. *Tetrahedron Lett.* **1993**, *34*, 2839).

(j) Benzenesulfonyl Oxaziridines

Derived from benzenesulfonimines, 2-sulfonyloxaziridines are stable, aprotic, and neutral oxidizing reagents. These reagents are prepared by oxidation of the parent sulfonimine with *m*-CPBA under biphasic conditions in the presence of a phase-transfer catalyst (Davis, F. A.; Stringer, O. D. *J. Org. Chem.* **1982**, *47*, 1774). 2-Sulfonyloxaziridines oxidize sulfides and disulfides to sulfoxides and thiolsulfinates, respectively, and hydroxylate carbanions (Davis, F. A.; Jenkins, R. H., Jr.; Yocklovich, S. G. *Tetrahedron Lett.* **1978**, 5171; Davis, F. A.; Manccinelli, P. A.; Balasubramanian, K.; Nadir, U. K. *J. Am. Chem. Soc.* **1979**, *101*, 1044). Alkenes are epoxidized by treatment with 2-sulfonyloxaziridines (Davis, F. A.; Abdul-Malik, N. F.; Awad, S. B.; Harakal, M. E. *Tetrahedron Lett.* **1981**, *22*, 917). This epoxidation is the basis for a method of α-hydroxylating ketones and esters that allows for the isolation of the intermediate α-siloxy epoxide. Treatment of a silyl enol ether with a 2-sulfonyloxaziridine affords an epoxide that can be isolated or hydrolyzed under mild conditions to give the α-hydroxy ketone or ester (Davis, F. A.; Sheppard, A. C. *J. Org. Chem.* **1987**, *52*, 954).

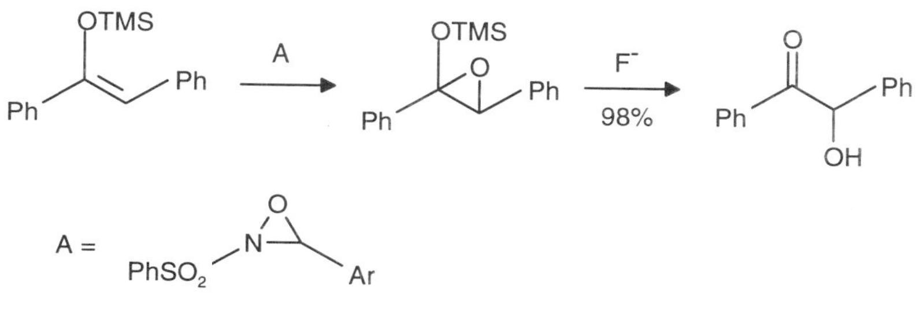

(k) Biological Activity of Benzenesulfonic Acids and Their Derivatives

Since their introduction in the 1930's, sulfonamides derived from sulfanilamide (4-aminobenzenesulfonamide) have been used extensively in the treatment and prevention of bacterial infections. Thousands of these 'sulfa drugs' were synthesized and many were introduced into human and veterinary medicine. These agents are active against both gram-positive and gram-negative bacteria. Although their importance as therapeutic agents has diminished with the advent of antibiotics, sulfa drugs remain clinically useful for the treatment of a variety of illnesses (including urinary tract infections) and in third world countries where the problems of storage and lack of medical personnel make the use of antibiotics difficult.

Sulfa drugs exert their action by inhibiting the utilization of 4-aminobenzoic acid, a necessary component of folic acid synthesis. These compounds impede the synthesis of folic acid and, therefore, are toxic to those bacteria that synthesize their own folic acid. The clinical importance of the sulfa drugs depends not only on their antibacterial properties, but on other factors such as solubility and metabolism. A wide range of pharmacokinetic properties are possible by varying the substitution pattern on the sulfonamide nitrogen. The best known of these active derivatives are sulfadiazine, sulfadimethoxine, sulfisomidine, sulfisoxazole, sulfamethizole, sulfamethoxazole, and sulfamethoxypyridazine (Table 1). For an excellent overview of sulfa drugs, see Foye, W. O. in *Encyclopedia of Chemical Technology* (*Kirk-Othmer*), 4th ed.; Kroschwitz, J. I., Ed.; Wiley: New York, 1992; Vol. 2, p. 876-93.

In addition to their antibacterial activity, sulfonamides also possess a variety of other medicinal properties. Tolbutamide and glyburide are typical

$$CH_3-\langle \rangle-SO_2NHCONH(CH_2)_3CH_3$$

Tolbutamide

$$\text{(2-OCH}_3\text{, 5-Cl-C}_6\text{H}_3\text{)}-CONHCH_2CH_2-\langle \rangle-SO_2NHCONH-\langle \rangle$$

Glyburide

Table 1

Common Sulfa Drugs

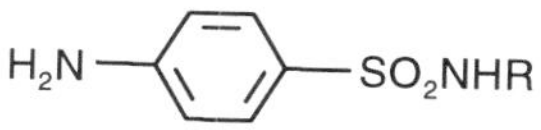

H₂N—⟨C₆H₄⟩—SO₂NHR

Name	R =
Sulfadiazine	2-pyrimidinyl
Sulfadimethoxine	2,6-dimethoxy-pyrimidin-4-yl
Sulfisomidine	2,6-dimethyl-pyrimidin-4-yl
Sulfisoxazole	3,4-dimethyl-isoxazol-5-yl
Sulfamethizole	5-methyl-1,3,4-thiadiazol-2-yl
Sulfamethoxazole	5-methyl-isoxazol-3-yl
Sulfamethoxypyridazine	6-methoxy-pyridazin-3-yl

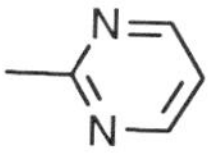

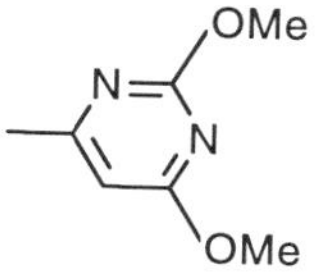

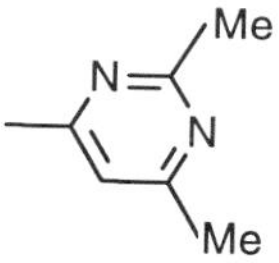

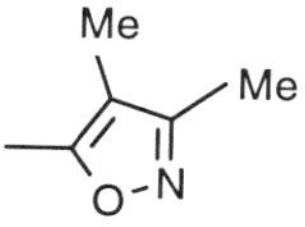

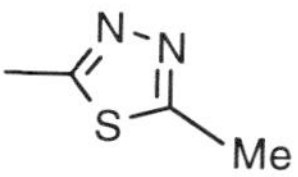

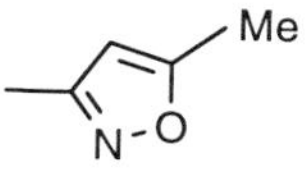

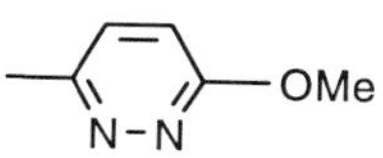

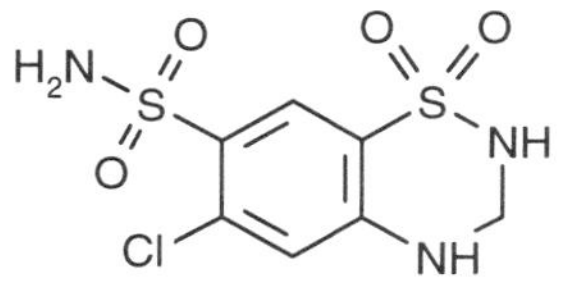

Hydrochlorothiazole

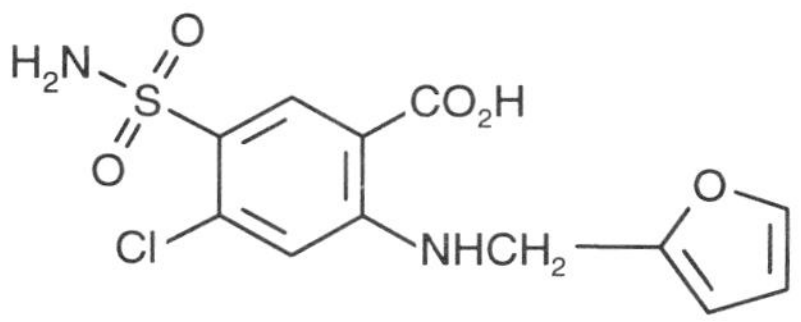

Furosemide

of the sulfonylureas, a subclass of sulfonamides, that are oral hypoglyceglycemic agents used for the treatment of non-insulin-dependent diabetes. Some sulfonamides inhibit carbonic anhydrase and have a profound diuretic effect. These derivatives include hydrochlorothiazide and furosemide and also inhibit the reabsorption of sodium chloride. Other clinically active sulfonamides include suclofenide (anticonvolsant), sulfadiazine (antimalarial) and sulpiride (antipsychotic).

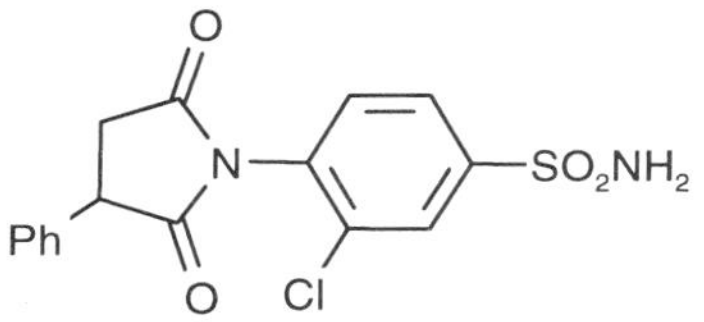

Suclofenide

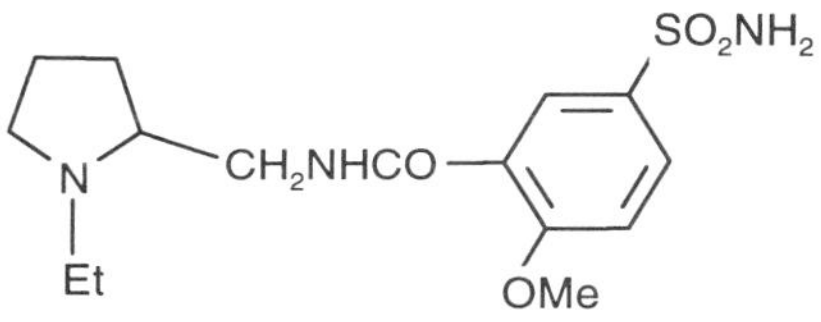

Sulpiride

N,N'-Bis(arylsulfonyl) hydrazines have demonstrated significant antineoplastic activity against a variety of cancers including the L1210 leukemia. The sulfonyl hydrazines break down *in vivo* to form a powerful alkylating agent (Shyam, K.; Cosby, L. A.; Sartorelli, A. C. *J. Med. Chem.* **1985**, *28*, 525).

Derivatives of sulfonamides have also proven to be potent herbicides. Sulfonylureas such as chlorsulfuron and sulfometuron methyl are highly active herbicides that inhibit acetolactate synthase, thereby disrupting branched amino acid biosynthesis (*Synthesis and Chemistry of Agrochemicals II*; Baker, D. R.; Fenyes, J. G.; Moberg, W. K., Eds.; ACS Symposium Series 443; American Chemical Society: Washington, DC, 1991; pp 16-119). This class of herbicides was pioneered by chemists at DuPont in the mid-1970's and controls weed species at application rates of only 2-75 grams/hectare. More recently, chemists in the laboratories of

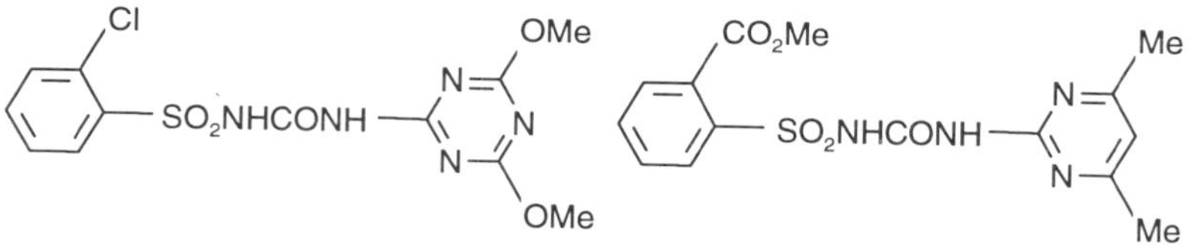

Chlorsulfuron

Sulfometuron Methyl

American Cyanamid have prepared sulfonyl carboxamides that inhibit acetohydroxy acid synthase and demonstrate herbicidal activity against broadleafed weeds (Alvarado, S. I.; Crews, A. D.; Wepplo, P. J.; Doehner, R. F.; Brady, T. E.; Gange, D. M.; Little, D. L. in *Synthesis and Chemistry of Agrochemicals III*, Baker, D. R.; Fenyes, J. G.; Steffens, J. J., Eds.; ACS Symposium Series 504; American Chemical Society: Washington, DC, 1992; pp 75-80).

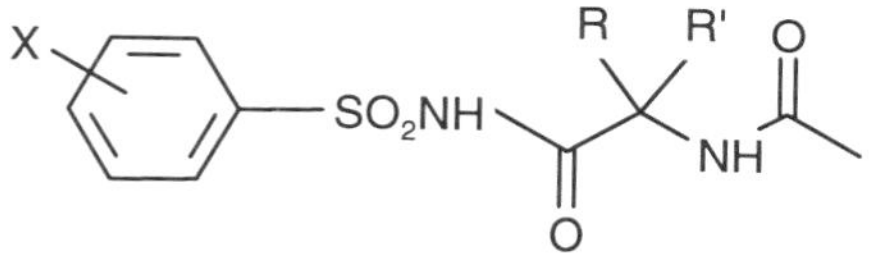

2. Benzenesulfinic Acids and Their Derivatives

Recent reviews on the chemistry of sulfinic acids and derivatives: Andersen, K. K. in *Comprehensive Organic Chemistry*; Barton, D. H. R.; Ollis, W. D., Eds.; Pergamon: Oxford, 1979; Vol. 3, Section 11.18; *The Chemistry of Sulphinic Acids, Esters and their Derivatives*; Patai, S., Ed., Wiley: Chichester, 1990.

(a) Benzenesulfinic Acids

(i) Preparation

Improvements in the existing methods of preparation of benzenesulfinic acids continue to be reported. Sulfones can be cleaved with strong base to form sulfinic acids. Treatment of thioxanthen-9-one 10,10-dioxides with dilute methanolic sodium methoxide at reflux affords benzenesulfinic acids via cleavage of the C-S bond of the more electrophilic aromatic ring. The resulting methoxy-substituted products are more stable than the corresponding phenols formed via caustic hydrolysis (Bennett, O. F.; Saluti, G.; Quinn, F. X. *Synth. Commun.* **1977**, 33). Similar cleavages have been reported with amines (Udre, V.; Lukevic, E.; Kemme, A.; Bleidelis, J. *Khim. Geterotsikl. Soedin.* **1980**, 320; *Chem. Abstr.* **1980**, *93*, 114414p). Partial oxidation of benzenethiols to benzenesulfinic acids has been achieved with 30% H_2O_2 at ambient temperature (Kamiyama, T.; Enomoto, S.; Inoue, M. *Chem. Pharm. Bull.* **1988**, *36*, 2652). The first examples of a Truce-Smiles rearrangement involving the migration of an *alkyl* group have been reported. Treatment of *o*-tolyl *tert*-butyl sulfone with *n*-butyllithium in THF affords *o*-neopentylbenzenesulfinic acid in 75-80% yield (Snyder, D. M.; Truce, W. E. *J. Am. Chem. Soc.* **1979**, *101*, 5432; Madaj, E. J.; Snyder, D. M.; Truce, W. E. *J. Am. Chem. Soc.* **1986**, *108*, 3466).

Benzenesulfonyl chlorides can be reduced to the corresponding sulfinic acids by treatment with sodium borohydride in THF at 0°C. Further reduction occurs at higher temperatures (Nose, A.; Kudo, T. *Chem. Pharm. Bull.* **1987**, *35*, 1770). The reaction of benzenesulfonyl halides with two equivalents of *p*-thiocresol in the presence of triethylamine at -76°C gives, after acidification, sulfinic acids in 64-92% yields and high purity. This method is often superior to conventional reduction with aqueous sodium sulfite (Lee, C.; Field, L. *Synthesis* **1990**, 391). Benzenesulfinic acids can be conveniently prepared by electrochemical cleavage of *N*-(phenylsulfonyl) phenylhydroxylamines, prepared *in situ* by the reaction of phenylhydroxylamines with benzenesulfonyl chlorides (Moinet, C.; Raoult, E. *Bull. Soc. Chim. Fr.* **1991**, 214).

(ii) Reactions

One of the principal reactions of benzenesulfinic acids is alkylation with an electrophilic carbon to form sulfones. An extensive review of this chemistry has been published (Schank, K. in *The Chemistry of Sulphones and*

Sulphoxides; Patai, S., Rappoport, Z., Stirling, C. J. M., Eds.; Wiley: Chichester, 1988; Chapter 7). Several improvements on the conversion of benzenesulfinic acids to sulfones using phase-transfer conditions have been reported. Treatment of sodium benzenesulfinates with alkyl halides in the presence of a catalytic amount of tetra-*n*-butylammonium bromide in either 1,2-dimethoxyethane or a 4:3:3 mixture of water:benzene:acetone affords sulfones in good to excellent yields (Wildeman, J.; van Leusen, A. M. *Synthesis* **1979**, 733; Crandall, J. K.; Pradat, C. *J. Org. Chem.* **1985**, *50*, 1327). Sulfones can be prepared *without solvent* by mixing the sodium benzenesulfinate and alkyl halide with a small amount of Aliquat 336 (Bram, G.; Loupy, A.; Roux-Schmitt, M. C.; Sansoulet, J.; Strzalko, T.; Seyden-Penne, J. *Synthesis* **1987**, 56). Benzenesulfinic acids or their sodium salts, pretreated with a macroreticular anion exchange resin containing quaternary ammonium groups, reacts with alkyl halides to form sulfones in high yields (Manescalchi, F.; Orena, M.; Savoia, D. *Synthesis* **1979**, 445). Finally, alkylation of benzenesulfinic acid with didodecyl prenylsulfonium perchlorate affords phenyl prenyl sulfone in 78% yield (Badet, B.; Julia, M.; Ramirez-Muñoz, M. *Synthesis* **1980**, 926).

Benzenesulfinic acids add to vinyl mercuric halides to form α,β-unsaturated sulfones ('vinyl sulfones') when irradiated at 350 nm (Hershberger, J.; Russell, G. A. *Synthesis* **1980**, 475). Vinyl sulfones can also be prepared by treatment of simple alkenes and dienes with benzenesulfinic acids and mercuric chloride ('sulfonomercuriation') followed by base-catalyzed eliminative demercuriation (Sas, W. *J. Chem. Soc., Chem. Commun.* **1984**, 862; Inomata, K.; Kobayashi, T.; Sasaoka, S.; Kinoshita, H.; Kotake, H. *Chem. Lett.* **1986**, 289; Andell, O. S.; Bäckvall, J.-E. *Tetrahedron Lett.* **1985**, *26*, 4555). Regioisomeric vinyl sulfones are formed by the reaction of alkenes with sodium sulfinates and iodine ('iodosulfonylation') and subsequent triethylamine-catalyzed dehydroiodination.

Allylic sulfones have been prepared *via* the palladium-catalyzed addition of benzenesulfinic acids to allyl acetates and allylic nitro compounds. In the case of allylic acetates, the structure of the product is dependent on the reaction time. The kinetic, more substituted product predominates if the reaction time is very short (<25 minutes) while the thermodynamically more stable isomer is the exclusive product if the reaction stirs overnight (Inomata, K.; Yamamoto, T.; Kotake, H. *Chem. Lett.* **1981**, 1357). In contrast, allylic nitro compounds react with sodium benzenesulfinate in DMF at 20-70°C in the presence of a catalytic amount of tetrakis(triphenylphosphine)palladium(0) to give the more substituted allylic sulfones (Tamura, R.; Hayashi, K.; Kakihana, M.; Tsuji, M.; Oda, D. *Tetrahedron Lett.* **1985**, *26*, 851; Ono, N.; Hamamoto, I.; Kawai, T.; Kaji, A.; Tamura, R.; Kakihana, M. *Bull. Chem. Soc. Jpn.* **1986**, *59, 405*).

R = CH=C(Me)$_2$, X = OAc   1 min   62 : 23

overnight   0 : 84

R = CO$_2$Me, X = NO$_2$ 7 days   82 : 18

Lewis acid-catalyzed additions of benzenesulfinic acids to protected aldehydes have been reported. Treatment of acetals with benzenesulfinic acids in the presence of boron trifluoride etherate gives α-sulfonyl ethers accompanied by minor amounts of sulfinate esters (Schank, K.; Schmitt, H.-G. *Chem. Ber.* **1977**, *110*, 3235). Synthetically useful 3-phenylsulfonyl phthalides can be prepared by the boron trifluoride-catalyzed reaction of benzenesulfinic acids with phthalaldehydic acids (Murty, K. V. S. N.; Pal, R.; Dutta, K.; Mal, D. *Synth. Commun.* **1990**, *20*, 1705).

Benzenesulfinic acids undergo Michael-type additions to alkynyl(phenyl)-iodonium salts. In methanol at 0°C, a quantitative yield of (Z)-(β-phenylsulfonylalkenyl)(phenyl)iodonium tetrafluoroborate is obtained. Treatment of this salt with triethylamine generates an alkylidenecarbene that undergoes a 1,5 C-H insertion reaction to afford 1-(phenylsulfonyl)cyclopentenes along with a small amount of rearranged alkynes (Ochiai, M.; Kunishima, M.; Tani, S.; Nagao, Y. *J. Am. Chem. Soc.* **1991**, *113*, 3135). With substrates incapable of undergoing insertion reactions, the alkynylsulfones are obtained in high yields exclusively (Tykwinski, R. R.; Williamson, B. L.; Fischer, D. R.; Stang, P. J.; Arif, A. M. *J. Org. Chem.* **1993**, *58,* 5235*).*

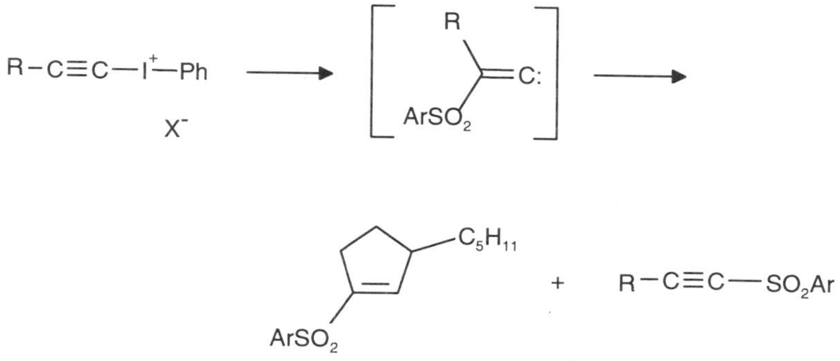

$$R-C\equiv C-\overset{+}{I}-Ph \quad \longrightarrow \quad \left[\begin{array}{c} R \\ ArSO_2 \end{array} \!\!\!\!\!\!\! \diagdown \!\!\!\! C\!\!=\!\!C\!: \right] \quad \longrightarrow$$
$$X^-$$

[cyclopentene with ArSO$_2$ and C$_5$H$_{11}$ substituents] + R−C≡C—SO$_2$Ar

R = n-C$_8$H$_{17}$ 80 : 20

R = t-C$_4$H$_9$ 0 : 100

The reaction of 2,3,3-trichloroacrylonitrile with sodium p-toluenesulfinate gives p-(tolylsulfonyl)acetonitrile as the only isolable product. In this extreme example of the nucleophilic addition of sulfinates to an activated alkene, the extensive substitution of all chlorines renders the double bond of the intermediate product susceptible to the further addition of water (Sepiol, J. J.; Sepiol, J. A.; Soulen, R. L. *J. Org. Chem.* **1984**, *49*, 1125).

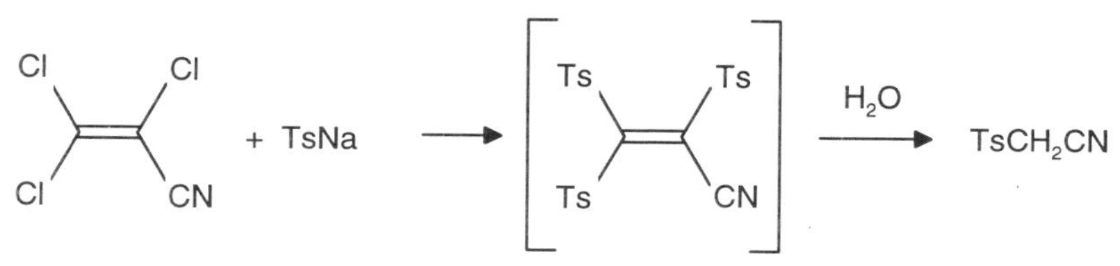

$$\underset{Cl}{\overset{Cl}{\diagdown}}\!C\!\!=\!\!C\!\underset{CN}{\overset{Cl}{\diagup}} \;+\; TsNa \;\longrightarrow\; \left[\underset{Ts}{\overset{Ts}{\diagdown}}\!C\!\!=\!\!C\!\underset{CN}{\overset{Ts}{\diagup}} \right] \;\overset{H_2O}{\longrightarrow}\; TsCH_2CN$$

The enzyme-mediated sulfonylation of resorcinol with benzenesulfinic acid and H$_2$O$_2$ has been reported in quantitative yield (Timofeeva, S. S.; Suslov, S. N.; Stom, D. I., USSR Patent 929 630, 1982; *Chem. Abstr.* **1982**, *97*, 144569t). An efficient method for preparing methylthiomethyl p-tolyl sulfone is the Pummerer rearrangement of dimethylsulfoxide with acetic anhydride followed by treatment of the resulting acetoxymethyl methyl sulfide with sodium p-toluenesulfinate in the presence of sodium acetate in acetic acid (Ogura, K.; Yahata, N.; Watanabe, J.; Takahashi, K.; Iida, H. *Bull. Chem. Soc. Jpn.* **1983**, *56*, 3543). Finally, attempted alkylation of isopropylidene 5-p-tosyloxymalonate with alkyl halides under phase transfer conditions affords, not the expected 5-alkylated products, but alkyl p-tolyl sulfones. This novel cleavage of a tosylate provides a synthetic equivalent to the tosyl anion (Xu, C.-L.; Chen, Z.-C. *Tetrahedron Lett.* **1991**, *32*, 2933).

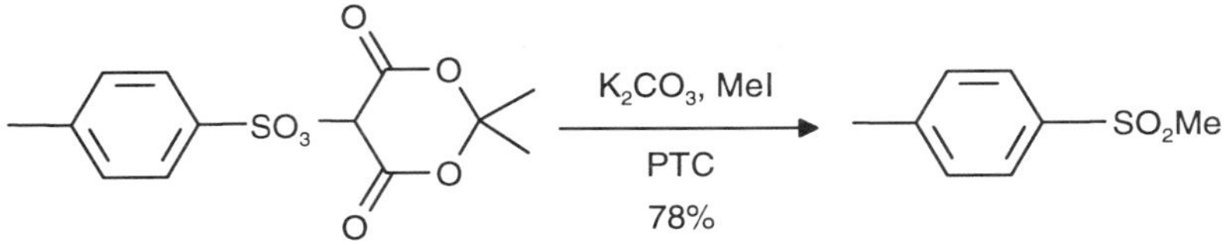

Aryl methyl sulfones can be prepared in excellent yields by the reaction of dimethyl methylphosphonate with sodium benzenesulfinates (Sutter, P.; Weis, C. D. *Phosphorus Sulfur* **1978**, *4*, 335). A convenient synthesis of (alkylsulfonyl)benzoic acids has been reported. Treatment of aqueous solutions of sodium sulfinylbenzoates, prepared by reduction of the corresponding bisacid chlorides with sodium sulfite, with α-halocarboxylic acids affords the sulfones in high yields. Alkylation of the carboxylate group does not occur (Brown, R. W. *J. Org. Chem.* **1991**, *56*, 4974). The reaction of p-toluenesulfinate anion with α-haloketones affords sulfones via competing ionic and radical mechanisms (Russell, G. A.; Ros, F. *J. Am. Chem. Soc.* **1985**, *107*, 2506).

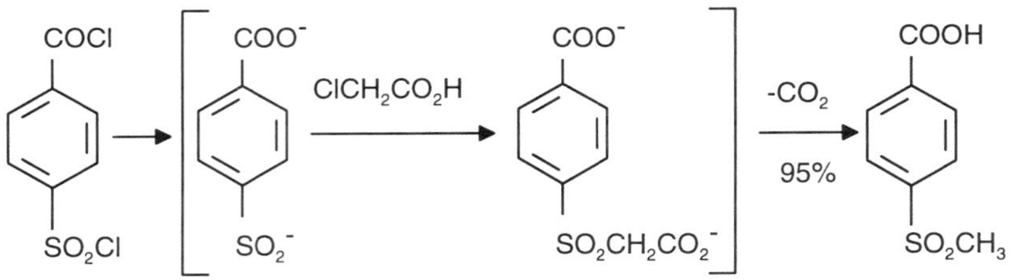

Benzenesulfinic acids react with a variety of reagents that activate the sulfinate moiety for further reaction. Treatment of benzenesulfinic acids with phenyl phosphorodichloridate and pyridine produces an intermediate that can react with alcohols, amines, and thiols to afford sulfinates, sulfinamides, and thiolsulfinates, respectively (Furukawa, M.; Ohkawara, T.; Noguchi, Y.; Isoda, M.; Hitoshi, T. *Synthesis* **1980**, 937). Diethyl azodicarboxylate-triphenylphosphine, 2-chloro-1-methylpyridinium iodide and γ-saccharine chloride are also effective coupling reagents (Furukawa,

M.; Ohkawara, T.; Noguchi, Y.; Nishikawa, M. *Synthesis* **1978**, 441; Furukawa, M.; Ohkawara, T.; Noguchi, Y.; Nishikawa, M.; Tomimatsu, M. *Chem. Pharm. Bull.* **1980**, *28*, 134). *p*-Toluenesulfinic acid reacts with phenyl phosphorodichloridate, diphenyl phosphorodichloridate, and 3-phthalimidooxy-1,2-benzisothiazole 1,1-dioxide to afford an activated intermediate that reacts further with Grignard reagents or enamines to give sulfoxides in low to moderate yields (Noguchi, Y.; Kurogi, K.; Sekioka, M.; Furukawa, M. *Bull. Chem. Soc. Jpn.* **1983**, *56*, 349). Chiral carbodiimides have been used to prepare optically active sulfinic acid derivatives of alcohols, thiols, and secondary amines (chiral at sulfur); however, the enantiomeric excesses of the products are ≤10% (Drabowicz, J.; Pacholczyk, M. *Phosphorus Sulfur* **1987**, *29*, 257). Alkyl sulfinates are the predominant product of the reaction of sulfinic acids with *O*-alkylisoureas. The ratio of sulfinate to sulfone is strongly dependent on the nature of the alkyl group on the isourea and the solvent. Secondary alkyl groups give exclusively alkyl sulfinates while primary alkyl groups lead to mixtures of sulfinates and sulfones (Kiełbasiński, P.; Żurawiński, R.; Drabowicz, J.; Mikołajczyk, M. *Tetrahedron* **1988**, *44*, 6687).

The facile reduction of benzenesulfinic acids to the disulfides upon treatment with thiols and TMSCl at ambient temperature has been reported (Oae, S.; Togo, H.; Numata, T.; Fujimori, K. *Chem. Lett.* **1980**, 1193). Sodium benzenesulfinates are readily converted to the corresponding thiocyanates *via* treatment with diethyl phosphorocyanidate in THF (Harusawa, S.; Shioiri, T. *Tetrahedron Lett.* **1982**, *23*, 447) or TMSCN in HMPA (Kagabu, S.; Maehara, M.; Sawahara, K.; Saito, K. *J. Chem. Soc., Chem. Commun.* **1988**, 1485).

Reaction of *p*-toluenesulfinic acid with methanol in the presence of NCS and dimethyl sulfide affords a mixture of methyl *p*-toluenesulfinate and methyl *p*-toluenesulfonate. Substitution of an amine for methanol gives the corresponding sulfinamides and sulfonamides. In all cases the sulfonic acid derivatives predominate (Furukawa, M.; Nishikawa, M.; Inaba, Y.; Noguchi, Y.; Okawara, T.; Hitoshi, T. *Chem. Pharm. Bull.* **1981**, *29*, 623). Oxidation of benzenesulfinic acids with dinitrogen tetroxide affords readily-separable mixtures of the corresponding sulfonyl nitrites and sulfonic acids. These sulfonyl nitrites are thermally unstable and decompose upon heating to give trisulfonylamine oxides (Oae, S.; Shinhama, K.; Kim, Y. H. *Tetrahedron lett.* **1979**, 3307). *N*-Chloroamines react with benzenesulfinic acids to give substituted sulfonamides (Nishikawa, M.; Inaba, Y.; Furukawa, M. *Chem. Pharm. Bull.* **1983**, *31*, 1374). The novel *S*-tosylsulfenyl chloride is formed by the reaction of p-toluenesulfinic acid with sulfur dichloride in dichloromethane (Kutateladze, A. G.; Beloglazkina, E. K.; Zyk, N. V.; Zefirov, N. S. *Izv. Akad. Nauk, Ser. Khim.* **1992**, 1217; *Chem. Abstr.* **1993**, *118*, 80575j). The addition of p-toluenesulfinic acid to the quinone of Rifamycin S gives an unusual O-sulfonyl derivative, not the expected Michael-type product (Taguchi, M.; Aikawa, N.; Yoshida, K.; Kitamura, M.; Tsukamoto, G. *Bull. Chem. Soc. Jpn.* **1988**, *61*, 2425). Finally, benzene-

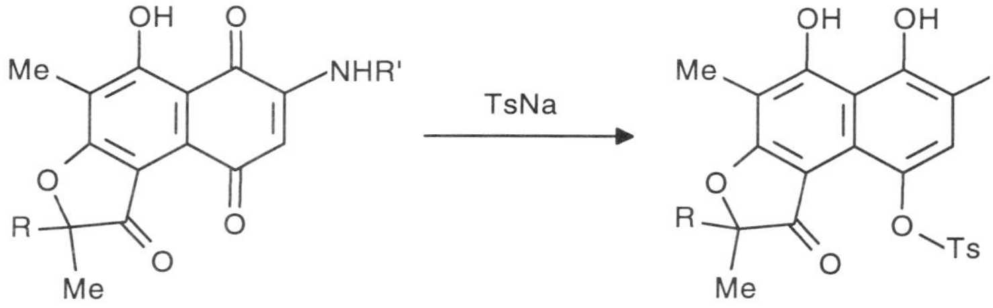

Rifamycin S

sulfinates form a self-assembled monolayer (SAMs) on gold. These SAMs are less stable than the corresponding monolayers prepared from aromatic thiols (Chadwick, J. E.; Myles, D. C.; Garrell, R. L. *J. Am. Chem. Soc.* **1993**, *115*, 10364).

(b) Benzenesulfinates

(i) Preparation

The most important method of preparing esters of benzenesulfinic acids involves reacting a sulfinyl chloride with an alcohol. Recent improvements on this transformation have been reported. Reaction of benzenesulfinyl chloride with neat alkoxytrimethylsilane at ambient temperature overnight affords the corresponding sulfinate in high yields (Harpp, D. N.; Friedlander, B. T.; Larsen, C.; Steliou, K.; Stockton, A. *J. Org. Chem.* **1978**, *43*, 3481). A variation of this method using alkyl chlorosulfites, hexamethyldisiloxane, and a catalytic amount of DMSO has been reported (Drabowicz, J. *Chem. Lett.* **1981**, 1753). Reduction of benzenesulfonyl chlorides with trimethyl phosphite in the presence of an alcohol and triethylamine gives the corresponding sulfinates in 48-92% yields *via* trapping of the initially generated sulfinyl chloride (Klunder, J. M.; Sharpless, K. B. *J. Org. Chem.* **1987**, *52*, 2598).

p-Toluenesulfinyl p-tolyl sulfone, readily prepared by treatment of sodium p-toluenesulfinate with sulfuric acid, reacts with hindered alcohols at ambient temperature to afford the p-toluenesulfinates in good yields (Boar, R. B.; Patel, A. C. *Synthesis* **1982**, 584). Stable, crystalline esters of p-toluenesulfinic acid can be prepared for a variety of alcohols using 1,1'-carbonyldiimidazole as a coupling agent (Lee, C.; Field, L. *Phosphorus Sulfur* **1989**, *45*, 35). Electrolysis of benzene-thiols and -disulfides in acetic acid containing sodium acetate and an alcohol gives high yields of the corresponding sulfinates (Nokami, J.; Fujita, Y.; Okawara, R. *Tetrahedron Lett.* **1979**, 3659).

Chiral sulfoxides are important substrates for asymmetric synthesis. The most practical technique for the preparation of enantiomerically pure sulfoxides is the reaction of Grignard reagents with resolved sulfinate esters of menthol. An improved large scale preparation of the (S)-(-)-menthyl p-toluenesulfinate has been reported in which the initial mixture of diastereomers is equilibrated with HCl to afford an 80% yield of the desired enantiomer (Solladié, G.; Hut, J.; Girardin, A. *Synthesis* **1987**, 173). p-Toluenesulfinate esters of *trans*-2-phenylcyclohexanol, however, can be prepared with greater kinetic selectivity (10:1 vs. 1.5:1), and both enantiomers of *trans*-2-phenylcyclohexanol are readily available (Whitesell, J. K.; Wong, M.-S. *J. Org. Chem.* **1991**, *56*, 4552). Treatment of the readily available (S)-(+)-N,N-diethyl p-toluenesulfinamide with an excess of an alcohol in the presence of 1.5 equivalents of boron trifluoride etherate at 0°C results in the formation of the (S)-(-)-sulfinates in excellent yields and stereospecificity (Hiroi, K.; Kitayama, R.; Sato, S. *Synthesis* **1983**, 1040). Optically active p-toluenethiolsulfinates can be prepared by the analogous reaction of optically active sulfinamides with thiols in the presence of trifluoroacetic acid (Drabowicz, J.; Mikołajczyk, M. *Tetrahedron Lett.* **1985**, *26*, 5703). Resolution of *tert*-butyl p-toluenethiolsulfinate via a β-cyclodextrin inclusion complex failed to give any significant enantiomeric enhancement (Mikołajczyk, M.; Drabowicz, J. *J. Am. Chem. Soc.* **1978**, *100*, 2510).

(ii) Reactions

The preparation of chiral sulfoxides *via* the reaction of Grignard reagents with chiral sulfinates is normally performed in ether solvent. Substitution of benzene as solvent leads to increased yield and stereospecificity and reduced formation of sulfide byproducts (Drabowicz, J.; Bujnicki, B.; Mikołajczyk, M. *J. Org. Chem.*, **1982**, *47*, 3325). Carbohydrates can be used in place of menthol as the chiral auxiliary in the synthesis of chiral sulfoxides (Ridley, D. D.; Smal, M. A. *J. Chem. Soc., Chem. Commun.* **1981**, 505).

β-Oxo sulfoxides have proven to be versatile synthetic intermediates in organic chemistry. A convenient method of preparation involves treatment of a silyl enol ether with benzenesulfinyl chloride in the presence of a stoichiometric amount of stannic chloride. Up to 94% yields of cyclic and acyclic products are realized (Meanwell, N. A.; Johnson, C. R. *Synthesis* **1982**, 283). A variation of this method has been used to prepare α-sulfinyl cycloalkanones with high stereospecificity. The silyl enol ether of cyclohexanone reacts with methyl (S)-p-toluenesulfinate in the presence of two equivalents of boron trifluoride etherate to afford (R)-2-(p-toluenesulfinyl)-cyclohexanone in 95% yield and 98% stereospecificity (Hiroi, K.; Matsuyama, N. *Chem. Lett.* **1986**, 65). This claim, however, has been disputed, and other authors report that the desired product can be prepared in >95% optical purity *via* the traditional reaction of (S)-(-)-menthyl p-toluenesulfinate and di(*iso*-propyl)magnesium bromide with cyclohexanone (Carreño, M. C.; García Ruano, J. L.; Rubio, A. *Tetrahedron Lett.* **1987**, *28*, 4861).

Chiral allylic sulfinates thermally rearrange to the corresponding γ-rearranged sulfones in good yields and stereospecificity in DMF at 90-100°C. Reactions in other solvents are less effective or fail completely (Hiroi, K.; Kitayama, R.; Sato, S. *J. Chem. Soc., Chem. Commun.* **1983**, 1470). This transformation may also be achieved under milder conditions using a palladium catalyst. Bulky substituents at the α-position of the allyl group, however, reduce the stereospecificity of the reaction significantly and give substantial amounts of the α-rearranged sulfones (Table 2) (Hiroi, K.; Kitayama, R.; Sato, S. *J. Chem. Soc., Chem. Commun.* **1984**, 303). Treatment of racemic allylic p-toluenesulfinates with tetrakis(triphenylphosphine)palladium in the presence of chiral phosphine ligands affords the corresponding optically active rearranged allylic sulfones in high optical yields. Again, bulky α-substituents lead to predominant formation of the α-rearranged sulfone (Hiroi, K.; Makino, K. *Chem. Lett.* **1986**, 617).

Stereochemical and kinetic studies of the acid-catalyzed alcoholysis of p-toluenesulfinates have been reported (Mikołajczyk, M.; Drabowicz, J.; Ślebocka-Tilk, H. *J. Am. Chem. Soc.* **1979**, *101*, 1302; Mikołajczyk, M. *Phosphorus Sulfur* **1986**, *27*, 31). Incubation of racemic benzenesulfinates with the microorganism *Rhodococcus equi* gives the corresponding sulfonates, leaving behind optically active sulfinates *via* an enantioselective

Table 2

Rearrangement of Allylic Sulfinates

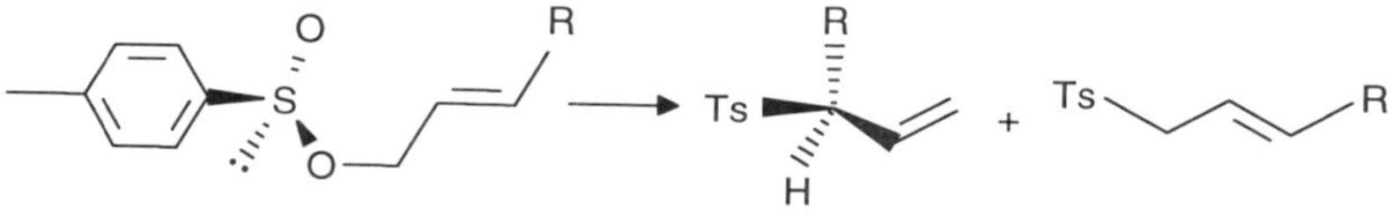

Substrate	Conditions	Yield	Ratio
R = Me	90°C, DMF	86%	100 : 0
	0°C, Pd(Ph$_3$P)$_4$	86%	80 : 20
R = C$_5$H$_{11}$	110°C, DMF	70%	100 : 0
	25°C, Pd(Ph$_3$P)$_4$	64%	30 : 70

oxidation. Substrates having longer alkyl chains as the alcohol component are more smoothly oxidized, and electron-donating substituents on the benzene ring accelerate the reaction (Kawasaki, T.; Watanabe, N.; Sugai, T.; Ohta, H. *Chem. Lett.* **1992**, 1611). Benzenesulfinates of benzoins decompose thermally or under basic catalysis to give α-diketones and benzenesulfenic acids (Müller, W.; Schank, K. *Chem. Ber.* **1978**, *111*, 2870).

Benzenethiolsulfinates undergo oxidation to give benzenesulfonic acids upon treatment with dinitrogen tetroxide (Oae, S.; Fukushima,, D.; Kim, Y. H. *Chem. Lett.* **1978**, 279) and benzenethiolsulfonates when treated with aqueous sodium periodate (Takata, T.; Kim, Y. H.; Oae, S. *Bull. Chem. Soc. Jpn.* **1981**, *54*, 1443). Treatment of benzenethiolsulfinates with alcohols in the presence of a catalytic amount of a halogen affords the corresponding benzenesulfinates *via* transesterification. Considerable amounts of symmetrical and unsymmetrical disulfides are also formed (Takata, T.; Oae, S. *Bull. Chem. Soc. Jpn.* **1982**, *55*, 3937). Methyl phenylsulfinyl thiocarbonate, readily prepared by oxidation of the corresponding sulfenyl thiocarbonate with *m*-CPBA, is a useful transfer reagent, giving phenylsulfinates upon treatment with alcohols (Harpp, D. N.; Granata, A. *Synthesis* **1978**, 782).

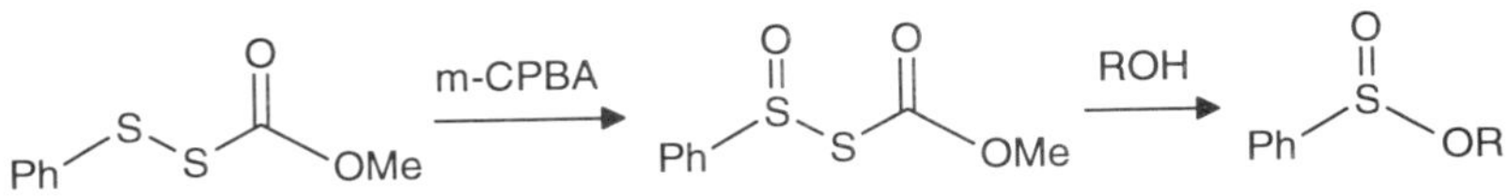

Ph-S-S-C(=O)-OMe →[m-CPBA] Ph-S(=O)-S-C(=O)-OMe →[ROH] Ph-S(=O)-OR

(c) Benzenesulfinyl Chlorides

Oxidation of benzene-thiols or -disulfides with sulfuryl chloride in acetic acid at low temperatures yields the corresponding sulfinyl chlorides. This method avoids the use of chlorine which may cause overoxidation (Youn, J.-H.; Herrmann, R. *Synthesis* **1987**, 72; Youn, J.-H.; Herrmann, R. *Tetrahedron Lett.* **1986**, *27*, 1493). Benzenesulfinyl chlorides are prepared in high yields by direct chlorosulfination of highly electron-rich benzenes with thionyl chloride (Bell, K. H. *Aust. J. Chem.* **1985**, *38*, 1209). The first synthesis of an aromatic sulfinic anhydride has been reported. Addition of benzene-*o*-disulfonhydroxyimide to ammonia followed by acidification with sulfuric acid affords benzene-*o*-disulfinic anhydride in 32% yield (Kice, J. L.; Liao, S. *J. Org. Chem.* **1981**, *46*, 2691).

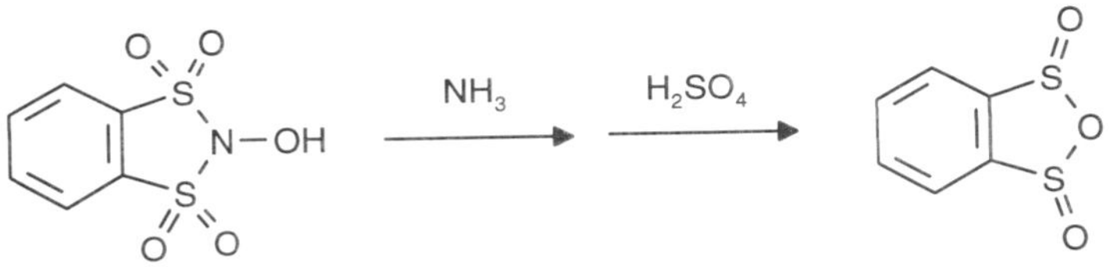

benzene-*o*-disulfonhydroxyimide →[NH₃] →[H₂SO₄] benzene-*o*-disulfinic anhydride

N,N-Dialkylhydroxyamines react with benzenesulfinyl chlorides at -70°C to give *O*-sulfinylated intermediates. At ambient temperatures, these intermediates undergo homolytic N-O bond cleavage, followed by rearrangement within the radical pair to afford benzenesulfonamides in 21-63% yields. Significant amounts of the corresponding imines and sulfinic acids are produced via escape of the radicals from the radical pair (Banks, M. R.; Hudson, R. F. *J. Chem. Soc., Perkin Trans. II* **1986**, 151). The *O*-sulfin-

$(PhCH_2)_2NOH$ →[PhSOCl, -70°C] $(PhCH_2)_2NOS(O)Ph$ →[25°C]

$(PhCH_2)_2NSO_2Ph$ + $PhCH_2N=CHPh$ + $PhSO_2H$

ylated intermediates formed in the reaction of N-methylhydroxamic acids with benzenesulfinyl chlorides can also be isolated below 0°C. These sulfinates decompose at ambient temperatures with simultaneous N-O and S-O bond fission to give N-acyl-N-methylsulfonamides and N-methyl-O-sulfonylhydroxamic acids by both in-cage and free pair radical recombination (Banks, M. R.; Hudson, R. F. *J. Chem. Soc., Perkin Trans. II* **1986**, 1211).

Ethylaluminum dichloride catalyzes the ene reaction of alkenes with benzenesulfinyl chlorides to give allylic sulfoxides (Snider, B. B. *J. Org. Chem.*

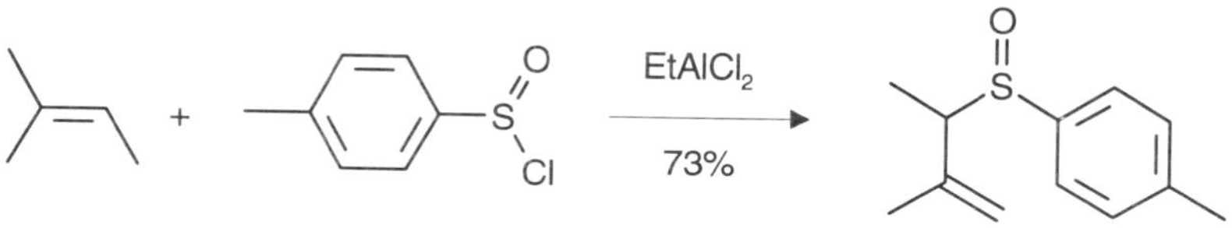

1981, *46*, 3155). Similarly, benzenesulfinyl chloride reacts with isoprene in the presence of silver tetrafluoroborate to afford an allylic sulfoxide. Benzenesulfinyl chloride and isoprene react under high pressure to give a chlorinated allylic sulfoxide (Moiseenkov, A. M.; Veselovsky, V. V.; Marakova, Z. G.; Zhulin, V. M.; Smit, W. A. *Tetrahedron Lett.* **1984**, *25*, 5929). p-Toluenesulfinyl chloride reacts with active methylene compounds in the presence of an amine base to yield α-(p-toluenethio)sulfones instead of the expected sulfoxides (Schank, K.; Buegler, S. *Sulfur Lett.* **1982**, *1*, 63).

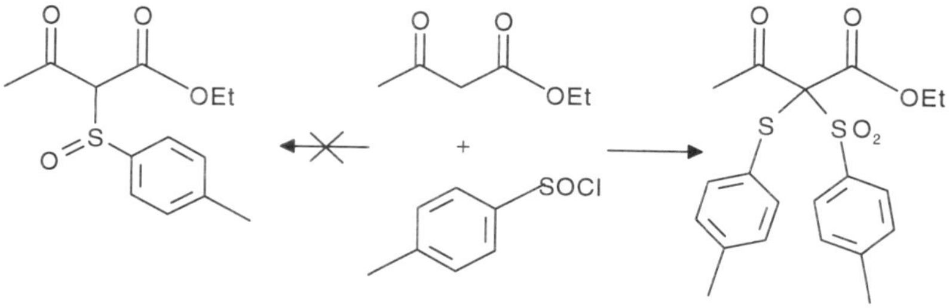

(d) Benzenesulfinamides

Treatment of 4-nitrophenyl phenyl sulfoxides and sulfoximides with elemental sulfur in liquid ammonia in an autoclave at 20-80°C gives substituted benzenesulfenamides with the sulfoximides providing higher yields. Bis(4-nitrophenyl)disulfide is formed as a byproduct (Sato, R.;

Chiba, S.; Takikawa, Y.; Takizawa, S.; Saito, M. *Chem. Lett.* **1983**, 535; Sato, R.; Saito, N.; Takikawa, Y.; Takizawa, S.; Saito, M. *Synthesis* **1983**, 1045). Diphenylcarbodiimides react with *N*-methyl-*N*-sulfinylmethanaminium tetrafluoroborate to give benzothiadiazine 1-oxides (Kresze, G.; Schwöbel, A.; Hatjiissaak, A.; Ackermann, K.; Minami, T. *Liebigs Ann.*

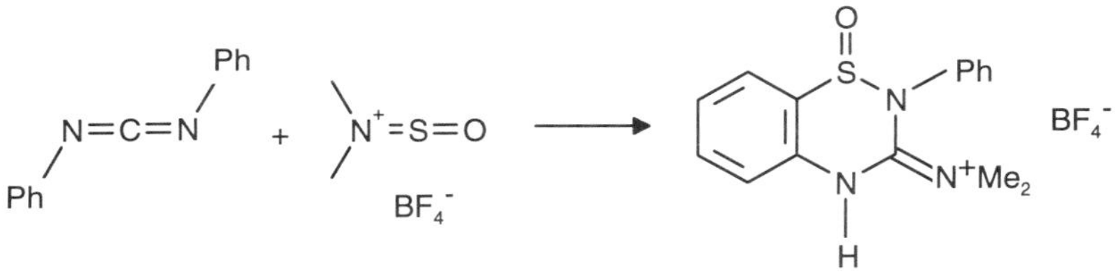

Chem. **1984**, 904). 2-Phenyl-4*H*-1,3-benzothiazin-4-ones undergo ring contraction upon treatment with aqueous sodium periodate, affording 1,2-benzisothiazol-3(2*H*)-one 1-oxides. The proposed mechanism involves opening of the thiazine ring to give an intermediate sulfenic acid (Szabó, J.; Szűcs, E.; Fodor, L.; Katócs, Á.; Bernáth, G. *Tetrahedron* **1988**, 44, 2985). Both *N*- and α-halogeno sulfoximides rearrange in the presence of base *via* a novel thiazirene intermediate to afford the corresponding *N*-sulfinylimines (Yoshida, T.; Naruto, S.; Uno, H.; Nishimura, H. *J. Chem. Soc., Chem. Commun.* **1982**, 106).

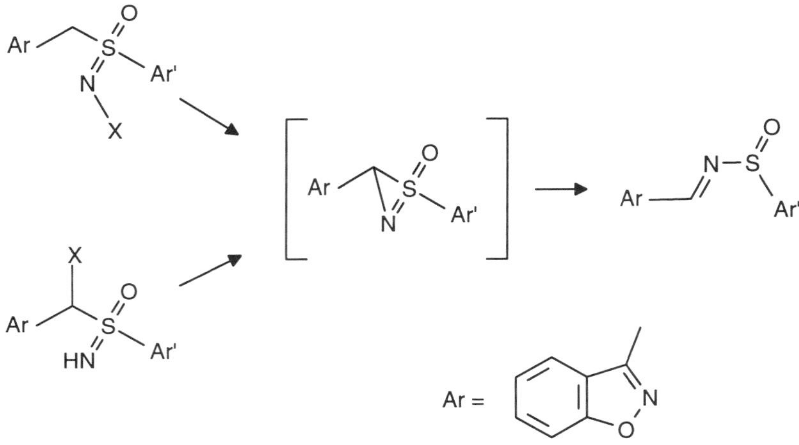

N-Phenyl benzenesulfinamides undergo *ortho*-lithiation with *n*-butyllithium in THF at low temperatures. Trapping of the lithiated intermediate with a variety of electrophiles provides a convenient route to a number of *ortho*-substituted benzenesulfinamides. Basic hydrolysis and desulfination with $HgCl_2$ or Raney nickel provides a convenient route to *meta*-substituted benzenes that might otherwise be difficult to prepare (Katritzky, A. R.; Lue, P. *J. Org. Chem.* **1990**, *55*, 74). Attempted *ortho*-lithiation of *N*-methyl-*N*-phenylbenzenesulfinamides with alkyllithium reagents gives only S-N bond cleavage resulting in the formation of *N*-methylaniline and the corresponding alkyl aryl sulfoxides. None of the expected rearrangement to an *ortho*-aminophenyl sulfoxide occurs. Rearrangement to mixtures of *ortho*- and *para*-sulfinyl substituted anilines does take place in the presence of HCl (in $CHCl_3$). The aniline ring must contain an activating group that directs ortho-para in electrophilic aromatic substitution reactions (*i.e.*, Me_2N, MeO, Me, etc.) and is located meta to the sulfinamido group for the rearrangement to occur (Andersen, K. K.; Malver, O. *J. Org. Chem.* **1983**, *48*, 4803).

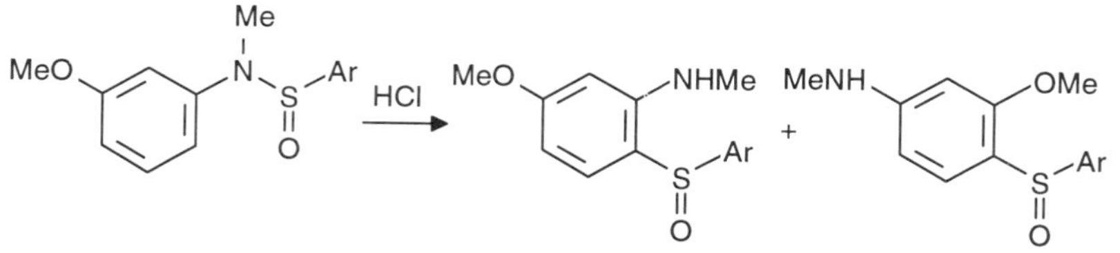

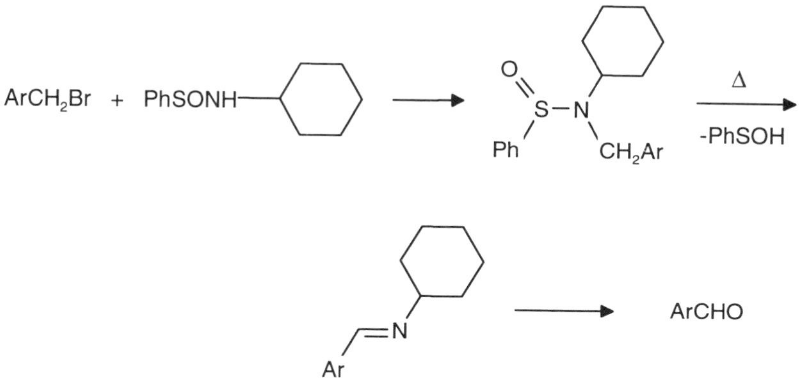

Thermolysis of *N*-allylic and benzylic benzenesulfinamides affords good yields of the corresponding imines. As allylic and benzylic halides readily alkylate secondary benzenesulfinamides, this reaction constitutes an effective means of converting those halides to the corresponding aldehydes (Trost, B. M.; Liu, G. *J. Org. Chem.* **1981**, *46*, 4617).

In the case of secondary and tertiary alcohols, the strong acid-catalyzed alcoholysis of chiral sulfinamides to sulfinates results in partial racemization. The use of boron trifluoride etherate as the acid catalyst affords optically active sulfinates in 90-100% stereoselectivity. Furthermore, the milder reaction conditions allow for the use of alcohols bearing acid-labile groups (Hiroi, K.; Kitayama, R.; Sato, S. *Synthesis* **1983**, 1040). The steric course of the acid-catalyzed alcoholysis of optically active *N,N*-diisopropyl *p*-toluenesulfinamide varies from predominant inversion to retention of configuration as the steric bulk of the alcohol increases. Methanol gives 69% inversion of configuration while cyclohexanol affords a sulfinate with 74% retention. Addition of silver perchlorate dramatically increases the degree of inversion (Mikołajczyk, M.; Drabowicz, J.; Bujnicki, B. *Tetrahedron Lett.* **1985**, *46*, 5699).

The preparation of benzenesulfonimidoyl chlorides by oxidation of sulfinamides with *tert*-butyl hypochlorite gives improved yields and is safer than the previous method involving chlorine (Johnson, C. R.; Wambsgans, A. *J. Org. Chem.* **1979**, *44*, 2278). Reaction of these sulfonimidoyl chlorides with fluoride ion constitutes the first synthesis of benzenesulfonimidoyl fluorides. Unlike the corresponding chlorides, benzenesulfonimidoyl

$$\underset{R}{\overset{O}{\underset{\|}{S}}}-NHR' \xrightarrow{t\text{-BuOCl}} R-\underset{\underset{NR'}{\|}}{\overset{O}{\underset{\|}{S}}}-Cl \xrightarrow{F^-} R-\underset{\underset{NR'}{\|}}{\overset{O}{\underset{\|}{S}}}-F \xrightarrow{MeLi} R-\underset{\underset{NR'}{\|}}{\overset{O}{\underset{\|}{S}}}-Me$$

fluorides react with primary alkyllithiums to give sulfoximines (Johnson, C. R.; Bis, K. G.; Cantillo, J. H.; Meanwell, N. A.; Reinhard, M. F. D.; Zeller, J. R.; Vonk, G. P. *J. Org. Chem.* **1983**, *48*, 1).

Chiral *N*-benzylidene *p*-toluenesulfinamides can be prepared by addition of alkyllithiums to benzonitriles followed by treatment with (*S*)-(-)-menthyl *p*-toluenesulfinate. The yields are higher than in the analogous reaction with Grignard reagents. Stereoselective addition of metal hydrides or allyl magnesium bromide to these benzylidene sulfinamides followed by acid-catalyzed methanolysis affords chiral amines in excellent yields and stereospecificity (Annunziata, R.; Cinquini, M.; Cozzi, F. *J. Chem. Soc., Perkin Trans. I* **1982**, 339; Hua, D. H.; Miao, S. W.; Chen, J. S.; Iguchi, S. *J. Org. Chem.* **1991**, *56*, 4).

Secondary sulfinamides bearing one hydrogen at the α-carbon react with acetic anhydride to form *N*-sulfenylimines and acetylimines *via* a Pummerer-type of rearrangement. The acetylimines are secondary reaction products derived from cleavage of the initially formed sulfenylimines. Both the yields and the ratio of sulfenyl- and acyl-imines increase with the electron-withdrawing power of the *para*-substituent on the aromatic ring (Isola, M.; Ciuffarin, E.; Sagramora, L.; Niccolai, C. *Tetrahedron Lett.* **1982**, *23*, 1381).

The first-order decomposition of benzenesulfinyl azide in the presence of various organic compounds has been studied. Only strong nucleophiles have any significant effect. Triphenyl phosphine dramatically increases the reaction rate. Thiols and amines decrease the rate and afford benzenethiolsulfinates and benzenesulfinamides, respectively (Maricich, T. J.; Angeletakis, C. N.; Mjanger, R. *J. Org. Chem.* **1984**, *49*, 1928; Maricich, T. J.; Angeletakis, C. N. *J. Org. Chem.* **1984**, *49*, 1931).

3. Benzenesulfenic Acids and Their Derivatives

Recent reviews on the chemistry of sulfenic acids and derivatives: Hogg, D. R. in *Comprehensive Organic Chemistry,* Barton, D. H. R.; Ollis, W. D., Eds.; Pergamon: Oxford, 1979, Vol. 3, Section 11.16; *The Chemistry of Sulphenic Acids and their Derivatives*, Patai, S., Ed., Wiley: Chichester, 1990.

(a) Benzenesulfenic Acids

The first instance of the isolation of a benzenesulfenic acid has been reported. Flash vacuum pyrolysis (FVP) of *tert*-butyl phenyl sulfoxides affords benzenesulfenic acids which can be trapped on a cold finger at -196°C. The corresponding benzenethiolsulfinates are isolated upon thawing of the condensate (Davis, F. A.; Jenkins, R. H., Jr.; Rizvi, S. Q. A.; Yocklovich, S. G. *J. Org. Chem.* **1981**, *46*, 3467). Infrared analysis of the clearly shows that benzenesulfenic acid exists in both the -OH and -SH forms (Davis, F. A.; Billmers, R. L. *J. Org. Chem.* **1985**, *50*, 2593). Benzenesulfenic acid is generated as a byproduct in the reaction of β-sulfinyl α,β-unsaturated ketones with butadienes *via* an ene reaction of the intermediate Diels-Alder adduct (Nishio, T.; Tokunaga, T.; Omote, Y. *Synth. Commun.* **1988**, *18*, 2083). Trimethylsilyl benzenesulfenates, masked equivalents of sulfenic acids, have been prepared in low yields by thermolysis of the corresponding *N*-benzylidene benzenesulfinamides with chlorotrimethylsilane and hexamethyldisilazane (Davis, F. A.; Rizvi, S. Q. A.; Ardecky, R.; Gosciniak, D. J.; Friedman, A. J.; Yocklovich, S. G. *J. Org. Chem.* **1980**, *45*, 1650).

Perfluorobenzenesulfenic acid, generated in high concentrations by FVP of *n*-butyl perfluorophenyl sulfoxide, self-condenses to afford perfluorodiphenyl disulfide and hydrogen peroxide. Decomposition of 4-nitrobenzenesulfenic acid under similar conditions gives mixtures of the corresponding disulfide, thiolsulfonate, and sulfinic acid (products of the decomposition of the initially-formed thiolsulfinate), but no hydrogen peroxide. The reason for this difference in reactivity is unclear, but partially attributed to the lower stability of the former sulfenic acid (Davis, F. A.; Jenkins, R. H., Jr. *J. Am. Chem. Soc.* **1980**, *102*, 7967). The importance of intermolecular hydrogen bonding towards lowering the energy of activation in the self-condensation of benzenesulfenic acids to give thiolsulfinates has been demonstrated by generating a series of 2-substituted benzenesulfenic acids in the presence of methyl propiolate as a trapping agent. Sulfenic acids capable of forming strong *intra*molecular bonds with the *ortho*-substituent react with the trapping agent to form the *trans*-vinyl sulfoxide. Acids that can not form intramolecular hydrogen bonds are too unstable to add to the methyl propiolate and instead form thiolsulfinates and further decomposition products (Scheme 3) (Davis, F. A.; Jenkins, L. A.; Billmers, R. L. *J. Org. Chem.* **1986**, *51*, 1033). Benzenesulfenic acids add regiospecifically to unactivated terminal alkynes to give vinyl sulfoxides in good yields (Jones, D. N.; Cottam, P. D.; Davies, J. *Tetrahedron Lett.* **1979**, 4977).

(b) Benzenesulfenates

A convenient method of preparing alkyl benzenesulfenates in good yields involves treatment of the corresponding sulfenyl chloride with an equivalent amount of a tri-*n*-butyltin alkoxide at ambient temperature (Armitage, D. A. *Synthesis* **1984**, 1042). Benzenesulfenates of (*E*)-4-hydroxy-2-alkenoates exhibit an interesting anisotropy. At ambient temperature, an electronic

Scheme 3

Hydrogen Bonding in Benzenesulfenic Acids

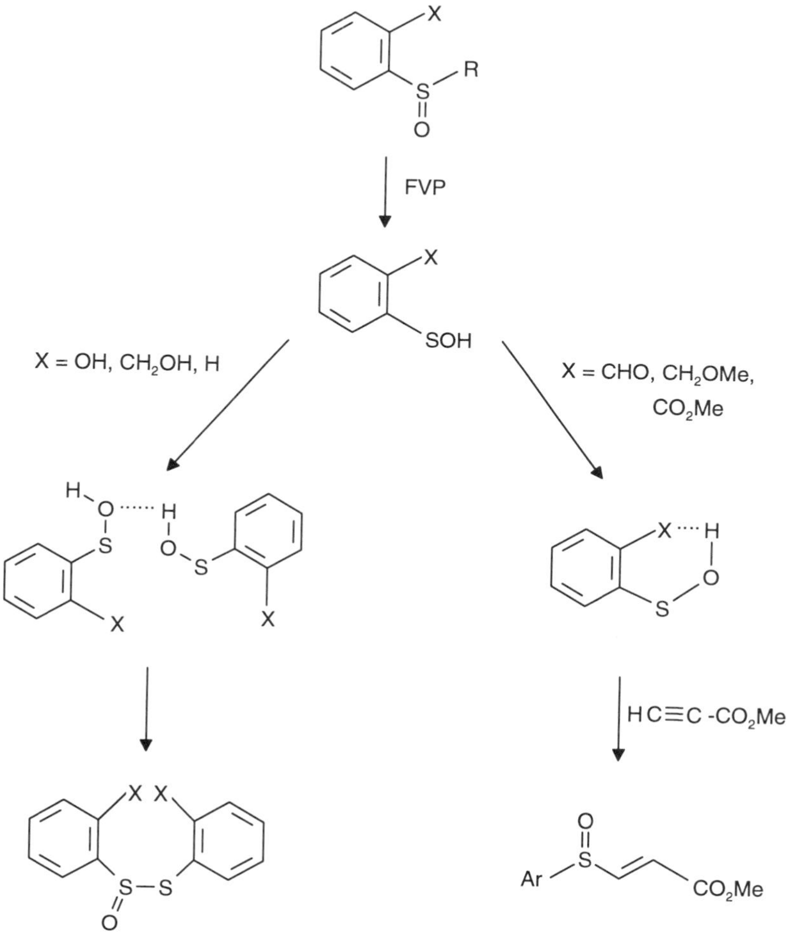

interaction between the lone pair electron on the sulfur atom and the π-cloud of the double bond is sufficiently strong that rotation about the C-O bond is inhibited, resulting in the formation of distinct isomers (Tanikaga,

R.; Kaji, A. *Chem. Lett.* **1988**, 677). The cyclic sulfenyl carboxylate 7-carboxy-3*H*-2,1-benzoxathiol-3-one is formed by heating either 2-(benzylsulfinyl)isophthalic acid or 4*H*-3,1-benzoxathiin-4-one 1-oxides (Walter, W.; Krishe, B.; Adiwidjaja, G.; Voß, J. *Chem. Ber.* **1978**, *111*, 1685; Krische, B.; Walter, W. *Chem. Ber.* **1983**, *116*, 1708).

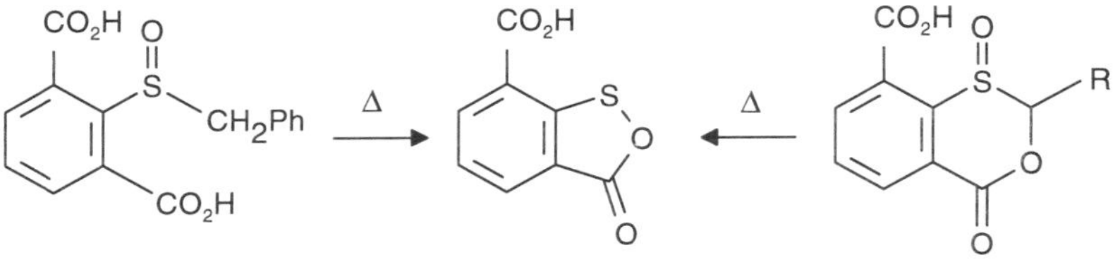

The acid-catalyzed hydrolysis of methyl *p*-toluenesulfenate has been studied. In the presence of a small amount of water, the reaction products are methyl *p*-toluenesulfinate, di-*p*-tolyl disulfide, and methanol; however, at higher concentrations of water, *p*-tolyl *p*-toluenethiolsulfonate, the disulfide, and methanol are formed. A mechanism involving the intermediacy of a sulfonium salt is proposed (Scheme 4) (Ciuffarin, E.; Gambrotta, S. Isola, M.; Senatore, L. *J. Chem. Soc., Perkin Trans. II* **1978**, 554). Recent kinetic evidence supports the initial formation of a hypervalent sulfur species (sulfuranide) as an intermediate in the acid hydrolysis of ethyl benzenesulfenate (Okuyama, T.; Nakamura, T.; Fueno, T. *J. Am. Chem. Soc.* **1990**, *112*, 9345). Attempted iodide catalysis of this reaction dramatically changes the product distribution, affording mainly diphenyl disulfide and only small amounts of the expected phenyl benzenethiolsulfinate (Okuyama, T.; Nakamura, T.; Fueno, T. *Tetrahedron Lett.* **1990**, *31*, 1017).

Propargyl sulfenates, prepared *in situ* by reaction of benzenesulfenyl chlorides with propargyl alcohols, undergo a [2,3]-sigmatropic shift to give

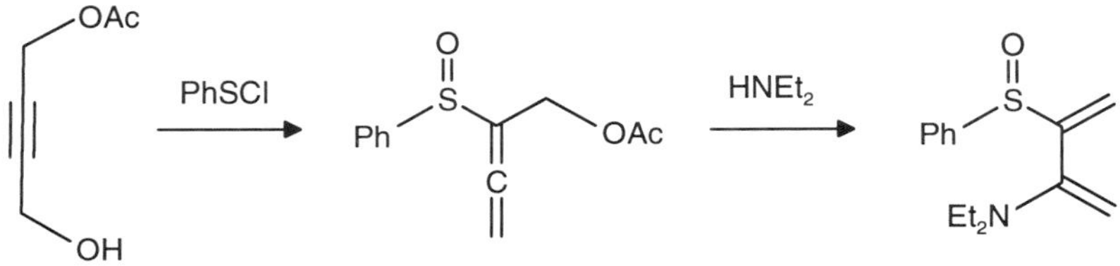

Scheme 4

Hydrolysis of Methyl p-Toluenesulfenate

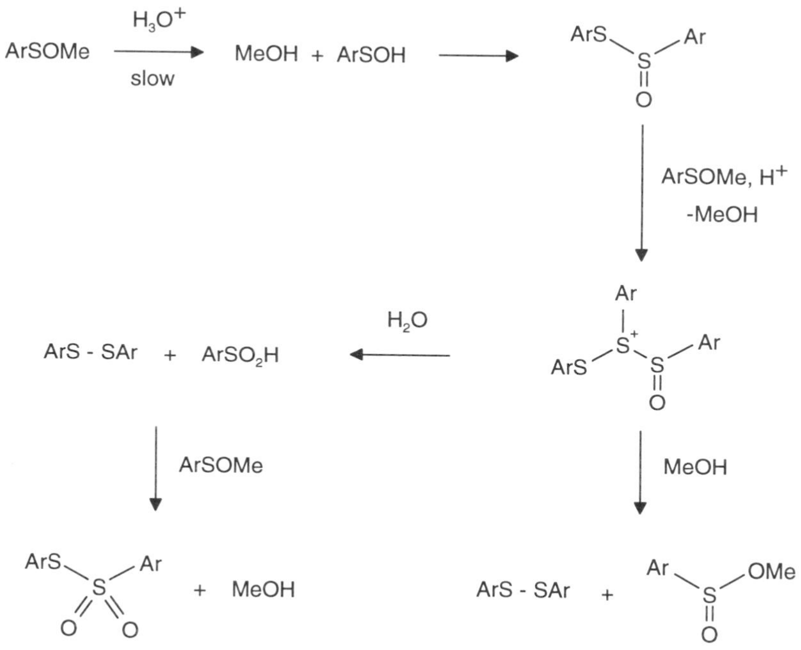

allenyl sulfoxides. Treatment of 4-acetoxybut-2-ynol with benzenesulfenyl chloride gives a 70% yield of the rearranged allenyl sulfoxide. When treated with two equivalents of diethylamine, this sulfoxide undergoes further reaction to give 3-(N,N-diethylamino)-2-(phenylsulfinyl)buta-1,3-diene in 93% yield (Bridges, A. J.; Fischer, J. W. *J. Chem. Soc., Chem. Commun.* **1982**, 665). These allenyl sulfoxides also undergo electrocyclic reactions with 1,3-dienes and nitrones (Reischl, W.; Okamura, W. H. *J. Am. Chem. Soc.* **1982**, *104*, 6115; Okamura, W. H.; Peter, R.; Reischl, W. *J. Am. Chem. Soc.* **1985**, *107*, 1034; Padwa, A.; Norman, B. H.; Perumattam, J. *Tetrahedron Lett.* **1989**, *30*, 663; Padwa, A.; Bullock, W. H.; Norman, B. H.; Perumattam, J. *J. Org. Chem.* **1991**, *56*, 4252).

Alkyl benzenesulfenates react with alkenes in the presence of a Lewis acid (TMSOTf, BF_3) to give episulfonium ions. These episulfonium ions undergo intramolecular cyclization with phenyl rings to form tetrahydronaphthalenes (Edstrom, E. D.; Livinghouse, T. *J. Am. Chem. Soc.* **1986**, *108*, 1334). Diterpenoids have been synthesized *via* a biomimetic polyene cyclization initiated by episulfonium ions generated in this manner (Edstrom, E. D.; Livinghouse, T. *J. Org. Chem.* **1987**, *52*, 949). If intramolecular attack is not available, nucleophilic attack by the alkoxy group of the original

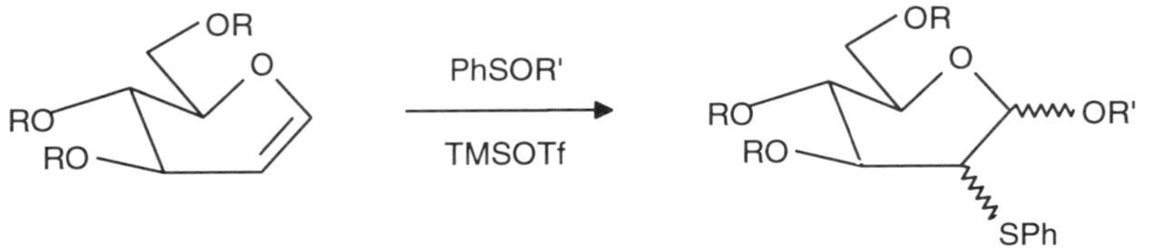

sulfenate occurs to give β-alkoxysulfides (Ito, Y.; Ogawa, T. *Tetrahedron Lett.* **1987**, *28*, 2723). Thioglycosides also react with alkyl benzenesulfenates and Lewis acids to afford *O*-glycosides *via* an intermediate sulfonium ion (Ito, Y.; Ogawa, T. *Tetrahedron Lett.* **1987**, *28*, 4701). Olefinic alcohols undergo intramolecular cyclizations upon treatment with benzenesulfenyl chloride to give tetrahydro-furans and -pyrans (Tuladhar, S. M.; Fallis, A. G. *Can. J. Chem.* **1987**, *65*, 1833).

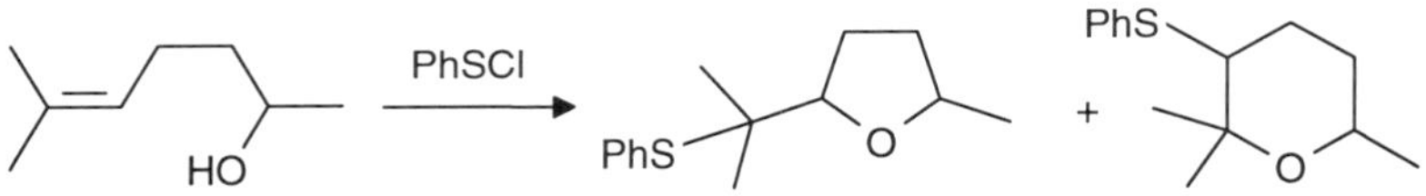

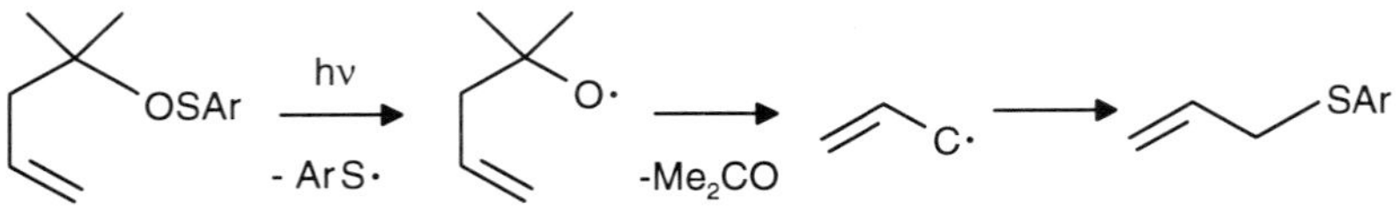

Irradiation of *tert*-alkyl 4-nitrobenzenesulfenates with >300-nm wavelength light in benzene solution results in the homolytic cleavage of the S-O bond. The alkoxy radicals thus formed undergo β-scission to produce carbon-centered radicals in essentially quantitative yields. These radicals can dimerize or, more likely, react with the aryl thiyl radical to form sulfides. Primary and secondary sulfenates react similarly, but undergo competitive disproportionation (Pasto, D. J.; L'Hermine, G. *J. Org. Chem.* **1990**, *55*, 5815). In the presence of oxygen, these primary and secondary alkyl sulfenates are oxidized to the corresponding sulfinates. Tertiary sulfenates do not react (Pasto, D. J.; Cottard, F.; Horgan, S. *J. Org. Chem.* **1993**, *58*, 4110). These alkoxy radicals can also be generated by treatment of alkyl benzenesulfenates with tributylstannane (Beckwith, A. L. J.; Hay, B. P.; Williams, G. M. *J. Chem. Soc., Chem. Commun.* **1989**, 1202). The asymmetric oxidation of sulfenates to sulfinates has been achieved using a modification of the Sharpless reagent (*t*-BuOOH, Ti(O-*i*-Pr)$_4$, diethyl (+)-tartrate, H$_2$O (1:1:2:1)). Enantiomeric excesses up to 36% are obtained (Nemecek, C.; Dunach, E.; Kagan, H. B. *Nouv. J. Chem.* **1986**, *10*, 761).

(c) Benzenesulfenyl Halides

Benzenesulfenyl bromide can be prepared from benzenethiol by reaction with *N*-bromosuccinimide (Bridges, A. J.; Fischer, J. W. *J. Org. Chem.* **1984**, *49*, 2954). Treatment of the acetate of benzenethiols with one equivalent of neat sulfuryl chloride affords the corresponding benzenesulfenyl chlorides in excellent yields. Excess sulfuryl chloride gives overoxidation to the sulfinyl chloride (Thea, S.; Cevasco, G. *Tetrahedron Lett.* **1988**, *29*, 2865).

$$ArSCOCH_3 + SO_2Cl_2 \longrightarrow ArSCl + CH_3COCl + SO_2$$

Reaction of benzenesulfenyl chlorides with (alkylthio)trimethylsilanes at 0°C affords unsymmetrical disulfides in good yields (Harpp, D. N.; Friedlander, B. T.; Larsen, C.; Steliou, K.; Stockton, A. *J. Org. Chem.* **1978**, *43*, 3481). An improved method of preparing unsymmetrical aryl sulfides has been reported. Equimolar amounts of benzenesulfenyl chlorides and the appropriate aromatic compounds react in nitroethane at ambient temperature within one hour to afford good to excellent yields of the corresponding sulfides. The mild conditions minimizes side-reactions and the products are easily isolated (Bottino, F.; Fradullo, R.; Pappalardo, S. *J. Org. Chem.* **1981**, *46*, 2793). Trimethylsilyl cyanide reacts with benzenesulfenyl chlorides to afford the corresponding thiocyanates in 81-96% yield (Lazukina, L. A.; Kukhar, V. P.; Romanov, G. V.; Khaskin, G. I.; Dubrinina, T. N.; Ofitserov, E. N.; Volkova, A. N.; Pudovik, A. N. *J. Gen. Chem. USSR (Engl. Transl.)* **1980**, *50*, 783; *Zh. Obshch. Khim.* **1980**, *50*, 985). Cyclic and acyclic α-diazoketones react with benzenesulfenyl chloride at ambient temperature to give α-chloro-α-(phenylthio)ketones. The adducts of cyclic ketones are not isolated, but undergo dehydrochlorination to 2-(phenylthio)-2-cycloalkenones upon treatment with triethylamine (McKervey, M. A.; Ratananukul, P. *Tetrahedron Lett.* **1983**, *24*, 117). Benzenesulfenyl chlorides react with *N*-chloro-amides and -ketimines to give *N*-substituted benzenesulfinimidoyl chlorides. In the case of ketimines, the sulfinimidoyl chlorides are unstable and eliminate chlorine to form *N*-(phenylsulfenyl)-ketimines *unless* the initial ketimine contains an α-trifluoromethyl group (Dubinina, T. N.; Levchenko, E. S.; Zabolotnaya, T. G. *J. Org. Chem. USSR (Engl. Transl.)* **1982**, *18*, 143; *Zh. Org. Khim.* **1982**, *18*, 162; Shermolovich, Y. G.; Talanov, V. S.; Pirozhenko, V. V.; Markovskii, L. N. *J. Org. Chem. USSR (Engl. Transl.)* **1982**, *18*, 2240; *Zh. Org. Khim.* **1982**, *18*, 2539).

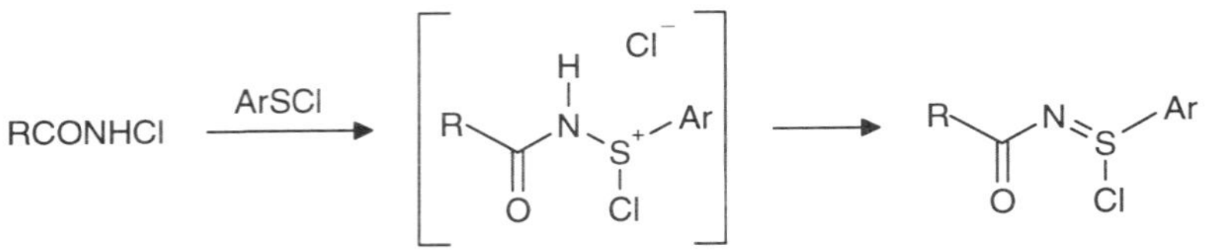

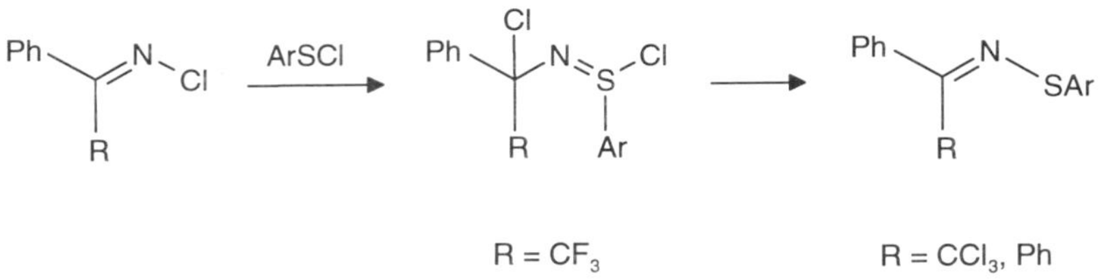

Recent applications of the cleavage of sulfides with benzenesulfenyl chloride have been reported. *tert*-Butyl thioethers of cysteine moieties in peptides can be selectively cleaved by treatment with 2-nitrobenzenesulfenyl chloride (Pastuszak, J. J.; Chimiak, A. *J. Org. Chem.* **1981**, *46*, 1868). Sequential addition of benzenesulfenyl chloride and diisopropylethylamine to thio-acetals and -ketals at -78°C affords the corresponding vinyl sulfides (Bartels, B.; Hunter, R.; Simon, C. D.; Tomlinson, G. D. *Tetrahedron Lett.* **1987**, *28*, 2985). Methionyl peptide bonds are cleaved by reaction with 2-nitrobenzenesulfenyl chloride (Galpin, I. J.; Hoyland, D. A. *Tetrahedron* **1985**, *41*, 895).

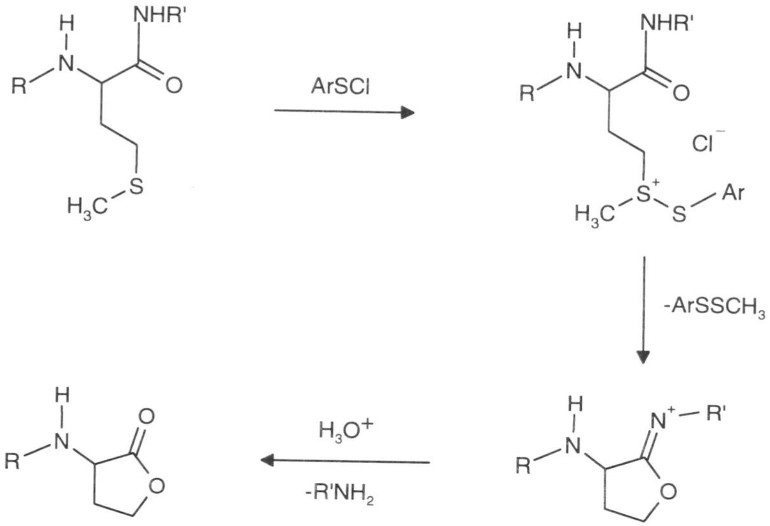

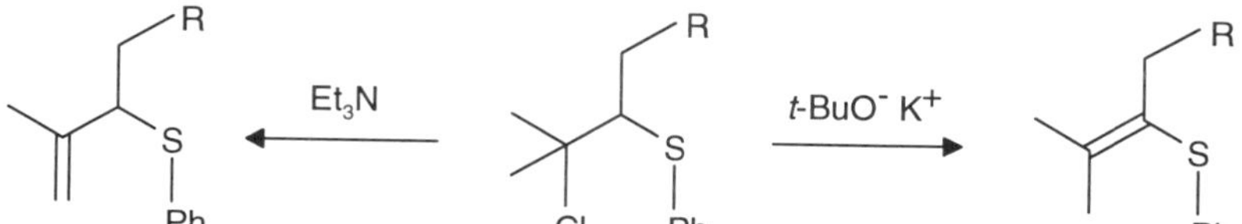

The addition of benzenesulfenyl chlorides to alkenes to afford β-chloroalkyl aryl sulfides is a well-established reaction. Base-catalyzed dehydrochlorination with triethylamine (DMF, 60°C) or potassium *tert*-butoxide (DMSO, 25°C) provides the corresponding allylic or vinyl sulfides, respectively (Masaki, Y.; Hashimoto, K.; Kaji, K. *Tetrahedron Lett.* **1978**, 4539; Masaki, Y.; Sakuma, K,; Kaji, K. *Chem. Lett.* **1979**, 1235). This methodolgy has been used in the synthesis of a chemically stable prostacyclin analogue (Bannai, K.; Toru, T.; Ōba, T.; Tanaka, T.; Okamura, N.; Watanabe, K.; Hazato, A.; Kurozumi, S. *Tetrahedron* **1983**, *39*, 3807). Treatment of electron-rich alkenes with benzenesulfenyl chloride and silver fluoride gives the analogous β-fluoroalkyl aryl sulfides (Purrington, S. T.; Correa, I. D. *J. Org. Chem.* **1986**, *51*, 1080). Sequential treatment of methyl vinyl ether with *p*-toluenesulfenyl chloride, titanium tetrachloride, and 1-methoxy-2-methyl-1-propene forms a cyclic sulfonium ion. Quenching of this reactive intermediate with an ethereal solution of allyl magnesium chloride affords a polyfunctional sulfide derived from three components in a one-pot

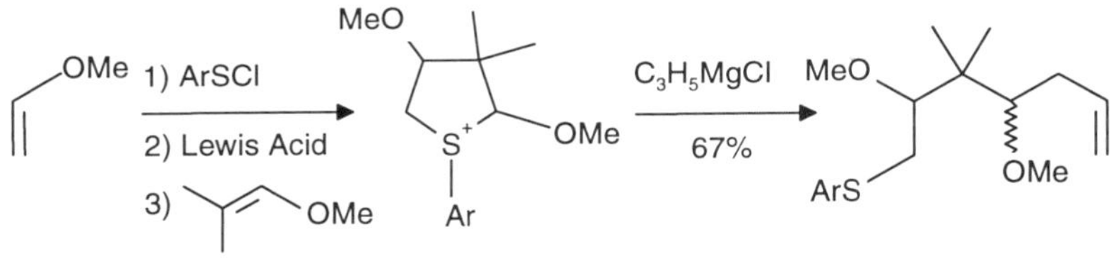

operation. An episulfonium ion is postulated as an intermediate (Smoliakova, I. P.; Smit, W. A.; Osinov, B. *Tetrahedron Lett.* **1991**, *32*, 2601). Sequential addition of benzenesulfenyl chloride and silver tetrafluoroborate to 1-alkenylcycloalkanols affords ring-expanded 2-(phenylsulfenylmethyl)-cycloalkanones (Kim, S.; Park, J. H. *Tetrahedron Lett.* **1989**, *30*, 6181). Continuing mechanistic studies of the addition reaction of benzenesulfenyl chlorides to alkenes have been reported (Smit, W. A.; Zefirov, N. S.; Bodrikov, I. V.; Krimer, M. Z. *Acc. Chem. Res.* **1979**, *12*, 282; Bodrikov, I. V.; Borisov, A. V.; Chumakov, L. V.; Zefirov, N. S.; Smit, W. A. *Tetrahedron Lett.* **1980**, *21*, 115; Bodrikov, I. V.; Borisov, A. V.; Smit, W. A.; Lutsenko, A. I. *Tetrahedron Lett.* **1984**, *25*, 4983).

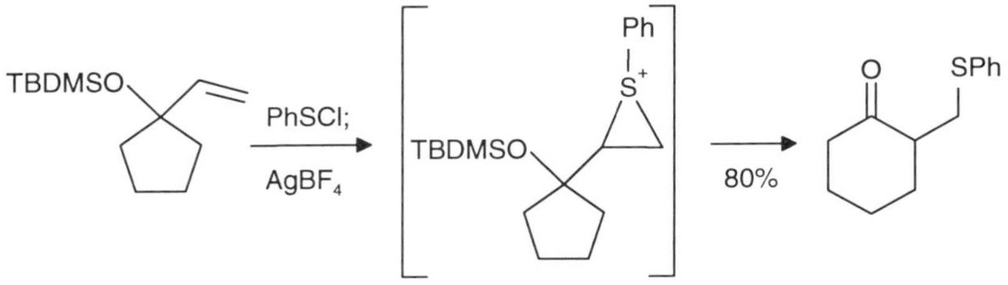

TBDMS = *t*-butyldimethylsilyl

(d) Benzenesulfenamides

The chemistry of sulfenamides has been reviewed recently (Craine, L.; Raban, M. *Chem. Rev.* **1989**, *89*, 689).

(i) Preparation

Diphenyl disulfide reacts with lithium dialkylamides in THF at low temperature to afford cleanly N,N-dialkyl benzenesulfenamides in good yields. The starting materials are relatively inexpensive and easily available (Ikehira, H.; Tanimoto, S. *Synthesis*, **1983**, 716). Treatment of 2-(methylsulfinyl)benzamides with thionyl chloride induces an intramolecular cyclization forming 2-alkyl- and 2-aryl-benzisothiazol-3(2H)-ones in excellent yields (Uchida, Y.; Kozuka, S. *J. Chem. Soc., Chem. Commun.* **1981**, 510). Alternately, these 2-alkyl- and aryl-benzisothiazol-3(2H)-ones may be synthesized by cyclization of 2-(bromosulfinyl)benzamides over activated basic alumina. Yields are excellent and the starting materials can be easily prepared by oxidation of the corresponding disulfides with bromine (Kamigata, N.; Hashimoto, S.; Kobayashi, M. *Org. Prep. Proced. Int.* **1983**, *15*, 315).

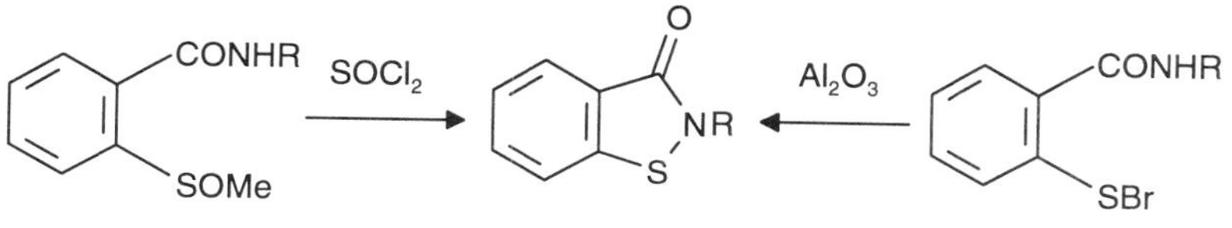

Benzothiete reacts with primary amines to give the ring-expanded 2,3-dihydrobenz[d]isothiazoles in good yields (Kanakarajan, K.; Meier, H. *Angew. Chem. Int. Ed. Engl.* **1984**, *23*, 244; *Angew. Chem.* **1984**, *96*, 220). Treatment of phenyl allyl sulfide with N-[(trifluoromethanesulfonyl)oxy]-carbamate affords N-allyl-N-(ethoxycarbonyl) benzenesulfenamide in 77% yield via a [2,3]-sigmatropic rearrangement of the intermediate sulfilimine

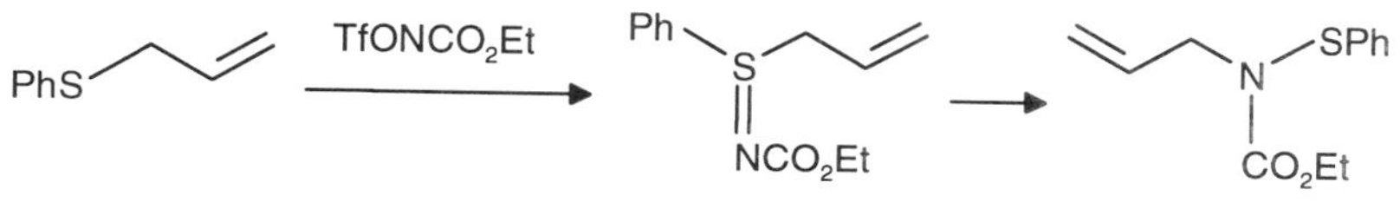

(Tamura, Y.; Ikeda, H.; Mukai, C.; Morita, I.; Ikeda, M. *J. Org. Chem.* **1981**, *46*, 1732). A simple method of preparing N-(phenylthio)succinimide, a useful sulfur transfer agent, from (phenylthio)tributylstannane and NBS has been reported (Harpp, D. N.; Aida, T.; DeCesare, J.; Tisnes, P.; Chan, T. H. *Synthesis* **1984**, 1037). The preparation and chemistry of various other sulfenamides as new sulfur transfer agents has been reported (Sosnovsky, G.; Krogh, J. A. *Liebigs Ann. Chem.* **1982**, 121; Romani, S.; Bovermann, G.; Moroder, L.; Wünsch, E. *Synthesis* **1985**, 512). Benzenesulfenylcarbamates and -ureas can be prepared by sequential treatment of benzenesulfenyl chlorides with silver cyanate and either an alcohol or a primary amine, respectively (Nagase, T.; Akama, K.; Ozaki, S. *Chem. Lett.* **1988**, 1385).

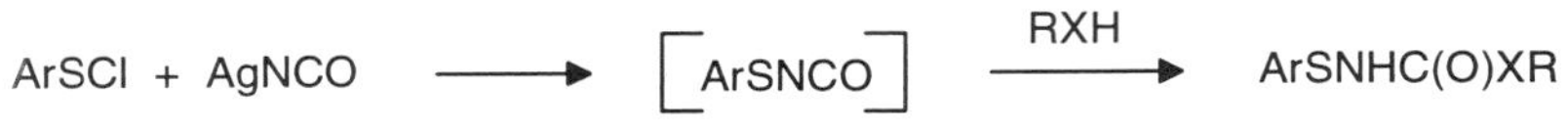

$$\text{ArSCl} + \text{AgNCO} \longrightarrow [\text{ArSNCO}] \xrightarrow{\text{RXH}} \text{ArSNHC(O)XR}$$

X = O or NH

Treatment of *N,N*-bis(trimethylsilyl) benzenesulfenamide with aldehydes or ketones in the presence of tetrabutylammonium fluoride provides a convenient method of preparation of benzenesulfenimines (Morimoto, T.; Nezu, Y.; Achiwa, K.; Sekiya, M. *J. Chem. Soc., Chem. Commun.* **1985**, 1584). Diaryl disulfides react with α-aminoalkanoates under electrolytic conditions to give benzenesulfenimines (Torii, S.; Tanaka, H.; Hamano, S.; Tada, N.; Nokami, J.; Sasaoka, M. *Chem. Lett.* **1984**, 1823). A variety of 2-nitrobenzenesulfenimines are readily available from the corresponding sulfenamides by electrochemical oxidation or by addition of stoichiometric amounts of a radical cation salt (Heyer, J.; Dapperheld, S.; Steckhan, E. *Chem. Ber.* **1988**, *121*, 1617). An interesting preparation of a sulfenimine has been reported as part of the synthesis of a cephalosporin derivative. Treatment of an α-aminolactam with three equivalents of *p*-toluenesulfenyl chloride affords the corresponding *protected and oxidized* sulfenimine in

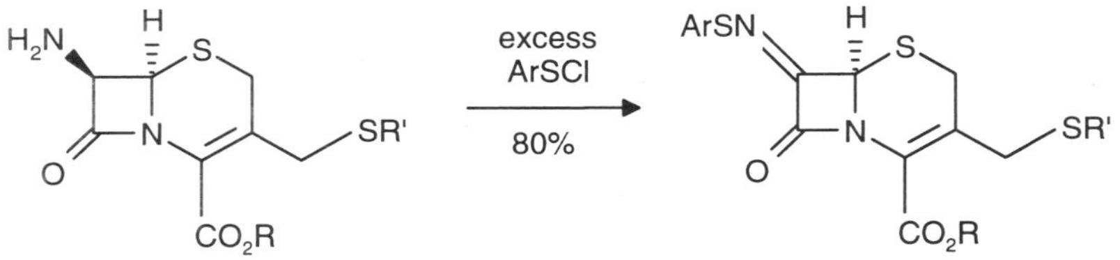

80% yield (Gordon, E. M.; Chang, H. W.; Cimarusti, C. M.; Toeplitz, B.; Gougoutas, J. Z. *J. Am. Chem. Soc.* **1980**, *102*, 1690). Benzenesulfenamides react with sodium *N*-chloro benzenesulfonamide to afford the corresponding *N*-(benzenesulfonyl) benzenesulfinamidines (Koval', I. V.; Oleinik,

T. G.; Kremlev, M. M. *J. Org. Chem. USSR (Engl. Transl.)* **1981**, *17*, 1938; Zh. Org. Khim. **1981**, *17*, 2174).

(ii) Reactions

Oxidation of *N,N*-dialkyl benzenesulfenamides with *N*-chlorosuccinimide in dichloromethane followed by hydrolysis of the intermediate sulfonium salt affords the corresponding sulfinamides in 66-85% yields (Haake, M.; Gebbing, H.; Benack, H. *Synthesis* **1979**, 97). Electrochemical oxidation of 2-nitrobenzenesulfenamides derived from secondary cyclic amines in methanol gives the corresponding sulfenates *via* an EC process (Sayo, H.; Yamada, Y.; Michida, T. *Chem. Pharm. Bull.* **1983**, *31*, 4530). The reaction of *N*-(4-nitrophenyl) benzenesulfenamides with thiols in the presence of boron trifluoride etherate provides an effective route to unsymmetrical

$$\text{ArSNH}-\text{C}_6\text{H}_4-\text{NO}_2 \xrightarrow[\text{BF}_3]{\text{RSH}} \text{ArSSR} + \text{H}_2\text{N}-\text{C}_6\text{H}_4-\text{NO}_2$$

disulfides (Benati, L.; Montevecchi, P. C.; Spagnolo, P. *Tetrahedron Lett.* **1986**, *27*, 1739). Dialkyl phosphites react with benzenesulfenamides to give phosphorothiolates in 67-99% yields (Torii, S.; Sayo, N.; Tanaka, H. *Chem. Lett.* **1980**, 695).

The asymmetric oxidation of sulfenamides to sulfinamides by a modification of the Sharpless reagent (*t*-BuOOH, Ti(O-*i*-Pr)$_4$, diethyl (+)-tartrate, H$_2$O (1:1:2:1)) has been achieved. Enantiomeric excceses up to 35% are obtained (Nemecek, C.; Dunach, E.; Kagan, H. B. *Nouv. J. Chem.* **1986**, *10*, 761). Asymmetric sulfenylations of 4-alkylcyclohexanones have been achieved in up to 50% enantiomeric excess with chiral sulfenamides in the presence of triethylamine hydrochloride (Hiroi, K.; Nishida, M.; Nakayama, A.; Nakazawa, K.; Fujii, E.; Sato, S. *Chemistry Lett.* **1979**, 969). Simple alcohols react with *N,N*-disubstituted benzenesulfenamides upon sequential treatment with *tert*-butyl hypochlorite and silver tetrafluoroborate to give alkoxy aryl dialkylaminosulfonium tetrafluoroborates (Haake, M.; Gebbing, H. *Synthesis* **1979**, 98).

Benzenesulfenanilides add to alkenes in the presence of boron trifluoride to give *trans*-β-(arylamino)sulfides (Benati, L.; Montevecchi, P. C.; Spagnolo, P. *Tetrahedron* **1986**, *42*, 1145). *N*-Alkyl and *N,N*-dialkyl benzenesulfenamides react in a similar fashion with various alkenes upon treatment with trifluoromethanesulfonic acid in dichloromethane; however, *in acetonitrile*, the intermediate episulfonium salt is intercepted by the solvent in a Ritter-type reaction, affording amidines as the final product (Brownbridge, P. *Tetrahedron Lett.* **1984**, *25*, 3759). Activated alkynes also react with benzenesulfenanilides in acetonitrile in the presence of boron trifluoride to give β-amidinovinyl sulfides (Benati, L.; Montevecchi, P. C.; Spagnolo, P. *J. Chem. Soc., Perkin Trans. I* **1989**, 1105). In acetic acid, the intermediate thiirenium ion reacts with the solvent to afford β-acetoxyvinyl sulfides (Benati, L.; Casarini, D.; Montevecchi, P. C.; Spagnolo, P. *J. Chem. Soc., Perkin Trans. I* **1989**, 1113). Oxidation of 2,4-dinitrobenzenesulfenamide with lead tetraacetate in the presence of electron-rich alkenes gives substituted *N*-(2,4-dinitrophenylsulfenyl)aziridines *via* the sulfenyl nitrene (Atkinson, R. S.; Judkins, B. D. *J. Chem. Soc., Perkin Trans. I* **1981**, 2615).

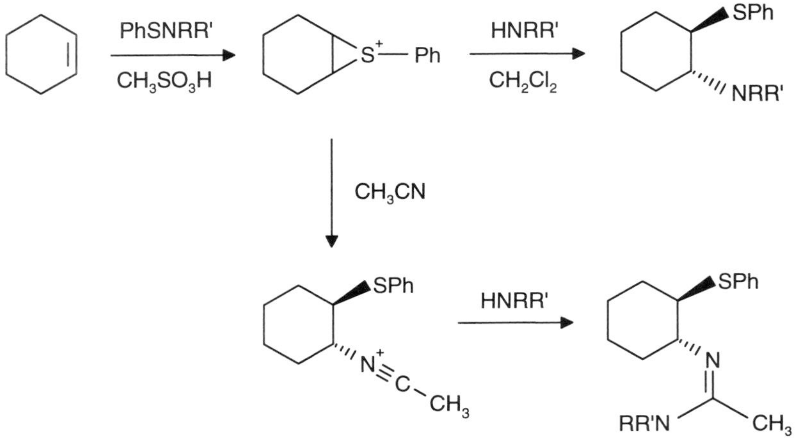

Alkyl- and aryl-lithium reagents add to the C=N bond of benzenesulfenimines. The resulting secondary and tertiary benzenesulfenamides can be isolated or carefully hydrolyzed without intermediate workup to afford primary amines (Davis, F. A.; Mancinelli, P. A. *J. Org. Chem.* **1977**, *42*,

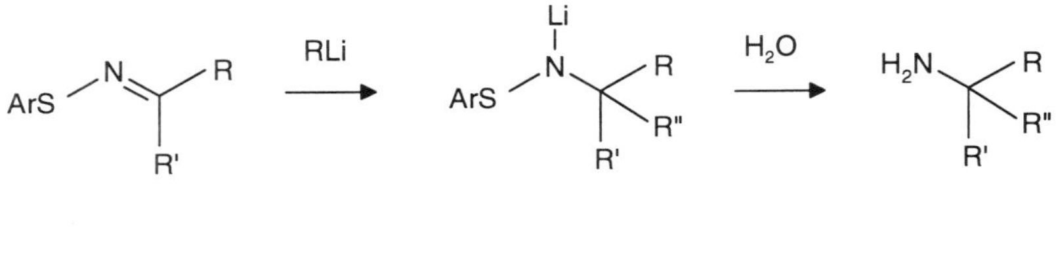

398). Treatment of benzenesulfenimines containing at least one β-hydrogen with lithium diisopropylamide generates a sulfenamide enolate equivalent (SEE) in high yield and with good regioselectivity. These carbanions can be trapped with electrophiles such as halides, carbonyl compounds,

$$\text{ArS}^{\diagdown}\text{N}\diagup\overset{R}{\underset{R'}{\diagdown}}\xrightarrow{\text{LDA}} \text{ArS}^{\diagdown}\text{N}\diagup\overset{R}{\underset{R'}{\diagdown}}^{-} \xrightarrow{E^+} \text{ArS}^{\diagdown}\text{N}\diagup\overset{R}{\underset{R'}{\diagdown}}\text{E}$$

and diaryl disulfides. Similar treatment of benzenesulfenimines derived from aldehydes results in cleavage to the corresponding nitrile and thiophenol (Davis, F. A.; Mancinelli, P. A. *J. Org. Chem.* **1978**, *43*, 1797). The method of quench affects the reaction pathway in the case of disulfides. Addition of the SEE to the diaryl disulfide gives the α-(phenylthio)sulfenimine in good yield; however, if the disulfide is added to the reaction mixture containing the SEE, the predominant product is a β-amino ketene thioacetal (Davis, F. A.; Mancinelli, P. A. *J. Org. Chem.* **1980**, *45*, 2597).

(e) Aryl Benzenethiolsulfonates

Aryl benzenethiolsulfonates (aryl benzenesulfenyl sulfones) are powerful sulfenylating agents; in some cases, they are superior to disulfides and sulfenyl halides (e.g., Trost, B. M.; Massiot, G. S. *J. Am. Chem. Soc.* **1977**, *99*, 4405). Symmetrical thiolsulfonates can be prepared in good yields by partial reduction of benzenesulfonyl chlorides with potassium iodide in anhydrous acetone containing a catalytic amount of pyridine (Palumbo, G.; Caputo, R. *Synthesis* **1981**, 888). Benzenethiolsulfonates can also be prepared from unsymmetrical thiolsulfinates in high yields by oxidation with sodium metaperiodate in polar solvents (Kim, Y. H.; Takata, T.; Oae, S. *Tetrahedron Lett.* **1978**, 2305). Sodium benzenesulfinates react with elemental sulfur in the presence of an amine or ammonia in a titanium autoclave at 20°C to afford sodium thiolsulfonates in essentially quantitative yields (Sato, R.; Goto, T.; Takikawa, Y.; Takizawa, S. *Synthesis* **1980**, 615). Further treatment with an alkyl halide and an anion exchange resin gives the corresponding alkyl benzenethiolsulfonates (Takano, S.; Hiroya, K.; Ogasawara, K. *Chem. Lett.* **1983**, 255). Reaction of *p*-toluenesulfonyl hydrazide with an excess of a benzenesulfonyl halide affords unsymmetrical thiolsulfonates along with an equimolar amount of diphenyl disulfide (Back, T. G.; Collins, S.; Krishna, M. V. *Can. J. Chem.* **1987**, *65*, 38). Finally, sodium benzenesulfinate reacts with *N*-(phenylthio)-1,4-benzoquinone imines in acetic acid to give thiolsulfonates (Kolesnikov, V. T.; Vid, L. V.; Kuz'menko, L. O. *J. Org. Chem. USSR (Engl. Trans.)* **1982**, *18*, 1906; *Zh. Org. Khim.* **1982**, *18*, 2163).

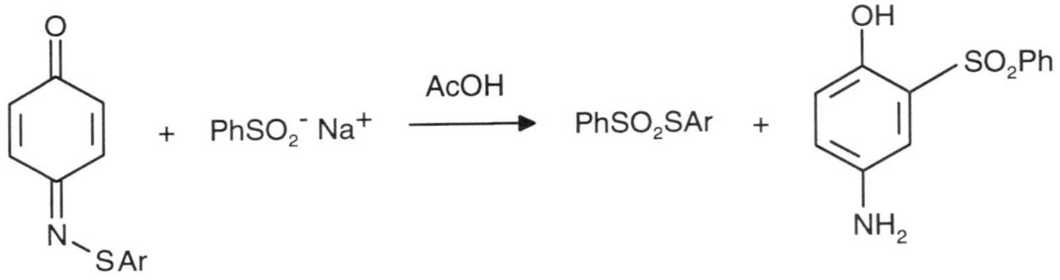

$$\text{quinone imine} + \text{PhSO}_2^- \text{Na}^+ \xrightarrow{\text{AcOH}} \text{PhSO}_2\text{SAr} + \text{aminophenol sulfone}$$

Aryl benzenethiolsulfonates react with benzenesulfenyl halides to form diaryl disulfides and benzenesulfonyl halides. Sulfenyl chlorides require more vigorous conditions for complete reaction (115°C, DMF) than the corresponding bromides (75°C, CH$_3$CN) (Lázár, J.; Vinkler, E. *Acta Chim. Acad. Sci. Hung.* **1980**, *105*, 171). Treatment of aryl benzenethiolsulfonates with (methylthio)trimethylsilane affords approximately equal amounts of the corresponding aryl methyl disulfides and trimethylsilyl benzenesulfinates (Capozzi, G.; Capperucci, A.; Degl'Innocenti, A.; Del Duce, R.; Menichetti, S. *Tetrahedron Lett.* **1989**, *30*, 2995). Organolithium compounds add to benzenethiolsulfonates *via* displacement of the benzenesulfinate moiety, providing aryl sulfides in excellent yields (Palumbo, G.; Ferreri, C.; D'Ambrosio, C.; Caputo, R. *Phosphorus Sulfur* **1984**, *19*, 235). Trimethylsilyl enol ethers of carbonyl compounds undergo a fluoride-mediated α-phenylsulfenylation upon treatment with phenyl benzenethiolsulfonate and tetrabutylammonium fluoride (Caputo, R.; Ferreri, C.; Palumbo, G. *Synthesis* **1989**, 464). Disulfides are smoothly prepared by reduction of the corresponding aryl benzenethiolsulfonates with chlorotrimethylsilane and sodium iodide (Palumbo, G.; Parrilli, M.; Neri, O.; Ferreri, C.; Caputo, R. *Tetrahedron Lett.* **1982**, *23*, 2391).

Chapter 7

MONONUCLEAR HYDROCARBONS CARRYING NUCLEAR SUBSTITUENTS CONTAINING SELENIUM, OR TELLURIUM

N. Furukawa and S. Sato

1. Introduction

Whereas organic selenium and tellurium compounds exhibit similar physical and chemical properties to the corresponding sulfur derivatives, the behavior of the related oxygen compounds is quite different. One reason for this is the fact that sulfur, selenium and tellurium have roughly the same electronegativities, but that of oxygen is much lower. Even so selenium and tellurium, together with sulfur, provide electron-accepting functional groups.

Another common feature of these three elements is that certain compounds e.g. thiols, selenols and tellurols; diselenides and ditellurides etc., can be synthesized by similar methods. These compounds can then be converted into a range of other derivatives as illustrated in Scheme 1.

Although there are many organic selenium and tellurium compounds, their chemical investigation has been restricted to those which are stable and only a few systematic studies have been performed which encompass the whole range of available types. In modern organic synthesis there is an intense demand for new and selective reagents. Many recently introduced reagents contain selenium, or tellurium and this has a dramatic effect on their chemical properties allowing them to be used to complement and extend the range of applications traditionally offered by sulfur bearing analogues. Coordination numbers vary in the Group 16 elements (chalcogenes) and, although oxygen commonly forms divalent and a few trivalent oxonium compounds, sulfur can increase its valency from two to six by expanding its valence shell from the normal octet to decet and even dodecet states.

Similar valence shell expansions also occur with selenium and tellurium affording highly coordinated (hypervalent) compounds. Sulfoxides and sulfones are normally stable molecules, but the corresponding tri- and tetra-coordinated compounds of selenium and tellurium are unstable. Selenoxides and telluroxides having β-protons, for example, undergo facile syn-eliminations to afford the corresponding olefins. Such reactions are of synthetic importance.

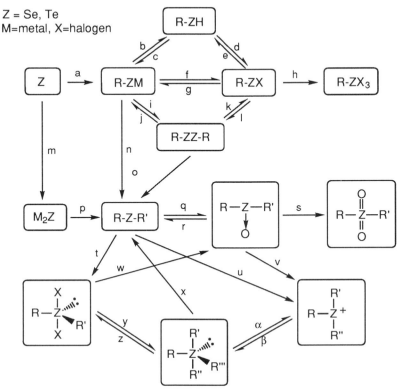

Scheme 1 Classification of Se and Te Compounds

a) RM(M=Li, MgBr); b) H$_2$O; c) MOH(M=Na, K); d) X$_2$; e) MH(M=Na, K); f) X$_2$; g) M(M=Li, Na, K); h) X$_2$; i) air/H$_2$O; j) M(Li, Mg); k) X$_2$; l) air/H$_2$O; m) MOH/Δ; n) RX; o, p) RM(M=Li, MgBr); q, s) mCPBA or NaIO$_4$; r) Na$_2$S or NH$_2$NH$_2$; t) X$_2$; u) RX(R=Alkyl); v) R'M(M=Li, MgBr); w) MOH(M=Na, K); x) Δ or hν; y) RM(M=Li, MgBr); z) X$_2$; α) HX; β) RM(M=Li, MgBr)

Another special feature of sulfur, selenium and tellurium compounds is their ability to form stable α-carbanions, which in turn react well with numerous electrophiles leading to a wide variety of usual products. In this area there is still room for investigation, although the problems of toxicity and smell mitigate against the casual use of compounds of this type.

This chapter briefly describes the monoarylic selenium and tellurium compounds of coordination number II to VI. Since the first edition covered the literature up until 1983, the present review mainly deals with developments in the field after this date, although some recourse to earlier material is necessary in order to accurately reflect recent progress. During this time several relevant books and review articles have been published. These include: Selenium: ("The chemistry of selenium and tellurium compounds", S. Patai and Z. Rappoport ed., Vol. 1 and 2 , John Wiley & Sons, New York, 1986 and 1987; D. Liotta, "Organoselenium chemistry", John Wiley & Sons, New York, 1987; C. Paulmier, "Selenium reagents and intermediates in organic synthesis", Pergamon press, Oxford, 1986; H. J. Reich, Acc. Chem. Res., 1979, **12**, 22; D. L. Clive, Tetrahedron, 1978, **34**, 1049; A. Krief, Tetrahedron, 1980, **36**, 2531). Tellurium: (S. Uemura, Synth. Org. Chem.,(in Japanese), 1983, **41**, 804; L. Engman, Acc. Chem. Res., 1985, **18**, 274; N. Petragnani and J. V. Commasseto, Synthesis, **1986**, 1).

Table 1. Physical Properties of Chalcogen Elements.

	X			
	O	S	Se	Te
electronegativity	3.5	2.5	2.4	2.1
electroaffinity	-7.28	-3.44	-4.21	-
ionization potential	13.61	10.36	9.75	9.01
pKa (aq. XH_2)	16	7.0	3.8	2.6
pKa (aq. XH)	-	12.9	11.0	11.0
bond length (Å)				
$C(sp^3)$-X	1.41	1.81	1.98	2.15
$C(sp^2)$=X	1.22	1.54	1.67	-
bond energy (kcal/mol)				
H-X	111	87	75	63
C-X	86	69	59	-
O-X	35	63	55	-
O=X	119	124.7	101	92

2. Se(II) and Te(II) Derivatives

(a) Arylselenols, diaryl diselenides and their derivatives

Arylselenols are the most common organic seleniums and these compounds are used as starting materials for the preparation of numerous other selenium derivatives. Benzeneselenol is commercially available and the synthesis of arylselenols, in general, simply requires a reaction between an aryl Grignard or aryllithium reagent and elemental selenium, followed by treatment of the product with acid. Benzeneselenol is a stronger acid than thiophenol or phenol: pKa C_6H_5SeH, 5.9; PhSH, 6.5; PhOH, 10.0 and it is a better nucleophile (G. Guani, C. D. Erba and D. Spinelli, Gazz. Chim. Ital., 1970, **100**, 184)). Arylselenols can be utilized in a similar manner to thiophenol and react with olefins, aldehydes and ketones yielding addition products and also selenoacetals or -ketals. Arylselenols, as well as arylthiols, undergo facile oxidation, even with oxygen, to afford the corresponding diaryl diselenides which are converted further to several arylselenenyl derivatives. Benzeneselenolate anion (PhSeM; M=Li, Na, K), which is efficiently prepared from diphenyl diselenide ((PhSe)$_2$) (**1**) with NaH or KH in DMF is a stronger nucleophile than thiolate anion and hence it reacts with alkyl halides (RX) by the S_N2 substitution process to give the corresponding selenides PhSeR (A. Krief, M. Trabelsi and W. Dumont, Synthesis, **1992**, 933). Similarly, treatment of (PhSe)$_2$ with Na in THF under irradiation by ultrasound gives PhSeNa, which reacts with sulfonates, halides and epoxides affording selenides (S. Ley, I. A. O'Neill and M. R. Caroline, Tetrahedron, 1986, **42**, 5363). Arylselenomagnesium halides e.g., p-TolSeMgBr, obtained by the reactions of arylmagnesiumhalides with elemental selenium, can also be utilized for production of alkyl aryl selenides in good yields. (K. K. Bhasin et al., Indian J. Chem., Sect. A, 1991, **30A**, 635). The reaction rates and equilibrium constants for the exchange reactions between aryl methyl selenides and the benzenelselenolate anion have been studied. These obey the Hammett equation i.e. for the rates (ρ^+= +1.1 ± 0.1) and the equilibria (ρ_{eq}= +2.9 ± 0.1) at 90°C in sulfolane. A comparison between these results and the analogous reactions of thiophenolate anion and aryl methyl sulfides reveals that the selenide anion reacts much faster than the sulfide anion (see equation 1) (E. S. Lewis, T. I. Yousaf and T. A. Douglas, J. Am. Chem. Soc., 1987, **109**, 2152; E. S. Lewis, M. L. McLaughlin and T. A. Douglas, ibid., 1979, 101, 417).

$$Ar-M-Me + PhM^- \underset{k^-}{\overset{k^+}{\rightleftharpoons}} ArM^- + Ph-M-Me \qquad (eq.\ 1)$$

$$[M = Se, S;\ k_{eq} = k^+ / k^-]$$

In contrast to arylselenols, several other dicoordinated arylselenium derivatives such as diphenyl diselenide (**1**) (equation 2), benzeneselenenyl halides (**2**) PhSeX (X=Cl, Br) (equation 3) (G. H. Schmidt and D. G. Garatt, J. Org. Chem., 1983, **48**, 4169), phenyl selenocyanate (**3**) (equation 4) (S. Tomoda, Y. Takeuchi and Y. Nomura, Chem. Lett., **1981**, 1069; idem, Synthesis, **1985**, 212), trimethylsilyl benzeneselenolate (**4**) (equation 5) (M. R. Detty and M. D. Seidler, J. Org. Chem., 1981, **46**, 1283), phenylseleno benzenesulfonate (**5**) (equation 6) (T. G. Back and S. Collins, Tetrahedron Lett., 1980, **21**, 2213; T. G. Back, S. Collins and M. V. Krishna, Can. J.

Chem., 1987, **65**, 38; R. A. Gancarz and J. L. Kice, J. Org. Chem., 1981, **46**, 4899) are well known as phenylselenenylating agents which are more stable thermally than benezeneselenol. Among these compounds diphenyl diselenide, benzeneselenenyl chloride and bromide are commercially available. Beside these divalent compounds, there are several other dicoordinated selenium derivatives which can be used as efficient selenenylating reagents. For example, N-phenylselenophthalimide (NPSP) (**6**) can be prepared from benzeneselenenyl chloride through reactions with potassium N-chlorophthalimide or diphenyldiselenide in combination with N-chlorophthalimide as shown in equation 7 (K. C. Nicolaou et al., J. Am. Chem. Soc., 1979, **101**, 3704; idem, Tetrahedron, 1985, **41**, 4835).

$$\text{PhSeH} \xrightarrow{[O]} \text{Ph-Se-Se-Ph} \quad \text{(eq. 2)}$$
$$\mathbf{1}$$

$$\text{Ar-Se-Se-Ar} + \text{SO}_2\text{Cl} \xrightarrow{\text{CCl}_4,\ \text{r.t.}} \text{ArSeCl} + \text{SO}_2 \quad \text{(eq. 3)}$$
$$\mathbf{2}$$
[Ar = p-X-C$_6$H$_4$; X=H, Cl, Br, F, CH$_3$, CF$_3$, OMe, NO$_2$]

$$\text{PhSeCl} + \underset{20\%\text{excess}}{\text{Me}_3\text{SiCN}} \xrightarrow{\text{THF/r.t.}} \text{PhSeCN} + \text{Me}_3\text{SiCl} \quad \text{(eq. 4)}$$
$$\mathbf{3}\ (96\%)$$

$$\text{Ph-Se-Se-Ph} \xrightarrow[\text{THF}]{\text{Na}} \text{PhSeNa} \xrightarrow{\text{ClSiR}_3}$$
$$\text{PhSeH} \xrightarrow{\text{n-BuLi}} \text{PhSeLi} \xrightarrow{\text{ClSiR}_3} \text{PhSeSiMe}_3 \quad \text{(eq. 5)}$$
$$\mathbf{4}$$
$$\text{PhSeH} + \text{HSiR}_3 \xrightarrow{\text{Ph(Ph}_3\text{P)}_3\text{Cl}}$$

$$\text{RSO}_2\text{NHNH}_2 + \text{PhSeO}_2\text{H} \xrightarrow{\text{CH}_2\text{Cl}_2} \text{PhSeSO}_2\text{R} \quad \text{(eq. 6)}$$
$$\mathbf{5}\ (83\sim100\%)$$
[R = Ph, p-Tol, Mes, MeOPh]

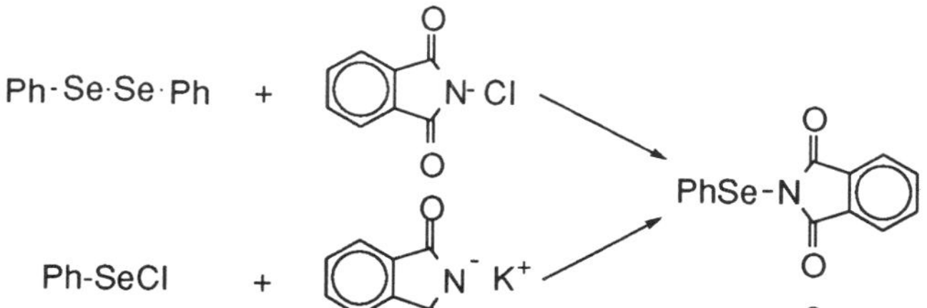

Ph-Se-Se-Ph + phthalimide-N-Cl ⟶ PhSe-N(phthalimide) (eq. 7)
Ph-SeCl + phthalimide-N⁻K⁺ ⟶ **6**

(b) Arylselenides

(i) Preparation
Normally, simple alkyl aryl or diaryl selenides (Ar-Se-R or Ar-Se-Ar') are prepared using either one of the following processes: (a) the substitution reactions of alkyl or aryl halides with appropriate selenolate anions in such solvents as THF or DMF in the presence of suitable catalysts. (b) The second standard procedure also involves the nucleophilic reactions of aryl or alkylselenenyl halides or diselenides with Grignard or organolithium reagents. Since selenols are readily oxidizable to diselenides, selenolate anions are generated from diselenides by reactions with appropriate reducing agents or selenophilic reagents. The selenolate anions thus generated are in general utilized in situ for further reactions with electrophiles such as alkyl, or sometimes, aryl halides. The following reagents are useful for the reduction of diaryldiselenides: sodium or potassium hydride in DMF or THF (equation 8) (A. Krief, M. Trabelsi and W. Dumont, Synthesis, **1992**, 933), $NaBH_4$ (K. K. Bhasin, A. Sandhu and R. D. Verma, Synth. React., Inorg. Met.-Org. Chem., 1988, **18**, 141), $NaBH_4$ in the presence of a phase transfer catalyst in EtOH at 5 °C and then with numerous alkyl halides, acyl chlorides, benzyl chlorides affording alkyl phenyl selenides in 76-100% yields (equation 9) (J. V. Weber et al., Synthesis, **1984**, 1044).

$$Ph\text{-}Se\text{-}Se\text{-}Ph \xrightarrow[DMF, THF]{MH\ (M = Na, K)} [RSeM] \xrightarrow{MeI} Ph\text{-}Se\text{-}Me \quad (5\sim88\%) \quad (eq.\ 8)$$

$$Ph\text{-}Se\text{-}Se\text{-}Ph \xrightarrow[resin/EtOH/5°C]{NaBH_4} PhSe^-Na^+ \xrightarrow{RX} Ph\text{-}Se\text{-}R \quad (76\text{-}100\%) \quad (eq.\ 9)$$

[RX= CH_2CHCH_2Br, AcCl, $PhCH_2Cl$]

$LiAlH_4$ has been used for the convenient ring opening and selenenylation of oxetane and related oxygen ring compounds. In the case of 2-methyloxetane this reagent gives 3-hydroxybutyl phenyl selenide (**7**) (equation 10) (K. Haraguchi et al., Chem. Lett., **1988**, 931).

$$\text{oxetane} + Ph\text{-}Se\text{-}Se\text{-}Ph \xrightarrow[dioxane/r.t.]{LiAlH_4} Ph\text{-}Se\text{-}(CH_2)_2\text{-}\underset{Me}{\overset{|}{CH}}\text{-}OH \quad (eq.\ 10)$$
7 (60%)

Diaryl diselenides are also reduced with metallic sodium under ultrasonic conditions in THF. Further treatment in situ with $(EtO)_2P(O)CH_2I$, for example, gives the Horner-Wittig type reagent, $(EtO)_2P(O)CH_2SePh$ (S. V.

Ley, I. A. O'Neill and C. M. R. Low, Tetrahedron, 1986, **42**, 5363). Thiourea dioxide (**8**) has been found to be a good reducing agent for both diaryl diselenides and diaryl ditellurides affording the corresponding aryl selenolate and tellurolate anions, which further react with alkyl halides to afford alkyl aryl selenides and tellurides (equation 11) (J. V. Commasseto et al., J. Organomet. Chem., 1987, **334**, 329). This reagent can also convert diaryl or alkyl aryl dichloroselenuranes and telluranes into the corresponding selenides and tellurides respectively (E. S. Lang and J. V. Commasseto, Synth. Commun., 1988, **18**, 301).

$$\text{Ar-M-M-Ar} \xrightarrow[\text{NH}_2 \quad \textbf{8}]{\text{HN=C-SO}_2\text{H}} [\text{Ar-MH}] \xrightarrow{\text{RX}} \text{Ar-M-R} \qquad (\text{eq. 11})$$

$$\left[\begin{array}{l} \text{M = Se, Te} \\ \text{Ar = Ph, p-MeOC}_6\text{H}_4\text{, p-Tol} \\ \text{RX = alkyl halide} \end{array} \right]$$

Diphenyl diselenide (**1**) reacts with Bu$_3$P in the presence of alkali to yield the benzeneselenolate anion which further reacts in situ in THF with alkyl halides, vinylic ketones and oxiranes to afford the corresponding selenides via a nucleophilic substitution or an addition (M. Sakakibara et al., Synthesis, **1992**, 377). Aryl halides react with MeSeLi in DMF via ipso-substitution and then undergo further nucleophilic substitution reactions with MeSeLi at the methyl carbon atom to generate lithium arylselenenates. These products react with alkyl halides, or ICN, to give ArSeR (R=alkyl, or CN) in high yields (M. Tiecco et al., J. Org. Chem., 1983, **48**, 4289). Similarly, the conversion of aryl halides to lithium arylselenolates or -tellurolates has been reported as a means of generating readily handleable selenium and tellurium anion sources (L. Engman and J. S. E. Hellberg, J. Organomet. Chem., 1985, **296**, 357). The copper(I) catalyzed reduction of diaryl diselenides and subsequent treatment with styryl, or ethynyl halides, in DMF and HMPA is also a convenient procedure for preparation of aryl olefinic, acetylenic and diaryl selenides in high yields (A. Braga et al., Tetrahedron Lett., 1993, **34**, 393; T. Ogawa, K. Hayami and H. Suzuki, Chem. Lett., **1989**, 769; N. V. Kondratenko et al., Synthesis, **1985**, 667). The reaction of diphenyl diselenide (**1**) with bis(triphenyltin)telluride (**9**) in the presence of fluoride ion generates the benzeneselenenate anion, which reacts in situ with alkyl halides to give alkyl phenyl selenides in good yields (equation 12) (C. J. Li and D. N. Harpp, Sulfur Lett., 1992, **15**, 155).

$$\text{Ph-Se-Se-Ph} + (\text{Ph}_3\text{Sn})_2\text{Te} + \text{F}^{\ominus} \xrightarrow{\text{RX}} \text{Ph-Se-R} \qquad (\text{eq. 12})$$
$$\quad \textbf{1} \qquad\qquad \textbf{9}$$
$$[\text{R = alkyl}]$$

Trimethylsilyl benzeneselenolate (**4**) and related trimethylsilyl derivatives are also utilized as modified selenenate anions (equation 13) (M. Yoshimatsu et al., Synlett., **1993**, 121; M. Sakakibara et al., ibid., **1992**, 965; N. S. Simpkins, Tetrahedron, 1991, **47**, 323; N. Miyoshi et al., Synthesis, **1988**, 175; S. Tomoda, Y. Takeuchi and Y. Nomura, ibid., **1985**, 212).

$$C_5H_{11}CH(OMe)_2 + Ph\text{-}Se\text{-}SiMe_3 \xrightarrow[-25\,°C]{Me_3SiOSO_2CF_3} Ph\text{-}Se\text{-}\underset{OMe}{CH}\text{-}C_5H_{11} \quad \text{(eq. 13)}$$
$$\text{(4)} \qquad\qquad 92\%$$

N-Phenylselenophthalimide (NPSP) (**6**) is a convenient selenenylating agent (T. Toru, S. Fujita and E. Maekawa, J. Chem. Soc., Chem. Commun., **1985**, 1082; T. G. Back and D. J. McPhee, J. Org. Chem., 1984, **49**, 3842). Allyl-, benzyl-, sec- and tert-alcohols can be converted into the corresponding selenides (**10**) upon simple treatment with aryl or alkyl selenols in the presence of $ZnCl_2$ or a strong acid such as H_2SO_4. The reactions are highly stereo- and regioselective (equation 14) and the reactivity is in the following order; allyl> tert-> sec-> prim- (M. Renard and L. Hevesi, Tetrahedron Lett., 1985, **26**, 1885; M. Clarembeau and A. Krief, ibid., 1984, **25**, 3625).

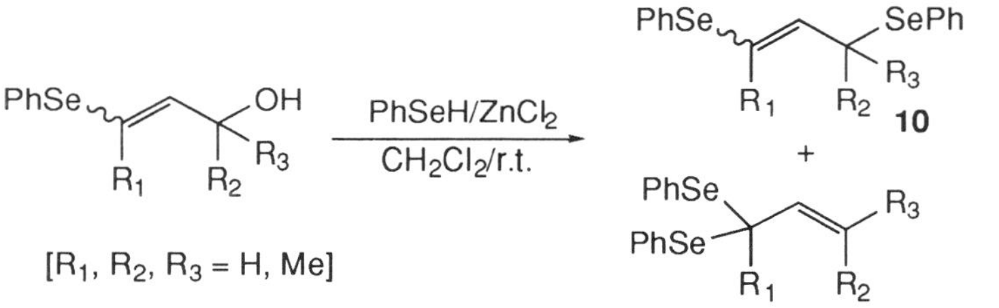

Diaryl (**12**) and alkyl aryl selenides are obtained through the reactions of aryl arylazo sulfones (**11**) with aryl, or alkyl selenolates, respectively. The necessary starting materials are formed from arene diazonium fluoroborates by treatment with sodium aryl sulfinates (M. J. Evers, L. E. Christiaens and M. J. Renson, J. Org. Chem., 1986, **51**, 5196). Similar reactions with aryl tellurides afford diaryl tellurides (**13**) (equation 15).

$$\text{Y-C}_6\text{H}_4\text{-NH}_2 \xrightarrow[\text{p-Tol-SO}_2\text{Na}]{\text{HNO}_2, \text{HBF}_4} \text{Y-C}_6\text{H}_4\text{-N}_2\text{SO}_2\text{-Tol-p} \quad \mathbf{11}$$

(eq. 15)

$$\mathbf{11} \xrightarrow[\text{r.t.}]{\text{CH}_3\text{CN}} \begin{cases} \text{RSeLi, R = Me, n-Bu} \rightarrow \text{Y-C}_6\text{H}_4\text{-SeR} \\ \text{PhSeLi} \rightarrow \text{Y-C}_6\text{H}_4\text{-SePh} \quad \mathbf{12} \\ \text{PhTeLi} \rightarrow \text{Y-C}_6\text{H}_4\text{-TePh} \quad \mathbf{13} \end{cases}$$

Se: 25-70 %
Te: 32-62 %

[Y = 2-COOCH$_3$, H, 3-Cl, 4-NO$_2$, 4-COCH$_3$, 4-Br]

(ii) Structures and reactions

Recently, the structures of various simple monoaryl selenides have been investigated using ^{1}H, ^{13}C and ^{77}Se NMR techniques. The chemical shifts of these nuclei can be correlated with the coordination state at the central selenium atom so that it is possible to distinguish between dicoordinate, tricoordinate and higher coordinations (W. Nakanishi et al., Phys. Org. Chem., 1990, **3**, 358). Long-range coupling between selenium and other atoms or ^{77}Se-^{77}Se homo-coupling are used for the determination of the conformation of aryl methyl selenides, or di-, tri-substituted bisselenides. ^{77}Se-^{77}Se Vicinal coupling reveals the stereochemical nature of addition reactions of diselenides to triple bonds (T. Schaefer and G. H. Penner, Can. J. Chem., 1988, **66**, 1641; I. Johannsen, L. Henriksen and H. Eggert, J. Org. Chem., 1986, **51**, 1657.). Electrochemical oxidation of diaryl and aryl methyl selenides has been performed anodically and the linear free energy relationship has been established between the oxidation potentials and the Hammett substituent constants (L. Engman et al., J. Chem. Soc., Perkin Trans. 2, **1992**, 1309).

The following three representative reactions of dicoordinated arylselenium compounds have been well documented. (1) Addition reactions of dicoordinated arylselenium compounds such as arylselenenyl halides and diaryl diselenides to olefins and acetylene derivatives. (2) Substitution of the arylselenenyl group by strong bases such as aryllithiums leading to carbanions, which can then be reacted with electrophiles. (3) Oxidation of selenides to selenoxides, followed by subsequent syn-elimination to provide numerous olefinic and acetylenic compounds. In the case of allyl selenides, oxidation is followed by [2,3]sigmatropic allylic rearrangement as a typical reaction. A combination of these reactions can be used for the production of numerous selenides bearing functional groups which makes it a very valuable synthetic procedure.

(1) Addition reactions

The conditions employed for addition reactions to alkenes are either ionic or radical in nature. In the ionic case either episelenenium cations (**14**), or episelenuranes (**15**) intermediates are involved, as shown in equation 16. The stereochemistry observed is usually anti-Markovnikov, but the regioselectivity depends on the nature of the substrates and conditions employed in the reactions (G. H. Schmidt and D. G. Garratt, Tetrahedron Lett., 1983, **24**, 5299; T. Ho. Pak and R. J. Kolt, Can. J. Chem., 1982, **60**, 663).

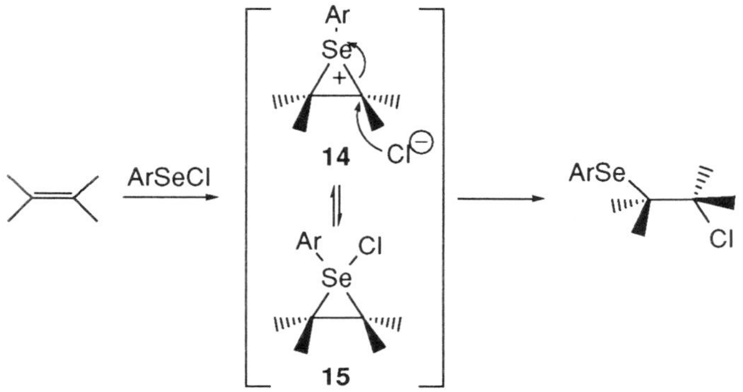

$$\text{(eq. 16)}$$

There are so many examples of addition reactions in the literature that only two typical illustrations necessary, i.e. those using the additions of PhSeSO$_2$Ph and PhSeBr to cyclohexene (equation 17). These lead to selenides (**16**) and (**17**) respectively, which when treated with oxidants afford the corresponding cyclohexene derivatives (**18**) and (**19**) via syn-elimination (T. G. Back and S. Collins, Tetrahedron Lett., 1980, **21**, 2215; P. G. Gassman, A. Miura and T. Miura, J. Org. Chem., 1982, **47**, 951).

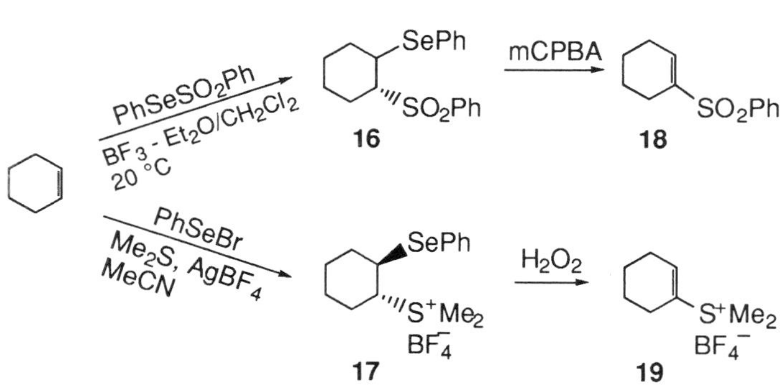

$$\text{(eq. 17)}$$

The addition of PhSeCl and PhSeBr to alkynes and allenes has been presented to proceed via both Markownikoff and anti-Markownikoff fashions and in the case of an allene the PhSe group adds to the central carbon atom resulting in the formation of vinylic selenides (**20**) (equation 18) (A. Toshimitsu, S. Uemura and M. Okano, J. Chem. Soc. Chem. Commun., **1982**, 965; D. G. Garratt, P. L. Beaulieu and V. M. Morisset, Can. J. Chem., 1981, **59**, 927; S. Halzy and L. Hevesi, Tetrahedron Lett., 1983, **24**, 2689.).

$$R_1\text{-CH=C=CH-}R_2 \xrightarrow{\text{PhSeCl}} R_1\text{-CH(Cl)-C(SePh)=CH-}R_2 + R_1\text{-CH=C(SePh)-CH(Cl)-}R_2 \quad (\text{eq. 18})$$

20

The preparations of vinylic selenides from acetylenic compounds and arylselenenyl halides reported before 1983 have been reviewed (J. V. Commasseto, J. Organomet. Chem., 1983, **253**, 131). The generation and addition of PhSeF (**21**) to olefinic double bond has been reported recently as shown in equation 19 (K. Uneyama and M. Kanai, Tetrahedron Lett., 1990, **31**, 3583; J. R. McCarthy, D. P. Matthews and C. L. Barney, Tetrahedron Lett., 1990, **31**, 973).

$$\text{Ph-Se-Se-Ph} + \text{XeF}_2 \xrightarrow[-20°C]{\text{CH}_2\text{Cl}_2} [\text{PhSeF}] \xrightarrow{\text{CH}_2=\text{CHCH}_2\text{CN}} \text{NC-CH}_2\text{-CHF-CH(SePh)-CH}_2\text{-CN} \quad (\text{eq. 19})$$

21 72 %

Direct addition of diphenyl diselenide (equation 20), to olefinic or acetylenic bonds under photo-irradiation have been performed. Such reactions proceeds via the initial formation of the PhSe• radical (A. Ogawa et al., Tetrahedron Lett., 1992, **33**, 1329; T. Masawaki et al., Chem. Lett., **1987**, 2407; A. Ogawa et al., Tetrahedron Lett., 1990, **31**, 5931). Similar reactions occur between benzenselenol and unsaturated compounds in the presence of oxygen.

$$R\text{-CH=C=CH}_2 + (\text{Ph-Se})_2 \xrightarrow[\text{CDCl}_3/15°C]{h\nu} R\text{-CH=C(SePh)-CH}_2\text{SePh} \quad (\text{eq. 20})$$

Under similar conditions PhCOSSePh and PhSeSO$_2$Ph both cleave to give the PhSe• radical, which can further react if olefins are present to give adducts of the type (**22**) and (**23**) (equation 21) (T. Toru et al., J. Chem. Soc., Perkin Trans. 1, **1989**, 1927; M. D. Bachi and E. Bosch, J. Org.

Chem., 1992, **57**, 4697; Y. H. Kang and J. L. Kice, J. Org. Chem., 1984, **49**, 1507; T. G. Back, S. Collins and R. G. Kerr, J. Org. Chem., 1983, **48**, 3077).

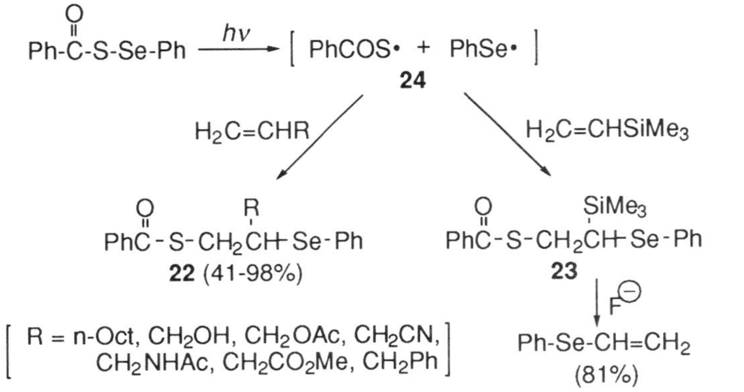

$$\text{Ph-C(=O)-S-Se-Ph} \xrightarrow{h\nu} [\text{PhCOS}\cdot + \text{PhSe}\cdot]$$
24

With $H_2C=CHR$:
$$\text{PhC(=O)-S-CH}_2\text{CH(R)-Se-Ph}$$
22 (41-98%)

[R = n-Oct, CH_2OH, CH_2OAc, CH_2CN, CH_2NHAc, CH_2CO_2Me, CH_2Ph]

With $H_2C=CHSiMe_3$:
$$\text{PhC(=O)-S-CH}_2\text{CH(SiMe}_3\text{)-Se-Ph}$$
23

$\downarrow F^{\ominus}$

Ph-Se-CH=CH_2 (81%)

(eq. 21)

Two new processes to generate directly the benzeneselenenyl cation(PhSe$^+$) (**25**) and the subsequent addition of this species to olefinic bonds have been reported. Benzeneselenenyl chloride reacts with silver salt of a strong acid such as CF_3SO_3H, affording PhSe$^+$SO$_3$CF$_3^-$. PhSe$^+$PF$_6^-$, PhSe$^+$SbF$_6^-$, PhSe$^+$SO$_3$Ar$^-$ are also prepared in a similar manner and when these salts are reacted with alkenoic acids lactones may form (equations 22). These products can be hydrolysed, or ring-opened in various ways. With alkynes similar additions of PhSe$^+$ occur leading to selenenylalkenes (equation 23) (W. P. Jackson, S. V. Ley and A. J. Whittle, J. Chem. Soc. Chem. Commun., **1980**, 1173; S. Murata and T. Suzuki, Chem. Lett., **1987**, 849, idem, Tetrahedron Lett., 1987, **28**, 4297, 4415; idem, Tetrahedron, 1989, **45**, 6819; T. G. Back and K. R. Murralidharam, Tetrahedron Lett., 1990, **31**, 1957; T. G. Back and M. V. Krishna, J. Org. Chem., 1988, **53**, 2533); T. G. Back and K. R. Muralidharan, ibid., 1991, **56**, 2781; M. Yoshida et al., Bull. Chem. Soc. Jpn., 1991, **64**, 416).

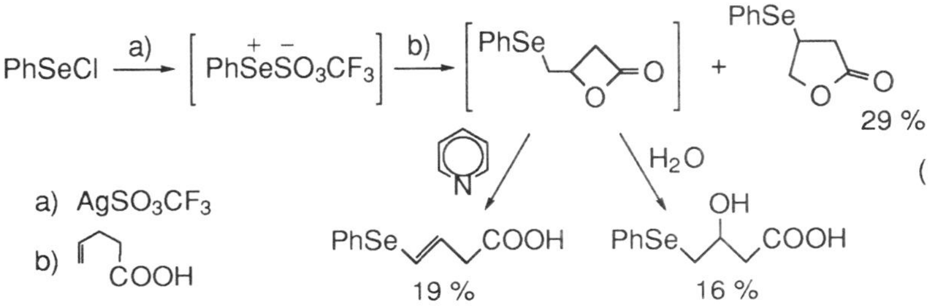

$$\text{PhSeCl} \xrightarrow{a)} [\text{PhSe}^+\text{SO}_3\text{CF}_3^-] \xrightarrow{b)} [\text{PhSe-}\square\text{-O}] + \text{PhSe-}\square\text{-O (29%)}$$

a) $AgSO_3CF_3$
b) $\square$-COOH

Via pyridine: PhSe-CH=CH-COOH (19 %)
Via H_2O: PhSe-CH$_2$-CH(OH)-COOH (16 %)

(eq. 22)

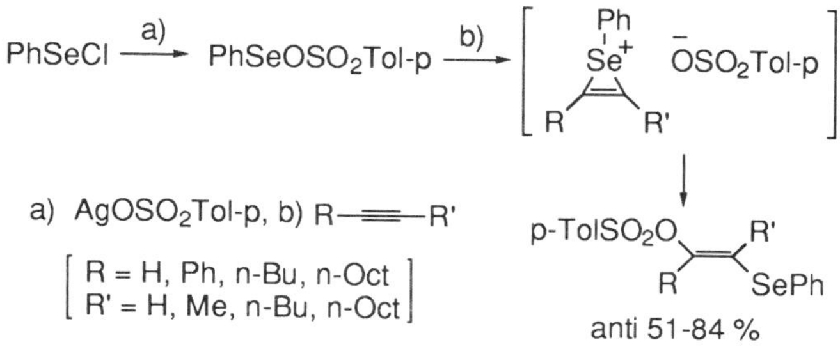

$$\text{PhSeCl} \xrightarrow{a)} \text{PhSeOSO}_2\text{Tol-p} \xrightarrow{b)} \left[\begin{array}{c} \overset{Ph}{\underset{R}{\overset{|}{Se}}}\overset{+}{} \quad \bar{O}SO_2\text{Tol-p} \\ \underset{R}{\diagdown}=\underset{R'}{\diagup} \end{array} \right]$$

(eq. 23)

a) AgOSO$_2$Tol-p, b) R≡R'

$\left[\begin{array}{l} R = H, Ph, n\text{-Bu}, n\text{-Oct} \\ R' = H, Me, n\text{-Bu}, n\text{-Oct} \end{array} \right]$

p-TolSO$_2$O, R'
 R   SePh

anti 51-84 %

The in situ generation of PhSe$^+$ (**25**) from diphenyl diselenide by treatment with ammonium persulfate ((NH$_4$)$_2^+$S$_2$O$_8^{2-}$) has been reported and the mechanism for the reaction has been demonstrated to proceed via an initial one electron-transfer oxidation from the diselenide to the persulfate anion as shown in equation 24. Once formed PhSe$^+$ can be reacted with a variety of alkenes as indicated in equation 25, and it is important to note how nucleophilic attack, either intramolecularly or intermolecularly, can be used to widen the scope and applicability of this procedure (M. Tiecco et al., Tetrahedron Lett., 1989, **30**, 1417; idem, Tetrahedron, 1988, **44**, 2261, 2273: idem, ibid., 1989, **45**, 6819; idem, J. Org. Chem., 1990, **55**, 429).

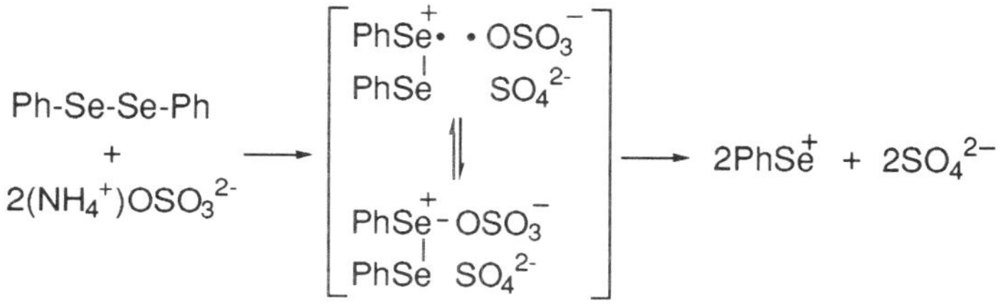

$$\text{Ph-Se-Se-Ph} + 2(\text{NH}_4^+)\text{OSO}_3^{2-} \longrightarrow \left[\begin{array}{l} \text{PhSe}^{+\bullet} \cdot \text{OSO}_3^- \\ \text{PhSe} \quad \text{SO}_4^{2-} \\ \quad \updownarrow \\ \text{PhSe}^+ - \text{OSO}_3^- \\ \text{PhSe} \quad \text{SO}_4^{2-} \end{array} \right] \longrightarrow 2\text{PhSe}^+ + 2\text{SO}_4^{2-}$$

(eq. 24)

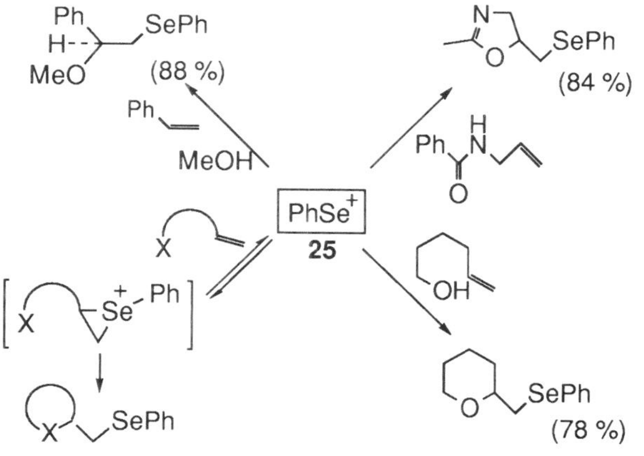

(eq. 25)

Products shown:
- Ph, H, MeO, SePh (88 %) from Ph=, MeOH
- oxazoline with SePh (84 %) from Ph-NH-allyl amide
- tetrahydropyran with SePh (78 %) from cyclohexenol (OH)
- cyclic X-SePh product from intramolecular attack via [X···Se$^+$-Ph] intermediate

A similar method can be used for the oxidation of active methylene compounds to dimethylacetals (equation 26) (M. Tiecco et al., J. Org. Chem., 1991, **56**, 5207). Phenyliodosodiacetate [PhI(OAc)$_2$] can also be used as an oxidizing agent for the generation of PhSe$^+$ (M. Tingoli et al., J. Org. Chem., 1991, **56**, 6809; idem, Synlett, **1993**, 211).

$$PhSe^+ + R\overset{O\quad O}{\underset{}{\diagdown\diagup}}R' \xrightarrow{MeOH} R\overset{O\quad O}{\underset{MeO\quad OMe}{\diagdown\diagup}}R' \qquad (eq.\ 26)$$

[R, R' = Me, Et, Ph, CO$_2$Me]   (15-75 %)

Other methods for the formation of PhSe$^+$ include the electrochemical oxidation of diphenyl diselenide and also its photolysis in the presence of a sensitizer. If these reactions are carried out in the presence of alkenols then cyclic ethers (**26**) are formed (equation 27) (K. Uneyama, M. Ono and S. Torii, Phosphorus and Sulfur, 1983, **16**, 6819; M. Lj. Mihailovic, S. Konstantinovic and R. Vukicevic, Tetrahedron, 1987, **28**, 4343, 6511; G. Pandey and B. B. V. S. Sekhar, J. Org. Chem., 1992, **57**, 4019; G, Pandey, B. B. V. S. Sekhar and U. T. Bhalerao, ibid., 1990, **112**, 5650).

$$Ph\text{-}Se\text{-}Se\text{-}Ph \xrightarrow[h\nu]{^1DCN^*} [Ph\text{-}Se\text{-}Se\text{-}Ph]^{+\bullet} \rightleftharpoons 2PhSe^+ \qquad (eq.\ 27)$$

PhSe$^+$ + (alkenol, n = 3, 4) → [PhSe intermediate] → **26** (60-72 %)

(2) Reactions of selenides with organolithium reagents (ionic and radical substitution of selenenyl group)

Normally selenides do not react with organolithium reagents, however in the case of selenides bearing active ligands such as allylic, acetylenic, or benzylic groups, C-Se bond cleavage takes place to afford the corresponding lithiated compounds. Notably, bisselenides such as selenoacetals or ketals react with organolithiums to undergo facile Se-Li exchange reactions. In such cases the remaining selenium atom can stabilize the α-carbanion formed. A comparison of pKa values of carbanions derived from several sulfides and selenides, shows that α-sulfenyl carbanions have lower pKa values than those of selenides (A. Krief, Tetrahedron, 1980, **36**, 2531; A. Krief, "The chemistry of organic selenium and tellurium compounds", ed. S. Patai, John Wiley & Sons, New York, 1987, Vol. 2, Ch. 17, p. 675; C. Paulmier, "Selenium reagents and intermediates in organic syntheses", Pergamon Press, Oxford, 1986, Ch. IX, p. 256). The pKa values of α-phenylthioacetophenone and α-benzeneselenenylacetophenone are 17.3 and 18.8 (in dimethyl sulfoxide) respectively, while those of bis diphenylthiomethane and its selenium analogue are 32 and 35, respectively (F. G. Bordwell et al., J. Org. Chem., 1977, **42**, 326). The structures of α-selenenyl carbanions have been studied intensively by Seebach et al. (D. Seebach and N. Peleties, Chem. Ber., 1972, **105**, 511; idem, ibid., 1975, **108**, 314). Normally the generation of α-selenenyl carbanions (**27**) from bisselenides or monoselenides bearing an α-proton requires either nucleophilic (or selenophilic alkyl lithium reagents), or strongly basic reagents, such as lithium diisopropylamide (LDA) (equation 28).

$$\underset{H}{\overset{Li}{\diagdown}}\overset{RLi}{\longleftarrow}\underset{H}{\overset{SeR}{\diagdown}}\overset{R'_2NLi}{\longrightarrow}\underset{Li}{\overset{SeR}{\diagdown}}\overset{RLi}{\longleftarrow}\underset{SeR'}{\overset{SeR'}{\diagdown}} \quad \text{(eq. 28)}$$

$$\textbf{27} \quad \begin{bmatrix} R = Bu, Ph \\ R' = i\text{-}Pr \end{bmatrix}$$

For the removal of a benzeneselenenyl group, butyllithium is usually employed this causeheterolytic fission of the C-Se bond and the formation of alkyl lithiums. Once generated these can be treated in situ with appropriate electrophiles. Typical reactions are illustrated in equations 29 and 30 (M. Clarembeau and A. Krief, Tetrahedron Lett., 1986, **27**, 1719, 1723; D. L. J. Clive, J. Chem. Soc. Chem. Commun., **1985**, 1205; T. Di. Gianberardino et al., Tetrahedron Lett., 1983, **24**, 3413).

$$PhSeCR^1R^2R^3 \xrightarrow[THF]{BuLi} R^2\underset{R^3}{\overset{R^1}{\diagdown}}CLi \xrightarrow{R^3\overset{O}{\diagdown}R^4} R^2\underset{R^3}{\overset{R^1}{\diagup}}\underset{H}{\overset{R^3}{\diagdown}}R^4 \quad \text{(eq. 29)}$$

$$\begin{bmatrix} R^1 = H, Me, Pr \\ R^2 = H, Me, Me_2CH \\ R^3 = Ph, o\text{-}ClC_6H_4, p\text{-}ClC_6H_4, p\text{-}FC_6H_4, p\text{-}MeOC_6H_4, PhCH_2 \end{bmatrix}$$

$$PhCH_2SePh \xrightarrow{BuLi} PhCH_2Li \xrightarrow{PhCHO} PhCH_2CH(OH)Ph \quad \text{(eq. 30)}$$
$$(86\%)$$

Bisselena-, bisthia- and selenathia-ketals of 3-hexanone have been prepared and a comparison of the stability of the corresponding carbanions formed, after treatment with butyllithium undertaken. The order of stability is as follows: PhS > PhSe > MeSe. This is paralleled by the selectivities and reactivities of these anions towards electrophiles (A. Krief, W. Dumont and B. E. Clarembeau, Tetrahedron, 1989, **45**, 2005, 2023). Conformational stability of selenenyl carbanions using 1,1-bisphenylselenenylketals of 4-t-butylcyclohexanone has been tested. These compounds are shown to afford the axial carbanions preferentially on treatment with n-BuLi and the stereochemical integrity of such reactions is retained when the anions are quenched with electrophiles (equation 31) (A. Krief et al., Tetrahedron Lett., 1989, **30**, 5635; A. Krief et al., ibid., 1991, **32**, 3231).

$$\text{(cyclohexane)}-\text{SeR}_1,\text{SeR}_2 \xrightarrow{\text{n-BuLi}}_{\text{THF}/-78°C} \left[\text{(cyclohexane)}-\text{Li},\text{SeR}_2 \right] \xrightarrow{\text{X-B}} \text{(cyclohexane)}-\text{X},\text{SeR}_2 \qquad \text{(eq. 31)}$$

$$\left[\begin{array}{l} R_1 = Ph, Me, CD_3, R_2 = Ph, Me \\ X\text{-}B = H\text{-}NH_3Cl, R_1Se\text{-}SeR_1 \; (R_1 = C_6D_5, Me, CD_3) \end{array} \right]$$

When there is a choice of aryltellurium, or arylselenium, groups the former is removed more readily. This is due to the lower electronegativity of tellurium. Thus when benzeselenenylbenzenetellurylmethane (PhSeCH$_2$TePh) is reacted with BuLi benzeselenenyllithium is the sole product (C. A. Brandt et al., J. Chem. Res., Synop., **1983**, 156). Monoaryl selenides bearing benzyl, allyl, propargyl and ethynyl groups undergo deselenenylation on treatment with organolithium reagents (equation 32) (M. Clarembeau and A. Krief, Tetrahedron Lett., 1985, **26**, 1093; A. L. Braga, J. V. Commasseto and N. Petragnani, Synthesis, **1984**, 240; K. Nishi et al., Chem. Pharm. Bull., 1992, **40**, 288).

$$R\text{—}\equiv\text{—}SeAr \xrightarrow[\text{THF}]{\text{n-BuLi}} \left[R\text{—}\equiv\text{—}Li \right] \xrightarrow{E^+} R\text{—}\equiv\text{—}E \qquad \text{(eq. 32)}$$
$$(51\text{-}78\%)$$

$$\left[\begin{array}{l} Ar = Ph, 3\text{-}CF_3\text{-}Ph, 4\text{-}Cl\text{-}Ph \\ E = H, Me, COCH_3, COOH, C(CH_3)_2OH \end{array} \right]$$

The removal of an arylselenenyl group can be performed using not only strong anionic species such as BuLi, but also radical, or cationic species in the presence of a radical initiator, or photolysis conditions as shown typically in equations 33, 34 and 35 (radical type reactions: V. H. Rewal et al., J. Org. Chem., 1991, **56**, 5245; P. Dowd and S. C. Choi, Tetrahedron, 1991, **47**, 4847; J. H. Byers, T. G. Gleason and K. S. Knight, J. Chem. Soc. Chem. Commun., **1991**, 354; J. H. Byers and Harper, Tetrahedron Lett., 1992, **33**, 6953; J. H. Gyers and B. C. Lane, ibid., 1990, **31**, 5697; D. P. Curran and G. Thoma, J. Am. Chem. Soc., 1992, **114**, 4436; T. K. Sakar, S. K. Ghosh and T. K. Satapathie, Tetrahedron, 1990, **46**, 1885; H. J. Reich and J. W. Ringer, J. Org. Chem., 1988, **53**, 455; D. L. Boger and R. J. Mathvink, J. Org. Chem., 1988, **53**, 3377, cationic type reactions: G. H. Schmidt and D. G. Garratt, Tetrahedron, 1985, **41**, 4787; C. G. Francisco et al., Tetrahedron Lett., 1984, **25**, 1621; A. M. Morrella and A. D. Ward, Tetrahedron Lett., 1984, **25**, 1197).

$$\underset{\text{SePh}}{\text{CH}_3\text{CO}\diagdown\diagup\text{CO}_2\text{CH}_3} + \diagdown(\text{CH}_2)_5\text{CH}_3 \xrightarrow[\text{42 hr}]{h\nu} \underset{\text{CH}_3\text{CO}}{\text{CH}_3\text{CO}_2 \quad \text{SePh}}\diagdown\diagup\text{C}_6\text{H}_{13} \quad \text{(eq. 33)}$$
(78 %)

$$\underset{[\text{X = CN, CO}_2\text{Et, COMe}]}{\text{PhSeCH}_2\text{X}} + \diagdown(\text{CH}_2)_5\text{CH}_3 \xrightarrow[\text{16-60 hr}]{h\nu} \text{X}\diagdown\underset{\text{SePh}}{\diagup}\text{C}_6\text{H}_{13} \quad \text{(eq. 34)}$$
(50-72 %)

$$\underset{\underset{\text{PhSe:}\longrightarrow\text{SeCl}}{|\qquad\quad|}}{\text{ClCH}_2\text{CH}_2 \quad \text{Ph}} \underset{\text{Path x}}{\rightleftharpoons} \underset{\underset{\text{PhSe}-\text{SePh}\;\text{Path y}}{|}}{\text{ClCH}_2\text{CH}_2 \overset{y\diagup\text{Cl}^-}{\underset{x}{(}}} \longrightarrow \underset{\text{ClCH}_2\text{CH}_2\text{Cl}}{\text{Ph-Se-Se-Ph} \atop +} \quad \text{(eq. 35)}$$

Aryl, alkyl, cycloalkyl and steroidal selenides are converted readily and quantitatively into the starting hydrocarbons on treatment with $NiCl_2/NaBH_4$ indicating a convenient process for removing the selenenyl group by the formation of a selenium-metal complex (T. G. Back et al., J. Org. Chem., 1988, **53**, 3815).

(3) Oxidation of selenides (syn-elimination, [2,3]- and [1,3]-sigmatropic rearrangement, solvolysis)

The oxidation of selenides should give the corresponding selenoxides, but with the exception of a few compounds, such as those with no β-protons (e.g. dimethyl, aryl methyl and diaryl selenoxides), these products are unstable. Selenoxides having at least one β-proton undergo syn-elimination quite rapidly to afford alkenes, or alkynes, even below 0 °C. This contrasts markedly with chemical behavior of sulfoxides which are thermally stable. Indeed, the propensity of selenoxides to undergo this reaction has been widely utilized in organic synthesis for the creation of numerous alkenes. Besides the facile syn-elimination of selenoxides, both allylic and propargylic selenoxides undergo [2,3] sigmatropic oxygen migration which converts selenoxides into allylic alcohols, allenyl derivatives and propargyl alcohols quite readily in high yields. These two characteristic reactions of organic selenium compounds have attracted most attention and consequently the oxidation of selenides is very important (see below).

The mechanism for the syn-elimination of selenoxides was studied by Kwart et al. and Reich et al. On the basis of kinetic investigations and isotope effects using allyl phenyl selenoxides, the rate determining step for the elimination is shown to be an initial β-proton abstraction by selenoxide oxygen atom with a concerted C-Se bond cleavage. The proton preferentially removed is, in the order: allylic > propargylic > methyl > methylene > methine (L. D. Kwart, A. G. Morgan and H. Kwart, J. Am. Chem. Soc., 1981, **103**, 1232; H. J. Reich and W. W. Willis, ibid., 1980, **102**, 5697). Furthermore, Kice reported that the rates of syn-elimination of the relatively stable compounds o-benzoylphenyl ethyl selenoxide (**28**) and o-nitrophenyl ethyl selenoxide (**29**) obey first order kinetics. The $\Delta H^{\neq}$ and $\Delta S^{\neq}$ values obtained are: (**28**), $\Delta H^{\neq}$=17.6 Kcal/mol, $\Delta S^{\neq}$=-12.9 e.u.: (**29**), $\Delta H^{\neq}$=19.5 Kcal/mol, $\Delta S^{\neq}$=-7.1 e.u. respectively. For the syn-elimination of phenyl n-propyl sulfoxide (**30**) the relevant data are: $\Delta H^{\neq}$=25-28 Kcal/mol, $\Delta S^{\neq}$=-11 ~ -16 e.u.. Thus the enthalpy of the syn-elimination of selenoxides is ca. 8 ~ 10 Kcal/mol less than that for the analogous thermal decomposition of sulfoxides. The entropy values of selenoxides are closely comparable to those for the pyrolytic decomposition of sulfoxides and hence the results are consistent with a mechanism proposed for syn-eliminations and summarized in Scheme 2. Here the role of the oxygen atom in deprotonating an adjacent carbon centre is emphasized and the geometry of the starting selenoxide is crucial in determining whether an alkyne (**A**), or an allene (**B**) is formed (S.-I. Kang and J. L. Kice, J. Org. Chem., 1985, **50**, 2968; D. W. Emerson and T. J. Korniski, ibid., 1969, **34**, 4115).

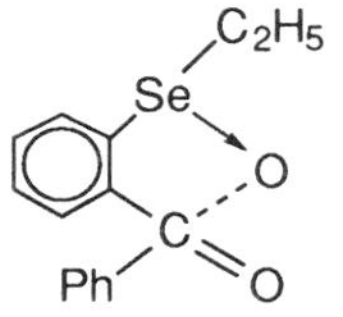

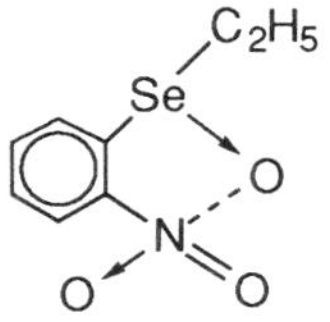

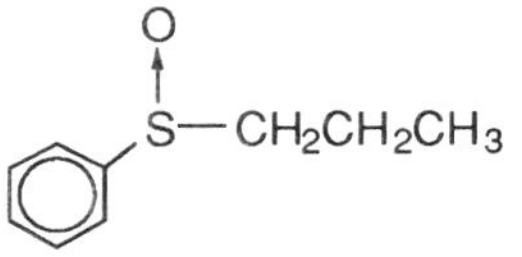

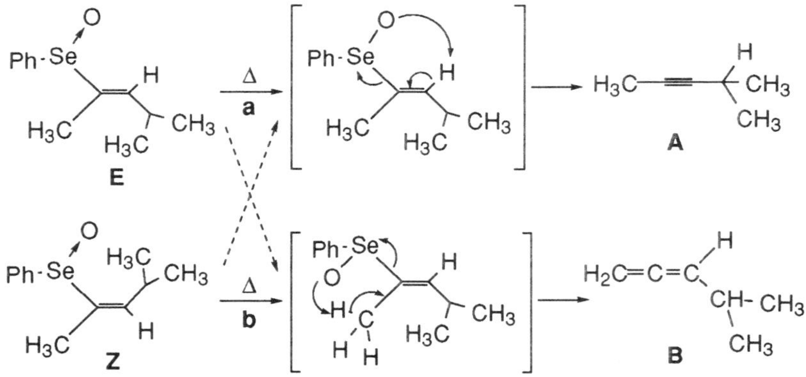

	A : B
E	85 : 15
Z	3 : 81

(Scheme 2)

There are numerous examples of syn-eliminations of selenoxides and it is necessary to summarize many results in the following account. Earlier, Back and his coworkers reported on the radical addition reactions of Se-Ph-p-toluene phenylselenenyl-4-toluenesulfonate (5) to acetylenes. These yields ethenes bearing both selenenyl and sulfonyl groups, which on oxidation with m-chloroperbenzoic acid (mCPBA) give the corresponding selenoxides. On heating these products decompose to afford ethynyl sulfones (31) (equation 36). Similarly, 1,3-butadiene can be converted by treatment of PhSeSO$_2$Ph, and subsequent treatment of the 1,2-adducts obtained with mCPBA into 2-sulfonyl substituted 1,3-dienes in high yields (T. G. Back, S. Collins and R. G. Kerr, J. Org. Chem., 1983, **48**, 3077; J. E. Baeckvall, C. Najera and M. Yus, Tetrahedron Lett., 1988, **29**, 1445).

$$\text{PhSeSO}_2\text{Tol-p} + \text{Ph-C} \equiv \text{C-H} \longrightarrow \begin{array}{c} \text{PhSe} \quad \text{H} \\ \diagup\!\!\!= \\ \text{Ph} \quad \text{SO}_2\text{Tol-p} \end{array}$$

5

$$\xrightarrow{\text{mCPBA}} \begin{array}{c} \text{O} \\ \uparrow \\ \text{PhSe} \quad \text{H} \\ \diagup\!\!\!= \\ \text{Ph} \quad \text{SO}_2\text{Tol-p} \end{array} \xrightarrow{\Delta} \text{Ph-C} \equiv \text{C-SO}_2\text{Ar}$$

(eq. 36)

31 (88-91 %)

There are several other papers dealing with the syn-elimination of selenoxides to give olefins and generally it is observed that the most acidic β-proton in the substrate is the one which is eliminated preferentially (V. J. Jephcote and E. J. Thomas, Tetrahedron Lett., 1985, **26**, 5327; idem, J. Chem. Soc., Perkin Trans. 1, **1991**, 429; S. Uemura et al., J. Chem. Soc. Chem. Commun., **1985**, 1037; R. S. Brown, S. C. Eyley and P. J. Parson, ibid., **1984**, 438). These reactions are rapid and when selenides bearing terminal sulfonyl group are oxidised under conditions likely to cause asymmetric induction, using either the Sharpless, or the Davis oxidation procedures no optically active selenoxides were obtained. Instead, syn-elimination occurred directly and allenes (e.g. **32**) were produced exhibiting a maximum 42 % optical activity (equation 37) (N. Komatsu et al., J. Org. Chem., 1993, **58**, 3697; idem, J. Chem. Soc. Chem. Commun., **1992**, 46).

As an aza-analogue of the syn-elimination process N-benzoylaminomethyl phenyl selenide (**33**) has been shown to undergo oxidation with either NaIO$_3$, or ozone, in methanol to give the corresponding selenone (**34**), or the methanolysis product N-benzoylaminomethyl methyl ether (**35**). The latter product is obtained via a syn-elimination of an intermediate selenoxide (equation 38) (B. P. Branchaud and P. Tsai; J. Org. Chem., 1987, **52**, 5475).

$$\underset{R}{\overset{H}{\diagup}}C=C\underset{SeAr}{\overset{CH_2SO_2Tol}{\diagdown}} \xrightarrow{i} \left[\underset{R}{\overset{H}{\diagup}}C=C\underset{*SeAr\downarrow O}{\overset{CH_2SO_2Tol}{\diagdown}} \right] \xrightarrow{ii} RCH=C=CHSO_2Tol\text{-}p$$
$$\textbf{32 (e.e.: 42 %)}$$

$$\left[\begin{array}{l} Ar = o\text{-}NO_2\text{-}Ph, 2,4\text{-}diNO_2\text{-}Ph, Ph, 2\text{-}Py \\ R = n\text{-}C_3H_7, n\text{-}C_7H_{15}, p\text{-}Tol\text{-}SO_2^- \end{array} \right] \quad \text{(eq. 37)}$$

i : oxidation

$$\left\{ \begin{array}{l} Ti(OiPr)_4 \\ (+)\text{-}DIPT \\ TBHT \end{array} \right. \left\{ \begin{array}{l} Ti(OiPr)_4 \\ (+)\text{-}DET \\ TBHT \end{array} \right. \left\{ \begin{array}{l} Ti(OiPr)_4 \\ (+)\text{-}DCHT \\ TBHT \end{array} \right.$$

[camphor-derived oxaziridine with Cl, Cl, NSO₃Ph] [camphor-derived sulfonyloxaziridine with Cl, Cl, S, N, O₂O]

ii : elimination   -[ArSeOH]

$$\underset{\textbf{33}}{Ph\overset{O}{\underset{H}{\overset{\|}{C}}}N\diagdown SePh} \xrightarrow{NaIO_4} \underset{\textbf{34}}{Ph\overset{O}{\underset{H}{\overset{\|}{C}}}N\diagdown SeO_2Ph}$$

$$\downarrow NaIO_4 \text{ or } O_3$$

$$\left[Ph\overset{O}{\underset{H}{\overset{\|}{C}}}N\underset{\downarrow O}{\diagdown}SePh \right] \xrightarrow{syn\text{-}Elim.} \left[Ph\overset{O}{\overset{\|}{C}}N=CH_2 \right] \xrightarrow{MeOH} \underset{\textbf{35 (89-98 %)}}{Ph\overset{O}{\underset{H}{\overset{\|}{C}}}N\diagdown OCH_3}$$

(eq. 38)

Oxidation of β-hydroxyethyl phenyl selenides in general gives a complex mixture of the products, however vinyl, aryl, acyl and iminyl substituted ether derivatives (**36**) give the corresponding enol ethers (equation 39) (M. Tiecco et al., J. Org. Chem., 1993, **58**, 1349.)

$$\underset{\underset{\textbf{36}}{OR'}}{RCHCH_2SePh} \xrightarrow{mCPBA} \left[R\text{-}\overset{H}{\underset{OR'}{\overset{|}{C}}}\text{-}CH_2\overset{O}{\overset{\uparrow}{Se}}\text{-}Ph \right] \longrightarrow \underset{R'O}{\overset{R}{\diagdown}}C=CH_2 \quad \text{(eq. 39)}$$
$$(52\text{-}93 \%)$$

(R = C_6H_{13}, Ph; R' = CH=CHX, aryl, COR, CH=NX)

Selenoxides having allylic or propargylic ligands are unstable and at elevated temperatures they undergo [2,3] sigmatropic oxygen migration to give allyl alcohol derivatives via the formation of allylselenenates. Allyl aryl sulfoxides are known to undergo similar [2,3] sigmatropic rearrangement to those of the corresponding sulfenates. The equilibrium constants and other data for the isomerization of the sulfoxide (**37**) and the selenoxide (**38**) have been compared (see Scheme 3 and Table 2). While the sulfoxide is more stable than the corresponding sulfenate, the opposite is true for the selenoxide which rearranges irreversibly into the isomeric selenate. Hydrolysis then affords an allylic alcohol (H. J. Reich, K. E. Yelm and S. Wollowitz, J. Am. Chem. Soc., 1983, **105**, 2503; H. J. Reich et al., ibid., 1981, **103**, 3112).

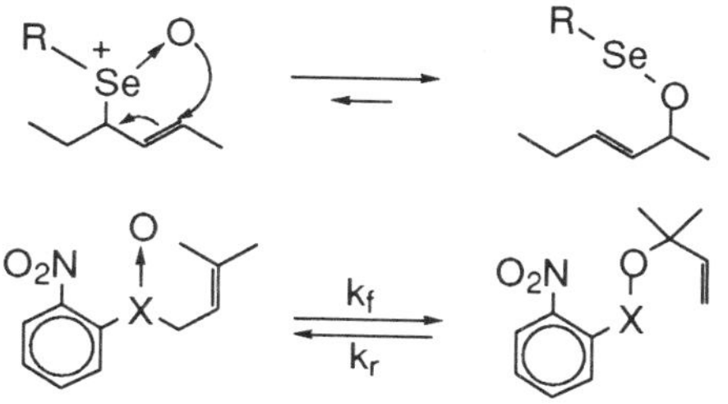

(Scheme 3)

37 (X = S)
38 (X = Se)

Table 2

	X	k_f/k_r	ΔG^o (Kcal/mol)
(37)	S	0.042	+1.5
(38)	Se	$10^{12.5}$	-11.0

There are many examples of the [2,3] sigmatropic rearrangement of selenoxides bearing allyl groups. Generally the reaction is utilized in combination with several other reactions and here a few representative reactions are depicted (see equations 40-42) (A. Lerouge and C. Paulmier, Tetrahedron Lett., 1984, **25**, 1983, 1987; K. M. Nsunda and L. Hevesi, Tetrahedron Lett., 1984, **25**, 4441; R. T. Lewis and W. R. Motherwell, Tetrahedron, 1992, **48**, 1465; idem, J. Chem. Soc. Chem. Commun., **1988**, 751).

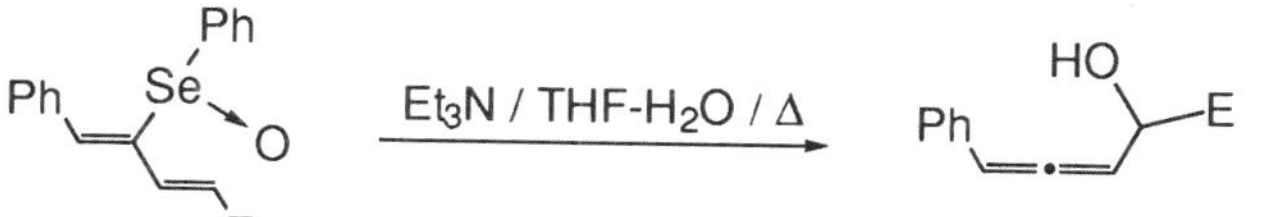

(eq. 40)

65 % (E = Ph)
72 % (E = CO_2Et)

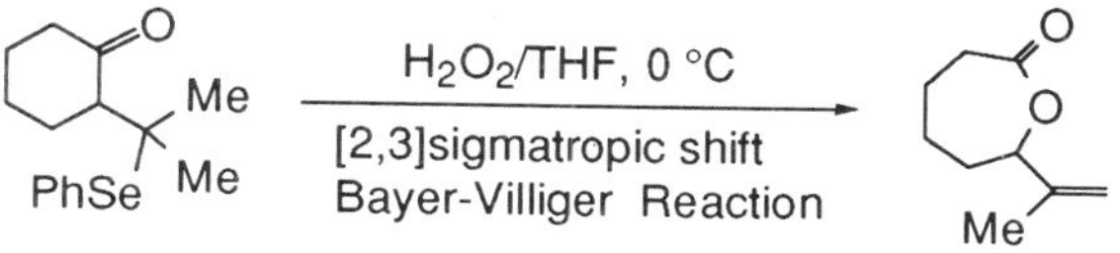

(eq. 41)

[2,3]sigmatropic shift
Bayer-Villiger Reaction

(62 %)

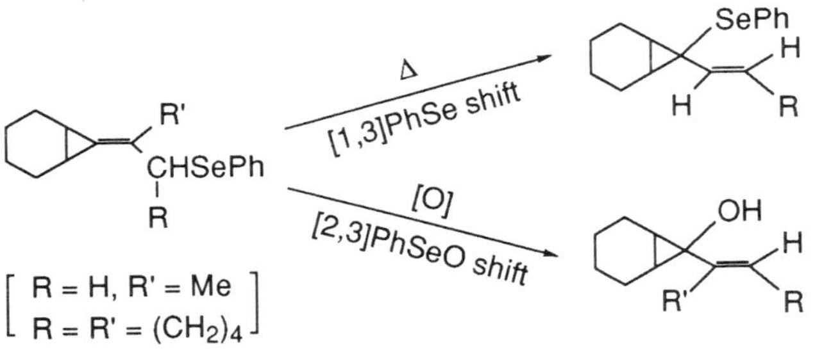

[Scheme for eq. 42 showing bicyclic alkene with CHSePh group undergoing [1,3]PhSe shift (Δ) to give vinyl SePh product, and [2,3]PhSeO shift ([O]) to give allylic alcohol]

$$\left[\begin{array}{l} R = H, R' = Me \\ R = R' = (CH_2)_4 \end{array} \right]$$

(eq. 42)

Interestingly, enzymatic oxidation of phenyl propargyl selenide (**39**) with cyclohexanone oxygenase gives rearranged allenyl phenyl selenenate which on hydrolysis and a [1,2] shift of the PhSe group affords (1-phenylselenenyl)acryl aldehyde (**40**) (equation 43) (J. A. Latham Jr. et al., J. Chem. Soc. Chem. Commun., **1986**, 528).

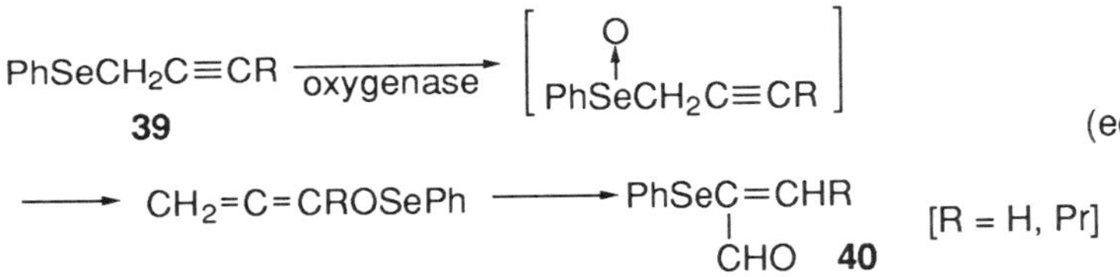

$$PhSeCH_2C\equiv CR \xrightarrow{oxygenase} \left[\overset{O}{\underset{\uparrow}{PhSe}}CH_2C\equiv CR \right]$$

$$\longrightarrow CH_2=C=CROSePh \longrightarrow \underset{CHO}{PhSeC=CHR} \quad [R = H, Pr]$$

40

(eq. 43)

The Davis procedure using an optically active oxaziridine and the Sharpless oxidation of aryl cinnamyl selenides both give optically active 1-phenyl-2-propen-1-ol in 10-65 % and 7-92 % e.e. via a [2,3] sigmatropic migration of the intermediate selenoxides (equation 44). (F. A. Davis, O. D. Stringer and J. P. McCauley, Tetrahedron Lett., 1983, **24**, 1213; N. Komatsu, Y. Nishibayashi and S. Uemura, Tetrahedron Lett., 1993, **34**, 2339).

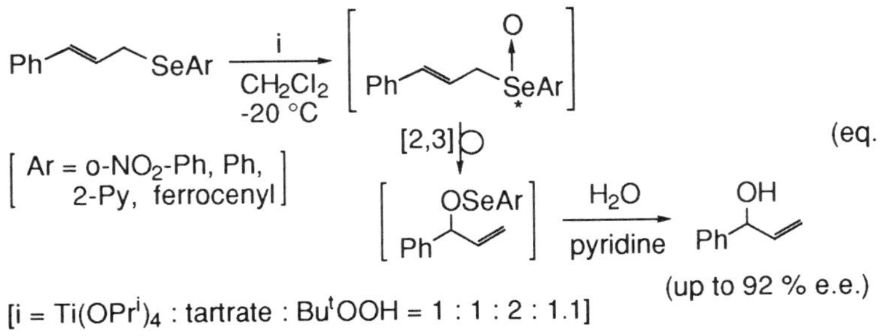

[Scheme for eq. 44: Ph-CH=CH-CH2-SeAr reacts with i/CH2Cl2/-20°C to give selenoxide intermediate, then [2,3] shift to allylic OSeAr, then H2O/pyridine to allylic alcohol]

$$\left[\begin{array}{l} Ar = o\text{-}NO_2\text{-}Ph, Ph, \\ 2\text{-}Py, ferrocenyl \end{array} \right]$$

(up to 92 % e.e.)

(eq. 44)

[i = Ti(OPr^i)$_4$: tartrate : Bu^{t}OOH = 1 : 1 : 2 : 1.1]

Similarly, selenilimines (**41**) generated on treatment of allyl aryl selenides and oxidants such as chloramine-T, or N-chlorosuccinimide

(NCS), in the presence of p-bromoaniline and triethylamine afford allylamines in high yield. The mechanisms of this type of reaction involve the initial formation of an unstable selenilimine (**41**), which undergoes a spontaneous [2,3] sigmatropic rearrangement to the allyl amine (equation 45) (R. G. Shea et al., J. Org. Chem., 1984, **49**, 3647; idem, ibid., 1985, **50**, 417; idem, ibid., 1986, **51**, 5243).

$$Ph-Se-CH_2-CH=CH-Ar \xrightarrow[MeOH/Et_3N]{NCS/p-BrC_6H_4NH_2} \left[\begin{array}{c} \text{selenilimine intermediate} \\ \textbf{41} \end{array} \right] \xrightarrow{[2,3]N \text{ shift}} CH_2=CH-CH(Ph)-NH-Ar \quad [Ar: p-BrC_6H_4] \quad (eq. 45)$$

If a five molar excess of an oxidising reagent, such as mCPBA in methanol, or THF-methanol, is used alkyl aryl selenides (and some selenoxides) undergo oxidation to the corresponding selenones, which are unstable and react immediately with methanol to afford alkyl methyl ethers (**42**) (equation 46). Selenides bearing an α-benzoyl group behave similarly. In the latter case oxidation gives rise to unidentified intermediates which may be selenones (e.g. **43**), although this is by no means certain. Ultimately, however, a phenyl group migration occurs giving rise to an ester (see equation 47). For β-hydroxyselenides oxidation is also followed by a shift of the phenyl group leading finally to a ketone (e.g. **44**) (equation 48) (S. Uemura and S. Fukusawa, J. Chem. Soc. Perkin Trans. 1, **1985**, 471; S. Uemura, K. Ohe and N. Sugita, J. Chem. Soc. Chem. Commun., **1988**, 111; M. Tiecco et al., Gazz. Chim. Italy., 1987, **117**, 423; S. Uemura et al., J. Chem. Soc., Perkin Trans. 1, **1990**, 907; S. Uemura, K. Ohe and N. Sugita, J. Chem. Soc. Perkin Trans. 1, **1990**, 1697; Y. Hirai et al., Phosphorus, Sulfur Silicon Relat. Elem., 1992, **67**, 173).

$$Ph\text{-}Se\text{-}R \xrightarrow[MeOH, r.t.]{excess\ mCPBA} \left[Ph\text{-}Se(=O)_2\text{-}R \right] \longrightarrow R\text{-}O\text{-}Me \quad (eq.\ 46)$$

$$\textbf{42}\ (97\%)$$

[R = Me(CH$_2$)$_n$-, n=12]

$$\text{PhSeCl} + \text{CH}_3\text{CH}_2\overset{\text{O}}{\underset{\|}{\text{C}}}\text{Ph} \longrightarrow \underset{\text{Me}}{\text{PhSeCHCPh}}\overset{\text{O}}{\underset{\|}{}} \xrightarrow[\text{MeOH}/25°\text{C}]{\text{mCPBA}}$$

$$\left[\underset{\text{O Me}}{\overset{\text{O O}}{\text{Ph-Se-CH-C-Ph}}} \right] \xrightarrow{\text{Ph migration}} \underset{\text{Ph}}{\text{CH}_3\text{CHCOCH}_3}\overset{\text{O}}{\underset{\|}{}} \quad \text{(eq. 47)}$$

43

$$\underset{\text{OH}}{\text{Ph}_2\text{CCH}_2\text{SePh}} \xrightarrow[\text{MeOH}]{\text{mCPBA / Ph migration}} \underset{\underset{\textbf{44}}{\text{O}}}{\text{PhCCH}_2\text{Ph}} \quad \text{(eq. 48)}$$

Interestingly, oxidation of 2-phenylethynyl phenyl selenide (**45**) with mCPBA yields neither the corresponding selenoxide nor the selenone. Instead the reduction product phenylacetylene (**46**) is formed. When oxygen is present coupling occurs and 1,4-diphenylbuta-1,3-diyne (**47**) is obtained in 86% yield, suggesting that the phenylethynyl radical, or its equivalent, is generated as an intermediate (equation 49) (J. V. Commasseto et al., Synth. Commun., 1986, **16**, 283; J. V. Commasseto et al., J. Chem. Soc. Chem. Commun., **1986**, 1067).

$$\text{PhC}\equiv\text{CSePh} \;\; \textbf{45} \quad \begin{array}{c} \xrightarrow[\text{THF, r.t.}]{\text{i) mCPBA, ii) 1N NaOH}} \text{PhC}\equiv\text{CH} + \text{PhSeO}_2\text{Na} \\ \textbf{46} \; (80\%) \\ \xrightarrow[\text{THF, r.t.}]{\text{mCPBA}} \text{PhC}\equiv\text{C-C}\equiv\text{CPh} \\ \textbf{47} \; (86\%) \end{array} \quad \text{(eq. 49)}$$

(c) Aryltellurols, ditellurides, tellurides and related derivatives

(i) Preparation

Aryltellurols are usually unstable and oxidizable and hence they are generated by the reaction of diaryl ditellurides with organometallic reagents and are used in situ for further reactions (equation 50) (N. Sonoda and A. Ogawa, "The chemistry of organic selenium and tellurium compounds", eds. S. Patai and Z. Rappoport, John Wiley & Sons, New York, 1986, Vol 1, Ch. 16, p. 650; L. Engman and J. S. E. Hellberg, J. Organomet. Chem., 1985, **296**, 357).

$$\text{PhLi} + \text{Te} \longrightarrow [\text{PhTeLi}] \xrightarrow{\text{RX}} \text{PhTeR} \qquad (\text{eq. 50})$$

There are several methods for preparation of alkyl aryl and diaryl tellurides. Diaryl ditellurides react with NaOH in the presence of quaternary ammonium salts as phase-transfer catalysts to generate aryltellurorate anions which react in situ with alkyl halides providing numerous alkyl aryl tellurides in 52-72 % yields (equation 51) (J. V. Commasseto, J. T. B. Ferreira and J. A. F. Val, J. Organomet. Chem., 1984, **277**, 261; A. Sandhu, K. K. Bhasin and R. D. Verma, Indian J. Chem., Sect.A 1990, **29A**, 1178).

$$\text{Ar-TeTe-Ar} + \text{NaOH} \xrightarrow[\substack{\diagdown \\ \diagup \text{N}^+\text{X}}]{} \begin{array}{c} 3[\text{ArTeNa}] \\ + \\ [\text{ArTeO}_2\text{Na}] \end{array} \xrightarrow{\text{RX}} \begin{array}{c} 3\text{Ar-Te-R} \\ (52\text{-}72\,\%) \end{array} \qquad (\text{eq. 51})$$

$$\left[\begin{array}{l} \text{Ar} = \text{Ph, p-Tol, 4-MeOC}_6\text{H}_4\text{, 4-EtOC}_6\text{H}_4\text{, 2-Naphth} \\ \text{R} = \text{Bu, Me}_2\text{CH(CH}_2)_2\text{, Me}_2\text{CHCH}_2\text{, PhCH}_2\text{CH}_2 \end{array}\right]$$

Aryl halides react with alkyl tellurorates and the alkyl aryl tellurides obtained further react with alkyl tellurorate anions to form aryl tellurorate anions. These anions on oxidation result in the formation of diaryl ditellurides while they give alkyl aryl tellurides by reactions with alkyl halides (V. A. Potapov, S. V. Amosova and P. A. Petrov, Tetrahedron Lett., 1992, **33**, 6515). Newly invented organometallic reagents such as samarium iodide (SmI$_2$) and alkylboranes (R$_3$B) also work as convenient reducing agents for diaryl ditellurides which give initially PhSmI$_2$ or PhTeR$_3$B and then, on treatment with alkyl halides or acyl halides, these intermediates are converted to alkyl aryl tellurides or acyltellurorates, respectively (Y. Yu Zhang and R. Lin, Synth. Commun., 1993, **23**, 189; M. Sekiguchi et al., Heteroatom Chem., 1991, **2**, 427; T. Abe et al., Chem. Lett., **1990**, 1671). Similarly, diaryl diselenides can be converted to the corresponding selenides in high yields (S. Fukuzawa et al., Heteroatom Chem., 1990, **1**, 491). Organometallic reagents of aryltellurides such as PhTeAl(i-Pr)$_2$, or PhTeSiMe$_3$ are also convenient reagents for the preparation of aryl tellurides (C. H. W. Jones and R. D. Sharma, J. Organomet. Chem., 1984, **268**, 113; K. Sasaki et al., Chem. Lett., **1991**, 415). There are several methods for the preparation of tellurides bearing functional groups. Aryl vinyl tellurides (**48**) are prepared from vinyl Grignard reagents and arylbromotellurides (equation 52) or substitution reaction of trans-2-phenylvinyl bromide (**49**) with sodium 4-methoxyphenyltellurorate stereospecifically (equation 53) (M. J. Dabdoub et al., J. Organomet. Chem., 1986, **308**, 211). Similarly, either a combination of ArTeLi with α-bromoketones, or PhTeI with the enolates of ketones, give high yields of the corresponding α-aryltelluroketones (T. Hiiro et al., Synthesis, **1987**, 1096;

L. A. III. Silks, J. D. Odom and R. B. Dunlap, Synth. Commun., 1991, **21**, 1105).

$$RCH=CR^1MgBr + R^2TeBr \xrightarrow{THF/C_6H_6} RCH=CR^1TeR^2$$
$$\textbf{48} \ (71\text{-}86 \%) \quad \text{(eq. 52)}$$

$[R = H, Ph; R^1 = H, Me; R^2 = 4\text{-MeOC}_6H_4, Bu]$

$$\underset{\textbf{49}}{\underset{H}{\overset{Ph}{>}}C=C\underset{Br}{\overset{H}{<}}} + CH_3O\text{-}\bigcirc\text{-TeNa} \longrightarrow \underset{H}{\overset{Ph}{>}}C=C\underset{Te}{\overset{H}{<}}\text{-}\bigcirc\text{-OCH}_3 \quad \text{(eq. 53)}$$
$$(86\%)$$

Radical type additions of alkyl aryl tellurides to acetylenic triple bonds in the presence of a catalytic amount of a radical initiator(AIBN) have been conducted to give stereoselectively E-ethenyl aryl tellurides (Li. B. Han et al., J. Am. Chem. Soc., 1992, **114**, 7591; idem, Phosphorus Sulfur and Silicon Relat. Elem., 1992, **67**, 243). Numerous preparations of diaryl tellurides, including (**50**) and (**51**), have been carried out. Standard procedures are the reactions of arenediazonium salts with KTeCN, or direct treatment of aryl halides with aryltellurorates (equations 54 and 55) (L. Engman, J. Org. Chem., 1983, **48**, 2920; T. Kemmitt and W. Levason, Organometallics, 1989, **8**, 1303; H. Suzuki and T. Nakamura, Synthesis, **1992**, 549).

$$Ar\text{-}NH_2 \xrightarrow{HNO_2} Ar\text{-}N_2^+BF_4^- \xrightarrow{KTeCN} Ar\text{-}Te\text{-}Ar \quad \text{(eq. 54)}$$

$$\bigcirc\underset{Br}{\overset{Br}{<}} + LiTeCH_3 \xrightarrow{THF} \bigcirc\underset{TeCH_3}{\overset{TeCH_3}{<}} \quad \text{(eq. 55)}$$

$$\bigcirc\underset{TePh}{\overset{TePh}{<}} \quad , \quad \bigcirc\underset{Y}{\overset{TeCH_3}{<}}$$
$$\textbf{50} \qquad\qquad \textbf{51}$$

$[Y = NMe_3, PMe_3, AsMe_3, SbMe_2, OMe, SMe, SeMe, Cl]$

When bis(2-phenylethynyl)telluride is treated with aryllithiums, an exchange reaction between the ethynyl group and an aryl group takes place to give diaryl tellurides in excellent yields. Although the reactions were followed by low temperature ^{125}Te-NMR, no ate complex type hypervalent intermediates were detected (L. Engman and S. D. Stern, Organometallics, 1993, **12**, 1445).

(ii) Reactions of dicoordinate aryltellurium compounds
Aryl tellurium compounds possess the similar reactivity and reactions to those of selenium compounds. The addition of aryltelluroates generated from diaryl ditellurides to triple bonds proceeds stereospecifically affording the Z-isomers (**52**) (equation 56) (S. M. Barros et al., Organometallics, 1989, **8**, 1661). The substitution reaction of butyl vinyl telluride (**53**) with butyllithium also proceeds stereospecifically to give the vinyllithium reagent (**54**) (equation 57).

$$RC \equiv CH + Te \xrightarrow[NaOH/EtOH-H_2O]{NaBH_4} (Z)\text{-}(RHC=CH)_2Te \quad \text{(eq. 56)}$$
$$\mathbf{52}$$

$$\underset{\mathbf{53}}{\underset{BuTe}{\overset{H}{>}}C=C\underset{Ph}{\overset{H}{<}}} \xrightarrow{BuLi} \left[\underset{\mathbf{54}}{\underset{Li}{\overset{H}{>}}C=C\underset{Ph}{\overset{H}{<}}}\right] \xrightarrow{PhCHO} \underset{Ph-\underset{H}{\overset{|}{C}}-OH}{\overset{H}{>}}C=C\underset{Ph}{\overset{H}{<}} \quad \text{(eq. 57)}$$

Acyl telluroates (**55**) are convenient sources for the generation of acyl, or aryltellurium radicals which can add to olefinic bonds. For example, photoirradiation of 1-naphthotelluroacetate gives both benzoyl and naphthyl tellurium radicals (C. Chen and D. Crich, Tetrahedron Lett., 1993, **34**, 1545; idem, J. Am. Chem. Soc., 1992, **114**, 8313). Telluroacetates [ArTe(O)OAc; Ar = Ph, p-MeOC$_6$H$_4$, 2-naphthyl] and aryltrihalotelluranes [p-MeOC$_6$H$_4$TeCl$_3$] undergo intramolecular addition to the double bond of pent-4-enol to give tetrahydrofuran derivatives (**56**) (equation 58) (J. V. Commasseto et al., Tetrahedron Lett., 1987, **28**, 5611; H. N. Xing et al., J. Org. Chem., 1989, **54**, 4391).

$$HO\text{-}(CH_2)_3\text{-}CH=CH_2 + ArTe(O)OAc \xrightarrow{CH_3COOH} \underset{\mathbf{56}\ (73\text{-}92\%)}{\text{[tetrahydrofuranyl-TeAr]}} \quad \text{(eq. 58)}$$
$$\mathbf{55}$$
[Ar = Ph, p-MeOC$_6$H$_4$, 2-Naphth]

Aryl vinyl tellurides react with dichlorocarbene, generated from CHCl$_3$ with NaOH under phase-transfer conditions, to give cyclopropanes with retention of the configuration of the double bond (X. Huang and S. Jiang, Synth. Commun., 1993, **23**, 431).

Oxidation of alkyl aryl tellurides with one, or two, equivalents of mCPBA in diethyl ether at room temperature affords the corresponding olefins via telluroxides (**57**) as intermediates. These reactions are stereospecific and are promoted if triethylamine is present (equation 59) (Y. Nishibayashi et al., J. Chem. Soc. Perkin Trans. 1, **1993**, 1133; S. Uemura, K. Ohe and S. Fukuzawa, Tetrahedron Lett., 1985, **26**, 895).

In the case of the oxidation of alkyl aryl tellurides using five molar excess of mCPBA in methanol the aryltellurium group is substituted with a methoxy group to give the substitution product (**58**) with concomitant formation of the corresponding olefin. In this reaction, using β-phenyl substituted tellurides, a 1.2-shift of the phenyl group is observed. Neither telluroxides nor tellurones were detected or isolated in the two reactions described above (S. Uemura and S. Fukuzawa, Tetrahedron Lett, 1983, **24**, 4347).

(eq. 59)

3. Se(III) and Te(III) Derivatives

Compounds belonging to this class are tricoordinated derivatives of Se and Te. Typical compounds are monooxides, monoimines, selenonium or telluronium salts and their ylides. Tricoordinated selenium and tellurium compounds bearing an oxygen or a nitrogen atom are unstable relative to their sulfur analogues and hence few stable species have been reported. Here a few typical examples of these compounds are illustrated.

(a) Selenoxides and telluroxides

As described in the previous chapter, both selenoxides and telluroxides are thermally unstable, except those without a β-proton, and undergo syn-elimination to afford olefins. As a result only a restricted number of representative compounds have been prepared. A few compounds which possess β-protons are known, but in such cases appropriate substituents stabilize the oxide by hydrogen-bonding or the group is protected by complexation.

Selenoxides are available from selenides by oxidation and a variety of oxidants including peroxides, ozone, $NaIO_4$ and iodosylbenzene dichloride have been used. Indeed the last reagent was once considered to be very efficient (M. Cinquini, S. Colonna and R. Givini, Chem. & Ind., **1969**, 1737). However, more recently the use of t-BuOCl-MeOH in the presence of pyridine is preferred especially for selenides having electron-withdrawing substituents. Here the products of oxidation are hydrolysed with alkali to release the selenoxide (equation 60) (M. Kobayashi, H. Ohkubo and T. Shimizu, Bull. Chem. Soc. Jpn., 1986, **59**, 503). Several examples of selenoxides reported are shown below (M. R. Detty, J. Org. Chem., 1980, **45**, 274).

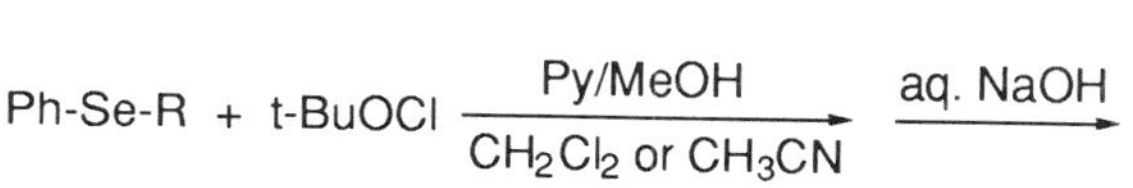

$$\text{Ph-Se-R} + \text{t-BuOCl} \xrightarrow[\text{CH}_2\text{Cl}_2 \text{ or CH}_3\text{CN}]{\text{Py/MeOH}} \xrightarrow{\text{aq. NaOH}} \text{Ph-Se(=O)-R} \quad (\text{eq. 60})$$

(b) Optically active selenoxides

Selenoxides are readily converted into the tetracoordinated dihydroxides (e.g. **59**) on contact with moisture (equation 61), hence optically active selenoxides once formed rapidly racemize. This is in marked contrast to sulfoxides which are much more stable. However, when moisture is excluded homochiral selenoxides can be handled satisfactorily and several procedures for the synthesis of such compounds have been reported.

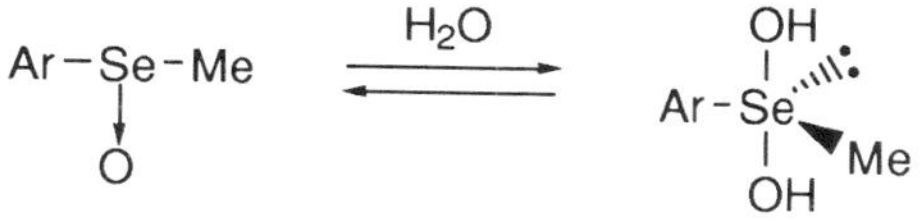

$$\text{Ar-Se(=O)-Me} \xrightleftharpoons{\text{H}_2\text{O}} \text{Ar-Se(OH)(OH)-Me} \quad (\text{eq. 61})$$

59

Racemic aryl methyl selenoxides can be resolved using 10-camphorsulfonic amide but the enantiomeric excesses are poor (F. A. Davis, J. M. Bilmers and O. D. Stringer, Tetrahedron Lett., 1983, **24**, 3191).

Chromatographic methods using chiral column have also been used, but again the results are not always satisfactory (T. Shimizu, M. Yoshida and M. Kamigata, Bull. Chem. Soc. Jpn., 1988, **61**, 3761; T. Shimizu and M. Kobayashi, J. Org. Chem., 1987, **52**, 3399; F. Toda and K. Mori, J. Chem. Soc., Chem. Commun., **1986**, 1357). These difficulties are partly avoided by asymmetric synthesis from selenides. Chiral oxidants such as the Sharpless oxidising system composed of tartarate-$TiCl_4$-t-BuOOH-H_2O can be used with modest success (M. Tiecco et al., Tetrahedron Lett., 1987, **28**,

3849; T. Shimizu, M. Kobayashi and N. Kamigata, Bull. Chem. Soc. Jpn., 1989, **62**, 2099). In early work the Davis chiral oxaziridine approach was not promising (F. Davis and A. C. Sheppard, Tetrahedron, 1989, **45**, 5703; F. A. Davis, O. D. Stringer and J. P. McCauley, ibid., 1985, **41**, 4747), but new reagents (+)- or (-)-3,3-dichloro-1,7,7-trimethylcamphor-N-p-toluenesulfonyl-2-oxaziridine (**60**) have been introduced and shown to give sulfoxides in excellent optical yields (F. A. Davis, R. T. Reddy, and W. Han, J. Am. Chem. Soc., 1992, **114**, 1428; F. A. Davis and T. Reddy, J. Org. Chem., 1992, **57**, 2599). The oxidation procedure is summarized in Scheme 4.

[Ar = 2,4,6-tri-i-Pr-C_6H_2-; R = Me, Et] (Scheme 4)

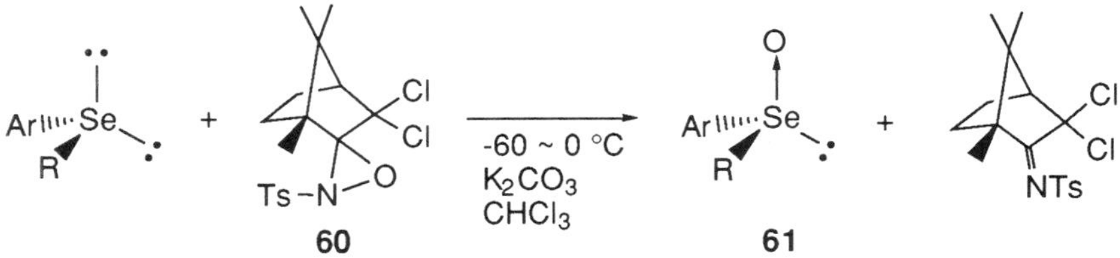

This procedure has been employed for the preparation of paracyclophane selenoxide (**62**) and also (S)-linalool (**63**) (equation 62) (H. J. Reich and K. E. Yelm, J. Org. Chem., 1991, **56**, 5672; H. J. Reich, E. E. Yelm. and S. Wollowitz, J. Am. Chem. Soc., 1983, **105**, 2503).

(eq. 62)

63 (83 % e.e.)

62

[X = Cl, Br]

64

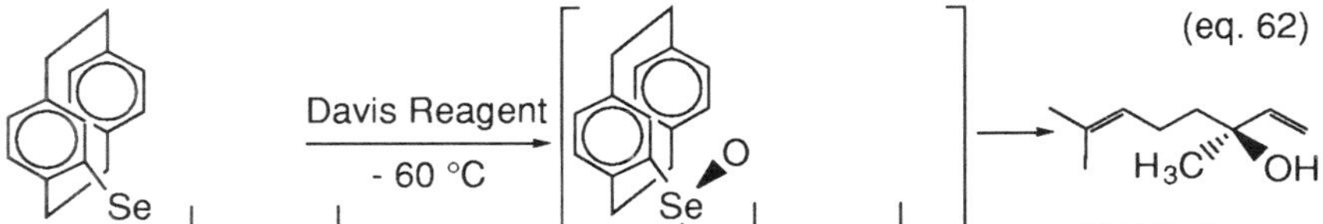

Telluroxides are also unstable and tellurides bearing β-protons undergo syn-elimination on oxidation. This has restricted the range of compounds and only diaryl and methyl aryl telluroxides have been prepared. Generally such compounds are obtained either from dihalodiaryltelluranes (**64**) by alkaline hydrolysis, or by the oxidation of tellurides with t-BuOOBu-t, or

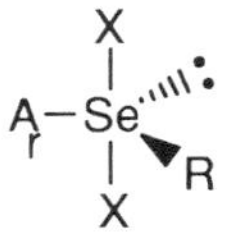

NaIO$_4$ (M. R. Detty, J. Org. Chem., 1980, **45**, 274; H. Lee and M. P. Cava, J. Chem. Soc. Chem. Commun., **1981**, 277; M. Akiba, K.-Y. J. Lakshmikantham and M. P. Cava, J. Org. Chem., 1984, **49**, 4819).

Di-(4-methoxyphenyl)telluroxide, pKa ~15, serves as a moderately strong base for aldolisation reaction, but is also an oxidant converting catechol into o-benzoquinone and thiols to disulfides (N. Petragnani and J. V. Commasseto, Synthesis, **1986**, 1).

(c) Selenoimines and telluroimines

Selenoimines and telluroimines are the aza-equivalents of selenoxides and telluroxides, respectively. Both are stable only when they contain no β-hydrogen atom (S. Oae, M. Fukumura and N. Furukawa, Phosphorus & Sulfur, 1981, **10**, 153; V. I. Naddaka, V. P. Krasnov, and V. I. Minkin, Zh. Obsch. Khim., 1983, **19**, 2632).

(d) Selenonium, telluronium salts and their ylides

Both selenonium and telluronium salts are not as popular as sulfonium salts which have been reviewed by Stirling, Patai and Rappoport ("The chemistry of sulfonium group", Vol. 1 and 2, eds. C. J. M. Stirling, S. Patai and Z. Rappoport, John Wiley & Sons, New York, 1981).

Triarylselenonium salts can be prepared according to the following methods (equations 63-66)) (W. Dumont, P. Bayet and A. Krief, Angew. Chem. Int. Ed., Engl., 1974, **13**, 274; Y. Iwama et al., Bull. Chem. Soc. Jpn., 1981, **54**, 2065).

Ar-H + SeCl$_2$ $\xrightarrow{AlCl_3}$ Ar$_3$Se$^+$Cl$^-$ (eq. 63)

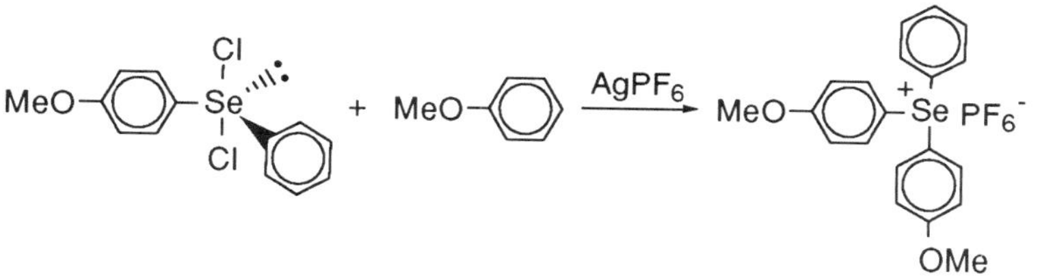

(eq. 64)

3 ArMgX + SeOX$_2$ ⟶ Ar$_3$Se$^+$ MgO$^-$ $\xrightarrow{HBr}$ Ar$_3$Se$^+$ Br$^-$ (eq. 65)

PhMgBr + Ph-Se(=O)-Ph ⟶ Ph$_3$SeBrMgO$^-$ $\xrightarrow{HBr}$ Ph$_3$Se$^+$Br$^-$ (eq. 66)

Alkylarylselenonium salts can be synthesised from alkyl aryl selenides by reactions with strong alkylating agents such as MeSO$_3$F, or using an alkylating agent in the presence of AgBF$_4$ or AgPF$_6$ (E. Maccarone and G.

Perrini, J. Chem. Soc. Perki Trans. 2, **1983**, 1605; P. G. Gassman, T. Miura and A. Mossman, J. Org. Chem., 1982, **47**, 954).

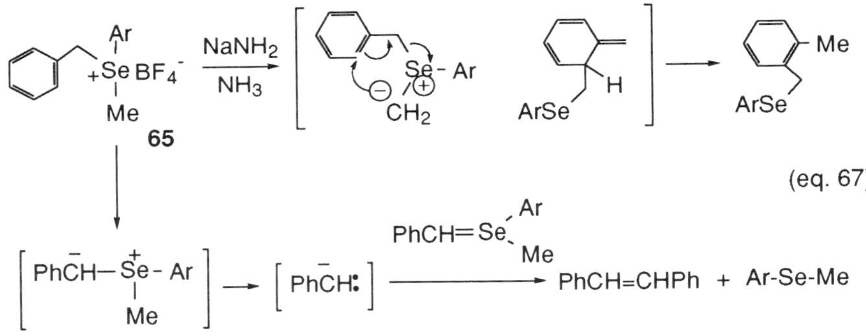

(eq. 67)

Arylbenzylmethylselenonium salts (**65**) are prepared either from trimethyloxonium tetrafluoroborate or $MeSO_3F$ with aryl benzyl selenides in 88-100 % yields. The selenonium salts (**65**) thus obtained undergo the Stevens type migration of the methyl group upon treatment with a strong base such as $NaNH_2$ in liq. NH_3 (equation 67).

Selenonium and sulfonium salts of benzothio and selenopyranium and related cyclic derivatives on treatment with alkali or alkaline earth metals undergo numerous interesting reactions. With magnesium metal, for example, they undergo C-S and C-Se bond cleavage, possibly through a one electron transfer process which involves an intermediate selenonium radical (**66**) (equation 68) (M. Hori et al., Tetrahedron Lett., 1989, **30**, 981;idem, J. Chem. Soc. Perkin Trans. 1, **1990,** 3017; idem, ibid., **1987**, 187.)

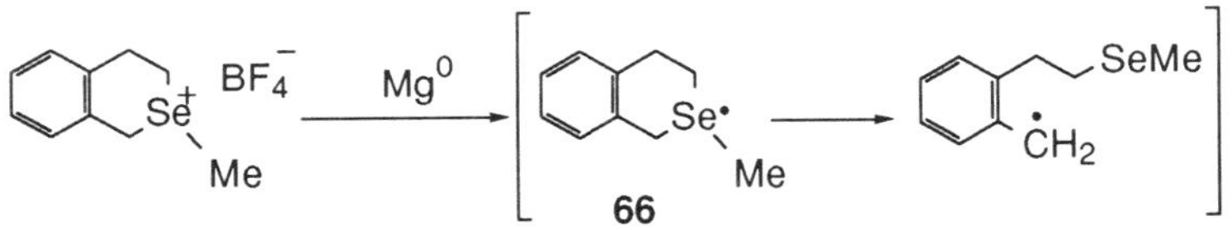

(eq. 68)

Few optically active selenonium salts have been reported. However, ethylmethylphenylselenonium perchlorate has been converted into the camphorsulfonate salt which on fractional crystallization gives both dextro and levo rotatory salts. Each salt can then be used as an optically active methyl group-transfer reagent, for example, in the preparation of 2-methoxycarbonyl-2-methylindanone (**67**) and 2-ethyl-2-methoxycarbonylindanone (**68**) (Scheme 5) (M. Kobayashi et al., Chem. Lett., **1986**, 2117).

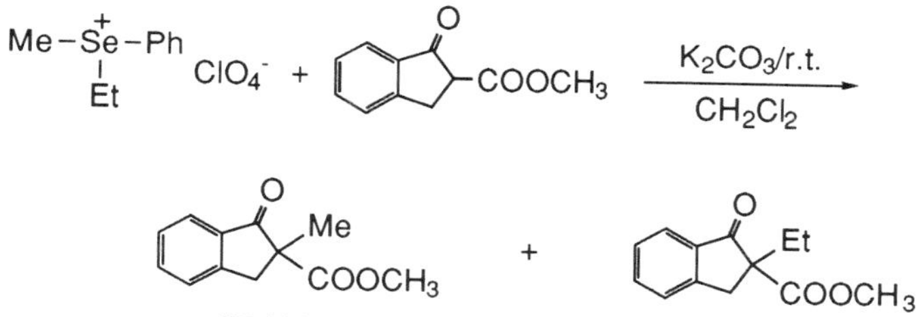

$$\underset{\text{Et}}{\text{Me}-\overset{+}{\text{Se}}-\text{Ph}} \text{ClO}_4^- \ + \ \text{[indanone]-COOCH}_3 \ \xrightarrow[\text{CH}_2\text{Cl}_2]{\text{K}_2\text{CO}_3/\text{r.t.}}$$

[indanone with Me and COOCH₃] + [indanone with Et and COOCH₃]

67 (41 %) **68** (36 %)

(Scheme 5)

The chemistry of telluronium salts is expanding, but only a few representatives have been documentated. Special interest in these salts centres on whether or not they contain hypervalent bonding, and this is controlled by the nature of the ligands attached to the central atom of the onium cation.

Diphenyl telluride reacts with methyl iodide at room temperature during two days to give a quantitative yield of diphenylmethyltelluronium iodide. This salt serves as a telluronium ylide on treatment with BuLi or PhLi, but unexpectedly when the ylide is reacted with aldehydes the products are sec-alcohols (equation 69) (Li-L. Shi, Z.-L. Zhou and Y.-Z. Huang, J. Chem. Soc., Perkin Trans. 1, 1990, 2847).

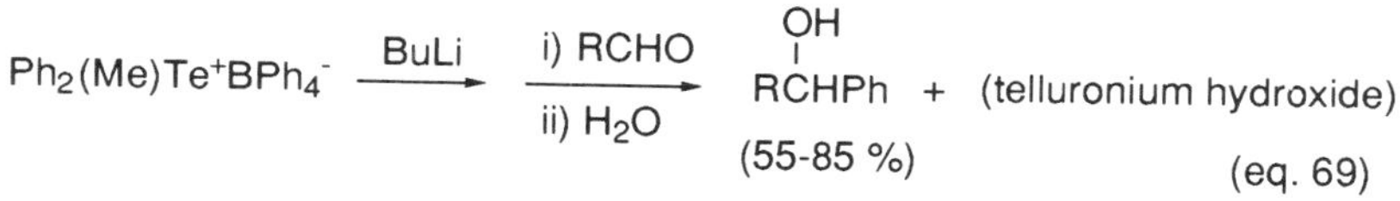

$$\text{Ph}_2(\text{Me})\text{Te}^+\text{BPh}_4^- \xrightarrow{\text{BuLi}} \xrightarrow[\text{ii) H}_2\text{O}]{\text{i) RCHO}} \underset{\text{(55-85 \%)}}{\text{RCHPh}} \ + \ \text{(telluronium hydroxide)}$$
$$\overset{\text{OH}}{|}$$

(eq. 69)

This may be interpreted as the formation of a tetracoordinated tellurane intermediate. However, if the strong base, lithium 2,2,6,6,-tetramethylpiperidide is used the ylide formed then reacts with suitable carbonyl compounds to afford oxiranes (Z. L. Zhou et al., Tetrahedron Lett. 1990, **31**, 7657). Stabilized ylides have also been reported (H. Osuka et al., Tetrahedron Lett., 1983, **24**, 2599, 5109.) and the exchange of counter anion in diphenylmethyltelluronium iodide has been conducted using silver salt of substituted benzoic acids to afford several telluronium benzoates. Telluronium benzoates contain a covalent bond between the tellurium atom and an oxygen atom of the carboxylate unit, an ionic bond is not present. This conclusion is supported by data from conductivity and ^{125}Te-NMR measurements and sets these compounds apart from most other onium salts (W. R. McWhinnie and J. Mallaki, Polyhedron, 1981, **1**, 13; N. S. Dance et al., J. Organomet. Chem., 1980, **198**, 131).

4. Se(IV) and Te(IV) compounds : Selenones and tellurones

Selenones, the analogues of sulfones can be prepared by the oxidation of selenides or selenoxides, however, whereas sulfones are stable, selenones are less so and the instability of selenoxides has restricted their use as starting materials. Despite this several oxidants have been recommended for the preparation of selenones from selenides, but the methods have not been optimized. In the case of methyl phenyl selenide oxidation with mCPBA in CH_2Cl_2 at 20 °C, CF_3CO_3H in CH_2Cl_2 at room temperature, or $KMnO_4$ in H_2O-CH_2Cl_2 at room temperature, all work sufficiently well to provide the selenone in 40-75 % yield (equation 70). These oxidants can be applied to other alkyl phenyl selenides, but the optimum conditions needed seem to be different in each case. For example, ethyl phenyl selenide affords the corresponding selenone in 82 % yield when treated with $KMnO_4$, but with mCPBA the yield drops to 14 %. Longer chain alkyl phenyl selenides also give the corresponding selenones, but success is restricted to those with a primary alkyl group such as n-decyl. Those with a secondary unit such as phenyl-dec-2-yl give many unidentified products, including some which result from syn-elimination. The structure of methyl phenyl selenone (**69**) has been determined by X-ray crystallography and this shows it to be almost identical in shape to methyl phenyl sulfone (A. Krief et al., J. Chem. Soc. Chem. Commun., **1985**, 569.).

$$\text{Me-Se-Ph} \xrightarrow{a)} \left[\begin{array}{c} O \\ \uparrow \\ \text{Me-Se-Ph} \end{array} \right] \xrightarrow{a)} \begin{array}{c} O \\ \| \\ \text{Me-Se-Ph} \\ \| \\ O \end{array} \quad \text{(eq. 70)}$$

a) O_3, mCPBA, $KMnO_4$ **69**

n-Decyl phenyl selenone undergoes substitution reactions with numerous nucleophiles, although the products are contaminated with phenylselenic acid. In a reaction with t-BuOK, followed by the addition of benzaldehyde this substrate forms 1-phenyl-2-n-nonyloxirane (**70**) in 74 % yield (equation 71). This behaviour is in marked contrast to that of the corresponding sulfone which fails to react under these conditions (A. Krief, W.Dumont and J-N. Denis, J. Chem. Soc. Chem. Commun., **1985**, 571; A. Krief, W. Dumont and J. L. Laboureur, Tetrahdron Lett., 1988, **26**, 3265; S. Uemura and S. Fukuzawa, Tetrahedron Lett., 1983, **24**, 4347).

$$\text{Ph-Se(=O)(=O)-n-decyl} \xrightarrow{a)} \left[\text{Ph-Se(=O)(=O)} \underset{K}{\overset{\text{nonyl-n}}{\diagup}} \right] \xrightarrow{b)} \left[\text{Ph-Se(=O)(=O)} \underset{\overset{\ominus}{O}}{\overset{\text{nonyl-n}}{\diagdown}} \underset{Ph}{\overset{H}{\diagup C \diagdown}} \right]$$

$$\longrightarrow \underset{\mathbf{70}\ (74\%)}{\overset{\text{n-nonyl} \diagdown \diagup \text{Ph}}{\underset{O}{\triangle} H}}$$

(eq. 71)

a) t-BuOK/THF, b) PhCHO

There are many reports showing the applications of the phenylseleninyl group in organic synthesis. For example, it acts as a strong activating group for α-deprotonation and as a good leaving group. A combination of these two effects is well illustrated in the synthesis of the oxirane (**70**) and the latter property is exemplified by the formation of oxetanes from phenylseleninylalkanols (see equation 72) (R. Ando, T. Sugawara and I. Kuwajima, J. Chem. Soc., Chem. Commun., **1983**, 1514; I. Kuwajima, R. Ando and T. Sugawara, Tetrahedron Lett., 1983, **24**, 4429; R. Ando et al., Bull. Chem. Soc. Jpn., 1984, **57**, 2897; T. Sugawara and I. Kuwajima, Tetrahedron Lett., 1985, **26**, 5571; M. Shimizu, R. Ando and I. Kuwajima, J. Org. Chem., 1984, **49**, 1230; M. Tiecco et al., Tetrahedron, 1986, **42**, 4889, 4897).

$$\text{Ph-Se(=O)(=O)} \diagdown \diagup \underset{OH}{\overset{R}{\diagup}} \xrightarrow{:B} \left[\text{Ph-Se(=O)(=O)} \diagdown \diagup \underset{O^{\ominus}}{\overset{R}{\diagup}} \right] \longrightarrow \underset{O}{\square}^R + \text{PhSeO}_2\text{H}$$

(eq. 72)

Tellurone is a little known compound, but bis(4-methoxyphenyl)tellurone has been prepared from the telluroxide by oxidation with $NaIO_4$ (yield 82 %). Although the structure of this compound has not been fully determined, it can be used as a mild oxidant which converts benzenethiol into diphenyl disulfide (yield 97 %) and hydroquinone into p-benzoquinone (yield 39%) (L. Engman and M. P. Cava, J. Chem. Soc. Chem. Commun., **1982**, 164).

5. Hypervalent compounds of Se(IV), Te(IV)

(a) General introduction and historical background of hypervalent compounds of chalcogens

Lewis' "Octet rule" has long been considered to be the major factor that governs the chemical bonding of simple molecules. However, in the past there have been reports of a few unusual compounds which do not obey the octet rule, for example, SCl_4, SF_4, SF_6, I_3^- etc. (A. Michaelis and O. Schifferdecker, Chem. Ber.,1873, **6**, 993; O. Ruff and A. Heizelmann, Z. Anorg. Allg. Chem.,1911, **72**, 63). The central atoms of these molecules are composed of elements below the third row and from the 14th to 17th groups of the periodic table. The characteristic bonding modes involve decets (E. A. Innes, I. G. Csizmadia and Y. Kanada, J. Mol. Structure (Theochem.), 1989, **186**, 1), or even dodecets of electrons (E. A. Robinson, J. Mol. Structure (Theochem.), 1989, **186**, 9); such compounds are considered as exceptions to the octet rule and called "unusual bonding molecules". A rational explanation for the unusual bonding was advanced first by Rundle (R. J. Hatch and R. E. Rundle, J. Am. Chem. Soc., 1951, **73**, 4321; R. E. Rundle, ibid., 1963, **85**, 112) and then by Musher (J. I. Musher, Angew. Chem., Int. Ed. Engl, 1969, **8**, 54). This invokes the concept of "hypervalent bonding", thus according to molecular orbital calculations, d-orbital participation is not important for hybridization, instead only 3s and 3p orbitals are needed to make decets and dodecets in unusually bonded molecules (M. M. L. Cheu and R. Hoffmann, J. Am. Chem. Soc., 1976, **98**, 1647; A. E. Reed and P. von R. Schleyer, J. Am. Chem. Soc., 1990, **112**, 1434; W. Kutzelnigg, Angew. Chem., Int. Ed. Engl., 1984, **23**, 272). The chemical bonding used in the decet is called three-center four-electron bonding, namely, the central atoms should have two types of chemical bonding, one a three-center four-electron bonding (or an axial bond) and the other three sp^2 hybridized orbitals orthogonal to each other (or an equatorial bond); hence, the molecular structure should be a trigonal bipyramid (TBP). This hypothetical molecular structure agrees well with the crystal structure of some unusually bonded molecules. The typical molecular orbitals for the TBP structure are illustrated in Figure 1, while the other unusual valency is a hexacoordinate square bipyramide (SP) which has three identical three-center four-electron bonds.

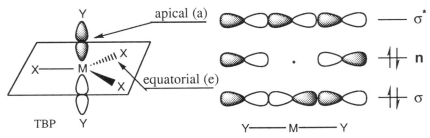

Fig. 1 Trigonal bipyramid structure and its MO

Since the pioneering work by Martin (J. C. Martin and R. Arhart, J. Am. Chem. Soc., 1971, **93**, 2339) and Kapovitz (I. Kapovitz and A.Kalman, J. Chem. Soc. Chem. Commun., **1971**, 649) on the first synthesis of hypervalent organic sulfur compounds, sulfuranes (**71**) and (**72**) (equations 73 and 74), sulfuranes and persulfuranes have been extensively and systematically investigated. (R. A. Hayes and J. C. Martin, "Organic Sulfur Chemistry", Elsevier, Amsterdam, 1985, Ch. 8, p. 408; J. C. Martin and E. F. Perozzi, Science, 1976, **191**, 154).

(eq. 73)

(eq. 74)

The chemical and physical properties of these compounds have been widely documented and according to Martin, molecular structures of hypervalent compounds, or the arrangement of the ligands around the central elements (TBP), obey the following restrictions; namely: 1) the Mutterties rule (E. L. Mutterties and R. A. Schunn, Quat. Rev., 1966, **20**, 245) which states that the more electronegative ligands occupy the axial positions while electropositive ligands must be at the equatorial positions. Lone pair electrons should be located at the equatorial positions, 2) five- and six-membered rings which stabilize the hypervalent compounds exist at both the axial and equatorial positions (R. A. Hayes and J. C. Martin, "Organic Sulfur Chemistry. Theoretical and Experimental Advances", Elsevier, Amsterdam, 1985, p. 408, Ch. 8). Martin has recommended a nomenclature system based on the formula N-M-L(AnBm), to describe hypervalent compounds, in such N is the number of valence electrons formally associated with the central atom M, and L indicates the number supplied by the ligands; A and B specify the ligands and n and m their numbers. For example, the sulfurane (**1**) is a member of the 10-S-4(C2O2) class (C. W. Perkins et al., J. Am. Chem. Soc., 1980, 102, 7753)

According to molecular orbital calculations (M. M. L. Chen and R. Hoffmann, J. Am. Chem. Soc., 1976, **98**, 1647) important factors to be noted for the preparation of hypervalent compounds are : 1) the smaller the electronegativity of the central element, the more stable the product, i.e., the three-center four-electron bonding must be stabilized; 2) as to the ligand, the more electronegative elements, or groups, stabilize the hypervalent bond; 3) lone pair electrons play an important role in the formation and stabilization of hypervalent (HV) bonds.

A combination of electropositive elements and electronegative ligands provides the most stable HV bonding, thus, in the case of chalcogenuranes, tellurium fulfills these requirements and gives the most stable HV molecules. Fluorine, chlorine and oxygen and groups containing these elements are appropriate ligands for hypervalent compounds. Valence bond calculations (A. Pross and S. S. Shaik, Acc. Chem. Res., 1983, **16**, 363; G. Sini, G. Ohanessian, P. C. Hiberty, and S. S. Shaik, J. Am. Chem. Soc., 1990, **112**, 1407) predict that the lower the bond energy the stronger is the HV bond. In fact telluranes are more stable than selenuranes. Sulfuranes are the least stable.

In the reactions of triphenylsulfonium chloride with phenyllithium, or phenylmagnesium bromide, the products obtained are biphenyl and diphenyl sulfide which are formed in quantitative yields. The mechanisms of these reactions have been investigated by Franzen and Mertz (V. Franzen and C. Mertz, Liebigs Ann. Chem., 1961, **643**, 24) who consider that a tetraphenylsulfurane is formed initially, but although both biphenyl and diphenyl sulfide were obtained, the sulfurane was not detected. A similar reaction between triphenylselenonium chloride and phenyl lithium affords biphenyl and diphenyl selenide, but again there was no evidence for tetraphenylselenurane. On the other hand, when diphenyldichlorotellurane, or triphenyltelluronium chloride, was the substrate tetraphenyltellurane could be obtained as a crystalline solid (G.Wittig, H. Fritz, Liebigs Ann. Chem., 1952, **577**, 39). This compound undergoes various substitution reactions with nucleophiles and also ligand couplings which lead to biphenyl and diphenyl telluride quantitatively . This is the first example of the isolation of a tellurane bearing four carbon ligands, but since that time more stable compounds of the type have been described. For example, bis(2,2'-biphenylylene)tellurane has been obtained by Hellwinkel (D. Hellwinkel, Annals of the New York Academy of Sciences, 1972, **192**, 158). Hellwinkel has also isolated bis(2,2'-biphenylylene)selenurane and recently Furukawa et al. (S. Ogawa et al., J. Chem. Soc., Chem. Commun., **1992**, 1141) have synthesized bis(2,2'-biphenylylene)sulfurane and have determined the crystal structure of this stable compound. Since the chemistry of sulfurane has been well documented by Martin, this review describes the HV compounds of tellurium and selenium. The article will be divided into two parts, 1) unstable HV compounds, such as ate-complexes of tellurium and selenium; 2) recent advancement in the study of the stable HV compounds of selenium and tellurium. Several review articles have been published on these subjects (D. L. Klayman and W. H. H. Günther, "Organic Selenium Compounds", John Wiley & Sons, New York, 1973; I. Hargittai and B. Rozsondai, "The Chemistry of Organic Selenium and Tellurium Compounds", ed. S. Patai and Z. Rappoport, Vol. 1, Ch. 3, John Wiley & Sons, New York, 1986; J. Bergmann and L. Engman, J. Sidén, ibid., Ch. 14; K. J. Irgolic, "The Organic Chemistry of Tellurium", Gordon and Breach, New York, 1974).

(b) Detection of unstable chalcogenuranes as intermediates

(i) Detection of tellurium and selenium ate-complexes [10-M-3(C3)] (M: Se, Te) using Li-Te and Li-Se exchange reactions

As shown in the previous section, dicoordinated tellurides and selenides are thermally stable. They react readily with electrophiles such as alkylating agents to give the corresponding onium salts, but they do not undergo substitution reactions with nucleophilic reagents. However, diaryl tellurides are known to react with Grignard reagents in the presence of Ni^{II} and Co^{II} salts as catalysts (S. Uemura, S. I. Fukuzawa, and S. R. Patil, J. Organomet. Chem., 1983, **243**, 9). The mechanism for this type of reaction may involve the formation of metal complexes (**73**) which react further with the Grignard reagent to undergo exchange of the ligands. Coupling then takes place to give mixtures of the products as shown in equation 75.

$$L_2MCl_2 \xrightarrow{RMgBr} L_2M\begin{smallmatrix}R\\R\end{smallmatrix} \xrightarrow[-R-R]{ArTeAr'} L_2M\begin{smallmatrix}TeAr\\Ar'\end{smallmatrix} \quad \mathbf{73}$$

[$L=R_3P$; M=Ni, Co]

$$\xrightarrow{ArTeAr'} \Big\downarrow \begin{array}{c} R'MgBr / \\ - ArTeMgBr \end{array} \quad (eq.\ 75)$$

$$L_2M \xleftarrow[-R'-Ar']{} L_2M\begin{smallmatrix}R'\\Ar'\end{smallmatrix}$$

In the case of diaryl sulfides, although they do not react directly with Grignard reagents, coupling reactions have been conducted in the presence of metallic salts (H. Takei et al., Chem. Lett., **1979**, 1447). Since the electronegativity of tellurium is 2.01, the compounds of tellurium may undergo nucleophilic reactions, or electron accepting reactions. In fact, recently Furukawa et al. and Reich et al. have found that diaryl tellurides undergo facile ligand exchange reactions on treatment with organolithium reagents. The organotellurium-organolithium exchange reactions proceed via the initial formation of $Ar_3Te^-Li^+$ telluranes [10-Te-3(C3)] (e.g. the ate-complex **74**) which are detected by examining the low temperature ^1H-, ^{13}C-, ^7Li- and ^{125}Te-NMR spectra. For example, in its ^{125}Te-NMR spectrum diphenyl telluride shows a peak at 670.4 ppm which shifts to upfield to 320.6 ppm when one equivalent of PhLi is added to the sample at -100 °C (S. Ogawa et al., Heteroatom Chem., 1992, **3**, 423). This chemical shift change indicates the unambiguous formation of a tellurane ate-complex as shown in equation 76 (H. J. Reich et al., J. Am. Chem. Soc., 1991, **113**, 414; idem, Phosphorus Sulfur Silicon, 1992, **67**, 83).

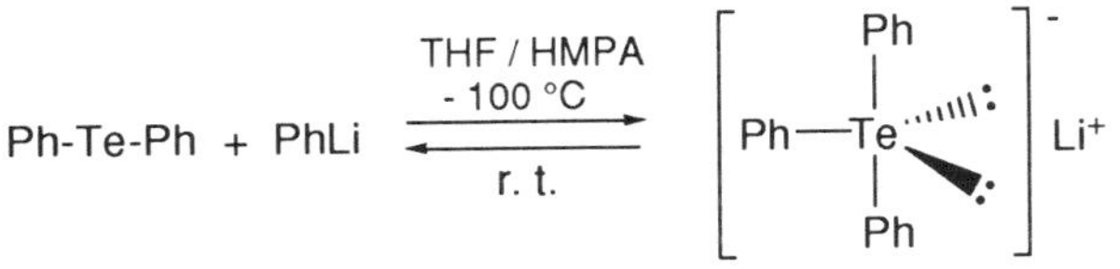

$$\text{Ph-Te-Ph} + \text{PhLi} \xrightleftharpoons[\text{r.t.}]{\text{THF / HMPA} \atop -100\,°\text{C}} \left[\begin{array}{c} \text{Ph} \\ | \\ \text{Ph—Te}\cdots \\ | \\ \text{Ph} \end{array} \right]^{-} \text{Li}^{+} \quad (\text{eq. 76})$$

74

Similarly, when phenyl pentafluorophenyl and 3,5-dichloro-2,4,6-trifluorophenyl tellurides were treated with pentafluorophenyllithium, the ate-complex (**75**) was observed by examining the low temperature ^{125}Te-, ^{19}F-, and ^{13}C-NMR spectra (S. Ogawa et al., Chem. Lett., **1992**, 2471). Interestingly, in these reactions, the ligand exchange clearly takes place at the apical position and pseudorotation is not observed (equation 77).

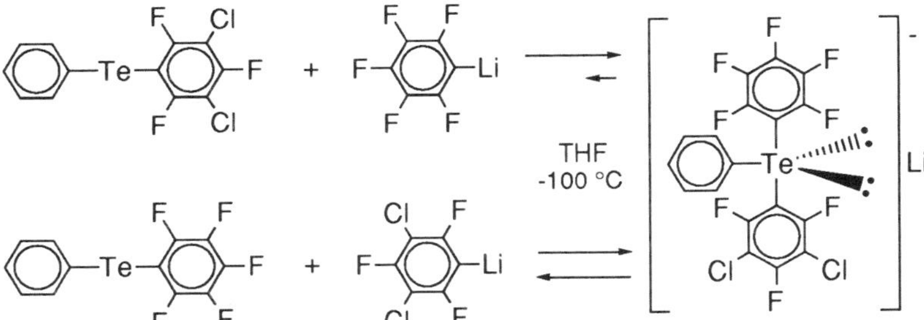

75 (eq. 77)

Neither the selenium nor the sulfur analogues of tellurium-ate complex (**75**) have been isolated or detected. However, variable temperature ^{77}Se-NMR spectroscopy has been used to prove the existence of selenium ate-complex (**76**), which is formed when 2-(2-bromophenyl)dibenzoselenophene is treated with butyllithium (equation 78) (H. J. Reich, B. O. Gudmundson, and D. R. Dykstra, J. Am. Chem. Soc., 1992, **114**, 7937).

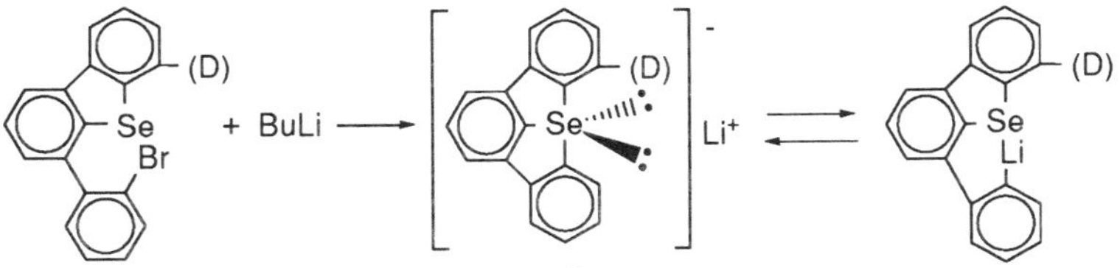

76

Masutomi and Furukawa (Y. Masutomi and N. Furukawa, Heteroatom Chem., 1994, in press) have also identified the selenium ate-complex (**77**) through ^{77}Se-NMR experiments. This complex is obtained by treating phenyl pentafluorophenyl selenide with pentafluorophenyllithium (equation 79).

$$\text{Ph-Se-C}_6\text{F}_5 + \text{F-C}_6\text{F}_4\text{-Li} \xrightleftharpoons[\text{-100 °C}]{\text{THF}} [\text{Ph-Se(C}_6\text{F}_5)_2]^- \text{Li}^+ \quad \text{(eq. 79)}$$

77

A similar complex is formed when 2,5-dichloro-2,4,6-trifluorophenyl phenyl selenide is reacted with pentafluorophenyllithium.

(ii) Organic synthesis via tellurium ate-complexes

Dicoordinated tellurium compounds undergo facile Te-Li exchange reactions via tellurium ate-complexes. Using this process, new organolithium reagents can be generated and applied for use in organic synthesis. Sonoda and his co-workers have extensively investigated the Te-Li metal exchange reactions using various tellurides such as telluroesters, tellurocarbamates and related compounds bearing many functional groups (T. Hiiro et al., Angew. Chem., Int. Ed. Engl., 1987, **26**, 1187; N. Sonoda et al., Organometallics, 1990, **9**, 1355; Synth. Commun, 1990, **20**, 1703; J. Am. Chem. Soc., 1990, **112**, 455). These authors report that the tellurium substrates on treatment with alkyllithium undergo facile Te-Li exchange reactions resulting in the formation of species such as acyl (**78**), carbamoyl (**79**), and thiocarbamoyl (**80**) anions. Anions of these types are difficult to prepare by other methods and they can be reacted in situ with electrophiles to afford various synthetically important products. For example, the acyl anions, generated from telluroesters by reacting them with butyllithium, combine with benzaldehyde to yield C-acylbenzyl alcohols (equation 80).

$$[\bar{\text{RCO}}] \quad [\text{RNH}\bar{\text{CO}}] \quad [\text{RNH}\bar{\text{CS}}]$$
$$\textbf{78} \quad\quad \textbf{79} \quad\quad\quad \textbf{80}$$

$$\text{RCOTeR'} + \text{BuLi} \xrightarrow{-78°\text{C}} [\bar{\text{RCO}}] \xrightarrow{\text{PhCHO}} \text{Ph-}\underset{\underset{\text{R}}{|}}{\overset{\overset{\text{OH}}{|}}{\text{C}}}\text{-H} \quad \text{(eq. 80)}$$
$$\textbf{78} \quad\quad\quad\quad \underset{\text{R}}{\overset{\text{C=O}}{}}$$

(iii) Ligand exchange and coupling reactions of chalcogenide oxides with organometallic reagents

Although sulfoxides are known to undergo simple ligand exchange reactions on treatment with Grignard, or organolithium reagents, there are only a few reports dealing with their ligand coupling reactions. One example is due to Andersen et al. (K. K. Andersen, R. L. Caret and I. K.-Nielsen, J. Am. Chem. Soc., 1974, **96**, 8026.; B. K. Ackerman et al., J. Org. Chem., 1974, **39**, 964) who describe the action of arylmagnesium bromides on diphenyl sulfoxide which produces the sulfuranes (**81**) (equation 81).

$$Ph-S(O)-Ph + ArMgBr \longrightarrow \left[\begin{array}{c} Ar \\ | \\ Ph-S-Ph \\ | \\ OMgBr \end{array} \right] \quad \text{(eq. 81)}$$
                                                                                          81

Substitution reactions of this type are generally considered to proceed via a simple S_N2 type process involving a hypervalent sulfurane, e.g. (**82**), as a transition state, rather than as an intermediate (equation 82) (J. P. Lockard, C. W. Schroeck and C. R. Johnson, Synthesis, **1973**, 485; N. Kunieda, J. Nokami and M. Kinoshita, Chem. Lett., **1977**, 244; T. Durst et al., Can. J. Chem., 1974, **52**, 76; M. Hojo et al., Synthesis, **1977**, 74; N. Furukawa, T. Shibutani and H. Fujihara, Tetrahedron Lett., 1987, **28**, 2727; C. Cardellicchio, V. Fiandanese and F. Naso, J. Org. Chem., 1992, **57**, 1718).

$$p\text{-Tol-}\overset{*}{S}(O)\text{-CH}_2R + R'Li \longrightarrow \left[\begin{array}{c} R' \\ | \\ p\text{-Tol}-S-OLi \\ | \\ CH_2R \end{array} \right] \longrightarrow p\text{-Tol-}\overset{*}{S}(O)\text{-R'} \quad \text{(eq. 82)}$$
                                                                             82   inversion

However, it has been found that thianthrene-5,5,10-trioxide reacts with butyllithium to afford the ring contracted dibenzothiophene-5,5-dioxide (**83**) in low yield (equation 83) (H. Gilman and D. R. Swayampatai, J. Am. Chem. Soc., 1955, **77**, 337; ibid., 1957, **79**, 208).

(eq. 83)

This reaction has been applied to synthesize various 1,9-disubstituted dibenzothiophenes and selenophene derivatives (**84**) from the reactions of 4,6-disubstituted thianthrene and selenanthrene monoxides with Grignard or alkyllithium reagents (equation 84). Such ring contraction processes may proceed via the formation of sulfuranes (**85**), or the corresponding selenuranes (**86**), as reaction intermediates (T. Kimura et al., Heteroatom Chem., 1993, **4**, 243).

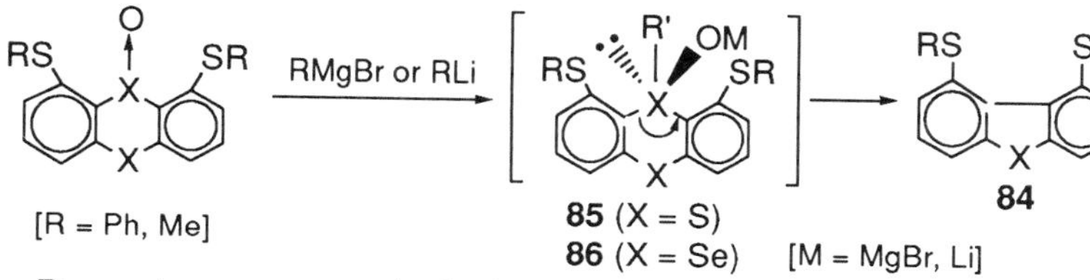

$$\text{[eq. 84]}$$

[R = Ph, Me]
85 (X = S)
86 (X = Se)   [M = MgBr, Li]

Recently unsymmetrical diaryl sulfoxides have been shown to undergo facile ligand exchange on treatment with aryllithiums. When all three aryl groups are different all three possible sulfoxides are produced via the disproportionation of the sulfurane (**87**) (equation 85) (S.-K. Chung and S. Sakamoto, J. Org. Chem., 1981, **46**, 4590; N. Furukawa et al., ibid., 1991, **56**, 6341).

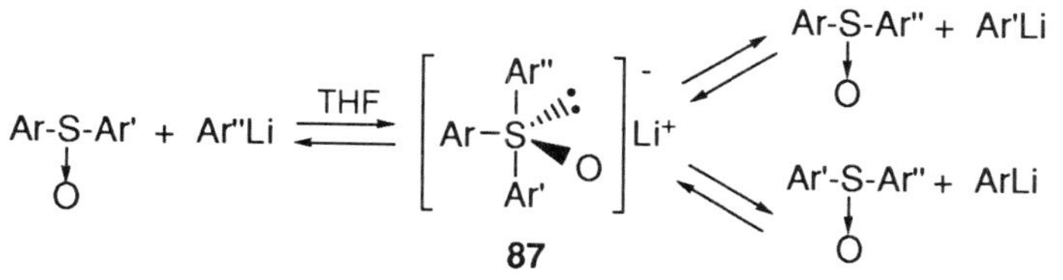

$$\text{(eq. 85)}$$

The reactions of sulfoxides are familiar, but those of selenoxides and telluroxides are less well known. Diphenyl selenoxide has been reacted with phenylmagnesium bromide to give triphenyl selenonium bromide which reacts with more phenylmagnesium bromide to afford tetraphenyl selenurane (**88**). This product then yields diphenyl selenide and biphenyl (equation 86). This result suggests a marked difference in reactivity between sulfoxides and selenoxides since Se-O bond scission must occur readily (Y. Iwata et al., Bull. Chem. Soc. Jpn., 1981, **54**, 2065).

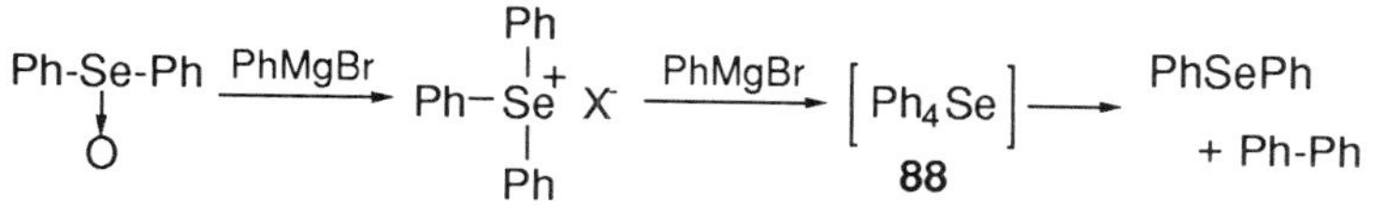

$$\text{(eq. 86)}$$

Recently, however, it has been proved that diaryl selenoxides react with organolithiums to give disproportionated selenoxides and selenides together with selenonium salts, through both C-Se and Se-O bond fissions acting competitively (see equation 87) (N. Furukawa et al., J. Org. Chem., 1991, **56**, 6341).

$$\text{Ph-Se-Tol-}p \xrightarrow[\text{THF}]{0.2 \text{ eq. BuLi}} \text{Ph-Se-Tol-}p + \text{Ph-Se-Ph} + p\text{-Tol-Se-Tol-}p \text{ (with O below each)} + \text{Ph-Se-Tol-}p + \text{Ph-Se-Ph} + p\text{-Tol-Se-Tol-}p \qquad (\text{eq. 87})$$

Diaryl telluroxides have been reacted with one molar equivalent phenyllithium to give triaryloxytelluranes (**89**) in very high yield (see equation 88). This type of reaction is not shown by sulfoxides or selenoxides (Y. Masutomi, N. Furukawa and T. Erata, Heteroatom Chem., to be published, 1994). Although the structures of the oxytelluranes have not been vigorously proved, on hydrolysis with dilute hydrochloric acid these products are converted into the corresponding telluronium salts (**90**) (equation 88).

$$\text{Ph-Te-Tol-}p \xrightarrow[\text{THF}]{\text{PhLi}} \left[p\text{-Tol}-\underset{\underset{\text{Ph}}{|}}{\overset{\overset{\text{Ph}}{|}}{\text{Te}}}\text{OLi} \right] \xrightarrow{\text{HX}} p\text{-Tol}-\underset{\underset{\text{Ph}}{|}}{\overset{\overset{\text{Ph}}{|}}{\text{Te}}}^{+} \text{X}^{-} \qquad (\text{eq. 88})$$

$$\qquad\qquad\qquad\qquad\qquad\qquad\text{89} \qquad\qquad\qquad\qquad \text{90}$$

The variability of reactivity towards organometallic reagents displayed by chalcogen oxides is accounted for in terms of the electronegativity and bond energies of M-O and M-C (M: S, Se, Te) as shown in Table 1, namely, in the case of sulfoxides the C-S bond energy is smaller than that of S-O bond and hence organometallic reagents attack preferentially at the C-S σ^*-bond rather than the S-O σ^*-bond. Meanwhile, in the case of telluroxides the Te-O bond energy is smaller than that of the Te-C bond, and telluroxides may give oxytelluranes (e.g. **89**) or telluronium lithium oxides.

(iv) Formation of tetraarylchalcogenuranes [10-M-4(C4)] (M: S, Se, Te) and ligand coupling reactions

The reactions of triarylsulfonium salts with organometallic reagents have long been investigated with the aim of detecting or isolating sulfuranes [10-S-4(C4)] (**91**) which lead to quantitative yields of ligand coupling products, namely, biaryls and diaryl sulfides (equation 89) (V. Franzen and C. Mertz, Liebigs Ann. Chem., 1961, **643**, 24; J. Bornstein, J. Schields and J. Supple, J. Org. Chem., 1967, **32**, 1499, Y. H. Khim and S. Oae, Bull. Chem. Soc. Jpn., 1969, **42**, 1968; B. M. Trost, Top. Curr. Chem., 1973, **41**, 1; M. Hori et al., Chem. Pharm. Bull., 1974, **22**, 2004, 2020).

$Ar_3S^+X^- + ArMgBr$ (or ArLi) $\longrightarrow [Ar_4S] \longrightarrow$ Ar-S-Ar + Ar-Ar (eq. 89)
                                            91

The formation of tetraaryl sulfuranes has been indirectly confirmed using the reactions of different combinations of triarylsulfonium salts and aryl Grignard reagents, or by examining the coupling modes of the substitution reaction using ^{14}C-labeled triphenyl sulfonium salts (**92**) (equation 90) (D. Harrington, J. Jacobus and K. Mislow, J. Chem. Soc. Chem. Commun., **1972**, 1079).

(eq. 90)

92   •: ^{14}C

Final proof of the participation of tetraarylsulfuranes was established by reacting pentafluorophenyllithium (**93**) with either pentafluorophenyl-trifluorosulfurane or sulfur tetrafluoride (see equation 91). The product is stable at 0 °C, but, so far, it cannot be isolated, however, its structure is secured through ^{19}F-NMR measurement. Above 0 °C it decomposes into di-(pentafluorophenyl)sulfide and decafluorobiphenyl (W. Sheppard, J. Am. Chem. Soc., 1971, **93**, 5597).

$3C_6F_5Li + C_6F_5SF_3$
$\qquad\qquad\qquad \xrightarrow{-80\,°C} [(C_6F_5)_4S] \xrightarrow{0\,°C} (C_6F_5)_2S + (C_6F_5)_2$ (eq. 91)
$4C_6F_5Li + SF_4$
$\qquad\qquad\qquad\quad\;\;$**93**

By way of contrast, tetraphenyltellurane (**94**) is a well known stable compound (see introduction) which is formed as shown in equation 92 (G. Wittig and H. Fritz, Liebigs Ann. Chem., 1952, **577**, 39). It can be isolated as a crystalline solid (I. C. S. Smith et al., Organometallics, 1982, **1**, 350).

$TeCl_4$ + 4PhLi
$\qquad\qquad\quad \xrightarrow{Et_2O}$ $\boxed{Ph_4Te}$ $\xrightarrow{\Delta}$ Ph-Te-Ph + Ph-Ph
$PhTeCl_3$ + 3PhLi $\qquad$ m.p. 105~108 °C $\xrightarrow{CHCl_3}$ Ph-CHCl$_2$ (eq. 92)
$\qquad\qquad\qquad\qquad\qquad\quad$**94**

The stable tellurane (**95**) and its selenium counterpart (**96**) have been obtained by Hellwinkel's group (see equation 93) (D. Hellwinkel and G. Fahrbach, Ann. Chem., 1968, **712**, 1; 1968, **715**, 68). Such compounds ring open when treated with acids and chloroalkanes, and react with *t*-butylthiol to afford biphenyl and di-*t*-butyl disulfide and also undergo extremely facile ligand exchange, or disproportionation (D. H. R. Barton, S. A. Glover and S. V. Ley, J. Chem. Soc. Chem. Commun., **1977**, 266; S. A. Glover, J. Chem. Soc. Perkin I, **1980**, 1338). Although the mechanism of the disproportionation reaction is uncertain, it seems possible that an ionic concerted pathway is followed, phenyl radicals are not involved.

$$\text{95} \xleftarrow[\text{Et}_2\text{O}]{\text{Te, Br, Br}} \text{[dilithiobiphenyl]} \xrightarrow[\text{Et}_2\text{O}]{\text{Se, NTs}} \text{96} \quad \text{(eq. 93)}$$

Quite recently Furukawa and co-workers (S. Ogawa et al., Tetrahedron Lett., 1991, **32**, 3179; 1992, **33**, 93) have succeeded in detecting tetraphenyl sulfurane (**91**) and tetraphenyl selenurane (**88**) by low temperature NMR spectroscopy. Tetraphenyl selenurane is stable at 0 °C and tetraphenyl sulfurane, long believed to be undetectable exists at -40 ~ -50 °C. These compounds can be formed either from the appropriate oxides, or triphenyl salts as indicated in equation 94.

$$\text{Ph-X-Ph} + 2\text{PhLi} \longrightarrow [\text{Ph}_4\text{X}] \longleftarrow \text{Ph}_3\text{X}^+\text{Y}^- + \text{PhLi} \quad \text{(eq. 94)}$$
$$\downarrow\text{O}$$
$$\text{91 (X = S)}$$
$$\text{88 (X = Se)}$$

All three tetraphenyl chalcogenuranes are thus recognized and it has been shown that the rates at which they undergo ligand coupling are in the order S > Se > Te, as expected from their relative stability (see Table 2) (S. Ogawa, S. Sato and N. Furukawa, Tetrahedron Lett., 1992, **33**, 7925). Although tetraphenyl chalcogenuranes give ample information on the thermally stable stability of the bonds involved, the quest for a sulfurane bearing four-carbon ligands was finally resolved when the sulfurane (**97**) was isolated from the reaction of dibenzothiophene S-oxide and o,o'-dilithiobiphenyl in the presence of trimethyl silyl triflate as an activator (equation 95) (S. Ogawa et al., J. Chem. Soc. Chem. Commun., **1992**, 1141). These authors determined the X-ray crystal structure of the sulfurane (**97**). The ORTEP drawing and some other physical parameters are shown in Figure 3. Both (**97**) and (**95**) [the structure of the selenurane (**96**) has not been determined as yet] are trigonal bipyramidal molecules in which apical and equatorial bonds can be distinguished.

Table 2. Activation Parameters of Ligand Coupling Reaction of Tetraphenyl Chalcogen Compounds. a)

Compd (Solvent)	k_1 (s^{-1})		Eact (kcal/mol)	$\Delta G^{\neq}_{298}$ (kcal/mol)	$\Delta H^{\neq}$ (kcal/mol)	$\Delta S^{\neq}$ (eu)
Ph$_4$S (THF-d$_8$)	2.48x10^{-4} (-67 °C)	1.22x10^{-4} (-72 °C)	10.9	17.5	10.5	-23.5
	5.00x10^{-5} (-77 °C)	3.25x10^{-5} (-82 °C)				
Ph$_4$Se (THF-d$_8$)	2.20x10^{-4} (0 °C)	1.30x10^{-4} (-5 °C)	21.3	20.4	21.3	3.1
	5.31x10^{-5} (-11 °C)	2.21x10^{-5} (-15 °C)				
Ph$_4$Te (Toluene-d$_8$)	3.91x10^{-4} (84 °C)	1.77x10^{-4} (74 °C)	29.0	26.9	28.4	5.2
	3.00x10^{-5} (63 °C)	8.26x10^{-6} (52 °C)				

a) Values shown are least-square treatments of Arrhenius and Eyring plots.

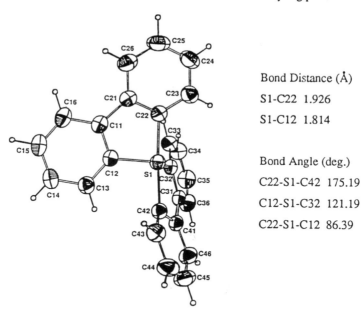

Bond Distance (Å)
S1-C22 1.926
S1-C12 1.814

Bond Angle (deg.)
C22-S1-C42 175.19
C12-S1-C32 121.19
C22-S1-C12 86.39

Figure 3 ORTEP drawing of **95**

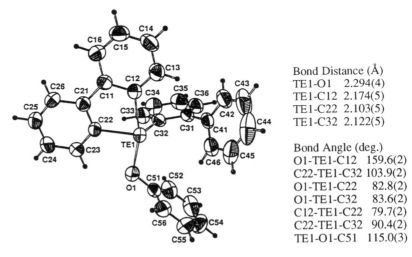

(eq. 95)

97

NMR studies conducted at room-temperature obscure this detail, however, and the two bond types are "averaged out" by pseudorotation. Fortunately this problem is resolved at lower temperatures and it is possible to use variable temperature NMR spectroscopy to compute the energy barriers for pseudorotation, or turnstile rotation. Thus for the tellurane (**95**) this value is 9 Kcal/mol and for the selenurane (**96**) it is 13 kcal/mol, showing that the stability and pseudorotational ability of the two compounds are opposed to one another (S. Ogawa et al, Tetrahedron Lett., 1992, **33**, 1915).

Chalcogenuranes of the type (**95, 96, 97**) react with hydroxyl compounds (e.g. phenol) to give ring opened products, which undergo ligand coupling or ipso-substitution (see equation 96) (N. Furukawa, Y. Matsunaga and S. Sato, Synlett., **1993**, 655; S. Sato and N. Furukawa, Chem. Lett., **1994**, 889). In the case of the tellurane the initial product (**98**, M=Te) is a crystalline solid, the X-ray crystal structure of which is shown in Figure 4. It has a slightly distorted trigonal bipyramidal geometry and it represents the first example of its class (S. Sato, N. Kondo and N. Furukawa, Organometallics, 1994, in press).

Bond Distance (Å)
TE1-O1   2.294(4)
TE1-C12 2.174(5)
TE1-C22 2.103(5)
TE1-C32 2.122(5)

Bond Angle (deg.)
O1-TE1-C12   159.6(2)
C22-TE1-C32 103.9(2)
O1-TE1-C22   82.8(2)
O1-TE1-C32   83.6(2)
C12-TE1-C22 79.7(2)
C22-TE1-C32 90.4(2)
TE1-O1-C51   115.0(3)

Figure 4. ORTEP drawing of **98** (M=Te)

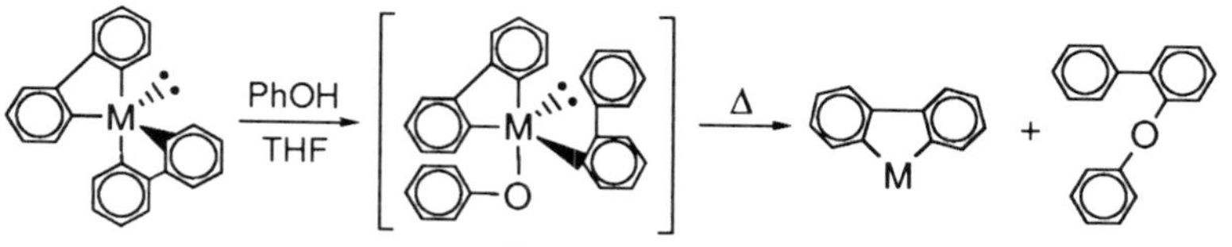

(c) Miscellaneous hypervalent compounds of sulfur, selenium and tellurium

Numerous examples of hypervalent compounds based upon selenium and tellurium have been reported. Chalcogenuranes of the type 10-M-4(C2X2) include dihalides of the constitution (**99**; M=S, Se, or Te). These can be prepared by mixing the appropriate diaryls with halogens (equation 97). Alternatively diaryltelluranes can be synthesized from arenes, or diarylmercury compounds, by treatment with tellurium tetrahalides (see equations 98 and 99) (N. W. Alcock, W. D. Harrison, J. Chem. Soc., Dalton Trans., **1983**, 251; C. S. Mancinelly, D. D. Titus and R. F. Ziolo, J. Org. Chem., 1977, **113**, 140).

$$\text{Ar-M-Ar} + X_2 \longrightarrow \text{Ar-M-Ar with } X, X \text{ (99)} \quad \text{[M = S, Se, Te]} \quad (eq.\ 97)$$

$$\text{ArH} + \text{TeCl}_4 \xrightarrow{\text{AlCl}_3} \text{Ar}_2\text{TeCl}_2 \quad (eq.\ 98)$$

$$\text{Ph-Hg-Ph} + \text{TeCl}_4 \longrightarrow \text{Ph}_2\text{TeCl}_2 \quad (eq.\ 99)$$

Fluorinated analogues can be formed from diaryls by reacting them with a range of fluorinating agents (see equations 100, 101 and 102) (K. J. Wynne, Inorg. Chem., 1970, **9**, 299; S. Herberg et al., Z. Anorg. Allg. Chem., 1982, **492**, 95; ibid., 1982, **494**, 151, 159). Other methods include the reactions of diaryltelluride monoxides with hydrogen fluoride (equation 103), or triaryltelluronium chlorides with hydrogen fluoride and silver oxide (equation 104) (I. D. Sadaekov et al., Zh. Obshch. Khim., 1976, **46**, 1660; 1977, **47**, 1305).

$$\text{Ar-Se-Ar} \xrightarrow{\text{AgF}_2} \text{Ar}_2\text{SeF}_2 \quad (eq.\ 100)$$

$$\text{Ph-Te-Ph} + \text{Ph}_2\text{XF}_2 \longrightarrow \text{Ph}_2\text{TeF}_2 + \text{Ph-X-Ph} \quad \text{(eq. 101)}$$
$$[X = S, Se]$$

$$\text{Ar-X-Ar} \xrightarrow{SF_4} \text{Ar}_2\text{XF}_2 \quad \text{(eq. 102)}$$
$$(X = Se, Te)$$

$$\underset{\downarrow O}{\text{Ar-Te-Ar}} \xrightarrow{HF} \text{Ar}_2\text{TeF}_2 \quad \text{(eq. 103)}$$

$$\text{Ar}_3\text{TeCl} \xrightarrow[HF]{Ag_2O} \text{Ar}_3\text{TeF} \quad \text{(eq. 104)}$$

Similarly the preparation of other diaryltelluranes and diarylselenuranes (e.g. **100** and **101**) can be achieved through the reactions of diaryltelluro- or -seleno-dihalides with isothiocyanate (equation 105) (C. S. Mancinelli, D. D. Titus and R. F. Ziolo, J. Organomet. Chem., 1977, **140**, 113) or thiocyanate (equation 106) (R. Paetzold and U. Lindner, Z. Anorg. Allg. Chem., 1967, **350**, 295), respectively.

$$\text{Ph}_2\text{TeCl}_2 \xrightarrow{NaNCS} \underset{\textbf{100}}{(\text{Ph}_2\text{TeNCS})_2\text{O}} \quad \text{(eq. 105)}$$

$$\text{Ar}_2\text{SeX}_2 \xrightarrow{NaSCN} \underset{\textbf{101}}{\text{Ar}_2\text{Se}(\text{SCN})_2} \quad \text{(eq. 106)}$$

Halogeno ligands can be replaced by hydroxyl groups and alkoxylated chalcogenuranes (e.g. **102**) have been made by treating dihalides with sodium alkoxide (see equation 107). Acyloxy analogues (**103**) are formed either from the appropriate oxides by the action of acid anhydrides (equation 108), or from the tellurides, or selenides, by treatment with hydrogen peroxide and a carboxylic acid in the presence of an ammonium salt (see equation 109) (M. Wieber and E. Kauzinger, J. Organomet. Chem., 1977, **129**, 339).

$$\text{R}_2\text{TeCl}_2 + \text{MeONa} \longrightarrow \underset{\textbf{102}}{\text{R}_2\text{Te}(\text{OMe})_2} \quad \text{(eq. 107)}$$
$$[R = Me, Ph; R' = Me, Et]$$

$$\text{R}_2\text{MO} + (\text{R'CO})_2\text{O} \longrightarrow \underset{\textbf{103}}{\text{R}_2\text{M}(\text{OCOR'})_2} \quad \text{(eq. 108)}$$
$$[M = Se, Te]$$

$$\text{R}_2\text{M} + \text{R'COOH} \xrightarrow[Et_3N^+\text{-R X}^-]{H_2O_2} \textbf{103} \quad \text{(eq. 109)}$$

Alkoxytelluranes are also formed by reacting triaryltelluronium halides with sodium alkoxides, and these products may be reacted further with thiols to afford thio derivatives (equation 110) (M. Wieber, E. Schmidt, Z. Anorg. Allg. Chem., 1989, **556**, 189; M. Wieber, S. Rohse, ibid., 1991, **592** 202). Other sulfur containing anions act as nucleophiles and combine with telluronium halides to give the corresponding telluranes (**104**). On heating these compounds fragment into diaryltellurides and arylthiolated derivatives (equation 111) (M. Wieber, E. Schmidt and C. H. Burschka, Z. Anorg. Allg. Chem., 1985, **525**, 127).

$$Ph_3Te^+Cl^- \xrightarrow{MeONa} Ph_3TeOMe \xrightarrow{R-SH} Ph_3TeSR + MeOH \quad (eq. 110)$$

$$Ph_3Te^+Cl^- \xrightarrow{RXNa} \underset{\mathbf{104}}{Ph_3TeXR} \xrightarrow{\Delta} Ph_2Te + RXPh \quad (eq. 111)$$

[X = S, S$_2$CO, S$_2$CS]

Attempts to synthesize dithiatelluranes (**105**), by reacting dialkoxytelluranes with dithiols fail, instead the only products isolated are polysulfides (**106**) and diphenyl telluride (equation 112) (M. Wieber and E. Kauzinger, J. Organomet. Chem., 1977, **129**, 339).

$$R_2Te(OEt)_2 + HS\text{-}X\text{-}SH \xrightarrow{/\!/} \underset{\mathbf{105}}{R_2Te\!\!\begin{array}{c}S\\ \diagdown\\ S\end{array}\!\!X}$$

$$\left[\begin{array}{l} R = Ph, Me \\ X = \text{-}(CH_2)_n\text{-} \ (n = 2, 3) \end{array}\right] \searrow \underset{\mathbf{106}}{\text{-}(S\text{-}X\text{-}S)_n\text{-}} + R_2Te \quad (eq. 112)$$

An interesting class of dicationic sulfuranes (**108**) and selenuranes (**110**) is available through the oxidation of the trisulfide (**107**) (equation 113) and its analogues (**109**, X=N, S, or Se) with nitrosonium phosphorus hexafluoride (equation 114). In the case of the triselenium product (**110** : X=Se) the ^{77}Se-NMR spectrum reveals a selenium-selenium spin-coupling constant J=200Hz (H. Fujihara et al., J. Am. Chem. Soc., 1992, **114**, 3117).

$$\underset{\mathbf{107}}{[107]} \xrightarrow{2NOPF_6^-} \underset{\mathbf{108}}{[108]} \ 2PF_6^- \quad (eq. 113)$$

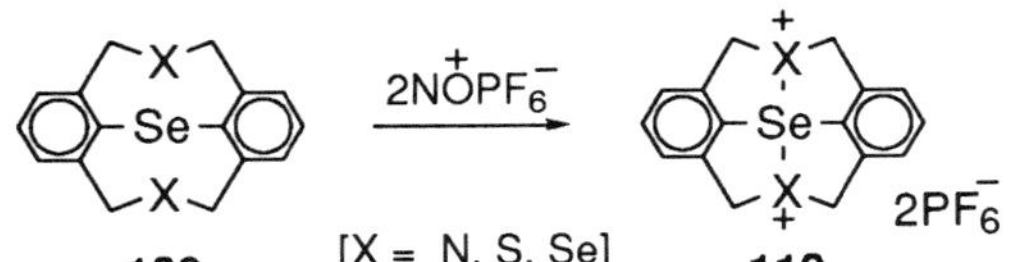

$$\text{109} \xrightarrow{2\text{NO}^+\text{PF}_6^-} \text{110} \quad [X = N, S, Se]$$

(eq. 114)

A new stable selenurane (**111**) has been prepared and determined its structure by X-ray diffraction analysis (equation 115) (H. Fujihara et al., J. Am. Chem. Soc., 1991, **113**, 6337).

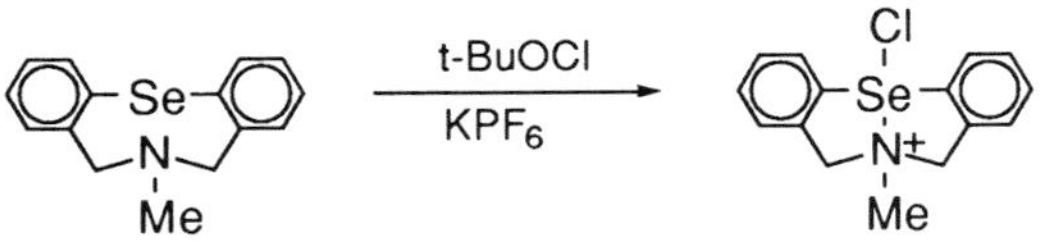

$$\xrightarrow{\text{t-BuOCl}, \text{KPF}_6} \text{111}$$

(eq. 115)

Guide to the Index

This index is constructed in a similar manner to the volume indexes of the first edition of the Chemistry of Carbon Compounds. However, to make the index easier to use, more descriptive entries have been made for the commonly occurring individual, and groups of chemicals.

The indexes cover primarily the chemical compounds mentioned in the text, and also include reactions and techniques, where named, and some sources of chemical compounds such as plant and animal species, oils, etc.

Chemical compounds have been indexed alphabetically under the names used by authors, editing being restricted to ensuring uniformity of entries under the same heading. In view of the alternative nomenclature that can often be used, a limited amount of cross-referencing has been done where it is considered to be helpful, but attention is particularly drawn to Convention 2 below.

For this and the succeeding volumes, the indexing conventions listed below have been adopted.

1. Alphabetisation

(a) A letter by letter alphabetical sequence is followed for entries, firstly for the main entry, followed by the descriptive entry.

(b) The following prefixes have not been counted for alphabetising:

n-	o-	as-	meso-	C-	E-
	m-	sym-	cis-	O-	Z-
	p-	gem-	trans-	N-	
	vic-			S-	
		lin-		Bz-	
				Py-	

Some prefixes and numbering have been omitted in the index, where they do not usefully contribute to the reference.

(c) The following prefixes have been alphabetised:

Allo	Epi	Neo
Anti	Hetero	Nor
Bis	Homo	Pseudo
Cyclo	Iso	

2. Cross references

In view of the many alternative trivial and systematic names for chemical compounds, the indexes should be searched under any alternative names which may be indicated in the main body of the text. Only a limited amount of cross-referencing has been carried out, where it is considered that it would be helpful to the user.

3. Derivatives

Simple derivatives are not normally indexed if they follow in the same short section of the text.

4. Collective and plural entries

In place of "– derivatives" the plural entry has normally been used. Plural entries have occasionally been used where compounds of the same name but differing numbering appear in the same section of the text.

5. Main entries

The main entry of the more common individual compounds is indicated by heavy type. Multiple entries, such as headings and sub-headings over several pages are shown by "–", e.g., 67–74, 137–139, etc.

Index

4-Acetamidobenzenesulfonyl azide, 432
2-Acetamido-6-hydroxy-3-iodo-7-methylnaphthoquinone, 233
Acetanilide, hydroxylation, 174
Acetone cyanohydrin nitrate, 167
Acetophenone, hydroxylation, 174
3-Acetoxy-2-(4-acetoxy-3,5-dimethoxyphenyl)-5,7-diacetylflav-3-ene, 369
4-Acetoxybenzofuran, 238
7-Acetoxybenzofuran, 238
3-Acetoxy-3[4,6-bis(dibenzyloxy)]-2,3,3a,8a-tetrahydro[2,3-b]benzofuran, 370
3-Acetoxy-2-bromo-5-methoxy-α,α-di(phenylthio)toluene, 354
5-Acetoxy-5-bromo-1,4-naphthoquinone, 261
2-Acetoxy-6-^{t}butylphenol, 361
5-Acetoxy-2,3-dihydro-2,2,7-trimethyl-4,6-di-^{i}propylbenzofuran, 362
3-(4-Acetoxy-3-methoxyphenyl)propyl acetate, mono-O-deacetylation, 345
6-Acetoxy-6-(3-methylenepenten-5-yl)-5-methylcyclohexa-2,4-diene, 256
10-Acetoxy-1,2,3,4-tetrahydro-9-hydroxyanthracene, 365
4-Acetoxythiochromane, ring-opening, 387, 388
β-Acetoxyvinyl sulfides, 466
9-Acetyl-acetamino-6,11-dihydroxy-7,8,9,10-5,12-naphthacene, 340
4-Acetylaminophenol, 326
2-(4-Acetylaminophenoxy)-2-methyl-1,3-benzodioxan-4-one, 326
3-(4-Acetylaminophenyl)-4-methyl-2-cyclohexenone, 267
1-Acetyl-1-arylsulfonyl-2-aroylhydrazines, 429
6-Acetyl-3,4-dihydro-7,8-dihydroxy-2-methyl-2H-benzofuran, 371
6-Acetyl-3,4-dihydro-7,8-dihydroxy-2-phenyl-2H-benzofuran, 371

2-Acetyl-1,4-dihydroxy-5,8-dihydronaphthalene, 338
Acetylenes, cyclotrimerisation, 119
3-Acetyl-4-hydroxybenzoate, 280
1-Acetyl-7-hydroxyindoline, synthesis from (1-acetylindolin-7-yl)thallium di(trifluoroacetate), 231
Acetylhypofluorite, 179
O-Acetylmorphine, reaction with N,N-dimethylformamide di-^{t}butyl acetal, 292
4-Acetylphenylfluorosulfonate, 285
2-Acetyl-4-phenyl-1-naphthols, 276
3-Acetyl-7-pyrrolidinylcoumarin, 328
4-Acetylveratrole, 149
Acylaminoketones, 156
Acyl anions, preparation from tellurium-ate complexes, 509
Acyl halides, reactions with benzyl bromide/palladium/zinc, 160
2-Acyl-3-hydroxycyclohex-2-enones, aromatisation, 336
6-Acyl-2-methoxynaphthalenes, 303
N-Acyl-N-methylsulfonamides, 449
Acyloxychalcogenuranes, 518, 519
2-Acylphenols, 275, 276
β-Acylpyrroles, synthesis from N-phenylsulfonylpyrrrole, 426
2-Acylresorcinols, 335, 336
Acyl sulfides, 408
1-Acyl-2-(4-tosyl)hydrazides reactions with sulfur halides, 427, 428
Adamantanediyl dication, 6
Aldehydes, synthesis from tosylhydrazones, 430
Alkanols, oxidation by bismuth(V) 132
O-Alkanols, phenylation, 132
Alkan-2-onyl aryl sulfides, 398
2-Alkanoyl-2,5-dimethyltetrahydrofurans, 331
Alkenes, epoxidation, 415
–, protonation, 139, 140

Alkenes (cont'd)
–, synthesis from thioesters, 390
–, – from thiolesters, 390
Alkenyl aryl selenides, 475
Alkenylphenols, cyclisation reactions, 265
N-Alkoxyaryl-N,N-dialkylamino-sulfonium tetrafluoroborates, 465
Alkoxychalcogenuranes, 518
Alkoxynitrocyclohexa-2,5-dienes, 204
3-Alkoxyphenols, 352
β-Alkoxypyranyl phenyl sulfides, 458
Alkylaminoarenes, synthesis by vicarious nucleophilic substitution, 85
Alkylarenes, protonation, 138
–, side chain rearrangement, 148
N-Alkylarenesulfonamides, 180
Alkyl aryl carbonates, methods of synthesis, 244
Alkyl aryl ethers, dealkoxylation, 298
–, dealkylation, 298, 299
Alkyl aryl selenides, methods of synthesis, 472
–, oxidation, 490–493
Alkylarylselenonium salts, methods of preparation, 499
Alkyl aryl sulfides, cleavage, 388, 389
–, methods of synthesis, 396
–, oxidation, 405, 406
–, uses in synthesis, 405–409
Alkyl aryl tellurides, oxidation, 493, 495, 496
–, reactions, 492
Alkyl benzenesulfenates, photolysis, 458, 459
–, reactions with alkenes, 457, 458
–, synthesis and reactions, 454, 455
Alkyl benzenesulfinates, methods of synthesis, 443
Alkyl benzyl sulfides, cleavage, 388
3-Alkylcatechols, 331
Alkyl 6-deoxykermesates, 339
N'-Alkyl-N-fluoro-1,4-diazonia-bicyclo[2.2.2]octane salts, 180
Alkyl 4-hydroxybenzoates, 280
Alkylhypochlorite, micelle-bound, 185
Alkylidenecyclopentano-cyclopentanones, methods of synthesis, 326

Alkyl 1-methyl-3,5,8-trihydroxy-anthraquinone-2-carboxylates, 339
Alkyl perbromides, 185
Alkylphenols, aldolisation type reactions, 263
–, methods of synthesis, 239
–, reactions, 262
2-Alkylphenols, synthesis from alkyl aryl ketones, 227
Alkyl phenyl ethers, methods of synthesis, 288
Alkyl phenyl selenides, oxidation, 502
5-Alkylresorcinols, 333
Alkyl selenoates, reactions with arene diazonium fluoroborates, 476
Alkylsulfonylbenzoic acids, 442
N-Alkylsulfonyl hydrazones, synthesis from tosylhydrazones/alkyl halides, 428
Alkyl 3-(sulfoxy)butanoates, 143
Alkyl tellurates, reactions with aryl halides, 492
Alkylthiols, synthesis from alkyl benzyl sulfides, 388
Alkylthiophenyl alkyl ethers, selective dealkylation, 320, 321
Alkyl 4-tolyl sulfones, 441, 442
Alkyl tosylates, methods of synthesis, 421
N-Alkyl-N'-tosylhydrazides, 427
Alkyl 3,5,8-trihydroxy-9,10-anthraquinone-2-carboxylates, 359, 360
Alkynes, cyclotrimerisation, 105–107
–, reactions with cyclobutadienes, 119, 120
–, – with haloalkenes/palladium, 120
–, – with vinylketones, 122
–, – with cobalt–divinyl ketone complexes, 106
–, trimerisation, 119
Alkynic tosylates, synthesis from alkynes, 420
Alkynyl aryl selenides, 475
Alkynyl aryl sulfones, methods of synthesis, 418
Alkynyl(phenyl)iodonium salts, synthesis from benzenesulfinic acids, 440

Alkynyl sulfones, 440, 441
Allenyl phenyl selenenate, 490
Allyl aryl selenides, 490
Allyl aryl sulfides, 411, 412
Allyl aryl sulfones, methods of synthesis, 417
–, synthesis from allyl arylsulfinates, 446
Allyl aryl sulfoxides, 488
2-Allyl-2,3-dihydro-2-vinylbenzofuran, 325
N-Allyl-N-(ethoxycarbonyl)benzenesulfenamide, 463
Allylic sulfones, synthesis from benzenesulfinic acids, 439, 440
Allyl 2-iodophenyl ether, reductive-deiodination, 202
3-Allyloxycarbonylchroman-4-ones, arylation, 129
Allyl selenates, 489
Allyl sulfinate, thermal rearrangement, 446
Allyl sulfoxides, synthesis from benzenesulfinyl chlorides/alkenes, 449
4-Allyl-2,4,6-trimethylcyclohexa-2,5-dienone, 256
Alternant hydrocarbons, 9
Amides, arylation, 130
B-Amidinovinyl sulfides, 466
Amines, arylation, 130, 132
Amino acids, synthesis via the Schöllkopf procedure, 145, 146
2-Aminoaryl benzenesulfonates, rearrangement, 422
4-Aminobenzenesulfonamide, sulfanilamide, 434
3-Aminobenzoic acid, iodination, 187
2-Amino-4-bis(isoxol-5-yl)methanes, 318
2-Amino-4-(4-bromophenyl)-3-cyano-7-hydroxy-4H-chromene, 350
α-Aminodialkyldiazenes, 428
4-Amino-1,3-dimethoxybenzene, carboxylation, 343
α-Aminoketones, 157
2-Amino-4-methoxyacetanilide, 326
2-Amino-6-(4-methoxyphenyl)pyrid-3-yl phenyl ketone, 308
Aminomethylation, arenes, 146

3-Aminophenol, O-sulfonylation, 247
Aminophenols, methods of preparation, 316–318
3-Aminophenols, formylation, 326, 327
7-Amino-4-trifluoromethylcoumarin, 328
3-Amino-2,4,6-triiodobenzoic acid, 187
Anilines, hydroxylation, 316, 317
–, $ortho$-lithiation, 87
–, synthesis from phenols, 248
–, – from tungsten carbenes/isonitriles, 108
Anisole, acylation, 153
–, methods of synthesis, 295, 296
–, nitrosation, 170
–, plumbylation, 128
–, sulfonation, 173
–, synthesis from phenol, 288
Anisole trichromium complexes, alkylation, 302
[18]-Annulene, 32
[22]-Annulene, 31, 32
[24]-Annulene, 31
[26]-Annulene, 31
[30]-Annulene, 32
[32]-Annulene, 32
[34]-Annulene, 32
[36]-Annulene, 32
[38]-Annulene, 32
Annulenediones, methanobridged, 33
Annulenes, 8–15
–, methanobridged, reduction, 24
[10]-Annulenes, 22
[12]-Annulenes, 26–31
[14]-Annulenes, 27–29
Anthracenocyclophanes, 51
Anthracyclinones, 156
Anthracyclins, 339
Anthranilic acids, methods of synthesis, 318
Anthraquinon-1-yl ethers, 295
9-Anthryl butyl sulfide, 300
Arbutin, 361
Arene–alkene cyclisations, 100
Arene biradicals, 21
Arene carboxylic anhydrides, synthesis from aryl halides, 93
Arene charge-transfer complexes, 167, 168

Arene chromium tricarbonyl anionic complexes, 90
Arene chromium tricarbonyl complexes, formation and reactions, 214–216
–, reactions with nucleophiles, 88–92
σ-Arene complexes, 19
Arenediazonium salts, photodediazoniation, 190
–, thermolyses, 128
Arene η^6-iron complexes, 91, 92
Arene η^6-manganese complexes, 91
Arenepyrosulfonic acids, 171
Arene radical cations, 76–79
Arenes, acetylation, 151
–, acylation, 147–154
–, alkylation, 139
–, – by allyl sulfones, 142
–, aminomethylation, 146
–, arylation, 129
–, asymmetric alkylation, 144, 145
–, benzylation, 139
–, Birch reduction, 109
–, bisacylation, 156
–, bromination, 183, 185
–, chlorination, 183
–, chlorosulfination, 448
–, cycloalkylation, 142, 144
–, synthesis by Diels–Alder reactions, 101
–, direct fluorination, 179
–, electron-transfer and electrophilic substitution, 74–79
–, synthesis from Fischer carbene complexes, 122
–, formylation, 160–163
–, Friedel–Crafts alkylation, 139
–, halogenation, 175–177, 204
–, halomethylation, 217–219
–, hydroxylation, 174, 175, 226–228
–, intramolecular acetylation, 157
–, iodination, 185
–, ipso-acylation, 157
–, ipso attack, 74
–, reactions with epoxides/Lewis acids, 143
–, Mannich reactions, 138
–, metal-complexation, 85–92
–, metallation, 128–137
–, microbial oxidation, 110

–, nitration, 164–170
–, – by single electron-transfer, 77–79
–, nitrosation, 170, 171
–, NMR spectroscopy, 115
–, nucleophilic substitution, 79–84
–, oxidative dearomatisation, 109
–, palladation, 134–136
–, perfluoroalkylation, 220–222
–, photoaddition reactions, 110
–, photochemistry, 110–114
–, photoreductions, 112, 113
–, plumbylation, 128–131
–, proto-dealkylation, 147
–, protonation, 137, 138
–, radical reactions, 92, 93
–, reactions with acyclic ethers, 143
–, – with carbon tetrafluoride/hydrogen fluoride, 221
–, – with osmium tetraoxide, 109
–, – with oxetanes, 143
–, – with perfluoroalkyl peroxides, 220
–, – with tetranitromethane, 76
–, substitution reactions, 73
–, sulfonation, 171–174
–, synthesis by cycloaddition reactions, 116
–, – from alkynes, 119
–, – from aryl tosylates, 423
–, – from dienes, 117
–, – from haloalkenynes, 120–122
–, thalliation, 128, 135, 136
–, trifluoromethylation, 221
–, Vilsmeier formylation, 138
Areneselenols, anodic oxidation, 389
Areneselenylmagnesium halides, 403
Arenesulfenyl stabilised carbanions, 405, 406
Arenethioesters, methods of synthesis, 389
Arenethiols, acidity, 379
–, acylation, 389
–, as nucleophiles, 379
–, desulfurisation, 391
–, general literature, 377, 378
–, oxidation, 391, 392
–, properties, 378–380
–, reactions, 391–394
–, synthesis from arenes, 384
–, – from arenesulfenyl halides, 387

–, – from arenesulfinic acids, 382
–, – from arenesulfonates, 382
–, – from arenesulfonic acids, 381
–, – from arenesulfonyl halides, 380, 381
–, – from aryl disulfides, 382
–, – from aryl halides, 386, 387
–, – from arylthiosulfonates, 381, 382
–, – from phenols, 382
Arenethionocarbamates, 382
Arenium ions, 74
–, chemical shifts, 115
Aroyl chlorides, decarbonylation, 188
Arsabenzenes, 38, 39
Arylacetonitriles, synthesis by vicarious nucleophilic substitution, 85
Aryl alkylthio carbonates, 244
Arylamine, nitrosation, 189
Arylamines, halogeno-deamination, 188, 189
–, *ipso*-nitrodeamination, 79
–, nitration, 168
–, nitrosation, 170
–, reactions with ᵗbutylthionitrite, 189
–, synthesis from aryllead tricarboxylates, 130
β-(Arylamino)sulfides, 465
Aryl anions, 20
–, synthesis from o-benzyne, 48
Aryl aryl sulfides, methods of synthesis, 396–404
Aryl azides, synthesis from aryllead tricarboxylates, 130
Aryl azo sulfones, 476
2-Aryl-1-(benzenesulfonyl)ethenes, 417
Aryl benzenethiolsulfonates, 467, 468
–, reactions with benzenesulfenyl halides, 468
3-Arylbenzofurans, synthesis from α-(aryloxy)acetophenones, 306
Arylbismuths, 129, 131
Arylboronic acids, 136
Aryl bromides, halogen exchange, 192
–, reductive-debromination, 200, 201
Aryl cations, 20
Aryl chlorides, chlorine–fluorine exchange, 191
–, reactions with trichlorofluoromethane/aluminium trichloride, 221, 222
–, synthesis from aryl bromides, 192
2-Aryl-2-chloro-1-(benzenesulfonyl)-ethanes, 417
Aryl cinnamyl selenides, oxidation, 490
Aryl coupling reactions, 95, 96
3-Aryl-1,2-cyclohexanediones, 423
Aryl diazo ethers, methods of synthesis, 319
2-Aryl-2,3-dihydrofurans, 95
Aryl *N,N*-dimethylaminomethyl ketones, 160
Aryldisulfides, synthesis from benzenesulfonic acids, 416
Aryl ethers, Birch reduction, 296
–, Diels–Alder cyclisations, 311
–, halogenation, 176, 300
–, sulfonation, 173
Aryl fluorides, fluoride ion displacements, 206
–, reduction by lithium aluminium hydride, 202, 203
Aryl fluorination reagents, 179–182
Arylglycines, 145
Aryl halides, cathodic reduction, 203
–, copper-assisted nucleophilic substitutions, 88
–, coupling reactions with alkylboranes, 136
–, electrophilic reactions, 204, 205
–, halogen exchange, 191
–, Heck reactions, 136
–, homolysis reactions, 92
–, reactions with copper(I) cyanide, 88
–, metal–halogen exchange, 192–198
–, metal complexes, 213
–, nucleophilic reactions, 205–211
–, photochemical reactions, 211–213
–, photolysis-carbonylation, 195, 196
–, reactions with alkali metal cyanides, 88
–, – with alkenes/palladium, 93, 94, 98
–, – with alkynes/palladium, 94
–, – with allyl alcohols, 94
–, – with allyltinalkyls, 96
–, – with dihydropyridines, 201, 202
–, – with tributyltin hydride, 202

Aryl halides, reactions (cont'd)
—, — with trimethylsilylketene acetals, 94
—, — with vinyltributyltins, 96
—, reductive-dehalogenation, 199–203
—, S_{RN} reactions, 82–84
—, Suzuki reactions, 96
—, tin mediated intramolecular cyclisations, 98
Aryl β-halogenoalkyl sulfides, 397
Arylhydroperoxides, hydrogenolysis, 231
N-Aryl-N-hydroxyamides, fluorination, 182
Aryl iodides, reactions with pseudohalides, 210
—, — with zinc, 192, 193
—, synthesis from aryl bromides, 192
Aryl ketols, synthesis from sulfonyloxy ketones, 422
Aryl ketones, synthesis from acyl halides/Grignard reagents, 158, 159
—, hydroxylation, 174
Aryllead triacetates, reactions with trimethylsilylenol ethers, 131
Arylmercury(II) reagents, 134
Aryl methoxymethyl ethers, 218
Aryl methyl disulfides, methods of synthesis, 468
Aryl methyl selenides, anionic exchange reactions, 472
Aryl methyl sulfones, methods of synthesis, 442
Aryl 4-nitrophenyl sulfoxides reactions with sulfur/ammonia, 449
Arylpalladium reagents, 134–136
Arylpalladiums, synthesis from arylthallium dicarboxylates, 136
Aryl perhydropyran-3-yl selenides, 399
Arylperoxysulfenates, 392
Arylperoxysulfinates, 392
N-Arylpyrrolidines, synthesis from enamines, 117, 118
Aryl radical anions, 20, 21
Aryl radical cations, 20
Arylselenenyl alkanones, deprotonation, 482
Arylselenenyl group, removal, 484

Arylselenenyl halides, addition reactions to alkenes/alkynes, 477–479
Aryl selenides, deprotonation, 478
Arylselenides, methods of synthesis, 474, 476
Arylselenium compounds, structural properties, 477
Aryl selenoate anions, 475, 476
Aryl selenoates, reactions with arene diazonium fluoroborates, 476
Arylselenols, oxidation, 472
Aryl sulfides, properties, 378
N-Arylsulfonamides, rearrangements, 425
2-Arylsulfonamidobenzophenones, 425
Aryl sulfones, synthesis from benzenesulfinic acids, 438, 439
Aryl sulfoxides, chiral, 445
—, methods of synthesis, 443
Aryl sulfides, general literature, 377
—, photolysis, 389, 390
Arylsulfinic acids, methods of synthesis, 391
Arylsulfones, 392
Arylsulfoxides, 392
Aryltelluroates, reactions with alkynes, 495
—, — with aryl halides, 494
Aryltelluroketones, 493
Aryl tellurate, 475
Aryltellurate anions, 492
5-Arylthianthrenium salts, 400
Arylthiocyanates, methods of synthesis, 443
Arylthiols, oxidation, 472
3-Arylthio-1-phenylindolin-2(3H)-ones, 408
Aryl tosylates, electrochemical reduction, 423
Aryltrialkylstannanes, 147
—, reactions with sulfur trioxide, 414
Aryl triflates, coupling reactions with organostannanes, 136
Aryl trifluoromethyl sulfides, 398
Aryl vinyl sulfides, 397, 409
Aryl vinyl sulfones, methods of synthesis, 418
Aryl vinyl tellurides, 492
—, reactions with dichlorocarbene, 495

Aryl xanthates, 245
Arynes, reactions with enolates, 126
–, rearrangement to benzocyclobutenes, 126
Atropisomeric thiols, 383
Aurentiacin, 155
Azaannulenes, 40
Azatriptycenes, photorearrangement, 113
Azlactones, 156, 158
Azulenes, 9, 25, 30, 31
Azulenocyclophanes, 50
1,3-Azuloquinones, 25, 26
Azupyrenes, 28, 29

Baeyer–Villiger reaction, 147, 174
Balz–Schiemann reaction, 179, 190
Beckmann rearrangement reaction, 147
Benylamines, lithiation, 346, 347
Benyl ethers, methods of synthesis, 293
5-Benyl-1,3,4-trihydroxy-2,6-diphenylbenzene, 367
Benzal chlorides, methods of synthesis, 222
Benzaldehyde, hydroxylation, 174
Benzaldehyde chromium tricarbonyl complex, lithiation/methylation, 91
Benzaldehydes, synthesis by vicarious nucleophilic substitution, 85
Benzaldimines, ortho-lithiation, 87
Benzalene, 43, 44
Benzamide, synthesis from benzene, 162
Benzamides, ortho-lithiation, 87, 88
–, synthesis from aryl halides, 93
Benzannulated annulenes, 9
Benzene, acetylation, 151
–, alkylation, 140, 142, 143
–, – by valerolactone, 143
–, automerisation, 43
–, hydroxylation, 227
–, photoaddition reactions, 110–112
–, photoisomerisation, 113
–, reactivity, 1
–, structure, 2–4
–, synthesis from Diels–Alder adducts, 15, 16
Benzene–arene adducts, 15

Benzene–benzvalene rearrangements, 43
Benzene dication, 21
Benzene 1,4-diradicals, 102
Benzene disulfides, oxidation, 448
Benzenedisulfides, reactions with alcohols, 445
Benzene-2-disulfinic anhydride, synthesis from benzene-2-disulfonohydroximide, 448
Benzene isomers, 43
Benzene manganese tricarbonyl complex, alkylation, 92
Benzenes, methods of synthesis, 116
α-Benzeneselenenylacetophenone, 482, 483
Benzeneselenenylbenzenetellurylmethane, deprotection, 484
Benzeneselenenyl cation, methods of generation, 480–482
Benzeneselenenyl group, base promoted removal, 480–483
Benzeneselenenyl halides, 472
Benzeneselenenyl radical, methods of generation, 479
Benzeneselenenyl salts, reactions with alkenes/alkynes, 480
Benzeneseleninic acid, as oxidant, 401
Benzeneselenol, methods of synthesis, 472
Benzene selenosulfonates, methods of synthesis, 418
Benzeneselenyl 4-toluenesulfonate, methods of synthesis, 422
Benzeneselinyl chloride, 176
Benzenesulfenamides, hydrolysis, 466
–, methods of synthesis, 462–465
–, reactions with alkenes, 465, 466
Benzenesulfenanilides, reactions with alkenes, 465, 466
Benzenesulfenates, chiral oxidation, 459
–, reactions, 454
Benzenesulfenic acids, addition to alkynes, 454
–, hydrogen bonding, 454
–, synthesis and isolation, 453, 454
–, – from benzenesulfinates, 447
Benzenesulfenimines, method of synthesis, 464

Benzenesulfenyl chloride, reaction with 4-acetoxybut-2-ynol, 457
Benzenesulfenyl chlorides, additions to alkenes, 461, 462
–, reactions with alkylthiotrimethylsilanes, 459
–, – with sulfides, 460
Benzenesulfenyl halides, methods of synthesis, 459
Benzenesulfenyl thiocyanates, 459
Benzenesulfinamides, alcoholysis, 452
–, alkylation, 451, 452
–, methods of synthesis, 442, 449–453
–, synthesis from benzenesulfinic acids, 443
–, – from benzenesulfinyl azide/amines, 453
–, thermolysis, 451
Benzenesulfinate esters, methods of synthesis, 444, 445
Benzenesulfinates, methods of synthesis, 442
Benzenesulfinic acids, additions to aldehydes, 440
–, alkylation, 438
–, methods of synthesis, 437, 438
–, reactions, 438–444
–, – with N-chloroamines, 443
–, reduction, 443
Benzenesulfinimidoyl chlorides, method of synthesis, 460
Benzenesulfinyl azide, decomposition, 453
Benzenesulfinyl chloride, reactions with alcohols, 444
–, – with alkoxytrimethylsilanes, 444
Benzenesulfinyl chlorides, methods of synthesis, 448, 449
–, reactions with N,N-dialkylhydroxylamines, 448
–, – with N-methylhydroxamic acids, 448, 449
Benzenesulfonamides, methods of synthesis, 424
–, *ortho*-lithiation, 424, 425
–, synthesis from benzenesulfinic acids, 443
–, – from benzenesulfinyl chlorides, 448

Benzene sulfonates, methods of synthesis, 420, 421
Benzenesulfonates, *ortho*-lithiation, 421, 422
Benzenesulfonic acids, alkylation, 421
–, biological importance, 434–437
–, *ortho*-lithiation, 415
–, reduction, 415, 416
–, synthesis from arenes, 413
–, – from arenes bearing sulfur-containing groups, 414
Benzenesulfonic peracids, synthesis from benzenesulfonyl imidazolides, 415
Benzenesulfonimidoyl halide, methods of synthesis, 452
Benzenesulfonyl anhydrides, methods of synthesis, 418, 419
–, reactions with enolates, 419
Benzenesulfonyl azides, methods of synthesis, 431
Benzenesulfonylazides, thermolysis, 432
N-(Benzenesulfonyl)benzenesulfinamidines, 464, 465
N-(Benzenesulfonyl)benzenesulfonamides, 424
Benzenesulfonyl chlorides, carbonylation, 418
–, coupling with alkenylstannanes, 418
–, reactions with alkenes, 417
–, – with styrenes, 417, 418
–, reduction, 438, 444
–, synthesis from benzenesulfonyl fluorides, 416, 417
Benzenesulfonyl fluoride, reactions with enolates, 417
Benzenesulfonyl halides, methods of synthesis, 416
–, reactions with thiophenols, 438
Benzenesulfonylhydrazines, reactions, 427–431
Benzenesulfonylhydrazones, addition reactions, 428, 429
–, methods of synthesis, 428
–, reactions, 427
–, – with alkyl lithiums, 429–431
–, reduction, 429
Benzenesulfonyl isocyanates, methods of synthesis, 432, 433

Benzenesulfonylnitrenes, 431, 432
Benzenesulfonyloxaziridines, use as oxidising agents, 433
2-Benzenesulfonyloxy-2-cyclohexenones, photolysis, 423
α-Benzenesulfonyloxy esters, 419
α-Benzenesulfonyloxy ketones, 419
Benzenesulfonyl peroxides, use as oxidants, 419, 420
N-(Benzenesulfonyl)phenylhydroxylamines, reduction, 438
Benzenethiols, oxidation, 438, 448
–, – to benzenesulfonic acids, 414
–, reaction with alcohols, 445
–, synthesis from benzenesulfonic acids, 415
Benzenethiol sulfinates, 453
Benzenethiolsulfinates, oxidation, 447
–, reactions with alcohols, 447
Benzenethiol sulfonates, 421
Benzenethiolsulfonates, methods of synthesis, 467
Benzene thiosulfinates, methods of synthesis, 442
Benzenium ions, 138
Benzenonium ion, 74
Benzfurans, ozoloysis, 275
1, 2-Benzisothiazol-3(2H)-one 1-oxides, 450
Benzisothiazol-3(2H)-ones, synthesis from 2-(bromosulfinyl)benzamides, 463
Benzo[3, 4]cyclobuta[1, 2-b]biphenylene, 119
Benzocyclo(n)alkenes, 49
Benzocyclobutadiene, 13
Benzocyclobutene, isomerism to styrene, 127
–, nitration, 166
Benzocyclobutenes, synthesis and reactions, 125
–, – from benzyne/alkenes, 47, 48
–, – from bromobenzene/enolates, 125, 126
Benzocyclobutenylidene, 127
Benzocyclopropene, stability and reactions, 123, 124
Benzocyclopropenes, methods of synthesis, 123, 124

Benzocyclopropenyl cations, 115
Benzofuran, synthesis from diphenyl ether, 306
Benzoic acids, halogenodecarboxylation, 188
Benzoin, benzene sulfinates, 447
Benzonorcaradienes, 24
Benzophenones, methods of synthesis, 149–161
Benzopyrazolidines, synthesis from azo compounds, 99
1, 4-Benzoquinone, reaction with sulfite ion, 414
Benzoquinones, methods of synthesis, 259
1, 2-Benzoquinones, synthesis from phenols, 260, 261
1, 4-Benzoquinones, methods of synthesis, 357
–, synthesis from methoxyarenes, 260
–, – from phenols, 260
1, 2-Benzoquinonimines, 258
Benzosultams, synthesis by vicarious nucleophilic substitution, 85
Benzothiadiazine 1-oxides, 450
Benzo[b]xanthone, 287
4H-3, 1-Benzoxathiin-4-one 1-oxides, 456
Benzoxazolin-2(3H)-one, 329
Benzoxepine-3, 5(2H, 5H)diones, 308
N-Benzoylaminomethyl phenylselenide, syn-elimination, 487
4-Benzoylbenzoyl chloride, synthesis from 4-hydroxybenzophenone, 247
Benzoyl cyanide, synthesis from iodobenzene, 93
1-Benzoyl-2, 2-dichlorocyclopropane 156
2-Benzoyl-4, 6-dimethoxybenzofuran, 163
N-Benzoyl-2, 4-dimethoxyphenylglycine, 130
N-Benzoyl-3-methanesulfonyloxyproline, 143
N-Benzoyl-2-methylalanines, 277
Benzpyrans, 98
Benzyl alcohols, fluorination, 188
Benzyl 4, 6-O-benzylidene-α-D-mannoside, O-phenylation, 132

Benzyl bromide, reactions with acyl halides, 160
Benzyl chlorides, methods of synthesis, 222
Benzylchlorobis(triphenylphosphine)palladium(II), catalyst in Friedel–Crafts reactions, 159
2-Benzyl-4-chlorophenol, 240, 322
2-Benzyl-5, 6-dimethoxyindan-1-one, 344
Benzyl halides, methods of synthesis, 216, 217
2-Benzyl-5-hydroxybenzofuran, 241
Benzyl 2-hydroxyphenyl ketone, 285
N-Benzylidene benzenesulfenamides, reactions with lithium reagents, 466
N-(Benzylidene)benzenesulfinamides, thermolysis, 454
N-Benzylidene-4-toluenesulfinamides, synthesis from benzonitriles, 452, 453
Benzyl iodides, 216
Benzyl 4-methoxyphenyl ether, O-debenzylation, 364
2-(6-Benzyloxy-3, 4-dihydro-3, 8-dimethylbenzopyran-2-yl) propionic acid, 362
2-Benzylphenol, 240
Benzyl phenyl ether, 289
2-(Benzylsulfinyl)isophthalic acid, 456
Benzyl 2, 4, 6-trihydroxyphenyl ketone, 368
Benzyne, methods of preparation, 46, 47
o-Benzyne, reaction with diazofluorene, 48
Benzyne–titanocene complexes, 45
Benzyne–zirconium complexes, 49
Benzynes, 45
–, synthesis from aryl halides, 84
o-Benzynes, adduct formation, 46
Bergman cyclisation, 1, 21, 102–105
BHT (2, 6-di-^{t}butyl-4-methylphenol), 270, 271
Biaryls, methods of synthesis, 319
–, synthesis by Suzuki coupling, 98
–, – from arylboronic acids, 136
–, – from aryl halides, 193–195
–, – from arylleads, 129
Bibismole, 37

Bicyclo[6.2.0]decapentene, heat of hydrogenation, 26
Bicyclopropenyl, 43
1, 1'-Binaphthyl-2, 2'-dithiol, chiral auxiliary, 383, 384
4, 4'-Binaphthyls, methods of synthesis, 310
Biphenyl, synthesis from chlorobenzene, 194
1, 1'-Biphenyl-2-carboxylate esters, 309
Biphenyls, photodechlorination, 211
Birch reduction, 286
–, arenes, 109
N,N'-Bis(arylsulfonyl)hydrazines, anticancer action, 436
Bisarynes, 46
Bis(4-bromophenyl)thione carbonate, 245
Bis(7-^{t}butylbenzofuran-2-on-3-ylidene), 269, 270
Bis(cycloocta-1, 5-diene)nickel(0), catalyst for cycloadditions, 120
Bis(dimethylamino)cyclobutenethiones, 403, 404
3, 4-Bis(2-ethylhexyloxy)benzaldehyde, 372
Bishomotropylium ion, 5
2, 2-Bis(4-hydroxy-3, 5-diphenylphenyl)propane, 267
1, 3-Bis(2-hydroxyethoxy)benzene, 347
3, 5-Bis[(2-hydroxy-2-methyl)ethyl]anisole, 305
1, 1-Bis(4-hydroxyphenyl)cyclohexane, reductive dehydroxylation, 249
1, 1-Bis(2-hydroxyphenyl)propane, 239
2, 2-Bis(4-hydroxyphenyl)propane, 240
3, 3-Bis(4-hydroxy-3-^{i}propylphenyl)phthalide, 267, 268
1, 6:9, 14-Bismethano[16]annulene, 31
Bis(2-methylphenoxy)methane, 289
Bismolyl anion, 37
4, 4'-Bis(4-nitrobenzoyl)biphenyl ether, 293
Bis(4-nitrophenyl) sulfide, 449, 450
Bis(phenol A), 240
Bis(2-phenylethynyl)telluride, 494
Bis(phenylselenenyl)acetals, conformations, 483, 484

Bis(tosyloxy)alkanes, synthesis from alkenes, 420
1,3-Bis(2,4,6-tribromphenyl)propan-2-ol, 291
Bis(trifluoroacetoxy)iodobenzene/iodine, 186
N,N-Bis(trimethylsilyl)benzenesulfenamide, reactions with carbonyl compounds, 464
2,6-Bis(trimethylsilyl)phenyl triflate, solvolysis, 20
Bis(trimethyl)silyl sulfate, 413
Bis[2,4,6]-tris(trifluoromethyl)phenyl diselenide, 386
Borabenzene-pyridin-1-ium complex, 38
Boracyclopropene, 36, 37
4-Bromoanisole, 301
1-Bromobenzocyclobutene, 125
–, synthesis from cycloheptatriene, 125
4-Bromo-2,6-di-^{t}butylphenol, 273
2-Bromochlorobenzene, reductive-debromination, 199
3-Bromochlorobenzene, reaction with thiols, 207
6-Bromo-8-chloromethyl-1,3-dioxane, 323
6-Bromoguiacol, 342
2-Bromo-4-hydroxycyclo-2,5-dienones, reductive aromatisation, 314
2-Bromomethoxyethylsulfonamido-polystyrene, 218
3-Bromo-4-methylphenol, 314
2-Bromomethyltoluene, 219
1-Bromo-2-naphthol, debromination, 323
4-Bromo-2-nitrophenol, 323
2-Bromophenol, 312
4-Bromophenol, 312
Bromophenols, methods of synthesis, 312
2-(2-Bromophenyl)dibenzoselenol, reaction with butyllithium, 508
2-Bromophenylethylamine, 196
3-Bromothiophenols, lithiation, 392, 393
Bucherer reaction, 230
Buckminsterfullerene, 2
–, aromaticity, 66–72
^{t}Butenylphenol, 284

5-But-2-enyl-2,3,4-trihydroxyacetophenone, 371
4-(But-3-onyl)-2-propylphenol, 264
5-Butoxy-1,2,3-trimethylbenzene, 296
2-^{t}Butylanisole, 342
4-^{t}Butylanisole, benzoylation, 155
Butyl aryl sulfoxide, ortho-lithiation, 87
3-^{t}Butyl-4-chlorophenol, 312
3-^{t}Butyl-1,5-dimethoxy-1,4-cyclohexadiene, 370
4-^{t}Butyl-2,6-dimethoxyphenol, dehydroxylation, 370
4-^{t}Butyl-2,6-dimethylphenol, 240
2-^{t}Butyl-5-methylanisole, 270
6-^{t}Butyl-2-methylphenol, dimerisation, 270
^{t}Butyl-1,3-oxazaspiro[5,5]undeca-2,7,10-trienone, 258
2-^{t}Butylphenol, reactions, 269–273
4-^{t}Butylphenol, reaction with glyceraldehyde acetonide, 264
–, – with tri-O-acetylglucal, 264
–, ring reduction, 263
4-^{t}Butyl phenyl ether, methods of synthesis, 289
6-Butyl-2-phenyl-4-propylphenol, 237
Butyl phenyl sulfide, 320
4-^{t}Butylstyrene, 305
Butyl vinyl telluride, 495
^{t}Butyrophenone, 159

Caesium fluorosulfate, 180
Calicheamicin, 104, 105
Calix[4]arene, 56
Calixarenes, 50, 55, 269
–, nitration, 169
Carbamoyl anions, preparation from tellurium-ate complexes, 509
1,2-Carbonyl group transformations, 430
6-Carboxy-1,3-benzodioxane, 287
3-(2-Carboxy-5-methoxyphenyl)-propionic acid, 309
Carcerands, 352
Carvacrol (5-^{t}propyl-2-methylphenol), 256
Catafusenes, 62
Catechol, disilyl ethers, 107
–, methods of synthesis, 330

Catechol (cont'd)
—, mono-O-acylation, 331, 332
Catechol acetals, acylation, 156
Catechols, mono-O-alkylation, 332
—, reactions, 340
—, sulfonation, 173
Chalcogen elements, physical properties, 413, 414
Chalcogenide oxides, ligand exchange reactions, 510–512
Chalcogenuranes, miscellaneous, 517–520
—, reactions with phenols, 516, 517
—, stability, 507
Chattaway reaction, 263
2-Chloroacylphenols, 275
2-Chloroanisole, 301
4-Chloroanisole, 300, 301
Chlorobenzene, synthesis from phenol, 248
Chlorobenzene chromium tricarbonyl complex, reactions, 215, 216
Chlorobenzenes, trifluoromethylation, 210
1-Chlorobenzocyclobutene, 125
3-Chlorobutanol, reaction with benzene, 143
4-Chloro-(N-butylamido)nitrobenzene, 294
3-Chloro-4,6-dihydroxy-2-methyl-5-(3-methylbut-enyl)benzaldehyde, 355, 356
4-Chloro-2,6-dimethylphenol, 312
6'-Chloro-3,3-dimethyl-3'-(1,2,4-triazol-1-yl)-spiro(tetrahydrofuryl)-2,2'-chromane, 321, 322
2-Chloro-3,5-dinitrobenzoic acid, 325
3-Chloro-4-ethoxycarbonyl-6-hydroxycoumarin, 363, 364
2-Chloro-6-fluorobenzonitrile, 191
3'-Chloro-2-hydroxy-5-methylbenzophenone, 276
6-Chloro-2-methoxycarbonylphenyl, 293
4-Chloro-5'-methoxy-3'-methylbiphenyl, 296, 297
2-Chloro-6-methoxymethyl-4-nitrophenol, 325
Chloromethylphenols, 312
N-Chloromorpholine, 185

2-Chloro-6-nitrobenzonitrile, fluorodenitration, 191
2-Chlorophenol, 312
Chlorophenols, methods of synthesis, 312
Chlorophenoxides, photochemical reactions with nucleophiles, 211, 212
1-(4-Chlorophenoxy)-3,3-dimethyl-1-(1,2,4-triazin-yl)butan-2-one, 292
2-(4-Chlorophenyl)chloroethane, reductive-monodechlorination, 200
α-Chloro-α-(phenylthio)ketones, 459, 460
Chlorosulfuron, 437
4-Chloro-2-tetrahydropyran-2-yl-1-naphthol, 240
2-Chloro-1,1,1-trifluoro-2-arylpropanes, 217
3-Chloro-4-(2,2,2-trifluoroethoxy)-nitrobenzene, 294
Chroman-4-one, synthesis from 2-hydroxyacetophenone, 285
Chromenopyrimidines, 283
6-Cinnamoyl-5,7-dihydroxychromane, 371
2-Cinnamoylphenol, 275
[7]-Circulene, 62, 65
Circulenes, 63–65
Claisen reaction, 283
Claisen rearrangements, 240
Codeine, O-demethylation, 342
[12]-Collarene, 64
Colletochlorin, 355
π-Complexes, 76
Copper(I) 2-[α-(N,N-dimethylamino)-ethyl]thiophenate, 379
Corannulene, 62, 65, 66, 68, 72
Coumarin, methods of synthesis, 284
Coumarins, synthesis from 3-aminophenols, 328
Cryptophanes, 56
1-Cyanobenzocyclobutene, 125
3-Cyano-7-(N,N-diethylamino)-coumarin, 328
2-Cyano-4,6-dimethoxybenzofuran, 163
Cyanoheterocylces, aromaticity, 39
3-Cyano-5-hydroxy-7-pentylcoumarin, 356

2-Cyano-6-hydroxy-2, 5, 7, 8-
tetramethylchromane, 358, 359
4-Cyanophenylazo phenyl sulfide, 319
[12]-Cyclacene, 64
Cyclacenes, 62–65
Cycloalkynes, 35
Cyclobutadiene, isomerism, 12
Cyclobutadiene dication, 13
Cyclobutadiene metal complexes, 13, 14
Cyclobutene-1, 2-diones, 36
Cyclo[n]carbons, 35, 36
1, 4-Cyclohexadiene, dehydrogenation, 18
Cyclohexadienes, synthesis from arenes, 112
1, 3-Cyclohexadienes, synthesis by Vollhardt cyclisations, 105
Cyclohexa-2, 5-dienones, halogenated, 257
Cyclohexadienones, methods of synthesis, 255–259
1, 2-Cyclohexanediones, 423
Cyclohexanone, tetraphenylation, 133
1, 2, 3-Cyclohexatriene, 43
Cyclohexenone oximes, aromatisation, 116
Cyclohex-2-enyl phenyl ether, 291, 292
Cyclohexen-3-yne, 43
3-Cyclohexyl-2, 4-dihydro-7-methoxy-2H-naphth[2, 1-e][1, 3]oxazine, 365, 366
3-Cyclohexyl-3, 4-dihydro-7-methoxy-2H-naphtho[2, 1-e][1, 3]oxazine, 306
Cyclohexyl 2-hydroxyphenyl ketone, 275
Cyclohexyl phenyl selenides, 478
1, 3-Cyclooctadiene-5-yne, 126
–, rearrangement to benzocyclobutene, 126
Cyclooctatetraene, protonation, 5
Cyclooctatetraene dication, 22
Cyclopenta[a]azulene, 30
Cyclopentadienyl anion, 14
($^5\eta$-Cyclopentadienyl)rhodium complexes, 119
[2.2.2.2.2.2](1, 2, 3, 4, 5, 6)cyclophane, 50
Cyclophanequinones, 51
Cyclophanes, 18, 49–55
–, bridged types, 51
–, electrical properties, 55

–, ring distortions, 51–54
Cyclopropa[4, 5]benzocyclobutene, 124, 125
Cycloproparenes, 12
Cyclopropene, iodination, 11
Cyclopropenium cation, 36
Cyclopropenium ions, 10
Cyclopropylarynes, 46
Cycolopentadienylcobalt–cyclobutadiene derivatives, 120
Cysteine ^{t}butyl sulfide, cleavage, 460

Dakin reaction, 283, 373
DAST (diethylaminosulfur trifluoride), 182
2-Decyl-1, 4-dihydroxynaphthalene, 238
2-Decyl-1-naphthol, 238
Decyloxybenzaldehyde, 283
Decyl phenyl selenone, reactions with nucleophiles, 502
Dehydro[8]annulenes, 22
Dehydroannulenes, 9
Dehydrocannabispiran, 356
Dendritic macromolecules, 352
Deoxyglycosides, method of synthesis, 399
Desoxybenzoins, synthesis from arylleads/silylenol ethers, 131
Dewar azulenes, 25
Dewar benzene, 43, 113
Dewar furans, 37, 113
Dewar pyridines, 113
Dewar thiophenes, 37, 113
4, 6-Diacetyl-2-iodoresorcinol, 349
Dialkylbenzenes, reaction with methoxycarbonylnitrene, 114
N,N-Dialkylbenzenesulfenamides, oxidation, 465
–, synthesis from diphenyl sulfide/ lithium dialkylamides, 463
N,N-Dialkyl 2-carboxybenzene-sulfonamides, 425
N,N-Dialkyl 2-chlorosulfonyl-benzamides, 425
Dialkyl hydroxyphenyldicarboxylates, 280, 281
2, 6-Dialkylphenols, oxidative coupling, 253, 254

Diaryl carbonates, methods of synthesis, 243
Diaryldilead diacetates, 129
Diaryl diselenides, addition reactions to alkenes/alkynes, 477
–, copper(I) reductions, 475, 476
–, methods of synthesis, 471
–, reduction, 474, 475
Diaryl disulfides, methods of synthesis, 391, 443, 468
Diaryl ditellurides, methods of synthesis, 476
Diaryl ditellurides reduction, 475
Diarylmethanes, 218, 219
2,3-Diarylphenols, synthesis from 2,6-diarylcyclohexanones, 235
Diaryl selenides, methods of synthesis, 476, 477
Diaryl sulfides, methods of synthesis, 319
Diaryl sulfones, 173
–, methods of synthesis, 476
Diarylsulfoxides, ligand exchange, 511
Diaryltelluranes, methods of synthesis, 517
Diaryl tellurides, 493–495
Diaryltellurides, reactions with hydrogen fluoride, 517, 518
Diaryl tellurides, reactions with sodium hydroxide, 492, 493
Diaryltellurides, synthesis from telluronium halides, 519
1,2-Diaryl-2-(2-tosylhydrazido)ethanols, reactions with formic acid, 427
2,2-Diaryl-1,1,1-trifluoroethanes, 141, 142
2,2-Diaryl-1,1,1-trifluoropropanes, 217
Diaza[22]annulenes, 40, 41
Diazo group transfer, 432
Dibenzo[a,e]cycloheptanone, 150
Dibenzo[1,4]dioxines, synthesis from catechol, 81
Dibenzoselenophenes, 511
Dibenzothiophenes, 511
Dibenzothiophene S-oxide, 514
5,6-Dibenzoyloxy-4',7-dihydroxyflavone, 228
Di-(O-benzyl)estradiol, dealkylation, 298

Dibenzynes, synthesis from tetrabromobenzenes, 59
1,2-Dibromobenzene, didebromocyclooligomerisation, 195
3,5-Dibromo-4-bromomethyl-2,6-dimethylphenol, 314
1,3-Dibromo-5,5-dimethylhydantoin, 177
2,4-Dibromo-3,6-dimethylphenol, nitration, 324
4,6-Dibromo-2-hydroxy-2,5-dimethyl-3,6-dinitrocyclohexanone, 324
2,6-Dibromo-4-methoxy-4-methylcyclohexa-2,5-dienone, 257
2,6-Dibromo-4-methylphenol, 257, 314
2,4-Dibromophenol, debromination, 323
1,4-Dibromo-2,3,5,6-tetrachlorobenzene, 198
2,6-Di-^{t}butyl-1,4-benzoquinone, 259
4,6-Di-^{t}butylbenzo-1,2-quinone, 272
2,6-Di-^{t}butylbenzo-1,4-quinone, 272
4,6-^{t}Dibutyl-3,1,3-benzothiadiazole, synthesis from 2,4-di-^{t}butyl-6-iodophenol, 270
3,5-Di-^{t}butylcatechol, aerial oxidation, 340
1,4-Di-^{t}butylcyclopropene, ^{1}H NMR spectrum, 11
1,4-Di-^{t}butyl-2,6-di(trimethylsilyl)-silabenzene, 37
2,6-Di-^{t}butyl-4-ethyl-4-hydroxy-cyclohexa-2,4-dienone, 273
2,6-Di-^{t}butyl-4-ethylphenol, 273
2,6-Di-^{t}butyl-6-formylphenyl, 273
2,6-Di-^{t}butyl-6-hydroperoxy-4-(2-methylpropanoyl)cyclohexa-2,4-dienone, 272
2,6-Di-^{t}butyl-4-methoxymethylphenol, 271
2,6-Di-^{t}butyl-4-methoxyphenyl ethers, in aldolisation reactions, 273
2,6-Di-^{t}butyl-4-methylphenol (BHT), 270
2,6-Di-^{t}butyl-4-methylphenyl 2-methoxybenzoate, demethylation, 299, 300
2,6-Di-^{t}butyl-4-(2-methylpropanoyl)-phenol, 272
2,6-Di-^{t}butylphenol, reactions, 269

–, – with paraformaldehyde, 270, 271
2, 6-Di-*t*butylphenol-*O*-glucoside, 271
2, 6-Di-*t*butylphenyl triflate, solvolysis, 20
2, 4-Dicarboxyphenol, 277, 278
1, 2-Dichlorobenzene, reductive-dechlorination, 200
2, 6-Dichloro-1, 4-benzoquinone, 259
7, 7-Dichlorobicyclo[4.1.0]hept-3-ene, dehydrochlorination, 123
Dichlorodidehydroannulenediones, methanobridged, 33, 34
2, 4-Dichloro-5-hydroxy-6-methyl benzoate, 312
Dichloromethylarenes, 85
2, 6-Dichloro-*N*-phenylaniline, 322
Di(4-chlorophenyl) sulfone, reaction with sodium thiophenate, 208
3, 5-Dichlorophenyl 2, 4, 6-trifluorophenyltelluride, reaction with pentafluorophenyllithium, 509
2, 5-Dichloro-2, 4, 6-trifluorophenyl phenyl selenide, reaction with pentafluorophenyllithium, 509
Dicyclopentadiene, reaction with 4-methoxyphenylmercury(II) chloride, 135
Di-(3, 5-di-*t*butyl-4-hydroxybenzyl) ether, 271
Dienylketenes, methods of generation, 123
2, 4-Diethoxycarbonyl-3, 5-dihydroxy-biphenyl, 335
1, 2-Di(ethoxycarbonyl)-5, 8-dihydroxy-naphthalene, 339
2, 4-Diethoxycarbonyl-5-methylphenol, 335
2, 6-Diethoxycarbonylphenol, 280
2-Diethylaminomethyl-4, 6-dimethyl-phenol, 265
3-(*N*,*N*-Diethylamino)phenol, 326
3-(*N*,*N*-Diethylamino)-2-phenylsulfinyl-1, 3-butadiene, 457
Diethylaminosulfur trifluoride (DAST), 182, 183
2, 2-Diethyl-2, 3-dihydro-5-methylbenzofuran, 265
2, 3-Diethyl-5-phenol, synthesis from 3-phenylcyclobutenone, 237

5, 6-Diethyl-3-phenyl-2-propylphenol, 237
S-(+)-*N*,*N*-Diethyl-4-toluene-sulfinamide, 445
Dieyne radicals, cyclisation, 19
1, 1-Difluoroalkenes, methods of synthesis, 423
1, 1-Difluoroalkyl ketones, methods of synthesis, 423
Difluorobenzenes, reactions with sodium methoxide, 208, 209
4, 4-Difluorocyclohexa-2, 5-dienone, 257
Difluorodiarylmethane, synthesis from diaryldithiolanes, 222
2, 6-Difluoro-3, 5-dimethoxy-4-trifluoromethylphenol, 209
2-(Difluoromethoxy)benzaldehyde, 283
5, 5'-Diformyl-2, 2'-dihydroxy-3, 3'-dimethoxybiphenyl, 375
Dihydric phenols, methods of synthesis, 330–340
1, 3-Dihydric phenols, oxidation, 365
2, 3-Dihydrobenz[*d*]isothiazoles, synthesis from benzothiete/amines, 463
Dihydrobenzofurans, 131
2, 3-Dihydro-1, 3-benzoxasilaole, 305
1, 4-Dihydro-2, 3-benzoxathiin, 125
1, 2-Dihydrocatechols, 110
2, 3-Dihydro-2, 3-cycloalkano-benzofurans, 266
6, 12-Dihydro-3, 9-dihydroxy-6, 12-diphenylbenzofurano[2, 3-*b*]benzofuran, 351
3, 4-Dihydro-5, 7-dihydroxy-4-phenyl-coumarin, 368
2, 3-Dihydro-5, 6-dimethoxy-2-(3, 4-dimethoxyphenyl)3-methylbenzofuran, 373
2, 3-Dihydro-2, 2-dimethylbenzofuran, 265
2, 3-Dihydro-2, 3-dimethylene-1, 4-dioxin, cycloaddition reactions, 347
2-(4, 5-Dihydro-4, 4-dimethyloxazol-2-yl)-1, 1'-binaphthyl, 310
5, 8-Dihydro-2-(2, 2-dimethyltetrahydro-furan-2-yl)-1, 4-naphthoquinone, 359
2, 6-Dihydro-2, 6-dioxo-3, 7-diphenyl-benzo[1, 2-*b*:4, 5-*b'*]bifuran, 362, 363

2,3-Dihydrofurans, arylation, 95
2,3-Dihydro-3-hydroxy-3-methyl-
  benzofuran, 266
2,3-Dihydro-6-hydroxy-2,2,3-
  trimethylbenzofuran, 351
3,4-Dihydro-6-hydroxy-
  2,2,4-trimethyl-7-(1,1,3,3-
  tetramethylbutyl)chromane, 358
1,3-Dihydroisobenzofurans, 124
Dihydroisocodeine, 100
4,4a-Dihydro-4a-methyl-9,10-
  anthraquinone, methyl group
  migration, 17
2,3-Dihydro-3-methylbenzofuran, 202
3,4-Dihydro-8-methylcoumarin,
  synthesis from ethyl acrylate, 266
2,3-Dihydro-3-methylidenebenzofuran,
  196, 197
3,6-Dihydro-2-methyl-4-
  trimethylsilylmethylanisole, 310
Dihydropentalenes, synthesis from
  benzocyclobutenes, 127
Dihydropyrenes, 20, 27, 28
2,3-Dihydro-2-vinylbenzofuran, 266
2,4-Dihydroxyacetophenone, 348
2,5-Dihydroxyacetophenone, 357
1,4-Dihydroxybenzenes, 357
–, methods of synthesis, 337
4,4'-Dihydroxybenzophenone, 276
2,2'-Dihydroxy-1,1'-binaphthyl, 254,
  255
3,4-Dihydroxybiphenyl, 330
Dihydroxy-5,5'-dimethoxybiphenyl, 254
3,4-Dihydroxy-4',5-dimethoxystilbene,
  372
4,4'-Dihydroxy-3,3'-dimethylbiphenyl,
  270
5,7-Dihydroxy-2,2-dimethylchroman-4-
  one, 368
3,10-Dihydroxymentha-5,11-dien-4,9-
  dione, 256, 257
4,5-Dihydroxy-3-methoxybenzaldehyde,
  373
2,2-Dihydroxy-5-(2-methoxycarbonyl-
  ethyl)biphenyl, 251
2,4-Dihydroxy-5-methylacetophenone,
  335
1,2-Dihydroxy-3-methylanthraquinone,
  331

7,10-Dihydroxy-2-methyl-6H-dibenzo-
  [b,d]pyran-6-one, 267
1,8-Dihydroxy-3-methylnaphthalene,
  285
Di(2-hydroxy-5-methylphenyl) sulfide,
  319
2,4-Dihydroxy-5-methylphthalide, 335,
  336
1,4-Dihydroxy-2-methylthiobenzene,
  358
2,3-Dihydroxy-1-napthaldehyde,
  O-methylation, 341
2,4-Dihydroxy-6-pentdecylbenzoic acid,
  335
1,4-Dihydroxyphenols, mono-O-
  acylation, 362
3,5-Dihydroxytoluene, 333
2,6-Diiodo-4-nitro-N-propylphenol, 324
Diltiazem, synthesis from 2-amino-
  thiophenol, 393
2,3-Dimethoxybenzaldehyde, mono-O-
  demethylation, 345
2,4-Dimethoxybenzaldehyde, 355
1,3-Dimethoxybenzene, 354
1,4-Dimethoxybenzene, C-acylation,
  363
–, asymmetric acylation, 153, 154
–, O-demethylation, 363
2,6-Dimethoxybenzenethiol, 385
4,5-Dimethoxybenzocyclobutene, 125
–, demethylation, 125, 126
4,4'-Dimethoxybenzophenone, 152
3,4-Dimethoxybenzoylcyanide, reaction
  with hydroxylamines, 344, 345
2-(3,4-Dimethoxybenzyl)indan-1-one,
  344
4,4'-Dimethoxybiphenyl, anodic
  dimethoxylation, 258
2,4-Dimethoxybromobenzene, 354, 355
3,4-Dimethoxybutyrophenone, 154
2,3-Dimethoxy-5-methyl-1,4-
  benzoquinone, 375
4,8-Dimethoxy-6-methyl-1-naphthol,
  oxidation, 259
5,8-Dimethoxy-1,4-naphthoquinone
  (naphthazarin dimethyl ether), 261
9,10-Dimethoxyphenanthrene, synthesis from 2-methoxyacetophenone,
  299

3,5-Dimethoxyphenol, C-alkylation, 369
3,5-Dimethoxyphenoxyacetophenone, 163
Di-(4-methoxyphenyl)tellurone, 503
Di(4-methoxyphenyl)telluroxide, oxidant and base, 499
3,5-Dimethoxytoluene, 333
3,5-Dimethylacetanilide, 116
N,N-Dimethylaminobenzylamines, palladation, 160
11-(N,N-Dimethylamino)coumarino[4,3-b]-1-benzopyrylium perchlorate, 328, 329
3-(N,N-Dimethyl)amino-6,7-dimethoxy-1-methylisoquinoline, 344
3-(N,N-Dimethylamino)phenol, 317
N,N-Dimethylaniline, halogenation, 176
N,N-Dimethylanilines, nitration, 168, 169
–, nitrosation, 170
2,3-Dimethylanisole, reaction with N-bromosuccinimide, 219, 220
Dimethyl 2-aryloxy-6-hydroxyphthalates, 281
1,4-Dimethylbenzene, halogenation, 176
7,7-Dimethyl-2(7H)-benzo[1,4,5-b]-bipyranone, 284
2,4-Dimethyl-1,3-benzodioxane, synthesis from phenol, 291
6,9-Dimethylbenzofuran-3-ol, 241
2,6-Dimethyl-1,4-benzoquinone, 254, 259
15,16-Dimethyldihydropyrene, 28
2,3-Dimethyl-1,4-dihydroxybenzene, 337
Dimethylgermylene, 17
2,6-Dimethylhalobenzene chromium tricarbonyl complexes, 215
2-(2,5-Dimethylhex-2-yl)phenol, 239
Dimethyl 2-hydroxy-4-(2-phenylethenyl)-1,3-benzenedicarboxylate, 281
Dimethyl 4-hydroxyphenylsuccinate, 252
2,6-Dimethylindanone, 149
Dimethyl isophthalate, nitration, 167
N,N-Dimethyl-4-methylbenzylamine, 147

1,4-Dimethyl-2-nitrobenzene, 168
N,N-Dimethyl-4-nitrosoaniline, 171
Dimethyl-(2,6-octadienyl)phenol, 239
2,4-Dimethylpentanoic acid, 273
2,6-Dimethylphenol, oxidation, 254
3,5-Dimethylphenol, synthesis from isophorone, 232
2,5-Dimethylphenyl 2-(N,N-dimethylamino)ethyl ether, 338
Dimethyl sulfoxide, Pummerer rearrangement, 441
N-(2,4-Dinitrophenylsulfenyl)-aziridines, 466
Di(perfluoroalkyl)benzenes, 223
3,3',5,5'-Diphenoquinones, 253
Diphenylamine, synthesis from aniline, 132
2,6-Diphenylbenzenethiol, 382
Diphenyl carbonate, 243, 244
Diphenyldichlorotellurane, reaction with phenyllithium, 506
Diphenyl diselenide, 472, 473
–, reaction with bis(triphenyltin)-telluride, 475
Diphenyl ether, 289
3,3-Diphenyl-3H-indole, 132
Diphenylmethane, synthesis from benzyl chloride, 139
1,2-Diphenylpropane, asymmetric synthesis, 145
Diphenylselenomethane, 483
Diphenylsulfoxide, reactions with arylmagnesium bromides, 510, 511
Diphenyltelluride, reaction with methyl iodide, 501
Diphenyltellurides, reactions, 506, 507
Di(4-phenylthiolphenyl) sulfone, 208
4,6-Diphenyl-2-thiolpyridinium iodide, 216
Diphenylthiomethane, 483
Diphosphirenium cations, 37
2,6-Dipropylphenol, synthesis from cyclohexanone/propanal, 232
Di-(propyn-2-yl)sulfoxide, reaction with propynol, 119
Ditellurides, methods of preparation, 493
7-(1,3-Dithianyl)-1-methylindole, 410
Dithiatelluranes, 519

Dithiatelluranes (cont'd)
—, oxidation, 519
1,6-Dithiocyanatocycloheptatriene, 124
1,4-Ditritiobenzene, β-decay, 20
4-Dodecylbenzenesulfonyl azide, 432
4-Dodecylphenol, 265
Duff reaction, 322
Durham route to arenes, 17
Dynemicin A, 104

Electrophilic addition-elimination reactions, 137
3-En-1,5-diynes, intramolecular cyclisation, 102
Enediyne antibiotics, 104
Enediynes, Bergman cyclisation, 21
Enols, O-phenylation, 133
Enynallenes, intramolecular cyclisation, 102
1,5-Enynes, synthesis from propargyl sulfides, 406, 407
Ephedrine analogues, 251
Episelenium cations, 478
Episelenuranes, 478
Episulfonium ions, 457
Epoxide, ring-opening, 399, 400
Estradiol, acetylation, 263
Estrone, reductive dehydroxylation, 249
Estrone methyl ether, 302, 303
3-Ethoxycarbonylcoumarin, 283, 284
3-Ethoxycarbonyl-7-(N-ethylamino)-6-methylcoumarin, 328
2-(1-Ethoxycarbonyl)-(1-hydroxyethyl)-phenol, 251, 252
3-Ethoxycarbonyl-2-iminobenzopyran, 283
4-Ethoxycarbonyl-2-phenyloxazol-5-one, arylation, 129, 130
4-Ethoxy-3-hydroxyprop-2-enylbenzene, 345
2-Ethoxynaphthalene, 290
4-Ethoxy-1,2-naphthoquinone, 254
2,3-Ethoxypropyl phenyl ether, 290
2-Ethoxytropylium salts, 5
2-[(N-Ethyl)amino-2-hydroxy-5-methylbenzoyl]benzoic acid, 327
3-[4-(N-Ethyl)amino-2-hydroxy-5-methylphenyl]isoindolinone, 327

Ethyl 7-^{t}butyl-2-methylbenzoxazole 5-carboxylate, 271
Ethyl cinnamate, synthesis from benzene, 135
Ethyl 3-cyano-2-hydroxy-4,6-diphenylbenzoate, 278
5-Ethyl-3,4-dimethylphenol, synthesis from 5-ethyloct-5-ene-2-yn-4-one, 233
Ethyl 4-ethoxyphenylglyoxylate, 155
Ethyl 6-ethyl-4-hydroxy-3-methylbenzoate, 280
Ethyl 2-hydroxy-2(2-hydroxy-4-methylphenyl)propionate, 263
O-Ethylisoeugenol, O-demethylation, 345
2-Ethyl-2-methoxycarbonylindanone, 500
Ethylmethylphenylselenonium camphorsulfonate, methyl group transfer, 500
Ethyl phenyl selenone, 502
Ethyl 1-thionaphthoxyacetate, 248
5-Ethynylpyrrolidin-2-one, 398

Fenton reaction, 174
Flavylium perchlorates, synthesis from aminophenols, 329
9-Fluorenyl cation, 14
Fluoroalkylarenes, industrial importance, 220
β-Fluoroalkyl aryl sulfides, 461
2-Fluoroanisole, 300
Fluorobenzenes, deprotonation, 198
—, ortho-lithiation, 87
4'-Fluoro-4-hydroxybenzphenone, 276
4-Fluoronitrobenzene, reaction with 4-aminobenzonitrile, 206
Fluoropentaiodobenzene dication, 22
2-Fluorophenol, 311
4-Fluorophenol, 311
4-Fluorophenyllead triacetate, reactions with arenes, 129
N-Fluoropyridone, 180, 182
N-Fluoropyridyl triflates, 181
Fluorosulfonylarenes, halogenodefluorosulfonylation, 190
2-Formylbenzenethiol, Schiff bases, 378
6-Formyl-3,4-dihydro-7,8-dihydroxy-2,2-dimethyl-2H-benzofuran, 372

4-Formyl-3, 5-dimethoxyphenoxy-
   acetonitrile, 163
Friedel–Crafts alkylation, 141–147
–, catalysts, 139
Friedel–Crafts catalysts, 149–161
Friedel–Crafts reaction, 139–161
Friedel–Crafts reactions, mediated by
   metals, 158–160
–, protection of amino groups, 158
Fries rearrangement, 275
[60]-Fullerene, 66–77
Fullerenes, 2, 66–72
–, adducts and complexes, 69, 70
–, Birch reduction, 71
–, electrophilic reactions, 70
–, radical reactions, 72
–, reactions with halogens, 71, 72
Fulleroids, 70
Fulvenes, 30, 31, 113
Furan, reaction with benzocyclopropene, 124
Furan adducts, synthesis from
   bisarynes, 46
Furanocyclophanes, 51
2-(Furan-3-yl)-4, 5-dimethoxy-2-
   methylcyclohexa-3, 5-dienone, 373
Furanyl phenyl sulfides, 458
Furosemide, 436
Fused benzenoids, 62–72

Gatterman–Koch reaction, 161
Gatterman reaction, 161
Glyburide, 434
Glycinyl cation equivalents, 145
Glycols, phenylation, 132
Glyoxylic acid esters, 155
Gomberg–Bachman–Hey reaction, 128
Graphite, 62
Graph theory, fused benzenoids, 62, 63
Group(IV) electrophiles, 138

Haegerman esters, 296
β-Haloalkylbenzenes, methods of
   synthesis, 217
π-Haloarene metal complexes, 213–216
Haloarenes, carbonylation, 243
Halodienes, carbometallation, 120
Halogenobenzenes, 178
–, methods of synthesis, 179–192

–, reviews, 178
Heck reaction, 93–98, 120–122, 134
Helicenes, 65
Hemicarcerands, 352
Hemispherands, 56, 57
Heptalene, 9
Heterannulenes, methanobridged, 40
Heterobenzenes, 37–42
Hexaazabenzene, 38
Hexabromobenzene, 197
Hexa-^{t}butylbenzene, 18
Hexachlorocyclohexadienone, 185
Hexadehydro[18]annulene, 35
Hexafluorobenzene, photochemical reaction with 1, 2-dichlorodifluoroethene,
   212, 213
–, reactions with nucleophiles, 209
Hexafluorobenzocyclobutanes, 213
Hexafluorocyclooctatetraenes, 213
Hexafluorocyclooctatrienes, 213
Hexahalogenobenzenes, 18
Hexaiodobenzene ditriflate, 21, 22
Hexakis(dimethylsilyl)benzene, 18
Hexakis(trimethylgermyl)benzene, 18
Hexakis(trimethylsilyl)benzene, 18
Hexakis(trimethylsilyl)benzene dianion,
   21
Hexamethylheptalene, 26
3-Hexanone, thia and selena acetals,
   483
Hexa-^{i}propylbenzene, 18
Hex-3-yne, cyclotrimerisation, 119
Homoanthracenes, 29, 30
Homoaromaticity, 5–8
Homofullerenes, 70
Homotropylium ion, 5
Hückel's rule, 2, 3
Hunsdiecker reaction, 188
Hydrochlorothiazole, 436
2-Hydroxyacetophenone, 275
Hydroxyacetophenones, 276
Hydroxyamines, synthesis from amines,
   420
6-Hydroxyanthranilic acid, 327
2-Hydroxyaraldehydes, 273, 274
4-Hydroxyaraldehydes, 274
N-(2-Hydroxyaryl)benzenesulfonates,
   422
4-Hydroxybenzaldehyde, 274

3-Hydroxybenzenesulfonic acid, 414
3-Hydroxybenzocyclobutene, synthesis from 3-formylbenzocyclobutene, 230
Hydroxybenzofurans, 241
4-Hydroxybenzoic acid, 277
Hydroxybenzoic acids, methods of synthesis, 277
2-Hydroxybenzonitrile, 281
2-Hydroxybenzophenone, 155
Hydroxybenzophenones, 276
3-Hydroxybenzo[a]pyrene, sulfonation, 247
3-(1-Hydroxybenzyl)phenol, 252
2-Hydroxy-1, 1'-binaphthyl, 255
3-Hydroxybutyl phenyl selenide, 474
1-Hydroxycarbazoles, method of synthesis, 318
Hydroxy carbonyl compounds, synthesis from silylenol ethers, 433
2-Hydroxy-4-cyanobiphenyl, 319
2-[2-Hydroxy-4-(N,N-diethylamino)-phenyl]-1, 4-naphthoquinone, 327
7-Hydroxy-2, 4-dihydroxyphenyl-2, 4, 4-trimethychromane, 350
4-Hydroxy-1, 5-dimethoxy-anthraquinone, 364
5-Hydroxy-4', 7-dimethoxyflavone, 370
2-Hydroxy-4-(N,N-dimethylamino)-benzylidene-4-nitroaniline, 327
4-Hydroxy-3, 5-dimethylbenzaldehyde, synthesis from 2, 4, 6-trimethylphenol, 261
6-Hydroxy-2, 4-dimethyl-1, 3-benzodioxine, 358
7-Hydroxy-2, 2-dimethyl-4-chromanone, 349
4-Hydroxy-1, 2-dimethyl-5-phenyl-benzimidazole, 234
4-Hydroxy-2, 5-dimethylphenyl 2-(N,N-dimethylamino)ethyl ether, 338
2-Hydroxy-3, 5-dinitrobenzoic acid, 325
1-Hydroxyethyl 4-methoxyphenyl ketone, 304
4-(2-Hydroxyethyl)phenol, 251
β-Hydroxyethyl phenyl selenides, oxidation, 488
2-Hydroxyhomotropylium ion, 5
6-Hydroxy-1-(4-hydroxyphenyl)-1, 3, 3-trimethylindane, 267

β-Hydroxylkanethiols, 387
2-Hydroxymandelic esters, synthesis by asymmetric alkylation, 145
4-Hydroxy-6-mercaptodibenzofuran, 384, 394
4-Hydroxy-6-mercaptophenoxythiin, 384, 385
2-Hydroxy-4-methoxyacetophenone, 347, 348
2-Hydroxy-5-methoxybenzaldehyde, 361
4-Hydroxy-3-methoxybenzaldehyde, 345
2-Hydroxy-4-methoxybenzophenone, 352, 353
3-Hydroxy-5-methoxybiphenyl, 333
2-Hydroxy4-methoxy-α-chloro-acetophenone, 353
3-Hydroxy-4-methoxy-6-methyl-benzaldehyde, 344
2-Hydroxy-3-methoxyphenazine N,N-dioxide, 374, 375
7-Hydroxy-2-methoxy-2, 4, 4-trimethyl-chromane, 349
2-Hydroxy-N-methylbenzamide, 282
4-Hydroxy-3-methylbenzoic acid, 277
3-(2-Hydroxy)-3-methylbenzoyl-2-(4-tolyl)isoindolinone, 277
7-Hydroxy-4-methylcoumarin, 350, 351
4-Hydroxymethylenedioxybenzene, 346
2-Hydroxy-4, 5-methylenedioxychalcone, 373
5-Hydroxy-6-methylindane, synthesis from 1, 6-heptadiyne, 236, 237
2-Hydroxymethyl-8-methoxy-1, 4-benzodioxane, 374
4-(1-Hydroxymethyl-1-methylallyloxy)-2, 3, 6-trimethylphenyl acetate, 362
3-Hydroxymethyl-1-naphthol, 233
4-Hydroxymethylphenol, selective O-acylation, 262
2-Hydroxymethylphenol (salicyl alcohol), 250
2-(4-Hydroxy-2-methylphenyl)ethanol, 333
1-Hydroxy-3-methylxanthone, 351
5-Hydroxy-1-naphthalenesulfonamide, 230
1-Hydroxynaphthalene-2-thiol, 384
1-Hydroxy-2-naphthoic acid, synthesis from 1-naphthol, 278

Hydroxynaphthoic acids, 277
2-Hydroxy-1-naphthoic acids, 278
2-Hydroxy-1,4-naphthoquinone, 365
4-Hydroxy-3-(2-nitrovinyl)benzoic acid, 286
6-Hydroxy-1-oxo-9-phenyl-1H-phenalene-7-carboxylic acid, 366
6-Hydroxy-2,2,5,7,8-pentamethylchromane, 359
2-Hydroxyphenol, dehydroxylation, 341
3-Hydroxyphenyl 2,4-dihydroxybenzoate, 348
2-Hydroxyphenylglyoxylic acid, 282
2-(4-Hydroxyphenyl)oxypropionic acid, 228
2-Hydroxyphthalic anhydride, 282
2-(β-Hydroxypropyl)phenol, synthesis from benzyl methyl ketone, 227
4-Hydroxy-α-pyrones, 116
4-Hydroxyquinizarin, 359
Hydroxyquinols, methods of synthesis, 366, 367
β-Hydroxytosylhydrazones, 429, 430
Hydroxy(tosyloxy)haloarenes, 419, 420
5-Hydroxy-2,2,4-trimethyl-2H-chromene, 350
4-Hydroxy-2,4,6-trimethylcyclohexa-2,5-dienone, 255, 256
1-Hydroxyxanthone, 351
Hypervalent compounds, structures, 504, 505
–, suitable ligands, 505, 506
–, theoretical considerations, 503, 504
– of selenium and tellurium, 503–520

Ibuprofen, 129
Imines, synthesis from hydrazines, 420
Iminophosphoranes, synthesis from benzenesulfonyl azides, 432
Indanones, synthesis via Friedel–Crafts reactions, 158
Indole, 3-phenylation, 132
Indoles, N-arylation, 130
Indolines, synthesis from bromophenylethylimines, 100
Interhalogen compounds, 176
Iodine dichloride, polymer-bound, 187
Iodine monofluoride, 186
4-Iodoanisole, 301

Iodoarenes, synthesis from benzenesulfonic acids, 415
Iodobenzene, palladation, 93
2-Iodoestradiol, 315, 324
4-Iodo-2-methylphenol, 315
4-Iodonitrobenzene, reaction with pyrrolidine, 206
4-Iodophenol, 315
Iodophenols, methods of preparation, 315
Iodophenols-β-cyclodextrin, complexation, 326
5-Iodovanillin, 342
$Ipso$-nitration, 168
Iptycenes, 58–61
Isocyanomethylarenes, synthesis by vicarious nucleophilic substitution, 85
Isoflavanoids, synthesis from chromanones, 129
Isopyrene, oxidation, 29
Isothymol, oxidation, 257

Kekulene, 64, 65
Ketones, synthesis from tosylhydrazones, 430
Knoevenagel reaction, 276, 283

β-Lactams, method of synthesis, 407
Lactones, synthesis from alkanoic acids, 420
Lasiodiplodin, 118
Lithium arylselenates, 475
Lithium aryltellurates, 475

Maesanin, 122
Magnetic anisotropy, 4
Mannich reaction, 146, 147
Meisenheimer complexes, 80–84
Menthyl 2-(2-hydroxybutylphenyl)-lactate, 263, 264
S-(–)-Menthyl 4-toluenesulfonyl-sulfinate, 445
4-Mercaptodibenzofuran, 384
Mesitol, synthesis from mesitylene, 175
Mesitylene, acylation, 148
–, halogenation, 177
–, hydroxylation, 175
–, nitration, 168
–, protonation, 138

[4]-Metacyclophane, 49
[2,2]-Metacyclophanes, 53
[2,m]-Metacyclophanes, chirality 54
[n]-Metacyclophanes, 49
[n, m]-Metacyclophanes, 49, 50
Metallocenecyclophanes, 50
Metallocene sandwich compounds, 15
Metalloproteins, 379, 380
1,5-Methano[10]annulene, isomerism, 24
1,6-Methano[10]annulene, 123
–, aromaticity, 22, 23
–, sulfonation, 172
Methano[12]annulenes, 27
1,6-Methanoaza[10]annulenes, 40
1,4,7-Methin[10]annulenes, 26
4-Methoxyacetophenone, 150, 152
2-Methoxybenzaldehyde, 282, 283
4-Methoxybenzenesulfonate, 301
2-Methoxybenzoic acid, Birch reduction, 286
–, O-demethylation, 309
2-Methoxy-1,4-benzoquinone, 355
6-Methoxybenzoxazolin-2-one, 353
Methoxybenzyl alcohols, polymerisation, 140, 141
3-(4-Methoxybenzylidene)-5-(4-methylphenyl)furan-2(3H)-one, 304
4-Methoxycarbazole, 297
3-Methoxycarbonyl-5-hydroxyphenyl ether, 293
2-Methoxycarbonyl-2-methylindane, 500
2-(4-Methoxycarbonylphenoxy)-ethylamine, 287
3-Methoxy-1-(2-chlorophenyl)propyne, 321
Methoxy-Dewar-azulene, 25
6-[α-(4-Methoxy)ethyl]-3,4-methylenedioxyphenol, 374
4-(1-Methoxyethyl)phenol, synthesis from 4-ethylphenol, 262
6-Methoxy-3-(4-methoxybenzyl)benzo-furan-2-one, 353
2-Methoxy-4-methylphenyl methyl carbonate, formylation, 344
2-Methoxynaphthalene, 290
–, benzoylation, 154

2-Methoxynaphthalene 1-sulfinyl-chloride, 302
7-Methoxy-1-oxoindan-4-carboxylic acid, 309
2-Methoxy-6-pentyl-1,4-benzoquinone (primin), 260
2-Methoxyphenol, methods of synthesis, 331
–, synthesis from 2-methoxy-benzaldehyde, 283
4-Methoxyphenol, C-alkylation, 361
–, O-alkylation, 360, 361
–, anodic oxidation, 360
–, oxidation, 254
1-(4-Methoxyphenyl) 2-oxocyclohexane-carboxylate, 128
2-(4-Methoxyphenyl)-4-phenylbutanoic acid, 361
4-Methoxyphenyl phenyl carbonate, 244
1-(2-Methoxyphenyl)-1-phenylethanol, 302
1-(4-Methoxyphenyl)phenyl ketone, 303
2-(4-Methoxyphenyl)propionate, 308
1-(4-Methoxyphenyl)propylamine, 303
4-Methoxyphthalonitrile, 294
4-Methoxystyrene, 310
Methyl 6-alkylsalicylates, 356, 357
Methyl 3-amino-5-hydroxybenzoate, 352
9-Methylamino-1-oxophenalene, 285, 286
2-(N-Methylamino)phenol, 317
2-Methylanisole, synthesis from 2-methylphenol, 288
4-Methylanisole, 288
Methyl benzenesulfonates, reactions with alkyl halides, 420, 421
2-Methylbenzenesulfonyl chloride, 248
4-Methylbenzoylacetone, 148
2-Methylbenzyl chlorides, flash vacuum pyrolyses, 125
Methyl 4-bromocinnamate, 136
1-(3-Methylbut-2-enyl)-2-naphthol, 239
Methyl 4-chloro-3-formylsalicylate, 322
Methyl 4-chlorophenoxy-2,2-dimethyl-acetate, 288
Methyl 2-(chlorosulfonyloxy)propionate, reaction with benzene, 143
Methyl cinnamate, synthesis from methyl acrylate, 135

Methyl [3, 4-dihydro-1, 4-benzoxazin-3(2H)-one-2-yl] acetate, 329
Methyl 2, 6-dihydroxybenzoate, 335, 336
Methyl 2-(2, 2-dimethyl-8-fluoro-1, 2-benzopyran-6-yl)propionate, 321
Methyldiphenyltelluronium iodide, 501
3, 4-Methylenedioxybenzaldehyde, 346
Methylenedioxybenzene, acetylation, 156
6, 7-Methylenedioxy-2H-benzopyran, 346
5, 6-Methylenedioxy-1-tetralone, 341
Methyl ethers, O-demethylation, 345
3-Methyl-4-ethoxycarbonylanisole, 296
Methyl 2-hydroxybenzoate, 287
Methyl 4-hydroxy-3, 6-dimethylbenzoate, 280
Methyl 3-hydroxy-5-methylbenzoate, 233
2-Methylindanone, 149
1-Methylindole, tricarbonyl chromium complex, 410
Methyl 4-methoxycarbonylcinnamate, 136
Methyl 3-methoxycarbonylcyclohexa-3, 5-dienolate, 279
4-Methyl-2-methylcyclohexanone, 311
5-Methyl-2-(6-methyl-5-hepten-2-yl)-phenyl acetate, 233
Methyl 5-methylsalicylate, 234, 279
3-Methyl-2-naphthol, 233
4-Methylnitrophenol, 316
Methyl orsellinate, 334
3-Methyl-1-oxa-2-azaspiro[4, 5]deca-2, 6, 9-trien-8-one, 257
2-Methyloxetane, reaction with benzene, 143
2-Methylphenol, synthesis from phenol, 239
–, – from toluene, 227
4-Methylphenol, oxidation, 274
–, reactions with glycals, 264
3-(4-Methylphenoxy)-N-phenylphthalimide, 311
3-Methyl-1-phenylindole, 132
2-Methyl-1-phenyl-1-octanone, 160
2-(4-Methylphenyl)-2-oxoethanal, 156
4-Methylphenyl 2-phenylacrylate, 243

Methyl 2-phenylpropionate, 143
Methyl phenyl selenide, oxidation, 502
Methyl phenyl selenone, 502
Methyl phenyl sulfide, acylation, 155
3-Methylphenyl trimethylsilyl ether, 289
4-Methylphenyl 2-(trimethylsilyl)-ethoxymethyl ether, 303
2-(3-Methylquinoxalin-2-yl)benzofuran, 284
3-Methylsalicylic acid, 277
Methyl sulfides, methods of synthesis, 403
N-Methyl-N-sulfinylmethanaminium tetrafluoroborate, 450
N-Methylsulfonyl-1, 4-benzazaquinone, protonation, 18
N-Methyl-S-sulfonylhydroxamic acids, 449
2-Methyltetrahydrofuran, reaction with benzene, 143
Methyl 5, 6, 7, 8-tetrahydro-3-methoxynaphthalene-2-carboxylate, 279
Methyl 2-thienyl tellurides, 403
4-Methylthiomethyl-5-(4-methoxyphenyl)pyrazole, 308
Methylthiomethyl 4-tolyl sulfone, 441
Methyl 4-toluenesulfenate, hydrolysis, 456, 457
α-[5-Methyl-2-(trimethylsilyl)ethoxymethyl]-α-phenylbenzyl alcohol, 303
Methyl vinyl ether, reaction with 4-toluenesulfenyl chloride, 461, 462
Mitsunobu reaction, zinc tosylates use in, 421
Molecular recognition systems, 380
Monoaryl selenides, spectra, 477
Monoarylselenium compounds, interconversion, 470
–, stability, 470, 471
Monoaryltellurium compounds, interconversion, 470
–, stability, 470
Monodeuteriopentafluorobenzene, 197
Monomeric lithium reagents, 385, 386
Muconic acids, synthesis from catechols, 261
Mycophenolic acid, 337
Myers reaction, 102, 103

Naphthalene, radical cation, 76, 77
–, synthesis from 1-methoxynaphthalene, 299
–, – from 2-naphthol, 249
Naphthalene-1,5-diol, oxidation 365
1,8-Naphthalene diradical, 103
Naphthalene-1,5-disulfonamide, Bucherer reaction, 230
Naphthalenes, polyfunctional derivatives, 338, 339
Naphthalenocyclophanes, 51
1,5-Naphthalenophanes, 54
Naphthazarin dimethyl ether, 261
1-Naphthol, esterification, 246
–, synthesis from 1-naphthyl benzoate, 228
–, – from 1-tetralone, 233
2-Naphthol, C-arylation, 133
–, synthesis from 2-naphthyl benzyl ether, 229
Naphthols, synthesis from Fischer carbene complexes/carbon monoxide, 109
–, – from aryldiazo ketones, 241
–, – from benzocyclohexanones, 233
–, – from naphthyl ethers, 240
1-Naphthols, synthesis from 1,2-aryldialdehydes, 238
Naphthoquinones, synthesis from benzyne, 101
1,4-Naphthoquinones, methods of synthesis, 261
1-Naphthotelluroacetate, photolysis, 495
1-Naphthyl benzoate, reaction with butylamine, 228
2-Naphthyl benzoate, 323
2-Naphthyl glucosides, 290
Naphth-2-ylmethanol, synthesis by tandem cyclisation, 103
Naphth-1-yl N-methylcarbamate, 246
Naphthyltellurium radical, 495
Neocarzinostatin, 104
Neoisostegane, 376
2-Neopentylbenzenesulfinic acid, 438
Neopentyl phenyl ether, 289
Newman–Kart reaction, 248, 416
Nitration, arenes, 164

Nitration reactions, mechanism, 167–170
Nitrite ion, scavenging agents, 190
Nitroanilines, synthesis by vicarious nucleophilic substitution, 85
4-Nitroanisole, 301
–, nuclear substitution reactions, 310
Nitroarenes, halogeno-denitration, 190
–, vicarious nucleophilic substitution, 84
Nitroarylethenes, 85
2-Nitrobenzene, trifluoromethylation, 210
Nitrobenzene cyclopentadienyl iron complex, 92
2-Nitrobenzenesulfenamides, anodic oxidation, 465
4-Nitrobenzenesulfenic acid, 454
4-Nitrobenzophenone, synthesis from 4-nitrobenzoyl chloride, 159
4-Nitrobiphenyl phenyl ether, 295
2-Nitroestrone, 316
Nitronium ion carriers, 164–168
4-Nitrophenol, hydroxyl group rotation, 225
Nitrophenols, methods of preparation, 315, 316
3-Nitrophenols, synthesis from 1,3-dinitroarenes, 230
2-Nitrophenylnitromethane, synthesis from nitrobenzene–cyclopentadienyl iron complex, 92
2-Nitrophenylsulfenyl chloride, reagent for sulfide group cleavage, 460
4-Nitrosoanisole, 170, 171
N-Nitrosobenzenesulfonamides, 424
2-Nitrotoluene, synthesis from penta-1,3-diene, 118, 119
4-Nitrotoluene, chlorination, 183
Nonadecatrycene, 59, 60
2-Nonyl-1-phenyloxirane, synthesis from decyl phenylselenone, 502
5-$'$Nonylsalicylaldehyde, 282

Octadehydro[24]annulene, 35, 36
Octet rule, 504
Octyl phenyl sulfide, 320
Oligoanthrylenes, 61, 62
Oligoarylenes, 61

Oligo[a,d]catafusenes, 63
Oligo-[1]-metacyclophanes, 55
Oligonaphthylenes, 61
Oligophenyls, 2
Olivetol, 334
Organoselenium compounds, stability, 470
Organotellurium–organolithium exchange, 507
Organotellurium compounds stability, 470
Orsellinic acids, 334
Orsellinic esters, 334
Oxetanes, ring-opening/selenenylation, 474
Oxidoannulenes, 70, 71
Oxiranes, synthesis from alkenes/benzenesulfonyloxaziridines, 433
–, – from phenylseleninylalkanols, 503
β-Oxo sulfoxides, methods of synthesis, 446

Paracyclene, 68
[4]-Paracyclophane, 49
[5]-Paracyclophane, 52
[6]-Paracyclophane, 52
[6]-Paracycloph-3-ene, 52
[n]-Paracyclophanes, 49
[n,m]-Paracyclophanes, 49
[n],[n]-Paracyclophanes, 54
[2.2]-Paracyclophanes, 53, 54
[2.2]Paracyclophanes, nitration, 166
[1.1.1.1.1.1]-Paracyclophanes, 50
Paracyclophane selenoxides, 497, 498
Pechmann reaction, 350
Peltogynols, 98, 99
Penta-O-acetylcatechin, bromination, 371
Pentachlorobenzene, reduction, 199
3-Pentadecylphenol, 265
Pentadeuteriophenol, 250
Pentafluorophenyl esters, 209
2-Pentafluorophenyl-1-fluoroethane, 213
Pentafluorophenyltelluride, reaction with pentafluorophenyllithium, 508
Pentafluorophenyltributylstannane, 197
Pentafuorophenyltrifluorosulfurane, 513
Pentalene, 9

Pentalene dianion, 25
1,3,5,6,6-Pentamethoxy-3-methyl-1,4-cyclohexadiene, 375
2,3,4,5,6-Pentamethylbenzene anion, 20
Pentaphenylbismuth, 133
Penta(trifluoromethyl)cyclopentadienyl anion, 14
Pentiptycenes, 59
Perbromodiphenyl ether, reductive-debromination, 201
Perfluorobenzene, photocycloaddition reactions, 212
Perfluorobenzenenium ions, 205
Perfluorobenzene oxide, 212
Perfluorobenzenesulfenic acid, 454
Perfluorobenzocycloalkanes, reactions with antimony pentafluoride, 223
Perfluorooxepin, 212
Perfluorotoluene, reactions with alcohols/sodium hydroxide, 209
Perfluoro-1,3,5-triphenylbenzene, 205
Perifusenes, 62
Perylene, 62
Petersen olefination reaction, 408, 409
Phenanthrene, synthesis from 2-methoxystilbene, 306
Phenanthridones, synthesis from N-arylbenzenesulfonamides, 425, 426
9-Phenanthrol, 241
Phenetole, acylation, 155
–, methods of synthesis, 289
Phenol, fluorination, 180, 181
–, iodination, 185
–, synthesis from allyloxybenzene, 228
–, – from benzoic acid, 229
–, – from cyclohexenones, 117, 118, 232
–, – from ethyl phenyl carbonate, 228
–, – from 1-(E,E-hexadienoyl)-3,5-dimethylpyrazole, 233
–, – from iodobenzene, 230
–, – from methoxybenzene, 228, 229
–, – from phenylboronic acids, 229
Phenol-aldehyde polymerisations, 250
Phenolates, as potential carbanions, 266
Phenolic acid, decarboxylation, 333, 334
Phenolic esters, methods of synthesis, 278
–, selective O-acylation, 364, 365

Phenolic ethers, methods of synthesis, 287–297
–, reactions, 297–309
–, selective O-dealkylation, 364
Phenols, acidities, 225
–, O-alkylation, 360
–, aminomethylation, 146
–, C-arylation, 133
–, O-arylation, 130
–, O- versus N-arylation, 133
–, as protection agents, 250
–, charge-transfer complexes, 176
–, chlorination, 184
–, cycloaddition reactions, 250
–, electronic spectra, 226
–, esterification, 241, 242, 262
–, general properties, 226
–, halogenation, 176, 177
–, hydrogenation, 262
–, hydroxyl group protection, 250
–, Michael reactions, 266
–, nitration, 79, 168
–, nitrosation, 170
–, oxidative coupling, 253–255
–, photoelectron spectra, 226
–, physical properties, 225
–, reactions with aryllead triacetates, 129
–, – with benzenesulfonyl halides, 416
–, – with cyclopentadienes, 337
–, – with dialkylthiocarbamoyl chlorides, 416
–, removal of hydroxyl group, 247, 248
–, synthesis from alkynes/vinylketene, 238
–, – from allyl aryl ethers, 240
–, – from araldehydes, 231
–, – from arenes, 226–228
–, – from arylhydroperoxides, 231
–, – from aryltriethoxysilanes, 231
–, – from azidomethylarenes, 229
–, – from cyclobut-2-enones, 101
–, – from cyclobutenones, 237
–, – from diazoketones/vinylketenes, 122
–, – from 2-exomethylenecyclobutenols, 101
–, – from Fischer carbene complexes/ alkynes, 107, 108

–, – from malonyl dichloride, 116, 117
–, – from 1-methoxyalkynes/ketones, 101, 102
–, – from phenolic esters, 228
–, – from phenolic ethers, 228, 229
–, tautomerism, 226
–, transcarbamylation, 245
–, transesterification, 245
2-Phenoxybut-3-enol, 290, 291
Phenoxyl radicals, 225
Phenyl acetate, hydroxylation, 174
–, photochemical rearrangement, 275
α-Phenylacetophenone, 482
2-Phenyl-4H-anthra[1, 2-b]pyran-4, 7, 12-trione, 286
Phenylazoalkenes, 428
N-Phenylbenzenesulfinamides, ortho-lithiation, 451
Phenyl benzenethiolsulfinate, 456
Phenyl benzoate, methods of synthesis, 242
2-Phenyl-4H-1, 3-benzothiazin-4-ones, ring contraction, 450
3-Phenylbutanol, 143
N-Phenylcarbazole, 132
Phenyl cations, 20
Phenyl cinnamate, 275
[3]-Phenylene, 58
[5]-Phenylene, 58
Phenylenes, 13
–, methods of synthesis, 57, 58
Phenyl ethers, ortho-lithiation, 87
2-Phenylethynyl phenyl selenides, oxidation, 493
Phenyl 4-fluorobenzoate, 242
Phenyl formates, methods of synthesis, 263
4-Phenylfuran-2(5H)-one, 135, 136
O-Phenylhydroxylamine, synthesis from mesitylsulfonylhydroxylamine, 247
8-Phenylmenthyl glyoxalate, 145
1-Phenyl-2-naphthol, 133
Phenyl nonanoate, transacylation, 329
2-(Phenyloxy)propionic acid, 4-hydroxylation, 228
Phenyl pentafluorophenyl selenide, reaction with pentafluorophenyl-lithium, 508
3-Phenylphenol, 235

2-Phenylphenyl acetate, 238
Phenyl 3-phenyl-3-oxopropionate, 242
Phenyl propargyl selenide, enzymic oxidation, 490
Phenyl 1-propynyl ether, 290
Phenyl pyanyl sulfides, 458
Phenyl radicals, 19, 20
Phenyl salicylate, 279
Phenylselenenylating agents, 472
Phenylseleninylalkanols, 503
N-Phenylselenophthalimide, 473, 476
α-Phenylseleno-β-silyloxyaldehydes, 402
β-Phenylselenyl enol tosylates, 422
2-Phenylselenenylphenols, 402
N-(Phenylsulfenyl)ketimines, 460
2-(Phenylsulfenylmethyl)cycloalkanones, 462
Phenylsulfenylnitrene, 17
1-(Phenylsulfonyl)cyclopentenes, 440
N-Phenylsulfonylhydrazones, synthesis from phenylhydrazones, 428
3-Phenylsulfonylphthalides, 440
Phenyl tetrazolyl ethers, reductive cleavage, 300
N-(Phenylthio)-1,4-benzoquinonimines, reactions with sodium benzenesulfinate, 468
Phenyl thiolates, *ortho*-lithiation, 87
2-Phenylthiophenol, synthesis from cyclohexanone, 400, 401
5-Phenylthiopyrrolidin-2-ones, 398
N-(Phenylthio)succinimide, 463
α-(Phenylthio)sulfenimines, 467
Phenyltributylstannane, catalyst in Friedel–Crafts reactions, 159
Phenyltrimethylstannane, 147
Phloroglucinol, C-acylation, 368, 369
Phloroglucinols, 116
–, formylation, 163
–, methods of synthesis, 366
Phosphabenzenes, 37, 38
Phosphaphenols, 38
Phosphinines, 37
Piperidino[3,4-b]thiochromone, 320
Piperitone, 262
Pivaloyl chloride, in Friedel–Crafts reactions, 154

Polycycles, synthesis by tandem cyclisations, 103
Polycyclic phenols, 241
Poly(2,6-diaryl-1,4-phenylene) oxides, 254
Polyenes, synthesis from benzvalene, 44
Polyhydric phenols, methods of synthesis, 366
Poly(paraphenylenevinylene)s, 61
Poly(1,4-phenylene sulfide), 211
Polyphenyls, 61
Polystyrene–divinylbenzene copolymers, 139
Porphyrin pigments, 41, 42
Porphyrins, diprotonated forms, 41
Prefulvene biradical, 43
Prestegane B, oxidative dimerisation, 342, 343
Primin (2-methoxy-6-pentyl-1,4-benzoquinone), 260
Prins reaction, 145
Prismane, 43
2-Propanoylphenol, 274
Propargyl benzenesulfenates, rearrangements, 456, 457
Propargyl sulfides, 406
Propene oxide, reaction with benzene, 143
2-^{i}Propoxyphenyl chloroformate, 341
3-(N-Propylamino)phenol, 317
^{i}Propylbenzene, direct plumbylation, 129
–, industrial synthesis, 139
5-^{i}Propyl-2-methylphenol (carvacrol), 256
Propylphenol, synthesis from 2-cyclopropylcyclohex-2-ene, 232
2-Propylphenol, reaction with but-3-en-2-one, 264
2-^{i}Propylphenol, 239
2-^{i}Propyl phenyl ether, 341
Pyranoquinolones, synthesis from aryl propynyl ethers, 307
Pyridinocyclophanes, 50
Pyrimidines, as masked malonaldehydes, 280
Pyrogallol, synthesis from 2-hydroxyisopthalaldehyde, 283
Pyrogallols, methods of synthesis, 366

Pyrogallol trimethyl ether, O-demethylation, 372
3-(Pyrrolidin-1-yl)phenols, methods of synthesis, 318
Pyrrolocyclophanes, 50

o-Quinodimethane, equilibration with benzocyclobutene, 126, 127
o-Quinodimethanes, 125
Quinones, reaction with sulfite ion, 414

Reimer–Tiemann reaction, 274
Reppe reaction, 119
Resorcinol, O-alkylation, 333
–, monomethyl ether, 352
–, sulfonation, 441
Resorcinols, reactions, 347
Rifamycin S, reaction with 4-toluenesulfinic acid, 444
Robinson reaction, 117
Rotenoids, synthesis, 309, 310

Salicyl alcohol (2-hydroxymethylphenol), 250
Salicylaldehyde, methods of synthesis, 274
Salicylic acid, iodination, 187
Sandmeyer reaction, 188
Schöllkopf reaction, 145
Selenenyl halides, reactions with Grignard reagents, 474
Selenides, Davis oxidation, 487
–, reduction with nickel(II) chloride/sodium borohydride, 485
–, oxidation, 485, 497
–, oxidative coupling, 491
–, Sharpless oxidation, 487, 489, 497, 498
Selenilimines, oxidation, 490, 491
Selenimines, 496, 499
Selenium-ate complexes, 508
Selenium based carbanions, 470
Selenoates, reactions with alkyl halides, 474
Selenols, methods of synthesis, 471, 472
–, reactions with benzeneselenide/zinc dichloride, 476
Selenones, 502, 503
Selenonium salts, 496, 499–501
–, reactions, 500
Selenoxides, 496–498
–, allylic, 488
–, homochiral, 497, 498
–, propargylic, 488, 489
–, rearrangements, 485–490
–, sigmatropic oxygen migration, 485
–, syn-elimination, 485–489
–, synthesis from selenides, 478
Selenurane dications, 519, 520
Selenuranes, reaction intermediates, 510
Semmler–Wolf reaction, 116
Sesamol, 346
Silabenzene, 37
Silacyclopentadienyl anion, 37
Silacyclopropenium cation, 36
Silatoluene, 37
O-Silylthioacetals, reactions with alkenes, 403
Smiles rearrangement, 248
Sodium benzenesulfinates, reaction with sulfur, 421
Sodium nonanoyloxybenzene, 329, 330
Spherands, 56
Spirochromanes, synthesis from resorcinol, 349
Spirocycloheptatrienes, 127
Stilbaindoles, 49
Stilbene, synthesis from styrene, 134
Stille coupling reactions, 95
Styrene, industrial synthesis, 139
Styrenes, microbial oxidation, 110
–, photoaddition reactions, 110
–, synthesis from aryl halides/alkenes, 93, 94
Suclofenide, 436
Sulfadiazine, 435
Sulfadimethoxine, 435
Sulfa drugs, 434–437
Sulfamethiazole, 435
Sulfamethoxazole, 435
Sulfamethoxypyridazine, 435
Sulfanilamide, 4-aminobenzenesulfonamide, 434
Sulfenamide enolate equivalents, 467
Sulfenamides, as sulfur-transfer reagents, 463, 464
–, asymmetric oxidation, 465

Sulfenimines, methods of synthesis, 464
Sulfenyl cation equivalents, 400
N-Sulfenylimines, 453
Sulfinamides, 443
–, Pummerer rearrangement, 453
Sulfinamidines, methods of synthesis, 464
N-Sulfinylamines, method of synthesis, 450
α-Sulfinylcycloalkanones, 446
Sulfisomine, 435
Sulfisoxazole, 435
Sulfometuron methyl, 437
Sulfonamides, biological importance, 434–437
Sulfonation reactions, mechanisms, 172
Sulfonomecuriation, 439
N-Sulfonyl-1H-azepines, 432
α-Sulfonyl ethers, synthesis from benzenesulfinic acids/acetals, 440
Sulfonyl nitrite, 443
Sulfonylureas, 436
Sulfoxides, synthesis from sulfides, 405
Sulfoximides, rearrangements, 450
Sulfoximines, synthesis from benzenesulfonimidoyl fluorides, 452
Sulfurane dications, 519
Sulfuranes, reaction intermediates, 510
Sulfur heterocycles, ring-opening, 387
Sulpiride, 436
Superoxide, oxidant for arenes bearing sulfur-containing groups, 414
Suzuki coupling reactions, 96–98

Taxol, 95
Telluranes, stability, 513
Tellurides, methods of preparation, 492–494
–, tellurium–lithium exchange, 509
Tellurium-ate complexes, 508, 509
–, uses in organic synthesis, 509
Tellurium based carbanions, 470
Telluroacetates, reactions with pent-4-enol, 495
Telluroimines, 496
Tellurols, methods of preparation, 492–494
Tellurones, 502

Telluronium benzoates, 501
Telluronium salts, 496
Telluroxides, 496–499
–, syn-elimination, 498, 499
1,2,3,5-Tetraacetoxybenzene, 375
3,4,5,6-Tetraacetoxy-2,7-diphenylisobenzofuran, 367, 368
Tetraarylsulfuranes, 512, 513
4′,5,6,7-Tetrabenzoyloxyflavone, selective hydrolysis, 228
2,3,5,6-Tetrabromo-4-methyl-4-nitrocyclohexa-2,5-dienone, 257
2,3,5,6-Tetrabromo-4-methylphenol, 257
Tetrabutylphosphonium hydrogendifluoride, 191
3,3′,5,5′-Tetra-^{t}butylstilbenequinone, 271, 272
1,2,3,5-Tetrachlorobenzene, reaction with octanethiol, 207
1,2,4,5-Tetrachlorobenzene, reaction with methoxide ion, 208
2,4,5,6-Tetrachlorophenol, 312
Tetracoordinated chalogen compounds, 470
Tetradehydroannulenediones, methanobridged, 33
Tetradehydromethano[18]annulene, 32
Tetradehydromethano[38]annulene, 32
2,3,5,6-Tetrafluoro-4-xylene, reaction with methyl fluoride/antimony pentafluoride, 204, 205
8,8,19,20-Tetrahydro-7H,18H-dibenzo[b,k]-1,7,10,16-tetraoxacyclooctandecane-6,10,17,21-tetrone, 346
2,3,4a,7-Tetrahydro-5,6-di(methoxycarbonyl)-8-methylbenzo-1,4-dioxine, 297
5,6,7,8-Tetrahydro-2-hydroxy-3-methoxynaphthalene, 231
1,2,3,4-Tetrahydro-6-hydroxynaphtho[2,3-f]quinoxaline-7,12-dione, 360
1,2,3,4-Tetrahydro-9-hydroxy-1-oxoanthracene, 276
1,2,3,4-Tetrahydroisoquinoline, N-phenylation, 132
1,2,3,4-Tetrahydroisoquinolin-1-one, 196

6, 7, 8, 9-Tetrahydro-4a, 7-methano-4aH-benzocyclohepten-2(5H)-one, 258
Tetrahydronaphthalenes, 457
5, 6, 7, 8-Tetrahydro-1-naphthol, 236
5, 5, 6', 6'-Tetrahydro-3, 3, 3', 3'-tetramethyl-1, 1-spiroindane, 340
5, 6, 7, 8-Tetrahydro-3, 4, 8-trimethyl-2-naphthol, 255
1, 2, 4, 5-Tetrakis(trimethylsilyl)benzene dianion, 21
Tetralins, trichloromethylation, 222, 223
α-Tetralone, 150
Tetralones, synthesis via Friedel–Crafts reactions, 158
Tetramercapto[1, 1, 1, 1]metacyclophane, 382, 383
4, 4', 5, 5'-Tetramethoxy-2, 2'-dimethylbiphenyl, 343
2, 3, 5, 6-Tetramethylbenzophenone 152
Tetramethyl-2-^{t}butylguanidine, 133
2, 2, 5, 8-Tetramethyl[1, 3]dioxolo[6, 7-d]chromane, 374
3, 3, 5, 5'-Tetramethyl-4, 4-diphenoquinone, 254
Tetramethyloctadehydrodihydroannulenediones, 34
1, 2, 4, 5-Tetraphenylbenzene, 197, 198
Tetraphenyl chalcogenuranes, 514–516
Tetraphenylcyclobutadiene, dipotassium salt, 14
Tetraphenylcyclopentadiene, cathodic reduction, 14, 15
Tetraphenylcyclopentadienyl anion, oxidative dimerisation, 15
Tetraphenylphosphonium hydrogendifluoride, 191
Tetraphenylselenurane, 506, 514
–, synthesis from diphenylselenoxide, 511
Tetraphenylsulfurane, 506, 514
Tetraphenyltellurane, 506, 514
Teuber reaction, 260
Thiannulenes, methanobridged, 40
Thianthrene monosulfoxides, reactions with butyllithium, 511
Thianthrene-5, 5, 10-troxide, 510
Thiatropylium ion, 5

Thioacetals, methods of synthesis, 409, 410
Thiocarbamoyl anions, preparation from tellurium-ate complexes, 509
Thioglycosides, reactions with alkyl benzenesulfenates, 458
Thiolating reagents, 387
Thiolesters, synthesis from Grignard reagents/phenyl chlorothioformate, 390
–, use in synthesis, 390, 391
Thiols, protection, 394, 395
Thiophenols, methods of synthesis, 319
Thioxanthen-9-one 10, 10-dioxides, 437
Thymol, oxidation, 257
–, oxidative coupling, 253
–, photochemical rearrangement, 262
Thymoquinone, 253
Thymylthymoquinone, 253
Tocopherylquinone, 260
Tolbutamide, 434
Toluene, acylation, 151
–, alkylation, 140
–, benzylation, 151
–, enzymic hydroxylation, 227
–, methylation, 141
–, thallation, 79
4-Toluenesulfinates, alcoholysis, 446
–, methods of synthesis, 445
4-Toluenesulfinyl chloride, reactions with enolates, 449
R-2-(4-Toluenesulfinyl)cyclohexanone, 446
N-(4-Toluenesulfonyloxy)-2-pyrrolidinone, rearrangement, 422
4-Toluenethiomethyl 4-toluene sulfones, 449
α-(4-Toluenethio)sulfones, 449
Tolyldiazomethanes, photolyses, 127
Tolyl diradicals, 102
S-Tolylsulfenyl chloride, 443, 444
4-(Tolylsulfonyl)acetonitrile, 441
Tosylates, inversion of stereochemistry, 420
N-Tosyl group, migration, 426
Tosylhydazides, 427
Tosylhydrazones, 427
3α-Tosyloxycholestane, 421
3-Tosyloxyphenoxylamine, 354

Transition metal–carbene complexes, 1, 2
Trans-4-*t*butyl-*cis*-cyclohexane-1, 2-diol, O-phenylation, 132
Triacetylenes, skipped, 120, 121
Trialkylstannyl benzenesulfonates, 414
1, 3, 5-Triamino-2, 4, 6-trinitrobenzene, 19
Triarylbismuthines, 131
Triaryllead tricarboxylates, 128–131
Triaryloxytelluranes, synthesis from diaryltelluranes, 512
Triarylselenonium salts, methods of synthesis, 499, 500
Triarylsulfonium salts reactions with organometallic reagents, 512–516
Triaryltelluronium chlorides, reactions with hydrogen fluoride, 517
2, 2, 2-Triaryl-1, 1, 1-trimethylfluoroethane, 148
2, 4, 5-Tribromo-3, 6-dimethylphenol, nitration, 324
3, 4, 6-Tribromo-2-hydroxy-2, 5-dimethyl-3, 6-dinitrocyclohexanone, 324
1, 2, 3-Trichlorobenzene, reaction with *i*propylthiol, 209
1, 2, 4-Trichlorobenzene, reductive-dechlorination, 199
2-(2, 2, 2-Trichloro-1-hydroxyethyl)-phenol, 251
Trichloromethylarenes, 221
3-Trichloromethyltoluene, synthesis from di(3-methylphenylmethyl) sulfide, 222
2, 4, 6-Trichlorophenol, 312
2, 4, 6-Trichlorophenyl acetate, 322
Tricoordinated chalogen compounds, 470
Tricyclopropylcyclopropenium ion, 10
Trifluoroacetylhypoiodite, 186, 187
Trifluoromethoxybenzene, 177
4-Trifluoromethylbenzoic anhydride, 152
1-Trifluoromethyl-3, 4-dihydro-6-methoxy-3-methylisoquinoline, 307
Trifluoromethylhypofluorite, 179
Trifluoromethyliodobenzene, reactions with enolates, 210, 211

2-(3-Trifluoromethylphenyl)-4-formyl-6-methoxyisoquinol-3-one, 307
Trihalogenomethylarenes, synthesis by vicarious nucleophilic substitution, 85
4, 5, 8-Trihydroxy-9, 10-anthraquinone, 359
Triiodocyclopropenium iodide, 11
2, 4, 6-Triiodophenol, 315
Trimethlphloroglucinol, aminoalkylation, 370
1, 3, 5-Trimethoxybenzene, lithiation, 86
2, 3, 4-Trimethoxybromobenzene, aryl coupling, 376
3, 4, 5-Trimethoxyphenyl methyl ether, 375
1, 3, 5-Trimethoxy-2, 4, 6-tri(propylthio)benzene, 86
2, 4, 6-Trimethylbenzaldehyde, 161
1, 2, 4-Trimethylbenzene, nitration, 167
1, 3, 5-Trimethylbenzene, 140
–, hydroxylation, 227
2, 3, 6-Trimethyl-1, 4-benzoquinone, 259, 357
1, 3, 5-Trimethylphenol, synthesis from 1, 3, 5-trimethylbenzene, 227
2, 3, 5-Trimethylphenol, synthesis from 2, 3-dimethylbut-3-enoyl chloride, 236
Trimethylsilyl arenesulfonates, 172
Trimethylsilyl benzeneselenolate, 476
Trimethylsilyl benzenesulfenates, as masked sulfenic acids, 454
Trimethylsilylbenzenesulfinates, 468
Trimethylsilyl benzyl ether, 293
Trimethylsilyl chlorosulfate, 414
2, 4, 7-Trimethyltropone, 261, 262
2, 4, 6-Trinitroanisole, reaction with methoxide ion, 82
1, 3, 5-Triphenylbenzene, reaction with sulfur, 385
2, 4, 6-Triphenylbenzenethiol, infrared spectrum, 380
Triphenylbenzenethiols, metal complexation, 386
Triphenylbismuth bis(carboxylates), 132
Triphenylbismuth carbonate, 133
Triphenylcyclopropenium ion, 12
Triphenylenes, methods of synthesis, 343

Triphenylethene, synthesis from 1,1-diphenylethene, 134
Triphenyl phosphate, methods of synthesis, 246
Triphenyl phosphite, methods of synthesis, 246
Triphenylselenonium bromide, reaction with phenylmagnesium bromide, 511
Triphenylselenonium chloride, reaction with phenyllithium, 506
Triphenylsulfonium chloride, reaction with phenyllithium, 506
Triphenyltelluronium chloride, reaction with phenyllithium, 506
Triphenylthiasilanol, 387
1,3,5-Triphosphabenzenes, 38
2,4,6-Tri-^{i}propylbenzenethiol, 385
Triptycene, synthesis from benzyne/anthracene, 58, 59
Triptycenes, photoisomerisations, 114
Triquinacene, 7
Tris(benzocyclobutadienato)benzene, 13
Tris(4-^{t}butylphenyl)thionophosphate, 246
1,3,5-Tris(cyclohexyl)-2,4,6-tris(mesyl)-1,3,5-tribora-2,4,6-triphosphabenzene, 38
1,3,5-Tris(dialkylamino)benzenes, 74
2,4,6-Tris(2,4-dihydroxyphenyl)-1,3,5-triazine, 348
2,4,6-Tris(^{i}propyl)phenylsulfonylazide, reactions with enolates, 432
Trisulfides, 380
Trisulfonylamine oxides, 443
Troponecyclophanes, 51
Truce–Smiles rearrangements, 438
Tyrosine, O-phosphorylation, 246

Ullmann reaction, 193
Umbellone, 262

Veratrole, acylation, 156
Vicarious nucleophilic substitution, 84, 85
Vilsmeier formylation reaction, 162, 163
Vilsmeier reaction, 155
Vinyl anion, generation and reactions, 429, 430
Vinyl carbamates, synthesis from vinyl isocyanate, 245
Vinylketenes, generation by photochemical Wolff rearrangement, 122, 123
Vinyl selenides, 479
Vinylstannanes, reactions with 4-chloro-2-cyclobutenones, 123
Vinyl sulfides, synthesis from tosylhydrazones, 430
Vinyl sulfonates, 417
Vinyl sulfones, synthesis from benzenesulfinic acids, 439
Vinyl sulfoxides, 454
Vollhardt cyclisation, 1, 105, 106

Wheland intermediates, 74, 75
Wolff rearrangement, photochemical, 122

Xenon difluoride, reactions with aryl derivatives, 179
Xylenes, benzoylation, 151
–, reductive alkylation, 141, 142
2,6-Xylenol, dimerisation, 267

Zinc tosylate, use in Mitsunobu reaction, 421
Zirconocenes, 45